Seeing Lithics

A Middle-Range Theory for Testing for Cultural Transmission in the Pleistocene

American School of Prehistoric Research Monograph Series

The American School of Prehistoric Research (ASPR) Monographs in Archaeology and Paleoanthropology present a series of documents covering a variety of subjects in the archaeology of the Old World (Eurasia, Africa, Australia, and Oceania). This series encompasses a broad range of subjects - from the early prehistory to the Neolithic Revolution in the Old World, and beyond including: hunter-gatherers to complex societies; the rise of agriculture; the emergence of urban societies; human physical morphology, evolution and adaptation, as well as; various technologies such as metallurgy, pottery production, tool making, and shelter construction. Additionally, the subjects of symbolism, religion, and art will be presented within the context of archaeological studies including mortuary practices and rock art. Volumes may be authored by one investigator, a team of investigators, or may be an edited collection of shorter articles by a number of different specialists working on related topics.

American School of Prehistoric Research (ASPR), Harvard University,
Peabody Museum, 11 Divinity Avenue, Cambridge, MA 02138, USA

For Geneviève Monnier Tostevin and Olivier Monnier Tostevin

(The reasons I care about cultural transmission)

SEEING LITHICS

A MIDDLE-RANGE THEORY FOR TESTING FOR CULTURAL TRANSMISSION IN THE PLEISTOCENE

Gilbert B. Tostevin

Oxbow Books
Oxford and Oakville

Published by Oxbow Books on behalf of the American School of Prehistoric Research.

Library of Congress Cataloging-in-Publication Data

Tostevin, Gilbert B.
Seeing lithics : a middle-range theory for testing for cultural transmission in the pleistocene / Gilbert B. Tostevin.
p. cm. -- (American school of prehistoric research monograph series)
Includes bibliographical references and index.
ISBN 978-1-84217-527-9
1. Paleolithic period. 2. Tools, Prehistoric. 3. Hunting and gathering societies. 4. Flintknapping. 5. Human beings--Migrations. 6. Pleistocene-Holocene boundary. I. Title.
GN771.T67 2013
930.1'2--dc23
2012042054

TYPESET AND PRINTED IN THE UNITED STATES OF AMERICA

Copyright Acknowledgments

Cover art: Nicole Tostevin, www.redheadart.com.

Figures 3.1 and 3.2 are reproduced from C. Carr, 1995, Building a Unified Theory of Artifact Design, in, *Style, Society, and Person: Archaeological and Ethnological Perspectives*, C. Carr & J. Neitzel, eds., pp. 151-170, New York, courtesy of Springer Press.

Figure 3.5 includes copyrighted photographs taken by Lars Smith ("Hadza man looking for game-Tanzania") and Stan Washburn ("San !Kung Woman with child at camp fire") provided courtesy of Anthro-Photo.

Figures 4.17 through 4.20 contain screen captures of the software program VD Flaker published in H. Dibble, S. McPherron, & B. Roth, 2003, *Virtual dig: a simulated archaeological excavation of a middle Paleolithic site in France, 2nd Edition*, reproduced courtesy of McGraw-Hill Higher Education, Mayfield Publishing Co.

Figures 4.33 and 4.34, from Donald O. Henry's article Correlations between Reduction Strategies and Settlement Patterns, are reproduced by permission of the American Anthropological Association from *Archeological Papers of the American Anthropological Association*, Volume 1, Issue 1, pages 140, 142, January 1989. Not for sale or further reproduction.

Figure 5.1 is reproduced in adapted form from figure 43, page 89, of Karel Valoch's 1993 Vedrovice V, eine Siedlung des Szeletien in Südmähren, in *Quartär* band 43/44:7-93, reprinted with permission courtesy of *Quartär* and the Hugo Obermaier-Gesellschaft.

Figures 5.2 and 5.3 are reproduced with permission from figure 61, p. 164, and figure 62, p. 165, of Karel Valoch's Die Erforschung der Kůlna Höhle 1961-1976. *Anthropos* 24, published in 1988, courtesy of the Anthropos Institute, Moravian Museum, Brno, Czech Republic.

Figures 5.8 and 5.27 are reprinted from figures 1 and 2 of J. Svoboda's 2000, Brno-Stránská skála (k.o. Slatina, okr. Brno-město), *Přehled výzkumů* 41:76-80, reprinted courtesy of *Přehled výzkumů* and the Institute of Archaeology Brno, Czech Republic.

Figure 5.10 is reprinted from figure 2 of P. Škrdla 1994, Rekonstrukce Paleolitických Technologií na Stránské skále, *Pravěk* 4.5-15, reprinted courtesy of *Pravěk*

Figures 5.11, 5.16, 5.17, 5.28, and 5.29 are reprinted from Jiři A. Svoboda and O. Bar-Yosef, Editors, *Stránská skála: Origins of the Upper Paleolithic in the Brno Basin*. American School of Prehistoric Research Bulletin 47. Dolní Věstonice Studies Volume 10. Copyright 2003 by the President and Fellows of Harvard College. Reprinted courtesy of the Peabody Museum of Archaeology & Ethnology, Harvard University.

Figure 5.34 is reprinted from figure 2.4, p. 9, of J. Svoboda's 1991, Stránská skála. Výsledky výzkumu v letech 1985-87, in, *Památky Archeologické* 82:5-47, reprinted courtesy of *Památky archeologické* and the Institute of Archaeology, Academy of Sciences of the Czech Republic, Prague.

Figure 7.2 is reprinted from *Journal of Archaeological Science*, Volume 38, by N. R. Rebollo, S. Weiner, F. Brock, L. Meignen, P. Goldberg, A. Belfer-Cohen, O. Bar-Yosef, and E. Boaretto, New radiocarbon dating of the transition from the Middle to the Upper Paleolithic in Kebara Cave, Israel, 2011, pp. 2424-2433, with permission from Elsevier.

Figure 7.3 is reprinted from *Journal of Archaeological Science*, Volume 23, by O. Bar-Yosef, M. Arnold, N. Mercier, A. Belfer-Cohen, P. Goldberg, R. Housley, H. Laville, L. Meignen, J. C. Vogel, B. Vandermeersch, The Dating of the Upper Paleolithic Layers in Kebara Cave, Mt. Carmel, 1996, pp. 297-306, with permission from Elsevier.

Figures 7.8, 7.15, and 7.21, are reprinted from Harold J. Hietala's, Boker Tachtit: Intralevel and Interlevel Spatial Analysis, in, *Prehistory and Paleoenvironments in the Central Negev, Israel*, Volume III The Avdat/Aqev Area, Part 3, A. E. Marks, ed., pp. 217-282, Dallas, courtesy of the Department of Anthropology, Southern Methodist University.

Figures 7.9 and 7.22 are reproduced from Philip W. Volkman's, Boker Tachtit: Core Reconstructions, in, *Prehistory and Paleoenvironments in the Central Negev, Israel*, Volume III The Avdat/Aqev Area, Part 3, A. E. Marks, ed., pp. 127-190, Dallas, courtesy of the Department of Anthropology, Southern Methodist University.

Figure 7.16 is reprinted from figure 14, page 63, of Petr Škrdla's 2003, Comparison of Boker Tachtit and Stránská skála MP/UP Transitional Industries, *Journal of the Israel Prehistoric Society* 33:33-69, reprinted courtesy of the Israel Prehistoric Society.

Contents

Preface

There is substantial debate over the extent to which the Middle to Upper Paleolithic transition and the dispersal of anatomically modern humans from Africa into Eurasia at the end of the Pleistocene were the result of the same process, related processes, or unrelated but coincident processes. The current debate shows a gap in archaeological method and theory for understanding how different cultural transmission processes create patterning in the material culture of foragers at the resolution of Paleolithic palimpsests. This research project attempts to bridge this gap with a middle-range theory connecting cultural transmission and dual inheritance theory with the archaeological study of flintknappers' flake-by-flake choices in the production of lithic assemblages. The project thus combines a new middle-range theory as well as a new approach to characterizing Paleolithic assemblages for systematic comparison of units of analysis appropriate to distinguishing forces of change in cultural evolution.

Using this new approach to asking culture historical questions of the Pleistocene record, this project tests the hypotheses of independent innovation versus cultural transmission as the dominant process(es) by which the Middle to Upper Paleolithic transition appeared across a stretch of western Eurasia. This is accomplished by reconstructing the diachronic patterns of technological behavior evidenced in 18 archaeological assemblages dated between 60,000 and 30,000 years before present in Central Europe, Eastern Europe, and the Levant.

The results of this research demonstrate the existence of two cultural transmission events across these three regions, one between 46–42,000 BP and another between 36–32,000 BP. The first event, spreading the "Bohunician Behavioral Package," appears first in the Levant at 47/46 kya, subsequently in Central Europe at 41 kya, and finally in Eastern Europe at 38 kya. This Bohunician Behavioral Package appears not to have greatly influenced the subsequent development of the Upper Paleolithic in any region, save for the Levant, due to the subsequent cultural transmission event which shifted hominin learning of flintknapping choices to the "Aurignacian Behavioral Package" by 36–32,000 BP. Both cultural transmission events are more likely the results of symmetric transmission of both genetic and cultural inheritance through mating networks than asymmetric transmission, i.e., the transmission of cultural inheritance only, although this is a probability statement, not an absolute judgment. These results have direct implications for the Danube Corridor Hypothesis for the route of entry of anatomically modern humans into Europe at the very beginning of the Upper Paleolithic. This study thus shows the utility of analyzing archaeological data according to units of analysis consistent with cultural transmission theory as well as the evolutionary analysis of material culture change.

All dates and notations to before present (BP) and thousands of years ago (kya) refer to uncalibrated radiocarbon years, unless otherwise indicated.

Acknowledgments

This research project was accomplished with the help of innumerable kind individuals and I will do my best to acknowledge as many as possible. I want first to thank my doctoral dissertation advisor, Ofer Bar-Yosef, for his help, humor, and encouragement during the long gestation of the original dissertation and his patience while waiting for this book. I have learned more about the pursuit of prehistory, and a good amount about life in general, from him than anyone in my career and I will count myself lucky if only a fraction of his insight has rubbed off on me during my studies with him. He is indeed a mentor worthy of the name.

Many individuals showed me great kindness and generosity during my research in the Near East and Europe, not only in providing access to their collections but often to their homes as well. Jiři Svoboda and Petr Škrdla of the Institute of Archaeology, Dolní Věstonice, Czech Republic, offered me not only free access to their material but their friendship as well and I heartily thank them for it. Anna Belfer-Cohen and Erella Hovers of the Hebrew University, Jerusalem, were equally generous with their collections and willingness to discuss prehistory at great length. I would also like to thank Karel Valoch and Petr Neruda of the Moravské zemské Muzeum, Brno, Czech Republic, for their help and kindness in allowing me to study their numerous collections, even when the timing of my visits was not especially convenient. Vitaly Usik, Yuri Demidenko, and Larissa Kulakovskaya of the Kiev Museum and the Ukrainian Academy of Science were also extremely generous with their time, energy, and friendship and made my visits to Ukraine very enjoyable. Alexandr Sytnik of the Institute of Ukrainian Studies, L'viv, Ukraine, similarly was of great assistance. I would also like to thank the Israel Antiquities Authority for access to their collections on several different occasions. The dissertation research which provided the raw data for the analyses presented in this book was supported by grants from the National Science Foundation (Grant #SBR-9714180) and the Mellon Foundation.

I am indebted to Harold Dibble, Greg Laden, Liliane Meignen, John J. Shea, and Martha Tappen for many hours of productive and thought-provoking discussions of lithic technology. Each of them was generous with their time and experience far beyond the call of duty and I feel extremely lucky to have been able to learn from such scholars.

I also would like to thank Matthew Hunstiger, Melissa Knight, and Sabrina Curran who, while students at the University of Minnesota, served as models (Hunstiger and Knight) and photographer (Curran) for the photographic sequence of an apprentice-knapper observing the etically-visible choices of the knapping process in Chapter 4. Their good humor at my heavy-handed application of Photoshop to their images is appreciated. The cover art for the book was far beyond me and so I am grateful for the wonderful help and expertise of my sister, Nicole Tostevin of www.redheadart.com.

I would also like to extend my great thanks to Joan O'Donnell of the Peabody Museum Press as well as the Board of the American School of Prehistoric Research at Harvard University, whose patience with the long process of revising this book was incredible. The benefits of the revisions, however, were worth the effort, particularly because of the important and insightful comments provided by two anonymous reviewers of the manuscript. I would also like to thank Wren Fournier, Publications Coordinator for the American School of Prehistoric Research, for all of her help with the editing and layout of the book.

Beyond all others, I owe my deepest gratitude, and indeed all of my happiness, to Gilliane Monnier, my wife and partner. Not only did she not give up hope on my finishing this project, even when I myself began to doubt, but she accompanied me on several trips, helped collect data, and even performed a medical rescue mission on a time. Her patience and encouragement, along with that of our children Geneviève Monnier Tostevin and Olivier Monnier Tostevin, made this work possible.

List of Figures and Tables

Figures

Tables

1

The Need for an Evolutionary Study of Culture History in the Pleistocene

Research into the anatomical and behavioral origins of modern humans, *Homo sapiens sapiens*, has flourished over the last three decades, with both fossil and genetic discoveries reinforcing the validity of some theories and disproving others. It is thus a productive period in which to work in paleoanthropology on a subject that clearly fascinates the public. Yet there are intriguing gaps in the method and theory of how we study the behavioral evolution of Pleistocene hominins. These gaps, which are as much artifacts of the history of our discipline as obstacles in the way of archaeological epistemology, suggest the need for a new method to ask what is an old question. How one asks the question, however, has the chance to change how we see the role of archaeology in testing paleoanthropological hypotheses.

1.1 The Paleoanthropological Issue

This is a relatively volatile period for research into the origins of anatomically modern humans. Recent discoveries of anatomically modern human fossils in locations and chronological settings, such as Herto at 160 kya in Ethiopia (Clark et al. 2003), have contributed to the strong support of earlier arguments for the recent African origin of anatomically modern humans (Stringer et al. 1984; Stringer and Andrews 1988; Klein 1995). Meanwhile, new discoveries in the identification of the historical basis of the variability within living human mitochondrial DNA (Cann et al. 1987; Wolpoff and Thorne 1991; Stoneking et al. 1993; Relethford 1998; Cann 2001; Templeton 2002; Forster 2004) as well as techniques for the extraction of paleoDNA from Neanderthal fossils, even more distantly related Late Pleistocene *Homo* fossils such as the "Denisovans", and Early Upper Paleolithic modern human fossils have contributed another independent data set (Krings et al. 1997; Pääbo et al. 2004; Serre et al. 2004; Krause et al. 2010; Green et al. 2010; Reich et al. 2010). Together, the fossil record (Lahr and Foley 1998; Harvati et al. 2004) and the genetic data sets have been interpreted to show not only a recent dispersal of modern humans from Africa but also only a miniscule genetic contribution of Neanderthals and their archaic cousins to the modern human genome (Green et al. 2010; Reich et al. 2010).

These studies stand in contrast to the growing complexity of the record in regions that we thought relatively well known. The discovery of a Gravettian anatomically modern child with derived Neanderthal features (Trinkaus and Zilhão 2002) in Portugal has lent greater support to the Assimilation Model, which argues for a recent African origin but with some admixture of Neanderthal genes to, at least, the initial population of moderns in Europe (Smith et al. 2005). Further, the diversity of anatomically modern cranial features at the very beginning of the Upper Paleolithic has grown substantially with the discovery of the Oase modern human remains (Trinkaus et al. 2003a,b; Zilhão et al. 2007). This variation between the Oase and the later Gravettian period modern humans further complicates the question of who was the hominin who first made the Aurignacian in Europe (Churchill and Smith 2000; Bailey et al. 2009). Further east,

Reich et al.'s (2010) identification of a DNA lineage in a finger bone dated between 50 and 30 kya at Denisova Cave (Siberia) that shares ancestry with Neanderthals but itself is an outgroup to the small genetic diversity within Neanderthals demonstrates that the Pleistocene hominin evolutionary tree was more diverse, even branching or bushy to use Bordes's (1950) term, than previously recognized. These discoveries, along with the surprise over the small-bodied *Homo floresiensis* in the Terminal Pleistocene of southeast Asia (Brown et al. 2004; Lahr and Foley 2004; Larson et al. 2007; Baab and McNulty 2009), remind us that we are only beginning to learn how much we still have to discover.

One of the most fascinating of the questions for which we do not yet have an answer is how dispersing modern humans interacted with Neanderthals and other descendants of *H. erectus*, *H. heidelbergensis*, etc. in Eurasia. While the large picture of the origins of anatomically modern humans and the fate of the Neanderthals is becoming clearer, the rarity of fossils and the even greater rarity of fossils from which paleoDNA can be sequenced limits our ability to characterize how, when, where, and under what adaptive contexts modern humans interacted with other Pleistocene *Homo* populations. Green et al.'s (2010) comparison of a complete Neanderthal genome derived from three individuals with the genomes of five living modern humans suggests that between 1-4% of the modern genome of non-African moderns derives from Neanderthals mating with modern humans between 80 and 50 kya. Yet the resolution of this conclusion cannot pinpoint under what adaptive context moderns and Neanderthals made contact on pair-bonding versus competitive terms. PaleoDNA is currently able to give us a continental-scale view of the question. Similarly, given the uncertain reaction to the Lagar Velho child fossil among some physical anthropologists, the paleontology of hominin skeletons may not be able to answer this question either any time soon. Yet without an answer to this proximate question, it will be increasingly difficult to evaluate which of the many hypotheses explaining the adaptive advantage of anatomically modern humans is actually more likely. Was the modern human adaptive advantage quantal speech (Lieberman et al. 2001), a new form of cognitive reasoning (Coolidge and Wynn 2005; Wynn and Coolidge 2010; Klein 1995), true projectile technology (Brooks et al. 2005; Shea 2006; Sisk and Shea 2009; Teyssandier et al. 2010), or the more secure pooling of subsistence risk during environmental fluctuations through long distance reciprocal altruistic relationships (Wiessner 1982) evidenced by raw material exchange (Roebroeks et al. 1988; Féblot-Augustins 1999; Adler et al. 2006; see also Foley and Gamble 2009)? Each of these hypotheses has its own analytical requirements for falsification and evaluation but each will also not make a convincing evolutionary argument without understanding the causal relationships between the proximate and ultimate mechanisms (*sensu* Mayr 1961) by which the actual adaptive advantage worked. How modern humans interacted with the other large-brained predator on the landscape, i.e., archaic *Homo* populations as a result of this adaptive advantage will likely form the causal link between the proximate and ultimate mechanisms by which *Homo sapiens sapiens* populated the planet. To what extent can archaeology help address this issue?

1.2 The Behavioral Issue

Archaeology is uniquely suited to studying the question of hominin social behavior (Gamble 1999; Gamble and Porr 2005). Our data set consists of the physical remains of hominin cultural performance, situated in contexts of time and space that can, with the appropriate methodology, inform on social structure and the evolution of technical acts through the Pleistocene. Archaeology has the further advantage of having large data sets from which to work, particularly in comparison with the paucity of fossils in the record. Yet archaeology to date, with some key exceptions (Gamble 1999; Hovers 2001), has been particularly bad at studying the evolution of cultural traits and social structure during the Pleistocene. This is partly due to methodological problems but also due to historical momentum in the discipline toward particular research questions. These research questions have not been particularly focused on anthropological theory or, for that matter, even the evolutionary understanding of cultural evolution itself.

This point is especially evident in the research into how anatomically modern humans dispersed into Eurasia, producing the possibility of encounters with biologically different populations such as the Neanderthals. As one of the most popular questions in paleoanthropology, research has focused on the most detailed record, that of western Europe, and the most debated issue, the Neanderthal Acculturation Hypothesis in which intrusive Aurignacian populations of modern humans are argued to have been responsible for Neanderthal populations changing their material culture behavior from one typed archaeologically as the Mousterian of Acheulian Tradition to one typed as the Châtelperronian (Klein 1973; Mellars 1989, 1993). Through a series of critiques and rebuttals (d'Errico et al. 1998, Mellars 1999; Zilhão and d'Errico 1999, 2003; d'Errico 2003; Conard and Bolus 2003; Bordes 2003; Mellars 2005, 2010; Gravina et al. 2005; Zilhão and Trinkaus 2007; Higham et al. 2010), the arguments and chronological data have been debated for both the independent evolution of the Châtelperronian and the Bow Wave Model (Mellars 1999, 2005), the latter being a long-distance version of the Acculturation Hypothesis. We have, however, only really learned about the chronostratigraphic *possibility* of contact and that only in southwestern France. We have yet to develop the methodological tools and theoretical bases for a *systematic* investigation into whether or not Châtelperronian artisans *actually did learn something* from Aurignacian artisans. Please note the emphasis on *systematic* here: there have been very thorough studies of both Aurignacian (Lebrun-Ricalens 1993, 2005; Schmider and Perpère 1995; Blades 2000; Bon 2002, 2005; Bon and Bodu 2002; Teyssandier and Liolios 2003; Teyssandier 2008; Teyssandier et al. 2010; Porraz et al. 2010) and Châtelperronian (Guilbaud 1993, 1996; Lévêque 1993; Pelegrin 1990, 1995; d'Errico et al 1998; Roussel 2011) lithic variability, as well as mobiliary art (White 1989, 1993a,b, 2007; Hublin et al. 1996; d'Errico et al. 1998). None of the methodologies utilized in these cases, however, have been created or structured from first principles to analyze evidence of shared learned behavior between these archaeological entities. With the growing consensus that there was no chronological overlap of Aurignacian moderns and Neanderthal Châtelperronian artisans within southwestern France (d'Errico et al. 1998; Zilhão and d'Errico 1999; J.-G. Bordes 2003; d'Errico et al. 2006 *contra* Gravina et al. 2005; Riel Salvatore et al. 2008 and Zilhão et al. 2008a,b *contra* Mellars and Gravina 2008), it is possible that more researchers will turn toward the question of how to determine how much cultural transmission occurred between hominin groups at various geographic and social distances.

This book begins this task by developing the middle-range theory to connect the study of the archaeological record with high-level theory questions derived from cultural transmission theory (Boyd and Richerson 1985; Durham 1991; Shennan 2003; Eerkens and Lipo 2007) as developed in both evolutionary biology and archaeology. This middle-range theory is needed for answering evolutionary questions, such as: can we use our current knowledge of culture historical processes modeled by proponents of cultural transmission theory to recognize features of the archaeological record that will identify the action of these processes in the Pleistocene? If so, can we use the reconstruction of cultural historical processes in the Pleistocene to map the dispersal of particular hominin morphotypes despite their rarity in the record? These are the questions specifically addressed by this research but their answers will lead to other questions for subsequent research: how important was cultural transmission between hominin morphotypes during the "replacement" process by which the modern human phenotype became the dominant form and our genetic variability became marked by the Pleistocene population bottleneck? What does it mean for the dual inheritance of human behavior (genetic and cultural) that Neanderthals and modern humans may have learned from each other? What portion of the behavioral change in the Pleistocene is a result of the selection for adaptations across the Eurasian and African continents versus the other three forces of evolution, i.e., the forces of historical contingency of mutation, genetic drift, and gene flow for biological evolution and innovation, cultural drift (isochrestic variation), and cultural transmission across space in cultural evolution?

These are all questions which the present research will bring closer to a resolution through the application of cultural transmission theory (Boyd and Richerson 1985; Durham 1991; Shennan 2003; Eerkens and Lipo 2007) to the study of Pleistocene hominin social interaction across the most obvious, but not necessarily well understood, phenomena of the Paleolithic: the Middle to Upper Paleolithic transition.

We know from the geographic progression of radiometric dates for the earliest Upper Paleolithic industries (Mellars 2006a), from the Levant at 47/46,000 BP (Marks 1983a; Bar-Yosef et al. 1996) to 43,000 BP in Southeastern Europe (Kozlowski and Ginter 1982), to 38,000 BP in western Europe (Richter et al. 2000; Rink et al. 1996a) that some process was responsible for the geographically-ordered appearance of these behaviors (Kozlowski 1990; Mellars 1996, 2006; Bar-Yosef et al. 1996; Bar-Yosef 2002; Kuhn 2003, 2004). But was the process the independent innovation of these changes under selective pressures common to all regions of western Eurasia, cultural transmission of ideas alone between regions, or the range expansion of hominins?

Between the smoking guns of the dating of the earliest Upper Paleolithic and the effective replacement of one hominin type by another evident in physical anthropology, the archeological record is just begging us to ask the question of which process or combination of processes best explains the *pattern* of change in material culture that occurred between 60 and 30 kya in western Eurasia.

In and of themselves, the obviousness of these Late Pleistocene smoking guns have not provided a better understanding of the Middle to Upper Paleolithic (MP-UP) transition as a historically contingent process/event. It is necessary first to determine *which* aspects of the behaviors which created the material culture changed and *how* or *by what processes* the behaviors changed before one can address the ultimate question of *why* the transition occurred. For just as no inter-regional phenomena may be studied without a prior understanding of intra-regional events, no processual explanation can be attained prior to historical description (Kroeber 1931:155). Pursuing this logic, the present research asks the following necessary questions of the archaeological record:

1) How do the flintknapping behaviors which created the material culture remains of the Middle to Upper Paleolithic transition pattern themselves across the landscape of Central Europe, Eastern Europe, and the Levant between 60 and 30 kya?

Using the results of this historical data question, the research addresses its processual correlate.

2) Is the pattern of these behavioral changes consistent with in situ change through independent innovation or cultural transmission between regions as the process by which the changes appeared?

These questions will be answered by deductively testing three hypotheses with their respective model predictions against the archaeological record.

This research proposes a new approach to the study of technological change in human evolution through a middle-range theory for asking culture historical questions of the Pleistocene archaeological record by means of both cultural transmission theory and a new methodology for characterizing stone tool assemblages. With this approach, it is hoped that we will be able better to understand the significance of the Pleistocene record as evidence of the historically-contingent processes of both biological as well as cultural evolution. Much of this progress will come from simply reorienting the archaeological view of culture history according to evolutionary principles. Culture history is not a question to ask in order to document the peregrinations of unknowable population groups for the sake of such a demonstration. The purpose of culture historical questions in Paleolithic archaeology must be the documentation of Pleistocene cultural transmission in the context of answering questions of the evolutionary significance of dual inheritance of cultural evolution versus biological evolution.

1.3 The Methodological Issue

This research analyzes how the Middle to Upper Paleolithic transition occurred between 60,000 BP and 30,000 BP in the three regions of western Eurasia pertinent to the issue. Lithic artifacts from 18 assemblages from Central Europe, Eastern Europe, and the Levant are analyzed using a methodology which incorporates numerous scholars' lithic attribute analyses to identify comparable behavioral units within the flintknapping process which produced each archaeological assemblage. These units consist of etically-recognized physical evidence of technical behaviors we know are the choice of artisans at particular periods during the lithic reduction sequence. The documentation of the cumulative tendencies of these technical choices constitutes the technological signature for each assemblage in a way that allows for evolutionary analysis of change in terms of frequency changes among variants through time. A comparison of the assemblage technological signatures as they appear between 60 and 30 kya in each region demonstrates regional diachronic patterns of change in these flintknapping behaviors. The comparison of these patterns of change within and between each region is then used to structure the production of, first, historical description and, second, processual explanation of the Middle to Upper Paleolithic transition.

This research tests the hypotheses of *in situ* behavioral change versus cultural transmission between regions as the dominant process by which the Middle to Upper Paleolithic transition appeared in each target region using an assessment of the goodness of fit between the inter-regional patterns of diachronic behavioral change and model expectations derived from a combination of archaeological (Carr 1995a,b; Sackett 1990; Andrefsky 1987; Lechtman 1977; Willey et al. 1956), history of science (Gille 1978; Basalla 1988), evolutionary (Boyd and Richerson 1985; Durham 1991; Shennan 2003; Eerkens and Lipo 2007), and social anthropological thought (Steward 1929; Kroeber 1931, 1940; Kluckhohn 1936; Barnett 1953; Rands and Riley 1958) on distinguishing different culture historical processes. Specifically, this research tests three hypotheses (below) concerning change in flintknapping behaviors between 60 and 30 kya in Central Europe, Eastern Europe, and the Levant.

Hypothesis 1 (Independent Innovation): flintknapping behaviors in a region changed due to *in situ* independent innovation without cultural transmission from neighboring regions. Any similarities in technological signatures are judged to be the result of adaptive convergence and/or cultural drift.

Hypothesis 2 (Cultural Transmission): flintknapping behaviors in a region changed due to the spread of behaviors from one region to another as a result of cultural transmission across space. This hypothesis is further subdivided according to the distinction between symmetric and asymmetric transmission of learned behaviors from cultural transmission theory (Boyd and Richerson 1985). This distinction between the selection of variants based on selection for people with the variant versus selection for the wider learning of the variant regardless of its fitness consequences helps clarify the analysis of the traditionally very muddy theoretical waters of the archaeological demonstration of population dispersal and behavioral diffusion based on artifactual variation.

Hypothesis 3 (Combination): flintknapping behaviors changed due to independent innovation in one or more regions with the subsequent transmission of the behaviors into adjacent regions.

The regions of Central Europe (specifically, the Czech Republic), Eastern Europe (western Ukraine), and the Levant (Israel) are used for two reasons. First, the archaeological record of these regions spans the apparent east to west geographic progression of radiometric dates for the appearance of the earliest Upper Paleolithic industries (Bischoff et al. 1989; Straus 1989, 1994; Kozlowski 1990; Mellars 1993; 1996; Bar-Yosef et al. 1996; Mellars et al. 1999). Second, a morphological similarity between several of the earliest Upper Paleolithic or transitional industries in these regions has already been noted. K. Valoch (1986, 1990a) first noticed the similarity between the industry of Brno-Bohunice (Valoch 1976), which became known as the Bohunician (Oliva 1979, 1984; Svoboda 1984), and the transitional industry of Boker Tachtit Level 1 (Marks 1983) in the southern Levant. Subsequently, scholars such as J. Kozłowski, Y. Demidenko, V. Usik, and P. Škrdla presented syntheses which argued that these similar industries represent one entity, a particular stage in the development of the transition (Kozłowski 1992, 2004; Ginter et al. 1996; Demidenko and Usik 1993a; Škrdla 1996). Before the present research, the meaning of these patterns had yet to be investigated.

Ideally, Anatolia and Southeastern Europe (the Balkans) would also be part of such a study. At the time of the primary data gathering for this project, however, as part of my doctoral dissertation research funded by a NSF doctoral dissertation improvement grant (#SBR-9714180), the new research into the Initial Upper Paleolithic in Anatolia (Kuhn et al. 1999, 2004; Kuhn 2003, 2004) had just begun. Similarly, the newly acquired archaeological assemblages from Temnata Cave in Bulgaria were not yet available for study. Western Europe, with its plethora of pertinent assemblages, was not selected in order to allow for more research effort to be devoted to Central and Eastern Europe which had not received as much inter-regional synthetic treatment (Gamble 1986:3, 393).

The sample of assemblages (Tables 2.1-2.3) was chosen to cover the range of time between 60 and 30 kya in order to include the regional variation before, during, and after the "transition" so that any changes would be placed in their proper context (Kuhn 1995:5). The assemblages were also chosen to represent the variation in known industrial types in each region. Most importantly, by analyzing all of the assemblages within this block of time using the same methodology, this research 1) avoids the common problem of analyzing incomparable regionally-specific archaeological systematics (as pointed out by Vermeersch 2001) as well as 2) avoiding the reification of analytical categories such as "Middle Paleolithic", "transitional", and "Upper Paleolithic".

1.4 Structure of the Book

The process by which one can answer the two questions proposed by this research takes the following form. Chapter 2 proposes a sample of the archaeological record of the Middle to Upper Paleolithic transition along the lines of objective criteria related to excavation quality, temporal resolution, and access to assemblages. A synthesis of the past and current epistemological and substantive problems facing the study of the culture historical processes through the analysis of artifactual variation follows in Chapter 3. As the title of this chapter, "Building Bridges from High-Level Theory: Part One of a Middle-Range Theory for Paleolithic Culture History", suggests, the chapter establishes the criteria for making the correct choice of cultural historical units of analysis given the intellectual goals of cultural transmission theory as applied to the Pleistocene. This is accomplished through the development of the concepts of social intimacy and taskscape visibility out of the building blocks of Behavioral Archaeology (Schiffer 1976, 1997; Schiffer and Skibo 1997), Carr's Unified Middle-Range Theory of Artifact Design (Carr 1995a; 1995b), the French *Technologie* School (Mauss 1936; Lemonnier 1992; Pelegrin 1990b), and the Americanist Organization of Technology approach (Nelson 1991).

Chapter 4, "Building Bridges from Low-Level Theory: Part Two of a Middle-Range Theory for Paleolithic Culture History", connects individual etic observations of the archaeological record, i.e., attributes on flakes and cores, to the processes of both learning and replicating flintknapping performances. This is accomplished through synthesizing what has been learned from the controlled study of flake fracture on glass cores that replicate prehistoric artifact morphologies using behavioral variables known to be under the control of the knapper (i.e., Dibble and Rezek 2009; Rezek et al. 2011). These variables are compared to what is physically observable and thus learnable during cultural transmission events to isolate a collection of lithic analytical techniques that satisfy the requirements of proxies for the choices of knappers during the process of creating stone tools. These are the most suitable units of analysis for answering cultural historical questions according to the social intimacy and taskscape visibility concepts derived from the high-level theory. Special attention is paid to creating tests of the lithic data that are behaviorally significant and readily comparable within and between different regions of western Eurasia. This chapter concludes with the assembly of five model expectations, for testing against the archaeological record, that predict how the intra-regional and inter-regional sequences of flintknapping behavior should appear given the three hypotheses proposed by this research.

The analysis of the lithic data collected during fieldwork in museums and institutes in all three regions between 1996 and 1999 begins with Chapter 5, "Regional Patterns of Behavior: Central Europe". Here the context of each assemblage in Central Europe is described along with the reconstruction of its technological signature according to the etic recognition of the cumulative results of knappers' choices at particular nodes in the flintknapping process. With each technological signature demonstrated, a systematic pair-wise comparison of assemblages as they appear through time in that region is used to produce an intra-regional sequence of change. This analysis is then duplicated for the next two regions, in Chapter 6 "Regional Patterns of Behavior: Eastern Europe" and Chapter 7 "Regional Patterns of Behavior: the Levant." The artifact drawings for each assemblage in the three regions are provided in Appendix A.

Chapter 8, "Inter-Regional Sequence of Behavioral Change", systematically compares the three regional sequences of change in flintknapping behavior and conducts pair-wise comparisons between assemblages which appear in a chronological progression in different regions. This chapter thus comprises the evaluation of the goodness of fit between the archaeological record and the five model expectations of the three hypotheses of the Middle to Upper Paleolithic transition.

Chapter 9, "Conclusions", assesses the significance of the results of the present research for the origins of the Upper Paleolithic as well as the debate over the fate of the Neanderthals and the origins of modern *Homo sapiens sapiens*. The methodological and theoretical contributions of this study are also evaluated in light of future research strategies intended to further our understanding of cultural evolutionary processes in the Pleistocene, in particular how and why the phenomenon of the Middle to Upper Paleolithic transition continues to represent one of the most significant events in human evolution.

1.5 Relationship to Previous Work

This book represents the development of a new method and theory for studying culture historical questions in the Pleistocene. The new approach taken here necessitates a particular analytical format. As a result, although this book uses the original artifact measurements I collected during my doctoral dissertation research (Tostevin 2000c), it is an entirely new theoretical perspective and analysis from that presented in Tostevin (2000a,b,c; 2003a). This book also includes data collected from the excavations at Stránská skála IIIc between 1997 and 1999, which were not included in my dissertation research but published elsewhere (Tostevin 2003b). As such, the presentation of this assemblage here is a broadening of the original sampling strategy of Tostevin (2000c) and also a reanalysis of this assemblage's data rather than a re-presentation of Tostevin (2003b). The development of the middle-range theory presented in Chapters 3 and 4 in fact necessitated significant changes to the technological signatures of each of the 18 assemblages included in Tostevin (2000c) and (2003b).

This reanalysis has allowed some corrections of analytical errors and misjudgments pointed out in critiques of the earlier approach (i.e., Marks 2003). This has undoubtedly improved the research and noticeably changed some of the results. The present study, however, is still constrained by the original limitations on sampling in place when I gathered the data between 1996 and 1999. Thus newer analytical measures such as retouch reduction intensity indices (Clarkson 2002; Eren et al. 2005; Clarkson and Hiscock 2008, 2009, 2011; Hiscock and Clarkson 2007, 2009) and cladistic or 3D landmark analyses (O'Brien et al. 2001; Darwent and O'Brien 2006; Lycett et al. 2006, 2010; Lycett 2007; Lycett and von Cramon-Taubadel 2008; Tostevin 2008) were not available for use in the present work. Similarly, assemblages (and regions) were not open to me in the late 1990s which have more recently become available for study, either because of politics or new discoveries in the archaeological record. It is hoped that whatever limitations of this research project are ascribable to the pragmatics of the sampling issue may be resolved in the future through the broadening of this research program, either through my own analyses or by others attracted to a quantitative approach to studying cultural evolution in the Pleistocene.

2

Sampling the Middle to Upper Paleolithic Transition

The Middle to Upper Paleolithic transition is a complex phenomenon in which multiple aspects of the archaeological record changed. Archaeologists have approached this record of change from many points of view, including the acceleration of culture change (Mellars 1989:366; Bar-Yosef and Kuhn 1999; Bar-Yosef 2002), use of mobiliary art (Leroi-Gourhan and Leroi-Gourhan 1965; White 1989, 2007), extensive alteration and use of bone and antler for tools (White 1982, 1989; Belfer-Cohen and Bar-Yosef 1981; Hahn 1972; Marshack 1972; Taborin 1977; Knecht 1991, 1993), investment in site architecture (Leroi-Gourhan 1976, Farizy 1988, 1990; Bánesz 1976), personal ornamentation (Leroi-Gourhan and Leroi-Gourhan 1965; Hahn 1972; White 1989, 2007; Kuhn and Stiner 2007), subsistence activities (Mellars 1973; White 1982; Chase 1987, 1989; Farizy and David 1992; Stiner and Kuhn 1992; Stiner 1995; Klein 1998; Adler et al. 2006; Kuhn and Stiner 2006), settlement size (White 1982; Mellars and French 2011), population density (Laville et al. 1980; Klein 1998; Bocquet-Appel and Demars 2000; Jöris et al. 2003; Mellars and French 2011), landscape use (White 1983; Kuhn 1995, 2004; Riel-Salvatore and Barton 2004; Riel-Salvatore et al. 2008), long distance contact (Gamble 1982; Taborin 1985, 1994; Roebroeks et al. 1988; Féblot-Augustins 1994), lithic technology (Hahn and Owen 1984; Pelegrin 1990a, 1995; Peyre 1992; Le Brun-Ricalens 1993; Guilbaud 1993; Guilbaud et al. 1994; Bar-Yosef and Kuhn 1999; Marks et al. 2001; Sarel 2004), and symbolism (White 1982, 1985; Chase and Dibble 1987; Chase 1991; Dibble 1989; Whallon 1989; Dibble and Chase 1990; Lindly and Clark 1990; Knecht et al. 2003; Byers 1994; Kuhn and Stiner 2007). The transition is thus a huge subject of research, with the result that the present project attempts to engage only an isolated aspect of the phenomenon. As such, no attempt is made here to be comprehensive in surveying the literature for a synthesis of the current view of the transition according to the perspectives listed above. The reader is directed to far better sources for thorough and recent surveys of the field, whether they be archaeological systematics (Sackett 1991; Groenen 1994; Monnier 2006), theoretical treatments (Hays and Thacker 2001; Zilhão and d'Errico 2003; Gamble 1999), or regional and topical syntheses (Marks 1990; Nitecki and Nitecki 1994; Svoboda et al. 1996; Mellars 1996; Goring-Morris and Belfer-Cohen 2003; Bar-Yosef 1994, 1998, 2002; Brantingham et al. 2004; Chabai et al. 2004, 2006; Shea and Bar-Yosef 2005; Mellars et al. 2007; Shea 2007, 2008a,b; Camps and Chauhan 2009).

The purpose of the present chapter is to outline the sampling strategy whereby specific lithic assemblages from the three regions under investigation are chosen for study. This task includes describing the known lithic variability in each region as well as its practical availability for the present study.

2.1 The Sampling Criteria

Collections from 18 assemblages from three

regions of western Eurasia (Tables 2.1-2.3) were studied to test the goodness of fit between the archaeological record and the model predictions of the three hypotheses of the Middle to Upper Paleolithic transition. The choices of which assemblages to include were based on five criteria:

1) the need to represent the known behavioral variability of each region, in terms of industrial types extant in the region within the chronological limits of the study;

2) the need to sample the variability between 60 and 30 kya, the time frame in which the hominin cultures were experiencing the effect of forces which caused the MP-UP transition, whatever those causes were. Assemblages earlier than 60 kya and later than 30 kya are not seen as directly informative on the transition as modeled in the research targeting these three specific regions of western Eurasia, as the transition in these regions all began long after 50 kya. If the project included Africa, however, there are abundant reasons to push the chronological beginning of the temporal sample as far back as 200 kya (McBrearty and Brooks 2000; Hensilwood and Marean 2002; Marean 2007, 2010; Shea 2006, 2008a);

3) the need to order the assemblages in each region chronologically with relative accuracy for the reconstruction of the sequence of change in behaviors within each region. This criterion requires a chosen assemblage to be chronometrically dated by radiocarbon, thermoluminescene, or electron spin resonance to a point in time, albeit likely with a large error range. Assemblages without dates of any kind were only included when stratigraphic correlation proved the assemblage's position in the chronologically ordered comparison between assemblages within the region;

4) the need for the debitage, core, and tools of an assemblage to have an anthropologically significant association in terms of the time depth of the archaeological palimpsest. This represents an evaluation of the integrity of the depositional context of each assemblage in terms of geological mixing of artifacts between otherwise temporally separated occupations. Old collections acquired with field methods likely to have mixed temporally distinct assemblages (e.g., Ksar 'Akil, Lebanon, and Kulychivka, Ukraine) were not selected due to this criterion; and,

5) the pragmatics of the access to extant collections in Czech, Israeli, and Ukrainian museums given limits on the availability of new collections, old collections never published but still in the hands of the original excavator, and the geopolitics of an immediately post-Cold War Europe. For site report citations for the description of these assemblages, please refer to Tables 2.1-2.3.

It must be stated that the "regions" within which this research was conducted are clearly artificial structures to facilitate analysis, since both cultural and genetic borders are porous (Barth 1969; Hudson 1977; Lefferts 1977; Green and Perlman 1985; Lightfoot and Martinez 195). The analytical definition of the regions as spatial associations of assemblages is, however, of an appropriate scale for investigating the affect of culture historical processes through time and space.

2.2 An Introduction to the Regional Sequences of the MP-UP Transition

2.2.1 *Western Europe*

Although Western Europe will not be included in the present study, it is important to discuss the industrial variability in the MP-UP transition in this region because it is here that the systematics which have so far been used were created. Three archaeological entities exist at the transition in

Western Europe: the late Mousterian, the Châtelperronian, and the Aurignacian. All three are currently best documented in the geographical arc between northern Spain, southern France, and southern Germany. The local Mousterian of Acheulian Type B, which lasted at least until 40 kya, has long been considered a direct ancestor of the Châtelperronian, based on the technological attributes of the blanks and the frequent overlap of the site distributions (Bordes 1958, 1972; Sonneville-Bordes 1960:488–89; see also Soressi 2002 and Soressi et al. 2008). While some researchers such as Lynch (1966) have considered the Châtelperronian as simply a late Mousterian industry with a few Upper Paleolithic *fossile directeurs* (specifically, the backed Châtelperronian knives), others have argued that the Châtelperronian is an acculturation effect of the arriving Aurignacian population upon a Mousterian industry (Harrold 1989; Mellars 1989, 2005; see also Wynn and Coolidge 2004). This interpretation was primarily due to three factors: the technological similarities between the Middle Paleolithic and the Châtelperronian (Harrold 1989), the discovery of Neanderthal fossils associated with Châtelperronian lithic industries at St.-Césaire (Lévêque and Vandermeersch 1980; Leroi-Gourhan 1984; Lévêque 1987; Leroyer 1987; Guilbaud 1993, 1996; Lévêque 1993; Lévêque et al. 1993) and La Grotte du Renne, Arcy-sur-Cure (Leroi-Gourhan and Leroi-Gourhan 1965; Farizy 1988, 1990; Bailey and Hublin 2006), and the initial interpretation of interstratification of Aurignacian and Châtelperronian assemblages at three sites: Roc de Combe, France (Bordes and Labrot 1967), Le Piage, France (Chapagne and Espitalié 1981), and El Pendo, Spain (González Echegaray 1980)(Harrold 1983, 1989; Leroyer 1984; Mellars 1989). Recently, the evidence for the interstratification of Aurignacian and Châtelperronian assemblages at these sites has suffered serious set backs, as taphonomic, dating problems, and excavator bias have been documented in each case (d'Errico et al. 1998; Zilhão and d'Errico 1999; J.-G. Bordes 2003). The last reported evidence for such interstratification, at Grotte des Fées de Châtelperron (Gravina et al. 2005) has also come under serious criticism (d'Errico et al. 2006; Zilhão 2008a,b *contra* Mellars and Gravina 2008). There are other viewpoints, however, that challenge the association of all Neanderthal fossils with Châtelperronian assemblages on taphonomic grounds (Bar-Yosef and Bordes 2010). With the contemporaneity of the Aurignacian and the Châtelperronian within southwestern France now in doubt, the acculturation hypothesis rests of the longer distance processes of the Bow Wave model, for instance from Aurignacian occupants of the Upper Danube Basin (Conard and Bolus 2003) or even further east. The consensus on the continuity of behaviors, or at least influence, between the Châtelperronian and later UP industries is also currently divided (Mellars 1989:353; Demars and Hublin 1989 *contra* Sonneville-Bordes 1960; d'Errico et al. 1998; Zilhão 2005).

Recent reanalysis of transitional assemblages, such as the Bachokirian, and Aurignacian assemblages has resulted in a clearer division of the Aurignacian technocomplex into two regionally specific phenomena (Teyssandier 2006, 2008; Mellars 2006b; Zilhão 2006): the Proto-Aurignacian (also "Archaic Aurignacian" or "Aurignacian 0" or Mellars's "Fumanian"; see Bon 2002; Porraz et al. 2010) in the western Mediterranean and the Early or Classic Aurignacian in temperate Europe (also "Aurignacian 1" as exemplified at sites such as Geißenklösterle in Germany, see Teyssandier and Liolios 2003 and Conard and Bolus 2003, and Willendorf II in Austria, see Nigst 2009, 2012). The Proto-Aurignacian is now perceived as the earlier, invasive archaeological culture that was potentially contemporaneous with the Châtelperronian and even involved in mutual acculturation (Zilhão 2006). The reciprocal nature of the acculturation argument presented by Zilhão is particularly welcome in that he recognizes and emphasizes the two-way nature of most culture contact phenomena. Interestingly, while the Early Aurignacian is now seen as autochthonous in Europe (Conard and Bolus 2003; Teyssandier 2008), the Proto-Aurignacian is now more commonly discussed (Meignen 2006; Zilhão 2006; Mellars 2006; Lebrun-Ricalens et al. 2009) as potentially deriving from the Initial Upper Paleolithic in the Levant, specifically the Early Ahmarian.

Despite the progress made in recognizing smaller scale groups of assemblages (i.e., Proto-Aurignacian assemblages as distinct from Early Aurignacian assemblages), most of the studies across the industrial-type units are still relatively incompatible, making a systematic test of the Ahmarian-to-Proto-Aurignacian Hypothesis just as methodologically fraught as the Neanderthal Acculturation Hypothesis has been. The technological studies of Châtelperronian assemblages, notably those of Lévêque et al. (1993) and Pelegrin (1990a, 1995), need to be extended to more assemblages and systematically compared to the Aurignacian technological studies completed thus far, notably Hahn and Owen (1984), Lebrun-Ricalens (1993, 2005), Bon (2002, 2005), Bon and Bodu (2002), Conard and Bolus (2003), Teyssandier and Liolios (2003), Teyssandier (2008), Nigst (2009), Teyssandier et al. (2010), and Porraz et al. (2010), if the relationships between the Châtelperronian, the Early Aurignacian, Proto-Aurignacian, and later industries are to be deciphered. The problem of analytical compatibility is paramount in this regard due to drastically different approaches evidenced between *chaîne opératoire* advocates such as Pelegrin and the Laplacian archaeologists such as Guilbaud and Lévêque. Roussel (2011) has begun this process for the Châtelperronian at la Grande-Roche-de-la-Plématrie à Quinçay, France, but his comparison was at the level of contrasting the *chaînes opératoires* at Quinçay with those reconstructed by Bon (2002) for the Proto-Aurignacian. Roussel's argument for comparison is convincing but it is not a meta-analysis of comparable data but a comparison of conclusions about the different technologies. With the hope that comparative studies become more frequent and access to western European MP–UP assemblages becomes more feasible than they have in the past (due to issues raised by Shea 2011a), the region is ripe for applying quantitative, multi-assemblage analyses such as advocated here.

2.2.2 *Central Europe: The Middle Danube*

The Central European record of the Middle Danube basin spanning the MP–UP transition is more complicated than that of Western Europe, mostly due to the presence of more archaeological entities of diverse character. The late Middle Paleolithic in Central Europe is primarily composed of the Eastern Micoquian, a Mousterian industry with a predominance of side-scrapers and very small distinctive handaxes, as found at Kůlna Cave, Layer 7a (Valoch 1967). Another Middle Paleolithic entity exists, however, in the form of the Szeletian, an industry with small blades and flakes, a mixture of Middle and Upper Paleolithic tool types, and distinctive small bifacial foliate points similar in technology to Micoquian handaxes (Valoch

1989). Numerous surface sites of Szeletian appearance are known from the Czech Republic and Hungary but only two systematically excavated, stratified Szeletian sites are currently known, Čertova pec Cave in Slovakia (Barta 1959, 1967, 1969, 1980; Vogel and Waterbolk 1964) and Vedrovice V in Moravia (Valoch 1984). Only the collection from Vedrovice V contains enough artifacts to warrant current study. Old excavated Szeletian assemblages are known from Hungary, where the industry was first identified at the turn of the 20th century, but the collection bias, not to mention poor stratigraphic control of the early excavations in the area, make inclusion of any old collections in this study unwise (Allsworth-Jones 1986; Svoboda and Síman 1989). Recent reanalysis by Nigst (2009, 2010, 2012) of layer Ah-2 from Willdendorf II has ascribed this assemblage, according to its technological variables, to the Szeletian as well. Although this assemblage was not available for study when the data collection was undertaken for the present research, it offers an opportunity for comparison in future work.

Apparently contemporaneous with the Szeletian in Moravia is the Bohunician, an industry with Levallois-like technology with a heavy blade component and a mixture of Middle and Upper Paleolithic tool types (Oliva 1984). Although Valoch (1976) originally labeled the industry at the Moravian type-site of Brno-Bohunice the *Szeletien de facies levallois*, the Bohunician is currently acknowledged as a distinct entity from the Szeletian (Svoboda 1984). The claim for contemporaneity between the industries is derived from overlapping radiocarbon dates from Vedrovice V, Brno-Bohunice, and the Bohunician assemblages of Stránská skála III and IIIa (Svoboda and Svobodová 1985; Svoboda 1987; Svoboda and Síman 1989; Svoboda and Škrdla 1995). While the number of AMS radiocarbon dates is low in Bohunician and Szeletian contexts (see Svoboda 2003c; Adams and Ringer 2004), both Szeletian and Bohunician industries are found within the same position within the same Last Interpleniglacial paleosol of southern Moravia, dating roughly between 42 and 38 kya. Paleosols are also infamously troublesome for the preservation of archaeological contexts due to the lack of sedimentation as well as the churning effect of bioturbation processes when the soil was forming (Waters 1991:94; Holliday and Goldberg 1993). Nevertheless, paleosols do provide a context for the formation of the archaeological record and these are the contexts currently accepted by Central European archaeologists for the Szeletian and Bohunician within Central Europe.

The two most recent chronometric dating programs that have helped clarify somewhat the temporal resolution of the Szeletian and Bohunician in Moravia merit some discussion. Richter et al. (2008) and Richter et al. (2009) report TL and OSL dating results, respectively, for the Bohunician assemblage recovered in the 2002 excavation at the type-site of Brno-Bohunice, Czech Republic. The weighted mean of 48.2 ± 1.9 ka BP_{TL} for 11 heated flint samples from Brno-Bohunice provides the first non-radiocarbon data on archeological material from the Bohunician. The OSL results in Richter et al. (2009) are corroborative of the TL results and are contextualized against the AMS radiocarbon dates, calibrated according to the CalPal Hulu model (Weninger et al. 2007; Weninger and Jöris 2008), from all available dated Bohunician assemblages.

Another chronometric dating program, Nejman et al. (2011), provides OSL dating results from multiple sites across the MP-UP transition in the Czech Republic and Slovakia. The results corroborate the existing chronological understanding of each of the sites save for the Szeletian site of Vedrovice V. Nejman et al. state:

> Based on the OSL results, the youngest possible age of the Vedrovice V assemblage is 45.1 ± 2.5 ka, because this is the OSL age of the overlying layer 4. The OSL age for the top of the artefact-bearing layer is 60.3 ± 3.4 ka and for the bottom of the layer 102.1 ± 7.0 ka... The OSL results are very useful and will be a basis for further dating projects in EUP studies, but it is important to realize that ^{14}C dates have the potential to provide dates for actual anthropogenic events (when the dated sample can be associated with human activity), so they are able date [sic] the actual human activities more precisely than OSL dates... If, however, the Vedrovice V Szeletian assemblage does date to between 60 and 100 ka as the OSL results suggest, then this antiquity would place it well into the Middle Palaeolithic (Nejman et al. 2011:1061-1062).

Given that OSL provides chronological information that equates more to geological contemporaneity (*sensu* Conard and Adler 1997) than anthropologically-meaningful contemporaneity (see also the discussion in Richter et al. 2009 of the significance of Dean 1978 in contrasting multiple dating methods), it is safer to refrain from discarding the Vedrovice V assemblage as falling outside of the temporal limits of this study. A geoarchaeological study of the paleosol in which the Vedrovice V assemblage was found is clearly needed to evaluate whether the ^{14}C dates or the OSL dates provide a better understanding of the temporal position of Vedrovice V. The possibility that Vedrovice V may predate rather than post-date the Kůlna Cave Layer 7a assemblage must, however, be kept in mind in the future.

While Valoch (1990a,b) and Neruda (2000) have argued that the Szeletian is derived from the local Micoquian (see Adams, Ringer 2004 for another view), it has been harder to explain the appearance of the Bohunician in the Middle Danube. Valoch (1986, 1990a) first noticed the typological and technological similarity between the industry of Brno-Bohunice (Valoch 1976, 1982) and the transitional industry of Boker Tachtit Level 1 (Marks 1983, Marks, Kaufman 1983) in the southern Levant, although he made no attempt at a systematic comparison. Subsequently, scholars such as Demidenko, Usik, Kozłowski, and Škrdla presented syntheses which argued that these similar industries represent one entity, a particular evolutionary stage in the development of the Middle to Upper Paleolithic transition across the regions (Demidenko and Usik 1993; Ginter et al. 1996; Škrdla 1996; Kozłowski 2000, 2004). These studies served as the inspiration for the present research.

Unfortunately, faunal preservation is extremely poor within open air sites in this period within Central Europe, thus preventing simultaneous studies of lithic and non-lithic operational sequences across the transition in this region. Cave sites such as Kůlna, however, do have sufficient faunal preservation and have been published in detail (Valoch 1967). Raw material studies have also been pursued within the region of Moravia and southern Poland/western Ukraine (Svoboda 1983, 1988; Kozlowski 1982a). The lithic economies of many of the assemblages across the MP-UP transition in Moravia have been studied by Nejman (2006), using the application of Kuhn's Mean Geometric Index of Reduction (1990b) and Clarkson's Index of Invasiveness (2002) to test for different patterns of economization through retouch. His contrast (Nejman 2009) of the Aurignacian from the Bohunician assemblages from Stránská skála by industrial type showed no differences in retouch intensity or invasiveness, leading him to conclude that the Bohunician and the Aurignacian hominins had similar landscape use patterns.

Sampling Central Europe: The Middle Danube Basin
In order to represent the behavioral variability between 60 and 50 kya, the Eastern Micoquian industry of Kůlna Cave, Layer 7a, has been included in this study. Other Middle Paleolithic assemblages in Central Europe which fall within the 60 to 30 kya range were excavated with a methodology which renders them incomparable for the analytical structure proposed here (see discussions in Allsworth-Jones 1986 and Svoboda and Síman 1989). Vedrovice V, the only well-excavated Szeletian site in Moravia, represents the Szeletian variation. The Bohunician is represented by the Stránská skála III Layer 5, Stránská skála IIIa, Layer 4, and Stránská skála IIIc assemblages. The Aurignacian assemblages from Stránská skála IIIa, Layer 3, and Stranská skálá IIa, Layer 4, represent the variability seen between 40 and 30 kya, as older excavations such as Istállóskö (Vertes 1955, 1961) and Mladeč (Szombathy 1925; Skutil 1938) are no longer comparable. Aurignacian sites from Lower Austria (Brandtner 1996; Nigst 2009, 2012) and Romania (Otte and Chirica 1993) were not selected due to time constraints and availability.

2.2.3 *Eastern Europe*

Chabai and Sitlivyi (1994) as well as Demidenko and Usik (1991, 1993b, 1993c) have made claims similar to Valoch's concerning the role of the Levalloisian blade industries as the earliest Upper Paleolithic in Eastern Europe. Demidenko and Usik focus their attention, however, on the *lame à crête* technique as the technological connection between the Middle and Upper Paleolithic, as it appears first within Middle Paleolithic contexts at Korolevo I, layer Ib. Further, they also associate the Near Eastern evidence at Boker Tachtit with that of the Central European sites of Korolevo I and II and Brno-Bohunice. The evidence from further east in the Crimea does not fit that of Central Europe. Anikovich (1992) claims a very late MP-UP transition in the Crimea, as late as 20 kya. Otte's reexcavation of Siurin I in the Crimea, the only reported early UP site in the region, has corroborated this date as no Aurignacian or any other early UP industry is really present. A shift from flake to blade technology with a *lame à crête* technique from the lowest level of Unit II to uppermost level (from 45-30 kya) at Kabazi II in the Crimea further supports a late and *in situ* transition between MP and UP in the region (Marks and Chabai 1995).

Sampling Eastern Europe: Western Ukraine
In order to represent the lithic variation present in Eastern Europe between 60 and 40 kya, the important Dnestr River Basin Mousterian industry of Molodova V, Layer 11, was chosen. Molodova I, Layer 4, would have been an equally viable option to choose to represent the Middle Paleolithic in this region. Slightly younger assemblages exist at Molodova V, specifically 10a-10b which predate the late OIS3 soil formation corresponding to 30 kya (Hoffecker 2002:144-145, 162-163). These assemblages, however, are too small for the initial development of this research but may be useful for later research in filling out the East European sample. The Early Upper Paleolithic "Molodova Culture" represented by layers 10-8, dating between 30-25 kya (Chernysh

Table 2.1 Central European sample assemblages

Site	Layer	Date (BP)	Dating Sample Reference	Curation and References
Kůlna Cave	7a Micoquian	50/53kya ± 5/6 kya	Samples 92202A-92233A	Moravské zemské Muzeum, Brno, Czech Republic. (Valoch 1967). This is an average value from 10 ESR dates with Early Uptake and Linear Uptake (Rink et al. 1996b).
Stránská skála IIIa	4 Bohunician	41,300 + 3,100 -2200	GrN 12606	Institute of Archaeology, Dolní Věstonice, Czech Republic (Svoboda & Svobodová 1985; Svoboda & Síman 1989; Svoboda 1983, 1991; Svoboda & Škrdla 1995; Svoboda 2003c: Table 2.1).
Stránská skála IIIc	Bohunician	38,300 ± 1,100 36,570 ± 940 36,350 ± 990 34,680 ± 820 34,530 ± 770 34,440 ± 720	AA-32058 AA-41476 AA-41478 AA-41480 AA-41477 AA-41475	Institute of Archaeology, Dolní Věstonice, Czech Republic (Svoboda 2003c: Table 2.1).
Stránská skála III	5 Bohunician	38,200 ± 1,100 38,500 + 1,400 - 1200	GrN 12297 GrN 12298	Institute of Archaeology, Dolní Věstonice, Czech Republic (Svoboda & Svobodová 1985; Svoboda & Síman 1989; Svoboda 1983, 1987; Svoboda & Škrdla 1995).
Stránská skála IIId	Bohunician	37,900 ± 1,100 37,270 ± 990 35,080 ± 830 34,530 + 830-740 35,320 + 320-300	AA-32059 AA-32060 AA-32061 GrA-11504 GrA-11808	Institute of Archaeology, Dolní Věstonice, Czech Republic (Svoboda 2003c: Table 2.1).
Vedrovice V	Szeletian	39,500 ± 1100 37,650 ± 550 37,600 ± 800 35,150 ± 650	GrN 12375 GrN 12374 GrN 15514 GrN 15513	Moravské zemské Muzeum, Brno, Czech Republic (Valoch 1984, 1993). GrN 17261 is not included due to noted root contamination (Valoch 1993:78), and GrN 19105 and 19106 are not included due to low carbon levels.
Stránská skála IIa	4 Aurignacian	32,350 ± 900	GrN 14829	Institute of Archaeology, Dolní Věstonice, Czech Republic (Svoboda & Svobodová 1985; Svoboda & Síman 1989; Svoboda 1991; Czudek et al. 1991; Svoboda & Škrdla 1995; Svoboda 2003c: Table 2.1).
Stránská skála IIIa	3 Aurignacian	30,980 ± 360	GrN 12605	Institute of Archaeology, Dolní Věstonice, CR. (Svoboda & Svobodová 1985; Svoboda & Síman 1989; Svoboda & Škrdla 1995; Svoboda 2003c: Table 2.1).

All dates are uncalibrated radiocarbon unless otherwise noted. Industrial affiliations are for descriptive purposes only.

1987:27-35) would also be interesting to study, but they postdate the end of the period currently being examined.

Unfortunately, the important Levalloisian Mousterian of Korolevo I, Layer Ib (Gladilin 1989a, 1989b; Gladilin and Demidenko 1989, 1990; Usik 1989; Demidenko and Usik 1993a) was unavailable at the time of the data collection for this research but it warrants a close examination when it becomes available for study. The period between 40 and 30 kya is represented by the bladey industries from Korolevo I, Layer Ia, and Korolevo II, Complex II.

The lower level of the site of Kulychivka (Savich 1975; Demidenko and Usik 1993b), specifically layer III-d, excavated in the late 1970s by V. P. Savych in western Ukraine is another important but difficult MP-UP site. The stratigraphic association of artifacts and stratum for this level unfortunately became too problematic after the premature death of its excavator for it to be included in this study. Re-analysis of the material by Stepanchuk and Cohen (2000-2001), however, portrays this assemblage as a Middle to Upper Paleolithic transitional entity pre-dating 30 kya. Labeling this the Kremenician Industry, they point to an absence of local antecedents in the Middle Paleolithic but also demonstrate analogies with the Bohunician of Moravia. Meignen et al. (2004) have also published a technological analysis of this assemblage.

With these three assemblages chosen, the western Ukrainian sample includes site locations on both sides of the Carpathian Mountains. Molodova V is located on the east side of the Carpathians while Korolevo is on the western side.

2.2.4 *The Levant*

The Middle to Upper Paleolithic transition in the Levant is perhaps the best documented, technologically, of all the regions included in this study. The site of Boker Tachtit (Marks 1983, 1990; Marks and Ferring 1988; Volkman and Kaufman 1983; Volkman 1983, 1989; Marks and Volkman 1983a, 1983b) represents a gradual transition in core technology from a Middle Paleolithic bidirectional blade technique to a unidirectional Upper Paleolithic blade technique around 40,000 BP that may reflect a technological shift in mobility (Marks 1988). No typological transition is present at Boker Tachtit as the earliest assemblage, Layer 1, contains only Upper Paleolithic tool types (Bar-Yosef 1994). The upper layer of the site represents an early form of Ahmarian, one of the two Upper Paleolithic entities of the Levant (Gilead 1991).

There are several Middle Paleolithic sites in Jordan, namely Tor Sabiha, Tor Faraj (Henry 1982, 1989, 1995b, 2003), and Ain Difla, and one in Israel, Rosh Ein Mor, which appear similar to Boker Tachtit (see Marks 2003), but this observation has never been systematically tested (but see Demidenko and Usik 1993a,b). There are also several Middle Paleolithic sites such as Kebara and Amud which possess industries quite different in typology and technology from anything seen at Boker Tachtit. These features of the Levantine record suggest less continuity across the Middle to Upper Paleolithic transition than first supposed (Bar-Yosef et al. 1996).

Chronologically, the second entity of the Upper Paleolithic in the Levant, the Levantine Aurignacian, appears in the region slightly before 30,000 BP (Belfer-Cohen and Bar-Yosef 1981; Bar-Yosef and Belfer-Cohen 1988). The rockshelter site of Ksar 'Akil, Lebanon, contains the longest sequence of Middle to Upper Paleolithic assemblages in the Near East, including many Levantine Aurignacian assemblages. The technological studies conducted to

Table 2.2 East European sample assemblages

Site	Layer	Date (BP)	Dating Sample Reference	Curation and References
Molodova V	11 MP	>40,300	GrN-4017	Historical Museum of L'viv, Ukraine. (Chernysh 1961, 1987; Ivanova & Chernysh 1965; Klein 1973).
Korolevo II	IIa Transitional	38,500 ±1,000	GIN-2774 (Moscow)	Institute of Zoology, Kiev, Ukraine. (Demidenko & Usik 1993a, 1993b; Gladilin 1989a, 1989b; Gladilin & Demidenko 1989, 1990; Usik 1989).
Korolevo I	Ia Transitional			Institute of Zoology, Kiev, Ukraine. (Gladilin 1989a, 1989b; Gladilin & Demidenko 1989, 1990; Usik 1989).

All dates are radiocarbon unless otherwise noted. Industrial affiliations are for descriptive purposes only.

date suggest a gradual technological transition between the Middle and Upper Paleolithic of a slightly different character from that of Boker Tachtit, with a unique typological signature at the transition (Azoury 1986; Ohnuma 1986; Bergman 1987; Ohnuma and Bergman 1990). Due to the poor stratigraphic control and unknown sample bias within the collections, the Ksar 'Akil assemblages will be used for reference but not as principle data for this study. The archaeological record of the Zagros Mountains, including the "Zagros Aurignacian" (Olszewski and Dibble 1994), falls outside of the Levant and warrants its own regional treatment.

The archaeological record from the Levant has produced a large number of synthetic works concerning both the issue of the origin of anatomically modern humans (Bar-Yosef 1989; 1994; Marks 1990, 1992; Clark and Lindly 1989a, 1989b; Shea and Bar-Yosef 2005) as well as the transition between the Middle and Upper Paleolithic (Marks 1990; Bar-Yosef 1994; Shea 2007). Despite recent syntheses, however, no consensus has been reached concerning Copeland's (1975) suggestion that the Near East may represent the possible source for the Aurignacian and the Upper Paleolithic which later reached Europe. The reasoning for her suggestion is primarily due to the ease of envisioning a transition from the Levantine Aurignacian assemblages of Ksar 'Akil and Abou Halka into a full European Aurignacian. The dates for the earliest Levantine Aurignacian (Bar-Yosef et al. 1996), however, are younger than its earliest appearance in Europe (Richter et al. 2000; Conard and Bolus 2003; Nigst 2009), discrediting this idea.

The recent research on core morphologies in sites spanning the transition in both the north and south Levant (Sarel and Ronen 2003; Sarel 2004) are offering new insights and may warrant future sampling from these localities to incorporate the variability Sarel (2004) recognizes.

Sampling the Southern Levant

In order to represent the industrial and behavioral variability in the Levant between 60 and 40 kya, the sequence of Middle and Upper Paleolithic assemblages from Kebara Cave was analyzed, with a focus on Unit VI for its late Tabun-B type industry, Units IV and III for Ahmarian assemblages, and Units II and I for their early Levantine Aurignacian assemblages. Unit V was not chosen due to the difficulties in delineating the stratigraphic boundaries between V and IV (Bar-Yosef et al. 1996). The important Middle Paleolithic

Table 2.3 Levantine sample assemblages

Site	Layer	Date (BP)	Dating Sample Reference	Curation and References
Kebara	Unit VI MP	48,300 ± 3,500 >48,000	Mean TL Gif-TAN-90029	Israel Antiquities Authority, Jerusalem. (Bar-Yosef et al. 1992; Meignen & Bar-Yosef 1988). The first date is a mean of TL dates (Valladas et al. 1987)
Boker Tachtit	1 Transitional	46,930 ± 2,420 47,280 ± 9,050 >45,570 >34,950	SMU-259 SMU-580 SMU-184 GY-3642	Israel Antiquities Authority, Jerusalem. (Marks 1983; Marks & Volkman 1983a, b; Volkman 1983, 1989)
Boker Tachtit	2 Transitional			Israel Antiquities Authority, Jerusalem. (Marks 1983; Marks & Volkman 1983a,b; Volkman 1983, 1989)
Kebara	Unit IV* Ahmarian	42,500 ± 1,800 42,100 ± 2,100	Pta-5002 Pta-4987	Hebrew Univ., Jerusalem. (Rebollo et al. 2011; Bar-Yosef et al. 1996; Bar-Yosef & Belfer-Cohen 1988)
Kebara	Unit III*	43,500 ± 2,200 >42,500 35,600 ± 1,600 >41,700 36,100 ± 1,100	OxA-3976 Gif-TAN-90037 OxA-1567 Gif-TAN-90168 Pta-4267	Hebrew Univ., Jerusalem. (Rebollo et al. 2011; Bar-Yosef et al. 1996; Bar-Yosef & Belfer-Cohen 1988)
Boker Tachtit	4 Transitional	35,055 ± 1,100	SMU-579	Israel Antiquities Authority, Jerusalem. (Marks 1983; Marks & Volkman 1983a, b; Volkman 1983, 1989)
Kebara	Unit II, Levantine Aurignacian	36,000 ± 1,600 34,300 ± 1,100	OxA-1230 Gif-TAN-90028	Hebrew Univ., Jerusalem. (Bar-Yosef et al. 1996; Bar-Yosef & Belfer-Cohen 1988).
Kebara	Unit I, Levantine Aurignacian	34,510 ± 740 32,300 ± 630	OxA-3974 Pta-4268	Hebrew Univ., Jerusalem. (Bar-Yosef et al. 1996; Bar-Yosef & Belfer-Cohen 1988)

All dates are uncalibrated radiocarbon unless otherwise noted. Industrial affiliations are for descriptive purposes only.

*Please see Rebollo et al. (2011) for the discussion and calibration of AMS dates with ABOx-SC pretreatment for this assemblage.

sites of Tor Faraj and Tor Sabiha (Henry 1982, 1992, 1995a, b, 2003) have not been included due to their dates beyond the 60 kya upper limit. Future research in this region should, however, include Tor Faraj as well as Ain Difla in the desertic regions of the Levant, as recommended by Marks (2003). The heavily reduced assemblage from Quneitra on the Golan Heights (Goren-Inbar 1990) would also be another significant contribution to the sample. It was excluded from the present study for practicality reasons, as the Kebara sequence was sufficiently large given the scale of this project. Quneitra, however, would increase the sample's coverage of environmental zones, a wise course for future work given the recent demonstration of differences in core reduction strategies between such ecological zones (Wallace and Shea 2006).

As no study of the MP-UP transition in the Levant would be complete without the inclusion of the transitional site of Boker Tachtit, this material was sampled through debitage analysis as well as synthesized through the published refittings of Volkman (1983, 1989). The extremely important assemblages of Ksar 'Akil Rockshelter, Lebanon, are used for reference and synthesis (i.e., Azoury 1986; Ohnuma 1986; Bergman 1987; Marks and Volkman 1986; Ohnuma and Bergman 1990) but not as principle data for this study due to the unknown stratigraphic control and sampling bias of the excavations (Copeland 1987). In the future, it may be possible to include these assemblages with suitable caveats on assemblage integrity.

2.2.5 Regions Not Sampled: Southeastern Europe and Anatolia

For the purposes of this study, Southeastern Europe constitutes the archaeological record of Bulgaria, a potentially important region for the transition between the Middle and Upper Paleolithic due to the earliest European dates, around 43,000 BP, for Aurignacian-like industries at Bacho Kiro and Temnata Caves (Garrod et al. 1939; Kozlowski 1979; Kozlowski and Ginter 1982). Both sites possess stratigraphic sequences from late Middle Paleolithic bladey Levalloiso-Mousterian levels through late Aurignacian industries but only Bacho Kiro was available for study at the time of the data gathering for this research. It will soon be possible, however, to add a sample of Temnata Cave assemblages to this research project in the near future, as they are now becoming available after the completion of the relevant site reports. At the time of the data gathering, however, this limited the variability that could be sampled in this region, preventing the inclusion of the Southeastern European archaeological record as a fourth study region for the present research. The cave of Samuilitsa II (Dzambazov 1959, 1967; Ivanova 1979; Sirakov 1979; Sirakov 1983) also possesses a bladey Levalloiso-Mousterian assemblage, although the stratigraphic context of the artifacts is doubtful enough to warrant its exclusion from the present study (Demidenko, personal communication 1994). Due to the lack of any literally "transitional" industry between the bladey Levalloisian industries and the full-fledged Bachokirian/Aurignacian, Kozłowski (1982, 1988, 2004) has expressed doubts about the continuity of MP to UP in Balkans and suggests Anatolia or the southern Near East as the origin for the Aurignacian influence into the Balkans.

Anatolia itself represents a region geographically well suited for this study but the paucity of available assemblages at the time of data collection prevented its inclusion. Although technically within the northern Mediterranean Levant rather than Anatolia *sensu stricto*, the research conducted at Üçağizli cave, Turkey (Kuhn et al. 1999, 2004, 2009; Kuhn 2003, 2004) nevertheless indicates that this region too will have much to contribute to future research on this question.

3

Building Bridges from High-Level Theory: Part One of a Middle-Range Theory for Paleolithic Culture History

This research proposes a new approach to the study of technological change in human evolution through both a new methodology for characterizing stone tool assemblages and a new middle-range theory for asking culture historical questions of the Pleistocene archaeological record. While the new method for characterizing stone tool assemblages is presented in the next chapter, the present chapter develops the middle-range theory necessary to specify which aspects of lithic material culture behavior reflect social processes involved in enculturation and cultural transmission, i.e., those processes appropriate for testing culture historical explanations in an evolutionary context. This middle-range theory is developed here in the context of a particular test case in order to demonstrate its applicability to data in practice as well as its theoretical basis. This test case, determining whether diffusion of stone tool manufacturing techniques occurred between hominins in three regions of western Eurasia between 60 and 30,000 years before present, is the focus of the second half of this book.

The theoretical approach to this research question is developed in two parts within this chapter. First, an evaluation of past theoretical approaches to using artifactual data to answer culture historical questions is presented in a roughly chronological framework. This evaluation considers both approaches to understanding cultural transmission processes as well as approaches to explaining artifact variability through time and space. These two issues have been highly interrelated in the development of archaeological method and theory and so are most successfully evaluated together; what an archaeologist attempts to reconstruct depends much upon what s/he believes to explain artifactual data. Throughout this discussion, each middle-range theory for connecting low-level archaeological observations (*sensu* Thomas 1998:66–94) with the high-level theory of social processes is contrasted to demonstrate commonalities and points of disagreement.

The successful perspectives as well as the cautionary tales of conducting culture historical research are then used in the second part of the chapter to build a new middle-range theory to connect observable variability in artifact manufacturing techniques with the different culture transmission processes by which foragers acquire their particular stone tool technologies. Specifically, this middle-range theory is used to select the appropriate units of analysis, i.e., flintknappers' decisions in the technological manufacturing process, to measure the effects of vertical (cultural inheritance) versus horizontal (cultural diffusion) transmission processes in the archaeological record. The specifics of how to measure and analyze these variables, i.e., low-level theory, is then detailed in the following Chapter 4.

In the final step in developing the theoretical approach to this research, three competing hypotheses are presented, at the end of Chapter 4, as high-level theory for explaining the diachronic change in lithic technology in the Middle Danube, Western Ukraine, and the Southern Levant between 60 and 30 kya. These hypotheses (cultural transmission, independent innovation, or both) are then operationalized into model expectations for how cultural transmission and independent innovation processes would pattern the lithic variables targeted by the middle-range theory for distinguishing vertical from horizontal transmission. These model expectations are then used in Chapters 5–8 to test the goodness of fit between the competing hypotheses and the archaeological record.

3.1 The Building Blocks: Past Theoretical Treatment of Culture Historical Processes and Artifact Variability

While the history of anthropology is beyond the purview of this work (see Harris 1968; Daniel 1975; Adams et al. 1978; Trigger 1989; Willey and Sabloff 1993; and Lyman et al. 1997), how culture historical research was conducted in the past is of direct relevance. In particular, previous efforts at explaining artifact variability in culture historical terms highlight epistemological and methodological problems to be solved by the present research. The following discussion, therefore, concentrates on relating the development of particular movements of archaeological thought, e.g., migrationism to the Processual Agenda, and how these movements contributed to four recent approaches to explaining artifact variability in the archaeological record. This section concludes with a systematic consideration of these four approaches: Carr's Theory of Artifact Design, the *chaîne opératoire* method of the *Technologie* school, the Behavioral Archaeology program, and the Organization of Technology.

3.1.1 Migrationism

Since the beginning of the discipline, anthropology has used the concept of contact between cultures as a means of explaining variation in cultural traits and behaviors through time and space. Within the North American context, this interest in viewing culture contact as the only causal agent was couched in European colonists' racist views of Native North Americans as incapable of inventing new or sophisticated cultural traits (Trigger 1985). Significantly complex archaeological remains such as the mounds of the Mississippian Culture were explained as evidence of migrations from the Old World or Central America (Willey and Sabloff 1993:22). The focus on migration as the explanation for cultural variability can also be seen as a simple carryover from the immigrant status of the Europeans themselves, especially given the importance of classical models of migrant expansion chronicled by Julius Caesar and the Old Testament. Regardless of the source of the bias, it was not until the end of the nineteenth century that anthropologists accepted the idea that Native Americans could be inventive enough to have created sophisticated archaeological remains (Thomas 1894). Until this realization was achieved, few methodological advances were made in studying culture contact.

This recognition of North American *in situ* development appeared at the same time as the views of Franz Boas and his students replaced the cultural evolutionism of Lewis Henry Morgan (Harris 1968:258–59). Boas's school of historical particularism argued that diffusion, not migration or deterministic evolutionary stages, played the largest part in shaping cultural change. This argument succeeded to a large degree due to E. B. Tylor's lead in separating culture, race, and language into separate variables (Kroeber and Kluckhohn 1952:15, 35), thereby removing the evidential support for many putative migrations (Adams et al. 1978). The Boasian school saw cultural diversity as the result of the idiosyncratic courses of historical contact between groups of cultures. Between this perspective and Boas's insistence on factual evaluation of explanations of cultural variability, anthropology began to construct analytical tools appropriate to the study of culture contact.

3.1.2 Boas and the Kulturkreislehre: The Rise of Diffusionism

Boas's factual emphasis brought North American anthropology closer to the type of culture history practiced in Old World ethnology. In particular, the diffusionist school of Boas found strong support in the Vienna school of ethnology and their concept of the Kulturkreislehre or 'culture circles perspective' (Graebner 1911; Schmidt and Poppers 1924; Van Bulck 1932; Schmidt 1939; for changes in this concept through time see Kluckhohn 1936; Heine-Geldern 1964; Andriolo 1979; and Braukämper 1979). As with Boas, the advocates of Kulturkreislehre held that a given cultural trait could be innovated only once, since humans are basically uninventive (Lowie 1937:178), and would spread from one group to the next, since humans are imitative and culture is contagious (Kluckhohn 1936). These two theoretical assertions provided the basis for reconstructing the history of cultures by tracing the diffusion of numerous cultural traits and objects between groups. In this type of study of culture contact, distribution maps of the ethnographic present were created for cultural elements of all media (material culture, economic behavior, social structure, and religious beliefs). The spatial and putative temporal distributions of these different data sets were then analyzed to define associations of elements which were too consistent to be random. These associations were treated as geographical cultural complexes or Kulturkreise which began in smaller regions and grew in size until they merged into the complexity that is the ethnographic present (Andriolo 1979). Culture history was then written based on the continuity of the geographical distribution of complexes, the quantitative assessment of elements shared between cultural complexes, and the parsimony of one innovation event per cultural element.

Three methodological achievements were made with the Kulturkreislehre concept. First, the criteria of explanation were explicit and quantitative. Second, the Kulturkreise or cultural complexes were not *a priori* assumptions, as the migration theories had been, but inferences constructed from the associations of data, at times with a good deal of self-reflection (Heine-Geldern 1964). Third, multiple rather than single types of data were used to construct the complexes (oral history, material culture, comparative religion, archaeological stratigraphy, etc.), thereby increasing the validity of the inferences (Whewell 1858:70, 85 96). The benefits of these three common sense principles were incorporated into the growing archaeological interest in culture history in the first half of the twentieth century, even as anthropology itself moved away from the pursuit of such questions. These principles also formed the basis of European culture history as exemplified by V. Gordon Childe (1929, 1951, 1956). Some of these principles will be discussed in the final section of Chapter 4 as contributions to the high-level theory for the study of the Middle to Upper Paleolithic transition.

Three methodological problems led to the failure of the Kulturkreislehre approach to studying culture contact. The most fundamental failing was the reliance on synchronic data to construct a diachronic explanation. While Kulturkreislehre advocates took advantage of available archaeological data to add time depth to their distribution studies (Kluckhohn 1936), the majority of their ethnographic data was inapplicable to diachronic analysis, regardless of how careful they were in minimizing recent acculturation in their data (Heine-Geldern 1964). This failing was central to the discipline of ethnology and cultural anthropology, although archaeological use of the Kulturkreislehre principles did not face this problem.

Second, the objectivity of the quantitative method was undermined by the subjective choice of which cultural elements to map as a means of defining cultural complexes (Kluckhohn 1936). The quantitative method was further weakened by the frequent interdependence of cultural elements used in the quantitative assessment of contact between complexes. In essence, Kulturkreislehre advocates could not avoid counting the same cultural elements more than once because their units were causally related. In the article which delivered the *coup de grâce* to the diffusionist and Kulturkreislehre schools of ethnology of the 1950s (see Heine-Geldern 1964; Andriolo 1979), Rands and Riley (1958) elucidated the phenomenon of "complex demand" to stress the absence of a need to resort to diffusion or migration to explain the discontinuous distribution of culture traits across cultural boundaries: "Once the nucleus of a particular trait-cluster is established [by independent innovation], the secondary traits will tend to develop sequentially from it, being channeled according to the particular needs of the complex" (1958:277). The similar principle of *tendance* has also been described by Leroi-Gourhan (1943:338). Without the recognition of complex demand in the elaboration of trait complexes, Rands and Riley argued, the acceptable probability of the independent innovation of one culture trait in two regions becomes the unacceptable probability of the independent innovation of a whole complex of related traits in each region, forcing the conclusion that diffusion has been active. Such bias due to the interdependence of units of analysis is known in the social sciences and statistics as Galton's Problem, from Sir Francis Galton's criticism of E. B. Tylor's (1889 (Moore 1961)) improper use of historically-related ethnic groups in his cross-cultural study of traits frequencies (Thomas 1986:446–456). Because Tylor did not use unrelated societies (independent events in the statistical sense) in constructing his tabulation of societies that had or did not have ritual in-law avoidance when compared to their pattern of postmarital residence, Galton pointed out that the interpretation of correlation of variables by Tylor was in fact erroneous and an artifact of his data. In this example it is possible to see how, between the subjectivity of the choice of which elements to map and their susceptibility to Galton's Problem, the Kulturkreislehre school lost its claim to objectivity.

The third failing of the Kulturkreislehre school was the claim that imitativeness and uninventiveness are human universals. This claim was never better than debatable. Many instances of cultural resistance to behavioral diffusion have been demonstrated in both the prehistoric (Dennell 1985; Price and Gebauer 1992) and the ethnographic record (Hodder 1982; 1985). Similarly, many technological innovations have appeared in isolation in different regions across the globe (Basalla 1988). Since the Kulturkreislehre school used these assumptions to define the criteria for the evaluation of a parsimonious argument, their conclusions were invalidated once these assumptions were questioned. By the time of the publication of the last great defense of the school (Schmidt 1939), anthropologists knew that the application of the Kulturkreislehre did not fulfill its theoretical claims and they turned their attention to more pressing issues such as acculturation, the relationship between culture and the individual, and cultural ecology.

3.1.3 *The Americanist Culture History School of Archaeology*

While an ethnological use of the Kulturkreislehre method was abandoned, the archaeological analysis of culture contact through distribution studies of artifact-traits and site-traits flourished between 1914–1960, in what is now known as the Classificatory-Historical Period of American archaeology (Willey and Sabloff 1993). Unlike the Kulturkreislehre, these distribution studies were diachronic, relying upon stratigraphy and seriation. Starting with the application of W. F. Libby's radiocarbon technique to archaeology in the 1950s, absolute dating increased the potential of diachronic studies.

The contributions of the archaeological method and theory of this period to the study of culture contact were summarized in the 1955 SAA symposium on culture contact (Willey et al. 1956, alternately Lathrap et al. 1956). The participants of the symposium categorized all culture contact scenarios into eight types, divided between two main variables. Contact situations were seen as falling into cases of site-unit intrusion, in which population movement or invasion was the causal agent, or trait-unit intrusion, in which behavioral diffusion without population movement was indicated. While the modern definitions of diffusion and migration place the latter as a subset of the former (Morrill and Manninen 1975; Kershaw 1978), diffusion/acculturation was used during the symposium as a separate entity from migration/invasion, applicable to artifacts rather than architecture, social structure, or other material behaviors. Four subcategories within the site unit and trait unit classes were envisioned based on the relative degree of contact and the degree of preservation or suppression of the local culture. Archaeological examples were given as representative of each of the eight contact situations but no theoretical reasoning was provided to explain why each archaeological example was interpreted to fit a particular contact situation (Lyman et al. 1997:195-96). Had Willey et al. (1956) drawn causal links between the process of culture contact and the resulting effect on the material culture of both parties, generalizable conclusions might have been produced. As it was, the symposium barely affected the method and theory of later culture contact studies since it only presented the participants' opinions of different archaeological case studies and not an elucidation of the manner by which contact processes produce different archaeological remains.

3.1.4 *The Fall of Culture History and the Rise of the Processual Agenda*

The above criticisms of Willey et al. (1956) were repeated in kind after 1960 by proponents of the New Archaeology as they attacked similar culture contact studies for their lack of explanatory power and theoretical basis (Binford 1962, 1965, and earlier, Taylor 1948). As Binford notes for the pursuit of culture history in general:

> It is evident that each culture trait tabulated in obtaining the ratio which measures degree of genealogical affinity must be evaluated to determine whether the similarity between traits arose as a function of lineal transmission, diffusion between cultural units, or independent development within each cultural unit. It is here that a basic, unsolved problem lies: How can archaeologists distinguish between homologous and analogous cultural similarities? (Binford 1968:8).

Binford's point strikes at the heart of the problem with culture history of that period: the unit of analysis was the same as that used by the Kulturkreislehre school, and neither school had developed a method of recognizing artifact variation due to functional or contextual requirements as distinct from variation due to cultural connections. While archaeology applied distribution studies in a truly diachronic fashion, the question of the suitability of the unit of analysis, the artifact-trait or site-unit, for identifying cultural relationships between groups had not been addressed since the original critique of Graebner (1911) and his successors. Graebner's two criteria for assessing culture contact were quantitative coincidence and artifact form. The former criterion had been advanced by Tylor (1896) but many cultural historians' lack of concern for unit interdependence resulted in susceptibility to the complications of Galton's Problem as mentioned above. The criterion of artifact form was adequately critiqued by Lowie, "The comparison of form can never do more than establish the identity of forms; that such identity is to be explained by genetic relationship is an hypothesis..." (1912:28). As such, culture historians could not answer Binford's question. Thus it was with irony that Binford's (1968) devastating critique of culture history used the same arguments which had humbled the Kulturkreislehre school decades before.

Lyman et al.'s (1997) analysis of the history of the American culture history school frequently reiterates Binford's point, i.e., the unquestioning reliance of culture historians on homologous similarities to recognize phyletic connections between archaeological occurrences in prehistory. Lyman et al. critique this reliance, however, not because they disagree that homology in material culture denotes phyletic connection between the producers of the material culture but because the culture historians failed to provide a theoretical, rather than commonsensical, reason for labeling the similarity as homologous rather than analogous. This failure to provide a theoretical basis for deciding between homology and analogy in artifacts of different temporal and spatial contexts provided the New Archaeologists the perfect tool with which to attack the culture historians and advocate exclusive research on analogy and artifact function.

The New Archaeology began to focus on analogous rather than homologous similarities in artifacts for two reasons. First, analogy in artifact form was seen as equivalent to functional requirements. By studying functional requirements, proponents of the Processual Agenda were closer to studying culture as the human "extrasomatic means of adaptation" (Binford 1965:205; White 1959:8) and their ultimate goal of creating generalizing laws of human behavior and adaptations through time. With the influence of Leslie White (1949;1959) and Julian Steward (1955), New Archaeologists saw greater utility in analogy than in homology. Second, the functional requirements of artifact form were better suited to be the default explanation for the artifact form's changes in frequency through time and space because they were theoretically disprovable. Through actualistic studies, it is possible to demonstrate, using an objective set of criteria, that a particular artifact form is less suitable or less efficient (in energy used, time spent, or raw material consumed) for a particular job than an alternative form. Conversely, if one assumes *a priori* that a particular artifact form is dictated by stylistic, functionally-neutral concerns, it is unlikely, if not impossible, to disprove this assumption. The hypothetico-deductive paradigm of Fritz and Plog (1970) thus favored questions of analogy rather than homology (Dunnell 1990:19).

As topics of inquiry in American archaeology changed with the momentum of the Processual Agenda (*sensu* Thomas 1998:23), historical questions came to be seen as less and less a part of the canon of archaeological science (Adams et al. 1978; Shennan 2000). In this, archaeology was following the larger trend in science to devalue fields of natural history such as geology and paleontology in favor of disciplines such as physics which produce more generalizable laws (Gould 1989:280–281). In directing archaeologists back to the objectives of anthropology, it was also obeying the social trend to configure academia into a more socially relevant component of American society (Thomas 1998:19–24). This trend was not wholly unproductive for culture history, however, as it allowed more research effort to be spent on creating the methodological and theoretical building blocks necessary for an improved understanding of culture change in all of its contextual richness.

First, while the sequence building aspect of the natural history tradition was deemphasized, the Processual Agenda encouraged the rise of geoarchaeology and thereby provided archaeologists with the field methods (Laville 1976; Courty et al. 1989) as well as theoretical tools (Butzer 1982; Schiffer 1987; Waters 1992) to understand how the archaeological record forms. Without understanding how the physical and anthropological components of site formation processes shape archaeological units of analysis, whether they be artifacts, assemblages, features, or regions, archaeologists could not hope to reconstruct sequences of culture change with any accuracy. Second, archaeological and anthropological studies began to explore the sources of artifact variability. The absence of an understanding of this topic had been the source of most of the methodological and theoretical shortcomings of the culture historical approach (Tschauner 1994:85). Researchers began to discover important cultural processes traditionally ignored by cultural historians, such as the stable mate of diffusion, the innovation process itself (see Renfrew 1978; Spratt 1982; Basalla 1988:26-134; Shennan 1989a; van der Leeuw and Torrence 1989; Schiffer 1993; 2005). The study of such processes in action in their systemic context, in the form of both ethnoarchaeological field work and actualistic studies, was one of the most successful contributions of the Processual Agenda to archaeology. Whereas Willey et al. (1956) failed to connect cultural processes with their resulting patterns of material culture, the Processual Agenda focused on developing this necessary middle-range theory (Binford 1977, 1978, 1983; Grayson 1986). In this way, the "reconstructionist" goals of the Processual Agenda (Tschauner 1994:88) have been instrumental in improving the modern archaeologist's ability to conduct research, including research on cultural transmission.

Actualistic studies and archaeometry, on one hand, and ethnoarchaeology on the other have thus been revolutionary for our understanding of the sources of artifact variability and the nature of the archaeological record. This has been partly a result of the close relationship between the low-level and middle-level theory building of these approaches with the high-level theory of the Processual Agenda (Thomas 1998:66–94). In particular, actualistic studies of artifact production and site formation processes have articulated well with the theoretical perspective of the Processual Agenda. But from the beginning, ethnoarchaeological efforts to build a more advanced *theoretical understanding* of "style", one of the most central analytical concepts in archaeology, had to face significant theoretical obstacles, but obstacles created in this case by the Processual Agenda itself.

3.1.5 Sources of Artifact Variability. Style Theory from Binford to Wobst to Wiessner

One of the most profound impacts of the Processual Agenda on American archaeology has been that of the dualistic opposition between style and function reified by Binford's original approach to situating archaeology within the sciences. Drawing on the concept of genetic drift from biology, Binford early on in the Processual movement defined style as that portion of artifact variability independent of functional and technological variation (Binford 1965:199-203; Binford and Binford 1966:245-246). In this conception, style encapsulated the residual artifact variability not accounted for by these other constraints (see also Dunnell 1978). Thus defined, style could vary through time and space independent of function as the process of cultural transmission randomly biased the continuity of stylistic elements between generations of artisans, just as genetic drift does independently of natural selection. This approach was highly suited to the Processual Agenda because it ordered research priorities according to its preference for ecological, systemic reconstruction of cultural systems over culture historical reconstructions. Given this ordering of priorities, stylistic variation could only be documented after all functional and technological variation had been thoroughly understood, a tall order for an indirect science. Thus in practice if not in theory, the Processual Agenda discouraged the investigation of style for the purposes of culture history, except for the most superficial, additive, and ornamental aspects of artifact variability, such as painted decoration on ceramic vessels.

The dualism of the style/function distinction in the Processual Agenda is most evident in the Bordes-Binford debate over the sources of variability between Bordes' Mousterian facies. From the patterning of assemblages into industrial taxa which resulted from the application of his typology (Bordes 1961a), Bordes maintained an interpretation of the facies as roughly synchronic learned traditions of flintknapping which etically marked particular populations (Bordes 1961b; Bordes and Bordes 1970). Binford, on the other hand, saw function as the default constraint on the form of the tools structuring the typological facies, with the result that the facies were more parsimoniously explained as adaptively distinct tool kits, suitable to different temporal and climatic conditions (Binford and Binford 1966; Binford 1973). Subsequent research has contextualized the straight debate between the Binfords and Bordes, through the documentation of chronological ordering among the facies (Mellars 1970; 1992), differential intensity of raw material utilization between the facies (Dibble 1987a; 1987b; Rolland and Dibble 1991; Dibble and Rolland 1992), differential effect of retouch on different blank forms (Collins 1970; Turq 1992; Hiscock and Clarkson 2009b), and the multifunctionality of Bordes's tool types as seen through similar use-wear traces (Beyries 1988). As a result, neither Bordes' nor the Binfords' position has stood the test of time. Regardless, the Bordes-Binford debate remains the touchstone for American conceptions of the analytical distinction between style and function. Along with the Ford-Spaulding debate (Ford 1952; Spaulding 1953), the Bordes-Binford debate is also a frequent focus for the distinction between etic and emic reality of archaeological artifact types (Dibble 1988; 1995b) and industrial taxa, although the relationship between what Bordes and the Binfords meant by the emic/etic distinction at the industrial level is often mutually misunderstood on both sides of the Atlantic (Sackett 1991:129). This issue will be discussed in further detail later in this chapter.

It is somewhat ironic that the intellectual offspring of the Processual Agenda, ethnoarchaeology, has contributed the most to the deconstruction of the dualistic notion of style and function. Martin Wobst's (1977) theoretical and ethnographic articulation of the role of material culture as a medium of communication of adaptively significant information was one of the first works to give style a function in the processual sense. Wobst also usefully provided predictions from his middle-range theory for the contexts in which material culture would broadcast information of social importance, such as group affiliation or membership, most notably in terms of the social and physical distance between the artisan and the intended audience. Some of these bridging arguments had been independently developed earlier by Friedrich (1970) but for the more express purpose of improving culture historical conceptions of style. Wobst, on the other hand, developed a more conscious middle-range theory, even if based on some questionable assumptions as to the cost of signaling (Dietler and Herbich 1989; 1998), and then applied the concepts to an ethnographic survey of Yugoslavian clothing styles and ethnic and religious boundaries. While Wobst's original argument was often misconstrued, according to his 1999 contemplation of the 1977 article, as advocating that material culture merely reflects social boundaries, the broader context of the argument placed material culture in a causal role: it helps create social meaning. "It is almost impossible to play the role of 'individual-consistently-different-from-others' for any length of time, or in any but the smallest population, without support from one's material shell (dress, body modifications, surrounding material accoutrements)" (Wobst 1999:121). In this way, Wobst's theory not only began to unite style with function but provided an integral role for material culture in the social mechanisms of the production of culture itself. In this regard, Wobst was joined by Hodder (1979b) whose ethnoarchaeological studies in East Africa demonstrated the existence of greater stylistic differences between more frequently interacting groups than between more distant groups, corroborating the view of style as a social mechanism for materially mediating the stress of boundary maintenance. Both Wobst and Hodder were working within an anthropology newly focused on the role of social boundaries as causal agents in the definition and realization of social units (Barth 1969a; Miller and Steffen 1977; Savage and Thompson 1979; Cohen 1978).

While Wobst (1977) in fact noted that multiple types of content could be communicated by material culture style, such as emotional state, identification (class affinity, social group affiliation, rank), authorship, ownership, prescription, proscription, and religious objectification, it is Polly Wiessner (1983) who is most often credited with expanding the active role of material culture style to include personal representation as well as group affiliation (see J. J. Clark 2001:6–22 for a fine summary of the development of style theory). Through her ethnoarchaeological field work among the Kalahari San, Wiessner perceived style in human behavior as an effort of the artisan to communicate her/his positive aspects to others with whom the artisan has reciprocal altruistic relationships (1984). She pointed out that, while the individual artisan could in some ways display creative individuality in the stylistic design of craftworks such as beaded headbands, the limited range of variability allowed to individuals within the ornamental conventions of these media suggested another function. Wiessner (1984) saw style as a method for individuals to use their effort at stylistic replication to communicate their quality, worth, and devotion to their exchange partners in the hxaro gift exchange system, a cultural mechanism for pooling risk with other foragers beyond the environmental catchment of the artisan's territory (1982). Wiessner labeled this role of encoding artifacts with personal information "assertive style," leaving the group affiliation role of material communication to "emblemic style" (1983:257–258). She noted, however, that specific elements of an artifact's style could serve both communicative functions. In her study of arrows among four San language groups, the !Kung, !Xo, G/wi, and Nharo (Ibid), Wiessner found that variability in the emblemic style of arrows was suppressed within groups that pooled risk in the hxaro exchange system but evident between these groups (see also Sinopoli 1991:64). These groups corresponded to language groups of a size (1,500–2,000 persons) larger than Wobst's maximal foraging band of 475 persons (1974, 1976). This larger than predicted size highlighted the contextual nature of the size at which meaningful cultural interaction and information exchange occurs. Although the ratio of assertive to emblemic style among these foragers was high compared to agriculturalists, the distinction between these functions of style allowed Wiessner both to corroborate and to extend Wobst' original conception of style as "active." In this, Wiessner supported Wobst in beginning the 'information exchange' school of style (Plog 1980; Pollack 1983; Gero 1989; Croes 1989; Hegmon 1992; Zedeño 1995; Emberling 1997; Duff 1998), a far richer development of Binford's "secondary functional variation" or the social use of an object (Binford 1965:206).

3.1.6 *Style Theory from Sackett to Lechtman*

The theoretical and ethnoarchaeological work of Binford, Wobst, and Wiessner provided a springboard for James Sackett's wider discussion of material culture style. Begun with the Bordes-Binford debate (Sackett 1973), Sackett's empirically derived arguments (1982, 1986a, 1990) provided a passive role to style, more akin to Binford's genetic drift model than to the active style of Wobst and Wiessner. In a long series of debates in the literature with both Wiessner and Binford, Sackett (1985, 1986a, 1986b, 1990) developed and clarified his concept of isochrestic variation (1982):

> In brief, I argue that the specific forms assumed by objects of material culture represent what amounts to series of specific choices made by artisans, whether conscious or not, from a broad spectrum of equally viable alternative ways of achieving the same end. I refer to such choices as constituting *isochrestic* (literally, 'equivalent in use') variation. The decisions artisans make among the isochrestic options potentially open to them are dictated in large part by the craft traditions in which they have been enculturated as members of social groups. The design notions peculiar to these traditions tend to be quite specific and consistently expressed at any given time and place, even though they are subject to revision in accordance with changes in patterns of social interaction among the artisans who carry on the traditions. Being thus socially bounded, isochrestic variation is idiomatic or diagnostic of ethnicity. And it is such variation that we perceive as style, even when it resides in form that for other purposes may be regarded as fully functional (Sackett 1986b:630, emphasis original).

Sackett's use of the term "ethnicity" in this quote and in his other publications has frequently led to criticisms of the application of the isochrestic and similar models to style in the distant past, specifically G. A. Clark and Lindley (1991) and G. A. Clark (1994). These scholars have targeted efforts to identify "identity-conscious social units" (Clark and Lindley 1991:578) as epistemologically problematic in situations in which the researcher cannot interview live informants. They are indeed quite correct that individual self-ascription to a group affiliation, one of the most well known anthropological definitions of an ethnic group (Barth 1969a, 1969b), is beyond a prehistorical archaeologist's resolution. Yet from Sackett's quote above in which the consciousness of the artisan's decisions has little to do with the source of the variation s/he instills in the object, it is clear that self-conscious social units are neither part nor parcel of the isochrestic model. The Barthian definition of an ethnic group has nothing to do with Sackett's use of the term, nor his argument concerning the spatial and temporal pattern of variation in artifact form as constituting both style and function.

In the above quote, it is clear that Sackett argued for an elimination of the style/function dualism characteristic of the Processual Agenda. The example he pursued to illustrate his point, a comparison of German, English, and French World War I rifles and helmets (1986b:630; 1990:38, 40), focused on the "ethnicity" of each rifle or helmet, despite their almost identical functional characteristics. His original synchronic example worked well for most archaeologists but has not been as successful with Paleolithic archaeologists. Adding a chronological component to his example helps to illustrate the relationship between style and function central to Sackett's isochrestic concept. Imagine three hunters pursuing their prey, one armed with a rifle, one with a crossbow, and one with a hafted Gravettian shouldered point. Each object is stylistic as diagnostic of a time and a place in terms of the frequency of its use at any given time and place in evolutionary history. This is a statistical statement, not an assertion or typologizing of time and space: most people currently use a rifle to hunt, while some enthusiasts still use crossbows, and only a few experimental archaeologists, such as Geneste and Plisson (1990), currently use Gravettian hunting tools. Conversely, 800 years ago, only the crossbow of the three would have been in use, and only the shouldered point would have been found in use some 25,000 years ago in particular areas of western Eurasia. It is thus possible to make probabilistic statements concerning the likelihood of finding any one of these objects in a given time and place. More importantly, each object is equally functional in that each is the tool with which the hunter kills her/his prey. In the large sense of observing where style resides (Sackett 1990), that aspect of a tool which is diagnostic of a time and place cannot be separated from its function since it is not a product of artifact form but part of the temporal and spatial contingency of the design process used to make the functionally useful object.

In reuniting style and function, Sackett' isochrestic model does not remove the utility of a cultural ecological (Winterhalder 1997), optimal foraging (Torrence 1983, 1989; Bleed 1986), or an organization of technology approach (Kelly 1988; Nelson 1991) to artifact variation through time. A rifle is still easier to use to kill an animal at a distance than a dart tipped with a Gravettian point. All things being equal, the former would be selected more often by a hunter given the two options; the differences in the selective advantage of attributes of both weapons are basic causes of stylistic change (Hill 1985; Braun 1995; Carr 1995a:166). Yet this does not negate the point that each tool is quantitatively diagnostic of its time and place as well as being functional. Rather, the isochrestic model helps identify what variables are available for differential selection by cultural adaptation to ecological constraints since it embeds the question in the time-space context of the material tradition in question: "...if both style and function are simultaneously present in artifactual form, how are we to tell when, and to what extent, the observed make-up of an assemblage reflects ethnicity and when, and to what extent, it reflects activity? The answer lies in the arenas of specificity and context. For style and function are obviously cross-cutting patterns, and whether artifactual variation presents itself in terms of its functional or stylistic aspect depends upon the level of resolution at which it is regarded and how the assemblage in which it appears is defined" (Sackett 1990:38). Depending upon the question being asked of the archaeological record, one may view how the rifle, crossbow, and Gravettian point are used in their respective ecological, systemic contexts as functional objects. Or, one may use the spatial-temporal patterning of these objects to help contextualize the archaeological record in order to understand the historical contingency of how hunting technology changed from the dart point to the crossbow to the rifle, as well as through intervening, not necessarily evolutionarily related hunting technologies.

In building off of Sackett's isochrestic model, Christopher Carr (1995a:166-167) presents three reasons for avoiding the dualistic conception of style and function when defining material style. The first is logical, holding that the "criteria of formal, spatial-temporal, and/or contextual patterning that defines an attribute as stylistic does not include or exclude logically whether the trait is essential to the artifact's utilitarian function. Thus, a design trait can be stylistic alone, functional alone, stylistic and function, or neither, from a strictly logical standpoint" (Carr 1995:166). The second reason for abandoning the distinction is quantitative and analytical. The attribute states of a functional object may be shaped by many causal constraints, even with an object as important to food procurement as a projectile point armature used to kill an animal at a distance. Carr uses the example of the multiple causal constraints on the design of the length and width dimensions of a projectile point to illustrate how the limitations on these dimensions are a result of utilitarian (ability to penetrate an animal's hide; the 'primary function' of Binford 1965:200), technological (raw material size constraints), and stylistic forces (the active style of border negotiation from Wobst, Wiessner, and Hodder; Figure 3.1). Carr's third reason is culture-historical, since "functional, spatially patterned variations in artifact design may also, through time, become a basis for social groups actively identifying or distinguishing themselves. Thus, functional, formal variations can come to participate in social processes and take on a role that is normally considered 'stylistic'. Whether that attribute is called functional or stylistic or both is a moot point" (Carr 1995:166-167).

In representing projectile points as an example, Carr's line B (1995a:163, fig. 6-4), reproduced in Figure 3.1, is reminiscent of Wiessner's 1983 study of San projectile points mentioned above. Wiessner's interviews with the makers of arrowheads in all four language groups showed that the variability diagnostic of the risk-sharing groups, according to the San, were not consciously created to be emblemic, despite the apparent patterning of the arrowheads' attributes as fulfilling a boundary maintenance function. San members stated that they "made things in a certain way because everything must be made some way and that was the way their parents did it" (Wiessner 1984:195). While Wiessner saw this as an unsatisfactory result to her study from the assertive style perspective, J. J. Clark concisely noted that "the San could not have presented a better response in support of Sackett's concept of isochrestic variation or Binford's 'cultural drift'. Because differences in point manufacture arose unconsciously through lack of contact, this variability cannot be easily explained and ultimately

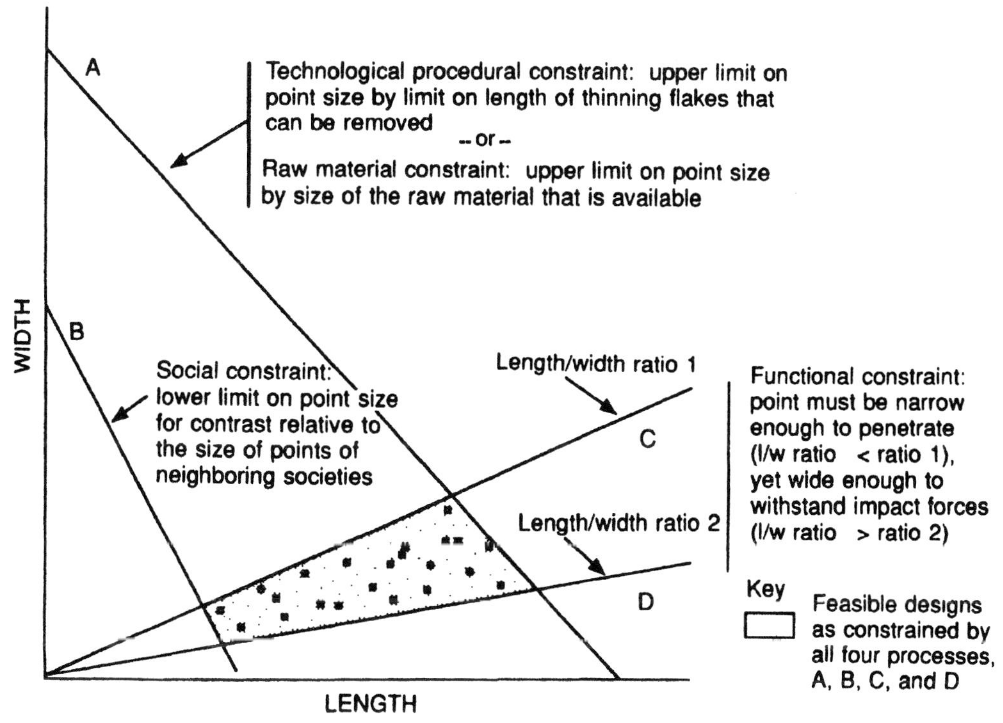

Figure 3.1 Style embedded within and manifested through the technology and function of the design variation of an hypothetical assemblage of projectile points (Carr 1995a:163-4, fig. 6-4, used by permission). The large dots within the shaded region represent the dimensions of individual projectile points.

could only be attributed to enculturative background" (J. J. Clark 2001:11). As the attributes of the arrowheads involved in the emblemic differentiation of the groups also provided the shape of the cutting edge of the armatures, Sackett's isochrestic variation can be seen to be clearly embedded in the functional traits of the tools, even in synchronic ethnographic contexts.

The equal efficacy of different tool forms used for the same function can be seen in other ethnoarchaeological studies of contemporary stone tool production, despite the fact that these examples reflect only a small subset of technologies utilized in the past (Gould et al. 1971; White and Thomas 1972; Gallagher 1977; Hayden 1979; Hayden and Nelson 1981; Torrence 1983; Burton 1984; Gould and Saggers 1985; White and Dibble 1986; J. E. Clark 1991; Roux et al. 1995; Watson 1995; Brandt et al. 1996; Stout 2002; Shott and Sillitoe 2004; Sillitoe and Hardy 2003; Weedman 2002a,b; see also Whittaker 2004). Weedman's ethnoarchaeological work among the Gamo of southern Ethiopia, who use hafted lithic end-scrapers to fletch hides, is a case in point. "The Gamo produce two different scraper forms (formal and informal), as a result of the presence of two different handle types (closed mastic handle and open non-mastic handle; Weedman 2000). Both of these handle and scraper types serve the uniform function of scraping cattle hides for bedding. Scraper and handle forms vary by social-group membership, rather than being related to differences in type of activity (for example hide type, efficiency, or final product) or access and distance to raw-material resources" (Weedman 2002a:733). As the morphological differences between the lithic components of the tool types are related to both the cutting edge and hafted edge, the variability is clearly isochrestic, with the emic distinction falling at the level of the handle, between the zucano handle of the Northern Gamo and the tutuma handle of the southern and central Gamo, rather than the "formal" versus "informal" endscraper itself. Thus despite Hoffman's (1989) argument to the contrary, stylistic variation is found in lithic tool forms but it is isochrestic style, not adjunct style or the "variation that is added on and supplemental to the utilitarian instrumental form" (Sackett 1991:33).

It must be pointed out that Sackett's isochrestic model embedded style in both technology and function. If style and function were dualistically opposed by the Processual Agenda, technology often formed a minor third member to a tripartite categorization, begun as a cross-cutting distinction of Binford's between technical and design dimensions (1965:206–207). In this and Sackett's discussion of the isochrestic model, technology became a locus for where stylistic and functional constraints were seen to operate. Binford's contingency table for systematizing the analysis of cultural variation (1965:207, tab. 2) clearly placed technical variation as influenced by both primary functional and secondary functional constraints, the latter being his term for active style as later developed by Wobst and Wiessner. For Binford, however, passive style such as his cultural drift model remained outside of technological variation (1962:220). For the functional argument, technology was thus considered as the combination of constraints represented by the physical properties of raw materials and the production procedures used to acquire and manipulate these raw materials, as it remains for most of the current Organization of Technology approach (Nelson 1991). Sackett's central point, however, stressed that most of isochrestic variation through time is to be found in differences in how objects are made rather than their final forms. As Sackett saw the style-function distinction as useless, the processes of material culture manufacture, i.e., technology, became the stage on which isochrestic variation played.

Sackett was not the only researcher during the height of the Processual movement to argue that technology, defined as the sequence of behaviors in the manufacture of material culture, was the locus of stylistic variation useful for culture-historical reconstruction. Heather Lecthman's "technological style" concept (1977; Lechtman and Merrill 1977) is similar to that of Sackett in conceiving of variation in the learning of methods of production between different social contexts as creating patterns of stylistic variation which can be used to recognize patterns of social context and interaction. As Wobst and Wiessner are said to have recognized "function in style" and Sackett "style in function," Lechtman's perspective recognized "style in technology." Lechtman's research was based solidly in the realm of archaeometry, utilizing microscopy for the study of prehistoric Andean metal alloys. Lechtman recognized commonalities among the Andean methods of manipulating various alloys, such as arsenical bronzes and the gold-copper alloy known as tumbaga, as distinct from the methods of manipulating metal objects in the Old World. In particular, methods of displaying rare elements on the surface of objects was found to be distinctly different between the Andean and European contexts. In the Andes, the rare element formed part of the alloy, and thus an intrinsic part of the whole object, but was brought to the surface of the object by hammering of the surface of the vessel. In Europe, the rare element was applied to the surface of vessel by gilding, as with gold leaf, rather than forming an integral part of the whole. As this concept of surface-as-part-of-the-whole pervaded both Andean metallurgy and Andean textiles, Lechtman saw this material culture patterning as evidence of a learned and thus culture-specific rule of production embedded with symbolic meaning.

While Lechtman argued that style was found in the production details of the manufacturing process as did Sackett, she emphasized the active communicative and symbolic role of such patterning: "Technologies are performances; they are communicative systems, and their styles are the symbols through which communication occurs" (Lechtman 1977:13). As with Wobst (1999), Lechtman saw the material assistance in human symbolic life as causal rather than an epiphenomenon of symbolic life. This is clearly evident in her recitation of the story of the dancer, Isadora Duncan, responding to a reporter's question as to the meaning of a dance: "If I could tell you what it meant, there would be no point in dancing it" (Lechtman 1977:13). In this way, Lechtman joined historical archaeologists such as Dietz (1965, 1977) and Glassie (1975) as well as later additions from Hodder (1982; 1990) in the search for fundamental cultural rules of material culture production, what Sackett calls the 'deep style' approach (1990:41).

3.1.7 The Nested Nature of Style Theory

The extensive body of literature debating style, function, and technology in the latter third of the twentieth century is evidence of debates waged because there was not sufficient realization that the different "types" of styles were in fact complementary. This realization was articulated most clearly by Carr and Neitzel (1995:4). Carr and Neitzel's argument for the "hierarchical relationship" or nesting of one conception of style within an other, or complementary to another, is represented spatially in Figure 3.2, almost as Russian matryoshka dolls nest within each other. The evidence of the nested structure of style theory, however, first became apparent through the debate between Wiessner (1985, 1990) and Sackett (1986a; 1990), a debate which has become an exemplar of productive communication between discussants. In his last engagement with this debate, Sackett recognized the nested relationships as integral to the definition of isochrestic variation: "…the choices involved in isochrestic behavior create the raw material of style, that style which informs upon ethnicity is an etic perception of the observer, and that style which mediates ethnicity is an emic phenomenon involving the operation of symbolic behavior upon the products of isochrestic choice. The notion of passive style differs from active style, then, in that it regards making choices and assigning meaning to those choices as two distinct kinds of behavior rather than one and the same behavior" (Sackett 1990:37). This is well illustrated in Sackett's thought example of an American tourist visiting Paris. Through the tourist's recognition of the active style apparent in a policeman's clothing as contrasted to the passive style of the furnishings of the tourist's hotel room (in particular, the bidet and the long sausage shaped pillow on the bed), the interplay between emic and etic meaning and the source vs. use of isochrestic variation is clearly visible.

Such thought examples helped clarify the theoretical approaches of the style proponents but did not illustrate how exactly active and passive style

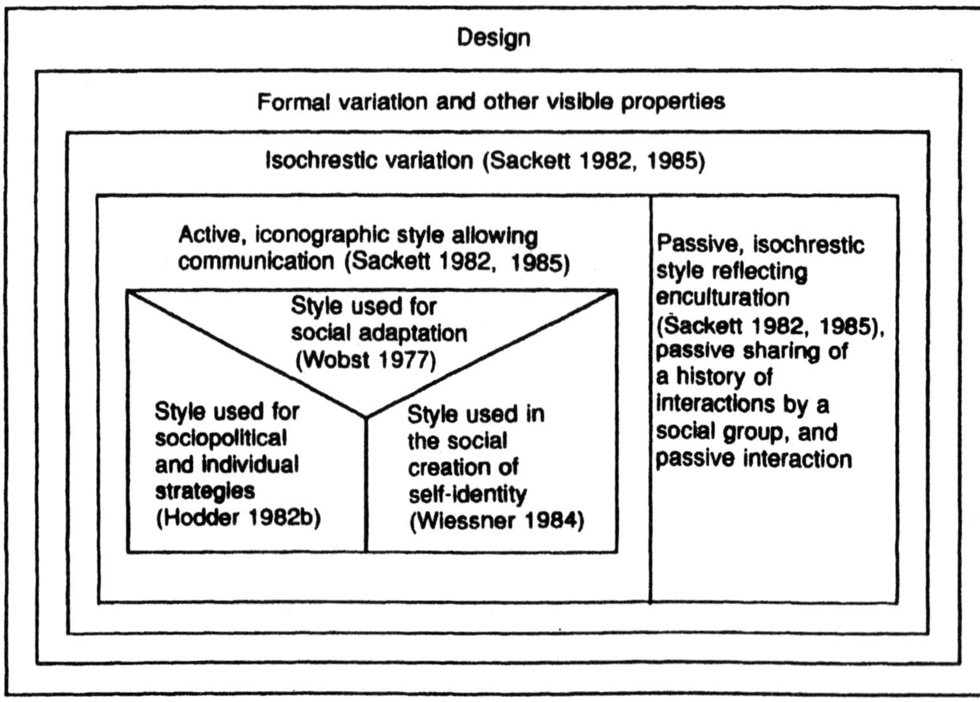

Figure 3.2 Carr's conception of artifact design as including morphology, physical properties, and hierarchically related, i.e., nested, concepts of style (Carr 1995a:158, fig. 6-1, used by permission).

worked together in creating artifact variability (see also Dietler and Herbich 1998; Hegmon 1998). One attempt at such an illustration was Carla Sinopoli's archaeological study of ethnographic examples of arrows collected by Powell from Numic speaking groups in the American Southwest (1991). Studying 172 arrows from the Kaibab, Moapa, and Deep Creek Gosiute bands within Powell's collection, Sinopoli showed that variables on the arrows were most distinctive between the geographically and linguistically closer groups, the Kaibab and Moapa Southern Paiute. This constituted an independent test of Hodder's findings in East Africa (1979b). Most interestingly, however, the attributes of the arrows which showed this marked difference were discrete, with "a limited number of unambiguous alternative forms available to each group" (Sinopoli 1991:73). As discrete, they were both easily visible and distinctive, following expectations for variables suitable for active style established earlier by Friedrick (1970), although she did not describe this as "active style." Sinopoli also demonstrated a second pattern of variability within the arrow sample: particular variables evidenced clinal variation between all three groups, showing an inverse correlation between stylistic similarity and distance between groups. These variables were continuous, less visible, and showed a pattern characteristic of isochrestic variability, rather than active style. As an excellent empirical illustration of how different types of variables operate differently according to their suitability for the different social processes targeted by different conceptions of style, Sinopoli's study suggests a direction for a unified approach to understanding artifact variability.

In the following section, four approaches are introduced which all attempt to the unify archaeological theory on explaining artifact variation, some from the perspective of artifact style and some not. This discussion is designed to illustrate the development of tools for culture historical reconstruction which can be utilized for answering the questions proposed by the present research. In describing these approaches, there is no intention to exclude other approaches from articulating with the method and theory being developed here. Rather, the intent is to highlight a few of the approaches which explicitly attempt to build a middle-range theory for understanding variation in artifact form through time and space. The four approaches chosen happen to be complementary and overlapping to a degree that makes their explicit inclusion more productive than some others. In particular, the four approaches can be bound together using Sackett's isochrestic model into a holistic approach to understanding artifact variation.

3.1.8 Approach I: Carr's Unified Middle-Range Theory of Artifact Design

Christopher Carr has proposed a reconceptualization of style theory into what he calls a "unified middle-range theory of artifact design" (1995a; 1995b) and, along with a team of colleagues, has applied it to a variety of material media in both archaeological and ethnographic contexts (Pryor and Carr 1995; Carr and Maslowski 1995; Neitzel 1995). This middle-range theory builds off of the recognition of the hierarchical relationships among current conceptions of style theory as well as a broader contextualization of the social processes which produce patterns of artifact form.

> The purpose of the unified theory of artifact design, as a middle-range theory, is to assign potential etic meanings to the design attributes of a class of artifacts. Possible etic meanings include processes and constraints that are technological, sociocultural, social psychological, personal behavioral, personal psychological, panhuman depth-psychological, and panhuman physiological in nature. This 'identification' process (Binford 1977) is important. The potential meanings that are assigned to attributes can be used as a basis for selecting for analyses a subset of attributes that are most likely to be relevant to the kinds of processes, constraints, and/or social units that are of interest. The probability of analytic concordance between the phenomenon of interest, selected data, and theoretical goals is thereby increased (Carr 1995b:246).

The breakthrough in Carr's approach is the recognition that the specifics of the physicality of a given material culture medium, say ceramics or lithics, dictates which observable attributes on artifacts can evidence specific social processes, be they active (and emic) signaling of group affiliation, active signaling of assertive style, or the etically-visible unconscious drift of isochrestic variation. Depending upon which question the archaeologist wishes to investigate, Carr's middle-range theory allows her/him to select which artifact attributes to study in order to gauge the effect of specific social processes, based not upon a statement of belief but on independent, *a priori* criteria for which attributes should evidence these processes based on their physical relation to the design process. Given the specificity required of culture historical reconstructions, and its absence in the methodology of previous approaches to the subject (G. A. Clark 1994), Carr's approach is of the highest utility for the present research.

To assign specific social processes to individual artifact attributes, Carr draws on ethnographic data in order to rank attributes by their relative position on three conceptual hierarchies: the decision sequence hierarchy, the production sequence hierarchy, and the visibility hierarchy. The decision sequence hierarchy is the order with which you must decide the attributes of the artifact you want to make, before any manufacture takes place. Earlier decisions limit the choices available for later decisions in the hierarchy. The production sequence hierarchy is the relative order of attributes dictated by the sequence in which you must create the physical attributes of the final product. The production sequence hierarchy is thus a sequence study approach as exemplified by Bleed (2001). The order of when an attribute-state must be decided on need not correspond positively with the order in which the attribute-state must be created during production. Demonstrating the subtlety of his approach, Carr provides examples of the relationship between the production and decision hierarchies for numerous material media, from ceramics to textiles to lithics, and how the physicality of the manufacturing process may establish positive or inverse relationships between these two hierarchies of attributes.

The third and most important hierarchy for Carr is the visibility hierarchy, the ranking of attributes by how visible they are both physically at different distances and relatively in different social contexts. In terms of the latter, the cultural restrictions on when, where, and with whom an artifact is made or used provides the level of its "contextual visibility." The ritual limitations on the use of Hopi kachina images is a clear example (Secakuku 1995). Another example is Dietler and Herbich's (1998:255-256) ethnoarchaeological study of the differential patterning of social interaction (including visibility) around artifact production versus artifact consumption. Drawing significantly from Wobst (1977), Wiessner (1983), Friedrick (1970), and Sackett (1990), the concept of contextual visibility of an attribute plays the dominant role in identifying what social processes could affect the variation in that attribute. The final identification of social processes working on a given attribute, however, is based on the visibility hierarchy as adjusted by each attribute's position on the production sequence hierarchy (which provides a temporal context to its visibility during manufacture) and its position in the decision sequence hierarchy. From this process of considering an attribute as situated in these three hierarchies, Carr's middle-range theory predicts what social processes can affect an attribute's variability. As an example, the choice of temper in a pot, whether it is shell or sand, can vary isochrestically with changes in the social environment in which potters learn their trade but cannot readily be used by potters for the broadcasting of active style in the same way as the shape of the pot, the latter being visible from a great distance and throughout the use-life of the artifact. Thus, the more visible the attribute on the final artifact, the larger the list of possible social processes which could work on its variability. Similarly, the earlier an attribute is produced during the manufacturing sequence, the more likely that attribute will be obscured by later manufacturing stages, if not removed altogether in the case of subtractive technologies, such as lithic reduction or wood carving. In such cases, the earlier an attribute is produced, the fewer the social processes which could act on that variability.

Carr's middle-range approach to linking social processes to artifact attributes has been applied to prehistoric ceramics (Lyons 2003), lithics (J. J. Clark 2001), basketry (Pryor and Carr 1995), domestic architecture (J. J. Clark 2001), and textiles (Carr and Maslowski 1995), from California to the Ohio Valley (see also LeMoine 2002). These applications of the theory have focused on varying social processes, from the action of individual artists' creativity vs. kin-based social learning (Pryor and Carr 1995), to social conformity at the level of regional interaction (Carr and Maslowski 1995). For the present research, the most important application of Carr's theory relates to isochrestic and unconscious enculturation in manufacturing techniques suitable for the archaeological demonstration of the migration of households (J. J. Clark 1991; Lyons 2003).

The most explicit independent evaluation of Carr's approach is J. J. Clark's (2001:14–22) test of the middle-range theory's predictions against a global survey of ethnographic and ethnohistoric studies. Sixty-one cases are used by J. J. Clark as a representative sample of the ethnographic and ethnohistoric present. Each case evaluates the success of particular material media in distinguishing neighboring groups (variously defined) when the distribution of the neighboring groups is known through non-artifactual evidence. As such, each case is an independent test of the theory's predictions for which material media and attributes are suitable for culture historical research, particularly relating to population movement or site-unit intrusions. Clark codes each example as "supportive" of the material medium's use for identifying a group, "inconclusive", or a "cautionary tale" against the use of that material medium. The results substantiate Carr's approach, with low visibility attributes (predicted to be subject to isochrestic, passive variation only) succeeding in at least 80% of the cases and high visibility attributes (predicted to be subject to passive and active style) failing as often as 60% of the time to map onto the material expression of passive enculturation needed to distinguish population groups. In concluding the results of the survey, Clark argues, "Unconscious and deeply embedded aspects of enculturation are often more stable through time than displayed identities such as ethnicity and more resistant to assimilation in mixed cultural settings generated by migration" (J. J. Clark 2001:10).

Importantly, Clark's units of analysis in the cross-cultural study consist of attributes combined with their position in the production sequence, not their physical visibility alone. Thus, ethnohistoric examples of attributes specific to non-decorative steps within ceramic manufacture were treated separately from examples of ceramic decorative techniques and designs visible on the final form. And significantly for the present research, flaked stone tool manufacture was treated separately from projectile point morphology. While the two case studies treating stone tool manufacture in the ethnographic present (Klingelhofer 1987; Wilkie 1996) support Carr's predictions, the studies treating projectile point morphology were evenly divided between supportive (Wiessner 1983; DeBoer 1990) and cautionary tales (Larick 1987; Lyons 1987). While the few lithic examples in the ethnographic and ethnohistoric present are not enough to substantiate the methodology on these examples alone, Clark's survey of examples of 16 media from Africa, the Americas, Europe, Asia, and the Pacific is substantial support for Carr's middle-range theory across different material media and classes of artifact types. The articulation of Carr's middle-range theory into the present theoretical approach to Pleistocene culture history is explained in section 3.2 of this chapter.

3.1.9 Approach II: Chaîne Opératoire and the French Technologie School

The inclusion of a production sequence hierarchy in Carr's middle-range theory finds a direct correlate in the *chaîne opératoire* concept employed by the *Technologie* School of French ethnography. This theoretical and methodological approach to understanding the manifestation of social behavior in material culture represents the second approach to a unified understanding of artifact variability directly related to the pursuit of culture historical research.

The current *Technologie* school in France, and now spreading throughout Western Europe, has its roots in the 1930s in the lectures of Marcel Mauss, notably his *Les techniques du corps* (1936), and in the 1940s-1960s work of the ethnologist and prehistorian, André Leroi-Gourhan (1943, 1945, 1964). Now represented by Pierre Lemonnier and Robert Cresswell's team of ethnologists at the Centre National de la Recherche Scientifique (CNRS) in Paris (entitled "Techniques et Culture"), this consortium of researchers utilizes a synthesis of Mauss' conception of body technique with Leroi-Gourhan's conception of the body technique applied to the objects of human action. For Mauss, the basic unit of human behavior, technique, is "an action which is effective and traditional" (Mauss 1936 [1979: 104, emphasis original]), meaning by tradition that which is learned as an established aspect of a culture. By "actes efficaces," Mauss stressed that a technique must have a role, i.e. be effective, in the social as well as the physical world. In his early theoretical works, *L'Homme et la Matière* (1943) and *Milieu et Techniques* (1945), Leroi-Gourhan stretched the concept of technique further to include objects created by *techniques du corps*, thus facilitating the modern perspective of the technological school of France.

In those two works, Leroi-Gourhan not only set up the theoretical basis for their subsequent investigations but also established their methodology. While Mauss' technique consisted of energy utilized through the action of the body, Leroi-Gourhan turned the body into a tool through which energy is applied to the physical world, thereby incorporating both material culture with the body as the tool or means of action as well as bringing the product of the action itself, physical or otherwise, within the domain of study. Only by conceiving of *le geste* or gesture as the articulation and unifying principle between the means and the energy of action was Leroi-Gourhan able to bring material culture under Mauss' conception of technique (Schlanger 1990). As a result, for the technological school, no tool is complete without the gesture used to put the tool into action. Thus, the school "embraces all aspects of the process of action upon matter, whether it is scratching one's nose, planting sweet potatoes, or making jumbo jets" (Lemonnier 1992:1).

Gesture also gives the ethnographers and archaeologists of the school a methodology for studying techniques as behaviors. This methodology, made concrete by the work of Haudricourt (1987) and later Pétrequin and Pétrequin (1994), begins with documenting the matter, energy, object or means of work, and the physical motion or gesture uniting all of these aspects in the production of one operation, for instance using a plow to till a field. By grouping sequential gestures, the ethnographer is able to document the *chaîne opératoire* or operational chain or sequence through which a technique is begun, implemented, and finished (e.g., the procurement of a plow and horse, the hitching of the horse to the plow, the operation of the plow, the planting, etc.). For the *Technologie* school, the *chaîne opératoire* is thus the methodological unit of data for documenting a technique and studying a culture.

The final theoretical element by which the *actes efficaces* are recognized as *actes traditonnels efficaces* is knowledge. To execute the production of a technique, whether a dance, the construction of a rocking chair, or the shaping and exploitation of a Upper Paleolithic blade core, a mental image of the technical process is required (Pelegrin 1990b). This type of knowledge is referred to as general knowledge or *connaissance*. The specific knowledge of how physically to execute a technique (knowing how to raise one's feet in the appropriate manner in a dance), referred to as know-how or *savoir-faire*, is distinct from the *connaissance* or *schéma opératoire* of a technique (knowing when to raise one's feet in a dance). *Savoir-faire* consists of practical knowledge which is acquired through observation, repetitive practice, or teaching within the social zone of frequent interactions, i.e. the social group in question.

The combination of *connaissance* and *savoir-faire* for all of the *chaînes opératoires* for all of the techniques of each member of a society constitutes the physical manifestation of the society's cultural information, the material expression of the society's thoughts (Lemonnier 1992: 82-85). Leroi-Gourhan calls this the *milieu technique*, while Lemonnier describes it as "the social representations of technologies" (Lemonnier 1992:79-103). Analytically, the *milieu technique* is approached through each technology by documenting the social context of the choices made by the producer/user among different options at each step within the operational sequence of the technology, whether it is the object's manufacturing process or its method of use. At each step or node in the process, the choice of a specific option is seen by the researcher as a potential window into the social reasoning of the artisan. As such, these social representations of technology, preserved in the materiality of that technology, comprise the shared, learned behavior so dearly sought by archaeologists trying to isolate meaningful population groups in space and time.

There are many similarities between the *chaîne opératoire* approach and the various concepts of style theory. In particular, Lemonnier (1989; 1992) shares with Glassie (1975), Deetz (1977), Lechtman (1977), and Hodder (1982; 1990) the search for a society's dominant social reasoning underlying the selection of technical options at many steps or nodes in the production sequences of different technologies. As such, the *chaîne opératoire* approach can be labeled "deep style" in Sackett's terms (1990:41). Similarly, the embeddedness of style in technology, particularly Sackett's emphasis on the technological choices of the artisan, correlates closely with the socially-meaningful choices of gestures and techniques for the *chaîne opératoire* school.

One difference is quite explicit in the approaches, however. For practitioners such as Lemonnier (1986; 1992:85-103; 1993) and Pétrequin (1993), a choice between two equally valid options in an operational chain is only of interest to the researcher if it can be demonstrated that the individuals involved in the technology know of each potential option. Thus, a choice is only informative of the social realm if the choice is emic, i.e., recognized by the artisan as a choice. For ethnographic research, this restricted view of the technological process is epistemologically valid, as an artisan can be questioned as to whether or not an alternative option at a given step is known within that group. The perspective, however, is not justified in all contexts, as it fails to recognize the historical contingency of technological knowledge, as exemplified by the isochrestic view of style. By ignoring the evolutionary ramifications of the use of one technological possibility versus another, regardless of whether or not it was a conscious choice, the etic understanding of technological processes and artifact variability is curtailed. As such, the failure of some *chaîne opératoire* practitioners to recognize the nested nature of technological style (*sensu* Carr and Neitzel 1995; Sackett 1990:37) limits the utility of the approach when investigating the historical contingency of cultural evolution (Tostevin 2006, 2011).

This problem is clearly evident in the differential success of the approach in ethnographic (Pétrequin and Pétrequin 1994) versus prehistoric contexts (Pétrequin 1993). In prehistoric contexts, neither of the two current methods for arguing for the emic availability of technical choices has been very successful, nor very epistemologically robust. Following the first type of argument, Pétrequin (1993) uses convoluted scenarios to argue that a technical option was emically known to a given group of artisans, despite the absence of its use within that group, because of the group's familiarity with another group which happened to use that technical option. While possible, this type of argument is only necessary if the variation in use of technical options is assumed to be conscious, active style, i.e. the intentional signaling of identity with that technical option. Wiessner (1983), however, has shown in ethnographic contexts that active, emblemic style is often unconscious, unintentional, and etic: San artisans were not conscious of making arrows whose style was diagnostic (emblemic) of their group. Nevertheless, they could select out their own arrowheads from a pool of several groups' arrows. More importantly, they reacted with fear and suspicion to the style of an unfamiliar group's arrows, clearly evidencing the past action of emblemic style in boundary maintenance.

The second form of argument found within the *chaîne opératoire* school for dealing with the perceived necessity to argue for the emic and conscious recognition of technical choices can be exemplified by Meignen (1994). In arguing that the use of blade technology in the Early Middle Paleolithic of the Levant served as a precursor to the use of blade technology in the Early Upper Paleolithic in this region, Meignen posits that all previous technological innovations would be consciously accessible to later artisans. This would make the switch to the technology of a distant past a matter of memory recall rather than the reinvention of the technology from scratch. While the social acceptance and proliferation of an infrequent example of a previous technology into a new dominant technology is possible and as an argument has merit (Renfrew 1978; Hovers 1998), the argument requires the demonstration of continuous enactment of the older technology until the appearance of the later technology (Tostevin 2003a). Without such continuous enactment by a few individuals within the society, technological memory alone would not preserve the rare options between generations. Even under the argument that the technological process itself continuously trained individuals in the enactment of all previous variations in the craft, ethnographic studies of material culture variation make this very unlikely. Artisans of one group are frequently surprised by technological choices made by contemporaries in non-adjacent groups despite the shared historical continuity and use of the technological process (Wiessner 1983). Given the expertise of both groups' artisans, their ignorance of the variation exhibited by their contemporaries in the same material medium fulfilling the same artifact function makes this argument unlikely.

The above discussion of how artisans must both receive cultural knowledge from the past as well as enact and create cultural behavior with each generation highlights another difference between some of the proponents of a *chaîne opératoire* approach and the style theory presented above. Specifically, Sackett's emphasis that isochrestic options "are subject to revision in accordance with changes in patterns of social interaction among the artisans who carry on the traditions" (1986b:630) places his conception of isochrestic variability beyond its predecessors in style theory and in contrast to much of the writings of Lemonnier. Sackett's emphasis removes the isochrestic model from the limitations of an essentialist, typological view of material style in which person to person transmission of manufacturing choices as cultural "norms" remains fixed indefinitely. This was one of the prime causes for the failure of the culture historical approach in American archaeology (Tschauner 1994). In contrast, the constant renegotiation of isochrestic variants within and between generations of artisans makes the isochrestic model more compatible with more recent social theory relying on the tension between the individual and her/his cultural learning, such as Giddens' structuration theory (Giddens 1979, 1984). As opposed to the Durkheimian view of social structure as outside of and imposed on the individual, Gidden's dualistic view of society gives inherited learning (structure) and individual negotiation of this learning through the potential to act (agency or the act of doing) equal roles in creating the social world (1984:162, 1979; Gamble 1999:38-39). In contrast, most *chaîne opératoire* practitioners, including Lemonnier (1992), either neglect to position themselves in relation to culture theory or through their studies demonstrate a strict Durkheimian and/or Structuralist view of social creation. In other words, a top-down approach to society is assumed.

This is in some ways understandable, given the Durkheimian origin of Mauss' original perspective on *actes traditonnels efficaces*. It is nevertheless surprising, however, since Leroi-Gourhan's emphasis on gesture, the tool in action, and the creation of social meaning through the movement of the artifact via the human body shows a clear complementarity in its units of analysis to Giddens as well as Bourdieu's theory of practice and habitus (1977, 1980; see the notes in Dietler and Herbich 1998:261-262). It is thus notable that ethnoarchaeologists such as Dietler and Herbich (1998) and archaeologists such as Hegmon (1998), Stark (1998; Stark et al. 1995, 1998), and Gamble (1999) have been more ready to integrate *chaîne opératoire* method and theory with Bourdieu's theory of practice and habitus than either ethnographers of the *Techniques et Culture* program (Mahias 1993; Gosselain 1998) or Paleolithic archaeologists whose work is otherwise fully situated in *chaîne opératoire* theory (Boëda et al. 1990).

This reliance of the traditional *Technologie* school on emic intentionality in material culture choices reflects a larger division of thought in anthropology and archaeology concerning how social behavior is produced in relation to social boundary distinctions. As noted above, Barth's (1969a) definition of ethnicity views the distinctive features of ethnic group identity (ethnic diacritica) as emically and consciously negotiated by the members of one group in opposition to the diacritica of a neighboring group. Barth's perspective has been adopted, developed, and pursued by both ethnographers (from Cohen 1978 to Comaroff 1987 and Comaroff and Comaroff 1992) and archaeologists (Upton 1996; Emberling 1997; Duff 1998). In contrast to this approach are ethnographers (Bentley 1987) and archaeologists (Shennan 1989b; Jones 1996; 1997; Lightfoot et al. 1998; Burmeister 2000) who argue that material culture variability sensitive to social identity such as ethnicity is instead to be understood best through the unconscious patterns of practice and Bourdieu's habitus concept. As Stone

(2003), J. J. Clark (2001), and C. Carr (1995b) have stressed, however, conscious vs. unconscious use of material culture in social relations depends upon the context of both the materiality of the variability and the form of social behavior. To rephrase Sackett (1990:37), unconscious style based in practice theory often serves as the raw material for the conscious manipulation of material culture variability by individuals for the purposes of emic, ethnic identification. Thus, depending upon whether you are trying to identify migrants through their distinctive, unconscious habitus or enculturating environment (J. J. Clark 2001:10) or trying to understand the contexts in which ethnicity as a social phenomenon was emphasized, suppressed, or ignored (Stone 2003), it is the context of the question and case rather than the "correctness" of the definition of culture which is important.

Despite the limitations of the current practice of some *Technologie* practitioners, the *chaîne opératoire* approach has been instrumental in reconfiguring our understanding of how the production process in different media (i.e., ceramic, lithic, textile, etc.) impacts the variation in final artifact form and how that relates to social behavior at various scales (White 1993a; Stark 1998). Much of the differences noted above are, of course, related more to which questions a researcher wishes to investigate through material culture variability than to methodological problems of the theory itself. In fact, as Gamble admirably notes in his articulation of how Paleolithic archaeology can pursue a bottom-up approach to social formation (1999:38–41):

> By uniting technical and social processes, Leroi-Gourhan proposed a single approach to the study of the Palaeolithic. He achieved this without once discussing culture as a concept. No longer, therefore, do we have to bemoan the lack of social data, as Childe did, because this period seems so impoverished in relevant materials compared to those which followed. Instead, as Leroi-Gourhan [1964] shows, we can fuse these concerns and study the long-term trajectory of social change by studying the evolution of techniques. Moreover, this fusion does away with the distinction between style and function in artifact analysis (White 1993a:xviii). Society does not just reside in the former as rules about identity and ethnicity but as social representations literally embodied in the acts of manufacture, use and discard (Gamble 1999:29).

Thus, despite its initial lack of concern for defining how culture and society are created on a daily and generational basis, the *chaîne opératoire* school provides appropriate units of analysis for exactly such a social approach to Paleolithic archaeology, i.e. behavioral choices at specific points in the operational sequence, regardless of their emic recognition. The late recognition of this perspective has had serious consequences, as White (1993) has noted:

> In American archaeology over the past twenty years, there has been significant concern with the concept of style which, while generally recognized as expressive of social identity, has been viewed in juxtaposition to function. Had these American scholars had access to this notion of operational sequences, this intellectual dead end might have been avoided (White 1993:xviii).

3.1.10 Approach III: Behavioral Archaeology and Behavioral Chain Analysis

Despite White's quote above, the concept of structuring the analysis of technology and artifacts from the point of view of artisans' choices during steps within manufacture and use sequences existed in American archaeology long before the 1990s popularity of Leroi-Gourhan's work in North America. This earlier antecedent is partly William Henry Holmes's (1894, 1897) demonstration of life history effects on stone tool typologies, as Shott (2003) has usefully argued, although the reduction sequence approach as developed in North America (Roberts 1935; Collins 1975; Bradley 1975; Callahan 1979; Flenniken 1985; Johnson andMorrow 1987; Amick and Mauldin 1989) has alternatively been argued (Tostevin 2006) to be analogous to the *chaîne opératoire* approach mostly in terms of the low-level theory common to all sequence studies (*sensu* Bleed 2001; Niebel 1988) rather than analogous in all middle-level theory methods or high-level theory goals. The more holistic antecedent is Michael Schiffer's behavioral chain analysis (Schiffer 1976:49–53; McGuire and Schiffer 1983). Developed as part of the analytical program entitled Behavioral Archaeology, behavioral chain analysis differs from the *Technologie* school in its strong emphasis on 'technological' and 'utilitarian' factors in determining artifact variation, rather than the latter's emphasis on the social context of technical choices, an observation which partly justifies White's comment above. Begun as an offshoot of the Processual Agenda, Behavioral Archaeology represents an attempt to unify the discipline of archaeology around "the study of the relationships between human behavior and material culture in all times and places" (Reid et al. 1975:864). Supported by a number of theorists over the years (i.e., J. J. Reid, M. Schiffer, W. Rathje, J. Skibo, and R. McGuire), the Behavioral Archaeology program adopts a framework which "is unabashedly behavioral, grounded in the concrete interactions that take place in the activities constituting the life histories of artifacts and people (e.g., Schiffer 1992, 1995a; Skibo et al. 1995)" (Schiffer 1997:28). As a consequence, its basic unit of analysis is the behavioral chain, the sequence of technical choices made and executed by the artisan in the context of either the manufacture or use of a technology. This is superficially similar to the *chaîne opératoire* concept in its observations and low-level theory (*sensu* Thomas 1998) describing artifacts as process, to the extent that some roughly equate one with the other (G. A. Clark 2005:381). Yet unlike the *chaîne opératoire* approach, the behavioral approach possesses an explicit middle-range theory designed to connect variation in behavioral chains with the consequences of the performance of each technical choice in the process:

> Each element in a given activity—whether person, artifact, or animal—has a set of interaction-specific capabilities known as performance characteristics (Braun 1983; O'Brien et al. 1994; Schiffer 1995b[1995]; Schiffer and Skibo 1987)... Thus, on the basis of feedback from performance, the artisan learns the consequences of specific technical choices. This brings us to the pivotal principle of our framework: any activity on an artifact's behavioral chain can, through feedback to the artisan about performance, lead to changes in the nature and sequence of technical choices. This principle can also be phrased in selectionist terms: behavioral chain activities are the 'immediate selective context' for technical choices (Schiffer 1996) (Schiffer and Skibo 1997:30–32, emphasis original).

This agrees in large part with Sackett's stress (1990:38) that technical choices are the contextual locus for human selection of options based on differential efficacy. Similarly, Carr also argues that technical choices in sequential activities "need not have the same 'selective advantage' with respect to a given milieu. Differences in the selective advantages of attribute states, or in changes in milieu that affect their selective advantages, are two basic causes of stylistic change (Hill 1985; Braun, Chapter 5)" (Carr 1995a:166).

The Behavioral Archaeology approach, however, uses the default assumption that all technical options possess slightly different performance characteristics and therefore "there are few true 'functionally equivalent' artifact types (Schiffer 1979)" (Schiffer 1996:644). This viewpoint appears at first glance to be in stark contrast to Sackett's isochrestic concept and Carr's arguments against the style-function dichotomy. Yet this contrast really depends on the scale at which differential performance characteristics are defined. Both Schiffer and Carr define differential efficacy at the level of the attribute chosen during one step within the behavioral chain, while Sackett's focus for differential efficacy is on the whole artifact, even though his argument for the source of artifact variability is at the level of the details of the technological process. As two beetles may appear different at higher magnification but exactly alike at low magnification, the different resolutions of the analyses of performance characteristics must be kept in mind if these viewpoints are to be understood. Thus, although the above quote can mistakenly be seen as reifying the style-function dichotomy, Schiffer and Skibo's (1997:28) recent reformulation of their approach to explaining artifact variability in fact argues that problematic analytical categories such as style vs. function can be discarded if one uses the Behavioral Archaeology program they espouse.

That being said, there is still a difference between the *chaîne opératoire* and the behavioral approach in what forces are perceived by the artisan as "performance characteristics." While theoretically part of Schiffer and Skibo's model, the performance of an attribute as an element of active style, working in the social world beyond its action on physical matter, is not the default assumption of Schiffer and Skibo's approach, whereas it is for Carr's middle-range theory. For Schiffer and Skibo, performance characteristics are more often related to the satisfactory fulfillment of utilitarian rather than social ends. While this is somewhat an overgeneralization, this puts Schiffer and Skibo's comment on Carr's middle-range theory in context: "Reaching a Herculean 108 pages of fine print, Carr's theory, despite having much merit, is in the end almost entirely about 'stylistic' phenomena, incapable of accommodating technological and utilitarian factors – as Carr (1995:229–230) himself comes close to admitting. On the other extreme is the authors' earlier formulation (Schiffer and Skibo 1987) that, although claiming complete generality, dwells on 'technological' and 'utilitarian' causes to the near exclusion of all others" (Schiffer 1997:28).

In this straightforward admission of the default assumptions of the two approaches, it is thus possible to see how they can be combined. Schiffer and Skibo's (1997) argument for the centrality of the theoretical relationships between elements' performances in activities and the artisan's subsequent choice of elements can be used as a model for conceptualizing when differential efficacy becomes visible for evaluation and selection by the artisan during the processes of manufacture and use. Carr's and Sackett's approaches do not attempt to model this important feature of the technological process since the variation to be explained is initially selected based on a rough equivalence in efficacy. The *chaîne opératoire* approach does, however, attempt to gauge how much an attribute or technical option can vary away from the morphology or gesture absolutely necessary for its stated function. While this is merely the attribution of the feature or gesture to a point on the sliding scale (*tendence*) between completely functionally determined (*premiers degrés du fait*) and completely independent of function (*derniers degrés du fait*) (Leroi-Gourhan 1943; 1945), it is a point to which Schiffer and Skibo's approach can forge a link to Lemonnier (1992; 1993) and others. Thus, the Behavioral Archaeology approach can be seen as theoretically complementary to Carr and Lemonnier at the point where the artisan chooses a given technical option in the behavioral chain or production sequence, which is also the point where the feedback on performance characteristics may influence this choice.

The fact that this complementarity has not been emphasized more strongly, despite Schiffer and Skibo's (1997:28) attempt to draw such connections, is somewhat a result of the structure of the research undertaken by behavioral archaeologists. Theoretically, whether one relies upon the default emphasis on utilitarian versus socially-mediated performance characteristics does depend, for Schiffer and Skibo as well as Carr and Lemonnier, on the contextual details of the specific technology and its articulation with the social and physical requirements of the society in question. The practicalities of the research endeavor, however, has a significant effect on this emphasis. The fact that experimental, laboratory-based studies of performance characteristics in archaeology focus on differential efficacy in physical rather than social interactions is understandable, if not necessarily justifiable, given the variables that are controllable in laboratory settings. It is easier to quantify the temperatures at which pots with different tempers crack than it is to measure the social impact on the artisan of selecting one temper over another, as Mahias (1993) has demonstrated through ethnographic research on pottery production techniques among different ethnic groups in India. Yet other laboratory-based disciplines, such as engineering, do not fix this bias into the research program. For instance, sequence studies of manufacture and use processes analogous to behavioral chain and *chaîne opératoire* analyses have been pursued in engineering since Frederick W. Taylor first started time and motion studies in 1881 (Niebel 1988). Yet these motion studies in engineering not only analyze physical aspects of body gestures in terms of maximums and minimums for performance (in units of time, energy, and material) but also evaluate the social effect of such performance on the artisan in the context in which s/he works. While there is thus more work to be done in Behavioral Archaeology to incorporate social filtering of artisans' perceptions of performance characteristics and technical choices, this is not the only branch of archaeology which biases utilitarian issues over social behavior in understanding artifact variability due to the inheritance of the Processual Agenda, as the discussion of the following approach demonstrates.

3.1.11 Approach IV: The Organization of Technology

The Organization of Technology approach to understanding artifact variability through time and space views technology as a strategy for fulfilling the functional needs of the society (Binford 1973; 1977; 1980). "Strategies are viewed as problem solving processes that are responsive to conditions created by the interplay between humans and their environment" (Nelson 1991:58). Begun through the ethnoarchaeological and theoretical work of Binford, this off-shoot of the Processual Agenda is one of the most productive in terms of its literature output. This is particularly true for the study of the effect of different mobility strategies on the organization of lithic technology (Kelly 1992), from the point of

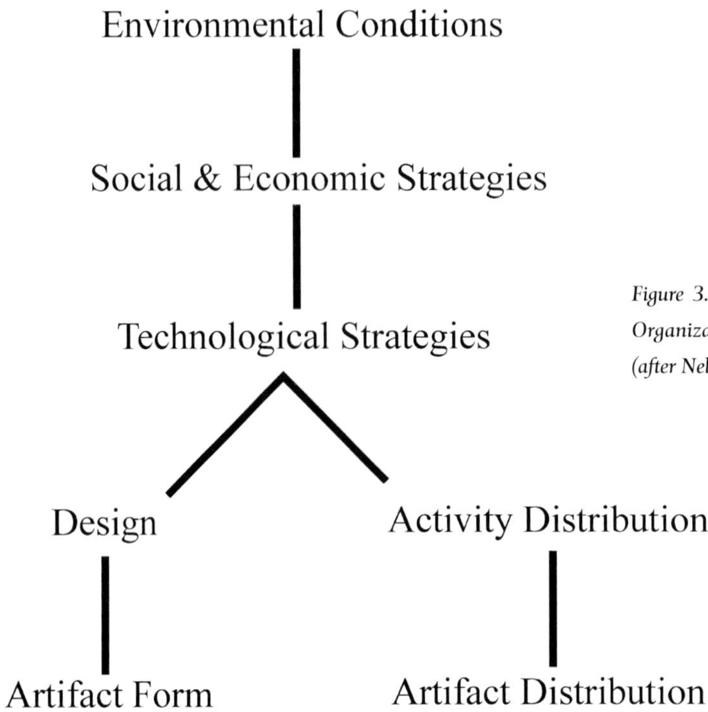

Figure 3.3 Levels of analysis within the Organization of Technology approach (after Nelson 1991:59, fig. 2.1).

view of expedient vs. formal core technology (Parry and Kelly 1987; Kelly 1988), raw material availability (Hayden 1989; Andrefsky 1994), and curation patterns (Binford 1978; 1979; Gramly 1980; Bamforth 1986; Shott 1986; 1989; Kuhn 1992; 1994; 1996; Morrow 1996). The diversity of hypotheses within this approach has been summarized and contextualized graphically by Nelson (1991:59, fig. 2.1), redrawn here as Figure 3.3. The relationships between these levels of analysis are the most significant contributions of this approach to explaining artifact variability. Although the Organization of Technology approach tends not to examine choices within technological sequences or the artisan's role in decision-making, the approach focuses on where and when specific activities within technological sequences are enacted on the landscape relative to the distribution of resources. In this, the Organization of Technology approach fills several gaps in our understanding of artifact production not explicitly addressed by the previous three approaches. First, the structure of the approach situates the design process as part of a technological strategy to fulfill social and economic needs which themselves are affected by environmental conditions. Second, how the design process is patterned on the landscape as artifacts produced by spatially-specific activities is a complexity little dealt with by Carr's theory of artifact design, the *Technologie* school, or Behavioral Archaeology. It is this latter innovation of the organization of technology on the landscape that will form the theoretical glue behind the middle-range theory developed in the second section of this chapter.

The ability of the Organization of Technology approach to test many of its hypotheses against archaeological data is partly due to its direct use of optimal foraging theory and its parent, evolutionary ecology (Krebs 1978; Smith 1983; Winterhalder and Smith 1981; Foley 1985; Winterhalder 1986; 1997; O'Connell 1995; Kelly 1995; Shennan 2002; for a comprehensive overview, see Krebs and Davies 1997). This body of theory,

> analyzes phenotypic variation in terms of "adaptive strategies" – that is, as a series of fitness-enhancing behavioral responses to different environmental states (assuming that these states have been recurrent within the evolutionary history of the organism's lineage and that the responses fall within its norm of reaction [Lewontin 1974:404; Stearns 1992:61-65]). This form of phenotypic response is thus construed to be based on a set of evolved "decision rules" (Krebs 1978)–genetically evolved cognitive mechanisms that guide development, learning, problem solving, and stimulus response (Boone and Smith 1998:S144).

Evolutionary ecology thus provides a conceptual tool for archaeologists to build expectations for how artisans would make adaptive choices in their creation of particular artifact forms for specific uses of a landscape's resources, whether stone as raw material for cutting edges or the seasonal availability of flora and fauna for subsistence. By emphasizing specific currencies to be optimized, maximized, or minimized during the acquisition and use of resources on the landscape, it is possible to propose patterns of mobility and artifact variation which can be tested against the archaeological record.

While the assumption that the design of artifacts is always subject to optimization, maximization, and minimization pressures is at times problematic (Torrence 1989a:2; Nelson 1991:61), there are principles which have been isolated which predict when such pressures would be most active during the use of lithic technology as a strategy for gaining access to landscape resources. These principles primarily relate to contexts in which technological performance is argued to have some effect on the reproductive success of the artisans, and thus engage the adaptive decision-making strategies which form the core of the optimal foraging approach. These principles focus heavily on hunting tools and subsistence practices. For instance, Bleed (1986) advances the principle of reliability, defined as the minimization of the amount of maintenance and repair work necessary during the time of use of the hunting tool. Reliability is particularly important in contexts in which the time of encounter with the resource is brief and cannot be interrupted by maintenance activities. Bleed points out that these contexts would favor technological organizations scheduled to resource availability. Blade core technology can be seen as following this principle, in that the significant core preparation (up to two hours of preparation before first blade removal, Madsen 2003), manufacturing time, and training (J. E. Clark 1987; Parry and Kelly 1987) precede the hunt while minimizing repair and replacement time during the hunt by means of the standardized products easily inserted into existing hafts (Nelson 1991).

Other principles of optimality in design include maintainability (Bleed 1986), the suitability of the tool for a variety of situations. This principle has been subdivided by Nelson (1991; see also Shott 1986) into flexibility, in which the form can be easily changed to suit a new task, and versatility, in which the form is already suitable for a set of different tasks. Transportability, often conceived in terms of potential cutting-edge utility versus tool weight (Kuhn 1994; 1996; Morrow 1996), has also been offered as an important principle for optimization due to the need to carry tools to specific points on the landscape (Binford 1979; Shott 1986). Likewise, conservation of raw material under differential conditions of raw material availability and quality also represents a significant argument for the fitting of artifact design to technological strategies to achieve social and economic ends (Bamforth 1986; Andrefsky 1994). While this discussion only begins to scratch the surface of the optimality principles offered to date (see the references cited within P. Carr's 1994 edited volume), it is important to point out that many of these principles can in fact represent concerns artisans would have considered "performance characteristics" (*sensu* Schiffer and Skibo 1997) of specific steps within a behavioral chain. This observation is only one of many points of articulation between the Behavioral Archaeology and Organization of Technology approaches.

While the Organization of Technology has been productive at the theoretical level, disagreement on which currency is optimized in a given technological or landscape context has slowed the creation of applicable middle-range theory from the first principles of optimal foraging theory. Energy in terms of kilocalories spent during production and/or use has been proposed as one currency (Camilli 1989, Morrow and Jefferies 1989, Jeske 1989, and Lurie 1989) while time spent during production and/or use has been offered as a competing currency (Foley 1985:229; Torrence 1989a; 1989b; Myers 1989; Boydston 1989). Similarly, raw material spent during production has also been suggested as highly important (Hayden 1989). Yet all of the proposed currencies lack a direct connection between artisan choice, stone tool form, and reproductive success, as Jochim points out:

> Ostensibly, evolutionary ecology, grounded in natural selection, is the theory underlying the work in much of this volume [Torrence 1989a]. If so, the role or weight of each factor affecting lithic technology should be determined by its relevance to reproductive success. Currently, however, we have not identified the selective penalties with the use of poor, local raw material versus higher quality, more expensive exotic stone, or with the carrying of a kilogram of stone rather than 500 grams... Rather, it is possible to argue, from first principles alone, for various, sometimes conflicting responses to the same situation. For example, an environmental deterioration causing decreasing overall resource abundance might be expected to encourage a hunter-gatherer group to respond by broadening their diet, increasing their mobility, or both (Jochim 1989:107–108, emphasis added).

Thus, from the extant theory alone, there is equifinality in the organization of technology as an adaptation to environmental conditions. The specificity of the context, as Jochim points out, is the only way to understand why a particular group adapts with one organization of technology while another group adapts differently to the same environmental change. This context is, in terms of Nelson's graphical representation of the Organization of Technology approach (Figure 3.3), the social and economic strategies of the society. Due to many factors, from its origins in the Processual Agenda to the perceived limitations of a lithic dataset, this is the level of analysis which has received the least amount of effort from this perspective. Yet it is the socially-mediated level of the organization of technology which must be explored before the approach can make accurate predictions of the archaeological record from first principles. Those concepts which do utilize the social side of technology, such as risk minimization (Wiessner 1982) and time spent in technological investment vs. subsistence handling time (Torrence 1989b; Bright et al. 2002), are in fact currently the most promising. As a consequence, it is apparent from the Organization of Technology literature that the theoretical equifinality in technological organization supports, rather than refutes as is often assumed (G. A. Clark 1994), Sackett's isochrestic view of technological variability through time. Adaptive responses can produce equifinal, isochrestic results, just as stylistic drift can (*sensu* Binford 1965:199–203).

3.1.12 *Causation in Artifact Design*

It is possible to rank the four approaches discussed above based on their default assumptions for how artifact variability relates to the artisan's concern for efficacy vs. her/his social-filtering of feedback on performance characteristics. Carr's approach views artisan choice as embedded within technological constraints (Figure 3.1) although the middle-range theory for assigning social processes to attributes depends upon these constraints first being fully understood. Given the fact that the specific function of the vast majority of artifacts in the Pleistocene archaeological record is unknown, this may not always be possible. For the *Technologie* approach, each point of choice in the technological sequence can be differently sensitive to functional requirements, a fact reflected in the relative number of available alternative choices. The Behavioral approach tends to document why artisans might choose among options based on physical rather than social performance characteristics, yet this is likely a bias of the laboratory rather than a theoretical necessity of the approach. The Organization of Technology approach is the most extreme of the four in its assumption that most technological decisions are shaped by utilitarian function, whether through artisan choice of 'optimal' options as perceived emically or differential reproductive success of artisans based on whether or not their choices are 'optimal' strategies for achieving Darwinian fitness, i.e. optimal from the etic point of view of the archaeologist. The repercussions of the strong default assumption of the Organization of Technology approach deserve further discussion.

The Organization of Technology approach as applied to the study of lithic technology is potentially very powerful because of its reliance on evolutionary theory to provide a causal mechanism for both the maintenance of stable strategies through time and the adoption of alternative strategies in other contexts. But the potential of evolutionary theory as applied to material culture in general has been hindered by the default assumption that the hypothetical relationship between one variable or optimized currency and the presumed reproductive success of individuals utilizing that variable alone explains the change in the whole organization of technology of a population through time. An examination of the application of optimal foraging theory to the behavior of living foragers should serve as a cautionary tale for assuming that an etically-recognized difference in the efficacy of a solitary variable suffices as a preferred explanation for technological change.

The Problem with Utilitarian Function as a Default Assumption for Artifact Form

Stanford's ethological studies of chimpanzees in Gombe, Tanzania (Stanford et al. 1994; Stanford 2001; 1999), have demonstrated that these chimpanzees consume less than 5% of their annual diet in the form of mammalian meat but spend substantial amounts of time, in particular instances weeks at a time, hunting red colobus monkeys and other prey. Because ethological studies of living societies allows multiple consequences of behavior to be viewed and integrated into understanding why a particular behavior is enacted by individuals, Stanford has been able to model why male chimpanzees spend so much time hunting (1998:189–209). Successful hunters rise in the male dominance hierarchy as a result of their use of hunting kills as coalition building events (Nishida et al. 1992). Through time, a successful hunter may replace otherwise larger and more dominant males as alpha male and thus father most of the group's progeny during his tenure at the top of the hierarchy. Further, meat from successful kills is frequently traded for copulations with females in estrus, to the extent that the number of swollen females in the chimpanzee party is the strongest predictor of when a hunt will occur, even holding all other variables constant (Stanford et al. 1994; see, however, Mitani and Watts 1999, Gilby 2006, Gilby et al. 2006, and Gilby et al. 2008 for conclusions from other study sites). If one looks only at the energetics of chimpanzee hunting, however, Stanford (1998:196) shows that no male chimpanzee at Gombe should ever hunt given the principles of optimal foraging theory. The risk to life and limb from prey counter-attacks as well as the energetic cost of the hunt itself are not compensated by the dietary contribution from hunting. Risk-free dietary alternatives exist year-round in the form of insects (McGrew 1992), fruit, and other vegetation, including oil palm and figs which provide eight times the calories per gram and more fat than monkey meat (Stanford 2001:127). These are in fact the constituents of most female chimpanzees' diets. This example thus shows how relying upon only one variable, in this case energetics in the ecology of hunting, incorrectly predicts what behavioral choices should be made by male chimps, i.e., to hunt or not to hunt, in order to optimize reproductive success (Rose 1999). Analogously, this should warn archaeologists that there is a chance of gross error in relying upon only one variable, whether it is tool transportability or another currency of tool energetics, as the performance characteristic considered by an artisan in choosing one technical option over another in the production of an artifact or the organization of lithic technology on a landscape. Even if the variable points to a logical utilitarian difference in the efficacy of an artifact's design, this assumption of the primacy of energetics could easily be a misleading proxy for reproductive success, as with the energetics of chimpanzee hunting.

We need not restrict the cautionary tale to non-human foragers. In the case of the Aché hunter-gatherers of Paraguay (Hill and Hawkes 1983), twenty years of field data have been gathered on this group's hunting ecology, food sharing, and strategies of reciprocal altruism. These data indicate that the social currency of reciprocity involving food and non-food items best explains the food sharing patterns resulting from successful hunts, not the straightforward attempt by hunters to maximize their intake of meat through improvements in their technological skill (Kaplan and Hill 1985). Yet male hunters are more likely to endeavor to be better hunters, not because they will earn the reward of more protein from their higher number of kills, but because they, their offspring, and relatives will be treated preferentially by other group members due to the hunter's generosity in sharing meat (Hawkes 1991). Thus Aché hunting, including the technological decisions made to improve hunting success rates, is a means of achieving the social strategies by which kin selection and reciprocal altruism are optimized, not necessarily the maximization of hunting returns for the purposes of dietary energetics alone.

Both Aché and chimpanzee hunting are adaptive due to the social consequences of hunting success rather than the dietary, utilitarian, or "functional" consequences of the meat procurement alone. In fact, both human and chimpanzee hunters utilize group-specific social filtering of purely functional concerns in making their adaptive choices related to hunting techniques and targeted prey (Stanford 2001:132-133; see also Boesche 1994). In Binford's terms (1965:206), the secondary function of the behaviors are thus the causal adaptive force, not the primary utilitarian function as is almost ubiquitously assumed in the application of optimal foraging theory to the organization of lithic technology. This shows on one hand that improvements to hunting success are indeed indirectly associated with higher reproductive success, the ultimate currency of optimal foraging theory and processual archaeology. On the other hand, however, these examples demonstrate that the social consequences of the hunter's success on other group members is what is directly associated with improving that hunter's reproductive success. Thus, to understand something as basic as how hunting success affects reproductive success, one must model the "social strategies" intervening between "environmental conditions" and "technological strategies" within Nelson's graphic representation of the Organization of Technology approach (Figure 3.3). This is exactly the aspect of the approach which has received the least amount of research effort because archaeologists trained since the Processual Agenda have used the default assumption, based on Binford (1965) and earlier Hawkes' ladder of inference (1954), that understanding utilitarian function must precede understanding social functions of material culture. McGuire and Hildebrandt's (2005) recent application of costly signaling theory to Middle Archaic Period hunters in the American Great Basin is the notable exception to this generalization, as it draws directly on social proximate mechanisms of hunters' reproductive success, such as Hawkes' (1991; Hawkes and Bird 2002) show-off hypothesis.

Proximate and Ultimate Evolutionary Mechanisms in Technological Change

While the contribution of meat to the diet of the Aché foragers is higher and thus more significant energetically than among Gombe chimpanzees (Stanford 2001), both groups of hunters use meat acquisition as a proximate mechanism for accessing an ultimate mechanism for achieving reproductive success. This distinction between proximate and ultimate mechanisms in evolutionary theory and the ethological study of behavior is a theoretical principle in behavioral biology but it has been long ignored in modern lithic studies. Given the fact that evolutionary theory is highly involved in causal explanations for artifact form, and thus explanations of historical changes in culturally transmitted behaviors related to artifact variation, the proximate vs. ultimate distinction in evolutionary theory deserves some clarification for the present research.

Ernest Mayr (1961, 1993) first advanced the concept of understanding biological causation as the interaction between two sets of mechanisms. Mechanisms related to the structural working of a behavior as both intrinsically performed and extrinsically stimulated are labeled by Mayr as "proximate." Those mechanisms related to how the consequences of the structural mechanisms are associated with achieving reproductive success are labeled by Mayr as "ultimate" (see Daly and Wilson 1983:15). A contemporary example of the distinction helps to understand his point.

Coitus in humans tends to be associated with pregnancy and reproductive success for both the male and female involved. This behavior can be understood, as well as its causal relationship to reproductive success, as the interaction of two mechanisms. Humans are driven, in part, to engage in coitus because of the intense pleasure which results from the high degree of enervation of the organs used in the behavior. The emotional and sensual pleasure associated with the behavior are some of the structural or proximate mechanisms by which individuals decide to engage in the behavior. Until the invention of contraceptives such as condoms, this was sufficient to cause those seeking the proximate mechanism also to achieve, whether they wanted it or not, the pregnancy and birth of offspring which constitute the ultimate mechanism by which reproductive success is incurred and the behavior selected for through evolutionary time. This perspective is useful because it requires the investigator to recognize that the behavior itself is not selected for as a genetic whole. Rather, genetic sequences for the individual proximate mechanisms of the enervation and hormonal feedback systems that provide pleasure to the actors are selected for due to the simple fact that those who have the proximate mechanisms are more likely to produce offspring, the ultimate mechanism. With the application of contraceptives, however, it is possible to disassociate or decouple the proximate and ultimate mechanisms, i.e., it is possible with contraceptives to pursue the proximate goals of emotional and sensual involvement between individuals without the likelihood of conception which provides the selective advantage of the behavior. In considering how the association can be broken, it is thus easy to see in situations without decoupling how selection works on the association between the mechanisms to produce the genetic (or cultural in other contexts) encoding of the proximate mechanisms.

This distinction between proximate and ultimate causation has been seminal in understanding causation in biology and philosophy (Alcock and Sherman 1994; Beatty 1994; Mayr 1994; Ariew 2003). The distinction is also closely mirrored in Tinbergen's list of four equally-valid ethological questions to explain a behavior (1963; Hauser 1997). While the complexities of the interaction between Mayr's and Tinbergen's concepts are more suitably addressed elsewhere, it should be made clear here how the proximate and ultimate distinction relates to understanding causes of technological variation. From the point of view of the relationship between artisan choice and the repercussions of the performance characteristics of the choice on the artisan, artifact forms and the organization of technology implied by the working of lithic tools on the landscape must be understood as proximate mechanisms meant to enable a social strategy to achieve reproductive success. The proximate mechanisms of lithic technology thus include both the mechanics whereby the cutting and piercing edge aids in procuring dietary necessities as well as the social consequences of artifact production and use on the artisan and user's efforts at optimizing direct Darwinian fitness (Daly and Wilson 1983:37-58), indirect Darwinian fitness (Hamilton 1964), and reciprocal altruism (Trivers 1971, 1985:361-394). As the examples of Aché and Gombe hunting illustrate, there are good reasons to avoid privileging the proximate mechanism of energetics over the proximate mechanisms of mating opportunities (Stanford 1998), kin selection (Sherman 1977), and reciprocal altruism (Kaplan and Hill 1985) in explaining a behavior. Thus, there are good reasons to avoid privileging functional energetics as the default assumption for the source of artifact design variation related to lithic technology.

This is not an argument for ignoring energetics or functional considerations in artifact design, only an attempt to highlight how important the social ramifications of behavior which results in stylistic variation can be in evolutionary ecology. For just as contraception decouples the proximate and ultimate mechanisms of human reproduction, it is fairly easy to demonstrate how energetic consequences of artifact design variation can be decoupled from the evolutionary ecology of artifact production. For instance, optimality modeling of lithic tool design frequently argues that the tool kit that minimizes time, effort, or raw material spent during particular aspects of the production, maintenance, and use of the tool kit should be the most selected through time, either by individuals acting "rationally," an anthropologically problematic assumption (Appadurai 1986; Torrence 1989a:2; Nelson 1991:61; Cosmides and Tooby 1992), or by natural selection favoring such artisans (Camilli 1989; Myers 1989; Boydston 1989; Hayden 1989; Rasic and Andrefsky 2001). But what if the tooling time and energy is actually being spent for two purposes, for the creation of the tool for use as well as the excuse for socialization and the investment of social meaning in the artifact for the purpose of demonstrating the worth of the artisan as a mate or exchange partner? Wiessner's ethnoarchaeology among the Kalahari San (1982, 1983, 1984, 2002) in fact demonstrates that time spent making artifacts, including arrowheads and arrows, is time spent conversing, gossiping, and socializing with band members, solidifying social bonds and intentions. At the same time as they are making functionally useful items for the hunt and other activities, they are also creating objects imbued with personal effort (i.e., assertive style) for gift exchange in the hxaro system of reciprocal altruism, a system vitally important for the San's mediation of subsistence risk through the social bonds of reciprocal altruism (Wiessner 1982, 2002). Multitasking and the amalgamation of material culture production and repair with socialization and other activities is well documented in many forager groups (e.g., Laden 1992). Which proximate mechanism for achieving reproductive success is to be privileged in understanding artifact production? Should production time be minimized (the time spent making a cutting edge) or maximized (the time spent demonstrating your quality to your hxaro partner)? The answer is clearly that the different causal proximate mechanisms must be optimized against each other but this is an obvious answer long ignored by the default assumption that artifact utilitarian function is paramount as the only evolutionary question worth asking. The fact that there is far less variation in involvement in the hxaro system among Kalahari bands (Wiessner 1982, 2002) than variation in the effective morphology of arrowheads among the San (Wiessner 1983) demonstrates where the selective force is strongest.

Given the simultaneous effect of these proximate mechanisms in many contexts (including various points in a hunter's life history strategy—Wiessner 2002), it is clear that technological principles that allow the optimization of both energetic and social proximate mechanisms in artifact design are important loci for the study of technological evolution. Bleed's (1986) reliability concept is a case in point, for as a "reliable" tool kit minimizes mid-hunt/forage tool maintenance time and thus minimizes lost subsistence opportunities, it also maximizes the in-camp tool production time which gives artisans the opportunity to converse and invest technological behavior with the social power of reciprocal altruism. Bleed's concept is thus more important than currently recognized as it is the social power of technological behavior to solidify social obligations of reciprocity and pair bonding that contributes, just as much as the food thereby acquired, to the association between tool production and the ultimate mechanisms for optimizing reproductive success.

Emic versus Etic Recognition of Artifact Design Variation

It is significant that Wiessner's informants emically recognized the importance of the social activity of technological behavior more than the importance of potential differences in the effectiveness of different tool forms. When asked about beaded headband or arrow production for hxaro partners, the San clearly expressed why they made gifts that emphasized the quality of the artisan (1982, 1984:204). When presented with arrows made by different linguistic groups, the morphological differences of the penetrating point were surprising to some artisans who were not familiar with the other group. However, "Because the G/wi and !Xo interviewed had contact with one another in their dry season, they could correctly identify each other's arrows. They attributed the stylistic differences to the fact that the arrows were made by different people who did things in a different way. Both the !Xo and the G/wi admitted that the G/wi point was probably a more effective design" (Wiessner 1983:269). It is telling that the !Xo viewed the G/wi design as more effective but did not consider the difference significant enough to adjust their design despite the long contact and emic recognition of the difference. This case is paralleled by the Australian foragers' refusal to adopt the bow and arrow from trading partners in neighboring New Guinea (Diamond 1997:298, 312), despite the argued superiority of this projectile technology compared to other hunting systems (Hughes 1998). The point reminds us that there are many social reasons for not adopting a better artifact morphology from an emically-recognized "other."

This observation points to a frequent assumption of the application of a single-variable approach to understanding artifact design. In many experimental assays of the efficacy differences between lithic technologies, such as cutting edge per volume by bifacial reduction versus bladelet core reduction (e.g., Rasic and Andrefsky 2001), there is rarely any explicit discussion of how the etically-demonstrated difference in the optimizable variable would have been observable to the artisan, not to mention significant enough to influence the artisan's choice between the possible trajectories. This assumption of a universal rationality to technological behavior stands in opposition to a wealth of documentation of the social filtering of such concerns among living humans.

The disciplines of history of science, the sociology of technology, and the anthropology of material culture specialize in studying how individuals, social groups, and societies negotiate the choice of production techniques and artifact forms from among the material culture possibilities. How such choices are made in different social and ecological contexts is informative for understanding how prehistoric culture change might occur. For instance, historians of science note the strong tendency throughout human history to choose certain technological solutions to specific problems not because they are more suitable or efficient but for non-functional, socially mediated reasons (Basalla 1988; Cowan 1985). Sociologists of technology and sociologists of knowledge also illustrate that there is never one human artifact which represents the most efficient tool for a particular function, primarily because different individuals and social groups have different perspectives and relationships to the technology (Pinch and Bijker 1987; Hughes

1987). While the literature produced by historians of science and sociologists of technology has focused more on subjective rather than objective reasons for choosing one technology over another due to their vested interest in explaining the illogical, these examples and those of a more broadly concerned anthropology of material culture (Miller 1985; Pfaffenberger 1992; Mahias 1993; Dobres and Hoffman 1994; Schiffer 1991; Schiffer et al. 1994; Dietler and Herbich 1989, 1998) make it clear that humans do not always choose artifacts based upon objective assessments of their efficacy for a particular function. This is true even in societies which possess the means of ascertaining the relative efficiency of different artifacts and which expressly believe in the rationality of human decision-making. Even modern rocket science (MacKenzie 1987) and profit-oriented capitalistic manufacturing corporations (Cowan 1985; Bijker 1987; Misa 2000) which produce technological solutions to functional problems identify entirely different aspects of the same technology for optimization and improvement. Technologies are even socially constructed (*sensu* Pinch and Bijker 1987) when life-threatening consequences exist for performance failure. Fallow (1985), for example, documents the clearly illogical and socially subjective choices made by the U. S. Army Ordinance Department in the 1960s in producing its newest infantry rifle, the M-16. The socially mediated "objectivity" of the ordinance specialists transformed the successful AR-15 infantry rifle prototype into the M-16, a rifle which failed six times faster than contemporary rifles, producing a death toll in the thousands. The M-16 case illustrates in particular the separation of the effects of artifact performance characteristics on the user from the decision-making processes of the producer.

As members of an industrial society have at times little influence on technological decisions, as in the case of the M-16 rifle, because of the separation between user and producer, one might presume that Pleistocene artisans who both produce and use the artifact would be more likely to adjust the technology based on inadequate performance characteristics. Thus they might privilege utilitarian function and rationality in technological behavior even more than modern industrialists. This argument stands in stark contrast to three issues. First, while Pleistocene artisans are more likely to adjust a technology based on inadequate performance characteristics, this does not mean that they would have ignored other social actors' perspectives on technology. Even chimpanzees make socially-mediated decisions related to technological behavior: chimpanzees only learn and replicate technological behaviors based on their evaluation of the appropriateness of the role model demonstrating the behavior (Boesch and Tomasello 1998). While chimpanzees are affected less than industrial humans by the influence of social groups on technological producers' decisions, Boesch and Tomasello's (1998) conclusions illustrate that performance feedback is categorized and privileged based on social relations in the living species closest behaviorally to the last common ancestor between the human and non-human ape lineages. Thus emic issues of social mediation are relevant, if moderately so, in Pleistocene contexts of technologies in which the user is also the producer.

Second, members of modern industrial societies are frequently immersed in the social ideology of technological progress in which rationality in technological choices is assumed (M. R. Smith 1994; M. L. Smith 1994). The observations of non-rational technological decisions by humans in industrial societies despite this ideological indoctrination should therefore warn archaeologists that prehistoric societies likely did not pursue technological decisions with a rationality unfiltered by emic social perspectives on technology. In fact, there is abundant evidence of a diversity of views of technology as a social prime mover among the world's forager and agrarian societies (Lemonnier 1992, 1993; Mahias 1993; Bulliet 1994; Perdue 1994).

Third, decades of ethnoarchaeological work with foragers and agrarian societies on many continents have shown that makers of traditional material culture do not distinguish in practice between the tripartite theoretical distinctions between style, function, and technology (Hodder 1979; Longacre 1981; Braithwaite 1982; Wiessner 1990; Childes 1991; Mahias 1993; Aronson et al. 1994; Stark 1995; Bulliet 1994; Dietler and Herbich 1989, 1998). In fact, these studies were the first to demonstrate that style is not without social function (Wobst 1977; Lechtman 1977; Wiessner 1983; Schiffer 1999) and analogous variation, i.e. utilitarian function, is not dictated solely by environmental forces. Functional constraints very rarely explain the huge variation in artifact forms or in the methods by which they are made (Lemonnier 1986, 1992). If they did, one would expect the production details of stone tool industries of the Mediterranean ecozones of California to resemble closely those of the Mediterranean ecozones of the Levant, an hypothesis which is clearly negated with even a brief examination of such industries (compare Pletka 2001, Arnold et al. 2001, and Preziosi 2001 with Henry 1995a). This equifinality of solutions to problems of technological organization given the same environment is not a result of insufficient archaeological resolution but due to the indirect nature of the effects of lithic tool form on the ultimate mechanisms for the reproductive success of the artisan.

Evolutionary Psychology and Material Culture

Artisans' social perspectives on how to make technological choices are more central to the application of evolutionary ecology to material culture than is currently recognized in archaeology. Evolutionary ecology is based on the premise that cognitive algorithms have been selected through evolutionary time to provide increased behavioral flexibility to the phenotypic adaptation (Boone and Smith 1998). These psychological mechanisms can emphasize adaptive, interpersonal social engagement through material culture as much as the optimization of constraining energetic variables in the technological process. Wiessner's (1984) discussion of the psychology of the investment of material culture in social relations bears this out. The discipline of evolutionary psychology, utilizing the same theoretical assumptions as evolutionary ecology, has in fact documented how human logic responds more accurately to evaluating social behavior and relationships than to abstract logic in the economic sense (Barkow et al. 1992). Cosmides and Tooby's (1992), building off of Wason (1966), presented informants with identical logical relationships in the form of statements, "If P then Q," and then asked them to identify which pieces of knowledge (P, not-P, Q, and not-Q) need to be discovered to falsify the statements. When the statements were worded in abstract terms, what Boyd and Silk (2000:587) term the "Clerical Problem," informants failed more than half of the time to choose the logically correct answer. When the statements were worded as conditional rules that expressed social contracts with a benefit and a cost, what Boyd and Silk (2000:588–589) term the "Bartender's Problem," 75% of the informants identified the logically correct response. These results were consistent even when familiarity with the types of questions was controlled in the analysis. From these and other experiments, evolutionary psychology has demonstrated that human reasoning is content-dependent, i.e., the subject matter that humans are asked to reason about seems to regulate how they reason (Barkow et al. 1992). Cosmides and Tooby's arguments that human reasoning is more accurate in cases in which there is a social "cheater" to identify follow directly from the same theory as evolutionary ecology. Accordingly, archaeologists must recognize that the cognitive algorithms that have been selected through evolutionary time to provide increased behavioral flexibility to the phenotypic adaptation (e.g., Smith and Winterhalder 1992; Bettinger et al. 1996; Boone and Smith 1998) relate to social behavior such as the material culture investment in social relations as much as to economizing behavior in artifact design. Thus, in the terms of Wiessner's ethnographic data, the stylistic investment in arrows for hxaro exchange is likely to be just as much a product of evolutionary psychological mechanisms as the desire to optimize the energy expenditure and dietary gain from the arrow's cutting edge.

Reinserting Social Function into the Evolutionary Ecology of Lithic Artifact Variation

This argument for reemphasizing the social context of technological activity in the processual understanding of behavioral change is meant to support, rather than to replace or critique, the evolutionary ecology approach in general. In fact, the behavioral ecology approach is powerful for the exact reason that it attempts to understand the evolution of social behavior and not just morphological biology (Daly and Wilson 1983; Boyd and Richerson 1985, 1996). Further it does not assume that every behavior is optimized according to evolutionary predictions; it only hypothesizes what a behavior would look like if it were optimized. The difference between the predicted optimal form and the observed form is in fact what makes the subject both fascinating and testable against the archaeological record (Shennan 2002:3). The reason for the necessity to reinsert the social context of technological behavior into the discussion of Pleistocene hominin evolutionary ecology is the narrowness with which some lithic analysts (e.g., Nelson 1991:61–62; G. A. Clark 1994) have taken the evolutionary ecology literature. The discussion of the organization of lithic technology frequently ignores the importance of the social context of technology, the role of cultural transmission processes in shaping what is available for selection, and the distinction between proximate and ultimate mechanisms in evolutionary theory. By acknowledging the adaptive aspect of the social side of forager technology (e.g., McGuire and Hildebrandt 2005), lithic analysts need not adopt a new perspective, only embrace the wider findings of evolutionary behavioral biology (Daly and Wilson 1983; Krebs and Davis 1997) from which the narrower field of optimal foraging theory was drawn.

3.1.13 Combining the Four Approaches to Explaining Artifact Variability

The previous section of this chapter has detailed why a theoretically-informed study of Pleistocene technological evolution must integrate into its approach both the likelihood of population-specific differences in the rationality of the emic evaluation of artifact performance characteristics as well as the existence of multiple social as well as energetic proximate mechanisms by which material culture contributes to an artisan's ultimate mechanisms for reproductive success. Together, these issues indicate that evolutionary explanations for technological change must avoid a strict style vs. function vs. technology distinction in dealing with the behavior of as highly social an animal as a human. In this, the four approaches to explaining artifact variability just reviewed are almost in unanimous agreement.

It is clear that there are other similarities among the concepts used by the different approaches that have been overlooked in the theoretical turf wars between the approaches. Carr (1995a; 1995b), Lemonnier (1992), and Schiffer and Skibo (1997) all argue against a straight style vs. function dichotomy in artifact design and in its place argue for a contextual evaluation of the physical relationships among individual attributes and/or technical choices within production sequences. Only the Organization of Technology approach implicitly reifies a style vs. function distinction (Nelson 1991:61-62) while explicitly including "social strategies" at the level of analysis between "environmental conditions" and "technological strategies" (Figure 3.3). Yet as the discussion of Mayr's proximate and ultimate mechanisms notes, the strength of the Organization of Technology's evolutionary theory necessitates the inclusion of both purely energetic function and social function in understanding why an artifact benefits its artisan and user.

Similarly, all four approaches treat material culture as a dynamic process in which the artisan chooses particular options at particular steps in the production sequence. Yet each approach offers its own advantage as a perspective on technological activities. The Organization of Technology approach is vital for its strong emphasis on technological activities as geographically specific and so contextualized against the available landscape resources, the mobility of the population, and the effect of environmental change on selective forces related to technological procedures. The Behavioral Archaeology approach is vital for its strong emphasis on the locus of artisan choice for the consideration of feedback from each choice's performance characteristics. The *chaîne opératoire* approach is vital for its emphasis on the embeddedness of tactical learning (*savoir-faire*) and strategic learning (*connaissance*) at the nodes of artisan choice in the production sequence. Finally, Carr's Middle-range Theory of Artifact Design, building off of Sackett, Wobst, and Wiessner, is unique for its predictions on what choices within these sequences would witness the effects of specific social processes, primarily through the visibility of those attributes produced by specific choices. Together these four approaches constitute mutually-supporting building blocks for the creation of a new middle-range theory for studying the evolution of technological behavior, including how that behavior is learned and enacted within and between groups of social actors.

All four of these approaches also recognize that artifact formal variation can reflect the effects of the life history of the artifact as much as the artisan's technical decisions in the initial creation of the artifact. In lithics, the effect of the history of resharpening and reduction on the morphology of tools and cores during their use-lives has become known as the Frison Effect (Frison 1968; Jelinek 1976; Dibble 1995b). While the following chapter details the identification of lithic technical choices as opposed to the Frison Effect, it is important to emphasize the distinction between artifact form and technical choices at this point in developing the present method and theory. For as Schiffer and Skibo note, "Regrettably, attributing variability to the artisan's knowledge and experience in the absence of historical or ethnoarchaeological evidence is fraught with difficulties, as generations of culture-historians discovered (Binford 1968). A common problem is that investigators, when invoking migration, diffusion, or influence as an 'explanation', often focus on formal properties, not on technical choices. However, the new analytical techniques and principles now available for inferring technical choices and locations of artifact manufacture, this difficulty is diminishing" (Schiffer and Skibo 1997:34).

Thus, through a healthy respect for (and a methodology to detect–see chapter 4) the degree to which artifact form responds to effects beyond artisan choice in the production sequence, it is possible to use a unification of the four approaches above to answer Binford's (1968:8) question to the normative culture historians, "How can archaeologists distinguish between homologous and analogous cultural similarities?" The answer is to study how artifacts are made, not artifacts' final forms. The four approaches above provide the units of analysis, i.e., technical choices in the production sequence, for such a study.

In all, this survey of sources of artifact variation and design has shown that some profitable principles remain after the frequent methodological and theoretical pitfalls are avoided. The remainder of this chapter uses these principles to propose a body of middle-range theory for asking a cultural historical question of the Middle to Upper Paleolithic transition.

3.2 Culture History as Evolutionary Process

The task of building an evolutionary approach to Pleistocene culture history requires multiple steps. The previous section of this chapter has accomplished one step, the introduction and synthesis of a body of method and theory for understanding the sources of artifactual variability in terms of the technical choices made by artisans during the process of creation and use of material culture. The present section of this chapter undertakes the next series of steps in this building process. This involves four tasks. First, it is necessary to understand the historical contingency of evolutionary processes, both biological and cultural, and how the documentation of the availability of particular variants for selection at a given time and place, i.e., culture history, is the first and irreplaceable step in the study of evolutionary change. This task includes the clarification of which theoretical course to steer among the multiple approaches to studying cultural evolution from a Darwinian perspective. Second, the mechanics of culture history must be modeled as consistent and understandable processes rather than random events which throw "evolutionarily stable" systems out of equilibrium. This task includes recognizing recent research on studying the movement of people and technological ideas as combinations of biological and cultural processes. Third, the current lack of understanding of how culture contact between foragers works as a process is addressed through the creation of a middle-range theory which uses cultural transmission theory and Carr's Unified Middle-range Theory of Artifact Design as the conceptual glues around the building blocks of the Organization of Technology, *chaîne opératoire*, and Behavioral Archaeology approaches. This middle-range theory provides predictions for where to look in the archaeological record of Pleistocene foragers for evidence of the enculturation and acculturation processes which drive culture historical mechanics. These predictions are described in the fourth task through the explication of the taskscape visibility concept.

3.2.1 Three of the Four Forces of Evolution

In the mid Twentieth century, theorists such as R. A. Fisher, J. B. S. Haldane, and S. Wright as well as biologists such as T. Dobzhansky, E. Mayr, and G. G. Smith combined population genetics and Darwinian theory to produce a powerful explanation of organic evolution (Boyd and Silk 2000:66;

Gould 2002:503-591). Under this "Modern Synthesis" of the 1950s, evolution came to be defined as a change in allele frequencies in a population through time. Natural selection, "the preservation of favourable variations and the rejection of injurious variations" (Darwin 1964[1859]:81; see also Lewontin 1970; Mayr 1982:ch.11), became one of four forces which can cause such a change. The other forces of evolution were recognized as: 1) mutation, the alteration of an individual's heritable genetic material through the change in one or more nucleotide base pairs; 2) genetic drift, the random bias in allele frequencies between generations in a population; and 3) gene flow, the contribution of alleles from one population to another through mating (Gould 2002:503-591). Since the Modern Synthesis, natural selection has thus been seen as the most important, but not the only significant, cause of evolutionary change. This latter point has been misunderstood and/or deemphasized in archaeology, to the determinant of the discipline, as Bettinger et al. (1996:141) and Shennan (2000) have pointed out.

An example will help illustrate this point. Once during a post-conference discussion with a distinguished Paleolithic archaeologist, I commented that there are many valid questions for archaeological research. The distinguished archaeologist responded that, "Yes, culture history is a valid question. It is just not interesting. It has nothing to do with human evolution." This type of statement can be found in the literature (e.g., G. A. Clark 1994, 2005) as well as in such anecdotal cases. On one hand, this view is understandable from the point of view of the Processual Agenda's critique of normative culture history and the fact that Paleolithic culture history has not yet advanced beyond the essentialist approach of the 1930s (e.g., Otte and Keeley 1990; Otte and Kozłowski 2003), a fact very clearly made by G. A. Clark and Lindly (Clark and Lindly 1991; Clark 1994) as well as Lyman et al. (1997). On the other hand, what is surprising is that this view reflects an erroneously exclusive equation of adaptation with evolution. The concept of evolution in the biological sciences since the Modern Synthesis clearly assigns three of the four forces of evolution to the processes of historical contingency. Put in terms of archaeological research, three of the four forces of change in cultural evolution are culture historical processes (mutation equating to historically-contingent cultural innovation, genetic drift equating to cultural drift and founder's effect, and gene flow as cultural transmission between populations), while only one of the four is adaptation resulting from natural selection (Boyd and Richerson 1985; Bettinger et al. 1996).

How Paleolithic archaeology developed a concept of evolution divergent from that in the biological sciences is best explained by the way in which the Processual Agenda espoused a view of evolution more in line with transformative change and Clementsian ecological systems theory than biological phylogenetic evolution and evolutionary ecology. The Processual Agenda saw culture as adaptation, specifically the "extra-somatic means of adaptation" (Binford 1965:205; White 1959:8), but in a systemic context, with sociocultural systems responding to a direct link between the environment and behavior. From this perspective,

> Evolution becomes the unfolding of a system or organism whose identity is preserved in the process. Binford is entirely within the Whitean-Spencerian tradition... This is a classical formulation of synchronic, ecological-functional proximate causality as opposed to evolutionary ultimate causality (cf. Pianka 1978:15 [Mayr 1961]). Explanations are couched in functional

terms, change is conceived of as a transformation of those comprehensive systems of which culture is but a subsystem (cf. Binford 1962:217), and equilibrium and disturbance to equilibrium are the forces which explain all change. This view of culture, analogous to mixing up evolution and succession in biological systems, is totally incompatible with Darwinian models of cultural evolution (Tschauner 1994:82).

So, given that the dominant paradigm in archaeology for the study of cultural evolution is incompatible with biology's synthesis of evolutionary processes, what body of theory is one to use to study the changes in Pleistocene hominin material culture behavior as evidence of cultural evolution?

In the place of the Whitean-Spencerian view of cultural evolution in which historical contingency is discounted, Tschauner (1994) makes an eloquent argument for the role of culture historical research in the study of human evolution within the dual inheritance approach to evolutionary theory. Begun by Cavalli-Sforza and Feldman (1981) and continued through the work of Boyd and Richerson (1985, 1987, 1996; Richerson and Boyd 1978, 2001), Durham (1990, 1991, 1992) and others (Soltis et al. 1995; Bettinger et al. 1996; Dennett 1996; Boesch and Tomasello 1998; Shennan and Steele 1999; Shennan 2000, 2003; Collard and Shennan 2000; Tehrani and Collard 2002), dual inheritance or cultural transmission theory makes the anthropological study of cultural evolution compatible with current biological evolutionary theory. Thus, Durham (1990) refers to this body of literature as evolutionary culture theory. For a number of reasons detailed below, this is the most suitable high-level theory (*sensu* Thomas 1998) for the evolutionary study of Pleistocene cultural behavior. Not the least of these reasons is that cultural transmission theory predicates the study of cultural evolution using units of analysis almost identical to those identified through the unification of Carr's (1995) middle-range theory, the Organization of Technology approach, the *chaîne opératoire* approach, and Behavioral Archaeology as detailed earlier in this chapter.

3.2.2 *Cultural Transmission Theory*

From the perspective of cultural transmission theory, culture is defined as "information capable of affecting individuals' phenotypes which they acquire from other conspecifics by teaching or imitation" (Boyd and Richerson 1985:33). Such transmission of information can be vertical, from parents, oblique, from related or unrelated individuals in an older generation, or horizontal, from conspecifics of the same generation (Cavalli-Sforza and Feldman 1981:54). Information passed through one of these modes can be modified by various forces of change, including 1) stochastic drift akin to Sackett's (1990) isochrestic variation (Bettinger et al. 1996:137); 2) guided variation in which individual trial-and-error learning improves upon the culturally inherited variant of a trait through the comparison of the trait's performance against the perceived goal (much like Schiffer's performance characteristics); and 3) biased transmission that affects the direction and/or fidelity of transmission through the learner's evaluation of features of the variant itself (direct bias), the frequency of the variant relative to other variants in the population (frequency-dependent bias), or the appropriateness or attractiveness of the role model displaying the variant (indirect bias) (see Boyd and Richerson 1985:chaps. 4–8; see also Boesch and Tomasello 1998). Direct bias is similar to guided variation in that the learner is making a decision on what variant of a trait to use, but the former is selecting from extant variation whereas the latter can generate new variation based on the performance variation of the inherited variant. In terms of the distinction between invention and innovation (Renfrew 1978), biased transmission takes an extant variant and makes a innovation out of it whereas guided variation produces an invention and works on its frequency to promulgate it as a group-wide innovation.

These forces modifying transmission, what Durham calls "descent with modification" using Darwin's term (Durham 1992:341), can be modeled through cost-benefit relationships that predict in which contexts one would expect certain transmission processes to be active (Shennan and Steele 1999). For instance, one can model when one would expect horizontal vs. vertical transmission, or what Moore (1994) calls "ethnogenesis" versus "phylogenesis." These models can then be used to test hypotheses against independent datasets, such as laboratory and field studies of cultural learning in enculturated (human-raised) versus wild chimpanzees (Boesch and Tomasello 1998), human ethnographic data (Soltis et al. 1995; McElreath et al. 2003; Collard and Tehrani 2005; Shennan and Collard 2005), and archaeological data (Bettinger et al. 1996; Bettinger and Eerkens 1999; Shennan 2000; Collard and Shennan 2000; O'Brien et al. 2001; Darwent and O'Brien 2006; Eerkens and Lipo 2007; Lycett and von Cramon-Taubadel 2008; Premo and Hublin 2009; Lycett 2010; Premo and Kuhn 2010).

The unit of analysis in cultural transmission theory is socially learned knowledge, regardless of the specific mechanism of learning or transmission. Thus this approach avoids much of the mechanistic arguments over the definition of cultural evolution as distinctly human or non-human, i.e., Tomasello (1994, 1999) versus McGrew (1992, 1994, 1998). Given the discussion of sources of artifact variability in the first half of this chapter, these units of social learning clearly include artisans' technical choices at nodes in the sequences of production and use of material culture. As Tschauner points out, these are culture historical units: "If our classifications are to measure cultural transmission and descent, they must be designed to establish the spatio-temporal continuity of cultural instruction sets. And this is what culture history was all about" (Tschauner 1994:87). This is in fact the question addressed by the present research: to what extent is there spatio-temporal continuity of flintknapping instruction sets between 60 and 30 kya in the Levant, Eastern Europe, and Central Europe?

The cultural instruction sets of individuals within a population are conceived of in cultural transmission theory as collections of different variables for which each individual may exhibit a different variant. The forces of selection, both biological and cultural (Durham 1992), thus work in tandem with historical contingency on these variants' relative replication advantages to determine the frequencies of these variants in the next generation. In contrast to the meme theory of Dawkins (1976; Blackmore 1999), Boyd and Richerson (1985) as well as Durham (1991, 1992) devote more attention to the modeling of selection pressures on both biological and cultural replication pathways to the next generation. Specifically, natural selection for one variant of learned behavior in a population occurs based on any replication advantage over the alternative variants (Durham 1992:343). This thus includes the greater replication of the learned behavior through the greater reproductive success of the behavior's bearers (symmetric transmission), as well as greater replication of the learned behavior through the greater frequency of its being learned by non-relatives, regardless of the possible lower

Darwinian fitness of its bearers (asymmetric transmission) (*sensu* Boyd and Richerson 1985:178-200). The dual inheritance approach also differs from Dawkins' concept of the meme as replicator of its own transmitted unit because the dual inheritance approach does not assume that cultural learning always produces an identical cultural instruction set in the mind of the learner. Cultural transmission theory thus avoids what some see as an essentialist argument in meme theory (Sperber 1996:100-118; Shennan 2003:46-47).

Cultural instruction sets have many analogs in the method and theory discussed in the first half of this chapter. Whether one conceives of a cultural instruction set as an individual's propriospect (Goodenough 1971:36; Wolcott 1991), structuration (Giddens 1979; 1984), or habitus (Bourdieu 1977, 1980) depends upon which "from-the-bottom-up" social theory one wishes to use in articulation with cultural transmission theory, as each is sufficiently complementary (but see the caveats of Hegmon 1998:269-270). What is required of any articulating social theory, however, is an avoidance of a strict, Durkheimian view of a superorganic culture forcing individuals into deterministic norms of behavior (Richerson and Boyd 2002; Gamble 1999:38–39).

For the present research, the variables of an individual's cultural instruction set are represented by decision nodes in the technological procedure of making stone tools, whereas the variants of each variable are represented by the different technical options an individual chooses to use at each node in the sequence. As detailed in the next chapter, one need not succumb to the trap of utilizing emic units of analysis, such as actual knowledge held in an individual's brain, as long as one uses etically observable evidence of the individuals' choices as they impact the physical form of material culture (Tostevin 2011). By characterizing the variation as well as central tendencies of these evidenced choices/variants within a population of individuals who contributed artifacts to a given archaeological assemblage, one can characterize the population-level cultural learning sets of a given time and place in an evolutionarily significant way. As Tschauner states:

> Such units would provide the appropriate basis for evolutionary studies: 'differences in the relative abundance of variant forms...in space and over time become the empirical basis on which the action of evolutionarily significant selective forces can be studied. For the Darwinist, change in diversity is evolution' (Rindos 1989:22, emphasis in original). As Shennan (1991:207) has argued, if cultural transmission is a process with its own properties, then culture history is a distinct phenomenon requiring examination in its own right. The study of cultural evolution does not end with a reconstruction of the 'genealogies' of cultures. But without such a reconstruction, it will never begin (Tschauner 1994:90).

The latter point, that the study of culture history is only the beginning of the process of studying cultural evolution, is important for several reasons. First, it reminds us to keep in mind the larger goal of explaining cultural evolution rather than just documenting it. Second, it reveals a direct and enlightening parallel with the distinct uses of cladistic and adaptationist explanations of biological evolution. Third, Tschauner's latter point reinforces his quotation of Rindos above to stress that cultural transmission theory characterizes cultural evolution as the selection of different variants through time rather than as the identification of period-specific, culturally-dictated norms. As a result, this approach to cultural evolution both avoids the fundamental essentialist error of normative culture history and unifies the study of cultural evolution with Darwinian biological evolution (Tschauner 1994).

The success of cultural transmission theory as well as its suitability as the basis for the present evolutionary study of Pleistocene cultural evolution can be seen when it is contrasted with alternative approaches to applying Darwinian principles to archaeology. First, cultural transmission theory represents a holistic rather than limited synthesis of modern evolutionary thought with models for how cultural evolution occurs. In contrast to the "Evolutionary Archaeology" or "Selectionist" school of Dunnell, O'Brien, Lyman and others (Dunnell 1980; 1989; Leonard and Jones 1987; O'Brien et al. 1994; and references within Teltser 1995b, O'Brien 1996, and O'Brien and Lyman 2000), the cultural transmission approach encapsulates the logic of the Modern Synthesis to recognize that the transmission of learned behavior between individuals differs from the transmission of genetic material in a biological system. Thus, cultural transmission theory recognizes that cultural evolution can behave quite differently through time compared to biological evolution. The Dunnell school, in contrast, fails to recognize this difference and so creates a rigid equation of artifact-phenotype-adaptation that does not equate with modern evolutionary theory. As Boone and Smith (1998) note in their defense of the evolutionary ecology approach against critiques from the Dunnell's school of evolutionary archaeology:

> ...given that evolutionary archaeology is positing cultural inheritance, its failure to pay attention to the effect of cultural transmission pathways (e.g., parental versus non-parental, generational versus peer-to-peer, one-to-many versus many-to-one) as well as secondary forces such as evolved preferences is puzzling. Though its proponents often cite the theoretical literature on cultural evolution (e.g., Boyd and Richerson 1985, Cavalli-Sforza and Feldman 1981, Durham 1991), they do not discuss the conclusion of this theory that cultural transmission might sometimes produce evolutionary trajectories that differ radically from those governing traits linked to Mendelian inheritance (Boone and Smith 1998:S144, emphasis original).

Cultural evolutionary processes are thus far more like viral genetics than eukaryote genetics, as Cavalli-Sforza and Feldman (1981:46–65) were the first to point out.

Cultural transmission theory's recognition of both the similarities and differences in the mechanics of cultural versus biological evolution is a result of its holistic grasp of evolutionary theory compared to the evolutionary archaeologists of Dunnell's school. As Boone and Smith (1998:S155) point out,

> Teltser's suggestion that evolutionary archaeology must develop new theory to deal with behavior because evolutionary biology "does not include the necessary terms to address behavioral phenomena" (1995a:3) would be true only if one ignored the explosion of theory and data developed in evolutionary behavioral ecology over the past three decades (e.g., Alcock 1993, Krebs and Davies 1997). Unfortunately, evolutionary archaeology seems generally to have done just that (Boone and Smith 1998:S155).

Thus, D. Hardesty's warning against limited "search and seizure missions" (1980:161) into other fields seems apt in the case of Dunnell's school of evolutionary archaeology. Not all of the products of Dunnell's evolutionary archaeological program have been this limited, however; research on computer simulations of stylistic drift (Neiman 1995), conceptual work on archaeological systematics (Lyman et al. 1997; O'Brien and Lyman 1999), and the application of cladistics to archaeology (O'Brien et al. 2001; O'Brien and Lyman 2003) are all positive contributions. It is unfortunate, therefore, that the idiosyncratic trajectory begun by Dunnell stands in such evident contradiction to evolutionary theory in biology and its application to paleontology. As Schiffer notes (1996:648-652), this has damaged the popularity of the use of Darwinian theory in archaeology and complicated the theoretical programs within the field. In contrast, Schiffer explicitly attempts to point out areas of clear intersection between the Behavioral program and a Darwinian evolutionary approach, including both the evolutionary archaeology program and cultural transmission theory. These intersections relate to where selection occurs (i.e., at the point an artisan chooses one technical option among many), what the artisan's selection (i.e., a proximate causality) is based on (i.e., performance characteristics), and how variation occurs (i.e., selection as well as historical contingency). In all, Schiffer's points demonstrate the appropriateness of cultural transmission theory for use with the Behavioral Archaeology program and the present research approach.

Compared to the alternative approaches as exemplified by Dunnell or Binford, cultural transmission theory integrates well with evolutionary ecology (Winterhalder and Smith 1981; O'Connell 1995; Kelly 1995; Krebs and Davies 1997) as well as evolutionary psychology (Cosmides and Tooby 1992), as discussed earlier in this chapter. As with both of these approaches, cultural transmission theory is careful to avoid the style vs. function dichotomy emphasized in other approaches (Bettinger et al. 1996:137). Likewise, Mayr's (1961) distinction between proximate and ultimate mechanisms by which selection affects variation is clearly reflected in the modeling of cultural and biological evolution in the dual inheritance approach as well as most evolutionary ecology studies (e.g., Boone and Smith 1998:S152; Bamforth 2002). In this, the agreement between evolutionary ecology and cultural transmission theory would be even better if lithic analysts' use of evolutionary ecology via the Organization of Technology approach incorporated more of the social as well as energetic proximate mechanisms for foragers to achieve reproductive success, as discussed above in relation to the progressive research of McGuire and Hildebrandt (2005).

In sum, cultural transmission theory provides a means of understanding the processes of transmission, modification, preservation, and loss of learned behavior, such as technical choices, in an evolutionary format (Premo and Hublin 2009; Premo and Kuhn 2010). From this body of theory, cultural evolution can be seen as both connected and distinct from biological evolution. Both, however, are united in being responsive to the historical contingency that provides the raw material for selection and incorporation of variants into the next generation, whether it be mutation, genetic drift, and gene flow in biological evolution, or innovation, isochrestism, and behavioral diffusion in cultural evolution. Culture history, therefore, is an integral part of any evolutionary explanation for the course of human evolution in the Pleistocene (Shennan 2003; Richerson and Boyd 2005; Boyd and Richerson 2005).

3.2.3 *The Processual Understanding of Culture Historical Mechanisms*

Having identified cultural transmission theory as the high-level source for generating hypotheses for the study of evolutionary culture history in the Pleistocene, it is now necessary to build archaeologically appropriate middle-level theories to connect the research questions and models with the observations available given the limits of the archaeological record (Binford 1977). Unfortunately, this task is challenging as a result of cultural historical mechanisms, such as culture contact, population migration, and behavioral diffusion, having only been the subject of processual inquiry in archaeology for the last few decades.

After the fall of the normative culture history paradigm (Lyman et al. 1997), the Processual Agenda suppressed the study of migrations and behavioral diffusion as unscientific subjects (Adams et al. 1978; Chapman and Hamerow 1997). Nevertheless, the Processual Agenda did benefit the study of culture history by leading archaeologists to search for processes resulting from one of the then-popular prime movers of culture change, population increase. This ultimately led to the understanding of phenomena such as migrations as processual explanations of culture change (Anthony 1990). These processes are now seen to have predictable patterns and recognizable effects on human cultures and the material culture they produce. Much of the advances in this area were accomplished with the help of other fields within the social sciences, such as geography, history, and sociology (Lee 1966; Kershaw 1978; Tilly 1978; Gmelch 1980; see references within Duke et al. 1978; Davis 1983; Anthony 1990, 1997; Rogers 1995; Golledge and Stimson 1997; Brettell and Hollified 2000). The processual role of return migration (Gmelch 1980; Traphagan 2000), migratory scouts and chain migration (Hudson 1977:20; MacDonald and MacDonald 1964), migration streams (Lee 1966:54; Lefferts 1977:39), gender-biased migration (Brown 1983; Jacobsen 1988; Moch 1992; Castles and Miller 1993; Burmeister 2000), age-biased migration (Leslie and Richardson 1961), apex families (Alvarez 1987:133–138), and frontier "middle-ground" societies (Van Kirk 1983; White 1991; Lightfoot and Martinez 1995; Merrell 1999) are all now recognized as integral to the archaeological study of human culture contact and migration (Kristiansen 1989; Cameron 1995; Lekson and Cameron 1995; Anthony 1997; Chapman and Hamerow 1997; Spielmann 1998; Lightfoot et al. 1998; Wells 1999; Burmeister 2000; J. J. Clark 2001; Lyons 2003).

Biology has also shaped the current understanding of cultural historical mechanisms as non-random processes of change. In particular, primatology and evolutionary ecology have equally contributed to understanding selective pressures that encourage individuals to seek other opportunities away from the place and social network in which they are currently residing. At the level of individuals, these processes include natal and breeding dispersion as a means for individuals to acquire mates and unclaimed resources, the proximate mechanisms (sensu Mayr 1961; Alcock and Sherman 1994) by which the ultimate mechanisms of avoiding both inbreeding and resource competition with relatives are achieved (Pusey 1992; Stenseth and Lidicker 1992). At the level of populations, these processes include the aggregate of individual actions through time, such as adaptive range expansion (Nowak 1975; Groves and Di Castri 1991), social group disintegration and formation through inter-group violence (Goodall 1986: 503-518; Cheney and Seyfarth 1987; Wilson and Wrangham 2003; Wilson et al. 2004), and shifts in population density as a result of changes in environment or ecological niche (Sinclair 1983; Bettinger and Baumhoff 1982; Shennan 2000). These processes are also recognized as having consequences for the health of the dispersing individual (Kaplan 1988; Isbell et al. 1990) as well as population-level effects on co-evolutionary relationships between hosts and parasites through disease transfer during dispersion and contact (Dobyns 1993; Cook 1998; Stojanowski 2004). It is thus possible to understand population movements and events of culture contact as both biological and cultural processes that have causes and consequences specific to both the population of origin and the population of destination.

The incorporation of the uniformitarian understanding of migration and contact processes today into the archaeological study of population movements in the past has produced more explicit and operational definitions of the concepts left undefined for so long by the normative culture historical school (e.g., Rouse 1958, 1986). This is particularly clear in the recent growth of a method and theory for objective research into migration processes in the prehistory of the American Southwest (Adams 1996; Ciolek-Torrello 1997; Reid 1997; Spielmann 1998; Stark et al. 1995). J. J. Clark (2001) has devised one of the more useful definitions of migration for archaeological analysis: "a long-term residential relocation beyond community boundaries by one or more discrete social units as the result of a perceived decrease in the benefits of remaining residentially stable or a perceived increase in the benefits of relocating to prospective destinations" (Clark 2001:2). This definition is designed to take advantage of several specific features: 1) it ignores "background noise" in terms of "movements by individuals such as marriage partners, traders, and itinerant craftsmen" (Clark 2001:2); 2) a group-level definition has a greater chance of being archaeologically observable compared to a definition based on the individual; and 3) the exclusion of seasonal and temporary movements makes it more likely that those movements included actually crossed social boundaries that would make the archaeological signature more visible.

The recognition of the nested nature of style (Carr and Neitzel 1995) as well as the recognition of its different uses in social practice (Hegmon 1992, 1998) have also produced far more subtle definitions of the social processes being targeted in archaeological studies of regional interaction (Hegmon 2000). Particularly useful in this regard is J. J. Clark's opposition of the different social processes which produce trait- or site-unit intrusions in the archaeological record, namely exchange, emulation, ethnicity, and enculturation (2001:6–10). Clark uses artifacts of non-local production to identify the exchange of goods at a distance (see also Renfrew 1975) whereas emulation of a distant population by a local population is recognized as the local production of an artifact or idea which was originally known from a socially distant origin. Although not treated as such by Clark, emulation theoretically could be the result of either stimulus diffusion of the idea of the object but not the technological procedure to produce it (Kroeber 1940) or the behavioral diffusion of the whole technology. Ethnicity is treated by Clark as the flexible, self-ascription or affiliation of an individual to a particular group membership, essentially as Barth (1969a,b) has defined it. In material expression, ethnicity produces stylistic similarity in objects made by people who may or may not be genetically related to the population of innovators of the original style.

J. J. Clark is clear in concluding that exchange, emulation, and ethnicity do not produce material culture evidence suitable for identifying the migration of households as defined above. Ethnic ascription is particularly problematic as a cultural reference for population demography, given the flexibility of the ascription and the rhizotic nature of its historical development, as Moore (1987, 1994) and Snow (1995, 1996) have pointed out. Instead, J. J. Clark focuses on the material evidence of the social process of enculturation as his target for the archaeological identification of population movement. Enculturation is seen as the process by which "specific ecological 'where-to-do's,' utilitarian 'how-to-do's,' and social 'what-to-do's' are passed from old to young through active instruction and passive imitation (Netting 1993:59, 63, 70-72)" (J. J. Clark 2001:9-10). Focusing on household-level cultural transmission, Clark comments that the household's "body of knowledge represents a shared tradition that can potentially be used to track the movements of associated households whether or not this collective consciously displays its identity. Unconscious and deeply embedded aspects of enculturation are often more stable through time than displayed identities such as ethnicity and more resistant to assimilation in mixed cultural settings generated by migration" (Ibid:10). Choosing enculturation as his social process of interest, Clark uses Carr's Unified Middle-Range Theory of Artifact Design (1995b), in particular the visibility and production sequence hierarchies, to predict that several types of not easily visible architectural details, including wall-fill construction techniques for pueblo buildings, are successful methods for identifying styles idiosyncratic to the enculturation of the household builder. P. D. Lyons' (2003) study of proto-historic migrations in the American Southwest similarly uses Carr's predictive theory, but in this case focusing on the geological origin of ceramic tempers to identify local production of ceramics in a non-local style. LeMoine uses similar reasoning in studying culture contact between Dorset and Thule Inuit (2002). Again, the visibility hierarchy serves to identify variability only responsive to enculturation.

For a case study on the other side of the globe, S. Burmeister (2000) has also developed a method and theory approach to studying migration from a very similar viewpoint. In particular, he distinguishes the patterning of immigrant material culture according to both the internal and the external culture sphere:

"The world of the immigrants is twofold. The public or external domain is the zone of contact with the society of the immigration area. In contrast to this is what I would like to call the internal domain. The two spheres of social life correspond to different practices in the use of material culture: adaptation in the realm of the public sphere, the external domain, and invariability in the realm of the private sphere, the internal domain. Although the distinction between the two domains is rather ideal, the habitus can be shown to have different effects in the two cases. It is in the external domain, the intersection of the social and the economic with the environment, that the habitus is confronted with change...Only with a shift of statuses within the family as a result of assimilation will the conditions that structure the internal domain be transformed over time. The habitus, a disposition rooted in the past, will show a stronger tendency toward tradition in the internal than in the external domain" (Burmeister 2000:542).

Burmeister uses analogies from New World historical archaeology to substantiate his theory of internal versus external cultural domains. Many of these ethnographic examples are in fact the same that gave rise to Carr's (1995b) theory, which subsequently informed Clark's study. While Burmeister does not explicitly utilize the principles of Carr's visibility hierarchy, the convergence of his theory with Carr's is undeniable, although Burmeister's arguments are situated more within social theory (i.e., Bourdieu 1977) than in processual middle-range theory, as with Carr. Nonetheless, each of these works utilizes the concept of social or contextual visibility limiting the social processes which can affect variation in material culture form and production sequences. A similar approach for the investigation of foraging societies is developed in the culmination of this chapter.

The last two decades of archaeological research have thus witnessed an improvement in the definition of terms and the operationalization of methods for identifying the known processes of culture history, namely migration and culture contact. The question remains, however: given the documentation of modern ethnographic and proto-historic processes of culture contact, migration, and diffusion, which of these processes would have been operating among Pleistocene foragers? It has been pointed out (Anthony 1990; G. A. Clark and Lindly 1991; Weiss 1988) that agriculturalists have longer distance short-term mobility due to their focal economy when compared to foragers with their diffuse economy (Cleland 1976). Simply, agriculturalists can store food and other resources for a longer residential move than foragers can. As a result, we must ask, how far does the uniformitarian principle apply? Attempts to answer this question have generated considerable debate. Resolving this debate requires a new means of modeling the material evidence of forager societies at the appropriate scale for archaeological analysis, with prehistoric processes and their resulting archaeological patterns clearly defined.

3.2.4 The Processual Understanding of Culture Contact among Foragers

Paleolithic archaeology, a subfield exclusively focused on foragers, never truly abandoned culture history as its central paradigm (Adams et al. 1978; Gamble 1993), despite the critiques of the Processual Agenda (Binford and Binford 1966; Binford 1973), the inappropriateness of ethnographic migrations of agriculturalists as analogies (Clark and Lindly 1991), and the, at times, clear lack of theoretical sophistication in the use of the paradigm in explaining human evolution (e.g., Otte and Keeley 1990). Nevertheless, arguments have been made for decades against the role, and even possibility, of culture historical processes affecting culture change in forager societies. Krantz (1976), in an extreme example of the marriage between nineteenth century evolutionism and twentieth century cultural materialism, has argued that forager populations could never successfully encroach upon the territory of another group due to the former's lack of territorial familiarity and the inability of any hunter-gatherer group to out-compete another once both have adapted behaviorally to their surroundings. Bettinger and Baumhoff (1982), however, have demonstrated just such a population expansion, in which a Numic-speaking forager group out-populated the Pre-Numic inhabitants of the Great Basin due to slight differences in their foraging versus collecting behavior (*sensu* Binford 1980).

If the hunter-gatherer subsistence base is the source for this hypothetical inability for population movements, primatology has many lessons to give us, since it also studies foragers, just ones that happen not to be human. Primatology provides numerous examples among chimpanzees (Goodall 1986: 503-518; Wilson and Wrangham 2003; Wilson et al. 2004) and other primates (Cheney and Seyfarth 1987) of intergroup hostilities leading to territorial expansions through group extinctions, forced assimilation of individuals from one group into a more dominant group, and population replacement. Beyond primates, Groves and Di Castri (1991) provides a series of studies documenting the invasion of Mediterranean flora and fauna into different habitats worldwide. Similarly, Nowak (1975) provides an exhaustive study of the causes of the twentieth century range expansion of twenty-eight

European mammal species, elucidating the mechanisms by which any animal, including humans, can encroach upon the territory of another. For instance, dispersal of the individual from its natal site to a more distant breeding location (Pusey 1992; Stenseth and Lidicker 1992) is a behavior as human as it is non-human; the majority of modern humans are in fact female-biased dispersed, i.e. with virilocal residence (Murdock 1967; van den Berghe 1979; Koenig 1989). This mechanism alone could afford for the spread of one group into another through mating networks. Exogamy has indeed recently been put forward as a mechanism for the spread of technological *connaissance* (Pelegrin 1995:267) in an argument analogous to the Deetz-Longacre Hypothesis for the effect of matrilineal versus patrilineal residence on stylistic variability (Longacre 1964; Deetz 1965; see Roe 1995:51-54). Based on these examples and the principles of evolutionary ecology (e.g., Foley 1984, 1987), it is more parsimonious to argue that hunter-gatherer populations can, rather than cannot, encroach upon other hunter-gatherer territories given sufficient adaptive incentive, usually in terms of changing environments, differential niche exploitation, and/or overpopulation (Sinclair 1983). Krantz's (1976) position, therefore, lacks support from both biological and anthropological research traditions.

While taking exception to Otte and Keeley's (1990) espousal of an Aurignacian migration as an explanation for the MP-UP transition, Clark and Lindly (1991) provide an exploration of the process of migration as viewed by the disciplines of evolutionary biology (Gauthreaux 1980; Foley 1987), geography (Lee 1966; Kershaw 1978; Gmelch 1980), genetics (Lasker and Mascie-Taylor 1988; Weiss 1988), and anthropology (Anthony 1990; Chang 1978; Duke et al. 1978; Rouse 1986). This is a very welcome discussion of a frequently undefined phenomenon (at least in anthropology) and G. A. Clark (1994) further clarifies the epistemological issue. In both papers, the authors are correct in criticizing the use of traditional systematics of retouched tool typologies to study population demography; as will be discussed below, retouched tool kits are not likely to be reflective of the enculturating environment of the artisans. Nevertheless, while Clark and Lindly correctly criticize Otte and Keeley's approach for these and other reasons, they fail to acknowledge the validity of population movement as a processual explanation for the study of prehistory, due primarily to an issue of scale.

The Problem of Scale and the Völkerwanderungen

Clark and Lindly criticize Otte and Keeley for following the continental peregrinations of lithic typological patterns which have no relation to "identity-conscious social units" (1991:578). As self-identification, a common definition of ethnic groups (Barth 1969a,b), is clearly beyond the resolution of any data set other than that of an oral interviewer, ethnic groups are indeed not appropriate targets for examining potential culture historical processes during the Paleolithic period. The inability to identify "ethnic" groups in the Paleolithic, however, does not negate the archaeologist's ability to partition the variation in material culture in order to define appropriate etic units of analysis of a larger scale (Adams and Adams 1991). The ethnic-scale definition of Paleolithic material culture patterning was recognized quite early by Childe (1951) as inappropriate for archaeological culture units. Except in specific cases such as Otte and Keeley (1990), there is thus no reason to critique the study of culture historical processes in the Paleolithic based on the use of ethnicity as a social process.

Ethnicity is thus somewhat of a red herring in this critique. As Sackett points out in response to similar arguments by Straus and Clark (1986):

> Although he is rarely cited, one seems to hear echoes of Lewis Binford in many of their [Straus and Clark] terms and much of their method. A Binfordian chord is surely struck by their gratuitous equation of 'cultural' with 'ethnic,' which serves to restrict a general term essential for anthropological discourse to a specific meaning for which a perfectly adequate word already exists (Sackett 1991:112).

There is thus no reason to discount culture historical processes because it is impossible to document ethnic groups in the Pleistocene since 1) there are other archaeologically identifiable social units which are behaviorally meaningful and yet not self-conscious, such as those used by J. J. Clark (2001) and P. D. Lyons (2003), suitable for the study of cultural transmission; and 2) retouched tool typologies are clearly not the most appropriate units to study in such cases, particularly given the Frison Effect and the predictions of Carr's middle-range theory for mobile tool kits (see below).

Other partitions of material culture behavior are possible for the Paleolithic since the variability exists and the archaeologist can choose what etic units to use to structure this variability. Just as other primates such as chimpanzees and orangutans produce material culture which varies isochrestically (*sensu* Sackett 1982, 1986) between geographically definable population groups (Goodall 1970; 1973; McGrew et al. 1979; McGrew 1992; 1994; 1998; Wrangham et al. 1994; van Schaik et al. 1999), parsimony favors the assumption that our last common ancestor as well as subsequent Late Pleistocene hominins produced material culture which is at least as varied as modern chimpanzees' and so appropriate for mapping learned behavior onto the Pleistocene landscape.

Tomasello et al. (1993) and Tomasello (1994, 1999) have argued that no other wild primate, including the chimpanzee, possesses the human learning abilities of imitation and collaborative learning which are required to achieve the high rate of the ratchet effect, the cumulative change in culturally transmitted information between generations, which makes human culture so unique. Subsequent collaboration between the laboratory exploration of captive chimpanzees and field observation of learning processes among wild individuals (Boesch and Tomasello 1998) has shown the contextual nature of such learning processes and reinforced the feedback mechanisms by which learning affects cognitive capacities (Donald 1991, 1998). Boesch and Tomasello thus corroborate many of Boyd and Richerson's (1985) concepts of conservatism in cultural transmission, including indirect biased transmission through the effect of role models and the frequency of the variant which is learned by different members of the next generation. Thus, while Late Pleistocene hominin foragers likely produced a slower ratchet effect than modern Holocene humans, the requirement for the study of culture history is the presence of material culture variation, not the rate of its change.

The population groups corresponding to archaeological entities, while certainly not self-identifying (Close 1989) and perhaps representing numbers of individuals larger than any of Wobst's groups (1974) as demonstrated in the San context by Wiessner (1983), are simply analytical units defined upon the basis of shared expressions of material culture behavior (the identification of behavior through artifact morphology is detailed in the next chapter). The investigation of the strength of cultural transmission in human evolution, through the measure of culture historical processes in the archaeological record, thus should be pursued based on analytical units identified by prehistorians for the purpose of specific inquiries, i.e., etic–not emic–material culture units (Adams and Adams 1991). For instance, the scientific relevance of distinguishing the presence or absence of bifaces across the Pleistocene landscape is both etic and of the same scale as distinguishing modern human learned behavior based on the use of chopsticks compared to fork and knife utensils, an analogy originally deriving from O. Bar-Yosef. This point is not meant to demean either the biface or the chopstick but to point out that the learned material culture behaviors involved in the creation, use, and evolutionary development of these objects is of the same resolution and neither have anything to do with the self-ascription to a social group with an internally and externally recognized identity (Barth 1969a,b). Such objects could be used emically to negotiate ethnicity in any given time and space (Sackett 1990:37), but the spatial and temporal distributions of the objects themselves do not equate to any one "ethnicity".

Clark and Lindly (1991) further stress the need to demonstrate the causes of a population dispersal, in terms of disadvantageous factors in the region of origin and advantageous factors in the destination region, before population movement for foragers or agriculturalists may be given as a valid interpretation. The approach taken by this research, however, sees causal questions such as why a set of events occurred to be beyond scientific reach without prior elucidation of what events occurred and how they occurred (Kroeber 1931; Lyman et al.1997:225). In this, Anthony (1990, 1997) provides a better methodology: the specifics of a population movement, diffusion, or regional independent innovation must be demonstrated through the patterning of the material culture record before the causes of these processes may be researched. If modern geographers and sociologists cannot agree on one cause of a historical migration happening before their eyes, it is unrealistic to expect prehistoric archaeologists to do better (Anthony 1997). Such criteria would prevent a recognized process of culture change from being acknowledged as ever having occurred in the past.

Paleontology (Whybrow 1992) and cultural transmission theory (Shennan 2003) equally argue for the demonstration of patterns before the explication of them. Yet the disadvantageous factors in the region of origin and advantageous factors in the destination region requested by Clark and Lindly are proximate causations used in the mind of the prehistoric potential migrant rather than ultimate causations for the creation of the regional distinctions. For Paleolithic foragers as discussed below, the directional (rather than circular) movement of the geographical center of a foraging range across the Pleistocene landscape is more a result of ultimate causation in terms of adaptive advantage than proximate decision making. This is particularly important when the adaptative advantage which allowed a range expansion may not be recognized because of the current limits of archaeological research techniques. The archaeological demonstration of cultural transmission across space, however, would signal that cultural or natural selection (*sensu* Boyd and Richerson 1985) was active in a given space and time, and thus new analytical techniques are required to investigate why such transmission across space has occurred.

As Gamble (1993) has noted, Clark and Lindly (1991) make a fundamental point in arguing that processes must be defined as operational concepts for archaeological study if they are to be afforded any explanatory power in reconstructing what happened in the past. In this regard, they are correct in arguing that historical migrations are unsuitable as definitions or analogies for Paleolithic dispersal since hunter-gatherers could not move their foraging territories as quickly as the historical migrations of populations using a focal, rather than diffuse, economy such as agriculture (Cleland 1976; Anthony 1990). Although not comparable to a völkerwanderungen such as the one Caesar observed during the Helvetii's one-year mass migration from Switzerland into western Gaul, Paleolithic hominins would have been able to adjust the geographical location of their foraging ranges via natal dispersal and breeding-age dispersal of individuals (Pusey 1992) over years, possibly thousands of years in duration, as long as the adaptive advantage remained. Such range expansions via natal dispersals have been shown for other members of the larger carnivore guild (Turner 1984, 1991, 1992). Such dispersals are best modeled as demic diffusion, the process by which an adaptive advantage (biological or technological) allows differential population growth at the expense of neighbors (Cavalli-Sforza et al. 1993; Ammerman and Cavalli-Sforza 1984; see also Fix 1999:168–182). Given the temporal scale of such range expansions, the distances and the difficulty of travel between regions (see Sapir 1916; Steward 1929) such as the Levant and Central Europe are, therefore, minor obstacles. We also know from modern ethnography of foragers that such societies are connected via artifact exchange and trade with other societies at great distances, as far away as 1,200 km (Mulvaney 1976, map 3; Murdock 1960). This is the scale of the phenomenon associated with the culture historical processes central to the hypotheses in this research.

The suitability of considering cultural transmission across the time scale of such demic diffusion processes has also been argued to be undemonstrated, if not unreasonable. As G. A. Clark notes in reviewing a series of European site reports, "While the patterns [of lithic reduction sequences] identified are undisputed, what is supposedly causing them to occur (social learning transmitted from one generation to the next over ca. 2000 generations) is epistemologically problematic" (G. A. Clark 2005:378). From the point of view of cultural transmission theory, however, there is no reason to presume change in cultural instruction sets over many, many generations beyond random drift from transmission errors, unless directional selection is active. In fact, conservative fidelity in technical learning such as frequency-dependent bias in transmission or "when in Rome, do as the Romans do" (Boyd and Richerson 1985:286), is strongly advantageous in spatially heterogeneous environments compared to trial-and-error evaluation of various alternatives. The stability of such technical instruction sets over many generations could thus be seen as stabilizing selection via either symmetric or asymmetric transmission. The alternative of complex flintknapping sequences being reinvented through trial-and-error learning in each generation simply due to an unknown environmental constraint or optimizing force seems more problematic than the transmission of such methods in contexts in which stabilizing selection for greater transmission fidelity has been documented (Boyd and Richerson 1985:152–157) The climatologically heterogeneous, even volatile, period of the Pleistocene (van Andel and Davies 2003) is just such an environment in which guided variation, i.e., independent trial-and error learning, represents too large a cost to the individual in terms of time, energy, and failed tasks compared to the adoption of the "tried" inherited cultural trait (Boyd and Richerson 1985:284-285). Selection for strong cultural transmission, weak guided variation, and a low potential for drift is thus predicted for spatially and temporally heterogeneous environments by both cultural transmission theory and Potts' (1998) concept of "variability selection", a specific example of natural selection in which methods of adaptation that buffer individuals' fitness across environmentally-caused resource fluctuations are favored. The study of modern human origins from the point of view of evolutionary geography (*sensu* Lahr and Foley 1998), in which population demography is correlated with climatic shifts, would further support this perspective.

Definition of Culture Historical Processes

Yet the question remains, what are the operational definitions of the present study? The approaches surveyed in this chapter, from Kroeber to cultural transmission theory, each contribute the necessary working units for constructing definitions of culture historical processes applicable to the study of Pleistocene foragers. I follow the standard practice in the social sciences of defining the largest culture historical process as the movement of ideas, concepts, and techniques from one geographical location to another by any means, with the process of moving ideas by moving people forming a subset of the larger process (Morrill and Manninen 1975; Kershaw 1978). Put in the terms of cultural transmission theory, this largest unit becomes Cultural Transmission Across Space. The movement of the ideas through the movement of the people rather than the spread of ideas by themselves is thus defined as Symmetric Transmission Across Space via the operation of demic diffusion. The transmission of variants of a cultural trait without the movement of people, i.e., in the absence of gene flow, is consequently defined as Asymmetric Transmission Across Space. As detailed in Table 3.1, these processes are defined according to their hierarchical relationships within cultural transmission theory. This is graphically represented in Figure 3.4.

Gamble (1992, 1993) has argued against the use of the term "diffusion" for the study of culture historical process, given its historical connections with an imperial biogeography. The point is well made but it must be recognized that most of our concepts in archaeology are equally history-laden. "Anthropology" is itself a colonialist term, for without the colonial perspective on "ethnographic others," anthropology would be subsumed within the discipline of history (Trigger 1985). Thus, the term diffusion is used, if cautiously, two times in the definitions in Table 3.1. First, the mechanism behind the symmetric transmission of cultural traits across the Pleistocene landscape is defined here according to the biological concept of demic diffusion as opposed to the asymmetric transmission mechanism of behavioral diffusion. Each instance of "diffusion" is thus defined in relation to a specific correlate in cultural transmission theory. Gamble's (1993) definitions of "dispersal" and "colonisation" are also useful in terms of the appropriate scale of the processes but they are not specific enough to the task at hand. In other words, the definitions must be made in terms of cultural transmission theory. With the intent of being as explicit as possible, Gamble's "dispersal" is thus incorporated as the proximate mechanism behind the larger process of demic diffusion, as the latter includes a causal force (demographic increase due to adaptive advantage) which is not contained in natal dispersal *sensu stricto*.

With these definitions of culture historical processes, it remains to structure their study by means of defining units of analysis that articulate with these processes in the archaeological record. This requires the development of another concept, social intimacy, in relation to which social

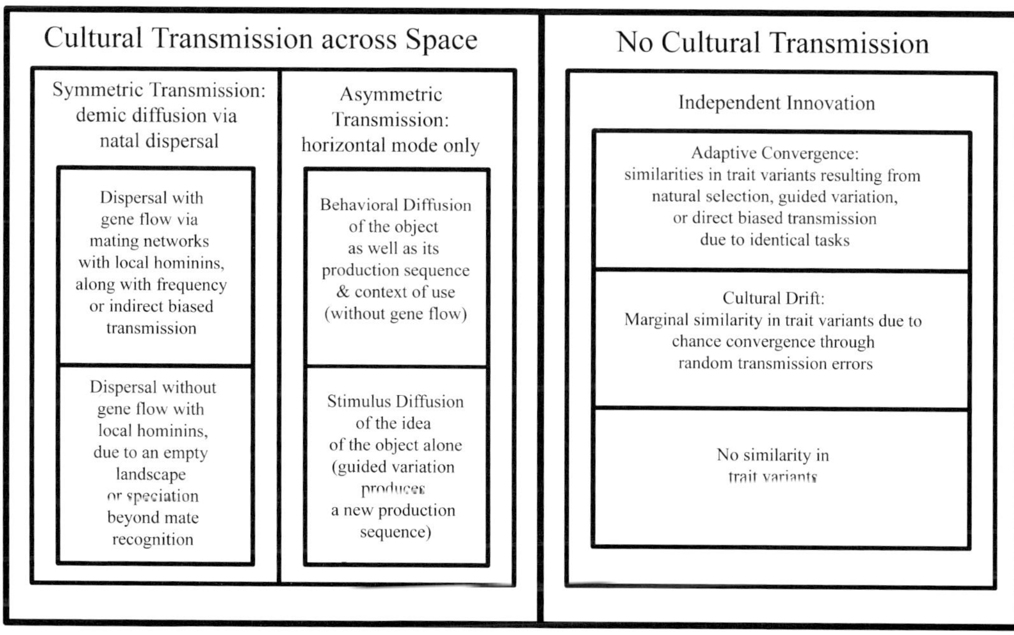

Figure 3.4 Nested structure of culture historical processes defined for foraging societies.

processes within cultural transmission theory produce material evidence appropriate for distinguishing between the nested concepts of symmetric transmission, asymmetric transmission, and independent innovation in Figure 3.4.

3.2.5 *Social Intimacy and the Taskscape Visibility Concept* [1]

In defining units of analysis within Pleistocene archaeology that will distinguish between the cultural historical processes defined above, it is important to consider how the units measured archaeologically as proxies for population units are produced socially. We need to know between what "units" or social groups cultural transmission is occurring. Although these units have nothing to do with self-ascribing ethnic groups or the oppositional force of active stylistic distinctions between them (e.g., Hodder 1979b, 1985), the task of defining population units in the present case is similar to Barth's conclusion that the critical focus for defining groups should be the boundary that defines the group, rather than the "cultural stuff" that it encloses (Barth 1969a:15). Thus, it is necessary to model Pleistocene social groups through the consideration of culture contact events between such groups as arenas of social processes.

The Late Pleistocene was a period in which potentially quite different social environments came into contact, the environments in which individuals were enculturated into how to lead their lives within a group. This enculturation, extending from birth to death (Boyd and Richerson 1985; Shennan and Steele 1999), includes how the individual learns and gains

Table 3.1 Definitions of key culture historical processes in terms of cultural transmission theory

Culture: "information capable of affecting individuals' phenotypes which they acquire from other conspecifics by teaching or imitation" (Boyd and Richerson 1985:33).

Cultural trait: an element of culture (as defined above) which is performed by an individual during a process which leaves material evidence of its performance in the archaeological record. For the present study, cultural traits are operationalized as necessary steps, choices, or nodes within operational sequences (i.e., behavioral chains, chaînes opératoire, or reduction sequences) for the production and use of lithic material culture for which individuals may perform one of a variety of options to fulfill the needs of the trait. Traits may be continuous or discontinuous in their expressed variability (Boyd and Richerson 1985:70-80). An individual's particular tendencies for variants within an operational sequence constitute her/his cultural instruction set or propriospect (*sensu* Wolcott 1991; Goodenough 1971:36) for that technology. The identification and measurements of such traits in lithic reduction sequences are detailed in the next chapter.

Cultural Transmission Across Space: the transmission of variants of cultural traits between generations across geographical space by either symmetric transmission or asymmetric transmission (Boyd and Richerson 1985:178-190). In traditional social science terms, this process subsumes transmission by the movement of people across the landscape, i.e., migration per Anthony (1990) as well as the transmission of cultural variants of a trait without the movement of people, i.e., behavioral diffusion (Kroeber 1940).

Symmetric Transmission Across Space: cultural and natural selection favor the transmission of the same variant of a cultural trait between generations through cultural transmission from parents or other older relatives to offspring. In traditional social science terms, this process includes the transmission of a cultural trait by the movement of people across the landscape. Demic diffusion via natal dispersal is defined as the operating force behind symmetric transmission across space in forager societies. Symmetric transmission can result from two forms of demic diffusion from an adjoining region: 1) one in which the variant of a cultural trait is maintained despite gene flow with the local inhabitants through frequency-dependent and/or indirect biased transmission (Boyd and Richerson 1985:153-157); and, 2) the transmission of a variant of a cultural trait from an adjoining region without the need of biased transmission due to the absence of gene flow in a landscape unoccupied by other hominins or occupied by a hominin population which has progressed along the speciation process beyond the Paterson point of recognizing members of the other population as suitable mates (Paterson 1985).

Demic Diffusion: the process by which an adaptive advantage (biological or cultural) allows differential population growth of bearers of a cultural trait at the expense of neighbors (Cavalli-Sforza et al. 1993; Ammerman and Cavalli-Sforza 1984; see also Fix 1999:168-182). Demic diffusion in forager societies is driven by natal and breeding-age dispersal of individuals seeking mates and unclaimed resources, the proximate mechanisms (*sensu* Mayr 1961; Alcock and Sherman 1994) by which the ultimate mechanisms of avoiding both inbreeding and resource competition with relatives are achieved (Pusey 1992; Stenseth and Lidicker 1992).

Asymmetric Transmission Across Space: "natural selection may favor one variant if the trait is transmitted by one system of inheritance and a different variant if the trait is transmitted by the other. For example, the traits favored in an adolescent prestige system with horizontal transmission [unrelated peer to peer] may be quite different from those favored in the vertical transmission system from parents" (Boyd and Richerson 1985:11). Within the context of Asymmetric Transmission Across Space, horizontal transmission includes the traditional social science concept of cultural or behavioral diffusion of a trait between unrelated individuals (no gene flow), whether as a full package of production sequence and social context of use or the stimulus diffusion of the idea of an object only.

Stimulus Diffusion: a subset of asymmetric transmission across space (behavioral diffusion), in which the idea of an object or cultural trait is transmitted without the operational sequence details needed to produce it or the exact social context of its use (Kroeber 1940). The operational sequence must be independently innovated through guided variation (Boyd and Richerson 1985:95-97).

Independent Innovation: the alternative to Cultural Transmission Across Space, in which both natural and cultural selection work to produce similar frequencies of variants of a cultural trait in two populations which have not engaged in cultural transmission. In biological terms, this is convergent evolution. Such an independent appearance of similar frequencies of variants in two populations can result from several forces: 1) adaptive convergence; and, 2) random cultural drift.

Adaptive Convergence: as a subset of independent innovation, adaptive convergence is the independent evolution of similar variants for the same cultural traits between isolated populations by one or more of three forces: 1) natural selection culling individuals from a population because of their use of a variant, who otherwise might have been available to transmit the unfit variant, leaving individuals who exhibit a more fit variant for the same adaptive task under selection in both population contexts (Boyd and Richerson 1985:174; see also Cavalli-Sforza and Feldman 1981); 2) guided variation in which individual trial-and-error learning improves upon the culturally inherited variant of a trait through the comparison of the trait's performance against the perceived goal (Boyd and Richerson 1985:174). Thus, as long as the goals of the task to which the variants are put are identical, guided variation in two isolated populations can produce convergence in the frequencies of variants. This is in fact the traditional social science definition of independent innovation; and, 3) direct biased transmission, the process by which extant variants of a trait are evaluated and adopted for later transmission to others based on their performance characteristics (Boyd and Richerson 1985:136-137). In the small population sizes of Pleistocene foragers, adaptive change such as this may be swamped by the inter-generational sampling error associated with cultural drift (Shennan 2000).

Cultural Drift: the accumulation of random errors in the cultural transmission of trait variants within operational sequences of two isolated populations which produces similar frequencies of variants between the populations. Akin to convergent evolution via genetic drift (Boyd and Richerson 1985; Cavalli-Sforza and Feldman 1981), as well as both Binford's residual style (1965) and Sackett's isochrestic style (1990) concepts (Bettinger et al. 1996), this force of convergence despite independent innovation is less likely than adaptive convergence since the chance of two operational sequences drifting in the same way is low. This assumes the traits within the operational sequences are functionally independent and that Rands and Riley's (1958) "complex demand" in not active (see below).

proficiency in a wealth of behaviors, such as the methods for exploiting a landscape's resources, the social uses of material culture, the formation of social relationships with kin and non-kin, the body techniques and operational sequences for material culture production (*sensu* Lemonnier 1992), and the articulation between technological choices and the organization of the society's tasks on the landscape (Nelson 1991). Enculturation here equates to Gabora's acquisition of the mind's worldview that acts as the replicator in cultural transmission theory (Gabora 2004:132–133). While this conception of enculturation agrees with Herskovits (1948:41) in presuming that more enculturation occurs earlier than later in life (particularly for unconscious body techniques), this definition differs from Herskovits's earlier use of the term in emphasizing the continuity of learning processes exhibited by non-human as well as human primate societies. Enculturation is thus a phenomenon shared by, rather than distinct from, non-human primates. In this sense, the enculturating environment discussed here is more akin to Donald's (1991, 1998) use of the term, as well as Boesch and Tomasello's (1998) and Kelly's (1995:153–6) discussions of social learning from role models, with the caveat that no two individuals experience the world or enculturation in exactly the same way (Goodenough 1971:36; Wolcott 1991).

In terms of cultural transmission theory, enculturation is the process by which an individual gains proficiency in that portion of her/his culturally inherited information that is the least influenced by guided variation and direct biased transmission. Indirect and frequency-dependent bias on the other hand are likely to contribute to the conservative transmission of an enculturating environment. As a result, enculturation tends to produce unconscious or unquestioned predispositions to details of production sequences and body techniques, for one must make "things in a certain way because everything must be made some way and that was the way their parents did it" (Wiessner 1984:195). For instance, as J. J. Clark's ethnographic evaluation of Carr's (1995b) Middle-Range Theory of Artifact Design shows (2001:14–22), attributes of artifacts which are produced early in the production sequence and/or have low physical or contextual visibility are the best material proxies for household-level groups. As most residential-level cultural transmission is parent to offspring according to the ethnographic surveys conducted by Shennan and Steele (1999) and Hewlett and Cavalli-Sforza (1986), enculturation has a strong, although not exclusive, association with symmetric transmission.

From recent ethnographic and ethnohistoric examples of culture contact, it is clear that such events are arenas for the interplay of numerous social as well as biological processes (Wiessner 1983; Lightfoot and Martinez 1995; Lightfoot et al. 1998; Wells 1999; Van Kirk 1983; Merrell 1999; White 1991; Fix 1999; Cook 1998). Regardless of whether the cases involve agriculturalists or foragers, these different processes are contingent upon the social proximity of the actors involved (Wobst 1977; Carr 1995a) and Palaeolithic archaeologists have yet to use this fact to our advantage in understanding which modes of cultural transmission were likely in different contexts of culture contact.

Drawing on the ethnoarchaeological work of Wiessner (1982, 1983, 1984) and Lee and DeVore (1976) among the !Kung San, the process of culture contact between hunter-gatherers can be understood to vary between events in which individuals have little exposure to each others' residential space, with meetings occurring at the

Figure 3.5 Spectrum of locations and social intimacy involved with culture contact among foragers. Left image: "Hadza man looking for game-Tanzania" taken by Lars Smith. Right image: "San !Kung Woman with child at camp fire" taken by Stan Washburn. Copyrighted photographs provided courtesy of Anthro-Photo.

edge of their respective ranges, to events in which marriage partners are exchanged and previously 'foreign' individuals become fixtures in each others' residential lives (Figure 3.5). Wiessner's work (1983) suggests that bow shot range is the likely distance for contact between unfamiliar individuals. As individuals become more familiar or *socially intimate* with each other (i.e., willing to engage in social interaction rather than aggressive territorial exclusion), interaction begins to take place within residential camps, where many individuals of diverse ages, *and particularly children*, are exposed to the stranger. This represents the access of the stranger to what Stanner (1965; Peterson 1986) termed the forager's estate, the core of her/his range. When socially distant individuals become familiar enough with each other to affect how they, *their respective offspring, and/or the offspring of their potential mating* learn about the world, their respective enculturating environments have begun to overlap. This overlap is the temporal, geographic, and social space in which acculturation occurs, defined here as the process of learning or cultural transmission from a socially distant individual. Two socially distant individuals are usually not genetically related (although there are exceptions), with the result that most cultural transmission which results from their contact, i.e., acculturation, is horizontal or asymmetric transmission. *Social distance* within an enculturating environment itself may be defined at various levels of intimacy, frequency of interaction, and form of interaction depending upon the question but may be taken here as encapsulating individuals within the "intimate", "effective" and possibly the "extended" networks discussed by Gamble (1999, 52-64; but see also Dunbar 2003; Milardo 1988; Hill and Dunbar 2003; Zhou et al. 2005). The social distance concept need not relate to a particular land tenure or territoriality system, as it is clear that hunter-gatherers have a diversity of social openness and land ownership strategies (Eerkens 1999; Kelly 1995:161-204; Peterson 1986; Dyson-Hudson and Smith 1978; Stanner 1965; Wilmsen 1973). Social distance is intentionally left widely defined and independent from a geographical definition for that reason. Due to the nature of Paleolithic palimpsests with potential millennial contributions of material culture to individual assemblages, it is unlikely that the relevant population sizes and social distances have remained the same in size or location for the duration of the formation of the assemblage. Thus, the actual range of social distance may be impossible to determine with Paleolithic data. For instance, the meaningful unit of social life would be constantly renegotiated as ecological and social systems adjust to climatological shifts that are amalgamated in the creation of one lithic assemblage (Gamble 1991, 1993, 1999). Moving beyond the above limitations, however, Hovers (2001) presents very thought-provoking arguments for how social units and territoriality might be modeled in Pleistocene hominins, including through the application of evolutionary culture theory to lithic variability. While not identical to the approach being built here, there are clear similarities which show a willingness on both of our parts to consider both the social and energetic proximate mechanisms behind forager social behavior.

Regardless of the absolute size or land tenure of the social entities engaged in contact, acculturation of both parties is usually the norm, rather than the one-way street presumed by colonialist literature (see examples within Fagan 1998). Yet there is variation in the degree to which enculturating environments are affected by contact, with some cases evidencing almost no acculturation. The contact between the Greenland Norse and Dorset Inuit is a clear example where cultural resistance to change as well as distinct ecological niches prevented the acculturation of at least the Norse, resulting in this culturally conservative group going extinct (McGovern 1981, 1994). Thus we should expect that Pleistocene examples of contact may evidence a range of effects, from significant to insignificant acculturation. Please refer to Table 3.2 for definitions of these concepts.

Where in this range of acculturation a particular case lies can be determined by an interesting relationship between where contact occurs and what can be learned between groups. Specifically, the differences in the location of contact, from unsocial encounters on the edge of a forager's range to social intimacy in the residential heart of the forager's encampment, result in differential amounts of exposure to information and behaviors of each group. Differential access to information thus produces the differential ability to learn and adopt from the other group (i.e., to be acculturated). This represents differential access to the enculturating environment. Thus, in a culture contact event between stone tool using hunter-gatherers, the amount of lithic technological knowledge learnable by each party depends upon where contact occurs relative to where lithic operational sequences are enacted on the social landscape.

Using the ethnoarchaeological work of Wobst (1977), Wiessner (1982, 1983, 1984), and Sinopoli (1991), the theoretical work of Gamble (1999:82-87) and Ingold (1993), and Carr's Middle-Range Theory of Artifact Design (1995a,b), the lithic core reduction sequence can be understood in two parts: 1) the sequence of behaviors which produced the blanks for use; and, 2) the sequence of behaviors of retouch application and edge rejuvenation of the curated, mobile tool kit. From our understanding of the organization of lithic technology relative to the demands of optimal foraging theory (Nelson 1991), blank production, i.e., how nodules are shaped into cores and reduced into blanks, occurs mostly at raw material sources and residential locations. Conversely, the knapping of

Table 3.2 Definitions of key terms in the middle-range theory of social intimacy and taskscape visibility

Enculturation: the process by which an individual gains proficiency in that portion of her/his culturally inherited information that is the least influenced by guided variation and direct biased transmission. Indirect and frequency-dependent bias on the other hand are likely to contribute to the conservative transmission of an enculturating environment, giving enculturation a strong but not exclusive association with symmetric transmission.

Enculturating environment: the learning environment provided by a population of hominins through which an individual is enculturated (as defined above). The enculturating environment can be seen in some ways to correspond with the structuration (Giddens 1984) and habitus (Bourdieu 1977) which influences an individual's cultural performance. The enculturating environment is thus made up of the physical performances of individuals in the group which are witnessed by the individual under enculturation and affect her/his adoption of performance techniques. This definition allows for a bottom-up approach to culture, in which individuals play an active role (Gamble 1999:33). Archaeologically, the enculturating environment is thus defined as the performances of those individuals whose flintknapping behaviors were enacted at a given spot on the landscape within the given time range of the formation of an archaeological assemblage (see Tostevin and Škrdla 2006 for the development of this argument). In other words, each archaeological assemblage represents one enculturating environment created by all of the hominins who have contributed material culture to its palimpsest.

Social Intimacy: the degree to which foragers from different enculturating environments are willing to engage in social interaction with each other rather than aggressive territorial exclusion.

Acculturation: the process of learning from a socially distant individual as a result of the overlap of two enculturating environments. As two socially distant individuals are usually not genetically related, acculturation tends to result from asymmetric cultural transmission.

Social Distance within an Enculturating Environment: the individuals within the "intimate," "effective," and possibly the "extended" networks discussed by Gamble (1999:52-64; but see also Dunbar 2003). The social distance concept is not specific to a particular land tenure or territoriality system.

Taskscape Visibility: the relationship between where, when, and with whom a cultural trait, such as a flintknapping behavior, is performed and the possible cultural transmission modes available for promulgating the trait into the next generation.

Stimulus Diffusion under the Taskscape Visibility concept for lithic technology: the transmission of the familiarity with specific tool kit morphologies without the transmission of the detailed, specific knowledge to produce these morphologies.

blanks from cores occurs less frequently on the pathways of the landscape due to the inefficiency of transporting a heavy core when a lightweight flake or retouched tool could provide the necessary cutting edge (Kuhn 1994), adaptive reliability (Bleed 1986), and functional flexibility (Shott 1986, Nelson 1991) for most anticipated tasks. That is not to say that cores and core-tools cannot be exported from raw material sources and residential locations, as this is clearly seen in their absences from refit data (Pitts and Roberts 1997: plate 27; Hallos 2004, 2005; Pope and Roberts 2005). Rather, in the contexts of the Middle to Upper Paleolithic transition in western Eurasia in which large core-tools (such as bifaces) are absent, there would have been few reasons to incorporate cores into the mobile toolkit, since they lack the tool association of bifaces (Kelly 1988). Also, their function as "sheathed" flakes (*sensu* Kuhn 1994:437) and items of an ad hoc provisioning of activities strategy (*sensu* Kuhn 2004:432–433) are more efficiently satisfied with a small set of flakes in the tool kit. Thus, in this temporal and geographic context, core reduction as a systemic activity would be rare on the pathways of the landscape. Instead, the tool kit carried into the landscape consists of blanks selected from the pool of debitage products made at the raw material and residential locations. Thus while blank production behaviors and initial tooling/hafting of the tool kit are enacted in these locations (and there the material evidence of these behaviors is thus deposited), only the selected blanks of the curated tool kit are transported onto the pathways of the landscape (see Close 2000 for an example of the demonstration of such artifact mobility via refits between sites). As a result, only the morphology of this tool kit would be visible on a hunter-gatherer during contact events at the edge of the foraging range, where she or he might encounter a social stranger. Compared to an individual who is allowed to visit at a residential site where a diversity of visible behaviors are enacted (including tools made for use only at the residential site), the stranger met on the pathways of the landscape would only be exposed to and thus only be able to learn from the curated tools of the forager. Compare what artifacts and behaviors are visible in the two images of foragers at different contact locations in Figure 3.5. A visitor to a hunting stand or retooling site, on the other hand, would only be exposed to the mobile tool kit and perhaps the retouch behaviors used to rejuvenate the cutting edges.

This opposition of behaviors enacted on the landscape versus those enacted at residential sites is analogous to Gamble's elegantly elucidated distinction between an *encounter* and a *gathering* (1999:68–71).

> Locales, rather than sites, are where interaction took place and is preserved. They take a number of forms. The starting point is the individual and her/his movement through their environment. This involves ambulatory perception (Gibson 1979) along paths and tracks. From this starting point we can begin to build a model of interaction....The *encounter* may involve an animal running across the hunter's path or another person coming into view... Both the encounter pattern of foragers and the intercept pattern of collectors (Binford 1980) can be described in these terms. The affordance of the encounter will lead to decisions about immediate and future action–eat the plant food, kill and carry the reindeer, make the spear, pick up the stone, greet the network partner, flee the stranger, remember and recount the information to others in your network. Encounters are information 'sound-bites' about the environmental track

along which the hominid is traveling (Gamble 1999:68–70, emphasis original).

A gathering, in contrast, is an “enduring locale” (Gamble 1999:70). The main difference between Gamble’s concepts and those advanced here is the present model’s specific predictions on what effect the difference in these two types of locales has on the potential acculturation of lithic technology which might result from culture contact in each context. Much of this predictive power comes from the ethnographically-derived middle-range theory of Carr’s (1995b) physical visibility hierarchy of artifact attributes. The present research’s predictions for the pattern produced by different culture historical processes, however, are not based only upon the physical visibility of different types of stone artifacts found in each location (cores and debitage vs. retouched tools) but also upon the context of whether or not a stranger has the opportunity *to witness the flintknapping behaviors* which produced a given tool in a specific locale. Flintknapping is a technological performance, matching the criteria used by anthropologists of performance (e.g., Beenan 2008) which can only be effectively taught and learned by observation of the body movements, the handling of the core and percussor, and the flake by flake strategizing on how to exploit core volumes. It does not necessarily require spoken instruction (although this does speed up skill acquisition), but it does require opportunities for lengthy visual observation (Ohnuma et al. 1997; Blackwell 2004). It also requires abundant opportunity for repetition and practice (Whittaker 1994:2-9; Waldorf 2006; Clarkson 2010:45). For example, if the social context of the contact between two individuals from different enculturating environments remains less than intimate and the contact event occurs away from residential and/or raw material locations, the visiting individual has little chance to observe and thus learn the early phase of the lithic core reduction sequence, i.e., the blank production behaviors. Given the equifinality of multiple manufacturing sequences leading to the same tool form (Sackett 1971, 1977, 1982, 1990; Lemonnier 1992) – a well recognized phenomenon in lithic technology (Marks and Kaufman 1983; Tixier 1984; Boëda 1995), it is fairly certain that the prehistoric visitor has little chance of reconstructing and thus copying the blank production behaviors of the other, even if the visitor is an excellent flintknapper. This is particularly true for a prehistoric knapper whose performance repertoire has been severely limited, compared to a modern knapper, by her/his own cultural tradition of knapping that dictates culturally-acceptable artifact forms and production sequences. As archaeologists, trained with the perverse desire to know how everything is made, and amateur knappers, living off of the sales of our regionally-varied artistic productions, we tend to forget that we have had the benefit of examining examples of knapped stone from many periods and parts of the globe. Prehistoric knappers did not have this global perspective, were likely discouraged from free innovation (to judge from Wiessner 1984:195), and so would have been less competent than a modern archaeologist in deducing the entirety of a core reduction sequence from a lone tool, even though they were likely better at their own production sequences than are modern knappers. Thus, while admittedly remaining an avenue for future exploration, it is safe to propose that even the best knapper is unlikely to be able to replicate the whole core reduction sequence from an item of the curated tool kit without the analysis of the debitage debris to use as inference. It is possible that a visitor could independently innovate some of the behaviors of blank production from examining a tool alone but it is unlikely that s/he would produce the same debitage-wide central tendencies for all of the behaviors in the process. A tool alone does not provide clues to all of the behavioral details of its production sequence, much as an idea does not include self-replication instructions as argued by Gabora (2004) for cultural transmission theory. Convergence through the independent innovation of behaviors within blank production is thus always a possibility, but not a high probability, as its likelihood depends greatly upon the size of the geotemporal context in which one is looking for convergence, as discussed later in Chapter 4.

In utilizing Carr's middle-range theory of artifact design, the context of the artifact's use is an important issue (Carr 1995b:185). In the case of the present middle-range theory, this context is explicitly temporal ("were you there when the flintknapping happened?") as well as geographic ("were you in the right place to see it?"). This contextual approach to tasks enacted on the landscape has been pursued in detail by Ingold (1993) through his concept of the taskscape.

> ...intuitively, the landscape seems to be what we see around us, whereas the taskscape is what we *hear*. To be seen, an object need do nothing itself, for the optic array that specifies its form to a viewer consists of light reflected off its outer surfaces. To be heard, on the other hand, an object must actively emit sounds or, through its movement, cause sound to be emitted by other objects with which it comes into contact (Ingold 1993:162, emphasis original).

Thus, the taskscape is defined and experienced, or *heard*, through the social activities and embodied information of technical acts (*sensu* Gamble 1999:71) conducted at given places and times in the environment. Given how similar this view of the taskscape is to the above discussion of the visibility of flintknapping behaviors at particular places and times on the landscape, it is fitting to combine Carr's and Ingold's terms to describe the context of when and where a technological act is learnable/transmittable (from the operational sequence, to the finished object, to its social use) as its *taskscape visibility* (Figure 3.6). By comparing the taskscape visibility of the two components of the lithic operational sequence, i.e. blank production and the mobile tool kit, it is possible to distinguish different degrees of contact and social intimacy between the knappers responsible for different archaeological assemblages. If contact has been intimate enough for periodic residential mixing of the groups, then the details of the blank production sequence could be copied between two groups. If contact has been less intimate, only the effective morphology of the tool kit could be copied. This is in essence the difference between behavioral diffusion and stimulus diffusion, as defined over 70 years ago by Kroeber (Table 3.2 and Figure 3.7). The taskscape visibility concept thus provides the theoretical concept to differentiate the flintknapping behaviors which constitute the enculturating environment (symmetric transmission within the group) from those behaviors that are susceptible to asymmetric transmission without social intimacy (via stimulus diffusion).

The Social Side of the Curation Concept

The curation concept has been employed by generations of archaeologists (Binford 1978; 1979; Gramly 1980; Bamforth 1986; Shott 1986, 1989; Kuhn 1991, 1992, 1994, 1996, 2004; Morrow 1996; see also references within Andrefsky 2008). Throughout this period, however, only optimality models driven by the principles of evolutionary ecology have been applied to this portion of the lithic operational sequence, despite clear problems with deciding which variable is being maximized (Torrence 1989) and how to deal with conflicting predictions from first principles (Jochim 1989). Yet ethnographic data has

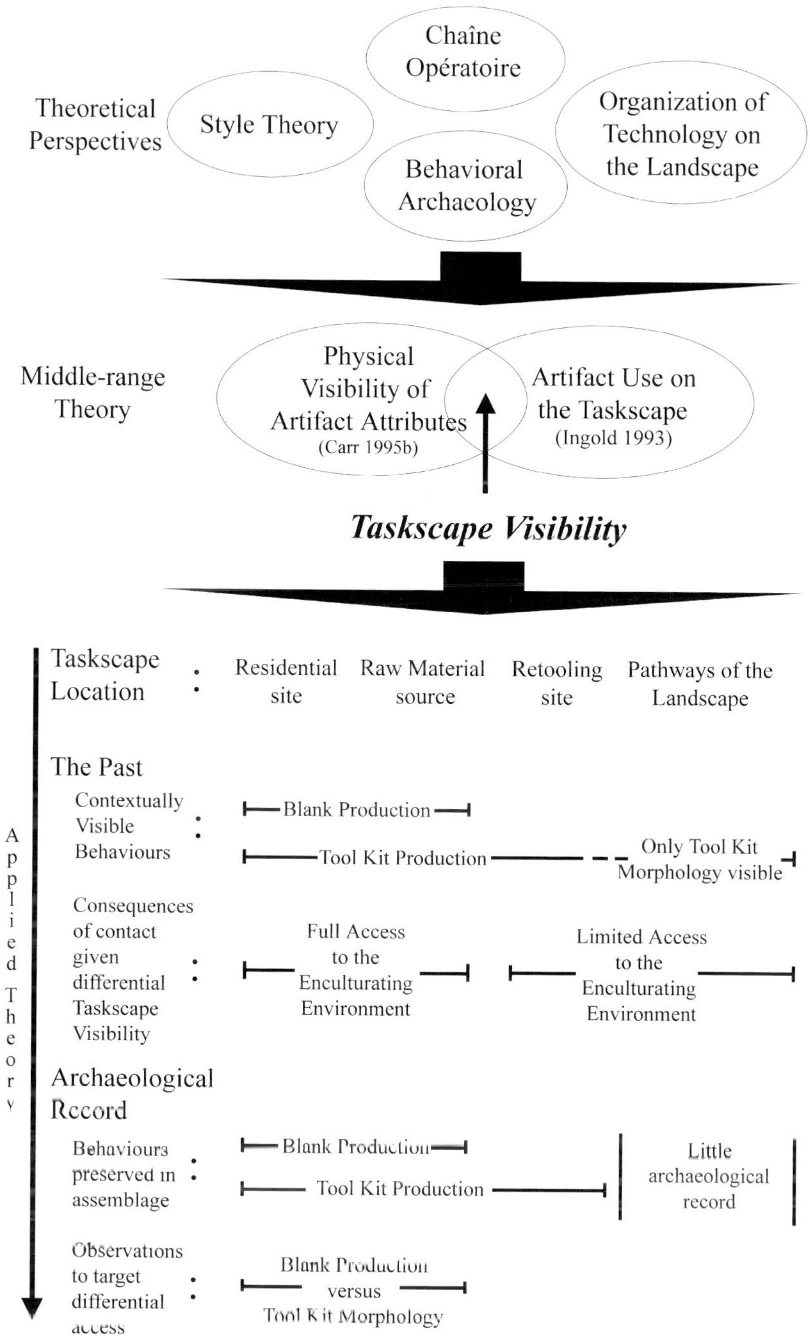

Figure 3.6 Development of the middle-range theory for the use of taskscape visibility to target particular aspects of the archaeological record for the differentiation between enculturation and acculturation among stone tool-using foragers.

long supported the view that the mobile tool kit is the most socially exposed aspect of a society's material culture (Wobst 1977), even if the production sequence details are not communicable through the tool kit. In fact, hunter-gatherers meeting on the pathways of the landscape are known to trade arrows (Wiessner 1983; Kelly 1995:164), with the result that arrowhead morphology is known to exhibit population specific isochrestic variability, emically and etically recognized (Sinopoli 1991), through materially reinforced social differentiation (Hodder 1985). Whallon's (2006) model for the communicative role of hunter-gatherer mobility networks also emphasizes the adaptive importance of sociality in mobility, if not explicitly the distinction between the curated tool kit and the enculturated blank production behaviors established by the taskscape visibility concept. The taskscape visibility concept, combining as it does the Organization of Technology approach with the style theory of social interaction (Figure 3.6), thus reflects the long neglected 'Social Side of the Curation Concept'. With the blank production behaviors serving as a proxy for an assemblage's enculturating environment, the tool kit morphology can serve as a proxy, following the logic of Carr (1995b), for both functional or task-specific needs on the landscape as well as possible social signaling through the communicative co-option of isochrestic variability. In fact, this is in many ways an attempt to fulfill Binford's plea (1982) for the alteration of the systematics of archaeological investigation in order to study 'the organizational relationships among places which were differentially used during the operation of past systems' (1982:5). The utility of the social intimacy and taskscape visibility concepts thus afford a reworking of the definitions of the nested relationships between culture historical processes identified for Pleistocene foragers, as reflected in Figure 3.7.

How to Choose the Appropriate Units of Analysis

Given the fact that this approach is based on investigating cultural transmission between hominins within the enculturating environment versus transmission during their contact events, the smallest units of analysis must be physical observations on artifacts which reflect learned behaviors related to how blanks are produced during flintknapping and how blanks are selected for retouch and incorporation into the mobile tool kit. As argued elsewhere (Tostevin 2000a, 2003a), these individual flintknapping behaviors should be measured as central tendencies and dispersions within an assemblage, rather than as anecdotal refits or more abstract (and thus epistemologically less secure) etic constructs of archaeologists, such as 'Levallois', 'desired end-products', and 'industrial types.' See Davidson and Noble (1993) for a critique of the "finished artifact fallacy" and Straus (2003) and Clark and Riel-Salvatore (2003) for the problem of the reality of industrial types limiting our ability to understand variability through time and space. While combining assemblages into industrial types, technocomplexes, and other more abstract entities has its appropriate function and utility for archaeological analysis (for some of these uses, see van Andel and Davies 2003), those appropriate tasks do not include measuring social intimacy or distinguishing between different modes of cultural transmission. The diminution of variability resulting from a typological approach to industrial complexes eliminates the specific data needed to answer a culture historical question (Tostevin 2003a; Tostevin and Škrdla 2006). In focusing on individual behaviors, the present approach has thus more in common with Schiffer's Behavioral Archaeology (1976, 1996) and Leroi-Gourhan's plea for a prehistory of techniques (1943) than with traditional

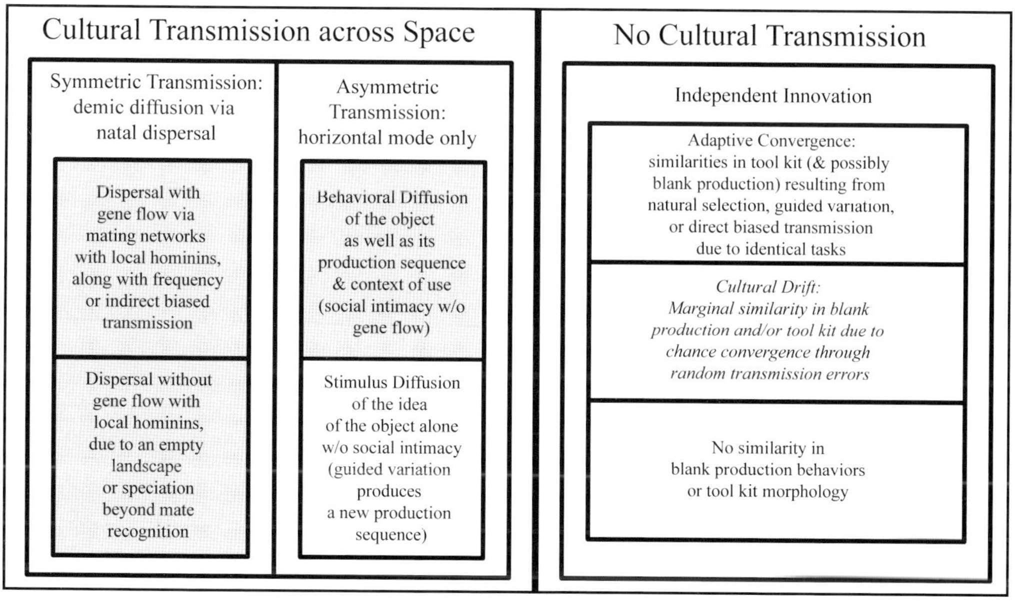

Figure 3.7 Nested structure of culture historical processes redefined through the social intimacy and taskscape visibility concepts. The shaded processes indicate strong similarity in blank production as reflecting similarity in enculturating environments between earlier and later archaeological assemblages. The italicized process indicates the random chance of a weak similarity in blank production not due to the overlap of enculturating environments.

Palaeolithic archaeology of either the American (Kuhn 1995) or European (Boëda et al. 1990) style (*sensu* Sackett 1991). Documenting the variability within individual flintknapping behaviors as cultural traits is also a requirement of the evolutionary approach to culture history, since cultural and natural selection work on variation rather than essential types.

The units of analysis advocated here are also non-traditional in regard to sampling the landscape. One might think that in opposing blank production behaviors with tool kit morphology one would contrast two different functional types of archaeological sites from two social entities, i.e., residential sites and 'pathway' sites from two industrial types. This is inadvisable for two reasons. First, this involves the culture historical question with epistemologically problematic units, i.e., retouch-defined industrial types. Second, 'pathway' sites, or encounters in Gamble's terminology, are archaeologically invisible (Gamble 1999:71): a contact encounter in which the tool kit is visible need not produce any discard into the local sedimentary environment. Thus raw material workshops and residential locations are the most suitable assemblage types for inclusion in Pleistocene culture historical research of this type. Specialized hunting sites, retooling sites, and sites distant from raw material require a slightly different treatment, as in Škrdla and Tostevin (2005) and Tostevin and Škrdla (2006).

The archaeological assemblage also needs to be defined in a particular way to maximize the potential of the taskscape visibility concept in relation to the concept of an enculturating environment. Each assemblage is defined as the association of flintknapping behaviors enacted at that spot on the landscape within the given time range of the assemblage's palimpsest (see Tostevin and Škrdla 2006 for the development of this argument). In other words, each assemblage represents one enculturating environment created by all of the hominins who have contributed material culture to its palimpsest. One consequence of this definition is that the represented hominins may include individuals who otherwise would be typed by archaeologists as being responsible for assemblages classified as different industrial types. On the other hand, this definition is both epistemologically justifiable and appropriate for the application of the taskscape visibility concept. The definition does require, however, the analyst to investigate the taphonomic integrity of the assemblage from the point of view of post-depositional alteration of artifactual content, after the closing of the original palimpsest. It also requires the analyst to use the association of artifacts that possesses the shortest temporal palimpsest justifiable from the geoarchaeological understanding of the site when defining the analytical "assemblage".

Given this definition, an assemblage from a residential location or raw material workshop can be compared to another such assemblage, according to the distinction in taskscape visibility: blank production behaviors in one assemblage are compared with those in another assemblage while the tool kit morphologies of the two assemblages are contrasted. The comparability of the assemblages, given the effects of differential raw material use on the morphology of the tool kit, are addressed in the next chapter. This pair-wise approach to comparing two assemblages according to their blank production behaviors and their tool kit morphologies is designed to test for continuity of these units through time and space, and thus test for the distinction between cultural transmission across space and the independent innovation of these units in each region. It is possible, however, to use the social intimacy and taskscape visibility concepts in other methodological configurations. Tostevin (2007) presents a separate methodology for a three-way comparison in which two enculturating environments (i.e., assemblages) are evaluated for their degree of overlap and subsequent contribution of culturally transmitted technical choices to a third assemblage. This three-way method of comparison is specific for acculturation hypotheses, such as the Bow Wave or Neanderthal Acculturation hypothesis (Mellars 1993, 2005; Klein 1973). It is also particularly useful in regions in which multiple phyla of industrial types have been hypothesized to exist. For the purposes of the present research, however, the pair-wise comparison methodology offers greater ease when dealing with three regions and 18 archaeological assemblages.

3.3 Bridging From the Other Bank: The Need to Model Knapper Technological Decisions from the Artifact Up

This chapter has only accomplished half of the task in building a middle-range theory that bridges the intellectual goals of studying cultural transmission in the Pleistocene, i.e. high-level theory, with the practical limitations of extracting meaningful information out of the available dataset, i.e., low-level theory. The following chapter, however, completes the construction of the middle-range theory by working from the level of individual flakes and cores, through the identification of the etically observable proxies on lithic remains that indicate which knapping decisions were learned and performed at given types of sites on the Pleistocene landscape.

4

Building Bridges from Low-Level Theory: Part Two of a Middle-Range Theory for Paleolithic Culture History

> No human investigation can be called real science
> if it cannot be demonstrated mathematically[1]
> - Leonardo da Vinci (1452-1519)

As the last chapter presents the high-level theoretical basis for the evolutionary study of cultural transmission during the Middle to Upper Paleolithic transition, the present chapter develops the complementary methodology that creates bridging arguments that connect low-level theory, i.e., the observations on artifacts, to the theoretical units of analysis, i.e., technological choices performed within the parameters of the social intimacy and taskscape visibility concepts identified in Chapter 3. This methodology consists of a body of lithic analyses designed to characterize the lithic reduction sequence of each assemblage studied. These analytical techniques will serve as proxies for the technological choices enacted within the knapping process, which themselves are the units of analysis for evaluating the goodness of fit between the five model expectations and the archaeological record of the three regions in question. For as Eerkens and Lipo state, "Although there are no agreed upon empirical units of CT [cultural transmission], we can build a unit for measuring CT" (2007:216). This chapter details why particular knapping behaviors are the best units of cultural transmission concerning this technological process and how these behaviors are recognized in particular lithic attributes for building the most suitable units of measurement of cultural transmission in the Pleistocene.

The construction of this list of lithic analytical techniques is far from simple, as the evolutionary approach to culture history presents several requirements for the standardization of the reconstruction of the large diversity of methods for producing cutting edges out of a block of stone. Each analytical technique for reconstructing a particular behavior within its procedural context must:

1) **serve as a proxy for the technological choices of the knappers in such a way as to characterize both central tendency and variability within the assemblage; and,**
2) **must be readily comparable between geographically-diverse assemblages according to the model expectations of the competing hypotheses.**

Behavioral Archaeology, the *Chaîne Opératoire* school, and the Organization of Technology approach, as detailed in the previous chapter, all identify technical choices made by artisans during the sequenced production of artifacts as the most appropriate units of analysis for fulfilling these two requirements. These approaches thus serve as middle- and high-level theory for this research. As noted earlier, the Organization of Technology approach is highly suitable for contextualizing how lithic variability is structured in relation to resources and processes on the landscape, as with the taskscape visibility concept from Chapter 3. The Behavioral Archaeology program treats technical choices within the sequence of artifact production as units suitable for evolutionary modeling (Schiffer 1996), such as with cultural transmission theory. Last but not least, the *Chaîne Opératoire* school conceives of each technical choice as a performance enacted in a given time and place, with the potential for individuals present to observe and thus learn both the conscious (*connaissance* or strategic) and unconscious (*savoir-faire* 'know-how' or tactical) knowledge of the process (Pelegrin 1990b). Through the study of ethnographic performance (Pétrequin and Pétrequin 1994; Stout 2002, 2005) and the cognitive, biomechanical, and neural-motor study of stone tool making (see references within Roux and Bril 2005), the *Chaîne Opératoire* approach thus contributes, along with other branches of archaeology, to my identification of "performances of technical choices" as cultural learning sets, the most suitable units of analysis for cultural transmission theory and the evolutionary study of Pleistocene culture history.

At the scale of low-level theory (i.e., the archaeologist's techniques of measurement and observation), however, *chaîne opératoire* practitioners frequently use a less quantitative and more emic-level approach to the demonstration of the details of a core reduction sequence (e.g., Boëda 1994). This is especially true when contrasted to the very etic and relatively quantitative approach of the attribute analyses of core reduction sequence studies in the Americanist tradition (e.g., Sackett 1966; Movius et al. 1968; Callahan 1979; Ohnuma 1986; Johnson and Morrow 1987; Bergman 1987; Amick and Mauldin 1989; Kuhn 1995; Bradbury and Carr 1999; Shott et al. 2000). While there are many similarities between these approaches (Bleed 2001; Shott 2003; Stark 1998), there are also aspects of the emic bias in *chaîne opératoire* theory (see Chapter 3) which penetrate from high-level theory to the level of observations and low-level theory (Tostevin 2006, 2011), making it necessary to choose between the approaches for the purposes of the present research.

4.1 Which Methodology: French, American, or Neither?

Leroi-Gourhan's concept of *milieu technique* and *chaîne opératoire* have had a significant impact on the pursuit of Pleistocene prehistory in France and elsewhere (Audouze 1999; Gamble 1999:32-97). For a time period from which only stone implements and a few animal bones are preserved, any theory and method which promises detailed description of actual sequences of behavior can be expected to be lauded from all sides. Significantly, the late Upper Paleolithic (Magdalenian) sites of Pincevent (Leroi-Gourhan and Brézillon 1983), Étiolles (Pigeot 1986), and Verberie (Audouze and Cahen 1984; Audouze 1994) which Leroi-Gourhan and his students used as case studies to develop the *chaîne opératoire* approach to Paleolithic archaeology are as close to fitting a Pompeii Premise as any Pleistocene sites known. Specifically, the numerous and extremely complete refittings of stone implements into the original nodules of flint used by the flintknappers allow precise description of the sequences of knapping operations from the beginning of knapping to the retouching of the blanks into tools. Due to these refittings, very little guesswork is needed to reconstruct the *connaissance/schéma opératoire* of the Magdalenian knappers and a few scholars have gone so far as to claim to have identified differences in *savoir-faire* among the refitted cores attributed to master-knappers and apprentices (e.g., Pigeot 1990; Karlin et al. 1993; Ploux and Karlin 1993). While many scholars are more reticent to identify individual flintknappers in the archaeological record (but see Van Peer 2007), the technological school is nevertheless theoretically very attractive for prehistorians.

There is a reason, however, why Paleolithic archaeologists should be cautious in adopting this particular methodology to reconstructing lithic technology for all questions: without refits, the validity of the interpretations frequently lies within the researcher's reputation and his or her personal expertise in inferring dynamic flintknapping operations from the static palimpsests of stone tool assemblages. Few individuals are trusted or feel sufficiently free from researcher bias to undertake such a research paradigm. A description of a technological study of a stone tool assemblage may make this point clear. In order to reconstruct the bodily gestures or physical hammer blows utilized by the prehistoric flintknappers to create a stone tool assemblage, i.e. in order to reconstruct the *milieu technique* of the group, the *Chaîne Opératoire* school relies upon recognizing the generalized (i.e., assemblage wide) step-by-step procedures of selection of the flint nodule, shaping of the core, removal of flakes as debitage from the cores, selection of blanks from the debitage for retouch, retouch of the flakes, use of the tools, and discard of the tools and debitage. This is accomplished frequently by refitting studies. Unfortunately, the opportunities to utilize refitting as a means of demonstrating an operational sequence are infrequent, mostly due to poor contextual preservation of a site, the original behavioral context of the site lacking refittable pieces, poor excavation or curation techniques, and/or a lack of laboratory time and space for such studies (see Goring-Morris et al. 1998, Schurmans 2007, and Laughlin and Kelly 2010 on the comparative study and applicability of refitting studies). Further, one or two totally refitted nodules may or may not be representative of the dominant *schéma opératoire* of the assemblage and so may misdirect the researcher's inferences. A method not reliant upon refitting studies is thus needed to augment extant refitting studies and thus to bring technological research to its theoretical potential.

Working on the Middle Paleolithic, a period in which sites typically contain few refittable sequences (Maastricht-Belvédère being a major exception – Roebroeks 1989; Schlanger 1996), Éric Boëda (1986, 1994) has championed an approach to technological studies which does not use refittings but a reading, *une lecture des schémas diacritiques*, of an assemblage's diagnostic pieces which allows one to infer the dynamic, step-by-step knapping behaviors within an assemblage (Boëda et al. 1990). Such *lectures* rely upon the researcher's personal knowledge of stone fracture mechanics and the trial replication of the assemblages to retrieve *les technique du corps* of the material culture. The inferences are not corroborated by an independent assessment of the validity of the observations in the published literature, except in a few cases (e.g., Geneste 1985, 1988). In the eyes of many North American archaeologists, this lack of transparency of the analytical process between observations and interpretations limits the impact of their conclusions and the general potential of the French technological approach.

As argued elsewhere (Tostevin 2003a:57), the striking disagreement between Boëda's (1988a) reconstruction of the flintknapping technology of the French Middle Paleolithic assemblage of Biache Saint-Vaast, Level IIA, and Dibble's (1995a) technological attribute analysis of the same assemblage demonstrates the problem of the reliability of *les lectures des schémas diacritiques*. Basing his conclusions upon an examination of the nuclei and debitage of this level of Biache Saint-Vaast, Boëda inferred that two systems of *recurrent* Levallois core reduction were present, Schema A being unidirectional and Schema B being bidirectional, in addition to a third, non-Levallois element. Boëda concluded that the two Levallois schemas were independently executed, i.e. that each core was reduced unidirectionally or bidirectionally but never both. Dibble analyzed the same assemblage and demonstrated strong correlations between debitage length and the dorsal scar patterns on the debitage. Longer debitage blanks are represented by a higher percentage of unidirectional dorsal scar patterns while shorter blanks are represented by a higher percentage of bidirectional, subradial, and radial scar patterns. A similar correlation was found between the amount of cortex on debitage blanks and dorsal scar pattern. As Dibble concluded:

> Now, while neither cortex nor size can allow us to state the exact point in the reduction sequence when a particular flake was removed, it is nonetheless clear that both of these measures do generally reflect the relative position. With both of them we can see that the relative proportions of the different scar pattern morphologies changes with respect to order of removal... (Dibble 1995a:104).

Dibble's attribute analysis thus shows that there is significant evidence for a relationship between Boëda's two schemas and stage of reduction of the cores. In contrast to Boëda's claims for the schemas' independence, the most parsimonious explanation of Dibble's analysis is that cores were reduced at the beginning of their use-lives using a unidirectional strategy (Schema A) but were reduced with a bidirectional strategy (Schema B) toward the end of their use-lives. Boëda's reliance upon a visual reading of the cores and debitage of the assemblage, without the corroboration of cumulative and recorded observations, did not enable him to recognize trends in the debitage that were easily revealed through attribute analysis. Other advocates of the *chaîne opératoire* approach to Paleolithic archaeology do not suffer from such a lack of analytical rigor, for instance Geneste (1985, 1988), Pelegrin (1990a, 1995), Soressi (2002), Roussel (2011), Soressi and Geneste (2011), and scholars fortunate enough to possess extensive refittings. Yet the methodological failings of the *lecture des schémas diacritiques* has tarnished the more rigorous applications of the *chaîne opératoire* approach and thus limited the true utility of the Mauss/Leroi-Gourhan high-level theory.

The example of Dibble, Boëda, and Biache Saint-Vaast shows how a quantitative attribute analysis according to the dynamic understanding of the reductive process is a more rigorous method for demonstrating a reduction sequence than a *lecture* of the artifacts which is not cross-checked with a quantitative demonstration of the *lecture*'s validity. But the quantification of the attribute analysis approach alone does not satisfy the requirements for the units of analysis of an evolutionary approach to culture history listed at the beginning of this chapter. Even with quantification and dynamic analyses, both the American and French approaches to lithic technology tend to categorize archaeological assemblages into typological units, such as Levallois (Boëda 1994), *discoïde* (Boëda 1993), blade (Hahn and Owen 1985), bipolar (Kuijt et al. 1995), bifacial (Parry and Kelly 1987), etc. Even some of the most theoretically and methodologically powerful experimental studies in the American approach (e.g., Bradbury and Carr 1999; Carr and Bradbury 2001) type reduction sequences in the same way as the least quantitative of *chaîne opératoire* practitioners, even though the former's demonstration of the existence of the reduction sequence is based on complex multivariate statistics. Instead, rather than suppressing the entirety of the variability in the assemblage into a type label, it is necessary to characterize each technical choice within the reduction sequence of the assemblage by means of its variability in the assemblage (Tostevin 2009). As Eerkens and Lipo (2007:239) argue for cultural transmission theory in archaeology in general, only through the measurement of the variability of traits in the sequence can one pursue an evolutionary study of how flintknapping behaviors change through time.

In addition to sharing the typological problem (see Tostevin 2011), both the American and French method of describing lithic reduction sequences frequently publish the results of their studies as a graphic illustration or drawing of the steps/stages of reduction through which the artifacts progress. In both cases, these stages of reduction are frequently similar: procurement of raw material, creation of one or more striking platforms, optional decortification, initial blank production, re-preparation of platform and debitage surfaces, late blank production, blank selection for tools, application of retouch, resharpening of tools, and discard of exhausted pieces. These stages of the core reduction sequence are commonsensical, based on the experience of modern flintknappers and examinations of the archaeological record. They are also of course arbitrary and etic. As such, these categories might be expected to serve as the units of analysis for the evaluation of the goodness of fit between the archaeological record and the five model expectations described at the end of this chapter. As noted at the beginning of this chapter, however, it is necessary for each behavioral step within the procedural context of flintknapping to have an analytical technique which 1) serves as a proxy for the technological choices of the knappers in such a way as to characterize both central tendency and variability within the assemblage; and 2) is readily comparable between assemblages according to the five model expectations. Without these two requirements, the current approach would quickly succumb to the failings exhibited by the *Kulturkreislehre* and early culture historical studies. These two requirements thus dictate that the flintknapping behaviors be divided into categories with respect to the quantitative method by which the behaviors are to be reconstructed, rather than by a commonsensical view alone. Unfortunately, the stage approach to reduction sequences is generally not defined in a way either to characterize the diversity within stages or to allow the accurate assignment of a prehistoric artifact to its actual reduction stage, as demonstrated with experimental knapping experiments (Shott 1996; Bradbury and Carr 1999) and the analysis of refitted archaeological assemblages (Shott et al. 2011). This makes these stages inappropriate for use in the present study.

Much of the inaccuracy of the stage approach to reduction sequences is the result of the nature of the models pursued in sequence studies. Bleed (2001:120–121) has identified two models common to both the American and French approaches. Bleed holds the *teleological* model of sequence studies to focus on the discovery of a predetermined process, often analogous to Deetz' (1967:43) mental template concept (e.g., Pelegrin 1990b). This predetermined process rarely can be characterized as anything but a type no matter how accurate the stages. In contrast, Bleed holds that the *evolutionary model* presents technological sequences as reactions to situations encountered in the sequence of production, use, and discard as part of tasks conducted in the environment. Studies following this situational approach, unfortunately, recognize responses at stages of reduction which may or may not be comparable between different reductions because different situational constraints and tasks may be present in each case. They are therefore not "evolutionary" in the context required by the present study.

While very useful for the study of reactions within the cultural context of one assemblage or closely related assemblages in time and space (e.g., Bleed 2011), the situational approach tends to produce incomparable results in the same way that the typological end-product concept of the teleological model does. A different concept of stage or node in sequenced behavior is required that recognizes where constraints in the technological process exist, i.e. points at which the artisan must choose a technical option, that are common to all of the assemblages under study. This new concept of nodes of technical choice must also, however, document the technical options as etic observations comparable between assemblages.

Thus, in contrasting the American and French approaches to the reconstruction of lithic reduction sequences, each approach has a useful aspect to contribute but neither alone is a sufficient methodology for the evolutionary study of Pleistocene culture history. Therefore, the answer to the question asked at the beginning of this section ("which methodology?") is *neither*.

4.2 What We Know the Knapper Chooses During the Flintknapping Process: The Argument for the Analysis of Variables by Domain

The history of anthropological studies of cultural transmission processes has shown that the objectivity of the quantitative method of early approaches, such as the *Kulturkreislehre*, was undermined by the subjective choice of which cultural elements to map as a means of defining cultural complexes (Kluckhohn 1936). In order to avoid the errors of the past, it is necessary to have *a priori* reasons for choosing particular culture historical units. The identification of such units is aided by both the taskscape visibility concept (Chapter 3) and experimental studies of what flintknappers control in the reduction process.

Given the fact that this research is based on determining which culture historical processes were responsible for the spatial and temporal continuity and/or discontinuity in *learned flintknapping behaviors* across the Middle to Upper Paleolithic transition, the smallest units of analysis must be physical observations on artifacts which reflect learned behaviors related to how blanks are produced during flintknapping and how blanks are selected for retouch and incorporation into the mobile tool kit. This requires the archaeologist to consider which flintknapping behaviors were in fact *learnable* through visual and auditory observation of the flintknapping process by prehistoric individuals becoming enculturated within that society. This task is called the *conceptual modeling of the observational learning of the flintknapping process* (Tostevin 2008). Beyond this task, however, the archaeologist is also required to know how those learnable behaviors affect the physical properties of the artifacts the archaeologist studies so that the etically-recognizable artifact properties (or attributes) can be used to recognize the behaviors that produced them, as argued by Van Peer (2007) and Bar-Yosef and Van Peer (2009). This task has long been the subject of research within experimental archaeology of both free-hand knappers and researchers into controlled flake fracture mechanics. We will find the appropriate units of analysis for the study of cultural transmission of flintknapping behaviors in those behaviors that are held in common between the conceptual modeling of the visual learning process and the experimental archaeology detailing the etically-observable effects of knapper choices. Significantly, analysts must abandon analytical approaches that use behaviors that are learnable but not recoverable etically by modern archaeologists as well as those behaviors which are wholly contingent or random, or reside at a level of abstraction beyond what is learned by the individual, or are not learned even if etically observable by the archaeologist.

4.2.1 Conceptual Modeling of the Observational Learning of the Flintknapping Process

In many ways, lithic analysts engaged in cultural historical reconstruction are currently in a position analogous to literary scholars of epic poetry who worked at the beginning of the twentieth century. Scholars who worked before the great innovations of Milman Parry and Albert Lord faced the daunting task of trying to understand how normal humans, albeit geniuses like Homer, could conceive, compose, memorize, and then perform oral poetry of such immense length as survive in transcribed texts such as the *Iliad* and the *Odyssey*. Through ethnographic research among living creators of Serbo-Croatian oral epic poetry in Yugoslavia in the 1930s, Milman Parry and Albert Lord recognized how the process of oral performance and the act, context, and mechanism of poetic creation allowed the performer to produce an oral piece of great length through the utilization of subtle structures of repetition. Parry's pivotal concept, developed further by Lord (1960, 2000) and subsequently further contextualized and augmented by John Miles Foley (1988), was that regular patterns of words would be used in particular metrical contexts to convey the same meanings. The poet learned how to readily satisfy the metrical requirements of the poetry using repetitive structures, such as the famously frequent phrase "rosy-fingered dawn" in Homer's *Iliad*. By learning in advance key repetitive elements that could be employed in particular contexts, a poet with a knowledge of the rough story line he wanted to convey could orally compose and perform immensely long poems without having to memorize each line in advance. They also emphasized that this mechanism made each performance subtly unique. This approach, now called Oral Theory, has allowed literary analysts to understand previously puzzling aspects of Homeric poetry as well as to bridge multiple theoretical problems in the study of how traditions are created, practiced, and replicated through time (Drout 2006).

I believe that the lessons learned from Oral Theory for understanding how poetic performers learned the structure of their oral poetic data can be usefully applied to the study of how the process of learning flintknapping by visual and auditory observation can structure both the forms of the artifacts produced and the means by which archaeologists reconstruct the behaviors that were both learned and performed in a given society. Consider the following scenario, a series of photographs of a knapping sequence undertaken in the Evolutionary Anthropology Laboratory at the University of Minnesota, Twin Cities Campus. In the photographs that follow, I highlight in particular the bodily gestures and knapping choices learnable by observers witnessing the performance at the time.

Figure 4.1 presents a seated knapper and her apprentice, the latter standing in order better to study her technological performance. Now, WHAT do we KNOW that the knapper needs to choose to do to remove a flake? Following her gaze in Figure 4.2, she examines the convexity of the dorsal surface of the core. She identifies a convexity to remove, in this case a convergent, pointed convexity (Figure 4.3). She then examines the platform (seen from the side in Figure 4.4) opposite this dorsal convexity in white. To remove this mass, she predicts the white line in Figure 4.5 as the fracture plane. In order to fracture the core there, however, she needs to strike with the platform thickness in Figure 4.6, equivalent to the theoretical platform thickness of Pelcin (1997c, 1998; Davis and Shea 1998) and the platform depth of Dibble and Pelcin (1995) and Dibble and Rezek (2009). She examines again the dorsal convexity (Figure 4.3) and realizes, by predicting the gray line in Figure 4.7 as the needed fracture plane, that she will need to remove more mass to strike off a convexity with this desired shape. This then requires a deeper platform thickness, seen in gray in Figure 4.8. Alternatively, she could make this change by altering the exterior platform angle (the white angle in Figure 4.9) instead of the platform thickness, as multiple controlled experiments have shown that external platform angle (EPA) and platform thickness (PT) together roughly predict the mass of the removal while the dorsal convexity contributes most significantly to giving specific shape to that mass (see below for more discussion). Thus, she may decide to alter the EPA through platform faceting (Figure 4.10) in order to change the EPA's relationship to PT such that a different mass is removed.

She is now ready to strike off the flake (Figure 4.11). But from the point of view of the observer learning the process (Figure 4.12), he can tell roughly where the knapper is gazing but not exactly what EPA and PT she is choosing emically. He can estimate the speed of the delivery of the strike from the position and gesture of the percussing arm (Figure 4.13). He can also estimate the angle of attack controlled by the arm and leg supporting the core (Figure 4.14). In contrast, the knapper has the full-body experience (Figure 4.15) of precisely controlling all of these variables in the split second it takes to deliver the blow and remove the flake. The observer has an etic viewpoint whereas the knapper has a fuller, emic viewpoint.

This difference in viewpoints is where new variation may be created in the transmission process between knapper and observer. While the observer has some notion of what variables are being chosen consciously or subconsciously through bodily performance (Figure 4.12) and he may practice later himself for hours or years to achieve a similarity between his products and those produced by the knapper, he never has her full-body emic experience of the original action. This conceptual distance allows errors of transmission to occur and new technological variation to be created during the process of learning. This inherent variability in the unit of cultural transmission is what makes cultural evolutionary processes so distinct from biological evolutionary processes. Boyd and Richerson (2000) artfully point this out in their critique of meme theory through an illustration of a figure handing a large cursive letter "A" to another figure who then hands a Courier-font "A" to the next figure who then passes on an at sign "@" to another figure outside of the illustration. The unit that is transmitted is itself shapeable, both behaviorally and morphologically, by the process by which it is learned. Other crafts performances may allow more or less fidelity in transmission between what is demonstrated and what is learned (as demonstrated by Eerkens and Lipo (2005) and Eerkens and Bettinger (2001)) but it is critical to the current project to understand where and how fidelity variation is created in the learning of even the most simple flintknapping operation, the striking off of one flake.

The conceptual distance between the knapper (Figure 4.15) and the learning observer (Figure 4.12) also allows the archaeologist to take an etic rather than emic perspective to the analysis of the archaeological record. With access only to the flake (Figure 4.16) produced by the blow and, if not reduced to shatter, perhaps the core as well, the emic choices are lost to the archaeologist just as most, if not all, of the emic choices of bodily performance are obscured from the learner by

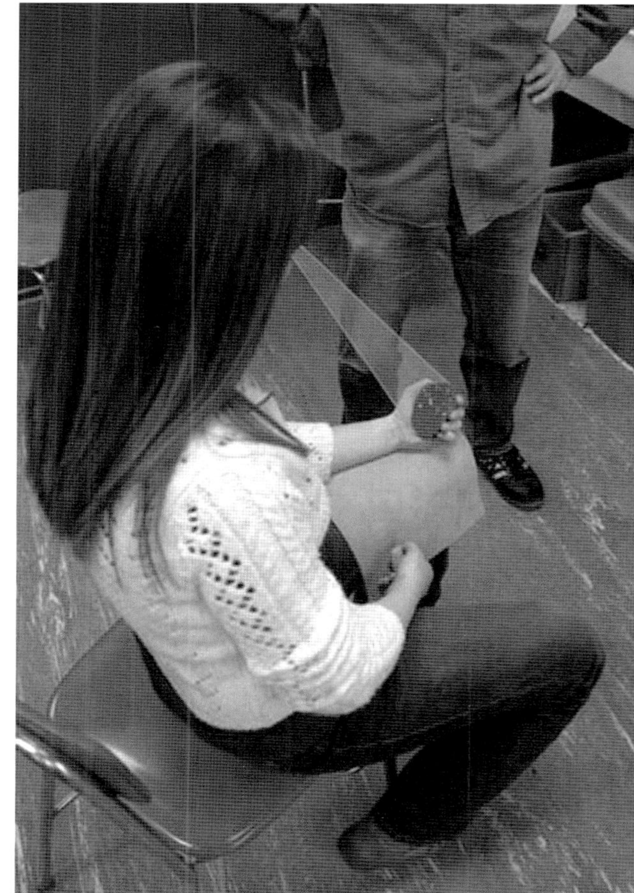

Figure 4.1 (left) Beginning of the scenario to model the observational learning of the flintknapping process. The seated knapper holds a core while her apprentice stands to watch her technological performance. Knapper, Melissa Knight; observer, Matthew Hunstiger; photographer, Sabrina Curran; Photoshop transformations of figures 4.1-4.16, Gilbert Tostevin. Figure 4.2 (right). The knapper examines the dorsal convexity of the core.

Figure 4.3 The knapper identifies a convergent-edged convexity, outlined in pale gray, for removal as a flake.

Figure 4.4 The knapper examines the lateral profile and curvature of the convexity identified for removal, shown in white.

Figure 4.5 The knapper predicts the fracture plane, in white, needed to remove the convexity.

Figure 4.6 The knapper predicts the platform thickness required to create the white fracture plane.

Figure 4.7 The knapper makes a second prediction for the location of the fracture plane, here in gray, needed to remove the pointed convexity from the core surface.

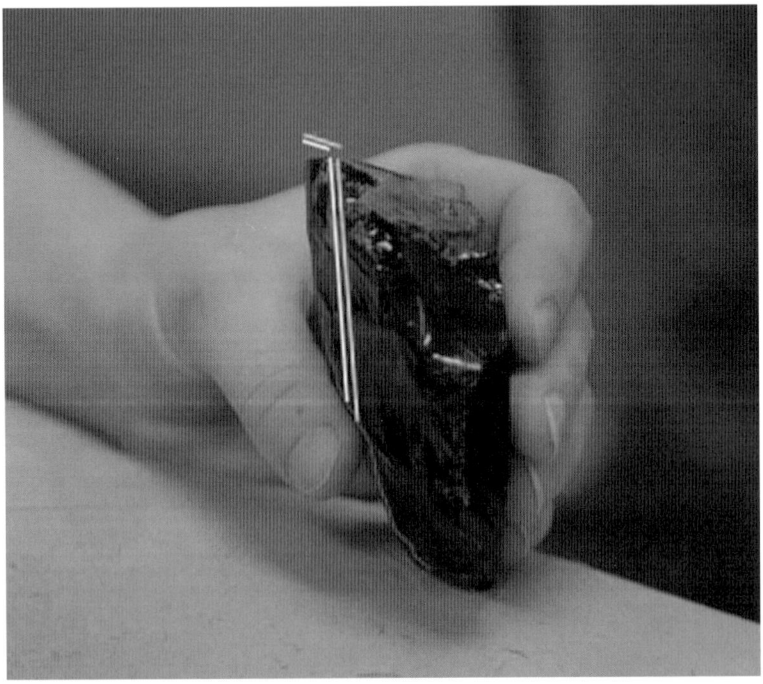

Figure 4.8 The knapper estimates a longer platform thickness, in gray, to match the deeper fracture plane needed to remove the pointed convexity from the core surface.

Figure 4.9 The knapper evaluates the exterior platform angle (EPA), here in white, needed for the flake removal.

Figure 4.10 The knapper facets the platform, in the direction of the white arrows, to change the exterior platform angle (EPA), so as to adjust the mass predicted at the platform for the flake removal.

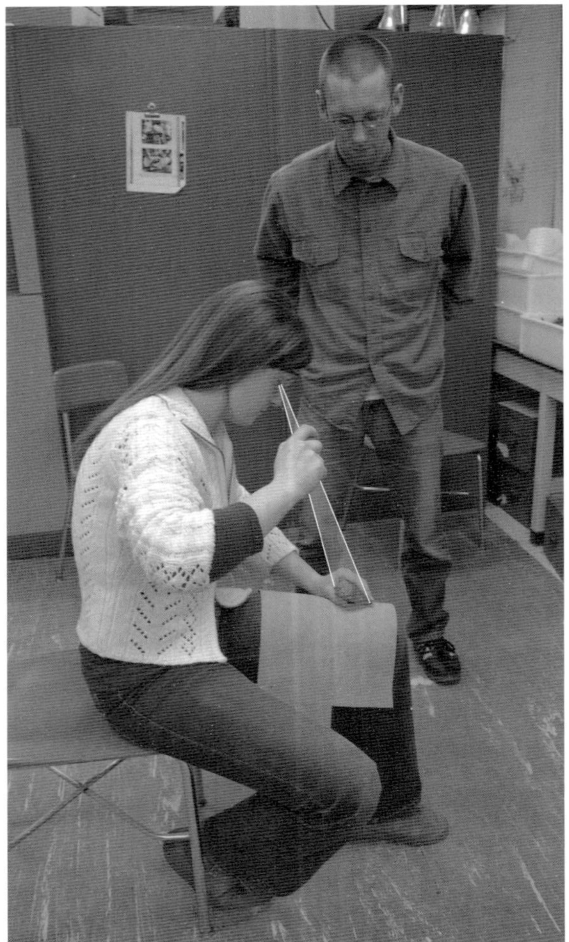

Figure 4.11 (left) The knapper is ready to strike off the flake, with her gaze focused on the platform and her arm ready to deliver the blow.

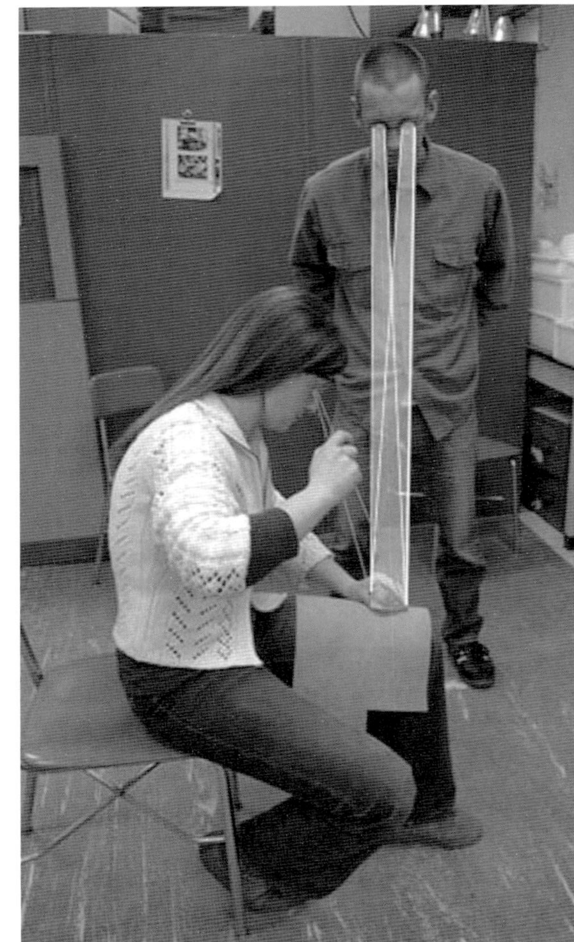

Figure 4.12 (right) From the perspective of the observer (represented by his lines of sight), he can tell where the knapper is looking in general, but not the specific exterior platform angle or platform thickness (at the point of percussion) chosen by the knapper before the strike.

Figure 4.13 (left) From the perspective of the observer, he can estimate the speed of the delivery of the knapper's strike from the position and gesture of the percussing arm, highlighted in this image.

Figure 4.14 (right) From the perspective of the observer, he can estimate the angle of attack controlled by the arm and leg supporting the core, highlighted here.

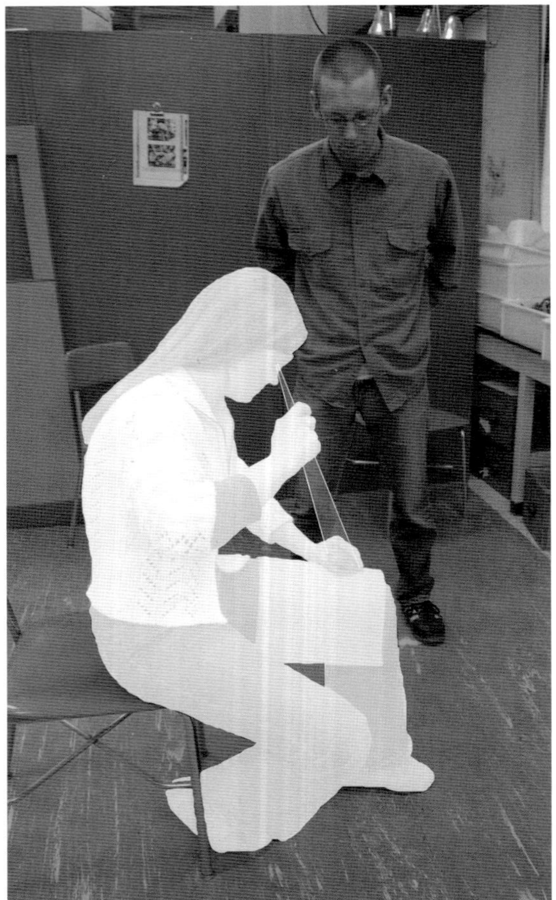

Figure 4.15 In contrast, the knapper has the full-body experience, highlighted here, of precisely controlling all of these variables in the split second it takes to deliver the blow and remove the flake.

Figure 4.16 With access only to the flake and the core that remain after the blow, the emic choices of the knapper are lost to both the prehistoric observer and the archaeologist, allowing the archaeologist to use the etically-recoverable variables on the flake as proxies for what the observer saw and learned during the transmission event.

the nature of the knapper's split-second performance of the blow. Yet both the observer and the archaeologist can pick up the flake to determine the etic results, i.e., the specific EPA, PT, platform preparation, and dorsal surface shape variables that were created by the blow, regardless of the fact that they cannot know the emic intention of the knapper. **The etic results, as a consequence, are measurable by the archaeologist and can serve as proxies for *what the observer saw as he learned*.**

The above conceptual scenario is clearly not ethnoarchaeology nor an actualistic study (the uniformitarian principles of using a 21st century North American to model the naive learning of flintknapping has its limitations, as Monnier (2007) showed in her critique of Bisson's (2001) actualistic study with "novice" knappers). Nor is it an experimental study of skill acquisition as usefully called for by Bamforth and Finlay (2008). The conceptual modeling above, however, does illustrate aspects of the process that are novel and significant to the application of cultural transmission theory to lithic data. The abundant examples of ethnoarchaeological studies of living knappers (Gould et al. 1971; White and Thomas 1972; Gallagher 1977; Hayden 1979; Hayden and Nelson 1981; Torrence 1983; Burton 1984; Gould and Saggers 1985; White and Dibble 1986; J. E. Clark 1991; Pétrequin and Pétrequin 1994; Roux et al. 1995; Watson 1995; Brandt et al. 1996; Stout 2002, 2005; Weedman 2000, 2002a,b; Shott and Sillitoe 2004; Sillitoe and Hardy 2003; Whittaker 2004; Roux and Bril 2005) have only hinted at these aspects of this very specific cultural transmission process. Yet the emic-richness of such studies might lead a reader to suppose that the conceptual distance between the knapper and the learning observer is not as separate as the above scenario presents, merely from the point of view that the knapper could verbally inform the observer of the variable states that she is choosing emically, as would any good teacher, rather than relying on the observer's poor etic viewpoint to capture those variable states. This point, however, merely illustrates that there will be variation between different cultural and temporal settings in the degree of fidelity of transmission based on differences in the context of the observational learning between the extreme of, on one end, imitation without the knapper's intent to educate (Thorpe's (1956) *local enhancement*) to the opposite extreme of full linguistically-aided teaching with the assistance of "scaffolding" (Greenfield 1984; Ferguson 2008), with all of the variations between (Boesch and Tomasello 1998). While I agree with Stout that the human mode of skill acquisition "has its foundations in the saturation of everyday social interactions with intersubjective awareness" (2002:719), the prehistory of lithic technology is likely to bridge the evolution of many modalities of socially-mediated observational learning and archaeologists should not bias our methodological approach in assuming particular learning and teaching processes that require an emically-recognizable level of intentionality. While any cultural tradition with a ratchet effect (Tomasello 1999) similar to that of modern humans implies that the learner had a theory of mind and intersubjective awareness (*sensu* Dunbar 2003), it is also the case that many modern human societies teach complicated skills without the use of language (de Beaune 2002) and even discourage questions from the observer during pedagogical demonstrations (Dr. Sven Haakanson, Jr., of the Alutiiq Museum, Alaska, personal comm. 2001). Lithic technology has also been shown to be learnable through non-linguistic imitation alone (Ohnuma et al. 1997). The conceptual modeling above thus represents the transmission process with the minimal degree of educational-intentionality from the point of view of the knapper watched by the observer and thus is amenable to further modeling to fit more complicated levels of educational-intentionality in future studies.

4.2.2 Lessons Learned from Experimental Archaeology

We must now determine which of the variables identified in the conceptual modeling of the observational learning of the flintknapping process are in fact etically observable to the modern lithic analyst. Fortunately, experimental archaeology helps answer this question, even more so than the mechanical engineering treatises that derive mathematical principles about glass and ceramic fracturing (e.g., Quinn 2007; see also Bertouille 1989). For blank production, both free-hand knapping experiments and controlled experiments have identified specific choices a flintknapper must make in order to produce blanks for use (Speth 1972; 1974; 1975; 1981; Bonnichsen 1977; Dibble and Whittaker 1981; Cotterell et al. 1985; Dibble and Pelcin 1995; Pelcin 1996, 1997a,b,c; Dibble and Rezek 2009; Rezek et al. 2011). These choices can be understood as being made in temporal clusters during the process of flintknapping, with some clusters of choices being made on a flake by flake basis, what may be called *tactical choices*, while others are made once or twice per core, which can be called *strategic choices*. These clusters of knapping choices, represented by artifact measurements, may be termed *flintknapping domains* (Tostevin 2000a,b,c, 2003a,b) and serve to structure the analysis of learned flintknapping behaviors.

The flake by flake variables and their two domains have received the most direct research. From free-hand experiments (Whittaker 1994), we know that the knapper who struck off a flake consciously chose or unconsciously determined through her or his body performance how far into the platform to strike (the platform depth or platform thickness), the angle of the exterior platform at that spot, and whether or not to alter that angle with platform preparation beforehand. Together these three variables constitute the ***platform maintenance domain*** as these choices are enacted together with the blow of the percussor. From controlled experiments into flake fracture mechanics, we know these variables to be independent choices of the knapper that together determine the mass of the resulting flake (Dibble and Whittaker 1981; Dibble and Pelcin 1995; Dibble 1997; Dibble and Rezek 2009). From free-hand experiments, we also know that the knapper chooses where to strike the core relative to the morphology of the dorsal surface. From Pelcin's (1997b) controlled experiments with the extremes of core surface morphologies (i.e., a high surface convexity equivalent to a burin spall vs. a low surface convexity of a flat glass plane), it has been shown that the morphology of the dorsal surface contributes significantly to the *shape* of the resulting flake but independently from the platform variables' determination of flake *mass*. Rezek et al. (2011) have corroborated Pelcin's results using the intermediate core surface convexities of glass cores that more faithfully replicate prehistoric core morphologies. While Rezek et al. showed a slight influence of core surface morphology on flake size compared to Pelcin's results, "the amount of flake variation produced when core surface morphology is held constant clearly shows in and of itself that the shape of a flake cannot be predicted by the geometry of the core" (2011:1357). In contrast, the platform variables are far more closely correlated with flake mass and size than is the case with core surface morphology. While our understanding of flake fracture mechanics is not advanced enough to enable platform variables alone to predict original flake mass with the accuracy needed to measure the intensity of mass lost due to subsequent retouch (Davis and Shea 1998; Shott et al. 2000; but this may soon change with the promising 3D approach in Clarkson and Hiscock 2011), all relevant controlled experiments (Pelcin 1997b; Dibble and Rezek 2009; Rezek et al. 2011) have concluded that platform variables are the most significant in determining flake size and mass, i.e., how much material is cut off of the core and at what angle to the core surface. In contrast, holding all platform variables constant and varying core surface morphology demonstrates that core surface morphology shapes the outline that the mass, determined at the platform, will take.[2] In other words, the choice of where to strike a core relative to the dorsal surface morphology reflects the knapper's choices relating to the flake's outline shape.

Considering these results of experimental archaeology, the knapper's choice of dorsal surface morphology may be characterized with five variables which together constitute the flake shape. These five variables reflect the longitudinal extent of the surface (the continuous variable of the length/width ratio of the flake), the vertical convexity of the mass to be removed (the continuous variable of the width/thickness ratio of the flake), the number of dorsal ridges defining the convexity (the ordinal variable of the flake's cross-section type), the longitudinal shape of the surface (the categorical variable of the flake's lateral edge type), and the curvature of the core surface (the categorical variable of the flake's profile type). Each of these physical features of the dorsal surface of the core are visible cues available to the flintknapper (*and the observer*) to choose the shape of the flake before the removal, as modeled in Figure 4.3. As each variable is in fact chosen simultaneously with the blow of the percussor, however, all five variables are combined within one temporal cluster, the ***dorsal surface convexity domain***.

While not all of the dorsal surface convexity domain variables used in this research were modeled by Pelcin (1997b) or Rezek et al. (2011), modeling even the interaction of only a few dorsal surface variables with the two continuous platform variables (platform thickness and exterior platform angle) is sufficient to simulate the production and morphology of conchoidal fracture. Dibble and McPherron's flintknapping simulation program named "VDflaker.exe," available as software included in Dibble et al. (2003), demonstrates the power of these few variables to simulate the production of conchoidal flakes. The simulation uses Dibble and Pelcin's (1995) mass predictor formula which, although it has been shown to be relatively weak in predicting the actual mass of archaeological specimens (Davis and Shea 1998; Shott et al. 2000; but see Clarkson and Hiscock 2011), is sufficiently accurate to produce flake morphologies more similar to archaeological artifacts than Pelcin's sheet-glass experimental products. The morphologies result from a few simple assumptions: the relationship between platform thickness and exterior platform angle from Dibble and Pelcin (1995), the establishment of the proximal-to-distal length of the bulb of percussion on the ventral surface as the width of the platform, and the gradual decrease in thickness from the end of the bulb until the feathered termination (Dibble personal comm. 2005). With these assumptions which are admittedly fudges of the actual fracture mechanics, it is possible to see how the alternation of one variable affects the morphology when the other variables are held constant. Figure 4.17 is a screen capture from VDFlaker which uses a high platform thickness with all other variables held constant in the at-start settings of the program. Figure 4.18, however, shows the same at-start settings but with a low platform thickness. The two resulting flakes are quite different in how much of the dorsal surface

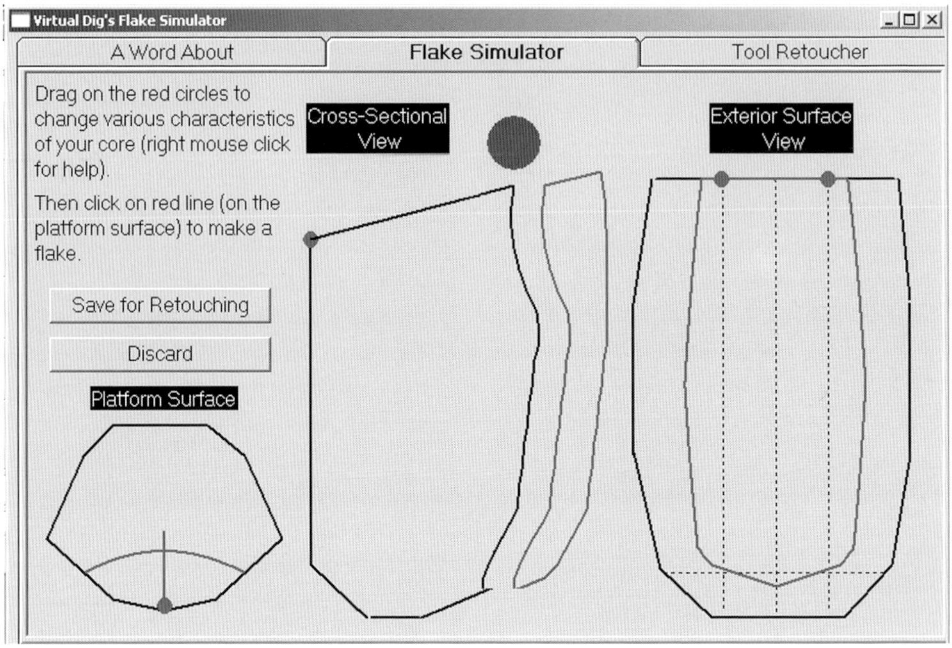

Figure 4.17 Screen capture from VDFlaker.exe (Dibble et al. 2003, used by permission), with a high platform thickness and all other variables with at-start values.

of the core they remove. Now compare Figures 4.17 and 4.19, both of which are identical in their parameters but for the lower exterior platform angle in 4.19. With the same dorsal surface of the core, the flakes are identical except for the lower exterior platform angle of 4.19 resulting in a shorter and thicker flake. Now, taking Figure 4.17 and changing the dorsal surface of the core by means of widening the two longitudinal ridges of the core as in Figure 4.20, one also increases the platform width because the fracture initiating at the point of impact must extend further to the sides until it can erupt down the dorsal face. Figure 4.20 also shows how changing the degree of the core's longitudinal convexity (in the VDFlaker program this is achieved by increasing the convexity of the distal end of the core), produces yet another flake morphology from the same platform thickness and exterior platform angle used in Figure 4.17.

The results of such a variable-by-variable alteration of flake morphologies illustrate the importance of the platform variables and the dorsal surface convexity in determining the size and shape of each flake the knapper chose to remove from the core. Dibble (1997) demonstrates these relationships statistically using both experimental and archaeological artifacts, by controlling for all but one variable at a time, much as done with Figures 4.17–4.20. Dibble et al. (2009) and Rezek et al. (2011) do the same with experimental flakes produced from glass cores replicating prehistoric morphologies. As it is certain that each flake was selected, consciously or not, for removal by the knapper at the measurable parameters of these variables, these etically-characterizable flake by

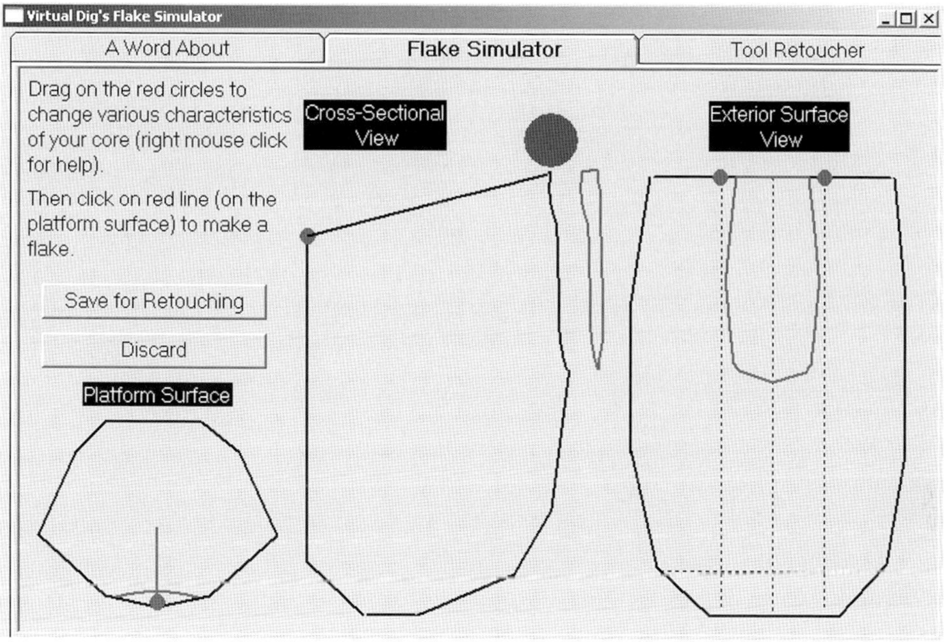

Figure 4.18 Screen capture from VDFlaker.exe (Dibble et al. 2003, used by permission), with a low platform thickness and all other variables with at-start values.

flake variables within these two domains, separated because of their functional independence (mass vs. shape), form the *tactical choices* of the lithic analyses in this research.

The clusters of decisions or domains which are made once or twice during a given core reduction have been subjected to far less quantification than the flake by flake decisions. They are in fact frequently represented by a typological approach to characterizing the whole core reduction, as in diagnostic *chaînes opératoires* such as *discoïde* or *Levallois préférentiel* (i.e., Boëda 1995), or unstandardized representations of refitting sequences. Instead of typing the whole assemblage as one reduction type versus another or using each assemblage as an anecdote, the evolutionary approach requires comparable choices within the strategic reduction of a core, both at the flake by flake level and at particular stages in a reduction. Of the stage related choices, two temporal/functional clusters of choices are recognized here as being comparable in all assemblages included in this research project: the *core modification domain*, which includes the choice of orientation of the raw material as a core and specific methods of repairing and maintaining convexities, and *direction of core exploitation*, which includes the dominant directions of debitage removal during both the early and late stages of core exploitation. The former domain is characterized by the presence of diagnostic flakes (such as core tablets and débordant removals) and the analysis of core attributes, whereas the latter domain is based on the evaluation of the correlation of debitage length with debitage dorsal scar patterns, as Dibble (1995a) did with Biache Saint-Vaast above. Each of

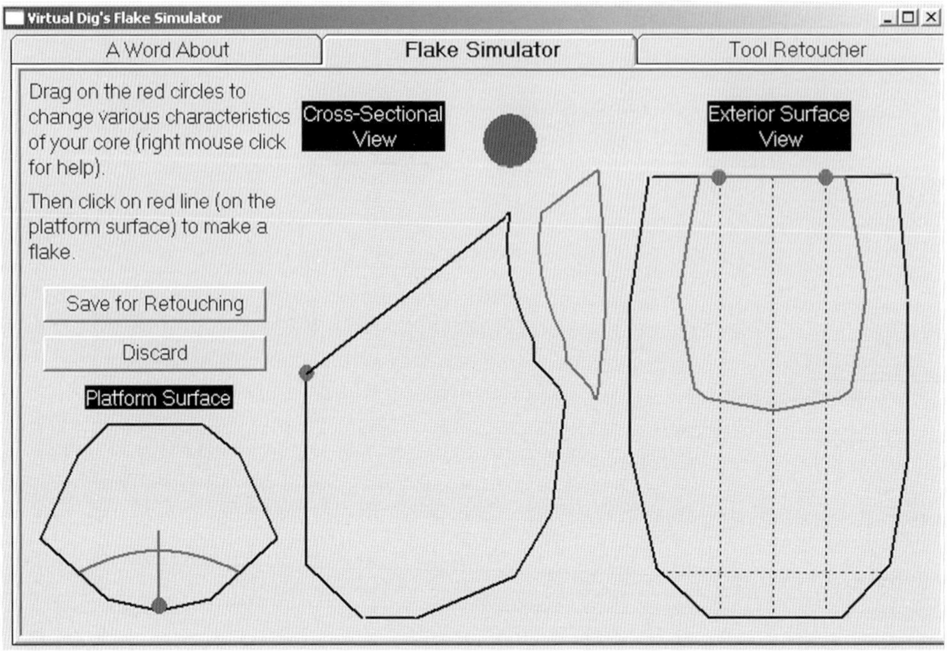

Figure 4.19 Screen capture from VDFlaker.exe (Dibble et al. 2003, used by permission), with parameters set as for Figure 4.1 but with a lower exterior platform angle.

these domains, their variables, and the analytical methods for their identification are defined in detail later in this chapter.

It is important to note that none of the attributes studied in this project are straight dimensions, such as blank width in millimeters. All continuous variables are non-dimensional ratios, so that local, situational constraints such as raw material availability differences between two locations on the landscape that could produce differences in artifact size despite identical behavioral operational sequences are not confounding the evaluation of the cultural transmission hypotheses. The only exception is platform thickness, which is measured as a scalar variable in millimeters. This variable is not seen as susceptible to this problem, given the small range of values typical for this variable.

As noted above, we must be wary choosing analytical variables that are not held in common between the conceptual modeling of what is observationally learnable and experimental archaeology's ability to recognize such variables from the study of lithic materials in the archaeological record. Unfortunately, not all of the significant tactical choices made by a knapper in removing a flake are currently etically-recoverable by lithic analysts. Both the conceptual modeling of the learning process as well as experimental replication identify several variables of importance that are not easily quantifiable or even recognizable on the resulting flakes. The first such variable is the angle of blow, which is known from free-hand knappers as important for positioning the fracture plane through the vector of the force of the blow (Pond

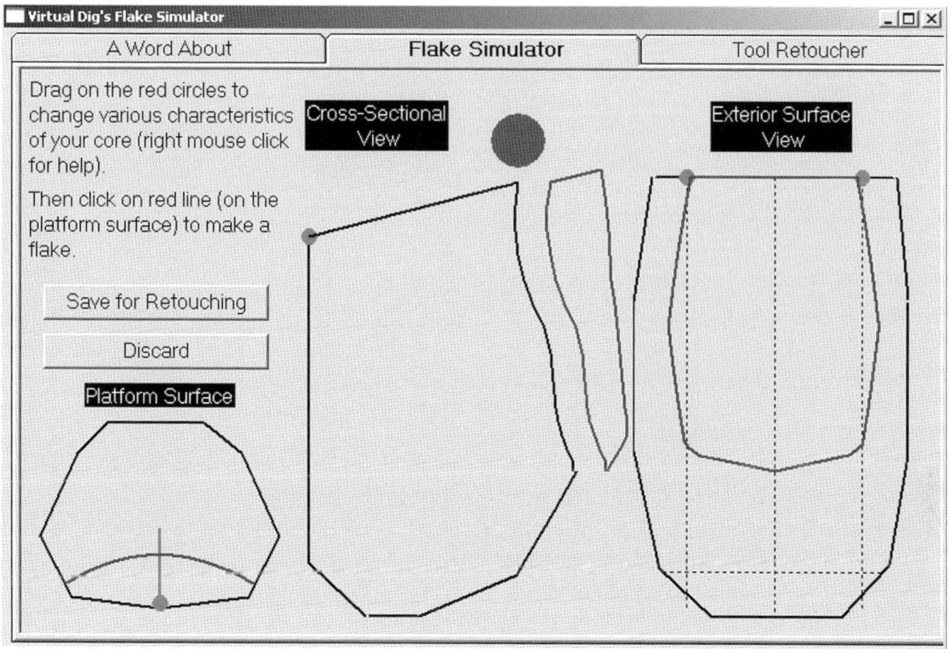

Figure 4.20 Screen capture from VDFlaker.exe (Dibble et al. 2003, used by permission), with a high platform thickness, wider dorsal ridges, more distal dorsal convexity, and at-start EPA value.

1930; Crabtree 1967, 1972; Whittaker 1994:91-98, 188-193). Despite the fact that angle of blow has been shown by controlled experiments (Speth 1975; Dibble and Rezek 2009:1950-1951) to have a significant effect on flake mass, particularly for direct as opposed to glancing blows, there is no generally recognized method of determining the angle of blow from the resulting flake morphology.

Similarly, the force of blow is known from free hand knapping (Whittaker 1994:91) and controlled experiments (Dibble and Rezek 2009:1951-1952) to be a key variable that a knapper must learn and apply differently when removing flakes of different masses. As controlled experiments have shown that "the amount of force required to remove a flake appears to be related only to the size of that flake" (Dibble and Rezek 2009:1951), it is a learned behavior whose variability is disguised by the fact that an insufficient force will not remove the flake whereas any force at or above the sufficient force, given its mass predicted by the raw material's hardness, the point of impact, and the exterior platform angle, will produce the same flake. This behavioral variable, even if moderately quantifiable with the characterization of the size of the ring crack at the point of percussion (Speth 1975), thus need not be included in the tactical choices studies here.

Of the strategic choices made by a flintknapper once or twice during the reduction of a core, there is one type of analytical characterization typically made by Paleolithic archaeologists which is not held in common between what is observationally learnable and what is recoverable etically by an archaeologist. Abstractions of the technological approach pursued within an assemblage to the level of a "typed" operational sequence or *chaîne opératoire*, such as *Levallois préférentiel*, are conflations of multiple variables at the tactical flake by flake level as well as multiple strategic rules for reducing core volumes by surfaces. As a result, the label of an assemblage as "being Levallois" does not equate to the variables as learned by the observer. Perhaps the rules or criteria of Levallois (*sensu* Boëda 1995) might be considered as individually learnable by observation but then those criteria, and not the type label of the whole assemblage being Levallois, would be the units of analysis to be applied to a given assemblage. An evolutionary approach to the study of cultural transmission would need to be able to reconstruct etically that one assemblage has criterion 1 but not criterion 6, for instance, so that variation as well as norm of behavior can be characterized. While an interesting possibility, such an undertaking faces enormous hurdles in bridging the gap between the emic descriptions of the Levallois criteria and the etic observations needed to demonstrate their existence in a given assemblage. As such, no current studies disarticulate the "type" abstraction of Levallois as a reduction method but merely document the abstraction's presence or absence in a given assemblage. Thus, as there is no understanding of variability within a type label (as argued in Tostevin 2011), an evolutionary approach to studying cultural transmission cannot use such abstractions. Thus, arguments that this approach is flawed because it may document similarity in blank production between two assemblages which are otherwise considered as different by other authors by the presence vs. absence of the Levallois reduction method, such as those made by Sitlivy and Zięba (2006:418), are in fact missing the point entirely.

4.2.3 Changes in the Number of Domains Necessitated by the Taskscape Visibility Concept from Chapter 3

The twelve variables within the four flintknapping domains introduced above represent the behavior by behavior approach to evaluating the degree of similarity and dissimilarity in blank production between assemblages. For blank production, the assemblage is composed of both complete debitage blanks (unretouched flakes) and the retouched tool kit. The retouched tool kit is included in, rather than excluded from, evaluating blank production since those pieces selected for retouch frequently represent the larger end of the variability actually created during blank production (Morrow 1996).

The analysis of the tool kit morphology on the other hand includes only those pieces with retouch in the same assemblage as well as those traditionally labeled as such despite a lack of retouch, e.g., Levallois products. The characterization of an assemblage's tool kit constitutes its own domain of choices enacted during selection of blanks for inclusion into the tool kit. Most of the variables reflected in the present characterization of tool kit morphology refer to the dorsal surface convexity variables which give form to the cutting edge of the retouched blanks in the assemblage. The attributes present on blanks available for inclusion in the tool kit are thus created initially during blank production. The central tendencies within a tool kit, however, can have as much to do with criteria for blank selection for subsequent retouch as the central tendencies within blank production before retouch. This may be the result of particular morphologies being selected out of the available blanks for use as retouched tools because of their preferred cutting edges or the way in which they bear the application of retouch through time. This is why characterizations of tool kits in previous analyses of this type (Tostevin 2000a,b,c, 2003a,b) have treated the identification of blank selection criteria (through the comparison of tendencies within the tool kit versus within the unretouched blanks) as the central concern for structuring the comparison of tool kit assemblages through time. In these earlier works which did not use the social intimacy and taskscape visibility concepts, blank selection criteria for tools formed a fifth domain in the demonstration of the technological signature for an assemblage. The present analysis, however, endeavors more to distinguish the dorsal surface convexity choices evidenced within the tool kits for direct comparisons between assemblages, rather than through the proxies of blank selection criteria. This is primarily due to the physical visibility of these attributes on the tools, visibility which affects which technical choices could be learned at particular degrees of social intimacy according to the taskscape visibility middle-range theory developed in Chapter 3. Blank selection criteria themselves, i.e., how different the tool kit is from the unretouched blanks for particular variables, are not visible features of the tool kit artifacts alone (they require comparison with the unretouched debitage) and therefore are unsuitable for investigating the contrast of the changes in blank production versus changes in tool kit variability according to the relationship between taskscape visibility and the social intimacy of the flintknappers responsible for each assemblage.

The analysis of the tool kit morphology includes the continuous characterization of the shape of the tools (length/width and width/thickness ratios) as well as the categorical characterizations of the dorsal surface convexity which have the most significant impact on the shape of the cutting edge itself, i.e. lateral edge type and profile type. Lateral edge type divides the variability of the dorsal-to-ventral surface interface into types which feasibly could affect the utility of the edge: convergent edges for slicing and piercing, parallel edges for a lateral haft and/or parallel motion of the tool against a worked material, etc.. Similarly, profile types such as straight, curved, or twisted could each affect the manner in which the cutting edge is employed in action, both for the ease of holding the flake in the hand (Tomka 2001) and the physics of the contact between the cutting edge and the direction of the flake's motion against another material. Cross-section type is not included in the characterization of the tool kit morphology since the number of ridges on the dorsal face seemingly has little to do with the efficacy of the cutting edge. While the dimensions of the cross-sectional area of a piece produced by the dorsal ridges relative to the ventral surface can pattern according to various functional categories, such as suitability for use as different types of armatures (Hughes 1998; Shea et al. 2001; Shea 2006; Sisk and Shea 2009), these dimensions are better characterized here through the continuous width/thickness variable rather than via the number of ridges on the surface, which is more relevant to the variable seen and chosen by the knapper for each removal during blank production.

The distal terminus type is perhaps the most relevant morphological variable for the suitability of a blank for particular functions, whether this is a pointed distal end required for a piercing gesture or a blunt end required for the application of a wider cutting edge to the material being worked. As such, a blank with the suitable distal terminus can be directly chosen from among the produced debitage to become a retouched tool based on the state of this attribute. The "distal terminus" type is included in the tool kit analysis but not in the blank production analysis. This relates to the difference between strategic and tactical choices made by the knapper, a distinction which has implications for the lithic analyst's ability to distinguish emic from etic operations of the knapper. Upon finding a retouched tool with a pointed distal end (pointed technologically before the application of retouch), the lithic analyst knows that the knapper chose this object with a particular distal terminus for subsequent retouch, although this etic observation does not substantiate any claim to whether or not the knapper emically chose the distal terminus over other considerations. The feature was visible on the piece, however, and so could have been a direct subject of selection, without the conflation of other variables.

Several scholars employing refitting methods of research have made the argument that the central goal of particular core reductions is the production of blanks with pointed distal termini. These arguments include Henry's (1995b, 2003) use of Demidenko and Usik's refittings at Tor Faraj, Marks and Kaufman (1983) for Boker Tachtit Levels 1-4, and Škrdla (2003c) for Stránská skála IIIa-4. These arguments are based, however, on inferences of the logic of the order of removals within refit sequences which are used to determine the desired end-product. All of these arguments, however, suffer from the epistemological danger of assuming an emic understanding of which products were intended and which were byproducts, simply because of the order of removals, as Sandgathe has pointed out for Levallois "preferential" debitage (2004). Without the study of use wear to document which pieces were utilized after blank production (i.e., expedient, non-retouch utilization) as well as presence/absence of retouch, the identification of emic "desired end-products," the subject of heated debates in the past (Ford 1952; Spaulding 1953), is too subjective for consistent use as a unit of analysis for comparison between assemblages which may or may not possess refit sequences.

I take a different approach to the etic demonstration of knapper choice in the present research. Instead of a supposition of the emic goal of the reduction sequence, I argue that a rigorous technological signature of an assemblage is created only through the identification of and quantification of the variability in attributes which are functional in the knapping process, learnable by an observer, and which were available as visible features for the knapper to choose at the given point in the process, i.e., during blank production and during the selection of blanks for the tool kit. With this in mind, we can return to the example of using the distal terminus as a variable for the tool kit but not blank production. While the distal end shape of a blank can be selected from a pile of debitage for use as a tool or retouched tool, the distal terminus as an individual attribute is not operationally chosen by the knapper before the blank is removed from the core. While the order of removal facilitates the production of a pointed morphology on the core surface, the actual core morphology which produces the distal terminus is more a result of the interaction between several variables than the result of the knapper's selection of one, individual feature on the surface of the core for removal. For instance, the distal terminus results from the shape of the convexity (as measured by lateral edge types), the directionality of core exploitation (for instance, using removals from an opposed platform to set up for the removal of a point from a Nubian 1 core - Van Peer 1992:40, Figure 21:2), and the relative vertical convexity being chosen for removal at the platform (as measured by the width/thickness ratio). Each of these variables must be proximately or immediately chosen by the knapper in order for the body technique of the hammer strike to remove the blank with the desired form. Thus, these variables are the ones being enacted in blank production, not the distal terminus shape itself. This is not to say that a flintknapper does not often regard the core morphology, decide the "type" of removal as pointed or not, and then strike to remove the desired shape. Bruce Bradley (1989) dramatically illustrates this power of prediction by the drawing of the edge of the future flake on the surface of the core in ink before he strikes the platform. The flintknapper's gestalt or power of prediction, however, does not change the fact that controlled fracture mechanics studies have shown that the behaviors actually enacted by the knapper through her/his body techniques that produce the blank are related to where to strike relative to a conflation of variables. These variables are already included in the characterization of blank production within an assemblage, making it unnecessary to include distal terminus itself within the analysis of blank production. Furthermore, while the strategic design of a knapper's manipulation of more proximate, tactical choices such as lateral edges, vertical convexities, and directional exploitation of the core may have been to produce a pointed blank at a given removal, the vagaries of the knapping process (i.e., raw material imperfections, errors in gestures, distractions around the knapper, etc., see Pelegrin 1990b) often alter the result of even the most experienced knapper's hammer strike. Thus, products which were strategically chosen to have pointed distal ends may have actually been produced with blunt, hinged, or overshot ends. In keeping with the intent of this study to examine what we know was under the control of the knapper on a flake by flake basis, only tactical choices of the knappers are used in characterizing the blank production tendencies in a given assemblage. As these tactical choices are what was actually observed by potential learners during the performance of the activity, the physical results of the technical choices constitute the units of analysis for the evaluation of assemblage similarities and social intimacy.

For the analysis of the variability evidenced in the tool kit morphology as opposed to the blank production, however, the distal terminus is a directly observable feature of each blank chosen for inclusion in the tool kit. There are no conflating variables: we know that the knapper chose this object with a particular distal terminus for subsequent retouch, although this etic observation does not substantiate any claim to whether or not the knapper emically chose the distal terminus over other considerations.

The last two of the seven variables used to characterize an assemblage's tool kit relate exclusively to the highly visible results of the application of retouch for the taskscape visibility concept, with one variable accounting for unique retouch types (bifacial, carinated, etc.) and another accounting for a general emphasis on lateral edge retouch (indicated by a majority of Middle Palaeolithic tools) versus distal retouch (indicated by a majority of Upper Palaeolithic tool types). This opposition between lateral and distal retouch is both a functional distinction and a highly visible feature of items in the tool kit.

4.3 To Make a Stone Soup: Description of the Technological Signature for Blank Production

Given the above reasoning for the existence of knapper technical choices by flintknapping domains, it is possible to compile analytical techniques to document each of these performed choices among the artifacts in an archaeological assemblage in a form suitable for comparison with the model predictions of the three hypotheses proposed by this project. The attribute analyses developed by many lithic scholars contributed to the identification of the central tendencies and variability within these knapping choices. These scholars include Baumler (1988), Bergman (1987), Bordes (1961), Crew (1975), Dibble (1995a), Dibble and Whittaker (1981), Geneste (1985), Henry (1989), Hours (1974), Kuhn (1990b; 1995), Meignen (1995; personal comm.), Movius et al. (1968), Ohnuma (1986), Pelcin (1996), Speth (1981), Van Peer (1992), and Volkman (1989). Most of the analyses are univariate tests and comparisons of pairs of flake and core attributes, often using the principles of dimensional change during reduction (Holmes 1919; Frison 1968; Newcomer 1971; Collins 1975; Jelinek 1976; Stahle and Dunn 1982; Henry 1989; Dibble 1987a) and cortical change during reduction (Sullivan and Rosen 1985; Geneste 1985; Mauldin and Amick 1989; Baumler 1989; Ahler 1989a; 1989b; Dibble 1995a). In almost all of the cases, the analyses were inspired by traditional attribute analyses, since attribute analysis has many benefits not represented by the other methods of studying stone tools, notably flexibility (resolution and amount of information) and generalizability (broadness of utility; Steffen et al. 1998:137, tab. 9.1). The heterogeneous origin of these attribute analyses makes a detailed discussion of their application and reasoning necessary.

The following discussion presents the four knapping domains of blank production, the domain of tool kit morphology, and the behavioral categories within each which are relevant to the present study. The number of behavioral nodes of choices within each domain ranges between two and seven. After the description of the possible behavioral variants which represent choices available to the flintknapper within a particular behavioral node, the attribute analyses used to identify which of those options was chosen by the flintknappers to produce a particular assemblage are presented. These characterizations of the cumulative behaviors used by the knappers for each step usually consist of a measurement of the central tendency and dispersion of the variability within the particular knapping step. It is necessary to present the variability within each step in terms of central tendencies and measures of dispersal, if possible, since one can find almost every knapping option used, if only to a small extent, in any given assemblage. This omnipresence of knapping options is a large source of potential bias in analyses which rely on a small number of refittings or anecdotal treatment of a few individual pieces without reference to the dominant tendency evidenced in the unrefit debitage, as argued in Tostevin (2003a). The characterizations of the flintknappers' choices for a particular behavioral step include both metric or continuous variables and discontinuous or categorical variables. Inferences based on available refittings are also used. After the definition of the domains and variables, I present the need and method of assessing the comparability of assemblages in terms of whether or not they share enough of their respective theoretical operational sequences (from initial raw material acquisition to exhaustion of the cores and retouched tools) to make a direct comparison meaningful.

4.3.1 Domain 1. Core Modification

The first domain in which the flintknapper has many equivalent options to choose from in order to produce cutting edges out of a block of stone is the modification of the raw material into a core. This domain includes two behavioral categories which are relevant to the present study of variation across the Middle to

Upper Paleolithic transition: core orientation and core management.

Domain 1. Step 1. Core Orientation

The orientation of the raw material as a core begins the exploitation for blanks suitable for immediate use or subsequent retouching. While the original form of the raw material, whether block, nodule, or core-on-flake, is determined by the nature of the raw material source and is often not etically demonstrable after extensive core modification, the orientation of that original form constitutes the first option to the flintknapper and is frequently identifiable on the larger core remnants. Some raw material forms affect nearly all of the options, as in pebble industries such as Tata (Vértes 1964). Yet even among potentially "restrictive" raw materials, it has been shown that technological variation is possible, as with the Levalloisian industries made on pebbles in the Italian Pontian (Kuhn 1995). Surveys of raw material sources for characterization of raw material package sizes and shapes are useful in considering this step (e.g., Baumler 1988; Rasic and Andrefsky 2001). Such studies, however, are not broadly applicable across different landscapes given the frequently small degree of overlap between the prehistoric landscape and the accessible modern landscape that allows us to access the archaeological record.

Among the assemblages studied in this research, the possible variants of core orientation include:

Longitudinal: The narrower and longer of the available surfaces is exploited for the majority of the debitage removals from the block.

Broad-face: The wider surface, usually with the largest surface area, is exploited for the majority of the debitage removals from the block.

Discoidal: Two surfaces, whose secant joint forms the circumference of the core, are alternately exploited by the flintknapper. Both or one surface is often conical in cross-section, resulting in most of the debitage removals terminating at the point of the cone. When both surfaces are used alternately as the debitage and platform surfaces for removals, this is termed a bifacial discoid orientation (Boëda 1993). When one surface comes to be used for the majority of the debitage and yet retains its conical cross-section, this is termed a unifacial discoid orientation (Jaubert et al. 1990; Jaubert 1993). Such a unifacial discoid orientation is most likely an exhausted form of the bifacial discoid orientation.

The attribute analyses used to identify which of the above variants were chosen by flintknappers in a particular assemblage involve both continuous and discontinuous variables and are numbered below:

1. Qualitative Examination of the Extant Complete Cores: While not the most rigorous of the analyses in this research protocol, the examination of core surface morphologies along the gradient of length of extant cores in the assemblage serves as a rough proxy for the tendency to knap raw material shapes according to the variants defined above. The longer cores can be taken as representing the core morphology earlier in reduction than cores on the shorter end of the length spectrum (Henry 1989). The presentation of illustrations of extant cores, showing the range of morphologies by size, is used to support the interpretation from the following analytical steps. In general, due to the problem of core exhaustion altering the morphology and surface relationships of extant cores in the record, the contribution of the examination of cores alone (e.g., Sarel and Ronen 2003; Sarel 2004) is limited in its utility for structuring the characterization of a lithic reduction sequence. Thus, in cases of core exhaustion, extant core morphology is devalued in relation to other aspects of the attribute analysis pointing to the tendencies within the assemblage for particular core orientations. Further, the lower frequency of complete cores compared to complete debitage in an assemblage makes a quantitative correlation of core form with length or cortex uninformative. Thus a gestalt impression of the core forms according to the three variants defined above against the length spectrum is used to suggest the use of one or more of these variants in the assemblage. It is hoped that analytical advances, such as Lycett et al. (2006) which appeared after this research project was begun, will improve our ability to analyze core morphologies quantitatively and thus improve on the method below without the use of typological categories as above.

2. Quantitative Definition of the Core "Debitage Surface(s)": To identify how the cores in a given assemblage were oriented according to the three variants above, it is necessary first to identify which specific surface on each core served as the "debitage surface." While most surfaces on Paleolithic cores show the scars of some past removals, usually only one or two surfaces were chosen to serve as the surface(s) from which most of the blanks were removed from that core. The other available surfaces served as platform surfaces or unretouched areas. This is a distinction which only pertains to the production of flakes and is not a distinction in the utility of flakes removed from either type of platform; flakes removed from a "platform surface" could be just as suitable for hominin use as a flake removed from the "debitage surface." Indeed, core tablets, a particular type of flake removed from the "platform surface" were preferentially chosen for tools in certain periods of the Paleolithic (Bergman 1987). To standardize and objectify the analyst's criteria used in deciding which of the available core surfaces fulfilled the role of the "debitage surface," one first counts the number of scars on each core surface. The surface with the higher number of flake scars is likely to be the "debitage surface." A minimum size of flake scars of one centimeter in any dimension standardizes such comparisons. Second, to corroborate this criterion, one determines which surface was last used to remove flakes. This is accomplished by counting the number of scars on each core surface which preserve the negative of the removal's bulb of percussion. The surface with the most, or dominant number, of negative bulbs of percussion represents the last surface from which flakes were removed, and thus is likely to represent the "debitage surface." Conversely, the dominant location(s) of core platform surfaces is determined by identifying the core surface(s) with the fewer scars and the fewer negatives of bulbs of percussion (Van Peer 1992: 23, fig. 12). The latter criterion cannot be used without some degree of interpretation, however, as exhausted cores may evidence the last removals from surfaces which did not produce the majority of the debitage during the use life of the core.

3. Definition of the Sector Location(s) of the Core Platform(s): Fortunately, this problem itself can be remedied by using a more objective criterion for determining the cumulative, assemblage-wide tendencies to place platforms in given locations. This is accomplished by arbitrarily dividing each core surface into four sectors (*sensu* Crew 1975) labeled A, B, C, and D clockwise around the core (see Figure 4.21). The choice of the first sector, A, on the first surface, Surface 1, is made based on the greatest number of bulbar negatives within that sector (this definition always results in sector A of Surface 1 being the richest in bulbar negatives in all cores). Each flake scar in the four sectors is then noted according to its direction of propagation, from the theoretical directions of A, B, C, and D as established by the orientation of the core sector A. This directional data by core section is then evaluated under the assumption that the greater the number of scars within a given sector that originated from the direction of that sector (for instance, scars from direction B within sector B), the more likely it is that the edge of that core sector served as a platform during the core reduction. This type of analysis is robust enough to determine accurately which sectors served as platforms during most of reduction, even if the last removals bias the appearance of bulbar negatives toward sectors which never served as platforms until the exhaustion of the cores.

Domain 1. Step 2. Core Management
Throughout the exploitation of a core, the flintknapper is faced with the need to continue the reduction of the core, i.e., to prevent the premature exhaustion of the core, through the strategic removal of flakes to adjust the convexities of the various core surfaces. These removals may represent unique reparation flakes or frequent and regular removals. Numerous options are available as choices for the flintknapper within this second behavioral category of Core Modification, "Core Management." Within the context of the present study, the options available for core management, which are not mutually exclusive, include:

Débordant Removals: The convexity of the "debitage surface" of a core may be perpetuated by removing flakes or blades along the two lateral edges of the core, thus leaving a convexity along the longitudinal center of the core (Meignen and

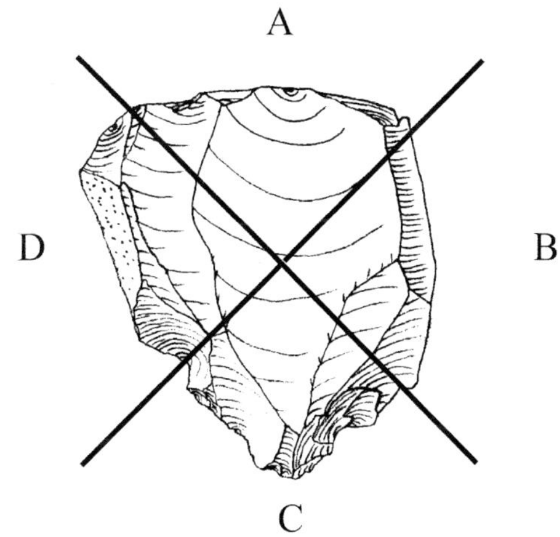

Figure 4.21 Definition of the Orientation of Core Sectors (after Crew 1975). Drawing by G. Tostevin.

Bar-Yosef 1988b). The isolated center of the core can then be removed, flattening the "debitage surface" again. Débordant removals are then necessary to begin the process again. This convexity management option may entail systematic and regular removals as above, or periodic reparations, as extant core morphologies and the presence of these débordant pieces in the assemblage can only indicate the use of the technique, not the degree of its use. Débordant removals are defined as flakes or blades with one lateral margin which removed the edge of the core. This lateral edge may be close to vertical and cortical, or more acute and composed of dorsal scars showing the removal of flakes, originating from a longitudinal ridge running down the length of the flake and terminating beyond the flake's lateral edge. These "cresting" removals can be toward one edge or both edges of the flake. Morphologically, débordant blades can overlap with crested blades used to begin or repair blade cores, save that the débordant removals tend to be twisted and/or turned in profile compared to crested blades.

Circumferential Removals: The convexity of the "debitage surface" of a core may be perpetuated by continually removing flakes from the whole circumference or periphery of a core. If such removals meet in the center of the "debitage surface" without leaving a mass too large to remove with subsequent removals, this management option allows the continuation of exploitation until exhaustion. The option is evidenced through the examination of extant cores, refits, and importantly the dorsal scar pattern of the debitage.

Frontal Crests: The convexity of the "debitage surface" of a core may be both created and perpetuated by establishing a central ridge or crest, through the removal of a line of flakes, down the long axis of the "debitage surface." This crest, which may be either bifacial or unifacial, serves as the arrise or *nervure-guide* (*sensu* Bordes 1961; Boëda 1986; Van Peer 1992) down which the force of a flintknapper's blow can propagate to remove a flake. This crested removal leaves two ridges, its lateral edges, which can serve as *nervure-guides* for subsequent removals. This technique may be used to commence exploitation of a core or to repair the longitudinal convexity of a reduction in progress. This option is evidenced through the examination of extant cores, refits, and importantly the identification of crested blades among the debitage.

Lateral Crests: This convexity management option utilizes the same principles as the frontal crest but is not created on the front of a "debitage surface" but behind one. By choosing to create two lateral crests, or *crête arrières* (Tixier 1984), on either side of a debitage surface, it is possible to narrow the "debitage surface" and thus manage its lateral convexity by means of removals from these lateral crests up the sides of the core. This option is evidenced through the identification of crested blades among the debitage but can only be differentiated from the use of front crests only through the documentation of its use on extant cores or refits.

Side Blade Removals: This convexity management option is relatively rare and represents an alteration of the reduction strategy from the use of a relatively broad debitage surface, which may be exhausted or irreparable, to the use of the lateral edge of this broad surface. In some ways, this option consists of removing numerous débordant flakes or blades from a surface situated on the side of the original surface. One form of this practice is to commence blade removals from the side surface after the termination of the exploitation of the original surface. Alternative forms distinct to the above example would be the continual shift from the flatter to the narrower side surface of the core and back again, as seen in Roussel's (2011) interpretation of the Châtelperronian assemblage at Quinçay. Refits and extant core morphologies allow the documentation of this option in practice.

4.3.2 Domain 2. Platform Maintenance

The knapping domain, "platform maintenance," includes the knappers' choices of how far into the platform to strike (the platform thickness), the angle of the exterior platform at that spot, and whether or not to alter that angle with platform preparation beforehand, for all complete flakes and complete platforms in the assemblage. As these choices are enacted together with the blow of the percussor for each flake, they are treated together within the same domain. The body techniques used in the delivery of the hammer blow which determines the point of impact, and thus these three variables, produce waves of force which move between

200 and 1,100 meters per second, depending on the material and the mode of application of force (Cotterell and Kamminga 1987). This process is thus too fast for complete conscious control and requires repeated exposure and practice in the reduction of cores to develop the tendencies one is expected to learn from the appropriate role models within a given society. The learned tendencies most likely do not reside at the conscious level or if they do they are likely to be incorrect. The fact that it has taken many controlled experiments to demonstrate that the intuition of the flintknapper is wrong, when their gestalt tells them that the type of percussor and the angle of attack are the most important variables in flake variation, means that it need not be obvious to the hominins who made the assemblages in question. An examination of D. C. Waldorf's illustration (1993:20, fig. 10) for the way to strike off a bifacial thinning flake shows a clear confounding of angle of attack with platform thickness, as well as exterior platform angle. Even though he is one of the most impressive knappers of the late 20th century, he incorrectly ascribes the results of this illustrated blow to angle of attack alone whereas controlled experiments have shown exterior platform angle and platform thickness to be the equally operative variables. Given the subconscious level of the body techniques used to produce these variable states, these variables are highly suitable for characterizing the enculturating environment of the hominins who created a particular assemblage.

Domain 2. Step 1. Debitage Platform Treatment
Platform treatment, whether prepared, abraded, or left unprepared, changes the exterior angle of the platform, and thus affects the resulting mass of the flake. Many treatment options are available (Inizan et al. 1992:79–82), including:

Platform Treatment Types

1. *Plain*: one scar serves as the platform surface.
2. *Cortical*: no treatment of the platform surface as the majority of the platform consists of cortex.
3. *Dihedral*: two scars serve as the platform surface.
4. *Facetted*: three or more regular scars, originating from the top of the platform, serve as the platform surface.
5. *Chapeau de gendarme*: a facetted platform which, when seen face-on, possesses a symmetrical profile of two ridges with a valley between (Inizan et al. 1992:80).
6. *Scarred*: three or more irregular scars, not necessarily originating from the top of the platform, serve as the platform surface.
7. *Punctiform*: a tiny point, of less than 2.0 mm, represents the whole platform surface.
8. *Linear*: a thin scar, of less than 2.0 mm in thickness, represents the whole platform surface.
9. *Removed*: platform surface removed by retouch sufficiently enough to prevent a clear identification.
10. *Broken/crushed*: platform surface damaged by postdepositional forces or during the removal of the piece, sufficient to prevent a clear identification. This is equivalent to "indeterminate" in other attributes.

For pair-wise comparisons between assemblages, the choice of platform treatment for a particular assemblage is represented by the relative frequency of blanks with "prepared" platforms (types 3–6) versus "unprepared" platforms (types 1–2, 7–8). The comparison is standardized by means of a G^2 likelihood ratio test (approximating the chi-square distribution; Sokal and Rohlf 1995; Sackett 1989) between these frequencies.

Domain 2. Step 2. Exterior Platform Angle
A flintknapper controls the exterior platform angle, the angle between the dorsal surface and

the platform surface, in order to affect, with the platform thickness and angle of attack, the mass or size of the flake to be struck off the core (Dibble and Pelcin 1995; Dibble and Rezek 2009). The assemblage-wide choices of treatment of the exterior platform angle are characterized for the present study by means of the central tendency statistic, the mean and standard deviation of the individual EPA measurements for intact blanks and proximal fragments.

Exterior Platform Angle (EPA): this measurement is taken with a goniometer immediately behind the point of percussion visible on the platform surface (Dibble and Bernard 1980). This measurement is the extant angle and thus, by means of beveling (the removal of small flakes behind the platform) and abrasion may measure higher than 90 degrees. The caliper method advocated by Dibble and Bernard (1980) is not used due to the greater ease of the gonionmeter method and the fact that the two methods have the same low inter-observer variance in the range of angles typically found in exterior platform angles at the Middle to Upper Paleolithic transition, i.e., >45° (compare Dibble and Bernard 1980:figs. 4–5).

Domain 2. Step 3. Platform Thickness

A flintknapper controls the platform thickness, the distance into the platform from the edge of the dorsal surface where the point of percussion falls, in order to affect, with the exterior platform angle, the mass or size of the flake to be struck off the core (Dibble and Pelcin 1995; Dibble and Rezek 2009). The assemblage-wide choices of the platform thickness are characterized for the present study by means of the central tendency statistic, the mean and standard deviation of the individual platform thickness measurements for intact blanks and proximal fragments.

Platform Thickness (PT): this measurement is taken with a digital caliper immediately at the point of percussion visible on the platform surface, perpendicular to the flake ventral surface, perpendicular to the platform width, and parallel to the surface of the platform (if the latter is flat). In other words, this length "is measured from the interior to the exterior surface of the platform at the point of percussion" (Dibble 1997:153). Please note that the terminology used here for "platform thickness" is in fact equivalent to Dibble and Pelcin's (1995:fig. 1) "platform depth" as well as Davis and Shea's (1999:fig. 2) "platform depth" and not to Dibble and Pelcin's (1995:fig. 1) "platform thickness" measurement of the thickness of the flake, when measured irrespective to the platform but perpendicular to the axis of flake propagation at the point of impact. If the point of percussion is not visible, the platform thickness is taken at the midpoint of the platform width.

4.3.3 Domain 3. Direction of Core Exploitation

The direction of core exploitation, the third behavioral domain used in this study, concerns the variability resulting from different options for removing flakes from single to multiple platform cores, creating debitage with particular dorsal scar patterns. The directional ripples on dorsal scars rarely affect the efficacy of the cutting edges, the tool shape for penetration, or the ability to hold/haft the tool. For these reasons, differences in how flakes were removed relative to core rotations between platforms serve as prime analytical proxies for isochrestic variation (*sensu* Sackett 1990) in how flintknappers are taught to reduce a core. This attribute thus has many uses for measuring the degree of social intimacy between groups of knappers based on whether or not they share these learned knapping behaviors. For each complete flake and tool in an assemblage, this choice is documented as follows:

Direction and Number of Scars per Blank Sector: In order to evaluate quantitatively as well as categorically each flake's evidence of direction of removal, one first counts the scars from the different cardinal directions relative to the orientation of the blank based on the platform. These counts are then used in the next attribute (*Dorsal Scar Pattern*) to produce a categorical type. To record this directional data, each blank is conceptually divided into four sectors (A to D; based on Crew 1975, Sirakov 1983; Meignen and Bar-Yosef 1988; Baumler 1988) by two perpendicular lines offset from the axis of propagation by 45° (Figure 4.22, based on Tostevin 2003, Figure 8.1, lower image). The number of scars from directions A (distal), B (right lateral), C (proximate), and D (left lateral) are noted using ripples and radial fissures or lances as indicators of scar directions. A dorsal scar must be at least 0.5 cm in any dimension to be counted. This data forms five attributes:

- *Direction A*: number of dorsal scars from direction A.
- *Direction B*: number of dorsal scars from direction B.
- *Direction C*: number of dorsal scars from direction C.
- *Direction D*: number of dorsal scars from direction D.
- *Direction U*: number of dorsal scars from an unknown direction.

Dorsal Scar Pattern: Using the observations of the number and direction of scars on the dorsal surface by sector to interpret the scar pattern for the whole piece, each blank is given a dorsal scar pattern type. In doing so, the direction of flake propagation is an extra "C" scar direction. Each blank falls into one of the following attribute-states (Volkman 1989; Meignen personal comm.):

Unidirectional: All scars come from the proximal part of the flake, sector C.

Unidirectional Convergent: All scars originate from the intersection of sectors B and C as well as C and D, converging toward the distal end of the piece. The variation in scar origins around the two intersections is approximately 30° (shown by the narrow black arrows around the wide

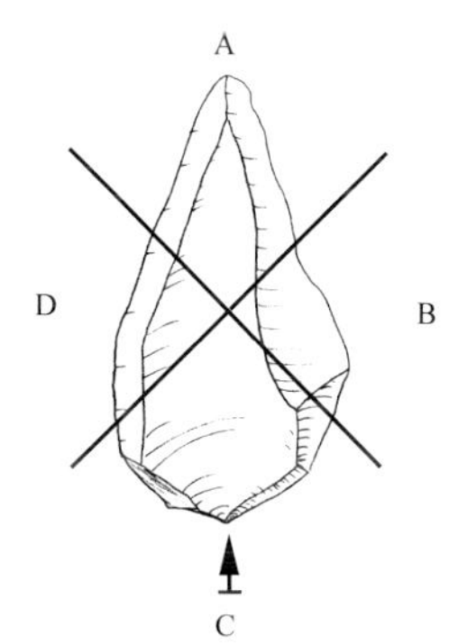

Figure 4.22 Definition of Sectors for the Directions of Dorsal Scars (after Crew 1975). Drawing by G. Tostevin.

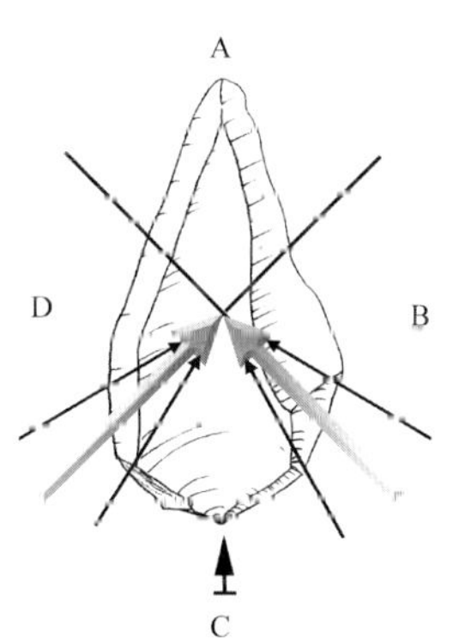

Figure 4.23 Definition of a unidirectional-convergent dorsal scar pattern.

shaded arrows in Figure 4.23). If scars are extant from both intersections, an additional scar for the proximal end (sector C) is still appropriate for this dorsal scar pattern. This category is subsumed within "unidirectional" in pair-wise assemblage comparisons.

Bidirectional: scars come from both the proximal and distal ends of the flake, sectors A and C. As the flake itself is from direction C, a bidirectional pattern can result from a blank with scars only from direction A.

Crossed: at least one scar originates from one of the lateral directions, sectors B and D. The flake may only have lateral scars or it may have some proximal scars as well.

Subcentripetal: scars originate from three sectors. This category may be combined with "centripetal" in pair-wise assemblage comparisons.

Centripetal: scars originate from all four sectors. This category may be combined with "subcentripetal" in pair-wise assemblage comparisons.

Unknown: the direction of one or more dorsal scars is unreadable. This category is removed from pair-wise assemblage comparisons.

If it is impossible to distinguish between two directions during the reading of a given scar and both of those directions are already accounted for by other scars, a specific interpretation other than "unknown" is warranted.

Theoretically, the dominant directional strategy used in each assemblage may be characterized by the relative frequencies of different dorsal scar patterns evidenced among the debitage. Dibble's (1995a) problem with Boëda's (1988a) reconstruction of level IIa from Biache Saint-Vaast, however, serves as a warning against the straight typology of either debitage or core scar patterns. The reduction strategy in an assemblage may drastically change as a core becomes exhausted. For this study, therefore, the dominant direction of core exploitation is determined using the principles of dimensional change during core reduction (Holmes 1919; Frison 1968; Newcomer 1971; Collins 1975; Jelinek 1976; Stahle and Dunn 1982; Henry 1989; Dibble 1987). As a core is reduced, it becomes shorter due to platform rejuvenation and the general subtractive nature of flintknapping. As the core gets shorter, so do the flakes struck off the cores. Thus, assuming that the general trend is for longer blanks to be produced earlier in core reduction and shorter blanks to be produced later, with the understanding that no particular blank can be absolutely placed in a stage of reduction based on its length, a correlation between debitage length and debitage dorsal scar pattern in an assemblage can indicate whether a change of directional strategy occurred during the continued reduction of the cores.

A similar correlation exists, according to experimental studies of bifacial technologies (Sullivan and Rosen 1985; Geneste 1985; Mauldin and Amick 1989; Baumler 1988; Ahler 1989a, 1989b), between debitage pieces with high versus low amounts of cortex and the relative position of the piece earlier and later in the process of core reduction. Relying on these experimental results, I used both cortex percent and debitage length in earlier versions of the characterization of this domain (Tostevin 2000a,b,c, 2003a,b) to correlate with debitage dorsal scar patterns to identify the direction of early exploitation (according to the cortex correlation) and late exploitation (the length correlation for non-cortical debitage). Replying to my characterization of Boker Tachtit Level 1 according to this early system (Tostevin 2003a), Marks (2003) pointed out that characterizing early core exploitation as "unidirectional" when the refits demonstrate a centripetal pattern represents a failed methodology. In hindsight, I must agree. The use of percent cortex on the dorsal surface of debitage is an unreliable proxy for identifying directionalities of decortification relative to the core's ultimate orientation, although it is certainly a credible proxy for the amount of early versus late core reduction present in an assemblage (Dibble 1995a; Baumler 1988; Odell 2004:126-127). The problem rests with the frequent tendency to rotate the core to exploit multiple platforms in decortification, as Volkman notes for Boker Tachtit Level 1 (1989:196-202), so that the dorsal scar patterns of the decortification flakes do not themselves indicate the orientation or rotation of the core itself. The experimental studies which provided the initial concept, being bifacial, should have warned me of the inappropriate use to which I put the variable.

Marks' critique necessitates a change in the methodology for characterizing early versus late exploitation of core as evaluated from data derived from debitage. It is still advisable, however, to base the characterization on debitage rather than another data source, such as infrequent refits or exhausted core forms, as the methodology necessitated by the high-level theory presented in the previous chapter also dictates the comparability of all units of analysis. Thus while refits may be more detailed in their reconstructions of behavior, they are inappropriate if their use produces distinctions with other assemblages simply because the latter lack the resolution provided by refits. Please refer to Chapter 7.2.3 for a complete reply to the critiques of Marks (2003), and Belfer-Cohen and Goring-Morris (2003).

The new system for characterizing early versus late exploitation uses only the correlation between debitage length and debitage dorsal scar patterns to suggest directional modalities of exploitation at these parts of the sequence. As core rotations during reduction are more standardized and the platforms more long-lived after most of the cortex has been removed, the artifacts used for such a correlation include only complete blanks (debitage and tools) with less than 60% of their dorsal surface, not considering in the platform, covered by cortex, according to the following definition:

Percentage Cortex: The dorsal surface of debitage blanks, excluding the platform, may be said to be covered in cortex, including old patinated and incipient fractures given the type and condition of the raw material, in the following proportions: 1) 0%, 2) 1-10%, 3) 11-40%, 4) 41-60%, 5) 61-90%, 6) 91-99%, 7) 100% (Dibble 1995a).

As noted in Tostevin (2003b), the number of dorsal scars on debitage blanks is an important consideration for the analysis of both steps within this domain. Flakes with few dorsal scars are less likely to evidence the more complex directional reduction options. For instance, a flake with two scars can only evidence unidirectional, bidirectional, crossed, or subcentripetal reduction directions (subcentripetal in this case is only possible if the direction of flake propagation is different from the two dorsal scars). Thus, a centripetal option cannot be evidenced on a flake with fewer than three dorsal scars. In order to avoid biasing the analyses of these two steps in favor of the simpler directional reduction strategies, only flakes with three or more dorsal scars are used to illustrate the change of dorsal scar pattern as debitage decreases in size and cortex decreases in percentage of dorsal surface covered.

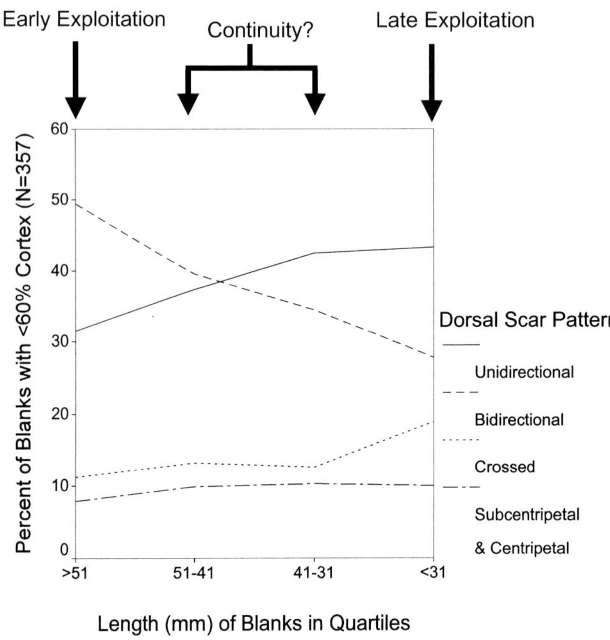

Figure 4.24 Example of a dorsal scar pattern by blank length in quartiles with indications where to characterize early exploitation, late exploitation, and potential continuity between these.

Domain 3. Step 1. Direction of Early Exploitation

The characterization of the central tendency in the direction of debitage removal early in core exploitation is based on the dorsal scar pattern within the blanks whose lengths fall into the largest quartile of the included blank population (see above). Thus, from the line graph of the length by dorsal scar pattern cross-tabulation, one examines the position of the lines for each dorsal scar pattern in the left most column, i.e., for the blanks that make up the longest 25% of the population. In the case of Figure 4.24, the most represented dorsal scar pattern for the blanks longer than 51 mm is "bidirectional," at around 49% of the population within that quartile. In a case in which two dorsal scar pattern lines are within values of 10% of the intra-quartile population of each other, both dorsal scar patterns are identified as the central tendency for this group of blanks. This "10% rule" standardizes the determination of the central tendency in directionality. In the case of Figure 4.25, slightly over 40% of the longest quartile of blanks evidence a bidirectional dorsal scar pattern while unidirectional blanks represent slightly fewer than 40% of the population. As the two values for bidirectional and unidirectional are within 10% of each other, they are both judged as the central tendency.

Domain 3. Step 2. Direction of Late Exploitation

Blank length in quartiles is correlated with dorsal scar pattern to identify the most numerous dorsal scar pattern (unidirectional, bidirectional, crossed, and subcentripetal together with centripetal) for late core exploitation in the fourth and thus shortest quartile of blanks. In the case

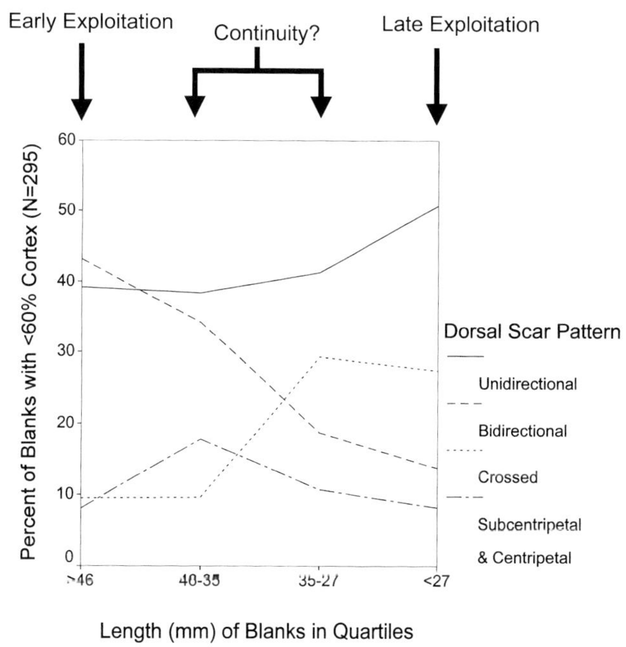

Figure 4.25 Example of a dorsal scar pattern by blank length in quartiles in which the 10% rule would apply to the characterization of early exploitation.

of Figure 4.24, unidirectional blanks represent over 40% of the blanks in the shortest quartile, and thus likely removed during late exploitation. As the bidirectional blanks fall below 30% and thus more than 10% of the population away from the unidirectional value, the 10% rule does not apply and late exploitation is classified as "unidirectional."

This method of characterizing the change in the tendency to rotate the core between different platforms is consistent and rigorous. It also is highly comparable between very disparate assemblages that lack refit sequences. Despite this comparability, it is also sensitive enough to corroborate extant refits, as can be seen in the comparison of the results of this method and the results of refit sequences from the same assemblages (see the discussions of Stránská skála IIIa-4 in Chapter 5 and Boker Tachtit Level 1 in Chapter 7). This is also the only true "sequence analysis" in this research protocol, distinguishing as it does between early and late stages of exploitation, etic categories though they are.

The method also allows for an evaluation of the relationship between the two characterizations of "early" and "late." The evaluation of the degree of continuity in the lines for the different dorsal scar directions between length quartiles provides a means of assessing the likelihood of the independence of the different directional approaches on different cores. As shown in Figure 4.24, the middle two quartiles should show a gradual change between the longest and shortest blanks if indeed the opposition is accurately capturing the difference between early and late reduction. Thus, in cases in which the middle two quartiles show a gradual change from the dominant direction in the largest quartile to another direction in the smallest quartile, as in Figure 4.24, this shows a change in directionality on the same cores. If the middle quartiles contrast drastically with the values on either end of the length continuum, it is likely that any distinction in directionality in the largest and smallest blanks represents two systems of directionality executed on two groups of cores. This method is thus more sensitive to behavioral variability in the pattern of core rotation across cores than admittedly more quantitative approaches (i.e., Clarkson et al. 2006) that sum the vectors of the scars on cores but only succeed in distinguishing random, radial, and laminar rotation patterns, all at the level of the exhausted core forms rather than throughout core reduction.

4.3.4 *Domain 4. Dorsal Surface Convexity*

While the interaction between the exterior platform angle and the platform thickness (and angle of attack) determines the mass of the resulting flake, the "dorsal surface convexity" domain includes the knapping steps which determine the shape of the flake as created by the *nervure-guide* or ridge pattern of the core exterior (*sensu* Bordes 1961; Boëda 1986; Van Peer 1992). The *nervure-guide* or ridge system is itself the result of the interaction between the scars from previous flake removals, as they shape the convexity of the core surface and so constitute the shape of the mass that is removed as the force of the flintknapper's hammer blow exits the core along the ridges between scars. A categorical attribute based on the pattern of the ridges themselves is currently inadvisable, as the extreme variability does not lend itself to easy characterization or inter-assemblage comparison. It is possible, however, to quantify the behavioral choices made by flintknappers in the course of controlling the ridge system through the following five attributes. These attributes are measured only on complete flakes and complete tools in order to improve the accuracy of the identification of the tendencies within the assemblage.

Domain 4. Step 1. Elongation of the Longitudinal Convexity: Length/Width Ratio

The ratio of blank length to width serves as an analytical proxy for flintknappers' behavioral tendencies to produce longitudinal ridge systems (for blade-like products) versus dispersed ridge systems (for flake products). In inter-assemblage comparisons, the mean and standard deviation of the length/width ratio is used in a student's t-test between the assemblages. This is a preferable method to reporting the Blade Index or *Index laminaire*, "the number of blades (flakes whose length is at least twice their width) divided by the total number of complete" pieces (Debénath and Dibble 1994:176). Although Belfer-Cohen and Goring-Morris (2003) have criticized the use of the length/width ratio in place of the Blade Index, the length/width ratio is clearly superior for the present purposes of this research, even if less frequently used in the literature and thus less comparable in meta-analyses. Unlike the Blade Index, the length/width ratio of complete blanks and tools allows the evaluation of both the central tendency and measure of dispersion of this relationship of elongation within the assemblage population. As Carr has argued, the structure of a variable "affects how the form of a distribution is characterized" according to "the number of alternative states that the attribute can assume. An attribute with a clinal distribution can be analytically constrained to and appear as a patchy-bounded or uniform-unbounded distribution if the attribute takes only two or a few states, as opposed to many rank or continuous-scale states" (Carr 1995b:241). Using a continuous-scale elongation variable allows one to identify more fine grain distinctions in the degree of elongation through time and space than would the ratio of blades to flakes, primarily because the latter ratio does not tell you how elongated the blades in fact are. The length/width ratio is calculated from the following attributes:

Debitage Length: length is measured with a digital caliper, as the maximum distance along the axis of flake propagation and thus perpendicular to the platform surface, between the top of the striking platform and the distal-most end of the piece, to the nearest tenth of a millimeter. In Andrefsky's illustration of different measuring techniques, this measurement is line "c" (Andrefsky 2005:fig. 5.8).

Debitage Width: width is measured at the mid-point of the above length, perpendicular to the axis of flake propagation. This is line b" in Andrefsky (2005:fig. 5.9).

These methods of measuring flake length and width, compared to maximal measurements irrespective of the axis of propagation (e.g., Debénath and Dibble 1994:fig.2.14c), are used here as they better reflect the dimensions of the core surface relative to the location of the core platform from which the flake was detached. As flake length in particular is used in analyzing change through time in directionality of core exploitation, this relationship between flake measurement and core orientation improves on the characterization of directionality discussed above.

Domain 4. Step 2. Shape of Convexity: Debitage Lateral Edges

The categorization of the lateral edge shapes evidenced in an assemblage serves as a behavioral proxy for flintknappers' tendencies to choose to strike along parallel, convergent, expanding, or diffuse ridge systems. The significance of different options utilized for this behavioral step in two assemblages is determined through a G^2 likelihood ratio test of the frequencies of these attribute-states.

Debitage Lateral Edge Types (adapted from Bergman 1987)

Converging: edges converge toward the distal end of the piece for the majority of the blank length, i.e., more than half of the length of the flake has lateral edges converging together at an angle greater than 30° from the axis of propagation.

Expanding: edges are divergent, expanding toward the distal end of the piece for the majority of the blank length, i.e., more than half of the length of the flake has edges expanding at an angle greater than 30° from the axis of propagation.

Ovoid: edges both expand and converge, producing rounded lateral edges, with roughly equal parts of the length of the flake having lateral edges expanding then converging.

Retouched: too much of the flake blank has been retouched away to determine accurately the lateral edge type.

Indeterminate: broken, or insufficient regularity to fit into typology.

Domain 4. Step 3. Curvature of Convexity: Profile

The categorization of the profile of debitage pieces evidenced in an assemblage serves as a behavioral proxy for the flintknappers' tendencies to exploit flat, curved, or twisted longitudinal core surfaces for debitage production. The significance of different options utilized for this behavioral step in two assemblages is determined through a G^2 likelihood ratio test. This variable is best seen by holding the platform up to the eye with the distal end pointing away from the eye, the dorsal surface up, and looking down the flake's propagation axis. Andrefsky's (1986) method for quantifying curvature in flake profiles would be an improvement on the typological approach used here. This is one of many modifications of the discrete variables used here which may allow the application of multivariate techniques which require continuous variables in future research along this line.

Debitage Profile Types (adapted from Bergman 1987)
Straight: This attribute-state depends upon a measure of the curvature of the blank profile, i.e., the distance between the midpoint of the length on the ventral surface of the piece to the top of the table surface on which the distal and proximal tips rest. For straight profiles, this measurement does not exceed 1/8 of the blank length. Also, "straight" is chosen when the distal half twists less than 45 degrees in either direction relative to the flat profile of the platform.

Curved: the curvature measurement exceeds 1/8 of the blank length. Also, "curved" requires that the distal half twists less than 45 degrees in either direction relative to the flat profile of the platform.

Concorde-type (sensu Meignen and Bar-Yosef 1991): the distal-most 1/3 of the blank length bends down at a striking angle. This attribute-state can be subsumed under "curved" in pairwise assemblage comparisons.

Twisted to left: distal tip of the piece is twisted at least 45 degrees to the left of a flat profile relative to the platform. The piece may be "curved" as well but "twisted" supersedes "curved" in the typology.

Twisted to right: distal tip of the piece is twisted at least 45 degrees to the right of a flat profile relative to the platform. This attribute-state can be subsumed with "twisted left" into a general "Twisted" category. The piece may be "curved" as well but "twisted" supersedes "curved" in the typology.

Retouched: Too much of the flake blank has been retouched away to determine accurately the profile type.

Unknown: broken, or insufficient regularity to fit into typology.

Domain 4. Step 4. Lateral Convexity: Cross-section
The categorization of the cross-section of debitage pieces evidenced in an assemblage serves as a behavioral proxy for the flintknappers' tendencies to utilize one versus two or more ridges as *nervures guides* in the propagation of debitage products. This attribute is assessed at the midpoint of the blank length. The significance of different options utilized for this behavioral step in two assemblages is determined through a G^2 likelihood ratio test. For pair-wise comparisons between assemblages, types 4 and 5 can be subsumed into an "other" category while types 1 and 2 are subsumed into a general "triangular" type.

Debitage Cross-section Types
(Meignen personal comm. 1995)
1. *Triangular*: effectively only one ridge present on the dorsal surface.
2. *Right-triangular*: one lateral edge is vertical in cross section and serves as the only dorsal ridge.
3. *Trapezoidal*: effectively two or more ridges present on the dorsal surface.
4. *Lenticular*: no dorsal ridges present. Cross-section convexity is similar to a lens.

5. *Domed*: multiple ridges form an extremely convex, almost hemispherical cross-section.
6. *Indeterminate*: too retouched, broken, or insufficient regularity to fit into typology.

Domain 4. Step 5. Vertical Convexity: the Width/Thickness Ratio

The ratio of blank width to thickness serves as a behavioral proxy for flintknappers' tendencies to exploit tall ridge systems versus flat or diffuse ridge systems. The vertical convexity of each flake as it comes off of a core can be understood through the metaphor of flakes being "peeled" from a core in thin sheets, as a potato is peeled, versus flakes being cut as thick wedges from the core, as a potato is cut for French fries, as Liliane Meignen once described it to me. This metaphor of peeling versus cutting flakes off during reduction serves as a model for how the width/ thickness ratio of each flake reflects the volumetric contrast in how a Levallois strategy removes flakes from the surface of the core while prismatic blade production cuts off a stick of volume with each removal. In this sense, the central tendency of the width/thickness ratio within an assemblage can be seen as a quantitative proxy for the volumetric conception of Boëda (1988b, 1994). This ratio requires the measurement of flake thickness, as defined below:

Debitage Thickness: thickness is measured with a digital caliper to the nearest tenth of a millimeter at the midpoint of the length as defined above, perpendicular to both the axis of flake propagation and blank width.

Marks, in a critique of this concept, comments, "Does anyone believe that the average width/thickness ratio of all debitage and tools within an assemblage '...quantifies the volumetric conception, *sensu* Boëda 1994)' (Tostevin [2003a:tab.6.1])?" (2003:264). I would respond simply with, should anyone believe Boëda's generalizations without quantification? What is the meaning of Boëda's volumetric conception of core reduction, or even the generalizations of Marks and Volkman (1983a) as derived from detailed refitting sequences, if the concepts and results cannot be put into action as comparative measures that allow one phenomenon to be analyzed in contrast to another? As Vermeersch (2001) has noted, the archaeological records of the Nile and the Levant during the Middle to Upper Paleolithic transition are incomparable and thus not understandable because of the lack of common archaeological systematics. As the quote from da Vinci at the beginning of this chapter emphasizes, we need measurements that different scientists can use consistently to study phenomena of interest; only then can we expect to build through time upon the work of single researchers. Put another way, it has been over two decades since Boëda's insightful and intuitive volumetric concept and no one has been able to operationalize it to study the way in which the pattern of the concept in the record informs on the evolution of human behavior. Similarly, it has been over two decades since Volkman's (1983, 1989) labor intensive refits and his introduction of a system for studying nearly complete refit sequences and yet no one has been able to gauge exactly how alike in comparable behavioral terms Boker Tachtit is to assemblages which lack refits. It is possible to generalize from Volkman's details (i.e., Marks and Volkman 1983b; Marks 1990) but without a quantitative means of assessing the accuracy of such historical generalizations/syntheses, they are difficult to build on scientifically. This is why it is so difficult to resolve drastically different historical syntheses from the same data, i.e., Marks (2003) versus Bar-Yosef (2003). The width/thickness ratio as a proxy for the vertical convexity removed on a flake by flake basis is offered here as one way of avoiding these limitations in comparability of technological reconstructions. Belief, for the culture materialist, develops when something shows itself useful in practice. The usefulness of the measure will be the test of its significance.

As noted in the discussion of Galton's problem in the previous chapter, it is necessary to use independent units of analysis in quantitative evaluations of similarity. With this in mind, the use of the width/thickness ratio results in the avoidance of the use of measures which are known to correlate significantly with either flake width or thickness. As flake width is correlated with the platform width of the flake (Dibble 1997), the latter measurement on the platform is avoided in characterizing the second domain, platform maintenance. While flake thickness is moderately correlated with platform thickness (R^2=0.28, n=3312, see below; see also Dibble 1997:165, fig. 9) but not with platform width (R^2=0.15, n=3380), the ratio of flake width over thickness lessens further this moderate correlation, making it more independent than dependent in terms of how it patterns on an assemblage-wide basis. These issues are explored further after the description of the analysis of the tool kit morphology for each assemblage.

4.4 Description of the Morphology of the Tool Kit

Tool kit morphology refers to the dorsal surface convexity variables which give the form of their cutting edge to the retouched blanks in the assemblage, as well as the diagnostic flakes without retouch traditionally categorized as tools - i.e., Levallois pieces and naturally backed knives. These are simultaneously the attributes of the tool kit which would be visible and thus learnable at a distance. Thus both functionality and visibility on the taskscape are modeled through the description of the morphology of an assemblage's tool kit. The attribute states within each categorical variable are typed with the intent to recognize the original form of the variable before the application of retouch. If retouch makes two attribute states equally likely as the original form of the blank, the type "retouched," "removed," or "indeterminate" (all equivalent for this purpose) is entered for that attribute. For continuous variables, the degree to which retouch reduction (the Frison Effect) has compromised the original blank form is treated in the Comparability of Assemblages section of this chapter.

As most of the variables included in this domain are described in detail in the dorsal surface convexity domain above, please refer to the earlier description when examining the following list of variables for the tool kit morphology description.

Step 1. Tool Laminarity: Length/Width Ratio. The elongation of blanks as described above.

Step 2. Tool Vertical Convexity: Width/Thickness ratio. A characterization of the cross-sectional geometry of the blank which affects the penetration potential of the tool as hand delivered or as an armature (Hughes 1998; Shea 2006; although this is not the tip cross-sectional area measurement used in these studies). As defined above.

Step 3. Shape of Tool Cutting Edges: Lateral Edge Type. As defined above.

Step 4. Shape of the Distal Terminus: As a characterization of the initial piercing ability, as opposed the ability to penetrate in depth as with the width/thickness ratio, the distal terminus is studied as follows:

Distal Terminus (adapted from Bergman 1987): The distal end of a flake is often a functional edge and this attribute represents selection for a particular type of edge.

1. *Blunt*: Dorsal view of the distal end of flake is blunt, i.e., the edges of the distal-most point of the blank form an angle >60° when a goniometer is placed on the distal termination. This attribute-state includes feathered terminations and vertical terminations.

2. *Pointed*: Dorsal view of the distal end of flake is pointed i.e., the edges of the distal-most point of the blank form an angle <60° when a goniometer is placed on the distal termination. This attribute-state includes feathered terminations and vertical terminations.

3. *Hinged* (Whittaker 1994:17).

4. *Overshot* (Whittaker 1994:19).

5. *Removed/Retouched*: too much of the flake blank has been retouched away to determine accurately the distal termination.

6. *Indeterminate*: broken, or insufficient regularity to fit into typology.

Step 5. Curvature of the Cutting Edge: Blank Profile: As defined above.

Step 6. Application of Unique Retouch. This behavioral category differentiates flintknappers' choices of different methods of applying retouch, by means of varying the angle of blow (i.e., for stepped as well as flat retouch) and the number of surfaces to which it is applied (i.e., bifacial retouch). The characterization of this choice is a qualitative operation to be sure as it relies upon the presence/absence of any of the possible retouch types in the assemblage. This is potentially meaningful, however, given the appearance of such different retouch applications as the flat bifacial retouch of the Central European Micoquian and Szeletian industries (Allsworth-Jones 1986, 1989) and the carinated retouch of Aurignacian and other industries (Le Brun Ricalens 2005; Belfer-Cohen and Bar-Yosef 2006; Bar-Yosef et al. 2006). Future research will hopefully improve upon this analytical proxy.

Step 7. Location of Retouch as Tool Types: Flintknappers' tendencies to choose particular locations for the application of retouch is gauged according to the types defined in Bordes (1961). Assemblages are designated as dominated by a majority of lateral edge retouched types, usually a majority of Middle Paleolithic tool types following Bordes (1961) as opposed to distally retouched types, usually a majority of Upper Paleolithic tool types. This distinction is made based on either published tool lists or my own typing of the tools in the artifact sample from the assemblage according to Bordes (1961). I also typed tools based on Hours' (1974) Levantine Upper Paleolithic and Epipaleolithic typology as a means of corroborating how the published type-lists relate to the distinction in distal and lateral retouch locations. This analytical step may be improved in the future, perhaps with a quantitative measure of tool kit diversity or coefficients of similarity such to the Brainerd-Robinson Coefficient (Odell 2004:116–120) or other methods (Shennan 1988:198; Stutz and Estabrook 2004). Given the serious limitations of the behavioral meaning that can be directly extracted with extant retouched tool typologies (Bisson 2000), however, the current low-power distinction is more behaviorally meaningful than a computationally-advanced treatment of a dubious type-list.

4.5 Evaluating the Statistical Independence of the Units of Analysis

Any quantitative method, such as counting similar behaviors in two assemblages, must use independent units so that the presence of one unit does not affect the appearance of another unit and bias the result (Rands and Riley 1958). Bias due to the interdependence of units of analysis, Galton's Problem mentioned above, is fortunately reduced in lithic technology by recognizing the results of fracture experiments that have shown that flintknapping steps related to platform treatment, external platform angle, platform thickness, dorsal scar ridge morphology, and subsequent placement of retouch are all functionally independent on a flake by flake basis (Speth 1972, 1974, 1975, 1981; Bonnichsen 1977; Dibble and Whittaker 1981; Cotterell et al. 1985; Dibble and Pelcin 1995; Pelcin 1996; Dibble and Rezek 2009; Rezek et al. 2011). This independence allows quantification of similar behaviors between assemblages to be used to test the competing hypotheses.

This assertion has been central to the discipline of Anthropology since its beginning (e.g., Tylor 1896:66; Kroeber 1931:151–152), although not always properly used. Rands and Riley's (1958) concept of "complex demand" (see Chapter 3) begs a question for the present study: to what degree does complex demand function in lithic operational sequences? Although the flake by flake attributes have been demonstrated to be independent, there has been no quantitative assessment by experimental research of the possibility that behavioral categories lose their independence when the knapper must sustain core exploitation. There is only so much variation possible in flintknapping given the constraints of fracture mechanics and the shrinking number of options available to the knapper as reduction continues, as noted by Odell and Henry (1989).

Nevertheless, there has been no controlled experiment to assess quantitatively the possibility that behavioral categories lose their independence when the knapper must sustain core exploitation. For instance, it is possible that the initial choice to exploit a narrow rather than a broad core face would influence the knapper to choose an elongated dorsal surface convexity (with a high length/width ratio) for the majority of subsequent removals in order to continue reducing the core. The example of the operational sequence of Molodova V Layer 11, however, illustrates that such an influence would not be absolute, as this assemblage exhibits relatively laminar/elongated blanks produced from broad core faces using multidirectional core reduction (see Chapter 6), a situation opposed to the above prediction. Thus, the interaction of complex demand and lithic technology must be assessed based on both general principles of lithic reduction and evidence from known operational sequences.

The data collected from the 18 archaeological assemblages in this study themselves can be used to test for the statistical strength of association between the variables used in this research. A evaluation of the strength of association between two variables at a time, tabulated into a matrix for all variables studied, is preferable in the present case compared to multivariate exploration of the variables that structure a multivariate space (such as principle component analysis). PCA is not favored in the present case because we are not interested in exploring what combination of variables explains particular percentages of the total population. Rather, the investigation of the statistical independence for dealing with Galton's Problem requires the specific variable by variable determination of how much of the variability of one variable is explained by another.

It is necessary to measure the strength of an association between variables separately from statistical significance, as it is the former which is informative as to how much one variable is going to affect the appearance of another rather than simply determining if it does affect the other variable. For nominal variables, the strength of association is measured through the deviation of the observed values from the expected values in a cross-tabulation, based on the marginal probabilities. This is a chi-square at least in form. Chi-square itself, however, is not a suitable method because its value is dependent upon both sample size and the strength of association. By dividing the chi-square value from a cross-tabulation of two variables by the sample size, one takes out this sample size component, leaving the statistic φ^2. Phi-squared is suitable only for 2x2 cross-tabulations, so the statistic is standardized as Cramér's V^2, which like Phi-squared, ranges from 0 indicating no relationship between variables, to 1, indicating a perfect relationship (Shennan 1997:115-117). Cramér's V^2 is thus used in the following matrix table (Table 4.1) to measure the strength of association between nominal variables.

Discontinuous variables that are rank ordered, such as platform preparation (which is either unprepared with 1 platform scar or prepared with 2 or more platform scars), as well as continuous variables that are broken into equal intervals by value of a variable can be treated as ordinal variables. The most appropriate measure of association in ordinal variables is Kendall's tau-c or τ_c, which varies between -1 and +1 for perfect negative and perfect positive associations, with 0 indicating no correlation between the variables (Shennan 1997:145-147). This statistic is used in comparisons between the ordinal variable "platform preparation" and continuous variables broken into ranked intervals in Table 4.1.

The measure of association most common for continuous variables is R^2, the coefficient of determination in regression analysis. The value of R^2 indicates the amount of the variation in the dependent variable that is accounted for by the independent variable. When multiplied by 100, R^2 equals the percentage of the variation accounted for in the dependent variable. As linear regression operates under the assumption of a normal distribution and several of the variables are not normal, the coefficients of determination were calculated from the regression of the log10 transformation of the non-normal variables. While the R^2 values from the regression of the continuous variables against each other are provided in Table 4.1, the scatterplots from which these values were regressed are useful in themselves in demonstrating the low degree of correlation within the data.

From the examination of Table 4.1 and Figures 4.26-4.31, it is clear that there is very little interdependence within the variables used in this study. The continuous variables most closely correlated are length/width and width/thickness, which is to be expected given the commonality of one variable in both ratios. Despite this commonality, the width/thickness ratio still only predicts 23% of the variation in the length/width ratio. Of the discontinuous variables, the Cramér's V^2 values are extremely small, with only the rare association controlling for more than 3% of the variability. The exception to this is the association between distal termination and lateral edge type, with a Cramér's V^2 value of 0.202. This is somewhat to be expected, given the correspondence between a pointed distal end and convergent lateral edges. This is the conflation of variable issue raised in the discussion of the role of distal termination earlier in this chapter. In addition to the *a priori* reasons for not treating distal termination as a variable directly chosen

Table 4.1 Statistical evaluation of the strength of association between continuous as well as discontinuous variables used in this research

	EPA†	PT*	Dorsal Scar Pattern	Length/ Width*	Lateral Edges	Profile	Cross-section	Width/ Thick.*	Distal Term.
Platform Prep.	Kendall τ_c= -0.08, n=3220	Kendall τ_c= 0.032, n=3311	Cramér's V^2= 0.003, n=3126	Kendall τ_c= -0.039, n=3676	Cramér's V^2= 0.003, n=3637	Cramér's V^2= 0.004, n=3648	Cramér's V^2= 0.002, n=3634	Kendall τ_c= .052, n=3676	Cramér's V^2= 0.000, n=2992
EPA†	X	R^2=0.11, n=3209	Cramér's V^2= 0.005, n=2732	R^2=0.00, n=3216	Cramér's V^2= 0.001, n=3185	Cramér's V^2= 0.003, n=3193	Cramér's V^2= 0.009, n=3178	R2=0.02, n=3216	Cramér's V^2= 0.002, n=2602
PT*	-	X	Cramér's V^2= 0.001, n=2793‡	R^2=0.07, n=3309	Cramér's V^2= 0.003, n=3264‡	Cramér's V^2= 0.011, n=3284	Cramér's V^2= 0.002, n=3261‡	R2=0.01, n=3309	Cramér's V^2= 0.000, n=2678‡
Dorsal Scar Pattern	-	-	X	Cramér's V^2= 0.010, n=3701‡	Cramér's V^2= 0.009, n=3667	Cramér's V^2= 0.005, n=3673	Cramér's V^2= 0.002, n=3662	Cramér's V^2= 0.004, n=3699‡	Cramér's V^2= 0.006, n=2997
Length/ Width*	-	-	-	X	Cramér's V^2= 0.029, n=4387‡	Cramér's V^2= 0.021, n=4398‡	Cramér's V^2= 0.006, n=4382‡	R2=0.23, n=4444	Cramér's V^2= 0.037, n=3620‡
Lateral Edges	-	-	-	-	X	Cramér's V^2= 0.005, n=4357	Cramér's V^2= 0.021, n=4348	Cramér's V^2= 0.020 n=4384‡	Cramér's V^2= 0.202, n=3581
Profile	-	-	-	-	-	X	Cramér's V^2= 0.007, n=4352	Cramér's V^2= 0.004, n=4397‡	Cramér's V^2= 0.003, n=3591
Cross-section	-	-	-	-	-	-	X	Cramér's V^2= 0.006, n=4381‡	Cramér's V^2= 0.028, n=3583

*Against another scalar variable, the log10 transformation of this variable is used for the calculation of R^2. Against discontinuous variables, this continuous variable is broken into four equal intervals of the population values, which is then treated as an ordinal variable against another ordinal variable or nominal against a nominal variable. In nominal vs. nominal cases in which the largest value interval is too sparsely populated to uphold the *chi*-square assumptions for expected cell counts for the calculation of Cramér's V^2, this largest value interval is removed before the calculation of Cramér's V^2, with the result indicated with the marker ‡.

†Against another scalar variable, no transformation is used for the calculation of R^2 due to the inappropriateness of this treatment of angle data. Against discontinuous variables, this continuous variable is broken into four equal intervals of the population, which is then treated as an ordinal variable against another ordinal variable or nominal against a nominal variable.

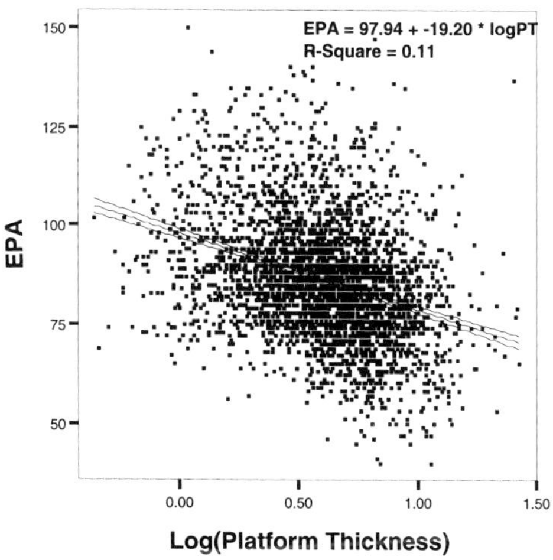

Figure 4.26 Scatterplot of exterior platform angle versus the log10 of platform thickness.

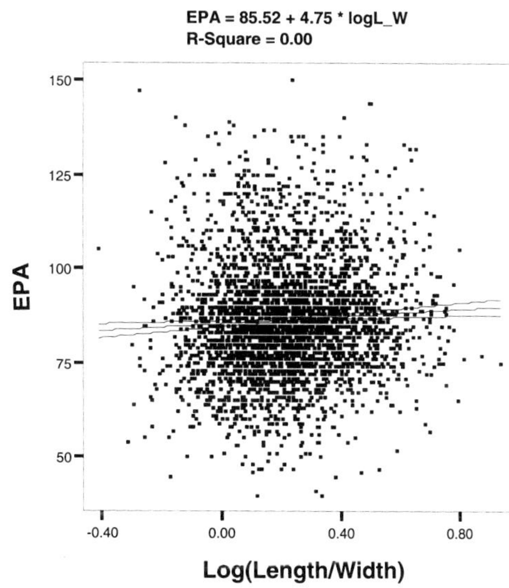

Figure 4.27 Scatterplot of exterior platform angle versus the log10 of the length/width ratio.

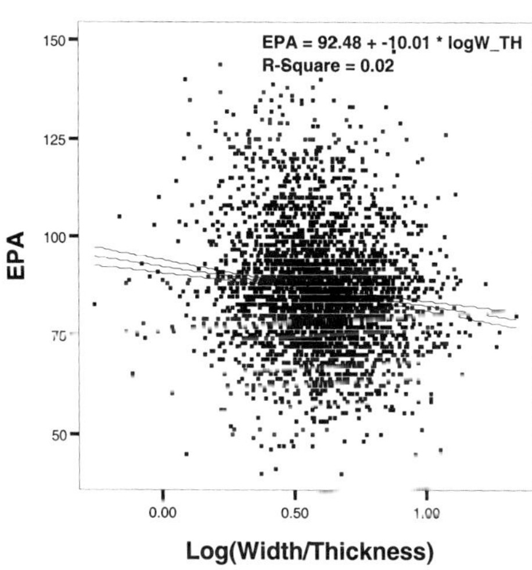

Figure 4.28 Scatterplot of exterior platform angle versus the log10 of the width/thickness ratio.

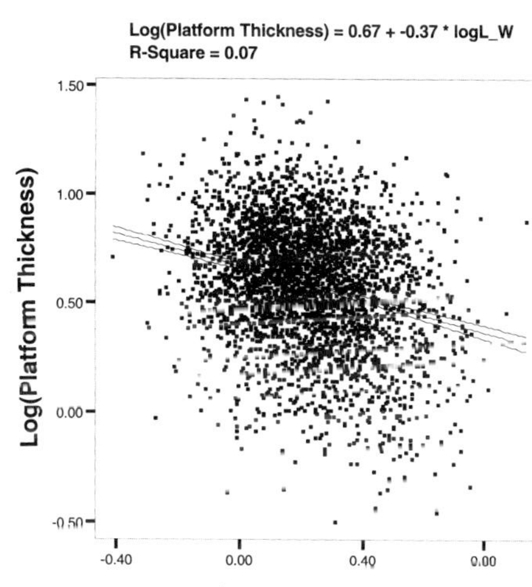

Figure 4.29 Scatterplot of log10 transformation of platform thickness versus the log10 of the length/width ratio.

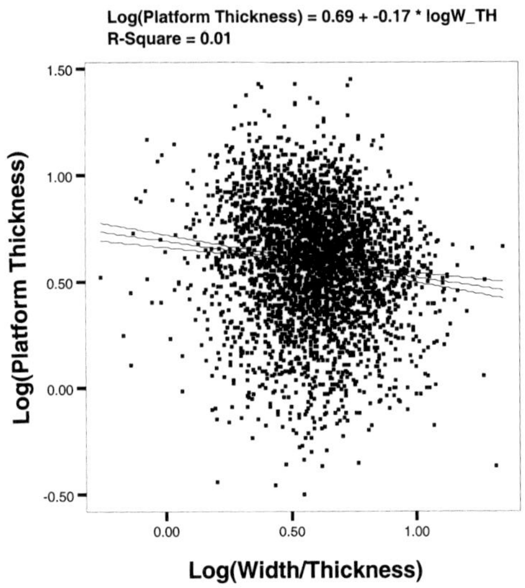

Figure 4.30 Scatterplot of log10 transformation of platform thickness versus the log10 of the width/thickness ratio.

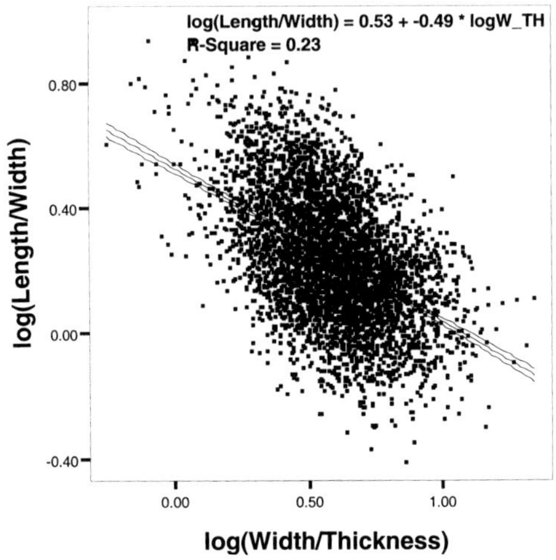

Figure 4.31 Scatterplot of log10 transformation of length/width ratio versus the log10 of the width/thickness ratio.

by the knapper's body technique, the statistical correlation between lateral edge and distal termination make it further unwise to include it with the blank production variables. As is, its inclusion among the tool kit morphology variables only affects 20% of the variability in lateral edge types and as the null hypothesis of this research is independent innovation for the appearance of tool kit morphologies, this slight interdependence contributes to the type II rather than type I error. Thus, contra Marks (2003:264), there is no statistical interdependence between the variables used in this research.

4.6 Evaluating the Comparability of Assemblages

As noted in Chapter 3, assemblages from raw material workshops and residential sites are preferable for this research compared to specialized hunting sites, retooling sites, or highly reduced assemblages due to raw material exhaustion. Specifically, in comparing the central tendencies of behaviors within the blank production of two assemblages, it is important to gauge how comparable the two assemblages are in terms of which portions of the operational sequence, from acquisition of raw material to discard of the exhausted cores, they happened to capture. It would be problematic to compare an assemblage which includes only the initial working of nodules into cores with an assemblage which includes extensive blank production down to the exhaustion of the cores. The central tendencies of the individual flintknapping behaviors within the four domains could be biased by this differential encapsulation of the operational sequences in each case. While some techniques have been proposed to solve this issue (Carr and Bradbury 2001; Wenzel and Shelley 2001; Dibble et al. 2005), they frequently rely on knowing the exact size and form of the original nodules of raw material, something particularly difficult in Pleistocene contexts in which whole raw material outcrops have been eliminated from the landscape by geological activity. The most clever of these approaches, Dibble et al.'s (2005) geometric method of measuring the difference between the observed surface area of cortex in an assemblage versus the expected surface area based on the geometric shape, size, and number of raw nodules used as cores, is attractive but requires too many assumptions about the raw material packages that cannot be controlled in the context of the present study.

The method used in the present study to gauge this comparability thus relies upon data from the assemblages themselves. Four analyses are conducted for each assemblage in order to approximate how comparable it is to the other assemblages. First, a ratio of non-cortical to cortical blanks (of the complete flakes and tools) is calculated for each assemblage and the entire 18 assemblages are graphed for this ratio in a histogram (Figure 4.32). It is possible now to situate any assemblage in relation to another. Specifically, a low value to the ratio of non-cortical to cortical blanks in an assemblage indicates the relative inclusion of more of the early than late portion of the operational sequence whereas a high value to this ratio indicates relatively more of the late portion of the sequence (Odell 2004:122-127; Mauldin and Amick 1989:70). A comparison of this ratio between assemblages does assume 1) relatively similar nodule sizes, as larger sized nodules will lead to more non-cortical flakes relative to the same degree of reduction in an assemblage with smaller nodules, as well as 2) relatively similar patterns of non-cortical blank export, as demonstrated by Dibble et al. (2005). Keeping these assumptions in mind, however, it is possible to use this ratio

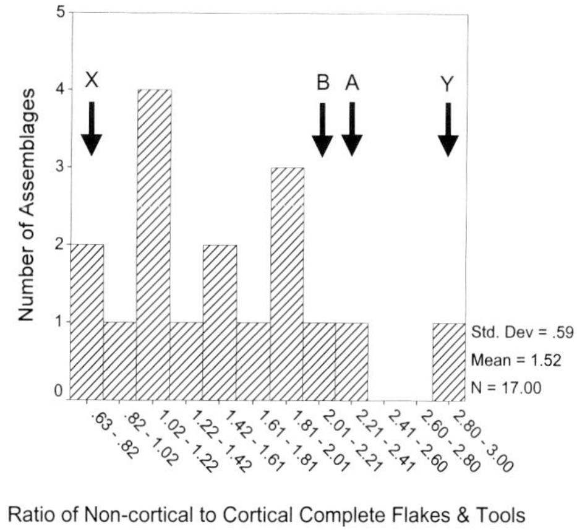

Figure 4.32 Example histogram of non-cortical to cortical blank ratios for a population of assemblages. With the values of four hypothetical assemblages are noted by the arrows, it is apparent that A and B are highly comparable, A and Y are moderately comparable, and X and Y are only marginally comparable.

as a cautious measure for ranking assemblages based on how much "early" working is being done on-site versus how much "field processing" has occurred before the cores reached the site. Thus, in the example provided in figure 4.32, assemblages found in column A and B would exhibit relatively similar portions of the core reduction sequence from raw material acquisition to core exhaustion. Specifically, they would both contain a fairly significant degree of field processing off-site followed by on-site core reduction, with less of the early portion of reduction sequence represented on-site. Conversely, assemblages X and Y would represent the extreme cases, with X possessing the proportion of cortical blanks indicative of a raw material workshop with initial decortification on-site and Y possessing the proportion of non-cortical blanks indicative of extensive field processing off-site.

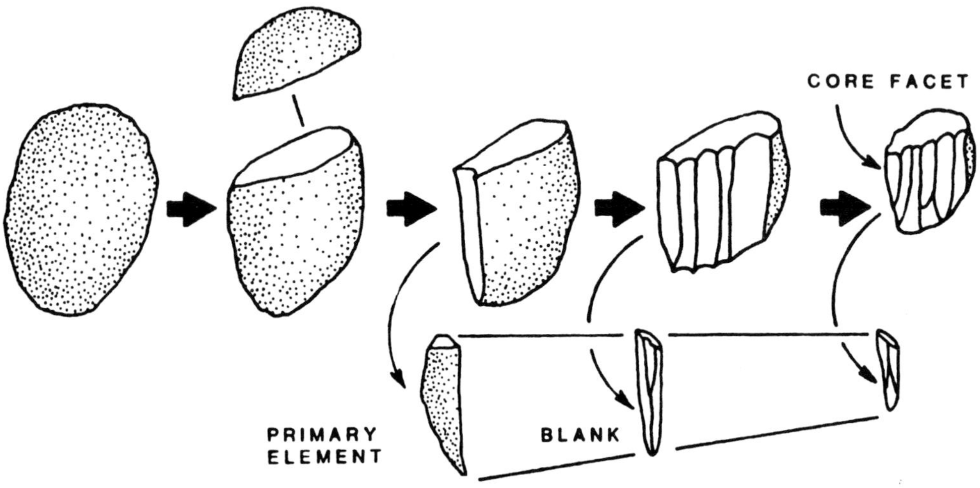

Figure 4.33 Change in the dimensions of blanks and core facets during core reduction (from Henry 1989: 140, fig. 1, used by permission).

Second, it is possible to evaluate how far toward exhaustion a set of cores were taken before being abandoned, i.e., how much of the very end of the reduction sequence is present, based on a comparison of an independent data set, the dimensional analysis of core facets versus complete debitage. In his 1989 paper, Don Henry argues that dimensional data can contribute greatly to the study of core reduction strategies and economic approaches to provisioning human activities with raw material for cutting edges. In particular, dimensional data allows one to gauge the intensity of core reduction in a given assemblage. By comparing the distribution of length to width ratios for all pieces of complete debitage to the distribution of these ratios for the last maximum flake scar on each core, one can determine whether or not the cores were worked down to the point where the last scar equals the length/width ratio of the smallest flakes in the assemblage (Figure 4.33). This comparison can even be improved if the debitage is separated into "primary elements" (in the present case meaning debitage with greater than 50% cortex) and blank elements (debitage with less than 50% cortex). If the distribution of the last maximum scars on cores (what Henry calls core facets) falls between the primary elements and the blanks (or at the top of the blank distribution; Figure 4.34A), the cores are not intensively reduced. In other words, the cores still have some use-life in them. On the other hand, if the core facets are as small as the smallest blank debitage, then little use-life remains to the cores (Figure 4.34B). This discussion does presume a minimum size for debitage blanks given the function to which they are to be applied. As Henry's analysis identifies if the very exhausted end of the reduction sequence has been included in the site formation processes which produced the assemblage, this scatterplot analysis is complementary to the consideration of the relative non-cortical to cortical blank ratio.

The third analysis used to address the comparability of assemblages relates to the effect of diverse levels of retouch application to the tool kit in different assemblages. The comparison of tool kit morphologies of two assemblages is as susceptible to bias by differential raw material exhaustion as is the comparison of blank production behaviors. Specifically, comparing a tool kit which has been retouched lightly against an extensively retouched tool kit could easily bias dimensional variables, such as the length/width and width/thickness ratio, to the extent that the blanks selected for the tool kits would look different when only resharpening was at issue. This has been shown through the extensive research on the Frison Effect (Frison 1968; Jelinek 1976; Dibble 1987a,b,1988, 1995b; Dibble and Roland 1992; Roland and Dibble 1990; McPherron 2000; Hiscock and Clarkson 2005, 2007). Different methods for evaluating the degree of resharpening relative to the original blank size have been proposed (Kuhn 1990b; Davis and Shea 1998; Clarkson 2002; Eren et al. 2005; Eren and Prendergast 2008; Clarkson and Hiscock 2008, 2011). While each method has much to offer a project of this type, the methods available before the data gathering phase of this research all had issues that made them problematic for studying the Middle to Upper Paleolithic transition, while others had yet to be invented. First, the flake-tool mass predictor evaluated by Davis and Shea (1998) was judged not yet accurate enough for application to the prehistoric record (Shott et al. 2000), although this may change in the future with the application of more 3D laser scanning (Clarkson and Hiscock 2011). Second, Kuhn's (1990b) Geometric Index of Unifacial Reduction (GIUR) was the only published reduction index at the time this research began but it is most appropriately used on side-scrapers and not end-scrapers

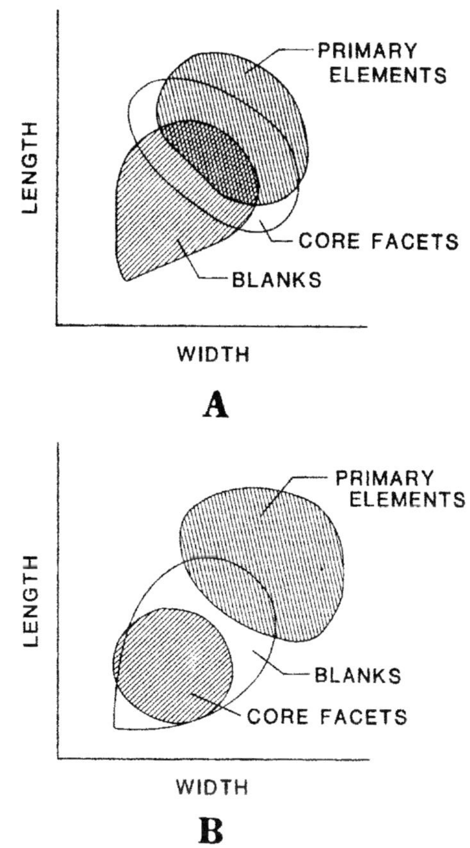

Figure 4.34 Henry's dimensional analysis of core reduction intensity: A) lightly reduced cores; B) heavily reduced cores (from Henry 1989:142, fig. 2, used by permission).

(Kuhn, personal communication), despite its illustration in use with the latter retouch type (i.e., Andrefsky 2005:fig. 7.27). While the index gradually reaches a maximum as side-scraper retouch continues, it far more quickly reaches a maximum as distal retouch is applied in the production of an end-scraper on a leptolithic blank. The application of this measure indiscriminately across both categories of retouched tools thus mixes different signals of reduction intensity and

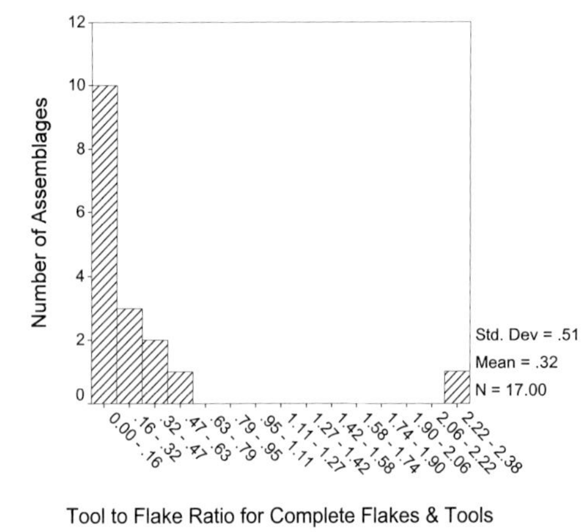

Figure 4.35 Example histogram of tool/flake ratios for a population of assemblages.

mass loss, each with their own rate of change, into the same measure. Clarkson and Hiscock's (2008) experimental test of the GIUR on flakes with both lateral and distal retouch which showed the success of GIUR to predict mass loss does not, unfortunately, address contexts frequent in western Eurasia, if infrequent in Clarkson and Hiscock's Australian study area, in which retouch is exclusively applied as lateral or distal but not both. As the distinction between lateral and distal retouch is strong within assemblages spanning the Middle to Upper Paleolithic transition in western Eurasia, this issue poses a distinct problem for the indiscriminate use of Kuhn's index for the assemblages studied in this project.

While the GIUR values of side-scrapers can be systematically compared exclusively to side-scrapers in different assemblages (and the GIUR of end-scrapers only to end-scrapers), this parsing of the tool kit produces low sample sizes in lightly retouched assemblages, such as raw material workshops, as well as a lack of comparability in some assemblages which are poor in distal retouched tools (many Middle Paleolithic assemblages) or poor in lateral retouched tools (many Early Upper Paleolithic assemblages). While beneficial findings can be produced in these contexts when the GIUR is applied alongside other measures of reduction such as Clarkson's (2002) Index of Invasiveness (see the discussion of Nejman 2009 in Chapter 5), it was judged at the time of the data gathering for this project (1996–1999) that Kuhn' index alone would not provide the measure of comparability of tool kit morphologies needed for the project. Of course, today lithic analysts have access to a new mix of reduction intensity measures that would, in an intelligent combination, serve as a means of evaluating the comparability of two assemblages' tool kits, specifically Clarkson's (2002) Index of Invasiveness, Eren et al.'s (2005) Estimated Reduction Percentage (Eren and Prendergast 2008), the improvements in how to measure the GIUR (Clarkson and Hiscock 2008; Hiscock and Clarkson 2007, 2009a; see also the debate between Eren and Sampson 2009 and Hiscock and Clarkson 2005, 2009a), and the more accurate predictions of original flake mass from 3D-measured platform variables (Clarkson and Hiscock 2011).

Although follow-up research along the lines of this project would do well to take advantage of this new mix of reduction intensity measures, these innovations in lithic analysis were unavailable when the initial data was gathered for this project. Thus, it was decided at the beginning of this project that the tool kit comparability must be evaluated based on other data within the assemblage, as with blank production comparability. This evaluation takes the form of the relative placement of an assemblage on the spectrum of values evidenced across a large sample of assemblages for the ratio of complete tools to complete flakes (Rolland 1981; Rolland and

Table 4.2 Predictions for comparability measures based on two example assemblages, one highly reduced (both in core reduction and tool reduction) and one lightly reduced

Variables	Highly Reduced	Lightly Reduced	Evaluation
Tools/Flake Ratio	High	Low	Location within histogram?
Non-cortical Debitage/ Cortical Debitage Ratio	High	Low	Location within histogram?
Henry's Dimensional Analysis	Little to no overlap in distributions	Appreciable overlap in distributions	Qualitative evaluation of the lengths & width scatterplot of blanks vs. the last maximum scar on cores
Proximity to Raw Material Source(s)	Distant	Close	Qualitative evaluation: adjacent/close/distant

Dibble 1990; Stiner and Kuhn 1992). A low value to this ratio indicates few flakes have been retouched, with the assumption that this correlates with low degree of retouch per flake. Conversely, a high value to this ratio indicates the retouch of many flakes, with the assumption that this also correlates with a high degree of retouch per flake. These assumptions are reasonable, given that retouch in most cases increases rather than decreases the edge angle and a low edge angle is the defining feature of "edge sharpness." As a result, an unretouched edge is almost always sharper than a retouched edge and the need for sharper rather than duller edges would imply that retouch would only be applied to greater degrees if unretouched flakes were scarce. Thus, in attempting to account for the differential working of the Frison Effect on two assemblages, it is possible to provide a relative comparison between assemblages in the context of the whole population in the study. In the case of the example histogram in figure 4.35, the left-most assemblages exhibit only rare retouch, while the right-most assemblages, and particularly the outlier, exhibit more retouched blanks. The comparison of these two extremes would require care in evaluating the degree to which differential reduction is responsible for any differences in variables based on morphological dimensions.

The fourth and final means of evaluating the comparability of two assemblages in terms of their differential use of raw material and the landscape is the reported proximity of a source or outcrop of knappable stone. For while sample assemblages were chosen for this project to represent as similar landscape roles as possible, there are instances in which an assemblage located close to a raw material source is compared with an assemblage in a site far from its raw material. Table 4.2 presents example predictions for the comparability measures of two assemblages with opposed patterns of retouched tool creation, core reduction intensity, and raw material proximity.

4.7 Procedures for the Pair-wise Comparison of Assemblages

For each pair-wise assemblage comparison, the lithic data collected from each assemblage is organized into a series of tables containing the individual analytical steps of the technological signature according to the four knapping domains defined for blank production behaviors (core modification, platform maintenance, direction of core exploitation, and the dorsal surface convexity domain; Table 4.3). A separate table is provided for the comparison of the tool kit morphologies between two assemblages (e.g., Table 4.4). The first column within these tables lists each analytical step within its behavioral domain. The next two columns contain a characterization of the cumulative behaviors used by the knappers at each decision node/step in the production of the two assemblages, respectively. These characterizations usually consist of measurements of the central tendency and dispersion of the variability within the particular knapping step. The number of artifacts and the frequencies for the various attribute-states are provided for categorical attributes. The number of artifacts and the mean and standard deviation are also provided for continuous variables. The fourth column contains a judgment of the significance of any difference between the assemblages' choice of option for each step in the operational sequence. This judgment is qualitative or quantitative, depending upon the nature of the type of lithic data available. Steps related to strategic domains, i.e., choices made once or twice per core reduction, often utilize data taken from cores that are qualitative, while steps related to tactical domains, with choices made with each flake or tool produced in the assemblage, are mostly quantitative. A "p" value in this column indicates the probability of obtaining values as different as these by the random sampling of one universal population of samples. Although a significance level of 5% is admittedly an arbitrary choice, this value has become traditional practice in the sciences and is used here for all statistical tests. The value of the relevant statistic, as given in the description of each knapping domain and its analytical steps above, is provided as well as the p-value, i.e., the student's t-value, the G^2 likelihood ratio, or Fisher's Exact Test (Sokal and Rohlf 1995) for 2x2 cross-tabulation tables. It should be noted that non-normally distributed variables are still tested with a student's t-test, despite its mathematical derivation from the assumption of normalcy, due to the robustness of the test against non-normalcy as demonstrated in Monte Carlo evaluations of the statistic (Sokal and Rohlf 1995).

To produce a quantitative measure of the difference between assemblages, one cannot simply sum up the number of operational steps in which a significant difference is judged to exist between the two options as this would bias the results through the interdependence of the units (as in Galton's Problem mentioned in Chapter 3). While the evaluation of the strength of association between the variables used in this analysis indicates that almost all of the variables are statistically independent, the *a priori* fact that each domain tends to include behaviors which are decided simultaneously by either conscious choice or unconscious bodily performance makes it reasonable to treat the variables within each domain as dependent upon each other but the domains as independent of each other. In order to quantify the pair-wise assemblage comparisons, therefore, the knapping steps in which significantly different options were used between assemblages are first summed within their specific flintknapping domain (i.e., within the categories of core modification, platform maintenance, direction of core exploitation, blank production, and tool manufacture) and divided by the total number of steps within that domain. The resulting numerical value of all four domains of blank production and 1 domain of tool kit morphology are then summed up to produce a measure ranging from zero (for assemblages with identical operational sequences) to four (for entirely different operational sequences). This procedure thus scales the measure of difference according to the variability

Table 4.3 Generalized comparison of the differences in blank production behaviors between two assemblages

Flintknapping Steps by Domain	Earlier Assemblage	Later Assemblage	Significant Difference?
Core Modification Domain (a strategic domain)			
Core Orientation	Longitudinal	Longitudinal	Qualitative distinction
Core Management	Frontal & lateral crests, core tablets	Frontal & lateral crests, core tablets	Qualitative distinction
Number of Changes/2 Steps			**n/2= 0–1.0**
Platform Maintenance Domain (a tactical domain)			
Platform Treatment	Unprepared: 76% Prepared: 24% n=354	Unprepared: 69% Prepared: 31% n=325	Ordinal scale, G^2 likelihood ratio test
External Platform Angle (degrees)	mean: 84.1, s.d.: 13.7, n=298	mean: 82.5, s.d.: 15.1, n=306	Ratio scale, t-test of means
Platform Thickness	mean: 4.09, s.d.: 2.93, n=302	mean: 4.80, s.d.: 2.52, n=304	Ratio scale, t-test of means
Number of Changes/3 Steps			**n/3= 0–1.0**
Direction of Core Exploitation Domain (a strategic domain)			
Direction of Early Exploitation	Unidirectional	Bidirectional & Unidirectional	Qualitative
Direction of Late Exploitation	Unidirectional	Unidirectional	Qualitative
Number of Changes/2 Steps			**n/2= 0–1.0**
Dorsal Surface Convexity System Domain (a tactical domain)			
Longitudinal Convexity: Length/Width Ratio	mean: 2.39, s.d.: 1.35, n=522	mean: 1.73, s.d.: 0.75, n=470	Ratio scale t-test of means
Shape of Convexity: Lateral Edges of Blanks	Parallel: 57% Convergent: 22% Expanding: 13% Ovoid: 8% n=472	Parallel: 57% Convergent: 17% Expanding: 16% Ovoid: 10% n=431	Nominal scale, G^2 likelihood ratio test
Curvature of Convexity: Profile of Blanks	Straight: 53% Curved: 28% Twisted: 19% n=501	Straight: 58% Curved: 24% Twisted: 18% n=430	Nominal scale, G^2 likelihood ratio test
Lateral Convexity: Cross-Section of Blanks	Triangular: 42% Trapezoidal: 53% Other: 6% n=493	Triangular: 46% Trapezoidal: 48% Other: 6% n=447	Nominal scale, G^2 likelihood ratio test
Vertical Convexity: Width/Thickness Ratio	mean: 3.76, s.d.: 2.00, n=522	mean: 3.45, s.d.: 1.51, n=470	Ratio scale, t-test of means
Number of Changes/5 Steps			**n/5= 0–1.0**
Total Measure of Difference in Blank Production			**0.0 - 4.0**

seen between these knapping domains and removes any effect of the simultaneity of the execution of the variables. For instance, in Table 5.68, the total number of significant differences between the behavioral flintknapping options used in the Kůlna Cave, layer 7a, assemblage and the assemblage from Stránská skála IIIa, level 4, is 9 out of a total of 12 comparisons. Using this

Table 4.4 Generalized comparison of the differences in tool kit morphology between two assemblages

Flintknapping Steps by Domain	Earlier Assemblage	Later Assemblage	Significant Difference?
Tool Kit Morphology Laminarity	mean: 2.35, s.d.: 1.50 n=101	mean: 1.73, s.d.: 0.78 n=52	Ratio scale, t-test of means
Width/Thickness	mean: 3.34, s.d.: 1.88 n=101	mean: 3.33, s.d.: 1.48 n=52	Ratio scale, t-test of means
Lateral Edges	Parallel: 59% Convergent:21% Expanding: 13% Ovoid: 7% n=80	Parallel: 48% Convergent: 13% Expanding: 17% Ovoid: 22% n=46	Nominal scale, G^2 likelihood ratio test
Distal Terminus	Blunt: 40% Pointed: 60% n=45	Blunt: 63% Pointed: 37% n=19	Nominal scale, G^2 likelihood ratio test
Profile	Straight: 56% Curved: 27% Twisted: 17% n=93	Straight: 63% Curved: 25% Twisted: 12% n=40	Nominal scale, G^2 likelihood ratio test
Unique Types of Retouch	Carinated retouch	Carinated retouch	Qualitative distinction
Tool Types	UP tools dominate	UP tools dominate	Qualitative distinction
Number of Changes/7 Steps			**n/7**
Total Measure of Difference in Tool Kit Morphology			**0.0-1.0**

proportion (9/12) would ignore the potential, although currently undemonstrated, effect of the simultaneity and Galton's Problem. Therefore, the measure of difference is weighted by the four behavioral domains, producing a value of 3.13 out of 4.0. The same procedure follows for tool kit morphology, only the number of differences is summed and divided by the number of analytical steps, i.e., 7, producing a maximum measure of difference as 1.0 with similarity indicated by 0.

It must be noted that, although some interdependence has been avoided with the above methodology, this study introduces its own bias in the structuring of pair-wise assemblage comparisons. Specifically, as the number of steps within each of the four domains of blank production differs, the steps in different domains are not weighted evenly. This situation is intentional, in that the number of steps within a domain reflects the potential for functional constraints to affect the choice of knapping options within that domain. Thus, tool kit morphology, with its 7 steps which directly affect the cutting edge of the tool, is the most influenced by functional utility and thus has the largest number of steps within the domain. Direction of core exploitation, on the other hand, has little or no affect on the functional utility of the resulting products. The only way directionality affects utility is the rate of raw material expenditure increases as you use and repair more platforms on the core (Marks 1988). Therefore, this domain has only two steps. This bias in the structure of the pair-wise comparisons should be kept in mind when evaluating the similarity/dissimilarity between assemblages traditionally typed together based on retouch typology alone.

4.8 Connecting the Bridging Arguments between High-level and Low-level Theory

Chapter 3 presented the beginnings of the middle-range theory to connect the evolutionary study of culture history with the archaeological record of Pleistocene hominins. The chapter presented bridging arguments from the basis of high-level theory, i.e., the intellectual objectives the researcher considered as the goals of the analysis. In this case, the bridging arguments began from the point of view of the high-level theory of dual inheritance modeling and cultural transmission theory that would allow the analyst to learn something new about *how*, therefore ultimately why, the phenomena of the Middle to Upper Paleolithic transition occurred in three regions of western Eurasia. Chapter 4 has presented the complementary bridging arguments, beginning with the low-level theory, for how to make observations of the archaeological record such that the units of the lithic analysis coincide with the expectations of the analytical objectives of the high-level theory. Specifically, by combining the conceptual modeling of the observational learning of the flintknapping process with the lessons lithic analysts have learned from controlled experiments in flake fracture mechanics, this chapter has identified the etic observations of the learnable and effective (*sensu* Mauss 1936, 1979) technological choices made by knappers during blank production and the creation of the mobile toolkit. Between the bridging arguments from each level of theory, it is hoped that a solid middle-range theory has been constructed that satisfies the needs for answering the paleoanthropological questions asked in this work. It remains to specify how the units identified in Chapter 4 are to be evaluated against the structure of the high-level theory established in Chapter 3. This is best pursued by defining the model expectations for these units of analysis and the intra-regional and inter-regional pair-wise comparisons that follow in Chapters 5–8 according to the competing hypotheses for how the Middle to Upper Paleolithic transition occurred, in terms of the action of cultural transmission, in these three regions of western Eurasia.

4.8.1 Cultural Transmission vs. Independent Innovation: The Model Predictions

This book represents a research project in which 18 archaeological assemblages dating between 60 and 30 kya in three regions of western Eurasia were analyzed in order to identify which flintknapping behaviors changed during the Middle to Upper Paleolithic transition and to compare the sequences of these diachronic behavioral changes within and between each region to determine how the transition occurred. Specifically, this research tests three hypotheses concerning change in flintknapping behaviors between 60 and 30 kya in Central Europe, Eastern Europe, and the Levant:

- Hypothesis 1 (Independent Innovation): flintknapping behaviors in a region changed due to *in situ* independent innovation without cultural transmission from neighboring regions. Any similarities in blank production and/or tool kit morphology are judged to be the result of adaptive convergence and/or cultural drift.
- Hypothesis 2 (Cultural Transmission): flintknapping behaviors in a region changed due to the spread of behaviors from one region to another as a result of cultural transmission across space. The more subtle distinction between symmetric transmission and asymmetric transmission forms alternative sub-hypotheses for each region:
- Hypothesis 2a (Symmetric Transmission): flintknapping behaviors changed due to the demic diffusion of hominins from one region

into another. This hypothesis does not distinguish between demic diffusion via gene flow through mating networks, the colonization of an empty landscape, or the intrusion of a population into a region with natives with whom they are unable to reproduce.

- Hypothesis 2b (Asymmetric Transmission): flintknapping behaviors changed due to the asymmetric transmission of flintknapping behaviors from one region into another without the dispersal of hominins or gene flow. Stimulus diffusion represents the process by which this sub-hypothesis functions, as noted below.
- Hypothesis 3 (Combination): flintknapping behaviors changed due to independent innovation in one or more regions with the subsequent cultural transmission of the behaviors into adjacent regions.

4.8.2 Expectations of the Three Hypotheses[3]

The three hypotheses described above are tested by assessing the goodness of fit between the archaeological record and a series of model expectations derived from the combination of the Behavioral Archaeology, *Chaîne Opératoire*, and Organization of Technology approaches to material culture as applied to culture historical processes defined from cultural transmission theory. The first step in this methodology, as explained earlier in Chapter 4, is the description of the technological signature for each assemblage according to the distinction between blank production behaviors (a proxy for enculturation) and the selection of the mobile tool kit (a proxy for functional needs and/or stylistic signaling). The second step is the comparison of the detailed technological signatures for assemblages that succeed each other through time within the same region, taking into account possible multi-phyla scenarios. Determining which flintknapping options in each portion of the technological signature appeared for the first time in a particular region or disappeared in later assemblages in that region produces an intra-regional sequence of behavioral change. The third step is accomplished by comparing which knapping options for each technological variable in the signature appeared and disappeared through time between adjacent regions, i.e., comparing the intra-regional sequences to produce an inter-regional sequence of change.

The final step in testing the hypotheses is the evaluation of the intra- and inter-regional sequences against the following five model expectations. Each expectation is based upon a theoretical principle which predicts how the intra-regional sequences and inter-regional sequences should appear given Hypothesis 1 (Innovation) or Hypothesis 2 (Cultural Transmission). As Hypothesis 3 (Combination) is the logical extension of Hypothesis 1 in some regions and Hypothesis 2 in others, it is not treated separately in the following description of the model expectations. This evaluation of the goodness of fit between the archaeological record of behavioral continuity in culturally transmitted traits can then be used to interpret which culture historical processes have been active in producing these behaviors in each region.

Expectation Number 1 (Antecedents: Continuity in Technical Choices)

This expectation asserts that "...any innovation is made up of preexisting components.... No innovation springs full-blown out of nothing; it must have antecedents..." (Barnett 1953:181; see also Eerkens and Lipo 2007:240). In terms of lithic technology, the presence or absence of specific flintknapping behaviors of a lithic operational sequence in a region can be used to judge the probability that a subsequent operational sequence appeared in that region through independent innovation or cultural transmission. This assertion is corroborated by social anthropology (Kluckhohn 1936; Kroeber 1940), history of science (Basalla 1988:49, 55), and archaeological theory (Ford 1952:330; Haury 1958; Deetz and Dethlefsen 1965; Dethlefsen and Deetz 1966; Willey et al. 1956:7; Andrefsky 1987:19; Renfrew 1978; Eerkens and Lipo 2007:240). This has been made explicit in Harrold's (1989:681) treatment of the MP-UP transition in western Europe as well as Hovers's (1998) and Tostevin's (2003a) in the Levant. From the point of view of cultural transmission theory, this expectation thus rests upon the premise that more than half of the forces of descent with modification in evolutionary culture theory are accounted for by biased transmission, which uses the extant variants in the population to shape the frequency of the variants in the next generation. In other words, biased transmission works on antecedents. As a result, to "trace the spread of a cultural instruction, we need to investigate where it develops locally and where it appears without an *in situ* developmental sequence" (Tschauner 1994:88).

Thus, the presence of antecedent elements for the technical choices in a given lithic operational sequence increases the likelihood that independent innovation was the mechanism by which the operational sequence appeared in that region, while decreasing the likelihood that cultural transmission across space played a role. While similar functional needs may lead geographically disparate flintknappers to select morphologically similar tool kits as a result of guided variation in each context, the phenomenon of equifinality common in core reduction (Marks and Kaufman 1983; Tixier 1984; Boëda 1995) makes the convergent evolution of exactly similar blank production behaviors unlikely. Thus antecedents of blank production behaviors can be seen as the continuity of an earlier enculturating environment whereas antecedents of tool kit morphologies within a region can be seen as continuity of tasks for which stone cutting edges were needed on the landscape.

Expectation Number 2 (Number of Changes in Technical Choices)

This expectation asserts that a few technical choices (within blank production and/or tool kit morphology) found in two assemblages in adjoining regions can be explained by independent innovation but as the number of such shared choices increases, so does the improbability of their independent recurrence (Taylor 1896; Steward 1929; Andrefsky 1987). Therefore, a quantitative measure of similar knapping behaviors between assemblages can be used to judge the parsimony of the competing hypotheses.

Any quantitative method, such as counting similar behaviors in two assemblages, must use independent units so that the presence of one unit does not affect the appearance of another unit and bias the result (Rands and Riley 1958). As Chapter 4 has been at pains to point out, the attribute proxies for the four domains of blank production and the one domain for toolkit morphology are sufficiently independent to avoid Galton's Problem. This independence allows quantification of similar behaviors between assemblages to be used to test the competing hypotheses. Thus, cultural transmission is more and more likely as the number of similar blank production choices or tool kit choices demonstrated between two assemblages increases. Similarly, fewer technical choices found to be similar between two assemblages makes independent innovation more likely.

Expectation Number 3 (Number of Domains Affected by Change)

This expectation asserts that cultural transmission results in many more types of material culture to be adopted together by the next generation than in instances of independent innovation (Steward 1929; Linton 1936:372; Lechtman 1977; Carr 1995b:249). This is due to the fact that although an instance of innovation in one material culture domain, ceramics for instance, may affect another material culture domain, such as metallurgy, it is unlikely to affect several independent domains simultaneously. On the other hand, cultural inheritance through vertical transmission from parent to offspring or horizontal transmission from peer to peer more frequently results in the learner adopting a whole package of cultural trait variants across material culture types. This is particularly true for the influence of indirect biased transmission: individuals "tend to acquire the clothing styles, pronunciation, and beliefs about the world that characterize the most successful individuals among potential models" (Boyd and Richerson 1985:243). Thus, one can use the number of independent material culture domains that show changes at the same time as an indicator of the strength of the cultural transmission process between regions.

While the Paleolithic record possesses only one major material culture domain, lithic technology, the principle of this expectation can nevertheless be applied by viewing independent flintknapping domains as analogous to independent material culture domains. To the extent that controlled fracture studies and the internal assessment of the independence of the variables allows (see above), the knapper's blank production choices can be meaningfully separated, from the point of view of what is learned, into four roughly independent domains. Two of these domains, platform maintenance and the dorsal surface convexity system, encapsulate the tactical choices faced by a knapper on a flake-by-flake basis, equivalent to the knowledge of the *savoir-faire* (*sensu* Pelegrin 1990b) in flaking within the given enculturating environment. The other two domains, core modification and the direction of core exploitation, represent the strategic choices faced by a knapper over the course of reducing each core, equivalent to the *connaissance* of the operational sequence within the given enculturating environment. The comparison of assemblages' blank production details by means of these four domains allows the quantification of the number of domains with similar knapping behaviors between two assemblages to be used to evaluate the competing hypotheses. Thus, this expectation predicts that the likelihood of the cultural transmission hypothesis increases, and the independent innovation hypothesis decreases, in proportion to the number of the four blank production domains which possess technical choices shared between assemblages through time and in adjoining regions.

Expectation Number 4 (Enculturation versus Stimulus Diffusion)

This expectation establishes the pattern by which the changes in technical choices within blank production and the curated tool kit between two assemblages can be interpreted as evidence for stimulus diffusion versus changes in enculturating environments. Following Carr (1995b) and the middle-range theory connecting social intimacy with taskscape visibility developed in Chapter 3, an instance of cultural transmission in which the individuals are not socially intimate results in the transmission of the knowledge of tool kit morphologies but not the blank production details into the next generation. The resulting pattern in the pair-wise assemblage comparison of similar tool kits but dissimilar blank production choices indicates the effect of one of several culture historical processes (Figure 3.7): stimulus diffusion or adaptive convergence. In both cases, the individuals responsible for the chronologically later assemblage choose cutting edge morphologies similar to those of the earlier assemblage for use on the landscape as part of their curated tool kit. This may be the result of the transmission from the individuals responsible for the earlier assemblage of the idea of such a tool (stimulus diffusion or asymmetrical transmission without social intimacy) or the independent innovation of the idea of such a tool morphology by means of guided variation to satisfy a task similar in both contexts. This latter case of convergence toward the same tool kit morphology can be distinguished from stimulus diffusion through the evaluation of Expectation 1. Specifically, the existence of antecedents for the tool kit morphology in the region of the later assemblage makes adaptive convergence the more parsimonious argument while the lack of antecedents for this tool kit morphology in the region makes stimulus diffusion from the earlier assemblage from a neighboring region more likely.

In another scenario, an instance of cultural transmission in which the individuals are socially intimate results in the transmission of both blank production details and tool kit choices into the next generation. This continuity in enculturating environments due to social intimacy may also indicate several culture historical processes (Figure 3.7): symmetric transmission via demic diffusion and gene flow, symmetric transmission without gene flow (see below), and, significantly, asymmetric transmission without gene flow but with enough social intimacy to allow the copying of knapping behaviors. The last process, which in traditional archaeological parlance could be called behavioral diffusion, is relatively unlikely in forager societies, as the likelihood of mating and gene flow between the societies in contact is correlated closely with the degree of social intimacy (Kelly 1995:261-92; Fix 1999; Van Kirk 1983). Foragers simply do not have the material means to spread cultural traits which require such social intimacy to observe without this proximity resulting in the formation of the mating networks that bind forager societies into environmentally-reflexible organizations. Most foragers do not have a term for an individual that does not place them in some kinship relationship to the others in the group (Gregory Laden, personal communication). Thus, gene flow is likely to have occurred to some degree. Put another way, foragers' diffuse economies (*sensu* Anthony 1990; Cleland 1976) make it unlikely for them to engage in the long distance culture contact via trade missions of craft specialists which would result in the transmission of a production sequence without significant gene flow. In contrast to the means available to agriculturalists, the mechanism for long distance culture contact among foragers is primarily demic diffusion via natal dispersal, thus necessitating both a high degree of social intimacy with neighbors and gene flow for a behavior to spread across a sizable landscape. All things being equal, therefore, evidence for very similar enculturating environments and thus strong social intimacy among foragers should indicate dispersal via demic diffusion and gene flow through time and space.

All things are not equal in Paleolithic archaeology, however. Two factors make it wise to be careful in proposing that all cases of similarity between enculturating environments indicate gene flow. First, other archaeological studies using the predictions of Carr's Middle-Range Theory of Artifact Design (1995b) to test for prehistoric migration (*sensu* J. J. Clark 2001) base their conclusions on multiple lines of evidence. These lines of evidence include either multiple types of material culture indicating the same pattern of enculturating environments moving across the landscape (ceramics, architectural wall design, room layout, etc., J. J. Clark 2001) or one type of material culture tied to chemical sourcing of raw materials (e.g., temper in Hopi ceramics, Lyons 2003). As these separate sources of data combine to indicate a particular significance, the validity of the significance increases through the "consilience of induction" (Whewell 1858:70, 85-96; Gould 1989:282). Thus studies relying upon multiple types of evidence have a greater change to be correct. As indicated above, Paleolithic archaeology primarily has only lithic technology in the context of the Middle to Paleolithc transition. Without consilience, therefore, there is a greater chance that the historical pattern of behavioral change is incorrectly correlated with a specific cultural transmission process. For this reason, our understanding of the Middle to Upper Paleolithic transition would be greatly advanced through the complementary application of this research's approach to testing for cultural transmission in other media, such as organic technology (Knecht 1991, 1993), non-conchoidal stone tools (de Beaune 2000, 2007) and mobiliary art (d'Errico et al. 1998; White 2007), although these material culture media present their own difficulties, being unevenly distributed across the transition.

The second reason to take care in equating all cases of similarity between enculturating environments with gene flow is the possibility that the two groups of hominins in contact have progressed along the speciation process (*sensu* Foley and Lahr 2005) beyond the Paterson point of recognizing members of the other population as suitable mates (Paterson 1985). In such a situation, the hominins may have been socially intimate enough to learn blank production details as well as the tool kit morphologies from each other despite not producing offspring together. Despite the arguments against such speciation among Late Pleistocene hominins between the Levant and Europe (e.g., Smith et al. 2005; Trinkaus and Zilhão 2002; Greene 2010), the arguments for such speciation (e.g., Tattersall 1998) make it unwise to discount this possibility completely. Thus, evidence of shared blank production details as well as shared tool kit morphologies cannot exclusively indicate gene flow and population dispersal. Depending upon the paleoanthropologist's conclusions in relation to these other research questions, the results of this research can be taken two ways. Figure 3.7 thus represents strong enculturation similarities as reflecting asymmetric transmission (except for stimulus diffusion) as well as symmetric transmission.

Despite this equifinality in the connection of transmission process to material evidence, the social intimacy and taskscape visibility concepts allow greater accuracy and rigor in identifying such processes than previous efforts. In particular, Tostevin (2000a,b,c) used only the principle of isochrestic variation or "technological style" to structure the interpretation of pair-wise assemblage comparisons. With the social intimacy and taskscape visibility middle-range theory, however, there is at least now a possibility of distinguishing cultural transmission of blank production behaviors as a separate set of processes from the stimulus diffusion of the tool kit alone. Thus, stimulus diffusion no longer represents a type II error (*sensu* Sokal and Rohlf 1969:158-159) in failing to falsify the null-hypothesis of independent innovation in a case when transmission of a type did occur.

Expectation Number 5 (Parsimony of the Inter-regional Sequence)

This expectation predicts that the multi-regional patterning of behavioral changes between 60 and 30 kya should document a continuous geographical and temporal progression for the appearance of individual behaviors between adjacent regions if cultural transmission were the mechanism for the appearance of the behavioral changes (Rouse 1986; Linton 1936:370; Meggers et al. 1965:157-78). In other words, a new knapping behavior or cluster of behaviors must appear first in one region, subsequently in an adjacent region, and only later in more distant regions. Conversely, it predicts that independent innovation is more probable given a temporally and geographically sporadic appearance of dissimilar behaviors on the multi-regional scale. For example, the operational sequences could show a non-centric pattern of appearance, without a point of origin and radiating pattern, as seems likely for the appearance of pastoralism in Africa (Harlan 1971; Marshall 1998). In the example of African pastoralism, the adoption of this mode of subsistence was not continent upon knowledge of the mode, as that knowledge was widespread before its adoption. This example provides a cautionary tale; it is only possible to document cultural transmission in material culture media in which it is impossible gradually to accrue the details necessary for its production.

This expectation rests upon the assumption that parsimony favors one locale of origin for a specific technical choice within the temporal and spatial limits of western Eurasia between 60-30 kya. This expectation thus relies upon both Expectation 1 (Antecedents) and Expectation 2 (Number of changes in technical choices). Diffusion is thus not indicated if the behaviors characteristic of the transmission phenomenon "jump" one region with a similar ecological situation to that of the donor region, only to appear in a more geographically-distant locale. Each region which receives behaviors from the source area(s) by cultural transmission must not only possess a diachronic behavioral sequence indicative of transmission but must exhibit the relevant behavioral changes in temporal order according to its geographic proximity to the source area, i.e., the behavioral changes must appear earliest in a recipient region which is closest to the source area and only later in more distant regions. This situation must be historically understandable in terms of the plausibility of the inter-regional temporal and geographic pattern of changes.

While modern cultural transmission events across space and even migrations often "jump" adjacent regions to appear in more distant regions by means of migration streams (Anthony 1990), no mechanism, such as pack animals or vehicles, for this type of movement of behaviors may be assumed to have existed in the Pleistocene. Such "jumps" in the Pleistocene are more likely artifacts of site formation and site recovery processes, a situation reminiscent of many archaeologists' favorite joke when explaining archaeological sampling of the past. In the dilemma of the drunk under the lamp-post, a policeman asks a drunk, who happens to be on his hands and knees under a lamp-post at night, "what are you looking for?" The drunk responds that he is looking for his lost keys which he dropped across the street. The policeman responds, "why are you looking here when you dropped them across the street?", to which the drunk answers "the light is better over here."

This dilemma explains why this expectation does not require one to identify an external (i.e., extra-regional) source for a new behavior before cultural transmission across space can be used as an explanation for its appearance, as others have argued (Rouse 1958; Trigger 1968:40–41; Adams et al. 1978). Following Snow (1996:795), I would argue that the nature of site formation processes (Schiffer 1987) and archaeological sampling of the Pleistocene landscape make such a requirement unreasonable. We simply have too limited an access to the material culture produced by Late Pleistocene hominins for us to believe, like the drunk under the lamp-post, that we have documentation of each innovation and transmission event exactly where it occurred. Each assemblage in each region is likely to fall at a given point in a space-time vector of change and therefore the beginning or end of the vector may not be represented. In addition, for the present research, only three regions were sampled, leaving many regions unsampled as possible sources of cultural transmission. The conclusions of the parsimony of the inter-regional sequence of change, therefore, represent hypotheses for subsequence research in these other regions.

The logic behind the expectation of a continuous inter-regional sequence from one point of origin has been weakened in past research by relatively crude units of analysis which have, by their very wide definitions and conceptualizations, incorporated the high likelihood of multiple innovations and regional origins. Functional retouched tool categories such as carinated technology are an example which have been argued to show cultural transmission across space (e.g., Otte and Kozłowski 2003) but which upon more detailed behavioral definitions of the production sequence (e.g., Le Brun-Ricalens 2005; Belfer-Cohen and Bar-Yosef 2006; and Bar-Yosef et al. 2006) have been shown to have several independent modalities which would increase the likelihood of each modality being innovated in a separate region (see also Zilhão and d'Errico 2003:344). Therefore, the satisfaction of this model expectation requires units of analysis of sufficient specificity that they can be defined as observable behaviors working as units of transmission between individuals. Technical choices the artisan must make at given nodes within a core reduction sequence are the more suitable units of analysis for this expectation compared to less detailed categories such as artifact dimensions. For instance, although debitage of blade dimensions exists from both the early Middle Paleolithic (Rust 1933; 1950; Garrod and Bates 1937; Copeland 1975; Jelinek 1989; Tuffreau et al. 1985; Révillion 1989; Nishiaki 1989; Révillion and Tuffreau 1994; Otte 1994; Ameloot-Van der Heijden 1994; Shimelmitz et al. 2011; see Conard 1990, Meignen 1994, and Bar-Yosef and Kuhn 1999 for summaries) as well as the early Upper Paleolithic, not all blanks with lengths twice their widths are the same (Tixier 1984; Pelegrin 1990; Meignen 1995). Until it is demonstrated that the blade industries from these temporally-disparate contexts were made with the same detailed knapping behaviors resulting from the artisans' sequenced technical choices, the cause for the appearance of blades in these two contexts must be sought separately. In the final analysis, it may be the case that the appearance of blade industries in both of these contexts was influenced by one hitherto unknown, ultimate cause. The proximate cause, whether cultural transmission across space or in situ innovation, could, however, still differ for each case (see Meignen 1995 for a different view). Thus, the temporal limits of this study are dictated by the logic that the processes which created the phenomenon of the Middle to Upper Paleolithic transition must be sought in the cultural processes of the hominins living during the context of that change (Hovers 1998; Bar-Yosef and Khun 1999), not in the context of hominins living 60,000 years earlier (Tostevin 2003a).

While not suitable for the identification of proximal level forces such as cultural transmission across space, it is in fact at the level of ultimate causation that broad generalizations concerning lithic technology, such as J. D. G. Clark's "modes" (1968) and the investigation of the paleogeography of these modes (e.g., Foley and Lahr 1997, 2003; Lahr and Foley 1998; Shea 2011b), may offer insights into why particular types of material culture may be more adaptively advantageous than others. Nevertheless, this is not the level of detail that will allow us to demonstrate the specific cultural transmission between one assemblage and another; modes are too generalized to demonstrate this level of dispersal of genes or cultural traits, particularly if the period of inquiry is within one mode or between two modes that are not easily distinguished (i.e., a Levallois-like blade technique as in the Bohunician – *sensu* Škrdla 2003c). This is the most significant problem with studies that rely on Clark's modes.

The logic of these temporal limits to the search for continuity and discontinuity in technical choices in flintknapping behavior are also reflected in the logic behind the spatial limits for this expectation. As Otte (1995) has argued, if one looks far enough, one is likely to find examples of a specific technique in places where it must be assumed that independent innovation has caused its appearance in the two locales. Otte's example of Levallois flakes being found in the Japanese Neolithic, when cultural continuity of technical behavior between the European Middle Paleolithic and this temporally and spatially distant context is highly unlikely, reminds us that the likelihood of independence innovation is spatially-dependent. Similar arguments (e.g., G. A. Clark 2000) may be made against the proposition of cultural continuity between the pressure flaking and bifacial techniques found in Clovis and Solutrean sites (Bradley and Stanford 2004). While the hypothetical Solutrean antecedent to Clovis in the New World is not beyond possibility, its demonstration, like that of any cultural transmission event tested in the present study, requires more of a quantitative evaluation of the 5 model expectations presented here than has currently been tried (e.g., Stanford and Bradley 2012). Nevertheless, the Solutrean-Clovis hypothesis will always be difficult to falsify because of the inability to test the intervening region, i.e., the North Atlantic, as Expectation 5 would dictate for a non-oceanic hypothesis.

Thus the present study focuses this expectation on the context of western Eurasia in this 30,000 year time slice, using the assumption that this is a small enough geographical and temporal context to make a single innovation of such detailed and often body technique-specific technical performances more likely than multiple innovations. This argument does not stand counter to d'Errico's (2003; d'Errico et al. 2003) argument for multiple innovations of behavioral modernity in several hominin lineages, as the units employed in those studies are relatively large scale (use of colorants, bone and shell modification, etc.) and the geographical context (South Africa to temperate Europe) and temporal context (200–30 kya) are extremely large. In that context and scale, one would expect multiple innovations. Yet within western Eurasia, there is a relatively strong consensus on the single innovation and resultant spread by stimulus diffusion of cereal agriculture (Harris 1996), block printing (Bulliet 1994), the Industrial Revolution (Bar-Yosef 2002), and other technical innovations which spread through western

Eurasia but were actually innovated further east (Needham 1954). These examples thus support the logic of the parsimony argument for the temporal and geographic context of this study.

4.8.3 *Evaluation of the Goodness of Fit*

Conceptually, the three hypotheses proposed in this chapter are fairly distinct. It must be admitted, however, that the evaluation of the goodness of fit between the archaeological record and the five model expectations described here is much less distinct. To put it bluntly, the space between the period at the end of the sentence in which the "p-value" is given and the beginning of the following sentence is the chasm of historical interpretation. The multiple model expectations and multiple units of analysis serve to limit the size of this chasm.

As the MP-UP transition is a historical event, unrepeated and unrepeatable in time, a truly deductive and unequivocal test of the fit between the archaeological record and the three competing hypotheses is nearly impossible to achieve. For, "How should scientists operate when they must try to explain the results of history, those inordinately complex events that can occur but once in detailed glory?...The issue of verification by repetition does not arise because we are trying to account for uniqueness of detail that cannot, both by laws of probability and time's arrow of irreversibility, occur together again" (Gould 1989:277-278).

Nevertheless, the body of method and theory which gave rise to the five model expectations provides the most objective and replicable evaluation of the goodness of fit between the three hypotheses and the archaeological data derived from the lithic analysis that is currently available. As the units of analysis for the model expectations are the details of technical choices within the operational sequences rather than the simple morphological shape of the final artifacts, the parsimony of each hypothesis is judged, not on an assumption of human uninventiveness as made by cultural historians and the advocates of the Kulturkreislehre before them, but on the evolutionary contingency of the details of behavioral changes in their temporal and spatial context.

5

Regional Patterns of Behavior: Central Europe

This chapter examines the details of the flintknapping behavior extant between 60,000 BP and 30,000 BP in Central Europe, as exemplified by the archaeological sequence in Moravia, Czech Republic. Each sample assemblage is described in order of chronological appearance in the region. The technological analysis of the lithic data collected between 1997 and 2000 is presented for each assemblage before the pairwise comparison of assemblages is conducted. The chapter concludes with a discussion of the significance of the intra-regional, Central European sequence of change for the cultural transmission across space and independent innovation hypotheses.

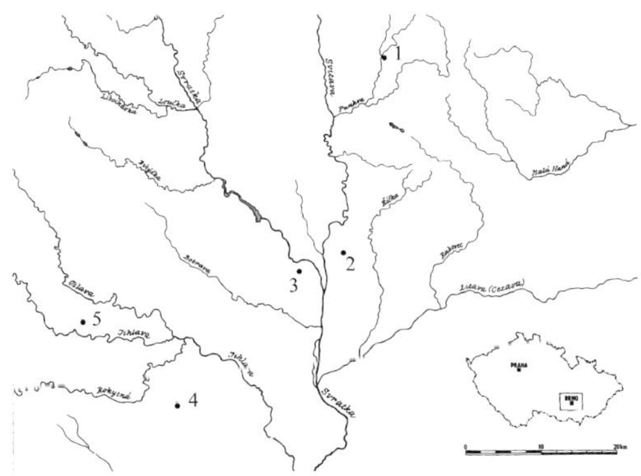

Figure 5.1 Map of Moravian archaeological sites. Key: 1) Kůlna Cave; 2) Stránská skála); 3) Bohunice; 4) Vedrovice V; 5) Mohelno. Insert of the Czech Republic for scale (adapted from Valoch 1993: 89, fig. 43, by permission of Quartär and the Hugo Obermaier-Gesellschaft).

5.1 Kůlna Cave, Layer 7a

5.1.1 Site Background

Kůlna is a large karstic cave in the Blansko/Sloup district of Moravia, north of Brno, Czech Republic. The site is situated 469 m above sea level and forms just one component of the large Moravian Karst system in the central highlands of Moravia. The site was first investigated in 1881 by J. Wankel, and again by M. Kříž between 1881 and 1886. J. Knies continued excavations from 1887 to 1913 but the cave lay scientifically dormant until the renewal of interest with the excavations conducted by Karel Valoch (1988) between 1961 and 1976. During World War II, the site was used by the German army as a munitions factory and the resulting disturbance of the archaeological deposits lessened the potential scientific contributions of the site. Nevertheless, Valoch's work at Kůlna represents the most careful excavation of a large and deeply stratified Middle and Upper Paleolithic sequence in the Czech Republic. With the earliest occupation, layer 14, argued to date to the penultimate glaciation (Rissian), the site contains both great chronological depth and industrial variability, with Taubachian (layer 11), Micoquian (layers 9b-6a), and Magdalenian (layers 6-5) occupations.

The layer of interest for the present study (see Figure 5.2), the final Micoquian occupation of Layer 7a, contained Neanderthal fossil remains, including a fragment of a maxillary (Kůlna I), a parietal bone (Kůlna II), and three deciduous molars (Kůlna III-V) (Jelinek 1988). A maximal ^{14}C date was procured during Valoch's excavations but has recently been superceded by a series of ten ESR dates which places Layer 7a at an average

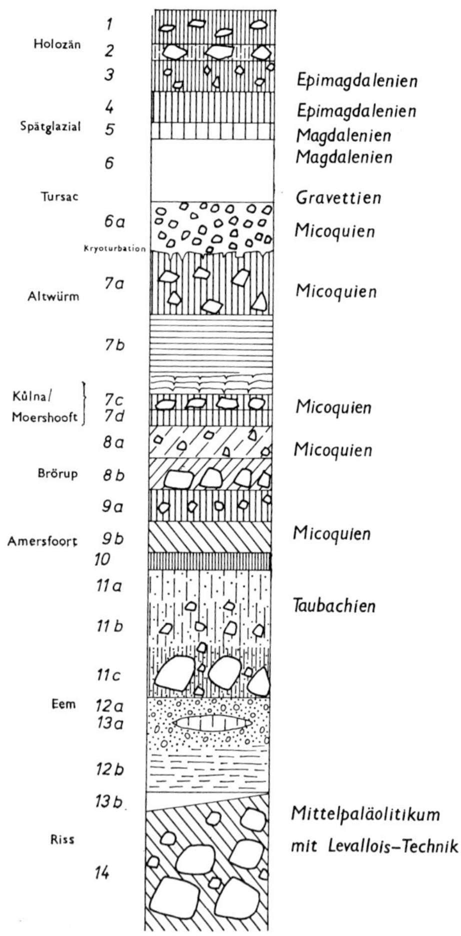

Figure 5.2 Kůlna Cave Stratigraphy (Valoch 1988, fig. 61, p. 164); used by permission.

date of 50/53kya ± 5/6 kya, early versus late uptake respectively (Rink et al. 1996b). The sedimentary matrix of this layer was described as 120 cm of "brown to dark-brown soil of a loessic origin... [and] contains scattered, quite big (up to 30 cm in a diameter), slightly rounded debris..." (Valoch 1988:310). Opravil (1988) used charcoal fragments to determine the presence of *Fraxinus excelsior* and *Pinus* species, as well as a single species of *Acer* cf. *Pseudoplatanus*, *Corylus*, *Picea excelsa*, and *Picea-Larix*. Musil's (1988) paleontological analysis documented the dominance of mammoth and reindeer among faunal species in Layers 7d–6a. Layer 6a, also Micoquian, was not available for study at the time of data collection. As Layer 7a is as rich in lithics, is associated with Neanderthal remains, and has been dated by several means, this assemblage was chosen to represent the earliest flintknapping signature in this region after 60 kya. Please consult Nejman and Clarkson (2008) for additional data on the distribution of retouch and angle of retouch in this layer.

5.1.2 Sampling

All of the intact artifacts from Sector D1, near the standing balk of the main trench in Kůlna Cave (see Figure 5.3), including debitage, tools, and cores, were studied. As the raw material from which the Kůlna artifacts were made varied in kind and quality, the value of certain technological attributes was lessened. For instance, due to the difficulty in seeing flake scar ripple marks on raw material of the Bořitov type of Cretaceous chert, the dorsal scar pattern on pieces of debitage was noticeably more difficult to identify, leading to the significantly larger percentage of Kůlna artifacts with "unknown" dorsal scar patterns compared to other sites in the region.

5.1.3 Blank Production

The reconstruction of the options chosen by the varied flintknappers whose cumulative efforts created the palimpsest of the Kůlna Layer 7a assemblage proceeds according to the twelve knapping behaviors or steps within the four domains of the flintknapping process.

Core Modification

The first domain, "core modification," concerns the orientation of the knapping surfaces of the

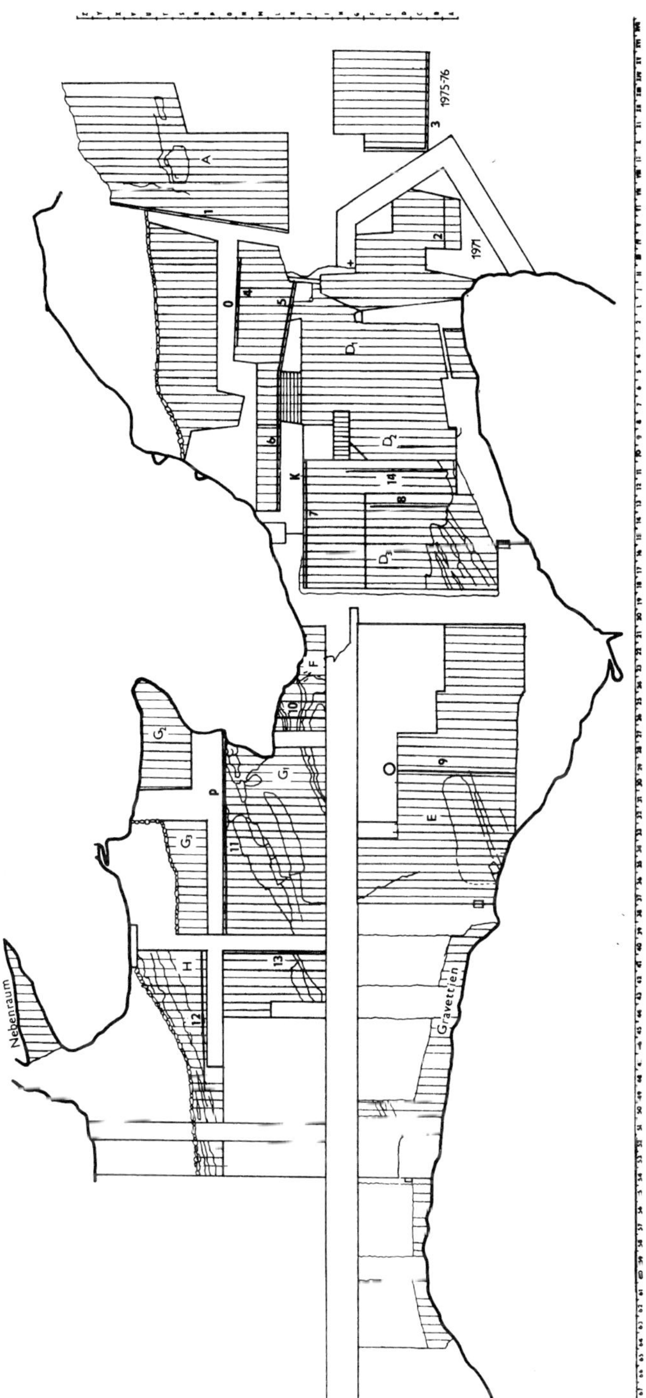

Figure 5.3 Kůlna Cave Site Plan (Valoch 1988, fig. 62, p. 165) Used by permission.

cores. The first step in this domain, "core orientation," is characterized by the distinction between different core forms, based upon an examination of extant core morphology (appendix Figures A1-A7) and an evaluation of where platforms were located during most of core exploitation. Table 5.1 presents the cumulative, assemblage-wide tendency for bulbar negatives to appear on given core surfaces ("Surface 1" and "Surface 2") and within given core sectors (A, B, C, and D). In the case of Kůlna 7a, it is clear that bulbar negatives are more frequent on Surface 1 than on Surface 2 and relatively evenly distributed among the former surface's four sectors. Table 5.2, showing the number of scars within each core sector which originated from the direction of that sector, corroborates the view that all four core sectors served as platforms during reduction. This figure also shows, however, that Surface 2 possesses almost as many scars in their sector of origin as Surface 1 and thus both surfaces may have served the role of debitage surfaces for blank production before the abandonment of the cores. This data, including the spatial orientation of the surfaces evident in the extant cores (see appendix Figures A1-A7) leads to the conclusion that the core orientation within Kůlna 7a is predominately a discoidal form, with a tendency later in reduction for a unifacial discoidal form. This is a different variant of the technology Boëda (1993) observed during his analysis of this assemblage but comparable to that seen at the Middle Paleolithic sites of La Borde and Mauran (Jaubert et al. 1990; Jaubert 1993). No refits have yet been made with artifacts from any layer of Kůlna Cave but Petr Neruda of the Moravské zemské Muzeum is currently attempting to solve this problem.

The second choice in this domain, "core management," includes the options which allow the knapper to continue reducing the core form to exhaustion through specific modifications of the surface convexities. In the case of Kůlna 7a, the constant refreshing of the convexity of the core surfaces was managed by exploiting all margins of the plane of intersection between the two core surfaces, with removals made at an acute angle to this plane of the intersection. This is adequately evidenced in the core morphologies (Figures A1-A7) and the removal of blanks from all sectors of the core margins, evidenced in Tables 5.1-5.2.

Platform Maintenance

The next knapping domain, "platform maintenance," includes the behaviors whereby the flintknapper modifies the platform surface and platform edge before each removal. These steps include platform treatment options, external platform angle, and platform thickness. Table 5.3 presents the frequencies of platform treatment types in this assemblage. Platform preparation was relative frequent, accounting of 48% of the debitage. Table 5.4 presents the central tendency statistics for the external platform angles and platform thickness in the Kůlna assemblage, indicating only moderately steep angles (a mean EPA of 83.9 degrees with a standard deviation of 14.8 degrees) and very thick platforms (a mean PT of 9.1 mm).

Direction of Core Exploitation

The next knapping domain, "direction of core exploitation," concerns the variability resulting from different options of removing flakes from single to multiple platform cores, creating debitage with particular dorsal scar patterns. The dominant strategy used in the assemblage was determined using the principles of dimensional change of debitage during core reduction (see Chapter 4). Table 5.5 presents the cross-tabulation of complete debitage and tools according to dorsal scar pattern and blank length (separated into quartiles so that one quarter of the sample

Table 5.1 Kůlna Layer 7a, percentage of intact cores (n=21) by number of bulbar negatives in each core sector (A, B, C, D) by surface (1, 2)

Number of bulbar negatives	Surf. 1 Sect. A	Surf. 1 Sect. B	Surf. 1 Sect. C	Surf. 1 Sect. D	Surf. 2 Sect. A	Surf. 2 Sect. B	Surf. 2 Sect. C	Surf. 2 Sect. D
0	19	19	19	19	62	52	52	43
1	38	62	48	57	33	48	19	48
2	33	14	19	14	5		19	5
3	5	5	14	10			5	5
4	0						5	
5	5							

Table 5.2 Kůlna Layer 7a, core sector preparation: percent of intact cores (n=21) by number of scars in sector of origin (A, B, C, D) by surface (1, 2)

Number of Scars	Surf. 1 Sect. A	Surf. 1 Sect. B	Surf. 1 Sect. C	Surf. 1 Sect. D	Surf 2 Sect. A	Surf. 2 Sect. B	Surf. 2 Sect. C	Surf. 2 Sect. D
0	10	43	29	38	29	38	29	38
1	14	33	29	14	19	38	29	10
2	38	10	29	33	33	19	14	33
3	24	5	14	10	10	5	14	14
4	10	5		0	10		5	5
5	5	0		5			5	
6		0					5	
7		5						

Table 5.3 Kůlna Layer 7a, platform types for complete and proximal fragments of flakes and tools

Platform type	Frequency	Percent
Plain	76	27.8
Cortical	17	6.2
Facetted	33	12.1
Scarred	38	13.9
Punctiform	1	.4
Linear	2	.7
Removed	26	9.5
Indeterminate	80	29.3
Total	273	100.0

Table 5.4 Kůlna Layer 7a, descriptive statistics for continuous platform variables of complete and proximal fragments of flakes and tools

	N	Minimum	Maximum	Mean	Standard Deviation
Exterior Platform Angle (degrees)	153	51.00	127.00	83.76	14.78
Platform Thickness (mm)	153	1.46	28.57	9.08	4.63

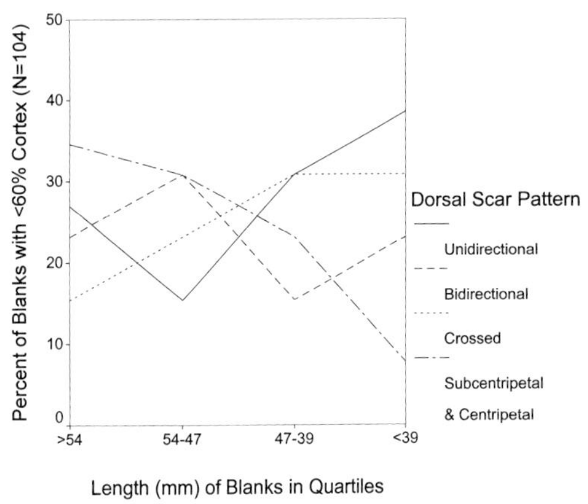

Figure 5.4 Kůlna Layer 7a dorsal scar pattern by blank length (mm) in quartiles cross-tabulation for complete flakes and tools (excluding pieces with a crested scar pattern, an indeterminate scar pattern, greater than 60% cortex, or fewer than three dorsal scars).

population lies within each of the four length ranges). By removing the debitage and tools with dorsal scar patterns noted as indeterminate and crested to clarify the pattern, it is possible to see (Table 5.6 and Figure 5.4) that the most numerous blanks in the first quartile of length possess dorsal scars of a centripetal pattern (34.6%), followed closely by a unidirectional pattern (26.9%). This is the signature of the first decision in this domain, the direction of early core reduction. The shortest blanks, those of the fourth quartile, possess a predominance of unidirectional patterns (38.5%) followed closely by crossed patterns (30.8%). This is the signature of the second decision in this domain, the direction of late core reduction. Examining the intervening quartiles, there is a continuous trend in the reduction of the frequency of centripetal, a continuous trend for the increase in frequency of crossed, and a fluctuation between unidirectional and bidirectional within the second quartile but a continuous increase in both patterns thereafter. These trends in the length-dorsal scar pattern suggests a consistent shift of direction of core exploitation during the reduction of the cores.

Dorsal Surface Convexity

The "dorsal surface convexity system" domain includes the knapping behaviors which encapsulate the cumulative choice of *nervure-guide* or ridge pattern(s) of the core exterior for the production of blanks. These choices are quantified by three categorical attributes (lateral edge type, profile type, and cross-section type) and two continuous variables (laminarity or length/width ratio and use of convexity or width/thickness ratio). Table 5.7 presents the counts and frequencies of lateral edge, profile, and cross-section types. Table 5.8, with the mean, range, and standard deviation for all blank dimensions, presents the longitudinal convexity (mean length/width ratio = 1.44) and vertical convexity (mean width/thickness ratio = 2.83) for the Kůlna complete tools and flakes. These variables indicate that the blanks in the assemblage are flake dominated, relatively thick in cross-section, and mostly trapezoidal.

5.1.4 Tool Kit Morphology

The first five characterizations of the tool kit morphology refer to the dorsal surface convexity variables which give form to the cutting edge to the retouched blanks in the assemblage, as well as the diagnostic flakes without retouch traditionally categorized as tools, i.e., Levallois pieces, although these types are absent for the Kůlna 7a collection. The Kůlna 7a tools are no more bladey than the debitage (mean length/width ration of 1.45) and equally thick in cross-section (mean width/thickness ration of 2.84; see Table 5.9). Even with a significant number of indeterminates due to the highly reduced nature of the tools, it is

Table 5.5 Kůlna Layer 7a dorsal scar pattern by blank length (mm) in quartiles cross-tabulation for complete flakes and tools (excluding flakes with greater than 60% cortex or fewer than three dorsal scars)

Dorsal Scar Pattern	>54	54-47	47-39	<39	Total
Unidirectional					
Count	7	4	8	10	29
% within Quartile	18.4%	13.3%	25.0%	30.3%	21.8%
Bidirectional					
Count	6	8	4	6	24
% within Quartile	15.8%	26.7%	12.5%	18.2%	18.0%
Crossed					
Count	4	6	8	8	26
% within Quartile	10.5%	20.0%	25.0%	24.2%	19.5%
Subcentripetal & Centripetal					
Count	9	8	6	2	25
% within Quartile	23.7%	26.7%	18.8%	6.1%	18.8%
Crested					
Count	9	3	3	5	20
% within Quartile	23.7%	10.0%	9.4%	15.2%	15.0%
Indeterminate					
Count	3	1	3	2	9
% within Quartile	7.9%	3.3%	9.4%	6.1%	6.8%
Total					
Count	38	30	32	33	133
% within Cortex	100%	100%	100%	100%	100%

Table 5.6 Kůlna Layer 7a dorsal scar pattern by blank length (mm) in quartiles cross-tabulation for complete flakes and tools (excluding pieces with a crested scar pattern, an indeterminate scar pattern, greater than 60% cortex, or fewer than three dorsal scars)

Dorsal Scar Pattern	>54	54-47	47-39	<39	Total
Unidirectional					
Count	7	4	8	10	29
% within Quartile	26.9%	15.4%	30.8%	38.5%	27.9%
Bidirectional					
Count	6	8	4	6	24
% within Quartile	23.1%	30.8%	15.4%	23.1%	23.1%
Crossed					
Count	4	6	8	8	26
% within Quartile	15.4%	23.1%	30.8%	30.8%	25.0%
Subcentripetal & Centripetal					
Count	9	8	6	2	25
% within Quartile	34.6%	30.8%	23.1%	7.7%	24.0%
Total					
Count	26	26	26	26	104
% within Cortex	100%	100%	100%	100%	100%

possible to see some differences between the blank production and tools in terms of discrete variables. Convergent lateral edges are more frequent (9.2% or 12.6% without the indeterminate pieces) among the debitage but much less so among the tools (4.3% or 6.8% without the indeterminate pieces). This tendency to avoid picking pointed blanks is corroborated by the higher frequency of blunt distal terminations than pointed.

The last two characterizations of the tool kit morphology refer to two variables describing the application of retouch to blanks: first, the application of distinctive and/or diagnostic forms of retouch and, second, the dominance or equality of Middle Paleolithic (lateral edge) versus Upper Paleolithic (distal and lateral edges) retouched tool types. In this case, Kůlna 7a exhibits the use of flat, bifacial retouch in the production of Micoquian handaxes, bifacial side scrapers, and bifacial knives. Middle Paleolithic tool types dominate in the application of retouch (Valoch 1988).

Intensity of Tool and Core Reduction

For the purposes of evaluating the comparability of the Kůlna 7a assemblage with other assemblages in this study, Table 5.11 presents three characterizations of the intensity of use of raw material within the collection. The location of most of the raw material from Kůlna 7a is close, 79% of which is Cretaceous chert/hornstone of the Bořitov type from the Moravian Karst region, around 10 km to the west. Local quartz from a radius of 10–15 km is found in the assemblage as well as some few examples of high quality flints and radiolarites from distant sources of up to 70 km and 230 km (Féblot-Augustin 1993:233; Svoboda et al. 1996:93). See Feblot-Augustin (1993; 1997) for a more detailed discussion of the raw material provenance of the Kůlna assemblages.

Table 5.7 Kůlna Layer 7a, frequencies of lateral edge, profile, and cross-section types for complete flakes and tools

Type	Frequency	Percent
Lateral edge		
Parallel	108	39.6
Convergent	25	9.2
Expanding	39	14.3
Ovoid	26	9.5
Indeterminate	75	27.5
Total	273	100.0
Profile		
Straight	143	52.4
Curved or Concord	50	18.3
Twisted	32	11.7
Indeterminate	48	17.6
Total	273	100.0
Cross-section		
Triangular	89	32.6
Trapezoidal	126	46.2
Other	16	5.9
Indeterminate	42	15.4
Total	273	100.0

Table 5.8 Kůlna Layer 7a, descriptive statistics for blank dimensions for complete flakes and tools

Blank Dimension	N	Range	Mean	Standard Deviation
Length	273	101.71	48.21	14.59
Width	273	100.54	35.38	11.50
Thickness	273	41.09	13.79	5.78
Length/Width Ratio	273	3.40	1.44	.49
Width/Thickness Ratio	273	8.65	2.83	1.06

Table 5.9 Kůlna Layer 7a, descriptive statistics for dimensions of complete tools

Blank Dimension	N	Range	Mean	Standard Deviation
Length/Width Ratio	162	3.40	1.45	.51
Width/Thickness Ratio	162	8.65	2.84	1.12

Table 5.10 Kůlna Layer 7a, frequencies of lateral edge, distal terminus, and profile types for complete tools

Type	Frequency	Percent
Lateral edge		
Parallel	57	35.2
Convergent	7	4.3
Expanding	24	14.8
Ovoid	15	9.3
Indeterminate	59	36.4
Total	162	100.0
Distal Terminus		
Blunt	48	29.6
Pointed	17	10.5
Hinged	20	12.3
Overshot	3	1.9
Removed	36	22.2
Broken	7	4.3
Indeterminate	31	19.1
Total	162	100.0
Profile		
Straight	85	52.5
Curved or Concord	25	15.4
Twisted	16	9.9
Indeterminate	36	22.2
Total	162	100.0

Table 5.11 Kůlna Layer 7a, intensity of core and tool reduction

Variables	Numerator	Denominator	Result
Tools/Flake Ratio	Complete Tools: 154	Complete Flakes: 76	2.03
Non-cortical Debitage/ Cortical Debitage Ratio	Non-cortical Complete Debitage: 125	Cortical Complete Debitage: 106	1.18
Henry's Dimensional Analysis	NA	NA	Almost no overlap, highly reduced cores

Certain aspects of the raw material economy at Kůlna Cave during the deposition of Layer 7a are distinctly different from most of the assemblages included in this research. The values calculated in Table 5.11 put the tool/flake ratio of Kůlna 7a as a distant outlier at the high end of the distribution of the other sample assemblages (Figure 5.5). With a tool/flake ratio over twice as high as the other assemblages, there is a remote possibility that an unwilling bias in excavation or curation towards the relative enrichment of the number of tools compared to debitage in the museum sample may be as much responsible for this value as the prehistoric raw material economy. Operating under the assumption that the value is relevant to the prehistoric situation, this data indicates that retouch was applied to many more blanks in Kůlna 7a than in other assemblages and therefore the variables based on relative tool shape dimensions (length/width ratio and width/ thickness ratio) may be biased when compared to less-reduced assemblages.

On the other hand, the value for the relative frequency of cortical blanks to non-cortical blanks indicates that the assemblage is not in a unique situation in terms of the degree to which raw material was worked from a nodular stage on-site (Figure 5.6). The assemblage falls in the low to middle-range of assemblages for this value, slightly closer to the raw material extraction end of the spectrum than the special function site end, in which most of the core reduction was pursued off-site. Given the minor degree of nodular working off-site, it is interesting that the Kůlna 7a assemblage shows a high intensity of reduction of its cores (Figure 5.7). From the dimensional analysis of core facets versus debitage lengths and widths proposed by Henry (1989), Kůlna 7a displays the vast majority of its core facets below the main distribution of the blanks, with 16 out of 27 facets found below the smallest debitage. This data may be slightly misleading as almost no debitage actually falls below 3.0cm in any dimension, indicating that the excavation or curational separation between the small finds collection and the debitage collection presented here is represented by a dimension larger than that used by the present study, 2.0 cm. Nevertheless, as 10 of these 16 facets are below 2.0 cm in size, it is clear that the cores are highly reduced in this assemblage when compared to the dimensions of the debitage, a situation close to core exhaustion (see Figure 4.34,B).

Between the cortical data, which indicates that a good portion of the early cortical-dominated portion of the core reduction sequence is present in the assemblage, and the intensity of core reduction data, which indicates that the very end of the reduction sequence is also present in the assemblage, it is clear that Kůlna 7a can be compared usefully with most assemblages, whether they are raw material extraction workshops or highly reduced activity sites. Both the intensity of tool reduction and the almost complete representation of the core reduction sequence in the site are

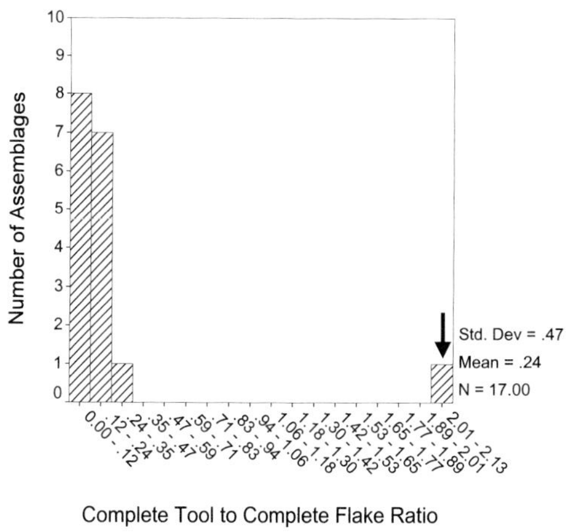

Figure 5.5 Kůlna Layer 7a, tool/flake ratio value relative to a histogram of all assemblages.

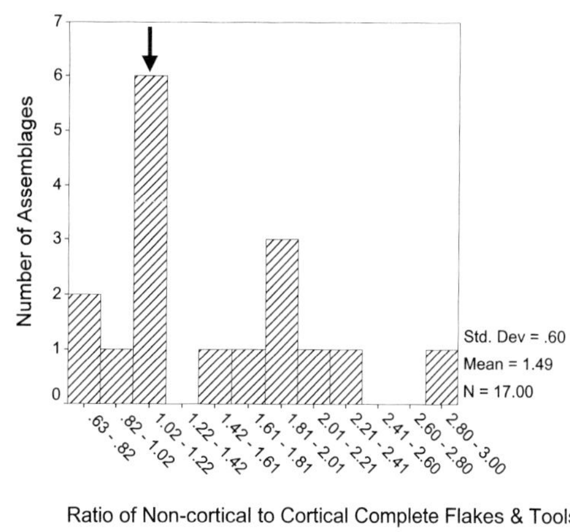

Figure 5.6 Kůlna Layer 7a, non-cortical/cortical flake ratio value relative to a histogram of all assemblages.

understandable in the context of this sheltered residential site in which presumably numerous types of activities took place. While the inclusiveness of the Kůlna 7a collection suggests that most activities found in other sites are likely also to be found in the Kůlna 7a assemblage, the degree to which the central tendencies at the assemblage level will reflect this commonality will have to be evaluated on a case by case basis.

5.2 Stránská skála IIIa Layer 4

5.2.1 Site Background

Located on the eastern outskirts of the modern city of Brno, Stránská skála is a hillside formed by a bedrock of Jurassic limestone overlain by Tertiary marine sands and Pleistocene loess deposits. Numerous Paleolithic localities have been discovered within the loess deposits of the hill as well as in the slope deposits on the western and northern cliff-faces. The western locality, labeled Stránská skála I, is the oldest site on the hillside, containing Middle Pleistocene paleontological deposits with a potential association with archaeological material (Musil and Valoch 1968; Valoch 1987). The other locality not situated directly on top of the Stránská skála hill is locality IV, an Epigravettian hunting site preserved in the slope sediments at the base of the north limestone cliff-face (Svoboda 1991; West 1996).

The locality of interest here is IIIa, situated in the middle to eastern half of the south facing slope of the hill. This locality is a multilayer site, possessing a lower Bohunician assemblage, Layer 4, and a superimposed Aurignacian assemblage, Layer 3 (Svoboda 1987). The sedimentary matrix consists of a soliflucted and redeposited sequence of paleosols, calcareous earths, and small gravels (Czudek et al. 1991; Svoboda et al. 1996). The pollen spectrum, although poor, indicates a cool environment for both occupations (Svobodová 1987). No fauna has been preserved in either assemblage.

The present assemblage, excavated in 1984 and categorized as Bohunician (Svoboda 1987), has been dated to 41,300 + 3,100/-2200 (GrN 12606), and lies within the last interpleniglacial

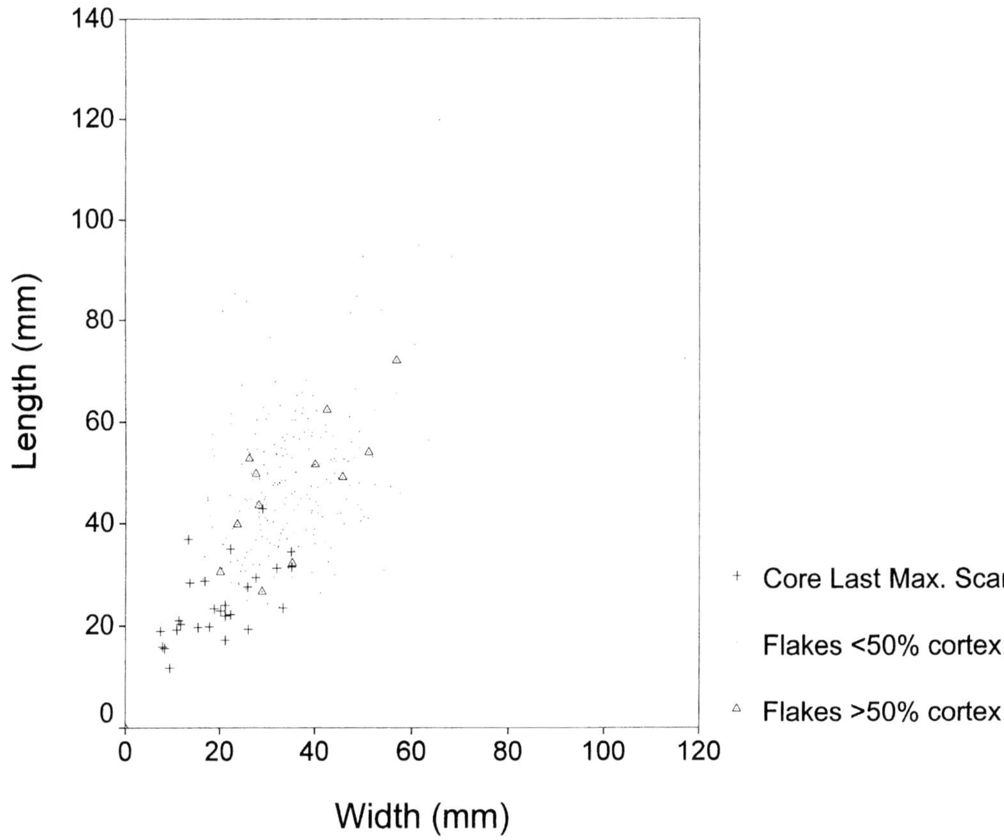

Figure 5.7 Kůlna Layer 7a according to Henry's core reduction intensity measure: length and width of last maximum core facets, highly cortical blanks, and blanks with little cortex.

paleosol (Svoboda et al. 1996:233). The sedimentary context of this layer is less affected by solifluction than the upper soil horizon which contains the Aurignacian assemblage.

Almost all artifacts in this assemblage are made on the local Stránská skála hornstone or chert, which outcrops from the Jurassic limestone at the current base of the hillside. The exact location of the Pleistocene outcrop in relation to the current hill is unknown due to the extensive sedimentation of loess during the Last Glacial Maximum but is argued to be extremely close (Svoboda 1983). *In situ* frost fracture is frequent in the unworked raw material when found it its Jurassic bedrock and therefore double patination is the norm in the knapped lithics. As the older of the two patinas is in essence a rind which predates the raw material's modification by hominins, this older patina was added to the normal chalky cortex when calculating the percentage of the dorsal surface area covered by cortex. This should be taken into consideration when studying this assemblage for the replicability of the data presented here. This should,

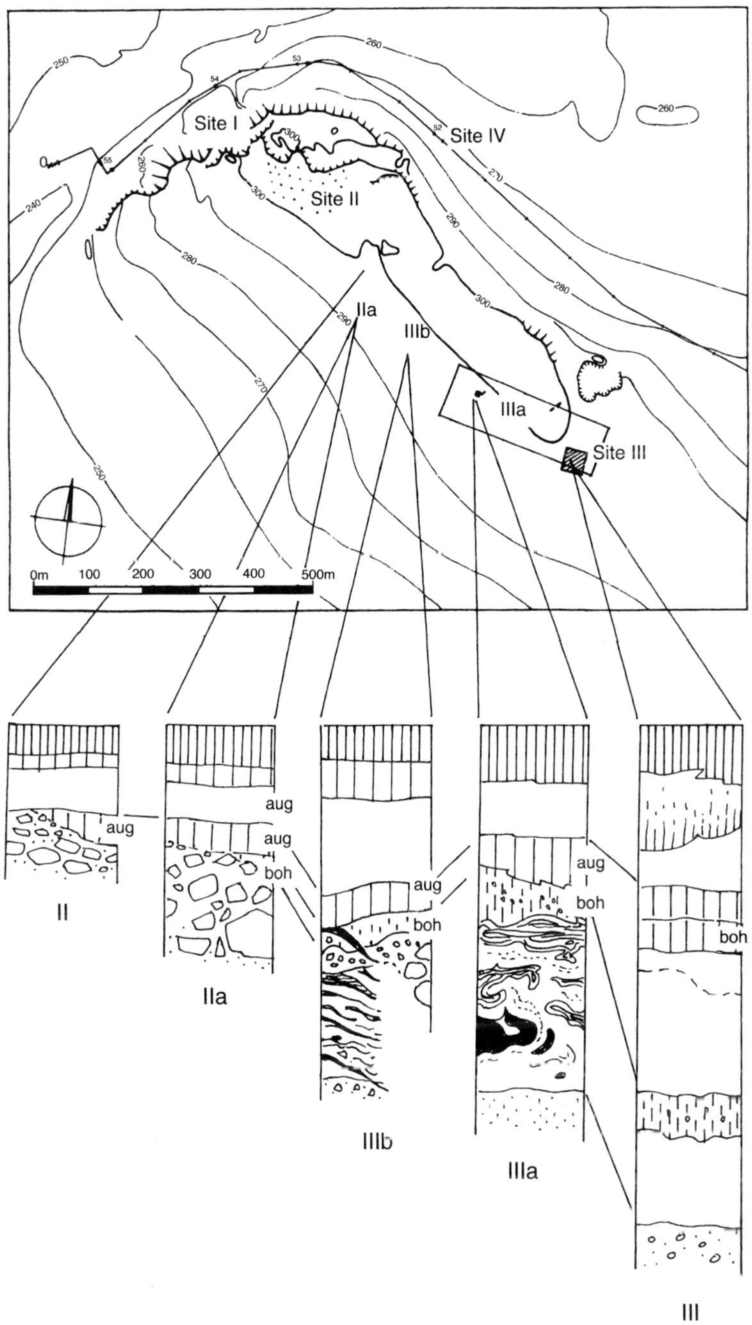

Figure 5.8 Map of Stránská skála hillside and the stratigraphic correlation between the main localities (Svoboda and Bar-Yosef 2003: fig. 1.8, p. 11, used by permission). The rectangle around IIIa and III refers to Figure 5.27.

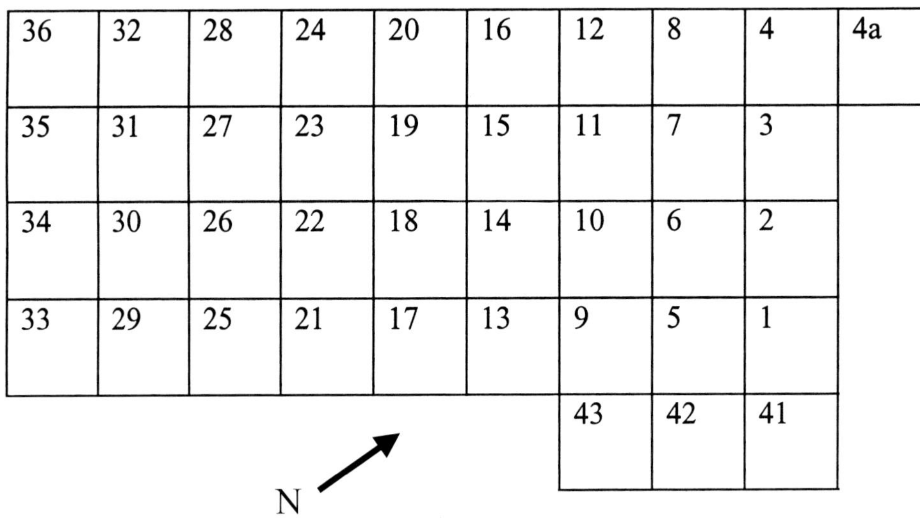

Figure 5.9 Stránská skála IIIa site plan.

however, not compromise the use of cortex percentages when comparing this assemblage to others in the region. All other assemblages excavated from the Stránská skála hillside were treated similarly in this regard.

5.2.2 Sampling

To maximize the accuracy of the technological study of both debitage and tools, all of the intact pieces from the meter squares with the largest number of artifacts from Stránská skála IIIa Layer 4 were studied. The square meters utilized in this study include: 2, 3, 4, 5, 6, 7, 8, 9, 10, 11, 12, 15, 16, 17, 18, 19, 20, 23, 24, 27, 28, 29, 41, 42, and 43. Only square meter associations are available for artifacts in this assemblage as three dimensional proveniencing was not conducted during the 1980s excavation of this site. The extensive refittings conducted by P. Škrdla (Škrdla 1994, 1996, 2003c; Svoboda and Škrdla 1995) were used to corroborate the attribute analysis of the debitage and cores.

5.2.3 Blank Production

Core Modification

Core modification in Stránská skála IIIa Layer 4 entails a longitudinal blade/flake core orientation, as evidenced in the identification of the narrow and long core face as the primary debitage surface in the assemblage (see appendix figures A8-A15). This identification is supported by Table 5.12, which indicates that Surface 1 strongly outweighs Surface 2 in number of bulbar negatives, suggesting a preference for exploiting one surface rather than two, at least as preserved in the final form of the cores. The longitudinal core orientation visible in the choice of Surface 1 is also corroborated by Table 5.13, which displays the dominance of Surface 1 over Surface 2 in preparation of core sectors. Sector A on Surface 1 displays the highest frequency of cores with removals from this direction, followed distantly by sector C, while sectors B and D (the lateral core margins) show only sporadic use, most likely the result of the preparation for

Table 5.12 Stránská skála IIIa Layer 4, percentage of intact cores (n=38) by number of bulbar negatives in each core sector (A, B, C, D) by surface (1, 2)

Number of Bulbar Negatives	Surf. 1 Sect. A	Surf. 1 Sect. B	Surf. 1 Sect. C	Surf. 1 Sect. D	Surf. 2 Sect. A	Surf. 2 Sect. B	Surf. 2 Sect. C	Surf. 2 Sect. D
0	8	37	42	42	71	84	66	92
1	29	37	18	37	16	5	11	5
2	37	24	39	13	8	8	21	3
3	18	0		5	5	3	3	
4	5	3		3				
5	0							
6	0							
7	0							
8	0							
9	3							

Table 5.13 Stránská skála IIIa Layer 4, core sector preparation: percent of intact cores (n=38) by number of scars in sector of origin (A, B, C, D) by surface (1, 2)

Number of Scars	Surf. 1 Sect. A	Surf. 1 Sect. B	Surf. 1 Sect. C	Surf. 1 Sect. D	Surf. 2 Sect. A	Surf. 2 Sect. B	Surf. 2 Sect. C	Surf. 2 Sect. D
0	13	71	50	74	37	74	53	74
1	16	16	3	13	21	11	18	16
2	18	8	32	8	24	5	16	0
3	34	5	13	3	11	3	5	11
4	11		0	0	5	8	5	
5	5		3	3	3		3	
6	0							
7	0							
8	0							
9	0							
10	0							
11	3							

débordant/lateral crested removals. On Surface 2, sectors A and C again show the most working but far less frequently than on Surface 1.

The second analytical step in this domain, core management, as evidenced through both the core morphology and P. Škrdla's refittings, began with the application of cresting removals to shape the longitudinal convexity of the core surface into a central ridge which was then struck off as a *lame à crête*. This is evidenced in the refit sequence number S08/84 (Škrdla 2003c:122-123). Core convexity management was also achieved in this assemblage through the removal of lateral débordant flakes/blades to reposition the dorsal convexity to the middle of debitage surface, when viewed from one of the narrow ends of the core. This appears to have been done after the surface had been overly flattened by the typically wide but thin removals characteristic of this assemblage (Figures 5.10-5.11). In particular instances, after the flattening of the front face of the core, blades were subsequently removed from the narrow side of the core, a 90 degree rotation to the right or left from the debitage surface when viewed from the platform. This is evidenced in refit

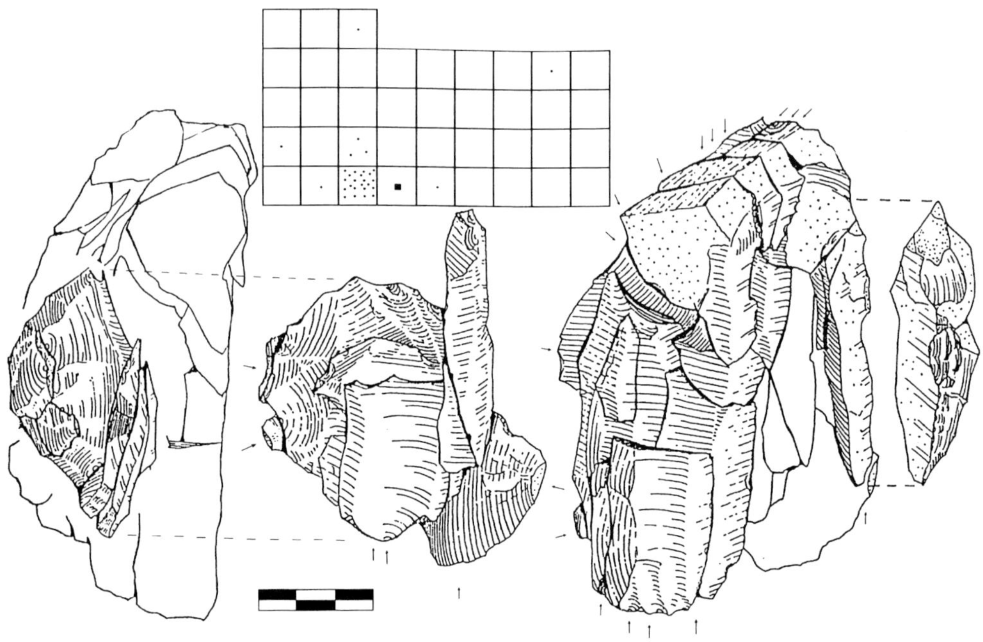

Figure 5.10 Refit core sequence (No. S01/84) from Stránská skála IIIa Layer 4. Grid shows spatial distribution of artifacts. Scale = 3cm. Drawing by J. A. Svoboda. Refit by P. Škrdla (from Škrdla 2003c:120, fig. 9.1, used by permission).

sequences such as number S03/84 published by Škrdla (2003c:131-132). This is a peculiarity of the Bohunician industry and others argued to be related to some degree (see Ginter et al. 1996).

Platform Maintenance

Platform maintenance includes a moderate amount of platform preparation, roughly 41% of the debitage when the indeterminate platforms are removed from the percentage (Table 5.14). Measurements of external platform angles and platform thickness (Table 5.15) indicate tendencies around 85 degrees and 5 mm respectively.

Direction of Core Exploitation

The analysis of the direction of core exploitation is presented in Tables 5.16-5.17. Figure 5.12 demonstrates that the dominant directional strategy during early core exploitation in the Stránská skála IIIa level 4 assemblage, i.e., the dorsal scar pattern within the largest length quartile, was bidirectional. Refit sequences corroborate the bidirectional reduction in this assemblage, for instance in refit sequence number S01/84 (Škrdla 2003c:120-122), which produced bidirectional debitage but left a core approaching a centripetal Levallois form at exhaustion (see Figures 5.10-5.11), as well as refit sequence number S08/84 (Škrdla 2003c:122-123). Conversely, the dorsal scar pattern of the pieces within the smallest length quartile of the blanks demonstrates a unidirectional strategy, characterizing the end of core exploitation. Taken together with the continuous pattern within the two middle quartiles, the

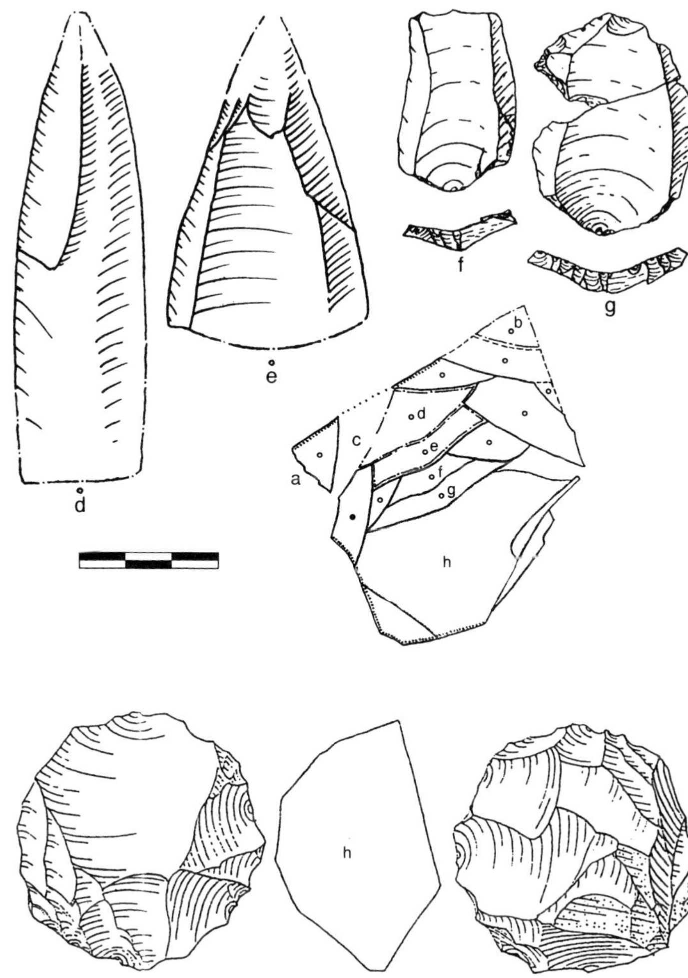

Figure 5.11 Products from refit core sequence (No. S01/84) from Stránská skála IIIa Layer 4. Scale = 3cm. Filled dots indicate a removal from the lower platform while an empty dot indicates a removal from the top platform. Drawings d and e represent negatives of artifacts missing from the refit sequence. Drawing by J. A. Svoboda. Refit by P. Škrdla (from Škrdla 2003: 121, fig. 9.2, used by permission).

extremes of the length-dorsal scar pattern cross tabulation demonstrate a strong change from bidirectional reduction at the beginning of core exploitation to unidirectional reduction at the end of core exploitation. The evidence from the refit data moderates this conclusion to some degree, as refit sequence numbers S07/84 and S06/84 (Škrdla 2003c:129–131) evidence the unidirectional reduction of "plate cores" or halved nodules, which possess a relatively short debitage length at the beginning of exploitation. Unidirectional pyramidal reduction sequences, S04/84 (Škrdla 2003c:132), and unidirectional non-pyramidal sequences, S03/84 (Škrdla 2003c:131–132), also exist. Given the combination of the refit data and the dorsal scar pattern by blank length data, it is probable that some unidirectional reduction of initially shorter core forms contributes to some degree to the pattern of bidirectional reduction changing to unidirectional reduction. The final form of the cores as seen in Tables 5.12–5.13, however, makes it clear that the

Table 5.14 Stránská skála IIIa Layer 4, platform types for complete and proximal fragments of flakes and tools

Platform type	Frequency	Percent
Plain	209	38.5
Cortical	33	6.1
Dihedral	9	1.7
Facetted	61	11.2
Chapeau	15	2.8
Scarred	103	19.0
Punctiform	6	1.1
Linear	12	2.2
Removed	8	1.5
Indeterminate	87	16.0
Total	543	100.0

Table 5.15 Stránská skála IIIa Layer 4, descriptive statistics for continuous platform variables of complete and proximal fragments of flakes and tools

	N	Minimum	Maximum	Mean	Standard Deviation
Exterior Platform Angle (degrees)	425	47.00	138.00	85.21	15.25
Platform Thickness (mm)	433	0.37	15.38	4.82	2.45

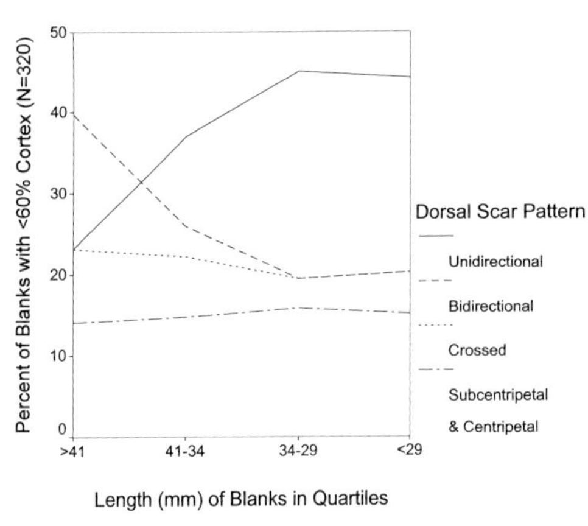

Figure 5.12 Stránská skála IIIa Layer 4, dorsal scar pattern by blank length (mm) in quartiles cross-tabulation for complete flakes and tools (excluding pieces with a crested scar pattern, an indeterminate scar pattern, greater than 60% cortex, or fewer than three dorsal scars).

two opposed platforms at the end of core exploitation were not equivalent in their degree of working: the sector A platforms on Surface 1 possessed far more bulbar negatives as well as non-bulbar removal scars from this platform than sector C, the opposed platform. This again suggests a pattern of bidirectional reduction at the beginning of core exploitation and a shift toward unidirectional reduction at the end of exploitation.

Additional refits also reinforce the inverse correlation between the length of the debitage surface and the progress of reduction. In refit sequence number 01219-77/84 (Škrdla 2003c:137–138) which lacks a core, a sequence of three Levallois point removals refit on top of each other. The staircase-like positions of their platforms relative to each other demonstrate how the debitage surface is shortened with the faceting of the platform after each removal.

Table 5.16 Stránská skála IIIa Layer 4, dorsal scar pattern by blank length (mm) in quartiles cross-tabulation for complete flakes and tools (excluding flakes with greater than 60% cortex or fewer than three dorsal scars)

Dorsal Scar Pattern	>41	41-34	34-29	<29	Total
Unidirectional					
Count	18	30	37	35	120
% within Quartile	17.6%	31.3%	35.9%	34.7%	29.9%
Bidirectional					
Count	31	21	16	16	84
% within Quartile	30.4%	21.9%	15.5%	15.8%	20.9%
Crossed					
Count	18	18	16	16	68
% within Quartile	17.6%	18.8%	15.5%	15.8%	16.9%
Subcentripetal & Centripetal					
Count	11	12	13	12	48
% within Quartile	10.8%	12.5%	12.6%	11.9%	11.9%
Crested					
Count	17	8	12	6	43
% within Quartile	16.7%	8.3%	11.7%	5.9%	10.7%
Indeterminate					
Count	7	7	9	16	39
% within Quartile	6.9%	7.3%	8.7%	15.8%	9.7%
Total					
Count	102	96	103	101	402
% within Cortex	100.0%	100.0%	100.0%	100.0%	100.0%

Table 5.17 Stránská skála IIIa Layer 4, dorsal scar pattern by blank length (mm) in quartiles cross-tabulation for complete flakes and tools (excluding pieces with a crested scar pattern, an indeterminate scar pattern, greater than 60% cortex, or fewer than three dorsal scars)

Dorsal Scar Pattern	>41	41-34	34-29	<29	Total
Unidirectional					
Count	18	30	37	35	120
% within Quartile	23.1%	37.0%	45.1%	44.3%	37.5%
Bidirectional					
Count	31	21	16	16	84
% within Quartile	39.7%	25.9%	19.5%	20.3%	26.3%
Crossed					
Count	18	18	16	16	68
% within Quartile	23.1%	22.2%	19.5%	20.3%	21.3%
Subcentripetal & Centripetal					
Count	11	12	13	12	48
% within Quartile	14.1%	14.8%	15.9%	15.2%	15.0%
Total					
Count	78	81	82	79	320
% within Cortex	100.0%	100.0%	100.0%	100.0%	100.0%

Table 5.18 Stránská skála IIIa Layer 4, frequencies of lateral edge, profile, and cross-section types for complete flakes and tools

Type	Frequency	Percent
Lateral edge		
Parallel	250	46.0
Convergent	121	22.3
Expanding	90	16.6
Ovoid	53	9.8
Indeterminate	29	5.3
Total	543	100.0
Profile		
Straight	278	51.2
Curved or Concord	152	28.0
Twisted	92	16.9
Indeterminate	21	3.9
Total	543	100.0
Cross-section		
Triangular	230	42.4
Trapezoidal	269	49.5
Other	28	5.2
Indeterminate	16	2.9
Total	543	100.0

Dorsal Surface Convexity

The analysis of the attributes serving as the proxies for the choices within the dorsal surface convexity domain are given in Tables 5.18–5.19. The ridge system used in the Stránská skála IIIa level 4 assemblage shows a propensity for relatively high longitudinal convexity (i.e., laminarity), thin products (width/thickness mean of 3.99), diffuse ridges (varied lateral edges), a bare majority of straight profiles, and slightly more trapezoidal than triangular cross-sections (i.e., following two rather than one ridge per detachment from the core).

5.2.4 Tool Kit Morphology

The first five characterizations of the tool kit morphology refer to the dorsal surface convexity variables which give form to the cutting edge of the retouched blanks in the assemblage, as well as the diagnostic flakes without retouch traditionally categorized as tools, i.e., the typological Levallois pieces frequently found in this assemblage. Possibly due to the inclusion of the typological unretouched Levallois points in this assemblage, the tool kit possesses a higher frequency of convergent lateral edges than the debitage as a whole, 31.6 % for the tools compared to 23.5% when the indeterminates are removed from the percentages. Otherwise, the tool kit morphology is fairly similar to that of the debitage.

The last two characterizations of the tool kit morphology refer to two variables describing the application of retouch to blanks: first, the application of distinctive and/or diagnostic forms of retouch and, second, the dominance or equality of Middle Paleolithic (lateral edge) versus Upper Paleolithic (distal and lateral edges) retouched tool types. In this case, Upper Paleolithic tool types dominate the tool kit (Svoboda 1987, 1991, 2003b). For the application of unique types of retouch, it must be remembered that no bifacial foliate points, nor bifacial thinning flakes, were recovered/recognized from this assemblage, contrary to the recovery and recognition of such pieces in another excavated Bohunician context at Brno-Bohunice 7.4 km to the west of Stránská skála (Tostevin and Škrdla 2003, 2006; Škrdla and Tostevin 2005). Under the logic of analyzing and comparing individual assemblages and not more generalized industrial types, bifacial retouch and foliate points cannot be ascribed to the Stránská skála IIIa level 4 assemblage.

It must also be pointed out that three atypical carinated endscrapers are reported for this assemblage (Svoboda 2003b:155, tab. 10.1). Given the atypical form, the small representation within the assemblage, and the likelihood of vertical m-ovement of these pieces down from the superimposed Aurignacian of layer 3 by the strong cryoturbation processes at this locality (as

Table 5.19 Stránská skála IIIa Layer 4, descriptive statistics for blank dimensions for complete flakes and tools

Blank Dimension	N	Range	Mean	Standard Deviation
Length	543	65.34	36.27	11.06
Width	543	64.29	23.15	8.45
Thickness	543	24.03	6.64	3.17
Length/Width Ratio	543	4.11	1.71	.67
Width/Thickness Ratio	543	18.35	3.99	1.82

Table 5.20 Stránská skála IIIa Layer 4, descriptive statistics for dimensions of complete tools

Blank Dimension	N	Range	Mean	Standard Deviation
Length/Width Ratio	67	2.25	1.58	0.50
Width/Thickness Ratio	67	10.05	4.28	2.14

shown by the contiguous vertical distribution of refits in Stránská skála IIIc), it was judged wiser to consider these pieces as insufficient evidence for a credible unique or diagnostic retouch for this assemblage. This decision admittedly opens up the possibility of a critique that the original industrial-type label of "Bohunician" rather than "Aurignacian" is being given more interpretive reality than the artifactual constitutents of the assemblage, or that the definition of an assemblage as "the enculturating environment created by all of the hominins who have contributed material culture to its palimpsest" (from Chapter 3) is being given only lip-service. Reification of the original industrial-type label, however, is not the aim in this decision. The three pieces in question simply do not have the morphology that would allow the retouch flakes removed from the working bit of the endscraper to serve as effective bladelets when compared to the numerous examples of carination elsewhere in western Eurasia (Lebrun-Ricalens 2005; Bar-Yosef et al. 2006). I personally would not type them as carinated atypical but I must in due diligence still report their original description as such. Further, relating to the question of how an

Table 5.21 Stránská skála IIIa Level 4, frequencies of lateral edge, distal terminus, and profile types for complete tools

Type	Frequency	Percent
Lateral edge		
Parallel	22	32.8
Convergent	18	26.9
Expanding	9	13.4
Ovoid	8	11.9
Indeterminate	10	14.9
Total	67	100.0
Distal Terminus		
Blunt	18	26.9
Pointed	18	26.9
Hinged	3	4.5
Overshot	2	3.0
Removed	21	31.3
Broken	2	3.0
Indeterminate	3	4.5
Total	67	100.0
Profile		
Straight	28	41.8
Curved or Concord	20	29.9
Twisted	14	20.9
Indeterminate	5	7.5
Total	67	100.0

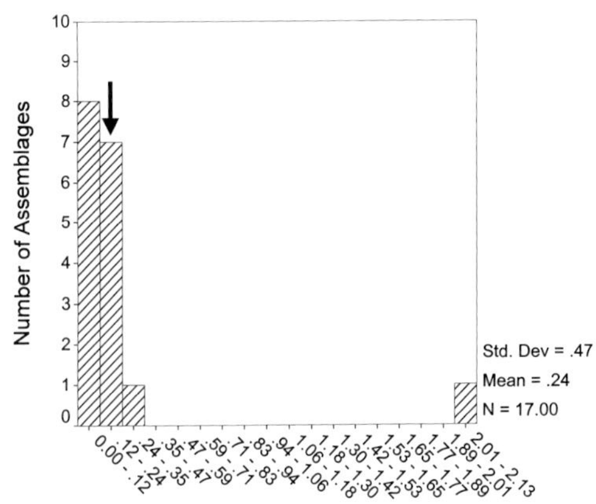

Figure 5.13 Stránská skála IIIa Layer 4, tool/flake ratio value relative to a histogram of all assemblages.

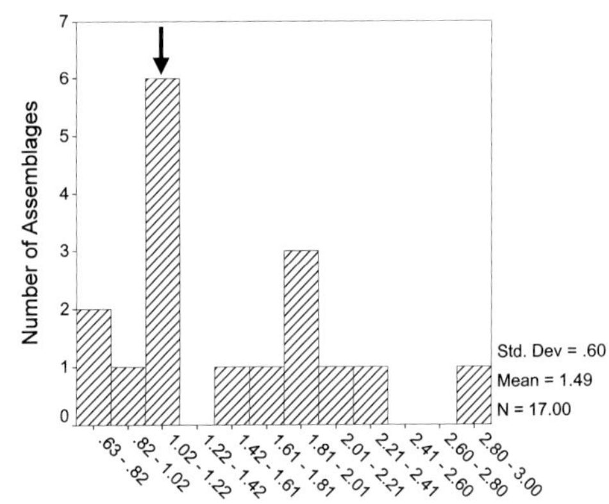

Figure 5.14 Stránská skála IIIa Layer 4, non-cortical/cortical flake ratio value relative to a histogram of all assemblages.

Table 5.22 Stránská skála IIIa Layer 4, intensity of core and tool reduction

Variables	Numerator	Denominator	Result
Tools/Flake Ratio	Complete Tools: 68	Complete Flakes: 437	0.16
Non-cortical Debitage/ Cortical Debitage Ratio	Non-cortical Complete Debitage: 260	Cortical Complete Debitage: 245	1.06
Henry's Dimensional Analysis	NA	NA	Core facets overlap with the low to middle of the blank distribution: slight core exhaustion

assemblage is defined, the definition provided in Chapter 3 pertains to the assemblage at the level of the original geoarchaeological context of the open palimpsest and does not limit one to evaluating post-depositional influences on the assemblage when they can be clearly described, as can be done at Stránská skála IIIa. Nevetheless, the way this research treats "unique types of retouch" is admittedly fairly essentializing, although not to the same degree as seen in Zilhão and d'Errico (2003:343-345), and this is a weakness of this analytical structure that future studies will hopefully avoid.

Intensity of Tool and Core Reduction

Given the situation of Stránská skála IIIa immediately adjacent to a raw material source, it is not surprising that in level 4 retouch was applied to blanks less frequently than most other assemblages (Table 5.22, Figure 5.13), nodules were worked from a fairly highly cortical stage on-site (Figure 5.14), and most of the last maximum facets on the cores overlap with the blanks in the middle to low end of their length and width distribution (Figure 5.15). While some of the refits show an exhausted form (Figure 5.11), as well as a few of the cores (6 out of 40) in Henry's dimensional

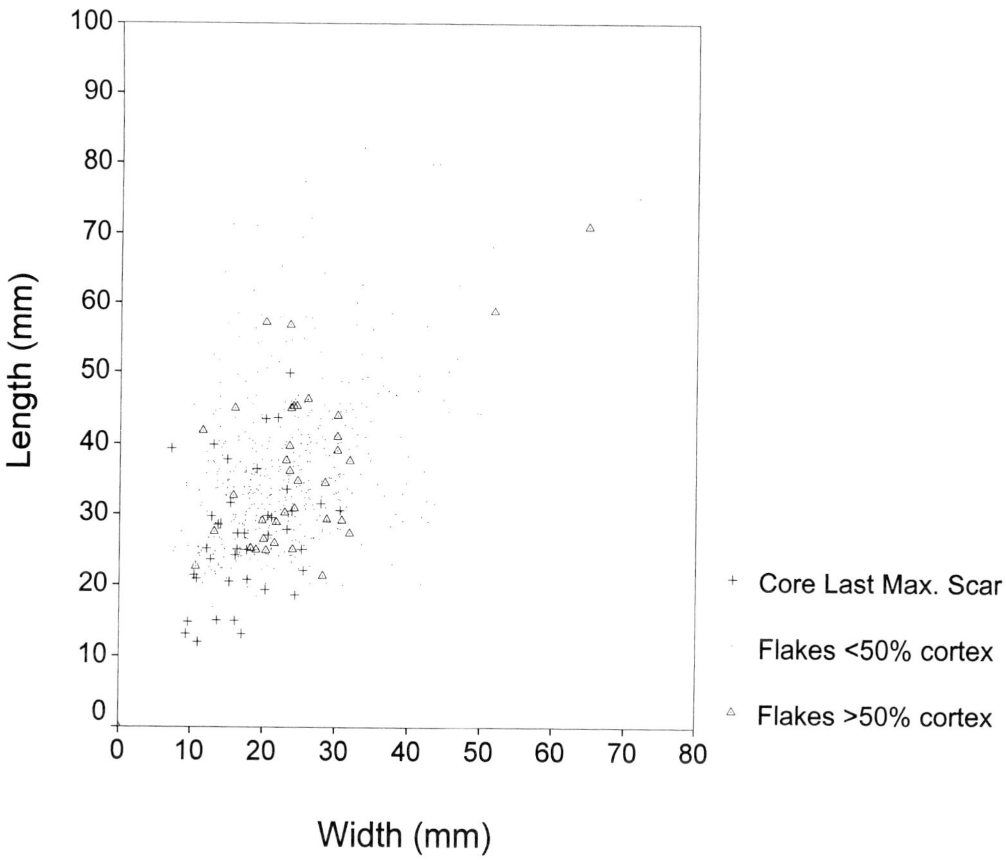

Figure 5.15 Stránská skála IIIa Layer 4 according to Henry's core reduction intensity measure: length and width of last maximum core facets, highly cortical blanks, and blanks with little cortex.

analysis, falling as they do below the 2.0 cm length cut off for the debitage analysis, this assemblage as a whole is not highly reduced in either its tools or its cores. This indicates that the early through at least the middle portion of the core reduction sequence is present in this assemblage.

5.3 Stránská skála III Layer 5

5.3.1 Site Background

Locality III of the Stránská skála hillside was discovered during rescue excavations of an overlying Chalcolithic Funnel Beaker Culture pit feature in 1981. The Paleolithic deposits, found in a stratigraphic setting very similar to that found in the later excavation of Stránská skála IIIa, were excavated by two teams, one led by Karel Valoch of the Moravian Museum (Valoch et al. 2000; Valoch 2003) and the other by Jiři Svoboda of the Institute of Archaeology, Brno (1987). Svoboda's excavated assemblage, later labeled "sub-site SSIII-2" by Svoboda and Valoch (2003), was chosen to represent the knapping behaviors at this

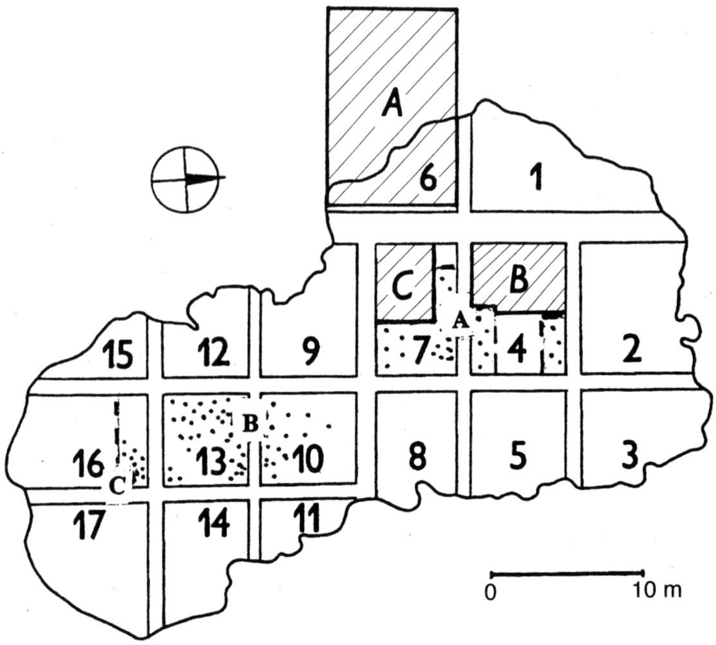

Figure 5.16 Stránská skála III site plan. Sectors and concentrations defined by Svoboda for the III-2 subsite. Hatched lines represent the III-1 subsite and concentrations excavated and defined by Valoch (from Svoboda and Valoch 2003:7, fig. 1.6, used by permission).

locality as Valoch's assemblage was not published at the time of the data collection presented here. Unlike locality IIIa described above, no Aurignacian industry was discovered in the upper of the two soil horizons. Only stratigraphic layer 5, soil 1, produced stone tools (see Figure 5.8). This was the first assemblage recognized as affiliated with Brno-Bohunice, the type-site for the industry which has become known as the Bohunician (Oliva 1984). The interest in this locality resulted in the discovery of the IIIa locality several years later. As with Stránská skála IIIa Layer 4, most of the artifacts in Stránská skála III layer 5 were made on the local hornstone. Significantly more of the retouched tool kit, however, is made on distant radiolarite materials, likely from western Slovakia (Svoboda 1987). The spatial distribution of the lithic finds at this locality formed three relatively distinct concentrations (A, B, and C), around hearth-like features, one of which produced radiocarbon dates of 38,200 ± 1,100 (GrN 12297) and 38,500 + 1,400/- 1200 (GrN 12298). Pieces of ochre were found in association with the hearths. Pollen analysis concluded the presence of a steppe landscape of *Pinus*, *Betula*, *Picea*, and *Alnus* species. After the 1997-1999 excavations conducted by Svoboda and Bar-Yosef (2003), the III locality was confirmed to extend further to the south, as encountered in the IIId locality. The IIId locality is, however, too small for the present technological study (Tostevin 2003b).

5.3.2 *Sampling*

The Bohunician assemblage from Stránská skála III, layer 5, is very rich in debitage but contains only a moderate number of retouched tools. For this reason, only one of the three main concentrations defined by Svoboda (1987) is necessary to provide an ample sample of debitage and cores for inter-assemblage comparative purposes. Concentration A of Svoboda's excavation area,

Table 5.23 Stránská skála III, percentage of intact cores (n=37) by number of bulbar negatives in each core sector (A, B, C, D) by surface (1, 2)

Number of bulbar negatives	Surf. 1 Sect. A	Surf. 1 Sect. B	Surf. 1 Sect. C	Surf. 1 Sect. D	Surf. 2 Sect. A	Surf. 2 Sect. B	Surf. 2 Sect. C	Surf. 2 Sect. D
0	5	49	35	41	46	78	57	76
1	35	35	32	32	38	22	24	16
2	43	14	27	16	8		8	3
3	14	3	5	11	8		8	5
4	3						3	

Table 5.24 Stránská skála III core sector preparation: percent of intact cores (n=37) by number of scars in sector of origin (A, B, C, D) by surface (1, 2)

Number of Scars	Surf. 1 Sect. A	Surf. 1 Sect. B	Surf. 1 Sect. C	Surf. 1 Sect. D	Surf. 2 Sect. A	Surf. 2 Sect. B	Surf. 2 Sect. C	Surf. 2 Sect. D
0	8	86	32	73	27	65	30	70
1	19	5	24	11	8	14	19	16
2	41	5	22	5	19	8	24	8
3	14	3	11	5	30	11	19	0
4	14		11	5	14	3	8	0
5	3				0			5
6	0				0			
7	3				3			

which includes Sectors 4, 5, and 7, was chosen for inclusion. To sample enough tools, however, all of the tools from all three of Svoboda's concentrations were studied. The extensive refittings conducted by Škrdla (1994, 1996, 2003c; Svoboda and Škrdla 1995) were used to corroborate the attribute analysis of the debitage and cores.

5.3.3 Blank Production

Core Modification

Core modification in Stránská skála III Layer 5 begins with a longitudinal blade/flake core orientation, as evidenced in the identification of the narrow core-face as the debitage surface (Appendix Figures A16-A17). This identification is supported by Table 5.23, which demonstrates that Surface 1 strongly outweighs Surface 2 in number of bulbar negatives, indicating a preference for exploiting one surface rather than two, at least as preserved in the final form of the cores. The longitudinal core orientation visible in the choice of Surface 1 is also corroborated by Table 5.24, which displays the dominance of the working of core sectors A and C on both core surfaces, with the latter surface serving as a platform for the other surface.

Core convexity management in Stránská skála III is achieved in similar ways to those in IIIa level 4, initially through the use of the *lame à crête* technique to shape the longitudinal convexity of the cores into a central ridge, as in refit sequence number S01/94 (Škrdla 2003c:123-124). Subsequent core convexity management utilized lateral débordant removals, as in refit sequence S02/94 (Figure 5.11). In particular instances, after the surficial removals from the flat surface of the core, late removal of blades proceeded from the narrow side of the core, situated 90 degrees to the previous surface when viewed from one of the narrow ends of the core. This is documented by refit sequence number S02/93 (Škrdla 2003c:127-128). Individual cores also show the

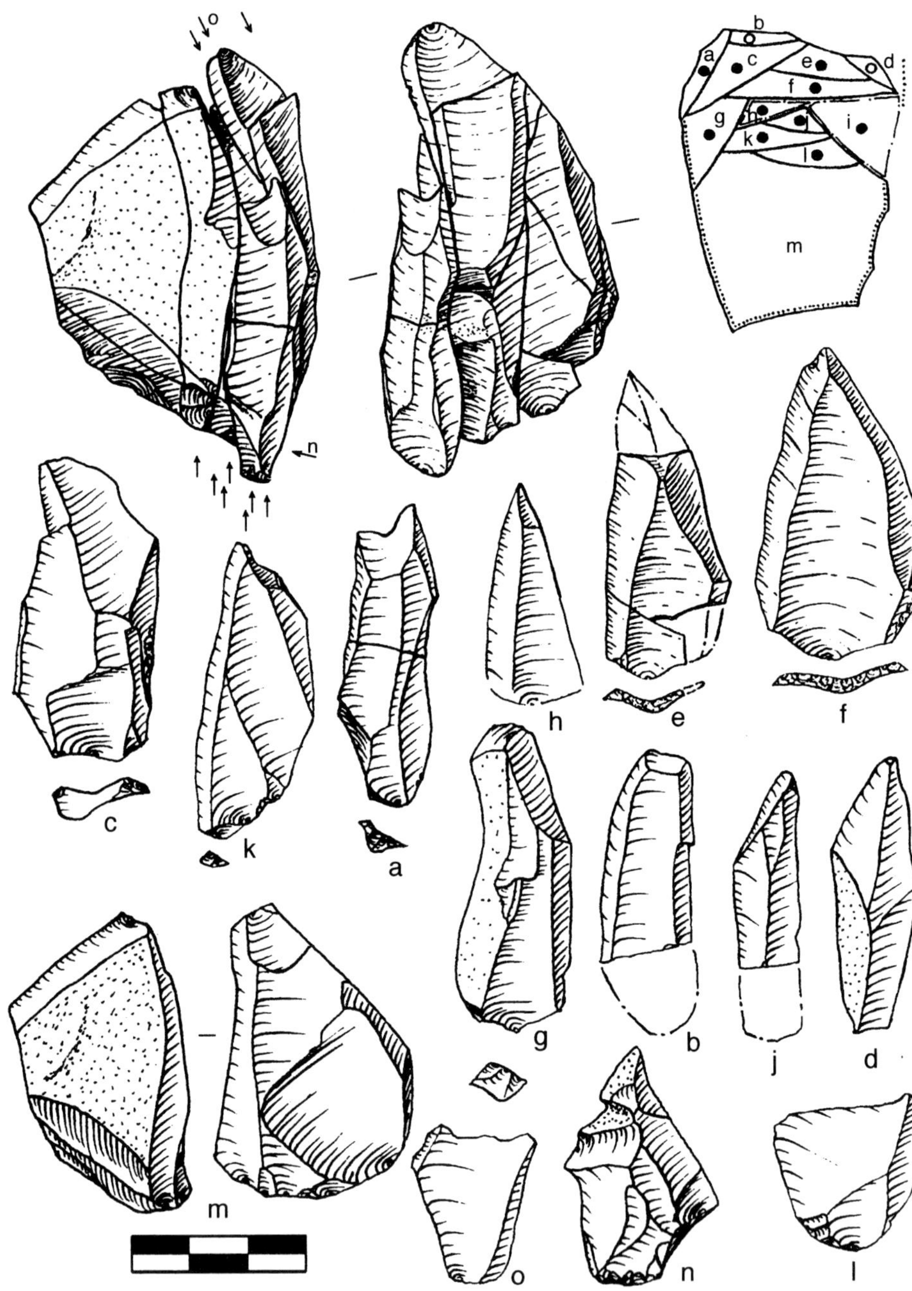

Figure 5.17 Refit core sequence (No. S02/94, top row) and products (bottom rows) from Stránská skála III Layer 5. Scale = 3cm. Filled dots indicate a removal from the lower platform while an empty dot indicates a removal from the top platform. Drawing by J. A. Svoboda. Refit by P. Škrdla (from Škrdla 2003c:126, fig. 9.6, used with permission).

Table 5.25 Stránská skála III, platform types for complete and proximal fragments of flakes and tools

Platform type	Frequency	Percent
Plain	159	37.8
Cortical	26	6.2
Dihedral	3	.7
Facetted	101	24.0
Chapeau	8	1.9
Scarred	55	13.1
Punctiform	4	1.0
Linear	11	2.6
Removed	1	.2
Indeterminate	53	12.6
Total	421	100.0

Table 5.26 Stránská skála III, descriptive statistics for continuous platform variables of complete and proximal fragments of flakes and tools

	N	Minimum	Maximum	Mean	Std. Deviation
Exterior Platform Angle (degrees)	340	51.00	137.00	86.36	14.04
Platform Thickness (mm)	344	0.64	12.66	4.42	2.16

tendency for narrow face reduction without prior Levallois or surficial working of the adjacent face, namely refit sequence numbers 57/82 and 66/82 (Škrdla 2003c:127-128). Both the débordant and side blade removals undoubtedly help reshape the lateral convexities of the cores in order to prolong exploitation for blanks.

Platform Maintenance

Platform maintenance includes a moderate amount of platform preparation, up to 45 % of the debitage when indeterminates are removed from the percentage (Table 5.25). Measurements of external platform angles and platform thickness (Table 5.26) indicate tendencies around 86 degrees and 4 mm respectively.

Direction of Core Exploitation

The analysis of the direction of core exploitation is presented in Tables 5.27-5.28. Figure 5.19 demonstrates that the dominant directional strategy during early core exploitation in the Stránská skála III level 5 assemblage, i.e., the most represented dorsal scar pattern within the longest length quartile, is bidirectional. Bidirectional reductions are seen in refit sequences such as S01/94 and S02/94 (Škrdla 2003c:123-124, 126-127). Conversely, the dorsal scar pattern of the majority of pieces within the smallest length quartile of the blanks is a unidirectional strategy. Taken together with the roughly continuous pattern within the two middle quartiles, the length-dorsal scar pattern cross-tabulation demonstrates a strong change from bidirectional reduction at the beginning of core exploitation to unidirectional reduction at the end of core exploitation.

The refit data can provide some added context to this result. Refit sequence number S03/93 (Škrdla 2003c:128-129) contains a core and debitage which was reduced from two opposed platforms. The surfaces exploited from each platform, however, were tilted relative to

Table 5.27 Stránská skála III, dorsal scar pattern by blank length (mm) in quartiles cross-tabulation for complete flakes and tools (excluding flakes with greater than 60% cortex or fewer than three dorsal scars)

Dorsal Scar Pattern	>44	44-34	34-27	<27	Total
Unidirectional					
Count	17	25	39	36	117
% within Quartile	20.2%	31.3%	51.3%	46.2%	36.8%
Bidirectional					
Count	28	23	16	12	79
% within Quartile	33.3%	28.8%	21.1%	15.4%	24.8%
Crossed					
Count	13	14	12	18	57
% within Quartile	15.5%	17.5%	15.8%	23.1%	17.9%
Subcentripetal & Centripetal					
Count	12	8	3	4	27
% within Quartile	14.3%	10.0%	3.9%	5.1%	8.5%
Crested					
Count	12	7	5	6	30
% within Quartile	14.3%	8.8%	6.6%	7.7%	9.4%
Indeterminate					
Count	2	3	1	2	8
% within Quartile	2.4%	3.8%	1.3%	2.6%	2.5%
Total					
Count	84	80	76	78	318
% within Cortex	100.0%	100.0%	100.0%	100.0%	100.0%

Table 5.28 Stránská skála III, dorsal scar pattern by blank length (mm) in quartiles cross-tabulation for complete flakes and tools (excluding pieces with a crested scar pattern, an indeterminate scar pattern, greater than 60% cortex, or fewer than three dorsal scars)

Dorsal Scar Pattern	>44	44-34	34-27	<27	Total
Unidirectional					
Count	17	25	39	36	117
% within Quartile	24.3%	35.7%	55.7%	51.4%	41.8%
Bidirectional					
Count	28	23	16	12	79
% within Quartile	40.0%	32.9%	22.9%	17.1%	28.2%
Crossed					
Count	13	14	12	18	57
% within Quartile	18.6%	20.0%	17.1%	25.7%	20.4%
Subcentripetal & Centripetal					
Count	12	8	3	4	27
% within Quartile	17.1%	11.4%	4.3%	5.7%	9.6%
Total					
Count	70	70	70	70	280
% within Cortex	100.0%	100.0%	100.0%	100.0%	100.0%

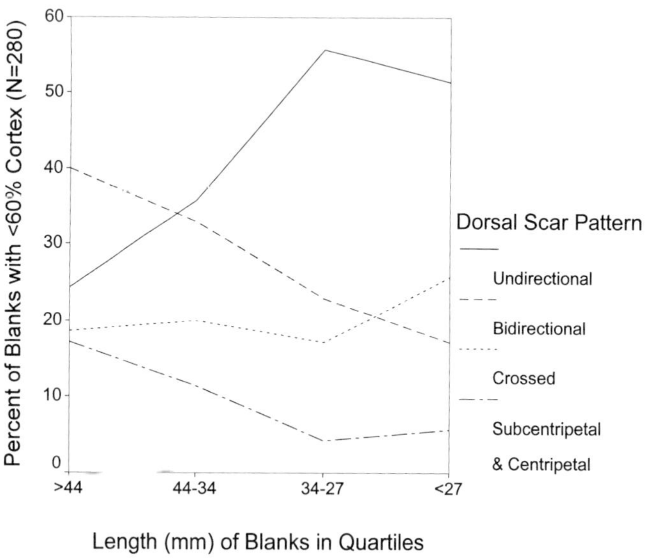

Figure 5.18 Stránská skála III, dorsal scar pattern by blank length (mm) in quartiles cross-tabulation for complete flakes and tools (excluding pieces with a crested scar pattern, an indeterminate scar pattern, greater than 60% cortex, or fewer than three dorsal scars).

each other so that the mutual debitage surface assumed an inverted "v" rather than a flat profile. This tilt increased during the course of reduction. As the tilt increased, the blanks began to fail to intersect the dorsal scars from the opposed platform. As a result, for the later and consequently shorter products, the dorsal scar pattern preserved on their dorsal surfaces was unidirectional rather than bidirectional. This refit sequence evidences one scenario by which the above length-dorsal scar pattern cross-tabulation can be produced. As few of the core remnants display such a two-angle debitage surface (i.e., they are usually quite flat), this is only one and not likely the dominant scenario for the production of this length-dorsal scar cross-tabulation. Strictly unidirectional refit sequences are known from this assemblage, as well. Actually a pair of sequences argued to derive from the same halved cortical nodule, sequences S04A/93 and S04B/93 (Škrdla 2003c:124–125) evidence unidirectional reduction from the platform of the large surface bifurcating the round nodule. This exploitation resulted in relatively acute exterior platform angles with extremely large platform thicknesses and very short blanks. These products do not match the majority of the debitage in terms of platform choices and as the blanks are highly cortical (and so not included in the length-dorsal scar pattern cross-tabulation if above 60% cortex), it is unlikely that reduction tendencies such as this produced the particular length-dorsal scar pattern cross-tabulation noted above.

Core refit sequences such as S02/94 (Škrdla 2003c:126–127, fig. 9.6) suggest that a continuous trend toward unidirectional removals late in the exploitation of bidirectional opposed cores is more likely the source for this cross-tabulation pattern. This refit sequence demonstrates bidirectional reduction from opposed platforms which produced long blanks with bidirectional dorsal scar patterns, the continued platform preparation between core rotations that progressively shortened the debitage surface, and unidirectional blanks removed late in the sequence (removals h, j, and l in Škrdla's fig. 9.6).

Table 5.29 Stránská skála III, frequencies of lateral edge, profile, and cross-section types for complete flakes and tools

Type	Frequency	Percent
Lateral edge		
Parallel	246	58.4
Convergent	74	17.6
Expanding	41	9.7
Ovoid	56	13.3
Indeterminate	4	1.0
Total	421	100.0
Profile		
Straight	197	46.8
Curved or Concord	123	29.2
Twisted	97	23.0
Indeterminate	4	1.0
Total	421	100.0
Cross-section		
Triangular	164	39.0
Trapezoidal	236	56.1
Other	20	4.8
Indeterminate	1	.2
Total	421	100.0

Table 5.30 Stránská skála III, descriptive statistics for blank dimensions for complete flakes and tools

Blank Dimension	N	Range	Mean	Standard Deviation
Length	421	84.13	36.47	12.36
Width	421	52.13	21.68	7.82
Thickness	421	18.52	6.24	3.05
Length/Width Ratio	421	5.63	1.83	.73
Width/Thickness Ratio	421	12.17	3.97	1.72

Corroborating the dorsal scar pattern and refit data, the core sector preparation and bulbar scar negatives (Figures 5.23–5.24) of the cores in this assemblage clearly demonstrate the extensive use of sectors A and C over the other sectors on both Surface 1 and Surface 2, the latter serving as the platform surface. As the cores themselves represent late exploitation, the relatively higher number of bulbar negatives and scars originating from sector A on Surface 1 compared to sector C, far beyond the definitional requirement to be considered sector "A", also indicates the tendency towards unidirectional reduction at the end of core exploitation.

Dorsal Surface Convexity

The analysis of the attributes serving as the proxies for the dorsal surface convexity domain are given in Tables 5.29–5.30. The ridge system used in the Stránská skála III layer 5 assemblage shows a propensity for only moderate laminarity (with a 1.83 mean, below the blade cut-off of 2.0), very thin/surficial products (width/thickness mean of 3.97) and longitudinal ridges (i.e., high frequency of parallel lateral edges).

5.3.4 *Tool Kit Morphology*

The first five characterizations of the tool kit morphology refer to the dorsal surface convexity variables which give form to the cutting edge of

Table 5.31 Stránská skála III, descriptive statistics for dimensions of complete tools

Blank Dimension	N	Range	Mean	Standard Deviation
Length/Width Ratio	30	3.41	2.02	0.91
Width/Thickness Ratio	30	5.73	4.29	1.57

Table 5.32 Stránská skála III, frequencies of lateral edge, distal terminus, and profile types for complete tools

Type	Frequency	Percent
Lateral edge		
Parallel	8	26.7
Convergent	15	50.0
Expanding	3	10.0
Ovoid	2	6.7
Indeterminate	2	6.7
Total	30	100.0
Distal Terminus		
Blunt	5	5.0
Pointed	19	63.3
Removed	6	20.0
Total	30	100.0
Profile		
Straight	15	50.0
Curved or Concord	8	26.7
Twisted	7	23.3
Total	30	100.0

the retouched blanks in the assemblage, as well as the diagnostic flakes without retouch traditionally categorized as tools, i.e., the typological Levallois pieces not infrequently found in this assemblage. As this assemblage includes more retouched points, both Levallois and non-Levallois, than the Bohunician assemblage from Stránská skála IIIa layer 4 above, the tool kit possesses a higher frequency of convergent lateral edges than the debitage as a whole, 53.6% for the tools compared to 17.7% when the indeterminates are removed from the percentages. The tools are also more laminar, actually surpassing the blade cut-off with a mean of 2.02, and more surficial in their width/thickness ratio (4.29 vs. 3.97).

The last two characterizations of the tool kit morphology refer to two variables describing the application of retouch to blanks: the application of distinctive forms of retouch and the relative dominance of Middle Paleolithic or Upper Paleolithic retouched tool types. It must be pointed out that two atypical carinated end-scrapers are reported for this assemblage (Svoboda 2003b:155, tab. 10.1). Given the atypical form and small representation within the assemblage, carinated retouch is judged not to represent a credible unique or diagnostic retouch for this assemblage. Thus, no unique type of retouch is accounted in the tool types in this assemblage and Upper Paleolithic tool types, primarily endscrapers, dominate the tool kit (Svoboda 1987, 1991, 2003b).

Intensity of Tool and Core Reduction

Given the situation that Stránská skála III is

Table 5.33 Stránská skála III, intensity of core and tool reduction

Variables	Numerator	Denominator	Result
Tools/Flake Ratio	Complete Tools: 30	Complete Flakes: 368	0.08
Non-cortical Debitage/ Cortical Debitage Ratio	Non-cortical Complete Debitage: 216	Cortical Complete Debitage: 182	1.19
Henry's Dimensional Analysis	NA	NA	Core facets overlap with the middle of the blank distribution: little core exhaustion

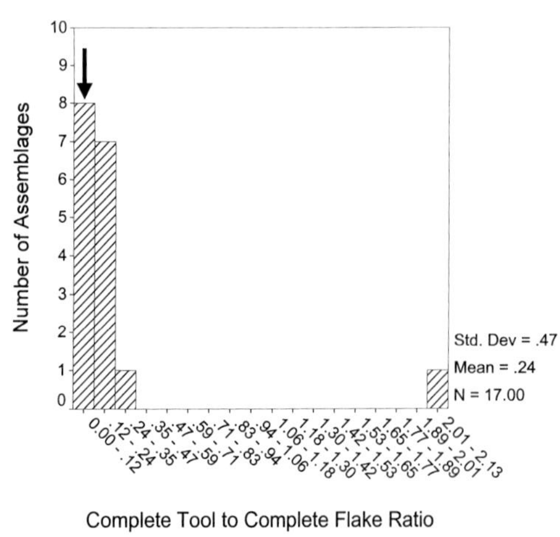

Figure 5.19 Stránská skála III, tool/flake ratio value relative to a histogram of all assemblages.

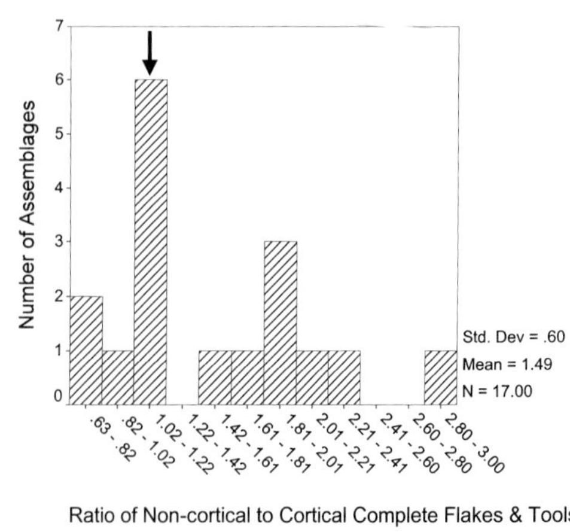

Figure 5.20 Stránská skála III, non-cortical/cortical flake ratio value relative to a histogram of all assemblages.

immediately adjacent to a raw material source, it is not surprising that retouch in this assemblage was applied to blanks less frequently than most other assemblages. The fact that the sampling of this assemblage incorporated all retouched tools from the excavation of the SSIII-2 subsite but only the debitage from one major concentration makes it even more telling that the sample demonstrated one of the lowest tool/flake ratios of all assemblages in this study, when such a sampling strategy theoretically could have biased the ratio significantly in the other direction (Table 5.33 and Figure 5.19). The non-cortical to cortical flake ratio, however, is quite similar to SSIIIa-4 above, possessing a frequency of cortical blanks indicative of significant working of cortical nodules on site (Figure 5.20). In addition to these indications of a workshop close to a raw material source, the III layer 5 assemblage shows no signs of core exhaustion, as the vast majority of the last maximum facets on the cores overlap with the middle of the blanks' length and width distribution (Figure 5.21). Few cores show facets that fall below the 2.0 cm length cut off for the debitage analysis (10 out of 48) and three of these show widths greater than lengths, indicating hinged removals which potentially ended exploitation

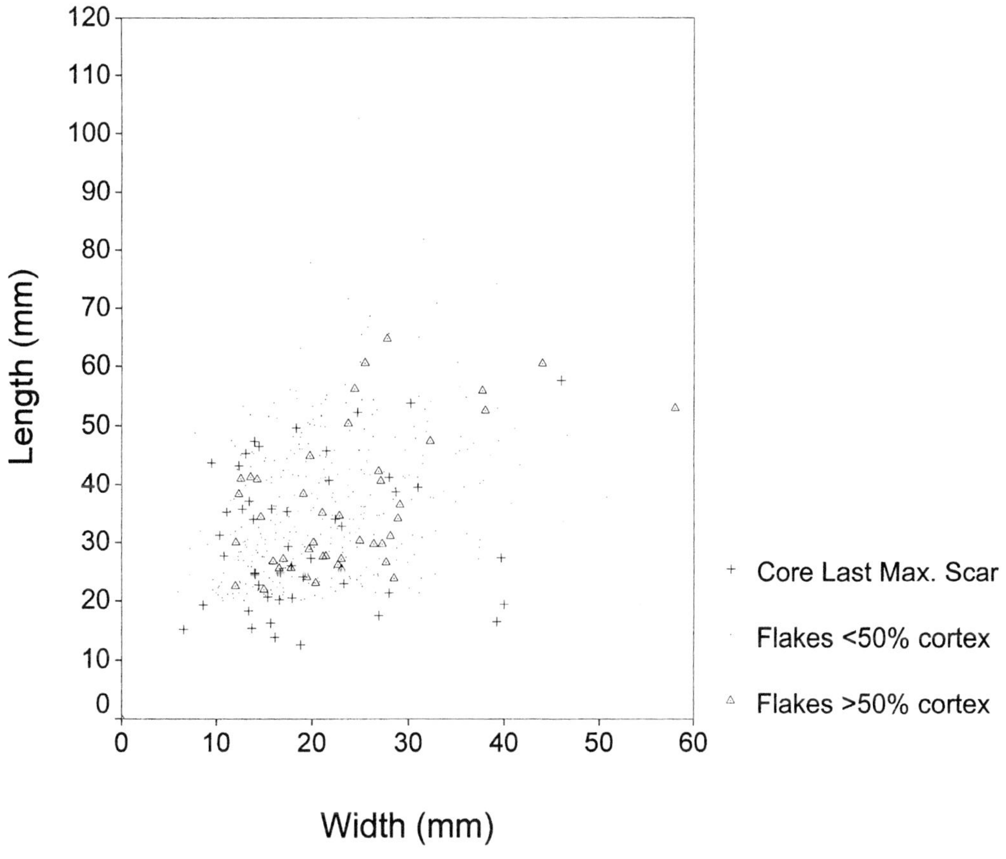

Figure 5.21 Stránská skála III according to Henry's core reduction intensity measure: length and width of last maximum core facets, highly cortical blanks, and blanks with little cortex.

prematurely. Therefore, this assemblage as a whole is extremely lightly reduced in both its tools and its cores. Given the presence of more retouched tools made on distant raw materials in this assemblage, compared to SSIIIa layer 4, (such as radiolarite from as much as 100 km away, either in western Slovakia or northern Austria, as well as less distant Krumlovsky-lès chert from as much as 30 km to the southwest), it is significant that these highly curated tools are not matched by any indication of other economization of raw materials. On the whole, the early to middle portion of the core reduction sequence is documented in this assemblage.

5.4 Vedrovice V

5.4.1 Site Background

One of many open-air Paleolithic sites located in the currently forested "Krumlovský-les" highland southwest of Brno, this assemblage from locality V is one of the most important Szeletian collections in Central Europe, as it is one of only a few stratified collections in the region excavated with reliable methods. Most Szeletian assemblages

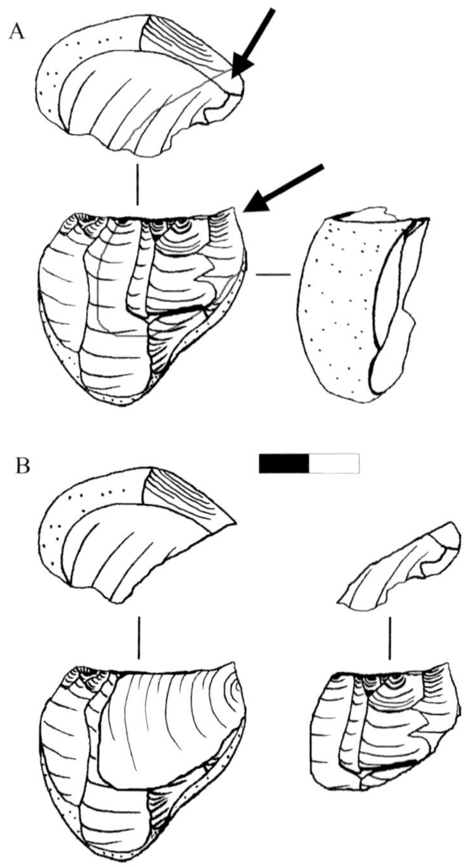

Figure 5.22 A hypothetical schematic/interpretive conclusion of Vedrovice V lateral core tablet technique: A) hypothetical example of a unidirectional core with a lateral core table predicted to remove the proximal half of the debitage surface. Scale 2 cm. Arrow indicates direction of removal; B) by-products of the lateral core tablet technique.

in the region are in fact surface collections from strata churned up by agricultural activities. As of 1997, the only other well excavated Szeletian collection in the Middle Danube region was that of Čertova pec Cave in Slovakia (Barta 1959, 1967, 1969, 1980; Vogel and Waterbolk 1964) but as it contains too few stone tools, it was not chosen for this technological study. This uniqueness of Vedrovice V has recently changed as B. Adams and A. Ringer (2004) have reexcavated and redated sections of Szeleta Cave in Hungary, the type-site for the Szeletian Industry, providing new assemblages for analysis. Along with these Hungarian assemblages, the site of Dzierżysław I, southern Poland, also now offers more of an opportunity to examine the details of Szeletian flintknapping behavior from defined stratigraphic contexts (Foltyn and Kozłowski 2003; Bluszcz et al. 1994). These assemblages now available for study will hopefully allow the future integration of data from multiple localities to help improve our understanding of the archaeological phenomena labeled the Szeletian.

Excavated by Karel Valoch (1984, 1993) in the early 1980s, the single occupation level within the last interpleniglacial paleosol at Vedrovice V produced a large number of stone artifacts, sufficient enough for technological studies of both debitage and tools. Unfortunately, few bones were preserved in the loessic sedimentary matrix of Vedrovice V. A series of radiocarbon samples from charcoal scatters (no features were defined during excavation) produced a range of conventional radiocarbon dates from 39,000 to 35,500 BP (Table 2.1). Adams and Ringer (2004:544) have argued that the series of radiocarbon dates from the occupation indicate that "Vedrovice V is conceivably contemporary with the Central European Middle Palaeolithic Micoquian, the Upper Palaeolithic Gravettian, or the "Epiaurignacian." This argument is based on undue consideration of charcoal samples with known problems for conventional dating, specifically, GrN 17261 (30,170 ±300 BP) with noted root contamination and GrN 19105 (>39,500 BP) and GrN 19106 (47,250 +3, 700/-2,500 BP) which contained insufficiently low carbon levels (Valoch 1993:78). Due to the problems with these three samples, only the dates

Table 5.34 Vedrovice V, percentage of intact cores (n=34) by number of bulbar negatives in each core sector (A, B, C, D) by surface (1, 2)

Number of bulbar negatives	Surf. 1 Sect. A	Surf. 1 Sect. B	Surf. 1 Sect. C	Surf. 1 Sect. D	Surf. 2 Sect. A	Surf. 2 Sect. B	Surf 2 Sect. C	Surf. 2 Sect. D
0	3	53	74	38	76	68	79	76
1	21	26	9	41	21	21	6	18
2	35	12	15	12	3	6	9	3
3	32	0	0	6		6	6	0
4	6	9	3	3				3
5	0							
6	0							
7	0							
8	3							

Table 5.35 Vedrovice V, core sector preparation: percent of intact cores (n=34) by number of scars in sector of origin (A, B, C, D) by surface (1, 2)

Number of bulbar negatives	Surf. 1 Sect. A	Surf. 1 Sect. B	Surf. 1 Sect. C	Surf. 1 Sect. D	Surf. 2 Sect. A	Surf. 2 Sect. B	Surf 2 Sect. C	Surf. 2 Sect. D
0	6	79	68	71	41	59	56	62
1	12	12	6	15	29	15	15	26
2	18	3	21	3	12	12	12	0
3	29	0	3	12	18	9	12	6
4	18	0	3			6	6	3
5	12	3						0
6	6	3						3

Table 5.36 Vedrovice V, platform types for complete and proximal fragments of flakes and tools

Platform Type	Frequency	Percent
Plain	167	30.8
Cortical	38	7.0
Dihedral	1	0.2
Facetted	60	11.1
Scarred	89	16.4
Punctiform	7	1.3
Linear	18	3.3
Removed	4	0.7
Indeterminate	158	29.2
Total	542	100.0

Table 5.37 Vedrovice V, descriptive statistics for continuous platform variables of complete and proximal fragments of flakes and tools

	N	Minimum	Maximum	Mean	Standard Deviation
Exterior Platform Angle (degrees)	342	40.00	150.00	89.03	17.36
Platform Thickness (mm)	359	0.31	27.22	5.01	3.64

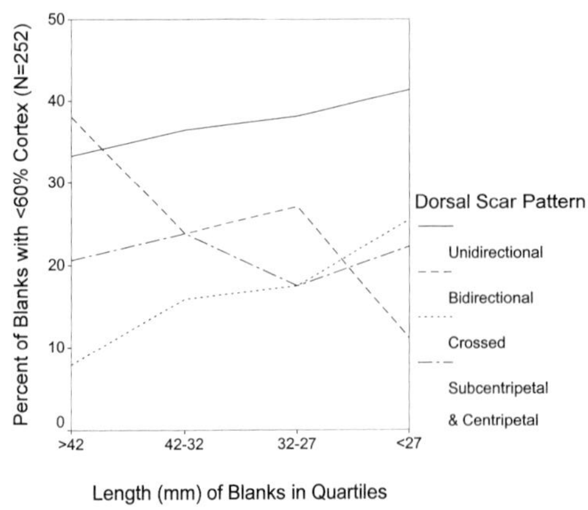

Figure 5.23 Vedrovice V, dorsal scar pattern by blank length (mm) in quartiles cross-tabulation for complete flakes and tools (excluding pieces with a crested scar pattern, an indeterminate scar pattern, greater than 60% cortex, or fewer than 3 dorsal scars).

presented in Table 2.1 are considered likely to represent the chronological position of this assemblage. Nevertheless, however, Adams and Ringer are correct that greater chronological resolution is desperately needed for all assemblages such as Vedrovice V which lack AMS dates as well as TL or ESR dates unaffected by the potential effect of the "Middle Paleolithic Dating Anomally" (Conard and Bolus 2003).

Almost all artifacts at Vedrovice V were made on the local Krumlovský les type raw material, a Jurassic chert in the form of very rolled river nodules with black cortex and bluish tinged interiors (as defined by A. Přichystal). A few pieces of radiolarite were also found.

5.4.2 Sampling

Although Vedrovice V was excavated using a square meter grid, the artifacts were not curated according to square. The normal sampling strategy of randomly selecting square meters for inclusion in the study was thus inappropriate. Consequently, the whole assemblage was studied without any sampling.

5.4.3 Blank Production

Core Modification

Core modification in Vedrovice V begins with a longitudinal blade/flake core orientation, as evidenced in the identification of the long core-face as the debitage surface (Appendix, Figures A18–A22). This identification is supported by Table 5.34, which demonstrates that Surface 1 strongly outweighs Surface 2 in number of bulbar negatives, indicating a preference for exploiting one surface rather than two at least at the end of core reduction. The longitudinal core orientation visible in the choice of Surface 1 is also corroborated by Table 5.35, which displays the dominance of the working of core sector A on Surface 1. Table 5.35 also demonstrates that, even though it was worked less than Surface 1, Surface 2 was worked evenly around all four sectors compared to the almost exclusive use of sector A on Surface 1. This is further confirmation that only one surface, Surface 1, served as the dominant debitage surface.

Core convexity management, as evidenced through both the core morphology and the attribute analysis of the debitage, was achieved through periodic changes of core orientation from one platform cores to crossed platform cores, producing lateral core tablets, and refreshing the convexity by using the original platform as a rejuvenating crested flake. As most of the cores in this assemblage were made on the small river cobbles of Krumlovský les type, it was easy for the knapper to use the cortical lateral edge of the core as a platform to strike off the junction of the old platform surface and the debitage surface. This interpretation is supported by the indications in Table 5.35

Table 5.38 Vedrovice V, dorsal scar pattern by blank length (mm) in quartiles cross-tabulation for complete flakes and tools (excluding flakes with greater than 60% cortex or fewer than 3 dorsal scars)

Dorsal Scar Pattern	>42	42-32	32-27	<27	Total
Unidirectional					
Count	21	23	24	26	94
% within Quartile	26.6%	27.7%	27.9%	31.3%	28.4%
Bidirectional					
Count	24	15	17	7	63
% within Quartile	30.4%	18.1%	19.8%	8.4%	19.0%
Crossed					
Count	5	10	11	16	42
% within Quartile	6.3%	12.0%	12.8%	19.3%	12.7%
Subcentripetal & Centripetal					
Count	13	15	11	14	53
% within Quartile	16.5%	18.1%	12.8%	16.9%	16.0%
Crested					
Count	8	10	7	7	32
% within Quartile	10.1%	12.0%	8.1%	8.4%	9.7%
Indeterminate					
Count	8	10	16	13	47
% within Quartile	10.1%	12.0%	18.6%	15.7%	14.2%
Total					
Count	79	83	86	83	331
% within Cortex	100.0%	100.0%	100.0%	100.0%	100.0%

Table 5.39 Vedrovice V, dorsal scar pattern by blank length (mm) in quartiles cross-tabulation for complete flakes and tools (excluding pieces with a crested scar pattern, an indeterminate scar pattern, greater than 60% cortex, or fewer than 3 dorsal scars)

Dorsal Scar Pattern	>42	42-32	32-27	<27	Total
Unidirectional					
Count	21	23	24	26	94
% within Quartile	33.3%	36.5%	38.1%	41.3%	37.3%
Bidirectional					
Count	24	15	17	7	63
% within Quartile	38.1%	23.8%	27.0%	11.1%	25.0%
Crossed					
Count	5	10	11	16	42
% within Quartile	7.9%	15.9%	17.5%	25.4%	16.7%
Subcentripetal & Centripetal					
Count	13	15	11	14	53
% within Quartile	20.6%	23.8%	17.5%	22.2%	21.0%
Total					
Count	63	63	63	63	252
% within Cortex	100.0%	100.0%	100.0%	100.0%	100.0%

Table 5.40 Vedrovice V, frequencies of lateral edge, profile, and cross-section types for complete flakes and tools

Type	Frequency	Percent
Lateral edge		
Parallel	176	32.5
Convergent	72	13.3
Expanding	141	26.0
Ovoid	62	11.4
Indeterminate	91	16.8
Total	542	100.0
Profile		
Straight	306	56.5
Curved or Concord	72	13.3
Twisted	87	16.1
Indeterminate	77	14.2
Total	542	100.0
Cross-section		
Triangular	258	47.6
Trapezoidal	135	24.9
Other	100	18.5
Indeterminate	49	9.0
Total	542	100.0

Table 5.41 Vedrovice V, descriptive statistics for blank dimensions for complete flakes and tools

Blank Dimension	N	Range	Mean	Standard Deviation
Length	542	97.90	37.27	14.39
Width	542	71.47	27.29	10.50
Thickness	542	36.87	7.69	5.13
Length/Width Ratio	542	4.20	1.47	.57
Width/Thickness Ratio	542	11.51	4.26	1.82

that Surface 1 core sectors B and D, the lateral edges of the core, were worked to a lesser degree than sector A but much more than the bottom of the core, sector C. This shows the prevalence of the tendency to use the lateral edges of the cores for late removals despite the dominance of the one-platform orientation.

Platform Maintenance

Platform maintenance includes less platform preparation than the assemblages discussed above, with 39.5% of the platforms prepared, excluding indeterminates in the calculation of the percentage (Table 5.36). Measurements of external platform angles and platform thickness (Table 5.37) indicate tendencies around 89 degrees and 5 mm respectively.

Direction of Core Exploitation

The analysis of the direction of core exploitation is presented in Tables 5.38–5.39. Figure 5.23 demonstrates that the direction of early core exploitation exercised in Vedrovice V was a mixture of a bidirectional strategy and an only slightly less frequent unidirectional strategy. Moving from the largest to the smallest length quartile, the unidirectional strategy gradually becomes the dominant direction by late core exploitation. This is corroborated by the form of the cores at discard, with the significant use of sector A on Surface 1 compared to all other sectors and surfaces. At the same time as the unidirectional exploitation becomes dominant, the frequencies of crossed and subcentripetal/centripetal scar patterns also increase, a feature of the data which matches the identification of lateral core tablets and changes of core orientation on the discarded cores. The only refit sequences from Vedrovice V evidence the unidirectional testing of small Krumlovsky-lès nodules after the core tablet-like removal of the top of the nodule (Valoch 1993:52, fig. 31:2, 5, 7).

Dorsal Surface Convexity

The analysis of the attributes serving as the proxies for the dorsal surface convexity domain are given in Tables 5.40 and 5.41. The ridge system used in the Vedrovice V assemblage shows a propensity for the low laminarity of flake production (length/width mean of 1.47), very thin products (width/thickness mean of 4.26) and straight, single ridge cross-sections.

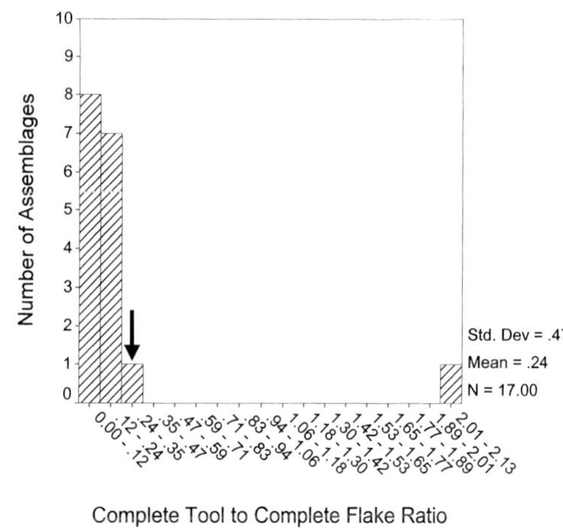

Figure 5.24 Vedrovice V, tool/flake ratio value relative to a histogram of all assemblages.

5.4.4 Tool Kit Morphology

The first five characterizations of the tool kit morphology refer to the dorsal surface convexity variables which give form to the cutting edge of the retouched blanks in the assemblage. Without a Levallois component, diagnostic flakes without retouch in this assemblage which are traditionally categorized as tools include only naturally backed knives. Bifacial foliate points are not included in the attribute analysis of the tool kit since, even if made on a flake blank, no technological features are preserved after the application of retouch. These pieces are, however, included in the last two characterizations of the tool kit morphology. The tool kit possesses a higher frequency of expanding lateral edges than the debitage as a whole, 45.9 % for the tools compared to 31.3 % when the indeterminates are removed from the percentages, with a concomitant decrease in ovoid and parallel lateral edges. Other categorical variables are similar to the debitage. Of the continuous variables, only the width/thickness of the tools appears to contrast with the debitage, but

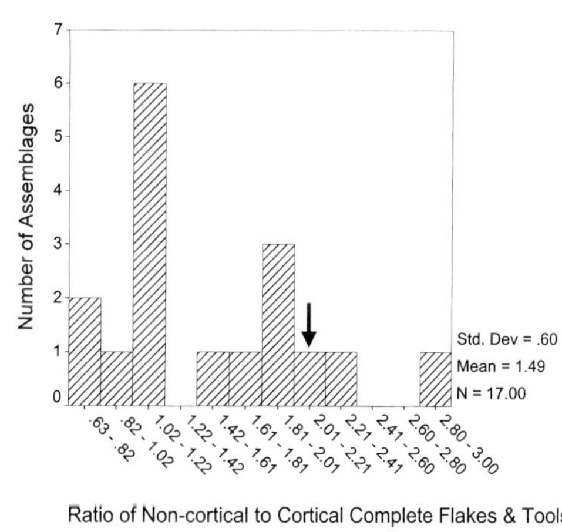

Figure 5.25 Vedrovice V, non-cortical/cortical flake ratio value relative to a histogram of all assemblages.

this may be a result of retouch on lateral edges decreasing width without concomitant decrease in thickness as much as blank selection.

For the last two characterizations of the tool kit morphology, the presence of flat bifacial retouch on the foliate points so famous in

Table 5.42 Vedrovice V, descriptive statistics for dimensions of complete tools

Blank Dimension	N	Range	Mean	Standard Deviation
Length/Width Ratio	115	2.64	1.51	0.52
Width/Thickness Ratio	115	7.03	3.59	1.56

Table 5.43 Vedrovice V, frequencies of lateral edge, distal terminus, and profile types for complete tools

Type	Frequency	Percent
Lateral edge		
Parallel	19	16.5
Convergent	7	6.1
Expanding	28	24.3
Ovoid	7	6.1
Indeterminate	54	47.0
Total	115	100.0
Distal Terminus		
Blunt	29	25.2
Pointed	7	6.1
Hinged	2	1.7
Overshot	2	1.7
Removed	1	.9
Broken	13	11.3
Indeterminate	61	53.0
Total	115	100.0
Profile		
Straight	48	41.7
Curved or Concord	10	8.7
Twisted	17	14.8
Indeterminate	40	34.8
Total	115	100.0

Table 5.44 Vedrovice V, intensity of core and tool reduction

Variables	Numerator	Denominator	Result
Tools/Flake Ratio	Complete Tools: 115	Complete Flakes: 358	0.32
Non-cortical Debitage/ Cortical Debitage Ratio	Non-cortical Complete Debitage: 317	Cortical Complete Debitage: 157	2.02
Henry's Dimensional Analysis	NA	NA	Core facets overlap with the low end of the blank distribution: moderate core exhaustion

Szeletian contexts represents the use of a unique type of retouch in Vedrovice V while Middle Paleolithic tool types dominate the tool kit (Valoch 1984).

Intensity of Tool and Core Reduction

Even though Vedrovice V is situated in the middle of a raw material procurement area (Svoboda 1983), that of the Krumlovsky-lès Jurassic chert, it is unknown if the raw material was procured immediately adjacent to the site due to changes in the landscape's topography since the Late Pleistocene. The assemblage has a tool/flake ratio of 0.32, falling above the mean for all of the

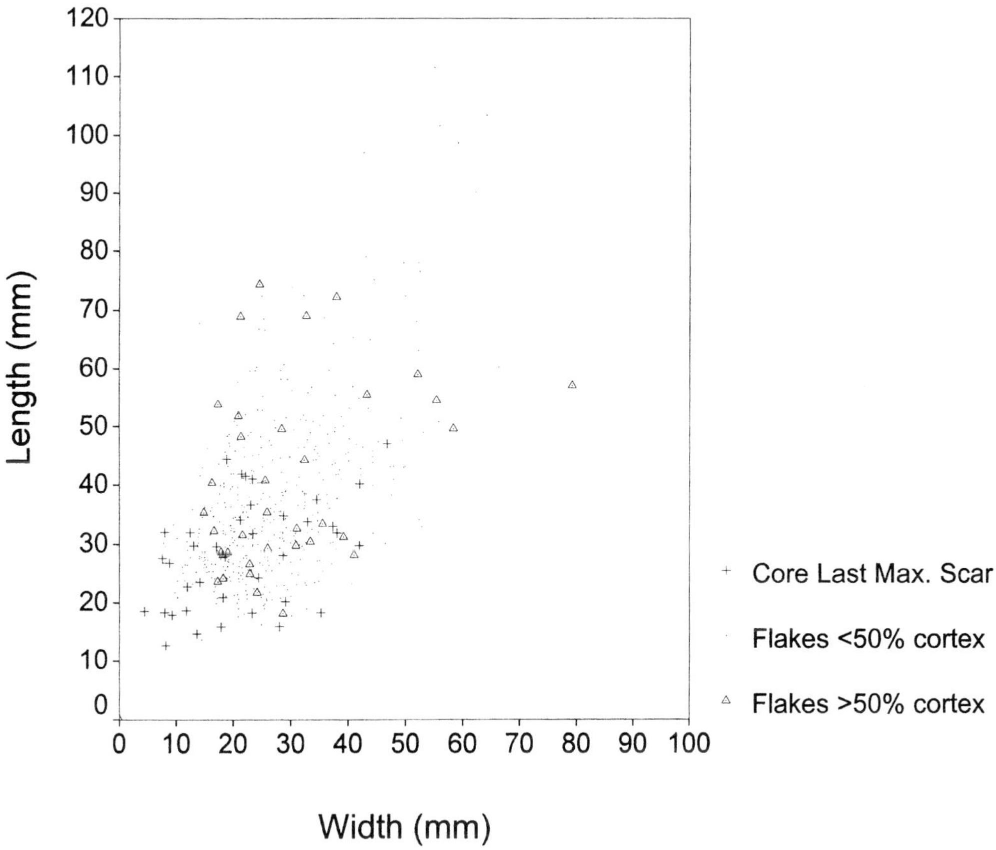

Figure 5.26 Vedrovice V according to Henry's core reduction intensity measure.

assemblages in this study, based on a distribution heavily weighted toward zero by a large number of lightly retouched assemblages (Table 5.44 and Figure 5.24). The frequency of retouch applied to blanks is thus indicative not of a raw material workshop but of an activity site in which the resharpening and shaping of blanks for use was more frequent than the previous two assemblages presented above. The non-cortical to cortical blank ratio (Figure 5.25) indicates that non-cortical blanks are twice as frequent as cortical ones, indicating some working of cortical nodules on-site but with a significant degree of reduction of the cores once the cortex was removed. This later conclusion derives from the fact that the cores' last maximum facets overlap with the lower end of the length and width distribution of the debitage (Figure 5.26). The fact that only seven out of 38 of the core facets fall below the 2.0 cm size limit for the debitage indicates that this assemblage is not utterly exhausted but significantly reduced, perhaps falling between Henry's two extreme examples, as presented in Figure 4.34. On the whole, the middle to late portions of the core reduction sequence are preserved in this assemblage, with less of the early

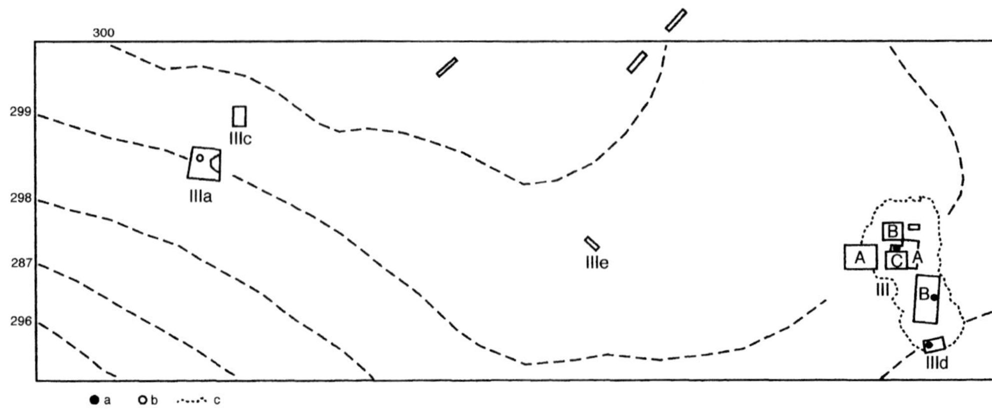

Figure 5.27 Detail of the localities on the Stránská skála III hillside, including IIIc. Key: a) Bohunician hearth; b) Aurignacian hearth; c) Neolithic and Chalcolithic features (from Svoboda and Bar-Yosef 2003: fig. 1.9, p. 12, used by permission).

raw material extraction behaviors likely contained in the assemblage.

5.5 Stránská skála IIIc Layer 5

5.5.1 Site Background

Locality IIIc of the Stránská skála hillside was excavated in 1997 and 1999 in an effort to locate a possible continuation of the dense occupational horizons uncovered at IIIa only a few meters to the south (Figure 5.27). This excavation was a joint venture between J. Svoboda of the Institute of Archaeology, Brno, and O. Bar-Yosef of the Department of Anthropology and Peabody Museum, Harvard University. This assemblage was published in Svoboda and Bar-Yosef (2003), including the typology (Svoboda 2003b), technological refitting (Škrdla 2003b; 2003c), spatial analysis (Monnier 2003), and attribute analysis (Tostevin 2003b). The technological study presented here reanalyzes the raw data which formed the basis for the conclusions presented in Tostevin (2003b), primarily the reworking of the direction of core exploitation domain and the addition of a new discussion of the tool kit morphology. Instances of redundancy with this previous publication will be noted in each circumstance.

The reasons for including this assemblage in the present study are three-fold. First, the mechanics of the present study differ from the previous publication enough to warrant its reworking here. Second, although two other assemblages ascribed to the Bohunician Industrial type have already been included in this work, the detailing of the blank production and tool kit morphology of a third example from a group of assemblages considered by many to possess a high degree of similarity allows an explicit testing of the principles and validity of the industrial type concept. This approach to multi-assemblage comparisons is somewhat analogous to that presented in Tostevin and Škrdla (2006) for new assemblage from Brno-Bohunice excavated in 2002.

Lastly, the chronological relationship of the IIIc assemblage to the other Bohunician assemblages is interesting. The AMS radiocarbon dates for this assemblage range between 38,300 ± 1,100 BP and 34,440 ± 720 BP, with an unweighted mean of 35.8 kya for the six dates listed in Figure 2.1. As one of the latest Bohunician assemblages despite its location within the same paleosol on the same hillside as IIIa layer 4 and III layer 5, its inclusion will allow the demonstration of any changes in technological organization within the Bohunician Industrial type through time, an aspect of cultural evolution usually obscured by the use of industrial affiliations in the analysis of assemblages.

In a review of the site monograph for this assemblage (Svoboda and Bar-Yosef 2003), Vishnyatsky (2006) comments on the original technological description by Tostevin (2003b): "it seems somewhat strange that the collection of Stránská skála IIIc is analyzed as if its homogeneity was beyond any doubt, whereas it is quite possible that it has an admixture of Aurignacian materials, as shown by G. Monnier [Monnier 2003] in Chapter 5 (p. 53)" (Vishnyatsky 2006:27). The possibility of an Aurignacian occupation in the vicinity of locus IIIc arose only as a result of the recognition of a clear Aurignacian occupation in the upper paleosol in locus IIIa, around 10 m to the south. In contrast to IIIa, locus IIIc however produced no diagnostic Aurignacian artifacts and only a few artifacts in the upper paleosol at all, any of which could have been moved there from the lower paleosol pedogenically (as Monnier suggested 2003:53). Thus there is no reason to assume any Aurignacian component at IIIc itself. As it is, however, there is no epistemologically valid means of characterizing an assemblage defined by geological association as "homogeneous" or not. As discussed at length by Tostevin and Škrdla (2006), an assemblage of artifacts in a paleosol context is merely "the material evidence of the portion of knapping behaviors captured and incorporated into the archaeological record at this particular locality during the length of the open palimpsest" (2006:38). Any other definition of the assemblage is riddled with unwarranted assumptions about etic abstractions such as industrial types. Thus, the assemblage from IIIc is treated as any other assemblage of artifacts.

5.5.2 *Sampling*

The Bohunician assemblage from Stránská skála IIIc is very rich in debitage but not as large as IIIa layer 4 or III. The whole assemblage was thus studied. G. Monnier participated in the attribute analysis of this assemblage. The refittings conducted by P. Škrdla (2003c) were used to corroborate the attribute analysis of the debitage and cores.

5.5.3 *Blank Production*

Core Modification

Core modification in Stránská skála IIIc begins with a longitudinal blade/flake core orientation, as evidenced in the identification of the narrow core-face as the debitage surface on extant cores (see Škrdla 2003b:69–71, figs. 7.11–7.16) and in the refit sequences, such as S22/00 (Škrdla 2003c:133–134). This identification is supported by Table 5.45, which demonstrates that Surface 1 strongly outweighs Surface 2 in number of bulbar negatives, indicating a preference for exploiting one surface rather than two. Note that this table presents bulbar scar negative data only by core surface, a simpler presentation than with the other assemblages in this study, but serving the same purpose. The longitudinal core orientation visible in the choice of Surface 1 is also corroborated by Table 5.46, which displays the dominance of the working of core sectors A and C on Surface 1 and of sector A on Surface 2, the latter sector presumably used as the platform surface.

The bulbar negative data on the extant cores is corroborated by refit sequences such as S25/00 (Škrdla 2003c:135, fig. 9.13). In this

Table 5.45 Stránská skála IIIc, percentage of intact cores (n=51) by number of bulbar negatives on each core surface (1, 2) (from Tostevin 2003b: 90, tab. 8.6)

Number of bulbar negatives	Core Surface 1	Core Surface 2
0	0	35
1	14	27
2	25	18
3	20	14
4	14	6
5	17	
6	2	
7	2	
8	4	
9	0	
10	2	

Table 5.46 Stránská skála IIIc, core sector preparation: percent of intact cores (n=51) by number of scars in sector of origin (A, B, C, D) by surface (1, 2) (from Tostevin 2003b: 90, tab. 8.7)

Number of Scars	Surf. 1 Sect. A	Surf. 1 Sect. B	Surf. 1 Sect. C	Surf. 1 Sect. D	Surf. 2 Sect. A	Surf. 2 Sect. B	Surf. 2 Sect. C	Surf. 2 Sect. D
0	0	78	49	84	43	76	84	80
1	14	18	8	6	16	14	4	10
2	10	0	12	4	25	4	6	4
3	14	4	12	2	8	4	4	2
4	19		7	4	6	2	0	2
5	23		2		0		2	2
6	10		2		2			
7	2		4					
8	2		2					
9	0		0					
>10	6		2					

sequence, a much-reduced bidirectional core is refitted to the 11 platform preparation and decortification flakes which set up the opposed platforms for bidirectional reduction. As reduction continued for many removals after the creation of these platforms, the bulbar negatives on Surface 2 are no longer extant, whereas the bulbar negatives on the reduced and shortened core are numerous.

Core convexity management is achieved initially through the use of cresting removals to shape the longitudinal convexity of the debitage surface into a centralized ridge for the removal of a *lame à crěte*. Refit sequences S20/00 (Škrdla 2003c:132-133), S24/00a and S24/00b (Škrdla 2003c:134-135), and S22/00 and S21/00 (Škrdla 2003c:133-134) all demonstrate the beginning of bidirectional reduction through the removal of a crested blade (see Figure 5.29). Lateral convexities were continually shifted to the center-line through the use of lateral débordant removals, struck at an angle between 45-70 degrees to the left or right of the flat debitage surface when viewed from one of the platforms.

Platform Maintenance

Platform maintenance includes a good amount of platform preparation, up to 42% of the debitage when indeterminates are removed from the percentage (Table 5.47). Measurements of external

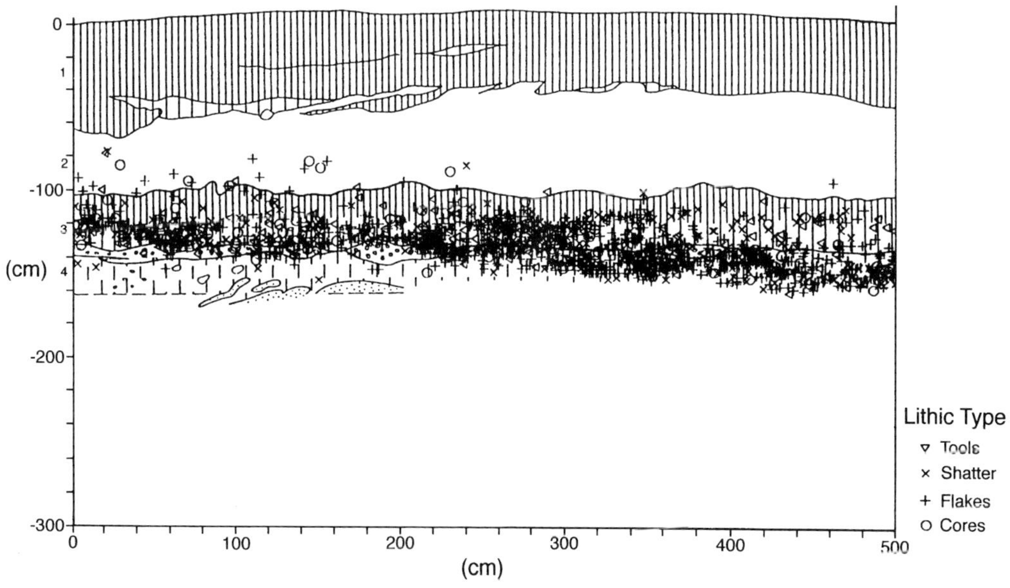

Figure 5.28 Stránská skála IIIc stratigraphic profile (from Svoboda 2003c: 19, fig. 2.1, used by permission).

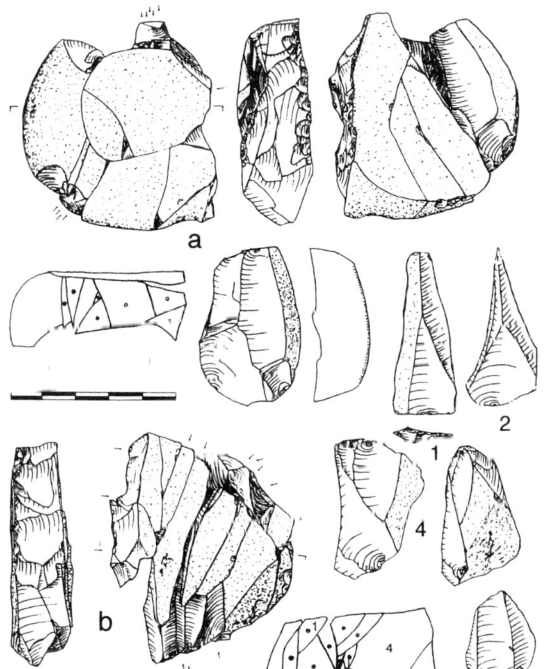

Figure 5.29 Refit core sequences from Stránská skála IIIc: a) No. S21/00; b) No S22/00. Scale = 3cm. Filled dots indicate a removal from the lower platform while an empty dot indicates a removal from the top platform. Refit by P. Škrdla (from Škrdla 2003c:133, fig. 9.12, used by permission).

Table 5.47 Stránská skála IIIc, platform types for complete and proximal fragments of flakes and tools (from Tostevin 2003b:90, tab. 8.8)

Platform type	Frequency	Percent
Plain	337	33.0
Cortical	78	7.6
Dihedral	69	6.8
Facetted	117	11.5
Chapeau	3	0.3
Scarred	134	13.1
Punctiform	8	0.8
Linear	21	2.1
Removed	9	0.9
Indeterminate	244	23.9
Total	1,020	100.0

Table 5.48 Stránská skála IIIc, descriptive statistics for continuous platform variables of complete and proximal fragments of flakes and tools (from Tostevin 2003b:90, tabs. 8.9 and 8.10)

	N	Minimum	Maximum	Mean	Standard Deviation
Exterior Platform Angle (degrees)	664	33.0	138.0	84.93	15.05
Platform Thickness (mm)	741	0.4	18.3	4.55	2.52

Table 5.49 Stránská skála IIIc, dorsal scar pattern by blank length (mm) in quartiles cross-tabulation for complete flakes and tools (excluding flakes with greater than 60% cortex or fewer than three dorsal scars). This table replaces the cross-tabulation in Tostevin (2003b:93, tab. 8.14)

Dorsal Scar Pattern	>47	47-37	37-29	<29	Total
Unidirectional					
Count	18	27	30	41	116
% within Quartile	13.3%	21.3%	21.7%	31.1%	21.8%
Bidirectional					
Count	32	24	27	12	95
% within Quartile	23.7%	18.9%	19.6%	9.1%	17.9%
Crossed					
Count	13	23	13	25	74
% within Quartile	9.6%	18.1%	9.4%	18.9%	13.9%
Subcentripetal & Centripetal					
Count	22	12	15	7	56
% within Quartile	16.3%	9.4%	10.9%	5.3%	10.5%
Crested					
Count	32	14	22	7	75
% within Quartile	23.7%	11.0%	15.9%	5.3%	14.1%
Indeterminate					
Count	18	27	31	40	116
% within Quartile	13.3%	21.3%	22.5%	30.3%	21.8%
Total					
Count	135	127	138	132	532
% within Cortex	100.0%	100.0%	100.0%	100.0%	100.0%

Table 5.50 Stránská skála IIIc, dorsal scar pattern by blank length (mm) in quartiles cross-tabulation for complete flakes and tools (excluding pieces with a crested scar pattern, an indeterminate scar pattern, greater than 60% cortex, or fewer than three dorsal scars). This table replaces the cross-tabulation in Tostevin (2003b:93, tab. 8.15)

Dorsal Scar Pattern	>47	47-37	37--9	<29	Total
Unidirectional					
Count	18	27	30	41	116
% within Quartile	21.2%	31.4%	35.3%	48.2%	34.0%
Bidirectional					
Count	32	24	27	12	95
% within Quartile	37.6%	27.9%	31.8%	14.1%	27.9%
Crossed					
Count	13	23	13	25	74
% within Quartile	15.3%	26.7%	15.3%	29.4%	21.7%
Subcentripetal & Centripetal					
Count	22	12	15	7	56
% within Quartile	25.9%	14.0%	17.6%	8.2%	16.4%
Total					
Count	85	86	85	85	341
% within Cortex	100.0%	100.0%	100.0%	100.0%	100.0%

platform angles and platform thickness (Table 5.48) indicate tendencies around 85 degrees and 5 mm respectively.

Direction of Core Exploitation

The analysis of the direction of core exploitation is presented in Tables 5.49-5.50. Figure 5.30 demonstrates that the dominant directional strategy during early core exploitation in the Stránská skála IIIc assemblage, i.e., the dorsal scar pattern within the largest length quartile, is bidirectional. Conversely, the dorsal scar pattern of the pieces within the smallest length quartile of the blanks demonstrates a unidirectional strategy at the end of core exploitation. The two middle quartiles exhibit an alternating frequency of crossed and bidirectional patterns, with a relatively continuous decrease in the sub/centripetal frequencies. Taken together, the length-dorsal scar pattern cross-tabulation demonstrates a change from bidirectional reduction at the beginning of core exploitation to unidirectional reduction at the end of core exploitation.

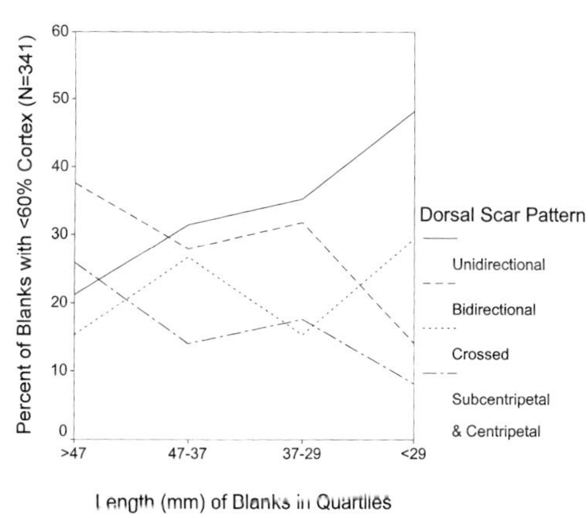

Figure 5.30 Stránská skála IIIc, dorsal scar pattern by blank length (mm) in quartiles cross-tabulation for complete flakes and tools (excluding pieces with a crested scar pattern, an indeterminate scar pattern, greater than 60% cortex, or fewer than three dorsal scars).

Table 5.51 Stránská skála IIIc, frequencies of lateral edge, profile, and cross-section types for complete flakes and tools (from Tostevin 2003b: 94, tab. 8.16)

Type	Frequency	Percent
Lateral edge		
Parallel	288	39.4
Convergent	114	15.6
Expanding	125	17.1
Ovoid	179	24.5
Indeterminate	25	3.4
Total	731	100.0
Profile		
Straight	411	56.2
Curved or Concord	184	25.5
Twisted	136	18.6
Total	731	100.0
Cross-section		
Triangular	293	40.1
Trapezoidal	341	46.6
Other	90	12.3
Indeterminate	7	1.0
Total	542	100.0

Table 5.52 Stránská skála IIIc, descriptive statistics for blank dimensions for complete flakes and tools (from Tostevin 2003b: 94, tab. 8.17)

Blank Dimension	N	Range	Mean	Standard Deviation
Length	731	80	37.96	13.38
Width	731	53	22.87	7.94
Thickness	731	32	6.36	3.30
Length/Width Ratio	731	4.73	1.82	0.80
Width/Thickness Ratio	731	12.01	4.15	1.84

Table 5.53 Stránská skála IIIc, descriptive statistics for dimensions of complete tools

Blank Dimension	N	Range	Mean	Standard Deviation
Length/Width Ratio	93	3.48	1.78	0.70
Width/Thickness Ratio	93	8.76	4.24	1.67

The refit sequences from this assemblage noted above, with the addition of S29/00 (Škrdla 2003c:136–137), clearly support the bidirectional tendencies within the assemblage. Only one refit sequence from a small cortical nodule of Stránská skála hornstone exhibits an exclusively unidirectional reduction, S26/00 (Škrdla 2003c:135–136). While the refit data does not provide any additional support of the change from bidirectional to unidirectional reduction seen in the attribute analysis, the core sector preparation data in Table 5.46 does. As the cores discarded at the end of their exploitation show a much greater propensity for removals to be found in sector A than in sector C, although not to the same degree as seen in Stránská skála IIIa layer 4, the assemblage-wide tendencies in the cores also suggest that unidirectional reduction dominated by late exploitation.

Table 5.54 Stránská skála IIIc, frequencies of lateral edge, distal terminus, and profile types for complete tools

Type	Frequency	Percent
Lateral edge		
Parallel	36	38.7
Convergent	16	17.2
Expanding	12	12.9
Ovoid	24	25.8
Indeterminate	5	5.4
Total	93	100.0
Distal Terminus		
Blunt	41	44.1
Pointed	28	30.1
Indeterminate	24	25.8
Total	93	100.0
Profile		
Straight	49	52.7
Curved or Concord	27	29.0
Twisted	17	18.3
Total	93	100.0

Table 5.55 Stránská skála IIIc, intensity of core and tool reduction

Variables	Numerator	Denominator	Result
Tools/Flake Ratio	Complete Tools: 93	Complete Flakes: 638	0.15
Non-cortical Debitage/ Cortical Debitage Ratio	Non-cortical Complete Debitage: 291	Cortical Complete Debitage: 440	0.66
Henry's Dimensional Analysis	NA	NA	Core facets overlap with the middle of the blank distribution: little core exhaustion

Dorsal Surface Convexity

The analysis of the attributes serving as the proxies for the steps within the dorsal surface convexity domain are given in Tables 5.51-5.52. The ridge system used in the Stránská skála IIIc assemblage shows a propensity for only a moderate laminarity (a mean of 1.82, below the 2.0 blade cut off), thin products (width/thickness mean of 4.15), straight profiles, and roughly longitudinal ridges (i.e., high frequencies of parallel and ovoid lateral edges) of most frequently two in number.

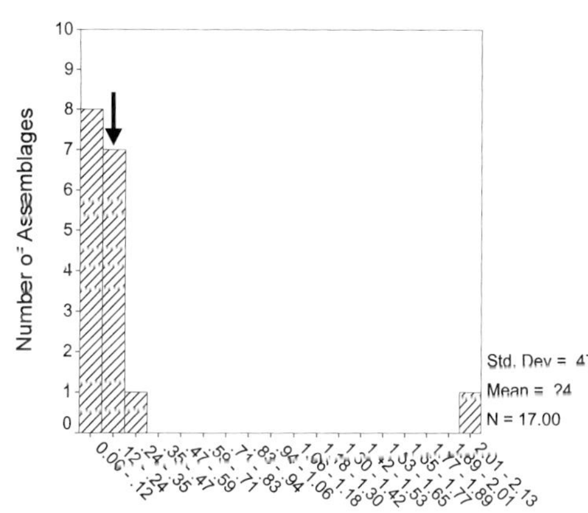

Figure 5.31 Stránská skála IIIc, tool/flake ratio value relative to a histogram of all assemblages.

5.5.4 Tool Kit Morphology

The first five characterizations of the tool kit morphology refer to the dorsal surface convexity variables which give form to the cutting edge of

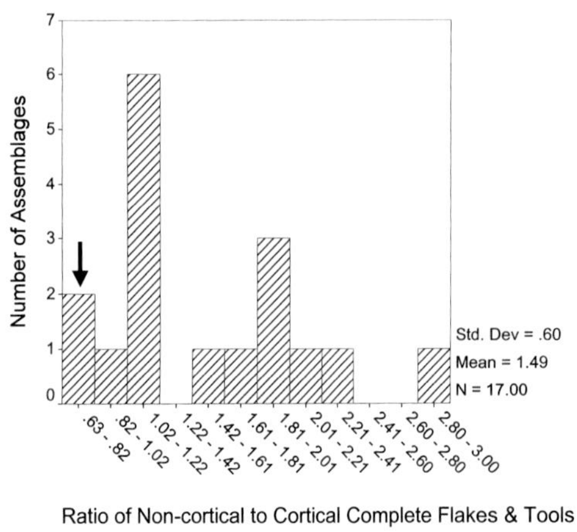

Figure 5.32 Stránská skála IIIc, non-cortical/cortical flake ratio value relative to a histogram of all assemblages.

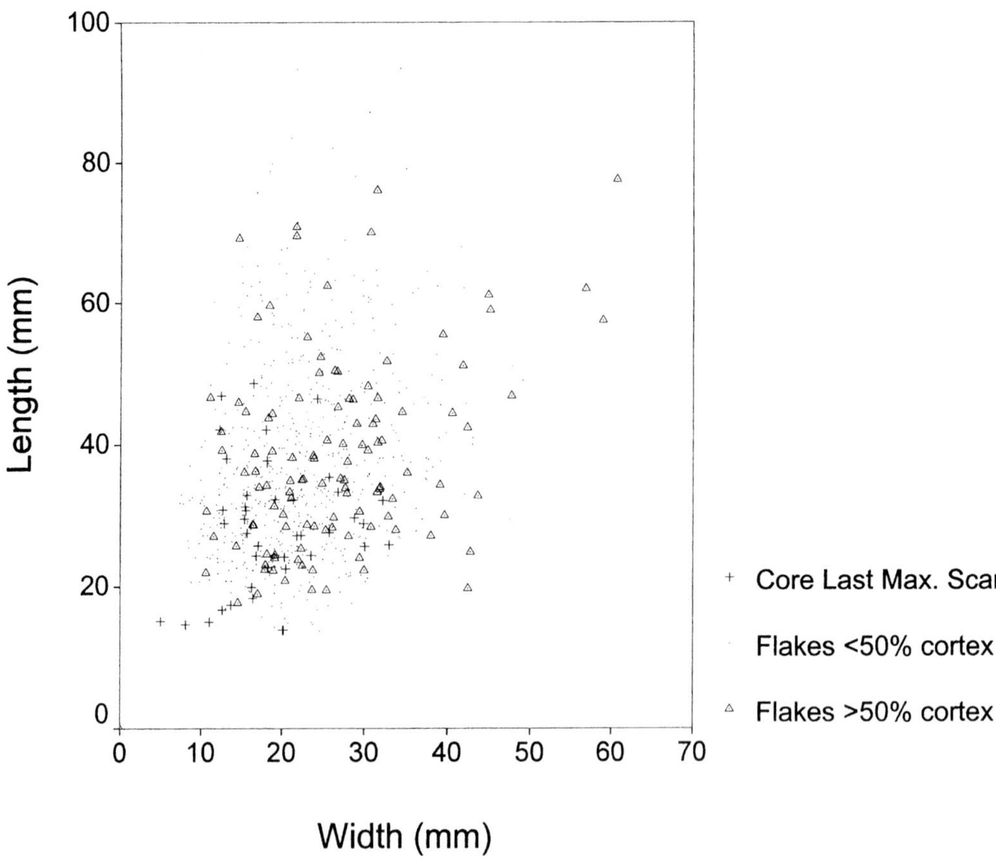

Figure 5.33 Stránská skála IIIc, according to Henry's core reduction intensity measure: length and width of last maximum core facets, highly cortical blanks, and blanks with little cortex.

the retouched blanks in the assemblage, including the unretouched Levallois component. Besides only moderately thinner but more flake proportioned supports, the tool kit differs little from the debitage.

For the last two characterizations of the tool kit morphology, no unique type of retouch is evidenced in the tool types in this assemblage and Upper Paleolithic tool types dominate the tool kit (Svoboda 1987, 1991, 2003b).

Intensity of Tool and Core Reduction

Given the situation that Stránská skála IIIc, as with III to the east and IIIa immediately to the south, is adjacent to a raw material source, it is not surprising that this assemblage possesses a fairly low tool to flake ratio (Table 5.55 and Figure 5.31). The non-cortical to cortical flake ratio also demonstrates a extremely low value, indicating the relatively high representation of cortical pieces in the debitage: significant working of nodules from a cortical stage occurred on site (Figure 5.32). In addition to these indications of a workshop close to a raw material source, the IIIc layer 5 assemblage shows no signs of core exhaustion, as the vast majority of the last maximum facets on the cores overlap with the blanks in the middle of their length and width distribution (Figure 5.33). Only a handful, six out of 41, of core facets fall below the 2.0 cm length limit for debitage. Therefore, this assemblage as a whole is extremely lightly reduced in both its tools and its cores, indicating that the very early to middle portions of the core reduction sequence are preserved in the assemblage.

5.6 Aurignacian Industries from Stránská skála IIa Layer 4 and Stránská skála IIIa Layer 3

5.6.1 Site Background

The IIa locality (Svoboda 1991; Czudek et al. 1991) on the Stránská skála hill contained two Aurignacian layers, a "Middle Aurignacian" dated to 32,350 ± 900 (GrN 14829, a cumulative charcoal sample) and an "Upper Aurignacian" without an associated radiometric date (see Figure 5.8). The former was deposited during a phase of loess sedimentation and pedogenesis, while the latter was deposited along with the last Würmian loess sedimentation in the region. For the present research, only the Middle Aurignacian assemblage is of import.

The sedimentary context of this Middle Aurignacian assemblage consists of a rubble redeposited by solifluction under extremely cold conditions. This is the cause of the extensive frost fracturing of a high proportion of the tools and debitage from this locality, making the technological study difficult. The size of the sample of artifacts made up for this difficulty to some degree, as this is the largest assemblage of this type excavated in the Czech Republic (Svoboda et al. 1996:231). However, in order to create a more complementary comparison of the Central European Aurignacian with other assemblages within and between regions, it was decided to augment the IIa Layer 4 sample by combining it with the smaller but less frost fractured Aurignacian assemblage from Stránská skála IIIa Layer 3, dated to 30,980 ± 360 (GrN 12605, from charcoal from a hearth). The potential risk of creating an overly diverse or unrepresentative sample assemblage through this combination is relatively minor, since the two localities are only 300 m apart on the hillside, the radiometric dates are close enough that they make a shorter palimpsest than that of other assemblages, and the stratigraphic position of each assemblage in the same continuous paleosol, or catena, is identical (Czudek et al. 1991), as seen in Figure 5.8.

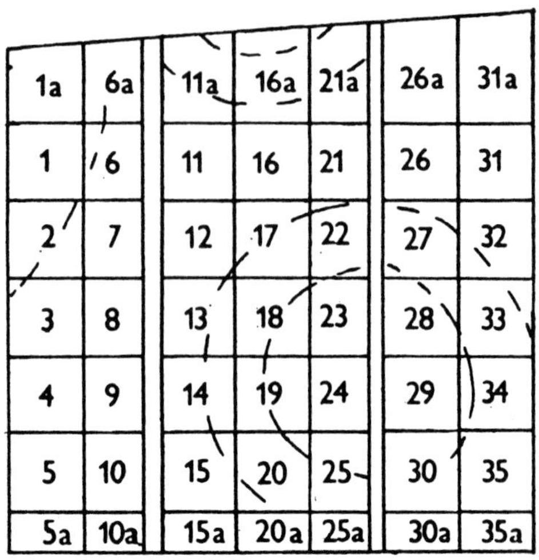

Figure 5.34 Stránská skála IIa site plan (from Svoboda 1991:9, fig. 2.4). Dashed lines indicate circular stone features in layer 6 likely produced by frost heave (Jarošová 2003:37, used by permission).

Table 5.56 Stránská skála IIa-4 and IIIa-3, percentage of intact cores (n=27) by number of bulbar negatives in each core sector (A, B, C, D) by surface (1, 2)

Number of bulbar negatives	Surf. 1 Sect. A	Surf. 1 Sect. B	Surf. 1 Sect. C	Surf. 1 Sect. D	Surf. 2 Sect. A	Surf. 2 Sect. B	Surf. 2 Sect. C	Surf. 2 Sect. D
0	15	52	67	67	59	89	78	78
1	19	30	26	26	26	4	11	15
2	41	11	4	4	7	4	11	4
3	11	4	4	4	7	0		0
4	7	0				0		4
5	4	0				4		
6	4	4						

5.6.2 Sampling

Due to the limited number of intact debitage and tools in both assemblages, all meter squares were studied as allowed by the state of curation of the assemblage. The squares studied in Stránská skála IIa-4 include: 1,2, 3, 4, 5, 5a, 6, 8, 9, 10a, 11, 11a, 12, 13, 14, 15, 16, 16a, 17, 17a, 18, 19, 20, 21, 22, 23, 24, 25, 25a, 26, 26a, 27, 28, 29, 31, 31a, 32, 33, 34, 35, and 35a (see Figure 5.34). To maximize the study of technological features on the relatively few retouched tools in the assemblage, all tools in the collection were studied.

The squares studied in Stránská skála IIIa-3 include: 1, 2, 3a, 4, 4a, 5, 6, 7, 8, 9a, 10, 11a, 12, 13, 14a, 15a, 16a, 18a, 19a, 20a, 23, 24a-28, 25a-29, 26, 27, 28, 30-32, 33, 41, 42, and 43 (see Figure 5.9). Debitage from squares 17, 21, 34, 35, and 36 were not found within the curated collections in the Institute of Archaeology, Dolní Věstonice, Czech Republic. To maximize the study of technological features on the relatively few retouched tools in the assemblage, all tools in the collection were studied.

5.6.3 Blank Production

Core Modification

Core modification in Stránská skála IIa-4 and IIIa-3 began with a longitudinal blade/flake core

Table 5.57 Stránská skála IIa-4 and IIIa-3, core sector preparation: percent of intact cores (n=27) by number of scars in sector of origin (A, B, C, D) by surface (1, 2)

Number of Scars	Surf. 1 Sect. A	Surf. 1 Sect. B	Surf. 1 Sect. C	Surf. 1 Sect. D	Surf. 2 Sect. A	Surf. 2 Sect. B	Surf. 2 Sect. C	Surf. 2 Sect. D
0	19	85	74	78	52	78	59	85
1	4	4	11	11	7	11	15	11
2	22	7	11	11	22	4	15	4
3	26	0	4		7	4	4	
4	11	0			11	0	4	
5	11	0				0	4	
6	4	0				0		
7	4	0				0		
8	0	4				4		

Table 5.58 Stránská skála IIa-4 and IIIa-3, platform types for complete and proximal fragments of flakes and tools

Platform type	Frequency	Percent
Plain	183	38.9
Cortical	21	4.5
Dihedral	1	.2
Facetted	24	5.1
Chapeau	1	.2
Scarred	76	16.2
Punctiform	6	1.3
Linear	13	2.8
Removed	21	4.5
Indeterminate	124	26.4
Total	470	100.0

form, as evidenced in the identification of the narrow core-face as the debitage surface (Appendix Figures A23–A24). This identification is supported by Table 5.56, which demonstrates that the longitudinally-oriented Surface 1 strongly outweighs Surface 2 in number of bulbar negatives, indicating a preference for exploiting one surface rather than two. The data in Table 5.57, from which it is possible to rank the sectors in order of degree of working, indicates that sector A on Surface 1 produced the most scars, followed by A on Surface 2, the would-be platform for the debitage surface. Sector C on Surface 2 has the third most number of scars from its edge, followed by Sector C on Surface 1, indicating that a minor amount of working from an opposed platform took place.

Core convexity management is achieved first through the use of frontal crests, visible in crested blade debitage, to control the frontal curvature of the core volume. The use of rear crests is visible in the extant core morphology (see Appendix, Figures A23–A24, also Škrdla 2003b:75, fig. 7.31) as well the minor frequencies of bulbar negatives and scars originating in sectors B and D on both surfaces. Rear crests served both to thin the backs of the cores and to constrain the lateral width of the debitage surface, which was frequently hemispherical when seen from the sector A platform, leaving the back of the core, Surface 2, flat and often unworked. The presence of core tablets (such as 01219-298 and 01219-294 of SSIIIa-3) in these assemblages indicates that this technique

Table 5.59 Stránská skála IIa-4 and IIIa-3, descriptive statistics for continuous platform variables of complete and proximal fragments of flakes and tools

	N	Minimum	Maximum	Mean	Standard Deviation
Exterior Platform Angle (degrees)	306	40.00	136.00	82.54	15.05
Platform Thickness (mm)	304	0.63	14.84	4.80	2.52

Table 5.60 Stránská skála IIa-4 and IIIa-3, dorsal scar pattern by blank length (mm) in quartiles cross-tabulation for complete flakes and tools (excluding flakes with greater than 60% cortex or fewer than three dorsal scars)

Dorsal Scar Pattern	>39	39-31	31-25	<25	Total
Unidirectional					
Count	17	27	23	35	102
% within Quartile	21.3%	34.6%	28.8%	52.2%	33.4%
Bidirectional					
Count	21	12	17	9	59
% within Quartile	26.3%	15.4%	21.3%	13.4%	19.3%
Crossed					
Count	13	12	8	7	40
% within Quartile	16.3%	15.4%	10.0%	10.4%	13.1%
Subcentripetal & Centripetal					
Count	5	5	8	5	23
% within Quartile	6.3%	6.4%	10.0%	7.5%	7.5%
Crested					
Count	13	11	10	0	34
% within Quartile	16.3%	14.1%	12.5%	0%	11.1%
Indeterminate					
Count	11	11	14	11	47
% within Quartile	13.8%	14.1%	17.5%	16.4%	15.4%
Total					
Count	80	78	80	67	305
% within Cortex	100.0%	100.0%	100.0%	100.0%	100.0%

is used to rejuvenate the top convexities, as well as the platform edges, on the cores.

Platform Maintenance

Platform maintenance includes a strong tendency to leave platforms unprepared, accounting for 69% of the debitage (Table 5.58). This data coincides with the identification of core tablets in the debitage. Measurements of external platform angles and platform thickness (Table 5.59) indicate tendencies around 83 degrees and 5 mm respectively.

Direction of Core Exploitation

The analysis of the direction of core exploitation is presented in Tables 5.60-5.61. Figure 5.35

Table 5.61 Stránská skála IIa-4 and IIIa-3, dorsal scar pattern by blank length (mm) in quartiles cross-tabulation for complete flakes and tools (excluding pieces with a crested scar pattern, an indeterminate scar pattern, greater than 60% cortex, or fewer than three dorsal scars)

Dorsal Scar Pattern	>39	39-31	31-25	<25	Total
Unidirectional					
Count	17	27	23	35	102
% within Quartile	30.4%	48.2%	41.1%	62.5%	45.5%
Bidirectional					
Count	21	12	17	9	59
% within Quartile	37.5%	21.4%	30.4%	16.1%	26.3%
Crossed					
Count	13	12	8	7	40
% within Quartile	23.2%	21.4%	14.3%	12.5%	17.9%
Subcentripetal & Centripetal					
Count	5	5	8	5	23
% within Quartile	8.9%	8.9%	14.3%	8.9%	10.3%
Total					
Count	56	56	56	56	224
% within Cortex	100.0%	100.0%	100.0%	100.0%	100.0%

Table 5.62 Stránská skála IIa-4 and IIIa-3, frequencies of lateral edge, profile, and cross-section types for complete flakes and tools

Type	Frequency	Percent
Lateral edge		
Parallel	246	52.3
Convergent	75	16.0
Expanding	68	14.5
Ovoid	42	8.9
Indeterminate	39	8.3
Total	470	100.0
Profile		
Straight	249	53.0
Curved or Concord	102	21.7
Twisted	79	16.8
Indeterminate	40	8.5
Total	470	100.0
Cross-section		
Triangular	206	43.8
Trapezoidal	214	45.5
Other	27	5.7
Indeterminate	23	4.9
Total	470	100.0

Table 5.63 Stránská skála IIa-4 and IIIa-3, descriptive statistics for blank dimensions for complete flakes and tools

Blank Dimension	N	Range	Mean	Standard Deviation
Length	470	71.63	33.47	10.23
Width	470	58.83	21.40	8.11
Thickness	470	32.88	7.08	3.65
Length/Width Ratio	470	5.97	1.73	.76
Width/Thickness Ratio	470	10.95	3.45	1.51

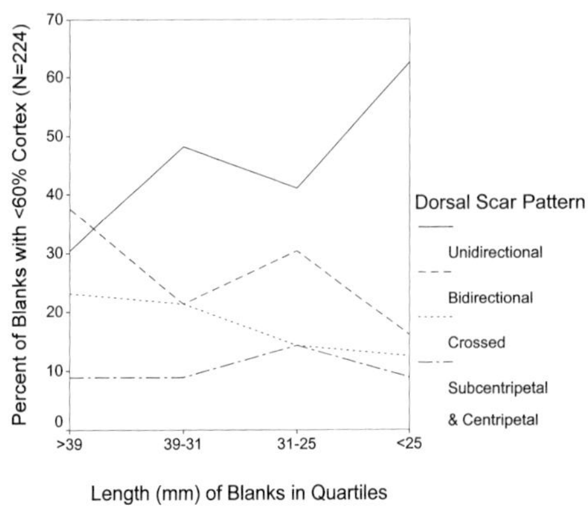

Figure 5.35 Stránská skála IIa-4 and IIIa-3, dorsal scar pattern by blank length (mm) in quartiles cross-tabulation for complete flakes and tools (excluding pieces with a crested scar pattern, an indeterminate scar pattern, greater than 60% cortex, or fewer than 3 dorsal scars).

demonstrates that the direction of early core exploitation exercised in these assemblages was a mixture of a bidirectional strategy and an only slightly less frequent unidirectional strategy. Moving from the largest to the smallest length quartile, the unidirectional strategy gradually increases, with one reversal, to become the pattern for the vast majority of the blanks by late core exploitation. Bidirectional scar patterns are already replaced by unidirectional patterns by the second longest quartile. Both the frequencies of bidirectional and crossed blanks continually decrease with blank length while subcentripetal and centripetal blanks remain fairly constant. The unidirectional tendency for the majority of the debitage when taking all length quartiles together (45.5% compared to the next highest pattern, bidirectional, at 26.3%) is corroborated by the fact that the highest frequency of bulbar negatives and scars originating from their sector of origin are found consistently in sector A of Surface 1 on the cores in both assemblages. The minor number of bulbar scars and originating scars in sector C on both surfaces can account for the presence of bidirectional reduction at the very beginning of core exploitation, interestingly from either Surface 1 or Surface 2 according to these data.

Dorsal Surface Convexity

The analysis of the attributes serving as the proxies for the dorsal surface convexity domain are given in Tables 5.62–5.63. The ridge system used in the Stránská skála Aurignacian assemblages shows a propensity for the low laminarity of a flake-based assemblage (length/width ratio of 1.73), thick products (width/thickness mean of 3.45) and longitudinal ridges (i.e., high frequency of parallel lateral edges).

5.6.4 Tool Manufacture

The first five characterizations of the tool kit morphology refer to the dorsal surface convexity variables which give form to the cutting edge of the retouched blanks in the assemblage; the assemblage does not include a Levallois component in this case. The tool kit differs little in length, width, thickness proportions from the debitage, although ovoid lateral edges (21.7 % vs. 9.7 % without indeterminates) and less twisted profiles (12.5 % vs. 18.4 % without indeterminates) are found among the tools than the debitage.

For the last two characterizations of the tool kit morphology, the presence of carinated tools in the tool kit counts for the use of a unique type of retouch while Upper Paleolithic tool types dominate the tool kit (Svoboda 1991, 2003b).

Intensity of Tool and Core Reduction

Given the situation that both Stránská skála IIIa and IIa are conveniently located near a raw material source, it is not surprising that this combined

Table 5.64 Stránská skála IIa-4 and IIIa-3, descriptive statistics for dimensions of complete tools

Blank Dimension	N	Range	Mean	Standard Deviation
Length/Width Ratio	52	3.67	1.73	0.78
Width/Thickness Ratio	52	7.22	3.33	1.48

Table 5.65 Stránská skála IIa-4 and IIIa-3, frequencies of lateral edge, distal terminus, and profile types for complete tools

Type	Frequency	Percent
Lateral edge		
Parallel	22	42.3
Convergent	6	11.5
Expanding	8	15.4
Ovoid	10	19.2
Indeterminate	6	11.5
Total	52	100.0
Distal Terminus		
Blunt	12	23.1
Pointed	7	13.5
Hinged	1	1.9
Overshot	3	5.8
Removed	10	19.2
Broken	2	3.8
Indeterminate	17	32.7
Total	52	100.0
Profile		
Straight	25	48.1
Curved or Concord	10	19.2
Twisted	5	9.6
Indeterminate	12	23.1
Total	52	100.0

assemblage possesses a fairly low tool to flake ratio (Table 5.66 and Figure 5.36). The non-cortical to cortical flake ratio demonstrates a value quite close to the mean of all of the assemblages, indicating an average proportion of cortical pieces in the debitage: cortical nodules were worked on site but only to a moderate degree (Figure 5.37). In fact, the non-cortical to cortical blank ratio of the Aurignacian assemblages is higher than any other assemblage on the Stránská skála hillside, indicating perhaps a different landscape utilization or mobility pattern in this time period. The dimensional analysis of the core facets relative to the debitage in the assemblages is also different from the preceding Bohunician assemblages on the hillside (Figure 5.38). While a few core facets are found in the high end of the blank distribution for length and width, most of them overlap the distribution of the smallest blanks. In addition, seven out of 37 core facets fall below the 2.0 cm length limit for debitage, indicating core exhaustion in these pieces. As a whole, this combined Aurignacian sample is only lightly reduced in its tool kit but moderately reduced in its cores, indicating the presence of the middle to late portions of the core reduction sequence in the assemblage.

Table 5.66 Stránská skála IIa-4 and IIIa-3, intensity of core and tool reduction

Variables	Numerator	Denominator	Result
Tools/Flake Ratio	Complete Tools: 52	Complete Flakes: 334	0.16
Non-cortical Debitage/ Cortical Debitage Ratio	Non-cortical Complete Debitage: 232	Cortical Complete Debitage: 154	1.51
Henry's Dimensional Analysis	NA	NA	Core facets overlap with the low end of the blank distribution: moderate core exhaustion

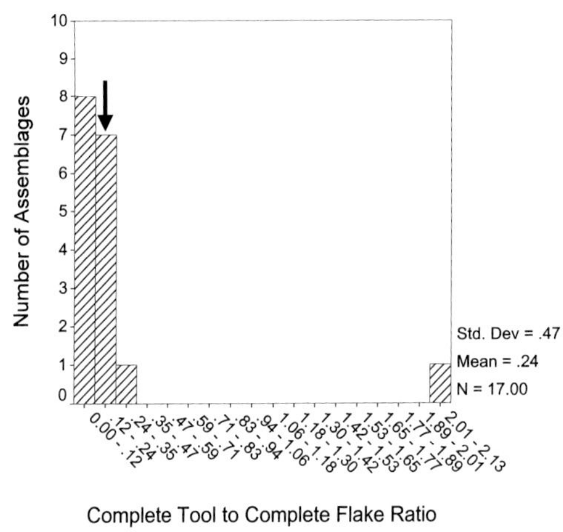

Figure 5.36 Stránská skála IIa-4 and IIIa-3, tool/flake ratio value relative to a histogram of all assemblages.

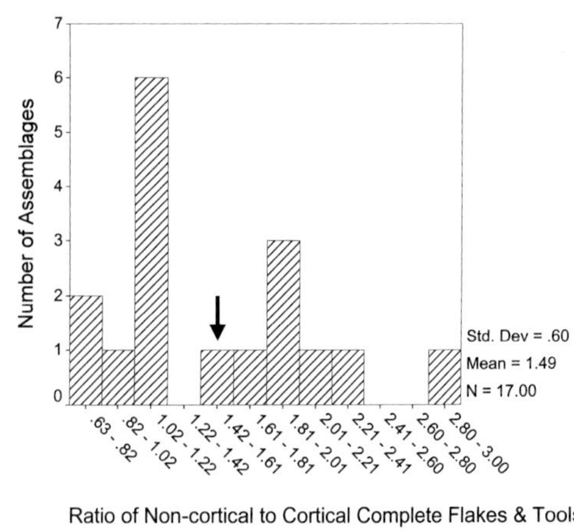

Figure 5.37 Stránská skála IIa-4 and IIIa-3, non-cortical/cortical flake ratio value relative to a histogram of all assemblages.

5.7 Comparison of Central European Flintknapping Behaviors

The construction of the Central European sequence of change between 60 and 30 kya is conducted on the basis of pair-wise comparisons between assemblages. Most of these assemblage pairs are chronologically related. Thus, the Central European sequence begins with the comparison of the earliest assemblage after 60 kya with the assemblage dated immediately thereafter. The next comparison pairs that second assemblage with its immediate successor, and so on till the latest assemblage between 60 and 30 kya has been examined. The reasoning for this procedure is discussed in Chapters 3 and 4 (see also Tostevin 2003a). In this chapter, however, several comparisons are made between assemblages which are not chronologically immediately successive, for instance between Kůlna Cave, Layer 7a, and Vedrovice V. For although the Bohunician assemblages from Stránská skála chronologically lie between them, the possibility

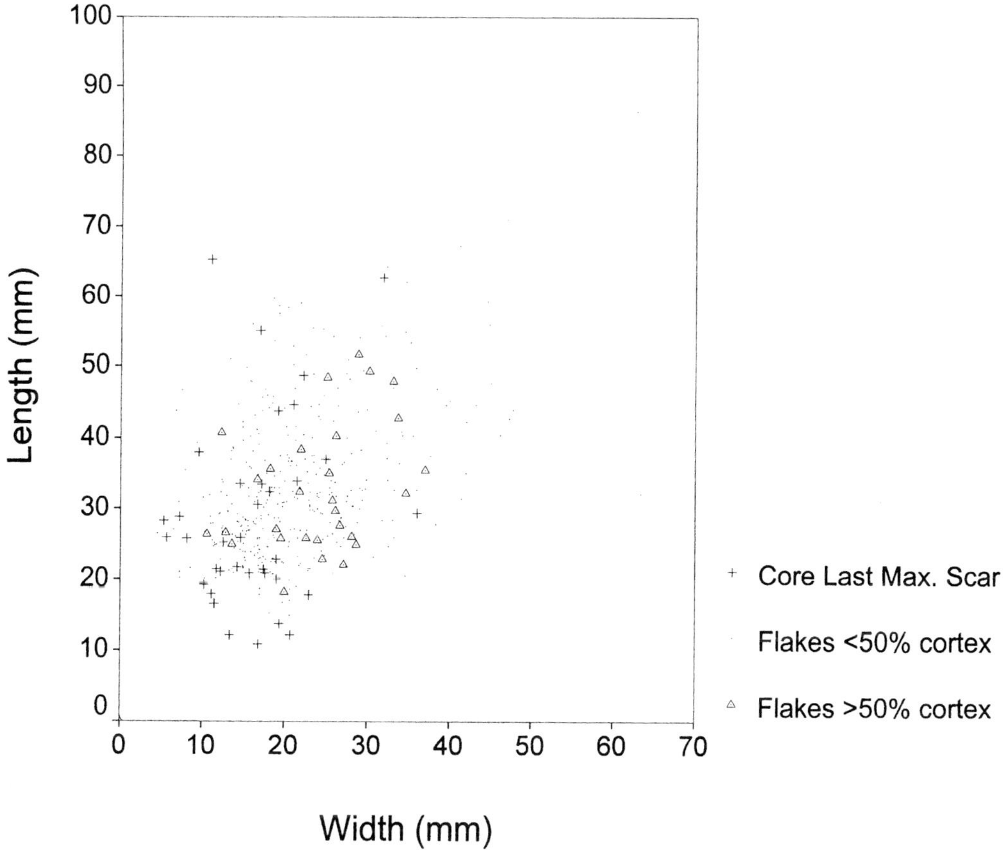

Figure 5.38 Stránská skála IIa-4 and IIIa-3, according to Henry's core reduction intensity measure: length and width of last maximum core facets, highly cortical blanks, and blanks with little cortex.

of a multiple phyla scenario makes it necessary to consider the technological relationships between the Szeletian of Vedrovice V and the Micoquian of Kůlna Cave, Layer 7a, as argued by Valoch (1990a, 1990b) and Neruda (2000).

5.7.1 Comparison of Kůlna Cave Layer 7a and Stránská skála IIIa Layer 4

The comparison between the Micoquian of Kůlna Cave layer 7a (Valoch 1967; 1988) and the oldest of the Bohunician assemblages, Stránská skála IIIa layer 4 (Svoboda and Svobodová 1985; Svoboda and Simán 1989; Svoboda 1983; 1991; Svoboda and Škrdla 1995; Škrdla 1996), is one of the most different in terms of landscape usage represented by each assemblage. Stránská skála IIIa is likely within a few hundred meters of the outcrop of hornstone used to produce the assemblage whereas most of the raw material used to create the 7a assemblage derives from 10 km away from Kůlna Cave. While the technological characteristics and nodule dimensions are similar

Table 5.67 Kůlna Cave Layer 7a and Stránská skála IIIa Level 4, comparability of the assemblages based on intensity of core and tool reduction

Variables	Kůlna Cave 7a	Stránská skála IIIa-4	Evaluation
Tools/Flake Ratio	2.03	0.16	Highly different
Non-cortical Debitage/ Cortical Debitage Ratio	1.18	1.06	Similar
Henry's Dimensional Analysis	Extreme core exhaustion	Slight core exhaustion	Highly different

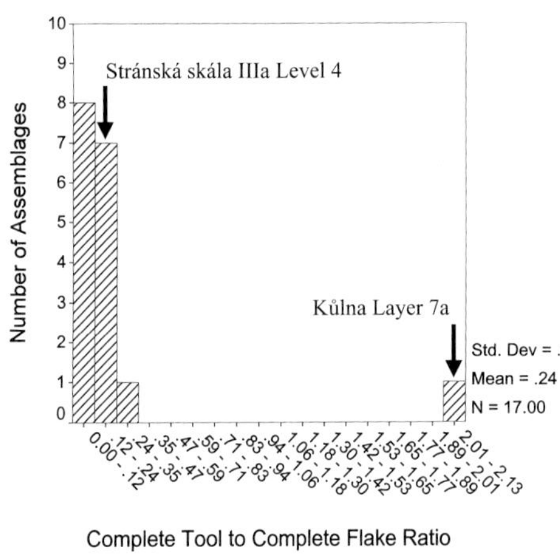

Figure 5.39 Kůlna Cave Layer 7a and Stránská skála IIIa Level 4 compared for tool/flake ratio values.

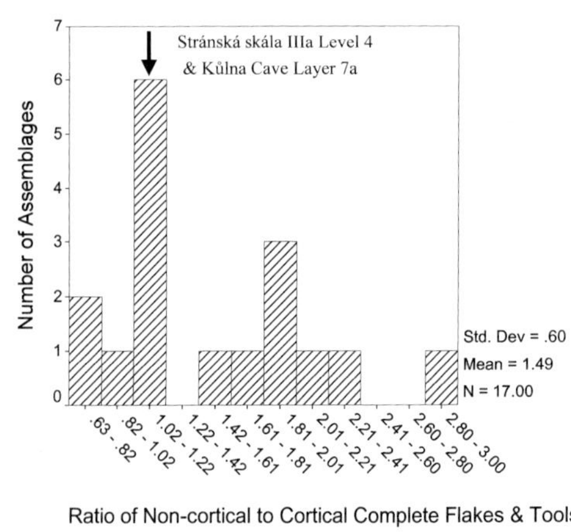

Figure 5.40 Kůlna Cave Layer 7a and Stránská skála IIIa Level 4 compared for non-cortical/cortical flake ratio values.

between the Stránská skála hornstone and the Bořitov type of Cretaceous spongolite chert used at Kůlna (Škrdla 2003a:41), the Micoquian made frequent use (13.92%) of quartz, quartzite, limestone, and other coarser raw materials not used at Stránská skála (Féblot-Augustin 1993:233). Thus, it is not surprising that the exploitation and discard of knapped stone at these two localities differ in several variables related to landscape use. The question of import is to what extent these differences affect the central tendencies and measures of dispersion reconstructed here for the different domains of flintknapping behavior at these two sites and time periods.

As indicated in Table 5.67 and Figure 5.39, both assemblages' frequencies of retouched to unretouched blanks in each assemblage are at the opposite extremes of the distribution for this study. While the excavation or curational history of the collection may contribute to the extreme position of Kůlna Cave layer 7a in this measure, caution is still warranted when comparing dimensional data, particularly with the tool kits, between these two assemblages. Kůlna's high frequency of tools per flake may indicate a greater application of retouch on each tool, thereby decreasing the length and width measurements of these tools relative to the tools in the Bohunician assemblage.

The categorical variables of lateral edge type, profile type, and distal end type are not necessarily biased by increased retouch intensity, as any application of retouch which compromises the interpretation of the attribute state is typed as "indeterminate" and removed from the comparison. While it is possible to eliminate the attributes dependent upon these dimensions, i.e., the length/width and width/thickness ratios, from the pair-wise assemblage comparison, this can increase rather than decrease the relative measure of difference between the assemblages, working against the null hypothesis. As a consequence, these attributes are left in the comparison but caution is warranted in evaluating how the greater retouch application in the Kůlna assemblage may affect the result.

Unlike the tool-flake ratio, both assemblages are similar in their non-cortical to cortical blank ratio (Figure 5.40), indicating that raw material was brought into the site in a similar cortical state. The difference, however, in the distances to the raw material sources, 10 km for most of the Kůlna artifacts and less than 1 km for Stránská skála, may indicate that other processes are also at work; all things being equal, Kůlna should possess less cortical relative to non-cortical debitage than Stránská skála IIIa. Further, Kůlna 7a clearly evidences much more core exhaustion than Stránská skála IIIa Layer 4. These factors should produce a higher non-cortical to cortical blank ratio, especially given the similar nodular sizes (and the potential flake per core production rate) between the Stránská skála hornstone and the Bořitov Cretaceous chert. The lower non-cortical to cortical blank ratio could, however, be partly affected by non-Cretaceous chert portion of the Micoquian assemblage included in this study, as these more local raw materials would bring more cortex into the assemblage.

In sum, the core reduction sequences preserved in these two assemblages are fairly comparable as both possess at least some early reduction and all of the middle portion of the core reduction sequence. The Bohunician assemblage theoretically lacks the behaviors utilized during the very late stages of core exhaustion witnessed in the Kůlna assemblage while the latter lacks the behavioral products of very early raw material shaping. Of the factors involved in a comparison between these assemblages, two are the most pertinent. First, it is likely that the dimensional data for Kůlna layer 7a is affected by its greater degree of retouch intensity. Second, despite the preservation in the Kůlna assemblage of more of the behaviors utilized at the end of the core reduction sequence, the few examples of exhausted cores at IIIa-4 (see Figure 5.11) do not resemble those in Kůlna layer 7a. This dissimilarity between the exhausted cores in each assemblage indicates that it is unlikely that the late reduction sequence behaviors at Kůlna would be similar to those at Stránská skála IIIa layer 4 if reduction had continued. The unknown effect of these possible differences on the central tendency statistics of the variables used in the comparison, however, should be kept in mind.

Blank Production

Table 5.68 presents the results of a comparison of blank production for the assemblages from Kůlna Cave, layer 7a and Stránská skála IIIa level 4. The first step in the core modification domain, core orientation, is characterized by the qualitative distinction between different core forms. In the case of Kůlna 7a, a predominately discoidal exploitation of two surfaces was evidenced in the majority of cores, although some of the exhausted examples show a unifacial discoidal approach. In the majority of the Stránská skála IIIa level 4 cores, a quite different orientation is evidenced, the preferential use of one surface on a longitudinal blade/flake core form. The second step in this

Table 5.68 Comparison of Blank Production for Kůlna Cave Layer 7a and Stránská skála IIIa Layer 4

Flintknapping Steps by Domain	Kůlna 7a	Stránská skála IIIa-4	Significant Change?
Core Modification			
Core Orientation	Discoidal use of two surfaces, unifacial at end	Preferential use of a Longitudinal Surface	Yes
Core Convexity Management	Secant surfaces exploited around the perimeter of the plane of intersection	Débordants, front crests, & side blade removals	Yes
Number of Changes/2 Steps			**2/2=1**
Platform Maintenance			
Platform Treatment	Unprepared: 58% Prepared: 42% n=167	Unprepared: 58% Prepared: 42 n=448	No p=.93, Fisher's Exact
External Platform Angle (degrees)	mean: 83.8, s.d.: 14.8, n=153	mean: 85.2, s.d.: 15.3, n=425	No p=.31, t=-1.019, df=576
Platform Thickness	mean: 9.08, s.d.: 4.63, n=153	mean: 4.8, s.d.: 2.5, n= 433	Yes p=.00, t=10.861*, df=183.069
Number of Changes/3 Steps			**1/3=0.33**
Direction of Core Exploitation			
Direction of Early Exploitation	Subcentripetal & Unidirectional	Bidirectional	Yes
Direction of Late Exploitation	Unidirectional & Crossed	Unidirectional	Yes
Number of Changes/2 Steps			**2/2=1**
Dorsal Surface Convexity System			
Laminarity: Length/Width Ratio	mean: 1.44, s.d.: 0.49, n=273	mean: 1.71, s.d.: 0.67, n=543	Yes p=.00, t=-6.551*, df=704.777
Lateral Edges of Blanks	Parallel: 54% Convergent: 23% Expanding: 20% Ovoid: 13% n=198	Parallel: 49% Convergent: 24% Expanding: 18% Ovoid: 10% n=514	Yes p=.01, G^2=11.496, df=3
Profile of Blanks	Straight: 64% Curved: 22% Twisted: 14% n=225	Straight: 53% Curved: 29% Twisted: 18% n=522	Yes p=.03, G^2=6.893, df=2
Cross-Section of Blanks	Triangular: 39% Trapezoidal: 54% Other: 7% n=231	Triangular: 44% Trapezoidal: 51% Other: 5% n=527	No p=.35, G^2=2.087, df=2
Convexity: Width/Thickness Ratio	mean: 2.83, s.d.: 1.06, n=273	mean:3.99, s.d.: 1.82, ,n=543	Yes p=.00, t=-11.435*, df=796.838
Number of Changes/5 Steps			**4/5=0.8**
Total Measure of Difference in Blank Production			**3.13**

*The t-test was calculated with equal variances not being assumed due to a significant F value on Levene's Test for Equality of Variances.

domain, core convexity management, includes the knapping options which allow the knapper to continue reducing the core form to exhaustion through specific modifications of the surface convexities. In the case of Kůlna 7a, convexity management proceeded by exploitation of the perimeter or circumference of the plane of intersection between two secant surfaces. This form of exploitation constantly refreshed the convexity of the surface. The Stránská skála IIIa level 4 debitage and cores evidenced management of the convexities through frontal crests, lateral débordant removals, and, in particular instances, the late removal of blades from the lateral, narrow side of the core, situated 90 degrees to the previous surface. A significant difference is qualitative ly judged to exist between the two options taken by the Kůlna Layer 7a and the Stránská skála IIIa level 4 knappers in this step.

The next knapping domain, "platform maintenance," includes the steps whereby the flintknapper modifies the platform surface and platform edge before each removal. The significance of the difference in the options exercised within the two assemblages is determined through a G^2 likelihood ratio test for platform treatment and t-tests for the other two steps. In the case of the Kůlna layer 7a and Stránská skála IIIa level 4 comparison, only mean platform thickness is significantly different, due to Kůlna's much thicker platforms. The frequency of platform treatment and the choice of exterior platform angles were not statistically different. The next knapping domain, direction of core exploitation, concerns the variability resulting from different options of removing flakes from single to multiple platform cores, creating debitage with particular dorsal scar patterns. A strong change from bidirectional reduction at the beginning of core exploitation to unidirectional reduction at the end of core exploitation is evidenced in the dorsal scar pattern of debitage from Stránská skála IIIa level 4 (Figure 5.12). For debitage from Kůlna layer 7a, the largest blanks are subcentripetal or unidirectional, while the smallest are unidirectional and crossed, suggesting a shift in strategy during the exploitation of the cores (Figure 5.4). Both of these directional options (for early and late exploitation) were judged to be significantly different between the Kůlna and Stránská skála IIIa assemblages.

The dorsal surface convexity domain includes the knapping steps which encapsulate the cumulative choice of *nervure-guide* or ridge pattern(s) of the core exterior for the production of blanks. The significance of different options for the three attributes, lateral edge type, profile type, and cross-section type, is determined through a G^2 likelihood ratio test. The significance of the convexity and laminarity measurements, both characterizations of the tendencies to remove particular volumes of stone through the choice of types of ridges to follow, is determined using a t-test. All of the behavioral choices exhibited in the two assemblages for these blank production steps were significantly different, save for the tendency to follow centralized versus dispersed dorsal ridge patterns, as measured by the cross-section attribute.

The final measure of difference between the blank production choices within Kůlna layer 7a and Stránská skála IIIa level 4, weighted by the four knapping domains, produces a value of 3.13 out of a possible maximum difference of 4.0, indicating that the blank production choices are distinctly different.

Tool Kit Morphology

The comparison of the morphology of the cutting edge of the tool kits of these two assemblages utilizes five categorical variables and two continuous variables. The latter are tested with a

Table 5.69 Comparison of Tool Kit Morphologies for Kůlna Cave Layer 7a and Stránská skála IIIa Layer 4

Tool Kit Morphology Variable	Kůlna 7a	Stránská skála IIIa-4	Significant Change?
Laminarity	mean: 1.45 s.d.: 0.51 n=162	mean: 1.58 s.d.: 0.50 n=67	No p=.09 t=-1.685, df=227
Width/Thickness	mean: 2.84 s.d.: 1.12 n=162	mean: 4.28 s.d.: 2.14 n=67	Yes p=.00 t=-5.214*, df=81.391
Lateral Edges	Parallel: 55% Convergent: 7% Expanding: 23% Ovoid: 15% n=103	Parallel: 39% Convergent: 31% Expanding: 16% Ovoid: 14% n=57	Yes p=.00 G^2=16.894, df=3
Distal Terminus	Blunt: 74% Pointed: 26% n=65	Blunt: 50% Pointed: 50% n=36 p=.03	Yes Fisher's Exact
Profile	Straight: 67% Curved: 20% Twisted: 13% n=126	Straight: 45% Curved: 32% Twisted: 23% n=62	Yes p=.01, G^2=8.577, df=2
Unique Types of Retouch	Flat bifacial retouch	Normal Retouch	Yes
Tool Types	MP tools dominate	UP tools dominate	Yes
Number of Changes/7 Steps			**6/7**
Total Measure of Difference in Tool Kit Morphology			**0.86**

*The t-test was calculated with equal variances not being assumed due to a significant F value on Levene's Test for Equality of Variances.

Table 5.70 Stránská skála IIIa Level 4 and Stránská skála III, comparability of the assemblages based on intensity of core and tool reduction

Variables	Stránská skála IIIa-4	Stránská skála III	Evaluation
Tools/Flake Ratio	0.16	0.08	Similar
Non-cortical Debitage/ Cortical Debitage Ratio	1.06	1.19	Similar
Henry's Dimensional Analysis	Slight core exhaustion	Little evidence of core exhaustion	Slight difference

t-test for the probability that the samples arose from the random sampling of the same population. Three of the categorical variables are tested with G^2 statistics and two with qualitative judgments of difference. Despite the concerns that the dimensions of the Kůlna tools would have been reduced to significantly different ratios, no difference is seen in the low laminarity of both assemblages' tools (it is theoretically possible that the Kůlna tools began even more laminar than those in the Stránská skála assemblage and were reduced to the current ratio but this is considered unlikely). The thick tools seen in the Kůlna assemblage, however, do statistically differ from the thin products of the Stránská skála assemblage. The tool kits also are statistically different in their lateral edges (parallel and expanding in Kůlna versus parallel and convergent in SSIIIa-4), distal terminations (Kůlna being less pointed), and profile types (Kůlna being far less twisted in profile).

The remaining two analytical steps within the comparison of tool kit morphology relate to qualitative issues of retouch. Only Kůlna 7a possesses a unique retouch type, flat bifacial retouch. The two assemblages also differ in the dominance of Middle Paleolithic tool types in the earlier assemblage and Upper Paleolithic tool types in the later one. Considering all seven analytical steps within the tool kit comparison, these assemblages exhibit a difference measure of six out of seven steps, or 0.86 out of a maximum of 1.0.

5.7.2 Comparison of Stránská skála IIIa Level 4 and Stránská skála III

The next chronological pair-wise comparison within the Central European region opposes two Bohunician assemblages from the Stránská skála hillside, Stránská skála IIIa Level 4 and Stránská skála III layer 5, dated to 41,300 + 3,100°2200 BP and 38,200 ± 1,100 BP respectively. Given their close proximity in terms of distance to the nearby raw material source, it is no surprise that the comparability measures are quite similar between these assemblages (Table 5.70). Both the tool to flake ratio (Figure 5.44) and the non-cortical to cortical blank ratio (Figure 5.44) show these two assemblages are identical to each other in relation to the other studied assemblages. A slight difference in the degree of core reduction is evidenced, with the older assemblage exhibiting slightly more reduction, but the difference is not great enough to argue that a significant portion of the core reduction sequence visible in IIIa layer 4 is missing from the less reduced III layer 5. The issue of comparison for the tool kit which is not dealt with by these variables is the greater frequency of distant raw material used for some of the retouched tools in III layer 5. Under the assumptions that greater distance from source to discard of the tool will produce a greater degree of resharpening in the form of retouch, it is possible that this minor raw material difference could play a factor in the comparison between the tool kits.

Blank Production

This comparison is presented in Table 5.71. The core modification domain begins with orientation of the core. In the majority of both the Stránská skála IIIa level 4 cores and the Stránská skála III cores, one surface of a longitudinal blade/flake core form is used. Core convexity management also shows a similar choice of options in the two assemblages, as the attribute analysis and existing refittings of individual cores from both assemblages indicate a management of the convexities through frontal crests, lateral débordant removals, and side blade removals. No significant differences are qualitatively judged to exist between

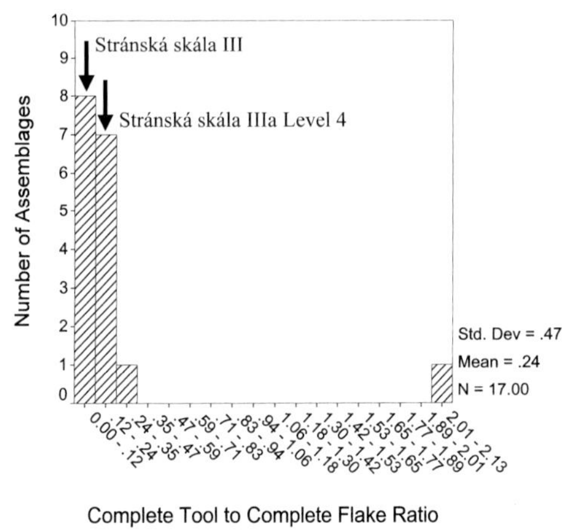

Figure 5.41 Stránská skála IIIa Level 4 and Stránská skála III compared for tool/flake ratio values.

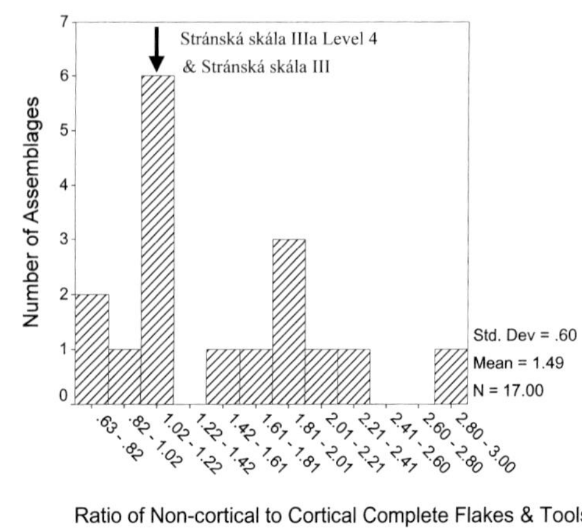

Figure 5.42 Stránská skála IIIa Level 4 and Stránská skála III compared for non-cortical/cortical flake ratio values.

these assemblages for this domain as the options taken by the two groups of flintknappers are the same for both analytical steps. The next knapping domain, platform maintenance, includes the steps whereby the flintknapper modifies the platform surface and platform edge before each removal. Of the three behavioral steps within this domain, only the cumulative choice of platform thickness is significantly different between the assemblages, producing a measure of difference for this domain of 1/3 or 0.33. This significant difference is due to the slightly thinner platforms of Stránská skála III.

The next knapping domain, direction of core exploitation, concerns the variability resulting from different options of removing flakes from single to multiple platform cores, creating debitage with particular dorsal scar patterns. As indicated in the discussion of Figures 5.12 and 5.18, both assemblages evidence the same tendencies to produce bidirectional debitage early in core exploitation and unidirectional debitage late in core exploitation. There is thus no significant difference within the direction of core exploitation domain between these assemblages.

Of the three categorical attributes within the convexity system domain, only frequencies of lateral edge types differ significantly between Stránská skála IIIa-4 and Stránská skála III, with the former possessing more convergent and expanding lateral edges. Of the two continuous variables in this domain, only the laminarity ratio differs significantly, due to the greater tendency for blade production at Stránská skála III. The direction of this statistically significant difference runs counter to the prediction that the tools made on distant raw materials in the latter assemblage would have been reduced in dimensions due to more retouch.

The cumulative measure of difference for the blank production between these two assemblages is 0.73 out of a maximum of 4.0, a difference resulting from only one platform variable and two dorsal surface variables. Thus, these two assemblages are extremely alike in blank production choices.

Table 5.71 Comparison of Blank Production for Stránská skála IIIa Level 4 and Stránská skála III

Flintknapping Steps by Domain	Stránská skála IIIa-4	Stránská skála III	Significant Difference?
Core Modification			
Core Orientation	Preferential use of a Longitudinal Surface	Preferential use of a Longitudinal Surface	No
Core Convexity Management	Débordants, front crests, & side blade removals	Débordants, front crests, & side blade removals	No
Number of Changes/2 Steps			**0/2=0**
Platform Maintenance			
Platform Treatment	Unprepared: 58% Prepared: 42% n=448	Unprepared: 55% Prepared: 45% n=367	No p=.31 Fisher's Exact
External Platform Angle (degrees)	mean: 85.2 s.d.: 15.3 n=425	mean:86.4 s.d.: 14.0 n=340	No p=.29 t=-1.070, df=763
Platform Thickness	mean: 4.8 s.d.: 2.5 n= 433	mean: 4.42 s.d.: 2.16 n=344	Yes p=.02 t=2.411, df=775
Number of Changes/3 Steps			**1/3=0.33**
Direction of Core Exploitation			
Direction of Early Exploitation	Bidirectional	Bidirectional	No
Direction of Late Exploitation	Unidirectional	Unidirectional	No
Number of Changes/2 Steps			**0/2=0**
Dorsal Surface Convexity System			
Longitudinal Convexity: Length/Width Ratio	mean: 1.71 s.d.: 0.67 n=543	mean: 1.83 s.d.: 0.73 n=421	Yes p=.01 t=-2.625* df=860.743
Shape of Convexity: Lateral Edges of Blanks	Parallel: 49% Convergent: 24% Expanding: 18% Ovoid: 10% n=514	Parallel: 59% Convergent: 18% Expanding: 10% Ovoid: 13% n=417	Yes p=.00 G^2=20.212 df=3
Curvature of Convexity: Profile of Blanks	Straight: 53% Curved: 29% Twisted: 18% n=522	Straight: 47% Curved: 30% Twisted: 23% n=417	No p=.07 G^2=5.311 df=2
Lateral Convexity: Cross-Section of Blanks	Triangular: 44% Trapezoidal: 51% Other: 5% n=527	Triangular: 39% Trapezoidal: 56% Other: 5% n=420	No p=.29 G^2=2.490 df=2
Vertical Convexity: Width/Thickness Ratio	mean: 3.99 s.d.: 1.82 n=543	mean: 3.97 s.d.: 1.72 n=421	No p=.87 t=0.166, df=962
Number of Changes/5 Steps			**2/5=0.4**
Total Measure of Difference in Blank Production			**0.73**

*The t-test was calculated with equal variances not being assumed due to a significant F value on Levene's Test for Equality of Variances.

Table 5.72 Comparison of Tool Kit Morphologies for Stránská skála IIIa Layer 4 and Stránská skála III

Tool Kit Morphology Variable	Stránská skála IIIa-4	Stránská skála III	Significant Difference?
Laminarity	mean: 1.58, s.d.: 0.50, n=67	mean: 2.02, s.d.: 0.91 n=30	Yes p=.02, t=-2.525*, df=37.083
Width/Thickness	mean: 4.28, s.d.: 2.14, n=67	mean: 4.29, s.d.: 1.56 n=30	No p=.98, t=-0.025, df=95
Lateral Edges	Parallel: 39% Convergent: 31% Expanding: 16% Ovoid: 14% n=57	Parallel: 29% Convergent: 54% Expanding: 11% Ovoid: 7% n=28	No p=.27, G^2=3.966, df=3†
Distal Terminus	Blunt: 50% Pointed: 50% n=36	Blunt: 21% Pointed: 79% n=24	Yes p=.03, Fisher's Exact
Profile	Straight: 45% Curved: 32% Twisted: 23% n=62	Straight: 50% Curved: 27% Twisted: 23% n=30	No p=.85, G^2=0.318, df=2
Unique Types of Retouch	Normal retouch	Normal retouch	No
Tool Types	UP tools dominate	UP tools dominate	No
Number of Changes/7 Steps			**2/7**
Total Measure of Difference in Tool Kit Morphology			**0.29**

*The t-test was calculated with equal variances not being assumed due to a significant F value on Levene's Test for Equality of Variances.

†Two cells (25%) have an expected count less than 5. The minimum expected count is 3.29.

Table 5.73 Stránská skála III and Vedrovice V, comparability of the assemblages based on intensity of core and tool reduction

Variables	Stránská skála III	Vedrovice V	Evaluation
Tools/Flake Ratio	0.08	0.32	Moderately different
Non-cortical Debitage/ Cortical Debitage Ratio	1.19	2.02	Moderately different
Henry's Dimensional Analysis	Little evidence of core exhaustion.	Moderate core exhaustion	Moderately different

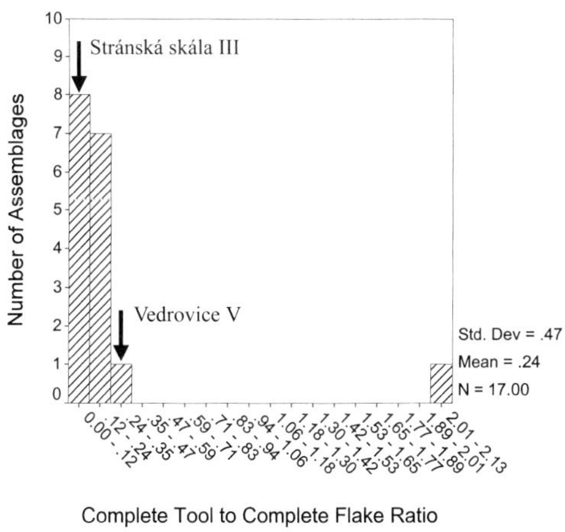

Figure 5.43 Stránská skála III and Vedrovice V compared for tool/flake ratio values.

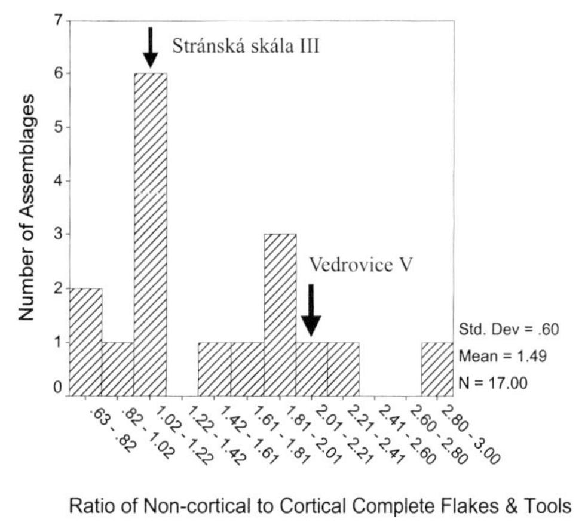

Figure 5.44 Stránská skála III and Vedrovice V compared for non-cortical/cortical flake ratio values.

Tool Kit Morphology

The comparison of the morphology of the cutting edge of the tool kits of these two assemblages produced two variables that were statistically different. The first, the laminarity of the tool kit, shows that the younger of the two assemblages has more bladey tools, a finding that again runs counter to the argument that a subset of these tools should be more reduced in dimensions by retouch and therefore less laminar. It is theoretically possible that retouch restricted to the lateral edges of the III layer 5 blanks would produce higher length/width ratios but as retouch was placed distally far more often in this assemblage, producing Upper Paleolithic types such as endscrapers and distally pointed blades, than laterally, which would produce more Middle Paleolithic tool types, this is deemed unlikely. The second difference in the tool kit variables, the distal terminus types, indicates that the younger assemblage possesses more pointed tools than the older assemblage. As noted above, both assemblages are dominated by Upper Paleolithic tool types, without the presence of any distinctive retouch type. In sum, the two tool kits evidence a difference value of only two out of seven or 0.29 out of 1.0. Together with the low measure of difference of the blank production, this data tends to corroborate the original allocation of these assemblages to the same industrial affiliation (i.e., Svoboda 1987).

5.7.3 Comparison of Stránská skála III and Vedrovice V

The next chronological pair-wise comparison within the Central European region opposes the later Bohunician assemblage from Stránská skála III layer 5, dated to 38,200 ± 1,100 BP, with the Szeletian assemblage from Vedrovice V, dated between 35,000 and 39,500 BP. Although situated in a raw material rich area, the Krumlovsky-lès region to the west of the Brno Basin where the Stránská skála sites are located, Vedrovice V indicates more behavioral distance from its raw material source than the III layer 5 assemblage (Table 5.73). Vedrovice

Table 5.74 Comparison of Blank Production for Stránská skála III and Vedrovice V

Flintknapping Steps by Domain	Stránská skála III	Vedrovice V	Significant Difference?
Core Modification			
Core Orientation	Preferential use of a Longitudinal Surface	Preferential use of a Longitudinal Surface	No
Core Convexity Management	Débordants, front crests, & side blade removals	Lateral core tablets, changes of orientation	Yes
Number of Changes/2 Steps			**1/2=0.5**
Platform Maintenance			
Platform Treatment	Unprepared: 55% Prepared: 45% n=367	Unprepared: 61% Prepared: 39% n=380	No p=.10, Fisher's Exact
External Platform Angle (degrees)	mean:86.4, s.d.: 14.0 n=340	mean:89.0, s.d.: 17.4, n=342,	Yes p=.03, t=-2.211*, df=652.999
Platform Thickness	mean: 4.42, s.d.: 2.16 n=344	mean: 5.01, s.d.: 3.64, n=359	Yes p=.01, t=-2.630*, df=587.872
Number of Changes/3 Steps			**2/3=0.67**
Direction of Core Exploitation			
Direction of Early Exploitation	Bidirectional	Bidirectional & Unidirectional	Yes
Direction of Late Exploitation	Unidirectional	Unidirectional	No
Number of Changes/2 Steps			**1/2=0.5**
Dorsal Surface Convexity System			
Longitudinal Convexity: Length/Width Ratio	mean: 1.83, s.d.: 0.73 n=421	mean: 1.47, s.d.: 0.57, n=542	Yes p=.00, t=-8.272*, df=779.614
Shape of Convexity: Lateral Edges of Blanks	Parallel: 59% Convergent: 18% Expanding: 10% Ovoid: 13% n=417	Parallel: 39% Convergent: 16% Expanding: 31% Ovoid: 14% n=451	Yes p=.00, G^2=68.776, df=3
Curvature of Convexity: Profile of Blanks	Straight: 47% Curved: 30% Twisted: 23% n=417	Straight: 66% Curved: 15% Twisted: 19% n=465	Yes p=.00, G^2=35.234, df=2
Lateral Convexity: Cross-Section of Blanks	Triangular: 39% Trapezoidal: 56% Other: 5% n=420	Triangular: 52% Trapezoidal: 27% Other: 20% n=493	Yes p=.00, G^2=101.339, df=2
Vertical Convexity: Width/Thickness Ratio	mean: 3.97, s.d.: 1.72 n=421	mean: 4.26, s.d.: 1.82, n=542	Yes p=.01, t=2.558*, df=925.101
Number of Changes/5 Steps			**5/5=1**
Total Measure of Difference in Blank Production			**2.67**

*The t-test was calculated with equal variances not being assumed due to a significant F value on Levene's Test for Equality of Variances.

Table 5.75 Comparison of Tool Kit Morphologies for Stránská skála III and Vedrovice V

Tool Kit Morphology Variable	Stránská skála III	Vedrovice V	Significant Difference?
Laminarity	mean: 2.02, s.d.: 0.91 n=30	mean: 1.51, s.d.: 0.52 n=115	Yes p=.00, t–2.939*, df=34.041
Width/Thickness	mean: 4.29, s.d.: 1.56 n=30	mean: 3.59, s.d.: 1.56 n=115	Yes p=.03, t=2.190, df=143
Lateral Edges	Parallel: 29% Convergent: 54% Expanding: 11% Ovoid: 7% n=28	Parallel: 31% Convergent: 12% Expanding: 46% Ovoid: 11% n=61\	Yes p=.00, G^2=21.263, df=3†
Distal Terminus	Blunt: 21% Pointed: 79% n=24	Blunt: 81% Pointed: 19% n=36 p=.00,	Yes Fisher's Exact
Profile	Straight: 50% Curved: 27% Twisted: 23% n=30	Straight: 64% Curved: 13% Twisted: 23% n=75	No p=.25, G^2=2.773, df=2
Unique Types of Retouch	Normal retouch	Flat bifacial retouch	Yes
Tool Types	UP tools dominate	MP tools dominate	Yes
Number of Changes/7 Steps			**6/7**
Total Measure of Difference in Tool Kit Morphology			**0.86**

*The t-test was calculated with equal variances not being assumed due to a significant F value on Levene's Test for Equality of Variances.

†Although 1 cell has an expected count less than 5, this accounts for only 12.5% of the cells and so does not invalidate the assumptions of a χ^2 test. The minimum expected count is 2.83.

has both more retouched tools in relation to its unretouched blanks (Figure 5.43) as well as more non-cortical blanks compared to cortical blanks (Figure 5.44) than in the III layer 5 assemblage. The former ratio indicates that the dimensions of the Szeletian tool kit may be more affected by a greater intensity of retouch while the latter ratio indicates that the Szeletian may have less of the beginning of its core reduction sequence than captured at III layer 5. The evidence of the core facets in relation to the dimensions of the debitage further indicates that Vedrovice V cores were reduced further than those at III layer 5. Together, these variables indicate that Stránská skála III layer 5 may capture more of the beginning and less of the end of the core reduction sequence as witnessed at Vedrovice V. Given the hypothetical close proximity to raw material in the Krumlovsky-lès region, it is possible that a difference in site function rather than distance to a raw material source is the cause for the differences in the variables summarized in Table 5.73. Nonetheless, enough of the core reduction sequence is common to these two assemblages to warrant a systematic comparison.

Table 5.76 Stránská skála IIIa-4 and Stránská skála IIIc, comparability of the assemblages based on intensity of core and tool reduction

Variables	Stránská skála IIIa-4	Stránská skála IIIc	Evaluation
Tools/Flake Ratio	0.16	0.15	Similar
Non-cortical Debitage/ Cortical Debitage Ratio	1.06	0.66	Similar
Henry's Dimensional Analysis	Slight core exhaustion.	Little evidence of core exhaustion.	Fairly Similar

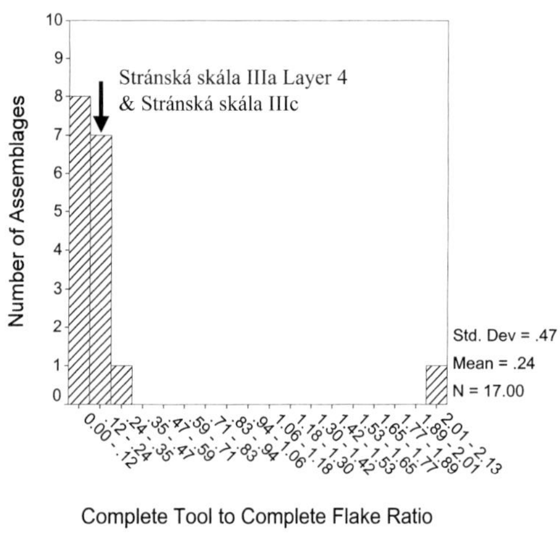

Figure 5.45 Stránská skála IIIa Level 4 and Stránská skála IIIc compared for tool/flake ratio values.

Blank Production

This comparison is presented in Table 5.74. For the first step within the first domain, the majority of both the Stránská skála III cores and the Vedrovice V cores evidence the preferential use of one surface on a longitudinal flake core form. The second step, core convexity management, exhibits different options in each assemblage. At Stránská skála III, a management of the convexities through frontal crests, lateral débordant removals, and lateral blade removals is indicated. At Vedrovice V, convexities were managed through changes of core orientation from one platform cores to crossed platform cores, producing lateral core tablets, and refreshing the convexity by using the original platform as a rejuvenating crested flake. Only in this second step, core management, do the two assemblages display significantly different options.

Of the three analytical steps within the platform maintenance domain, only the cumulative choice of platform treatment (prepared versus unprepared) is similar between these two assemblages. The external platform angles of debitage from Vedrovice V are significantly larger and the platforms thicker than in Stránská skála III. This produces a measure of difference for this domain of 2/3 or 0.66.

The next knapping domain, direction of core exploitation, concerns the variability resulting from different options of removing flakes from single to multiple platform cores, creating debitage with particular dorsal scar patterns. As indicated in the discussion of Figures 5.18 and 5.23, the assemblages differ in the dominant dorsal scar patterns in early core exploitation, specifically the high frequency of unidirectional as well as bidirectional blanks at Vedrovice V. In late core exploitation, both assemblages are dominated by unidirectional scar patterns. This produces a difference of 0.5 for this domain.

All five behavioral categories within the surface convexity domain differ between the two assemblages. This is the result of Vedrovice V's flake-based blank production, variable lateral edges, straighter profiles, and thinner but more triangular use of convexities. This produces a difference measure of 1. For the whole of blank production taken together, the measure of difference between these assemblages is 2.67 out of a maximum of 4.0, indicating that they are more dissimilar than they are similar.

Tool Kit Morphology

Of the variables which characterize the tool kit morphology of these two assemblages, all but one show statistically significant differences. Only the profile of the tools is similar in both assemblages. The Bohunician assemblage possesses tools of a higher laminarity, thinner cross-sections (a reversal of the differences between the debitage samples), more pointed distal ends, and more convergent lateral edges. The test of the last variable, lateral edge, contains one cell with an expected count of fewer than five observations (the ovoid cell for SSIII layer 5), making its significance problematic but for the clear distinction in the convergent and expanding frequencies between the assemblages. The lack of bifacial retouch in the Bohunician site and the dominance of Middle Paleolithic retouched tool types in the Szeletian site also contribute to the final difference measure between the tool kits of 0.86 out of 1.0. This indicates that the tool kits are very different in terms of their cutting edge morphologies.

5.7.4 *Comparison of Stránská skála IIIa Layer 4 and Stránská skála IIIc*

The next comparison does not chronologically follow from the previous comparison but it warranted as an assay of the variability within the Bohunician Industrial type through time. As the assemblage from Stránská skála IIIc has provided dates younger than those at IIIa layer 4 but is only ten or so meters to the north of the older excavation, it is intriguing to evaluate the degree of difference in their blank production and tool kit morphology variables.

In terms of comparability, the tool/flake ratios (Table 5.76 and Figure 5.45) are almost identical, both at the low end of retouched tool production among the study assemblages. The non-cortical to cortical blank ratio (Figure 5.46), however, shows IIIc as possessing slightly more cortical pieces relative to non-cortical pieces than IIIa layer 4. In this way, IIIc witnesses more of the initial working of the cores and less of the extensive reduction of them than IIIa layer 4, although the latter is clearly not without its portion of early core exploitation to judge by its position relative to the other assemblages for this measure, not to mention the evidence of the refits from this assemblage. This interpretation is corroborated by the almost complete absence of core exhaustion evidenced at IIIc compared to a slight indication of such at IIIa layer 4. Together, these comparability variables indicate that IIIc and IIIa layer 4 are quite comparable as both are

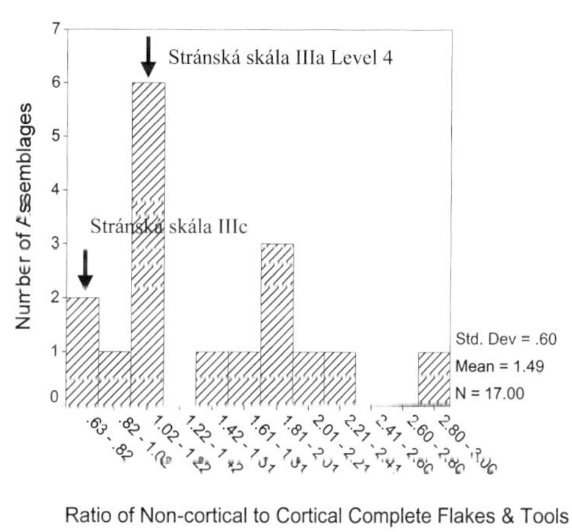

Figure 5.46 Stránská skála IIIa Level 4 and Stránská skála IIIc compared for non-cortical/cortical flake ratio values.

Table 5.77 Comparison of Blank Production for Stránská skála IIIa-4 and Stránská skála IIIc

Flintknapping Steps by Domain	Stránská skála IIIa-4	Stránská skála IIIc	Significant Difference?
Core Modification			
Core Orientation	Preferential use of a Longitudinal Surface	Preferential use of a Longitudinal Surface	No
Core Convexity Management	Débordants, front crests, & side blade removals	Débordants, front crests, & side blade removals	No
Number of Changes/2 Steps			**0/2=0**
Platform Maintenance			
Platform Treatment	Unprepared: 58% Prepared: 42% n=448	Unprepared: 58% Prepared: 42% n=767	No p=1.00, Fisher's Exact
External Platform Angle (degrees)	mean: 85.2, s.d.: 15.3, n=425	mean: 84.9, s.d.: 15.0 n=664	No p=.77, t=-0.297, df=1087
Platform Thickness	mean: 4.8, s.d.: 2.5, n= 433	mean: 4.6, s.d.: 2.5, n=741	No p=.07, t=-1.808, df=1172
Number of Changes/3 Steps			**0/3=0**
Direction of Core Exploitation			
Direction of Early Exploitation	Bidirectional	Bidirectional	No
Direction of Late Exploitation	Unidirectional	Unidirectional	No
Number of Changes/2 Steps			**0/2=0.0**
Dorsal Surface Convexity System			
Longitudinal Convexity: Length/Width Ratio	mean: 1.71, s.d.: 0.67, n=502	mean: 1.82, s.d.: 0.80 n=731	Yes p=.01, t=2.456*, df=1181.171
Shape of Convexity: Lateral Edges of Blanks	Parallel: 49% Convergent: 24% Expanding: 17% Ovoid: 11% n=489	Parallel: 41% Convergent: 16% Expanding: 18% Ovoid: 25% n=706	Yes p=.00, G^2=47.086, df=3
Curvature of Convexity: Profile of Blanks	Straight: 54% Curved: 29% Twisted: 17% n=496	Straight: 56% Curved: 25% Twisted: 19% n=731	No p=.28, G^2=2.573, df=2
Lateral Convexity: Cross-Section of Blanks	Triangular: 45% Trapezoidal: 50% Other: 5% n=495	Triangular: 41% Trapezoidal: 47% Other: 12% n=724	Yes p=.00, G^2=19.056, df=2
Vertical Convexity: Width/Thickness Ratio	mean: 4.03, s.d.: 1.83, n=502	mean: 4.15, s.d.: 1.83 n=731	No p=.28, t=1.080, df=1231
Number of Changes/5 Steps			**3/5=0.6**
Total Measure of Difference in Blank Production			**0.60**

*The t-test was calculated with equal variances not being assumed due to a significant F value on Levene's Test for Equality of Variances.

likely raw material workshops with core reduction behaviors overlapping for most the sequence, even if IIIc may lack some of the behaviors resulting from late core reduction seen in IIIa layer 4.

Blank Production

As presented in Table 5.77, in the majority of both the Stránská skála IIIa layer 4 cores and the Stránská skála IIIc cores, one surface of a longitudinal blade/flake core form is used preferentially over the other. Core convexity management also shows a similar choice of options in the two assemblages, as the attribute analysis and existing refittings of individual cores from both assemblages indicate a management of the convexities through frontal crests, lateral débordant removals, and side blade removals, although this last feature

Table 5.78 Comparison of Tool Kit Morphologies for Stránská skála IIIa-4 and Stránská skála IIIc

Tool Kit Morphology Variable	Stránská skála IIIa-4	Stránská skála IIIc	Significant Difference?
Laminarity	mean: 1.58 s.d.: 0.50 n=67	mean: 1.78 s.d.: 0.70 n=93	Yes p=.03 t=-2.148* df=157.955
Width/Thickness	mean: 4.28 s.d.: 2.14 n=67	mean: 4.24 s.d.: 1.67 n=93	No p=.89 t=0.139, df=158
Lateral Edges	Parallel: 39% Convergent: 31% Expanding: 16% Ovoid: 14% n=57	Parallel: 41% Convergent: 18% Expanding: 14% Ovoid: 27% n=88	No p=.13 G^2=5.654, df=3
Distal Terminus	Blunt: 50% Pointed: 50% n=36	Blunt: 59% Pointed: 41% n=69 p=.41	No Fisher's Exact
Profile	Straight: 45% Curved: 32% Twisted: 23% n=62	Straight: 53% Curved: 29% Twisted: 18% n=93	No p=.64, G^2=0.896, df=2
Unique Types of Retouch	Normal retouch	Normal retouch	No
Tool Types	UP tools dominate	UP tools dominate	No
Number of Changes/7Steps			**1/7**
Total Measure of Difference in Tool Kit Morphology			**0.14**

*The t-test was calculated with equal variances not being assumed due to a significant F value on Levene's Test for Equality of Variances.

Table 5.79 Stránská skála III and Stránská skála IIIc, comparability of the assemblages based on intensity of core and tool reduction

Variables	Stránská skála III	Stránská skála IIIc	Evaluation
Tools/Flake Ratio	0.08	0.15	Similar
Non-cortical Debitage/ Cortical Debitage Ratio	1.19	0.66	Fairly Similar
Henry's Dimensional Analysis	Little evidence of core exhaustion	Little evidence of core exhaustion	Similar

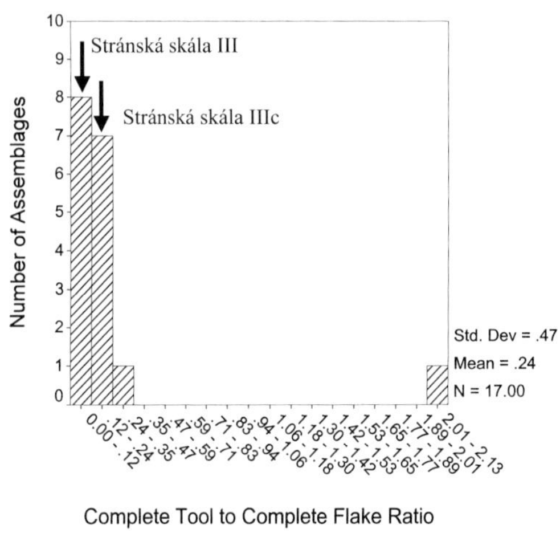

Figure 5.47Stránská skála III and Stránská skála IIIc compared for tool/flake ratio values.

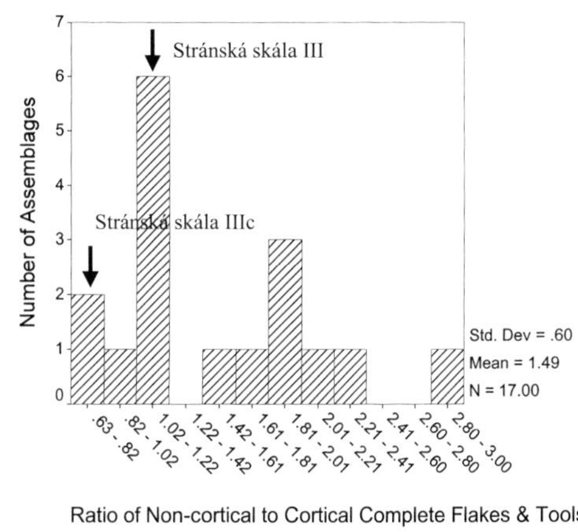

Figure 5.48 Stránská skála III and Stránská skála IIIc compared for non-cortical/cortical flake ratio values.

is not as evident in IIIc. No significant differences are qualitatively judged to exist between these assemblages for this domain as the options taken by the two groups of flintknappers are the same for both analytical steps. Of the three behavioral steps within the platform maintenance domain, none show any statistical difference between the assemblages. Likewise, the directionality of core exploitation shows no qualitative differences to exist between early or late core exploitation. That being said, the debitage in IIIc does show longer quartiles and a slower increase in the frequency of unidirectional dorsal scar patterns when compared to IIIa layer 4 (unidirectional patterns are barely more frequent in the second and third quartiles in IIIc compared to a drastic difference by the second quartile in IIIa layer 4), all agreeing with the interpretation that IIIc represents more of the initial core reduction sequence compared to IIIa layer 4. Of the five behavioral categories within the surface convexity domain, the slightly higher laminarity of IIIc marks it as statistically different from the older assemblage while the increase in "other" cross-sections as well as ovoid lateral edges in IIIc contribute to the statistical difference in these two variables. All told, blank production differs by 0.60 out of a maximum of 4.0 for the comparison between IIIc and IIIa layer 4.

Tool Kit Morphology
Of the seven variables considered for the morphology of the tool kit, only the higher laminarity of the few tools in the IIIc assemblage are statistically different from the tools in IIIa layer 4 (Table 5.78). Neither of these tool kits is actually blade dominated but IIIc is slightly more elongated than IIIa layer 4. This results in a difference measure of only 0.14 out of 1.0.

5.7.5 Comparison of Stránská skála III Layer 5 and Stránská skála IIIc

The next pair of assemblages may be considered a chronological comparison but it also serves to elucidate differences in blank production and tool kit morphology across multiple assemblages ascribed to the same industrial complex, the Bohunician.

In terms of comparability, the tool/flake ratios (Table 5.79 and Figure 5.47) are very similar, both at the low end of retouched tool production among the study assemblages. The non-cortical to cortical blank ratios (Figure 5.48), however, show IIIc as possessing even more cortical pieces relative to non-cortical pieces than III layer 5. In this way, IIIc witnesses more of the initial working of the cores and less of the exploitation of the cores compared to III layer 5, although the latter is clearly not without its portion of early core exploitation to judge by its position relative to the other assemblages for this measure, not to mention the evidence of the refits from this assemblage. This interpretation is corroborated by the almost complete absence of core exhaustion evidenced at IIIc compared to a few instances of core facets falling below the distribution of the smallest debitage in III layer 5. Together, these comparability variables indicate that IIIc and III layer 5 are extremely comparable as both are likely raw material workshops in which the behaviors of core reduction overlapped for most of the sequence, even if IIIc possesses even more of the initial decortification stage and/or less exploitation of the cores than see in III layer 5.

Blank Production
As presented in Table 5.80, in the majority of both the Stránská skála III layer 5 cores and the Stránská skála IIIc cores, one surface of a longitudinal blade/flake core form is used preferentially over the other. Core convexity management also shows a similar choice of options in the two assemblages, as the attribute analysis and existing refittings of individual cores from both assemblages indicate a management of the convexities through frontal crests, lateral débordant removals, and side blade removals, although this last feature is not as evident in IIIc. No significant differences are qualitatively judged to exist between these assemblages for this domain as the options taken by the two groups of flintknappers are the same for both analytical steps. Of the three behavioral steps within the platform maintenance domain, none show any statistical difference between the assemblages. Likewise, the directionality of core exploitation shows no qualitative differences to exist between early or late core exploitation. As with IIIa layer 5, the debitage from IIIc shows a slower increase in unidirectional scar patterns by length quartile when compared to III layer 5. However, as one might expect given the even closer similarity in the degree of core reduction between IIIc and III layer 5 compared to IIIa layer 4, both IIIc and III layer 5 show only marginally more unidirectional blanks in the second quartile. By the third quartile, however, III

Table 5.80 Comparison of Blank Production for Stránská skála III and Stránská skála IIIc

Flintknapping Steps by Domain	Stránská skála III	Stránská skála IIIc	Significant Difference?
Core Modification			
Core Orientation	Preferential use of a Longitudinal Surface	Preferential use of a Longitudinal Surface	No
Core Convexity Management	Débordants, front crests, & side blade removals	Débordants, front crests, & side blade removals	No
Number of Changes/2 Steps			**0/2=0**
Platform Maintenance Platform Treatment	Unprepared: 55% Prepared: 45% n=367	Unprepared: 58% Prepared: 42% n=767	No p=.31 Fisher's Exact
External Platform Angle (degrees)	mean: 86.4 s.d.: 14.0 n=340	mean: 84.9 s.d.: 15.0 n=664	No p=.15 t=-1.454*, df=1002
Platform Thickness	mean: 4.42 s.d.: 2.16 n=344	mean: 4.6 s.d.: 2.5 n= 741	No p=.37 t=0.892*, df=768.839
Number of Changes/3 Steps			**0/3=0**
Direction of Core Exploitation			
Direction of Early Exploitation	Bidirectional	Bidirectional	No
Direction of Late Exploitation	Unidirectional	Unidirectional	No
Number of Changes/2 Steps			**0/2=0.0**
Dorsal Surface Convexity System			
Longitudinal Convexity: Length/Width Ratio	mean: 1.83 s.d.: 0.74 n=397	mean: 1.82 s.d.: 0.80 n=731	No p=.72 t=-0.358, df=1126
Shape of Convexity: Lateral Edges of Blanks	Parallel: 59% Convergent: 17% Expanding: 10% Ovoid: 14% n=395	Parallel: 41% Convergent: 16% Expanding: 18% Ovoid: 25% n=706	Yes p=.00 G^2=43.870, df=3
Curvature of Convexity: Profile of Blanks	Straight: 47% Curved: 30% Twisted: 23% n=396	Straight: 56% Curved: 25% Twisted: 19% n=731	Yes p=.01 G^2=8.916, df=2
Lateral Convexity: Cross-Section of Blanks	Triangular: 40% Trapezoidal: 55% Other: 5% n=397	Triangular: 41% Trapezoidal: 47% Other: 12% n=724	Yes p=.00 G^2=20.580, df=2
Vertical Convexity: Width/Thickness Ratio	mean: 3.98 s.d.: 1.73 n=397	mean: 4.15 s.d.: 1.84 n=731	No p=.15 t=1.449, df=1126
Number of Changes/5 Steps			**3/5=0.6**
Total Measure of Difference in Blank Production			**0.60**

*The t-test was calculated with equal variances not being assumed due to a significant F value on Levene's Test for Equality of Variances.

Table 5.81 Comparison of Tool Kit Morphologies for Stránská skála III and Stránská skála IIIc

Tool Kit Morphology Variable	Stránská skála III	Stránská skála IIIc	Significant Difference?
Laminarity	mean: 1.78 s.d.: 0.70 n=93	mean: 2.02 s.d.: 0.91 n=30	No p=.13 t=-1.513, df=121
Width/Thickness	mean: 4.24 s.d.: 1.67 n=93	mean: 4.29 s.d.: 1.57 n=30	No p=.88, t=-0.152, df=121
Lateral Edges	Parallel: 41% Convergent: 18% Expanding: 14% Ovoid: 27% n=88	Parallel: 29% Convergent: 54% Expanding: 11% Ovoid: 7% n=28	Yes p=.00, G^2=14.437, df=3*
Distal Terminus	Blunt: 59% Pointed: 41% n=69	Blunt: 21% Pointed: 79% n=24	Yes p=.00, Fisher's Exact
Profile	Straight: 53% Curved: 29% Twisted: 18% n=93	Straight: 50% Curved: 27% Twisted: 23% n=30	No p=.83, G^2=0.362, df=2
Unique Types of Retouch	Normal retouch	Normal retouch	No
Tool Types	UP tools dominate	UP tools dominate	No
Number of Changes/7 Steps			**2/7**
Total Measure of Difference in Tool Kit Morphology			**0.29**

*Although 1 cell has an expected count less than 5, this accounts for only 12.5% of the cells and so does not invalidate the assumptions of a $\chi 2$ test. The minimum expected count is 3.62.

Table 5.82 Stránská skála IIIc and Vedrovice V, comparability of the assemblages based on intensity of core and tool reduction

Variables	Stránská skála IIIc	Vedrovice V	Evaluation
Tools/Flake Ratio	0.15	0.32	Moderately different
Non cortical Debitage/ Cortical Debitage Ratio	0.66	2.02	Moderately different
Henry's Dimensional Analysis	Little evidence of core exhaustion.	Moderate core exhaustion	Moderately different

layer 5 evidences many more unidirectional blanks while IIIc waits until the last quartile to evidence the clear dominance of unidirectional over bidirectional exploitation. When considering the absolute length of blanks dividing the first and second quartile in each assemblage (47 mm for IIIc, 44 mm for III, and 41 mm for IIIa layer 4), it is clear that the assemblage with more of the beginning of the core reduction sequence possesses the longer blanks. This all agrees with the relative ordering of IIIc, III layer 5, and finally IIIa layer 4 in terms of decreasing representation of the earliest portion of the core reduction sequence. It also indicates that the

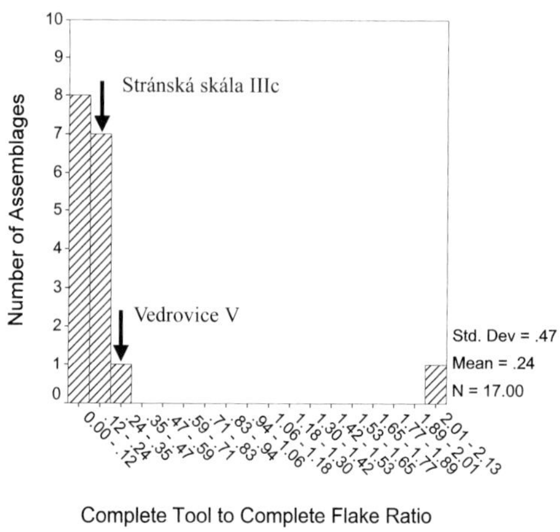

Figure 5.49 *Stránská skála IIIc and Vedrovice V compared for tool/flake ratio values.*

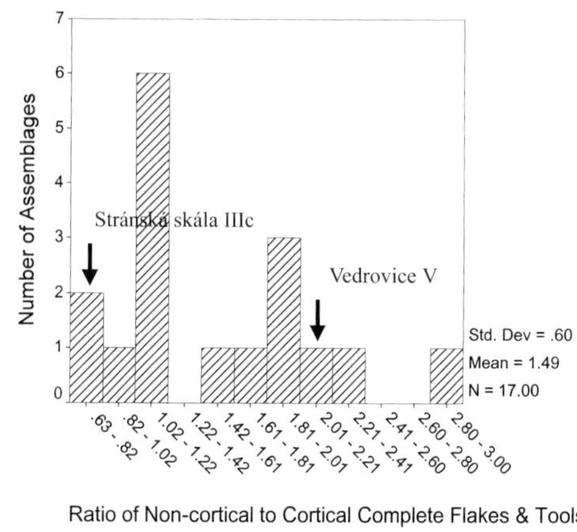

Figure 5.50 *Stránská skála IIIc and Vedrovice V compared for non-cortical/cortical flake ratio values.*

relative speed (from first to fourth quartile) by which the bidirectional scar patterns are replaced by unidirectional patterns is related to how much of the earliest working of the raw material is included in the assemblage.

Of the five behavioral categories within the surface convexity domain, only lateral edge type, profile type, and cross-section type are statistically significantly different between the assemblages. All told, blank production differs by 0.60 out of a maximum of 4.0 for the comparison between IIIc and III layer 5, the same value (although based on different distinctions in blank production) seen between IIIc and IIIa layer 4. These assemblages are thus far more similar than dissimilar.

Tool Kit Morphology

Of the seven variables considered for the morphology of the tool kit, only two variables resulted in statistically distinct values (Table 5.81), the higher frequency of ovoid lateral edges in IIIc compared to the high frequency of convergent edges in III layer 5 as well as the high frequency of pointed distal ends in III layer 5. These two differences result in a difference measure of 0.29 out of 1.0.

5.7.6 Comparison of Stránská skála IIIc and Vedrovice V

The next comparison opposes a late Bohunician, dated between 38,300 ± 1,100 BP and 34,440 ± 720 BP, with the Szeletian assemblage from Vedrovice V, dated between 39,500 ± 1100 and 35,150 ± 650 BP. Found in the same paleosol but approximately 30 km apart, the radiometric dates for these assemblages make them contemporary. As with the comparison with Stránská skála III layer 5, Vedrovice V indicates more behavioral distance from its raw material source than the IIIc assemblage (Table 5.82). Vedrovice has both more retouched tools in relation to its unretouched blanks (Figure 5.49) as well as more non-cortical blanks compared to cortical

Table 5.83 Comparison of Blank Production for Stránská skála IIIc and Vedrovice V

Flintknapping Steps by Domain	Stránská skála IIIc	Vedrovice V	Significant Difference?
Core Modification			
Core Orientation	Preferential use of a Longitudinal Surface	Preferential use of a Longitudinal Surface	No
Core Convexity Management	Débordants, front crests, & side blade removals	Lateral core tablets, changes of orientation	Yes
Number of Changes/2 Steps			**1/2=0.5**
Platform Maintenance			
Platform Treatment	Unprepared: 58% Prepared: 42% n=767	Unprepared: 61% Prepared: 40% n=380	No p=.41, Fisher's Exact
External Platform Angle (degrees)	mean: 84.9, s.d.: 15.0, n=664	mean: 89.0, s.d.: 17.4, n=341	Yes p=.00, t=-3.708*, df=605.825
Platform Thickness	mean: 4.55, s.d.: 2.52, n= 741	mean: 5.00, s.d.: 3.64, ,n=358	Yes p=.04, t=-2.120*, df=527.666
Number of Changes/3 Steps			**2/3=0.67**
Direction of Core Exploitation			
Direction of Early Exploitation	Bidirectional	Bidirectional & Unidirectional	Yes
Direction of Late Exploitation	Unidirectional	Unidirectional	No
Number of Changes/2 Steps			**1/2=0.5**
Dorsal Surface Convexity System			
Longitudinal Convexity: Length/Width Ratio	mean: 1.82, s.d.: 0.80, n=731	mean: 1.50, s.d.: 0.59, n=473	Yes p=.00, t=8.024*, df=1184.317
Shape of Convexity: Lateral Edges of Blanks	Parallel: 41% Convergent: 16% Expanding: 18% Ovoid: 25% n=706	Parallel: 39% Convergent: 15% Expanding: 31% Ovoid: 14% n=410	Yes p=.00, G^2=35.253, df=3
Curvature of Convexity: Profile of Blanks	Straight: 56% Curved: 25% Twisted: 19% n=73	Straight: 64% Curved: 16% Twisted: 20% n=419	Yes p=.00, G^2=14.617, df=2
Lateral Convexity: Cross-Section of Blanks	Triangular: 41% Trapezoidal: 47% Other: 12% n=724	Triangular: 50% Trapezoidal: 29% Other: 21% n=431	Yes p=.00, G^2=41.527, df=2
Vertical Convexity: Width/Thickness Ratio	mean: 4.15, s.d.: 1.84, n=731	mean: 4.26 s.d.: 1.84, n=473	No p=.32, t=-0.992, df=1008.648
Number of Changes/5 Steps			**4/5=0.8**
Total Measure of Difference in Blank Production			**2.47**

*The t-test was calculated with equal variances not being assumed due to a significant F value on Levene's Test for Equality of Variances.

Table 5.84 Comparison of Tool Kit Morphologies for Stránská skála IIIc and Vedrovice V

Tool Kit Morphology Variable	Stránská skála IIIc	Vedrovice V	Significant Difference?
Laminarity	mean: 1.78, s.d.: 0.70 n=93	mean: 1.51, s.d.: 0.52 n=115	Yes p=.00, t=3.035*, df=164.009
Width/Thickness	mean: 4.24, s.d.: 1.67 n=93	mean: 3.59, s.d.: 1.56 n=115	Yes p=.00, t=2.885, df=206
Lateral Edges	Parallel: 41% Convergent: 18% Expanding: 14% Ovoid: 27% n=88	Parallel: 31% Convergent: 12% Expanding: 46% Ovoid: 11% Parallel: 31% n=61	Yes p=.00, G^2=20.479, df=3
Distal Terminus	Blunt: 59% Pointed: 41% n=69	Blunt: 81% Pointed: 19% n=36	Yes p=.03, Fisher's Exact
Profile	Straight: 53% Curved: 29% Twisted: 18% n=93	Straight: 64% Curved: 13% Twisted: 23% n=75	Yes p=.05, G^2=6.190, df=2
Unique Types of Retouch	Normal retouch	Flat bifacial retouch	Yes
Tool Types	UP tools dominate	MP tools dominate	Yes
Number of Changes/7 Steps			**7/7**
Total Measure of Difference in Tool Kit Morphology			**1.0**

*The t-test was calculated with equal variances not being assumed due to a significant F value on Levene's Test for Equality of Variances.

blanks (Figure 5.50) than in the IIIc assemblage. The former ratio indicates that the dimensions of the Szeletian tool kit may be more affected by a greater intensity of retouch while the latter ratio indicates that the Szeletian may have less of the beginning of its core reduction sequence than captured at IIIc. The evidence of the core facets in relation to the dimensions of the debitage further indicates that Vedrovice V cores were reduced further than those at IIIc. Together, these variables indicate that Stránská skála IIIc may capture more of the beginning and less of the end of the core reduction sequence as witnessed at Vedrovice V. Given the hypothetical close proximity to raw material in the Krumlovsky-lès region, it is possible that a difference in site function rather than distance to a raw material source is the cause for the differences in the variables summarized in Table 5.82. These differences, however, are not significant enough to problematize the comparison of the assemblages, as most of the core reduction sequence is shared by both assemblages.

Blank Production

For the first step within the first domain, the

Table 5.85 Vedrovice V and Stránská skála IIa-4 and IIIa-3, comparability of the assemblages based on intensity of core and tool reduction

Variables	Vedrovice V	Stránská skála Aurignacian	Evaluation
Tools/Flake Ratio	0.32	0.16	Moderately different
Non-cortical Debitage/ Cortical Debitage Ratio	2.02	1.51	Slightly different
Henry's Dimensional Analysis	Moderate core exhaustion	Less moderate core exhaustion	Slightly different

majority of both the Stránská skála IIIc cores and the Vedrovice V cores evidence the preferential use of one surface on a longitudinal flake core form. The second step, core convexity management, exhibits different options in each assemblage. At Stránská skála IIIc, a management of the convexities through frontal crests, lateral débordant removals, and, to a moderate degree, lateral blade removals is indicated. At Vedrovice V, convexities were managed through changes of core orientation from one platform cores to crossed platform cores, producing lateral core tablets, and refreshing the convexity by using the original platform as a rejuvenating crested flake. Only in this second step, core management, do the two assemblages display significantly different options.

Of the three analytical steps within the platform maintenance domain, only the cumulative choice of platform treatment (prepared versus unprepared) is similar between these two assemblages. The external platform angles of debitage from Vedrovice V are significantly steeper and the platforms thicker than in Stránská skála IIIc. This produces a measure of difference for this domain of 2/3 or 0.66.

The next knapping domain, direction of core exploitation, concerns the variability resulting from different options of removing flakes from single to multiple platform cores, creating debitage with particular dorsal scar patterns. As indicated in the discussion of Figures 5.23 and 5.30, the assemblages differ in the dominant dorsal scar patterns in early core exploitation, specifically the high frequency of unidirectional as well as bidirectional blanks at Vedrovice V. In late core exploitation, both assemblages are dominated by unidirectional scar patterns. This produces a difference of 0.5 for this domain.

All but one of the five behavioral categories within the surface convexity domain differ between the two assemblages. This is the result of Vedrovice V's flake-based blank production, variable lateral edges, straighter profiles, and more triangular use of convexities. Only in the thinness of the convexities exploited (width/ thickness ratios) did the assemblages match. For the entirety of the blank production behaviors, the measure of difference between these assemblages is 2.47 out of a maximum of 4.0, indicating that they are more dissimilar than they are similar.

Tool Kit Morphology

Of the variables which characterize the tool kit morphology of these two assemblages, all show statistically significant differences. The Bohunician assemblage possesses tools of a higher laminarity (although not blade-dominated), thinner cross-sections, more pointed distal ends, more curved profiles, and more parallel and ovoid lateral edges than the Szeletian assemblage. The lack of bifacial retouch in the Bohunician site and

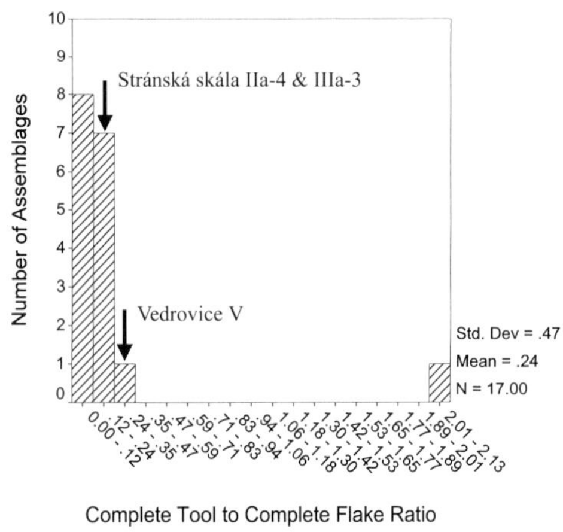

Figure 5.51 Vedrovice V and Stránská skála IIa-4 and IIIa-3 compared for tool/flake ratio values.

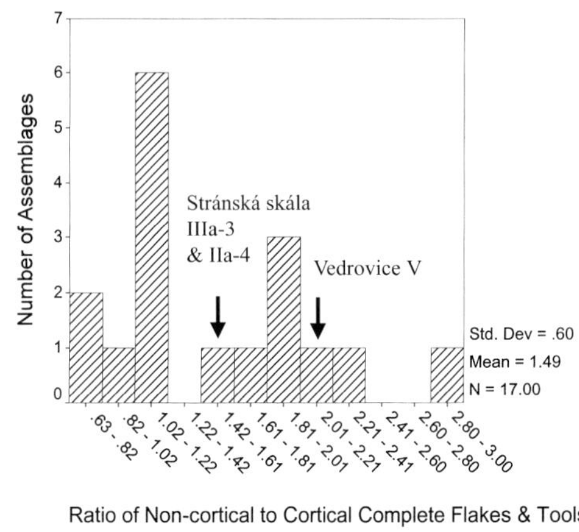

Figure 5.52 Vedrovice V and Stránská skála IIa-4 and IIIa-3 compared for non-cortical/cortical flake ratio values.

the dominance of Middle Paleolithic retouched tool types in the Szeletian site also contribute to the final maximum difference measure between the tool kits of 1.0 out of 1.0. This indicates that the tool kits are as different as measurable given the current analytical structure.

5.7.7 Vedrovice V and the Aurignacian of Stránská skála IIIa-3 and IIa-4

The next chronological pair-wise comparison within the Central European region opposes the Szeletian assemblage from Vedrovice V, dated between 39,500 ± 1100 and 35,150 ± 650 BP, with the Aurignacian assemblages from the Stránská skála hillside, Stránská skála IIIa-3 and IIa-4, dated to 30,980 ± 360 and 32,350 ± 900 BP, respectively. As with the comparison with the Bohunician assemblages at Stránská skála, Vedrovice V indicates more behavioral distance from its raw material source than the Aurignacian assemblage (Table 5.85). Vedrovice has both more retouched tools in relation to its unretouched blanks (Figure 5.51) as well as more non-cortical blanks compared to cortical blanks (Figure 5.52) than in the Aurignacian assemblage. The former ratio indicates that the dimensions of the Szeletian tool kit may be more affected by a greater intensity of retouch while the latter ratio indicates that the Szeletian may have less of the beginning of its core reduction sequence than captured in the Aurignacian samples. The evidence of the core facets in relation to the dimensions of the debitage further indicates that Vedrovice V cores were reduced slightly more than those in the Aurignacian assemblage. For instance, even though both assemblages show core facets at the low end of the blank distribution, there are few core facets in the Szeletian assemblage which overlap with the upper end of the blank distribution, unlike the Aurignacian assemblage. Together, these variables indicate that the Stránská skála Aurignacian may capture more of the beginning and perhaps less of the end of the core reduction sequence than witnessed at Vedrovice V. The comparability of Vedrovice V is

Table 5.86 Comparison of Blank Production for Vedrovice V and Stránská skála IIa-4 and IIIa-3

Flintknapping Steps by Domain	Vedrovice V	Stránská skála IIIa-3 & IIa-4	Significant Difference?
Core Modification			
Core Orientation	Preferential use of a Longitudinal Surface	Preferential use of a Longitudinal Surface	No
Core Convexity Management	Lateral core tablets, changes of orientation	Frontal & lateral crests, core tablets	Yes
Number of Changes/2 Steps			**1/2=0.5**
Platform Maintenance			
Platform Treatment	Unprepared: 61% Prepared: 39% n=380	Unprepared: 69% Prepared: 31% n=325	Yes p=.03 Fisher's Exact
External Platform Angle (degrees)	mean: 89.0 s.d.: 17.4 n=342	mean: 82.5 s.d.: 15.1 n=306	Yes p=.00 t=5.095*, df=645.367
Platform Thickness	mean: 5.01 s.d.: 3.64 n=359	mean: 4.80 s.d.: 2.52 n=304	No p=.39 t=0.862*, df=636.759
Number of Changes/3 Steps			**2/3=0.67**
Direction of Core Exploitation			
Direction of Early Exploitation	Bidirectional & Unidirectional	Bidirectional & Unidirectional	No
Direction of Late Exploitation	Unidirectional	Unidirectional	No
Number of Changes/2 Steps			**0/2=0.0**
Dorsal Surface Convexity System			
Longitudinal Convexity: Length/Width Ratio	mean: 1.47 s.d.: 0.57 n=542	mean: 1.73 s.d.: 0.75 n=470	Yes p=.00 t=-6.137*, df=866.069
Shape of Convexity: Lateral Edges of Blanks	Parallel: 39% Convergent: 16% Expanding: 31% Ovoid: 14% n=451	Parallel: 57% Convergent: 17% Expanding: 16% Ovoid: 10% n=431	Yes p=.00 G^2=41.186, df=3
Curvature of Convexity: Profile of Blanks	Straight: 66% Curved: 15% Twisted: 19% n=465	Straight: 58% Curved: 24% Twisted: 18% n=430	Yes p=.01 G^2=10.079, df=2
Lateral Convexity: Cross-Section of Blanks	Triangular: 52% Trapezoidal: 27% Other: 20% n=493	Triangular: 46% Trapezoidal: 48% Other: 6% n=447	Yes p=.00 G^2=66.271, df=2
Vertical Convexity: Width/Thickness Ratio	mean: 4.26 s.d.: 1.82 n=542	mean: 3.45 s.d.: 1.51 n=470	Yes p=.00 t=7.731*, df=1007.809
Number of Changes/5 Steps			**5/5=1**
Total Measure of Difference in Blank Production			**2.17**

*The t-test was calculated with equal variances not being assumed due to a significant F value on Levene's Test for Equality of Variances.

Table 5.87 Comparison of Tool Kit Morphologies for Vedrovice V and Stránská skála IIa-4 and IIIa-3

Tool Kit Morphology Variable	Vedrovice V	Stránská skála IIIa-3 & IIa-4	Significant Difference?
Laminarity	mean: 1.51 s.d.: 0.52 n=115	mean: 1.73 s.d.: 0.78 n=52	No p=.07 t=-1.824*, df=72.086
Width/Thickness	mean: 3.59 s.d.: 1.56 n=115	mean: 3.33 s.d.: 1.48 n=52	No p=.31 t=1.016, df=165
Lateral Edges	Parallel: 31% Convergent: 12% Expanding: 46% Ovoid: 11% n=61	Parallel: 48% Convergent: 13% Expanding: 17% Ovoid: 22% n=46	Yes p=.02, G^2=10.487, df=3
Distal Terminus	Blunt: 81% Pointed: 19% n=36	Blunt: 63% Pointed: 37% n=19	No p=.20 Fisher's Exact
Profile	Straight: 64% Curved: 13% Twisted: 23% n=75	Straight: 63% Curved: 25% Twisted: 12% n=40	No p=.18 G^2=3.465, df=2
Unique Types of Retouch	Flat bifacial retouch	Carinated retouch	Yes
Tool Types	MP tools dominate	UP tools dominate	Yes
Number of Changes/7 Steps			**3/7**
Total Measure of Difference in Tool Kit Morphology			**0.43**

*The t-test was calculated with equal variances not being assumed due to a significant F value on Levene's Test for Equality of Variances.

Table 5.88 Stránská skála III layer 5 and Stránská skála IIa-4 and IIIa-3, comparability of the assemblages based on intensity of core and tool reduction

Variables	Stránská skála III	Stránská skála Aurignacian	Evaluation
Tools/Flake Ratio	0.08	0.16	Similar
Non-cortical Debitage/ Cortical Debitage Ratio	1.19	1.51	Slightly different
Henry's Dimensional Analysis	Little evidence of core exhaustion	Moderate core exhaustion	Different

better with the Aurignacian than with the Bohunician assemblages at Stránská skála, however, as more of the core reduction sequence is held in common in each assemblage. Given the hypothetical close proximity to raw material in the Krumlovsky-lès region, it is possible that a difference in site function rather than distance to a raw material source is the cause for the differences in the variables summarized in Table 5.85.

Blank Production

Blank production is compared in Table 5.86. Within the core modification domain, the majority of both the Vedrovice V cores and the

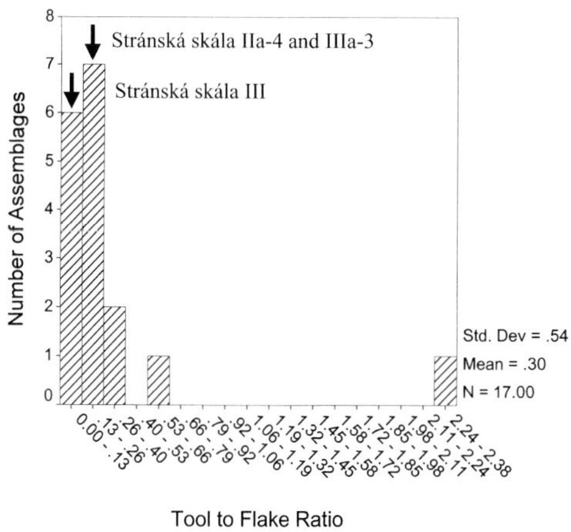

Figure 5.53 Stránská skála III layer 5 and Stránská skála IIa-4 and IIIa-3 compared for tool/flake ratio values.

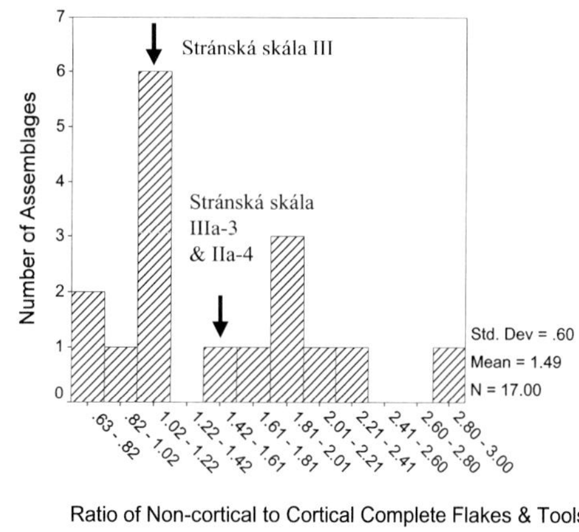

Figure 5.54 Stránská skála III layer 5 and Stránská skála IIa 4 and IIIa-3 compared for non-cortical/cortical flake ratio values.

Stránská skála Aurignacian cores possess the preferential use of one surface on a longitudinal core orientation. Different options were used, however, in the second step in this domain, core convexity management. In the assemblage from Vedrovice V, convexities were managed through consistent changes of core orientation from one platform cores to crossed platform cores, producing lateral core tablets, and refreshing the convexity by using the original platform as a rejuvenating crested flake. In the Aurignacian assemblages, cores exhibit the use of both frontal crests and rear crests to control the volume of the core. Core tablets are also used to rejuvenate surfaces. Only in this second step, core convexity management, do the two assemblages display significantly different options.

Of the three behavioral steps within the platform maintenance domain, only the cumulative choice of platform thickness is similar between these two assemblages. The platforms are more prepared and the external platform angles of debitage from Vedrovice V are significantly larger than in the Aurignacian assemblages. This produces a measure of difference for this domain of 2/3 or 0.66.

As indicated in the discussion of Figures 5.23 and 5.35 for the direction of core exploitation domain, both assemblages evidence the same tendencies to produce bidirectional and unidirectional debitage in early core exploitation while producing mostly unidirectional debitage at the end of core exploitation. This results in no difference for this domain.

Within the surface convexity domain, all five behavioral categories differ between the two assemblages. This is the result of Vedrovice V's less laminar blank production, variable lateral edges, straighter profiles, and thinner and more triangular use of convexity. This produces a measure of difference of 1.

The final measure of difference between the blank production choices of the Stránská skála Aurignacian and Vedrovice V assemblages, weighted by the four knapping domains, produces a value of 2.17 out of a possible maximum difference of 4.0, indicating that the details of blank production are slightly more distinct than similar.

Tool Kit Morphology

Of the dorsal surface convexity variables which characterize the tool kit morphology of these two assemblages, all but one fail to show a statistically significant difference. Only the lateral edge types are different between the assemblages, primarily due to relative frequencies of expanding edges in Vedrovice V and ovoid edges in the Aurignacian. This is an interesting result given the fact that the debitage dorsal surface convexity variables all showed statistically significant differences. The lack of bifacial retouch in the Aurignacian assemblage and the dominance of Middle Paleolithic retouched tool types in the Szeletian site also contribute to the final difference measure between the tool kits of 0.43 out of 1.0. This indicates that the tool kits are more similar than different.

5.7.8 Stránská skála III and the Aurignacian of Stránská skála IIIa-3 and IIa-4

The next pair-wise comparison is not only chronological but also necessary, given the contemporaneity between the later Bohunician industries and the Szeletian industry in Moravia. Either type could serve as an ancestor to the subsequent industry in the region, the Aurignacian. Thus, the present comparison opposes the later Bohunician assemblage from Stránská skála III layer 5, dated to 38,200 ± 1,100 BP, with the Aurignacian assemblages from the Stránská skála hillside, Stránská skála IIIa-3 and IIa-4, dated to 30,980 ± 360 and 32,350 ± 900 BP, respectively.

Given their close proximity in terms of distance to the nearby raw material source, it is no surprise that the comparability measures are quite similar between these assemblages (Table 5.88). The tool to flake ratio (Figure 5.53) shows these two assemblages to be extremely similar, while the non-cortical to cortical blank ratio (Figure 5.54) shows only a slight difference in the degree to which cortical nodules were begun on site. While the Bohunician assemblage tends toward the values associated with raw material workshops close to the outcrop, the Aurignacian assemblage is at the mean of all assemblages, making a raw material workshop designation less warranted. This difference in the beginning of the core reduction sequence is also reflected in a slight difference in the degree to which cores were reduced toward exhaustion. More of the core facets in the Aurignacian assemblage fall below or at the bottom of the blank distribution than in the Bohunician assemblage. While these differences would suggest that the Aurignacian assemblage captured a little more of the end of the core reduction sequence than the Bohunician assemblage, there are other processes which could also be involved. The fact that both assemblages are so similar in their distance to raw material makes the lack of numerous refit sequences in the Aurignacian suggestive, as compared to those found in the Bohunician assemblage. Whether a result of fewer consecutive knapping events which left refittable sequences or the removal of refittable blanks from the site by curation, the lack of debitage sequence refits in the Aurignacian may indicate that it is a different type of a palimpsest of activities than III layer 5, resulting from different mobility strategies or site function. That being said, these two assemblages are more comparable than other possible pairs of assemblages represented in the study as they share most of the core reduction sequence in common. Further, it is unlikely that retouch intensity will have any effect on the comparability of the dimensional data for either blank production or tool kit morphology. Nejman's (2009) study of the retouched tools from all Stránská skála Aurignacian and Bohuician assemblages showed no difference in Kuhn's Mean Geometric Index of Reduction (1990b) or in the patterns Clarkson's Index of Invasiveness (2002), indicating that tools were not differently reduced, according to these measures.

Blank Production

Within the first domain, core modification, the majority of both the Stránská skála III cores and the Aurignacian cores demonstrate the preferential use of one surface with a longitudinal blade/flake core orientation (Table 5.89). Core convexity management, however, differs between the two assemblages. In the assemblage from Stránská skála III, the management of convexities through frontal crests, lateral débordant removals, and side blade removals is indicated. In the Aurignacian assemblages from Stránská skála, convexities were managed through the application of frontal and rear crests as well as core tablets.

The next knapping domain, platform maintenance, includes the steps whereby the flintknapper modifies the platform surface and platform edge before each removal. All three behavioral steps within this domain show different options chosen in the two assemblages. In general, the Aurignacian platforms are less prepared, lower in EPA, and larger in PT than the Bohunician. This produces a measure of difference for this domain of 1.

The direction of core exploitation domain shows dissimilar options chosen during the beginning of core exploitation, with the Aurignacian showing a co-dominance of bidirectional and unidirectional scar patterns. By late core exploitation, however, the assemblages demonstrate similar central tendencies to produce unidirectional blanks. In section 5.7.5 above in which the direction of core exploitation evidenced in all three Bohunician assemblages was discussed, the hypothesis was presented that the relative absence in a given assemblage of the earliest working of cores increased the speed by which unidirectional blanks replaced bidirectional blanks as the dominant reduction strategy by length quartile. This hypothesis is suggestive when considering a combination of data related to the Stránská skála Aurignacian: the minor dominance of bidirectional over unidirectional patterns in the first quartile of blank length, the large dominance of unidirectional over bidirectional patterns by the second quartile, the smaller size of debitage in general as seen in the smaller length division between the first and second quartiles compared to the Bohunician assemblages (only 39 mm), and the higher noncortical to cortical blank ratio compared to all of the Bohunician assemblages. These data suggest that, given the above hypothesis, the relative difference in the direction of reduction at the beginning of core exploitation between the Aurignacian and the Bohunician from the Stránská skála hillside may have more to do with the degree to which the Aurignacian sample assemblages lack the earliest portion of the core reduction sequence than to discrete differences in hominin choices for the whole reduction sequence. While the Aurignacian and Bohunician data support this hypothesis and this interpretation, it is wiser to keep the evaluation of a distinct difference in the direction of reduction during early core exploitation between these assemblages in order to preserve the same degree of analytical resolution between regions for the larger study at hand. As other regional sequences were not sampled as thoroughly as the Central European one, with multiple assemblages studied

Table 5.89 Comparison of Blank Production for Stránská skála III Layer 5 and Stránská skála IIa-4 and IIIa-3

Flintknapping Steps by Domain	Stránská skála III	Stránská skála IIIa-3 & IIa-4	Significant Difference?
Core Modification			
Core Orientation	Preferential use of a Longitudinal Surface	Preferential use of a Longitudinal Surface	No
Core Convexity Management	Débordants, front crests, & side blade removals	Frontal & rear crests, core tablets	Yes
Number of Changes/2 Steps			**1/2=0.5**
Platform Maintenance			
Platform Treatment	Unprepared: 55% Prepared: 45% n=367	Unprepared: 69% Prepared: 31% n=325	Yes p=.00 Fisher's Exact
External Platform Angle (degrees)	mean: 86.4 s.d.: 14.0 n=340	mean: 82.5 s.d.: 15.1 n=306	Yes p=.00 t=3.333, df=644
Platform Thickness	mean: 4.42 s.d.: 2.16 n=344	mean: 4.80 s.d.: 2.52 n=304	Yes, p=.04 t=-2.067*, df=601.663
Number of Changes/3 Steps			**3/3=1.0**
Direction of Core Exploitation			
Direction of Early Exploitation	Bidirectional	Bidirectional & Unidirectional	Yes
Direction of Late Exploitation	Unidirectional	Unidirectional	No
Number of Changes/2 Steps			**1/2=0.5**
Dorsal Surface Convexity System			
Longitudinal Convexity: Length/Width Ratio	mean: 1.83 s.d.: 0.73 n=421	mean: 1.73 s.d.: 0.75 n=470	No p=.06 t=1.920, df=889
Shape of Convexity: Lateral Edges of Blanks	Parallel: 59% Convergent: 18% Expanding: 10% Ovoid: 13% n=417	Parallel: 57% Convergent: 17% Expanding: 16% Ovoid: 10% n=431	Yes p=.04, G^2=8.541, df=3
Curvature of Convexity: Profile of Blanks	Straight: 47% Curved: 30% Twisted: 23% n=417	Straight: 58% Curved: 24% Twisted: 18% n=430	Yes p=.01, G^2=9.684, df=2
Lateral Convexity: Cross-Section of Blanks	Triangular: 39% Trapezoidal: 56% Other: 5% n=420	Triangular: 46% Trapezoidal: 48% Other: 6% n=447	Yes p=.05, G^2=6.059, df=2
Vertical Convexity: Width/Thickness Ratio	mean: 3.97, s.d.: 1.72 n=421	mean: 3.45, s.d.: 1.51, n=470	Yes p=.00, t=4.768, df=889
Number of Changes/5 Steps			**4/5=0.8**
Total Measure of Difference in Blank Production			**2.80**

*The t-test was calculated with equal variances not being assumed due to a significant F value on Levene's Test for Equality of Variances.

for the same industrial type, there is no comparable data to apply similar hypotheses as above to distinctions in the blank length to dorsal scar pattern cross-tabulation. As such data may be collected in the near future, it is likely that a more subtle use of a combination of reduction sequence representation with the length/scar pattern cross-tabulation will soon be possible.

Returning to the comparison of the Stránská skála Aurignacian with the Bohunician of III layer 5, all but one of the five behavioral categories within the surface convexity domain differ between the assemblages. Only the degree of laminarity demonstrates a similar choice of knapping options between the Bohunician and the Aurignacian, both for elongated flake blanks. The differences in the dorsal surface convexity domain lie mainly in the Aurignacian's blanks possessing more triangular and thick cross-sections, straighter profiles, and slightly more expanding lateral edges. This produces a measure of difference of 4/5 or 0.80.

The final measure of difference between the blank production choices of the Stránská skála III layer 5 and the Stránská skála Aurignacian assemblages, weighted by the four knapping domains, produces a value of 2.80 out of a possible maximum difference of 4.0, indicating that the details of blank production are more distinct than similar. These differences are found mostly in the treatment of platforms and the exploitation of different dorsal surface convexities.

Tool Kit Morphology

Of the dorsal surface convexity variables which characterize the tool kit morphology of these two assemblages, four out of the seven show a statistically significant difference. As with the blank production comparison, the width/thickness ratio and lateral edge type frequencies differ significantly, the latter's divergent frequencies clearly different despite the violation of the test's assumptions with two cells with expected counts of fewer than 5. The distal terminus types also differ significantly, with opposite tendencies to blunt and pointed blanks. The lack of a credible carinated tool kit in the Bohunician assemblage (see above) produces the last significant difference, resulting in the final difference measure between the tool kits of 0.57 out of 1.0. This indicates that the tool kits are more distinct than different.

5.7.9 Kůlna Cave Layer 7a and Vedrovice V

The last pair-wise assemblage comparison in the Central European sequence does not follow chronologically from the preceding comparison. Since the Szeletian and Bohunician industries appear to be relatively contemporaneous, with the possibility that the Bohunician is slightly older as well as younger given the current dates, it is necessary to gauge the difference in knapping options taken between the Micoquian of Kůlna Cave, layer 7a, and the Szeletian of Vedrovice V. This is especially pertinent given the numerous claims for genetic relationships between these two industrial types (Cervinka 1927; Prošek 1953; Kadic 1915; Valoch 1976; Allsworth-Jones 1986; 1989; Neruda 2000).

As indicated in Table 5.91 and Figure 5.55, the differences in the tool to flake ratio between these assemblages is extreme. Even though Vedrovice V is relatively retouched compared to most of the assemblages in the study, the fact that Kůlna lies to the extreme right of the distribution of assemblages indicates significantly more application of retouch in this assemblage. While the excavation or curational history of the collection may contribute to the extreme position of Kůlna Cave layer 7a in this measure, caution is still warranted when comparing dimensional data, particularly with the tool kits, between these two assemblages. Kůlna's high frequency

Table 5.90 Comparison of Tool Kit Morphologies for Stránská skála III Layer 5 and Stránská skála IIa-4 and IIIa-3

Tool Kit Morphology Variable	Stránská skála III	Stránská skála IIIa-3 & IIa-4	Significant Difference?
Laminarity	mean: 2.02, s.d.: 0.91 n=30	mean: 1.73, s.d.: 0.78 n=52	No p=.13, t=1.539, df=80
Width/Thickness	mean: 4.29, s.d.: 1.57 n=30	mean: 3.33, s.d.: 1.48 n=52	Yes p=.01, t=2.775, df=80
Lateral Edges	Parallel: 29% Convergent: 54% Expanding: 11% Ovoid: 7% n=28	Parallel: 48% Convergent: 13% Expanding: 17% Ovoid: 22% n=46	Yes p=.00, G^2=14.537, df=3*
Distal Terminus	Blunt: 21% Pointed: 79% n=24	Blunt: 63% Pointed: 37% n=19	Yes p=.01, Fisher's Exact
Profile	Straight: 50% Curved: 27% Twisted: 23% n=30	Straight: 63% Curved: 25% Twisted: 12% n=40	No p=.44, G^2=1.651, df=2
Unique Types of Retouch	Normal retouch	Carinated retouch	Yes
Tool Types	UP tools dominate	UP tools dominate	No
Number of Changes/7 Steps			**4/7**
Total Measure of Difference in Tool Kit Morphology			**0.57**

*Two cells have an expected count less than 5 (25% of the cells). The minimum expected count is 4.16.

of tools per flake may indicate a greater application of retouch on each tool, decreasing the length and width measurements of these tools relative to the tools in the Bohunician assemblage. While it is possible to eliminate the attributes dependent upon these dimensions, i.e., the length/width and width/thickness ratios, from the pair-wise assemblage comparison, this can increase rather than decrease the relative measure of difference between the assemblages. As a consequence, these attributes are left in the comparison but caution is warranted in evaluating how the greater retouch application in the Kůlna assemblage may affect the result.

Unlike the tool-flake ratio, both assemblages are moderate in their non-cortical to cortical blank ratio (Figure 5.56) compared to the other sample assemblages, indicating that raw material was brought into the site in a roughly similar cortical state. The Kůlna assemblage, however, contains more cortical pieces relative to non-cortical pieces than the Szeletian assemblage, indicating more early core reduction was captured in this assemblage. The fact that the Kůlna cores demonstrate a far greater degree of

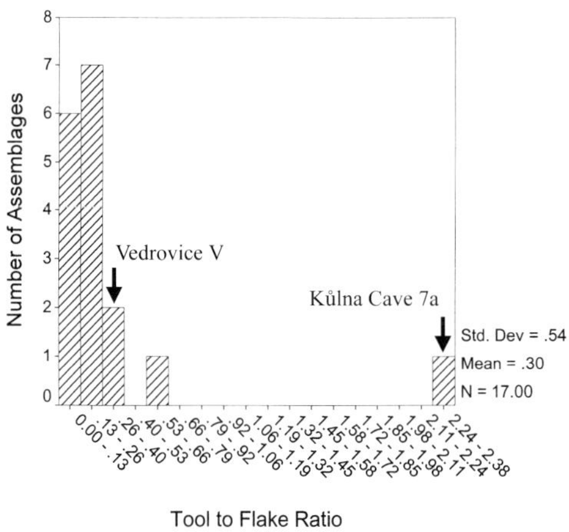

Figure 5.55 Kůlna Cave Layer 7a and Vedrovice V compared for tool/flake ratio values.

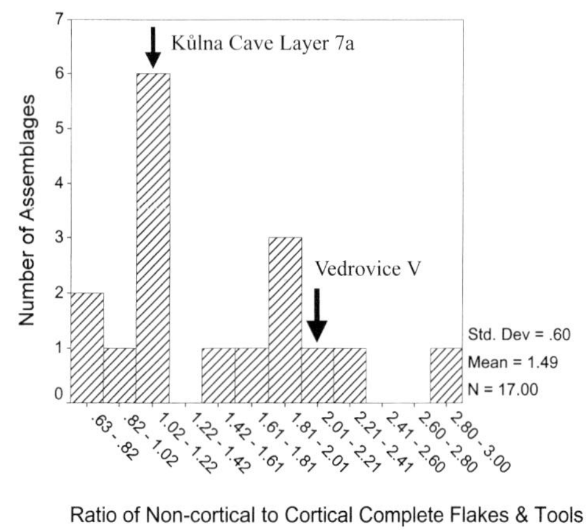

Figure 5.56 Kůlna Cave Layer 7a and Vedrovice V compared for non-cortical/cortical flake ratio values.

Table 5.91 Kůlna Cave Layer 7a and Vedrovice V, comparability of the assemblages based on intensity of core and tool reduction

Variables	Kůlna Cave 7a	Vedrovice V	Evaluation
Tools/Flake Ratio	2.03	0.32	Extremely different
Non-cortical Debitage/ Cortical Debitage Ratio	1.18	2.02	Slightly different
Henry's Dimensional Analysis	Extreme core exhaustion	Moderate core Moderate core	Different

exhaustion than the Szeletian cores, however, suggests that other factors may have a significant effect on this combination of variables, notably the number of blanks which could be removed from a given nodule size and the differences in site function implied by the open-air and large cave contexts of the Szeletian and Micoquian assemblages, respectively. This question requires further research in the direction pursued by Wenzel and Shelley (2001), that of controlling for the original nodule size by prospecting in the original outcrops utilized in prehistory. This approach has merit for Holocene sites such as explored by Wenzel and Shelley but may be hindered in its application to Late Pleistocene contexts by uncertainty in the exact exposed outcrops utilized by the hominins compared to the outcrops visible to archaeologists and geologists today. This is clearly the case, even in such well surveyed geological landscapes as Moravia.

In the meantime, these two assemblages are considered comparable despite the above differences, since many of the knapping behaviors preserved in these assemblages clearly occurred during the same portions of the core reduction sequence. In addition, the likelihood that the

Table 5.92 Comparison of Blank Production for Kůlna Cave Layer 7a and Vedrovice V

Flintknapping Steps by Domain	Kůlna 7a	Vedrovice V	Significant Difference?
Core Modification			
Core Orientation	Discoidal use of two surfaces, unifacial at end	Preferential use of a Longitudinal Surface	Yes
Core Convexity Management	Secant surfaces exploited around the perimeter of the plane of intersection	Lateral core tablets, changes of orientation	Yes
Number of Changes/2 Steps			**2/2=1**
Platform Maintenance			
Platform Treatment	Unprepared: 58% Prepared: 42% n=167	Unprepared: 61% Prepared: 39% n=380	No p=.51, Fisher's Exact
External Platform Angle (degrees)	mean: 83.8, s.d.: 14.8, n=153	mean: 89.0, s.d.: 17.4, n=342	Yes p=.00, t=-3.471*, df=339.948
Platform Thickness	mean: 9.08, s.d.: 4.63, n=153	mean: 5.01, s.d.: 3.64, n=359	Yes p=.00, t=9.692*, df=235.603
Number of Changes/3 Steps			**2/3=0.67**
Direction of Core Exploitation			
Direction of Early Exploitation	Subcentripetal & Unidirectional	Bidirectional & Unidirectional	Yes
Direction of Late Exploitation	Unidirectional & Crossed	Unidirectional	Yes
Number of Changes/2 Steps			**2/2=1**
Dorsal Surface Convexity System			
Longitudinal Convexity: Length/Width Ratio	mean: 1.44, s.d.: 0.49, n=273	mean: 1.47, s.d.: 0.57, n=542	No p=.40, t=-0.849*, df=623.660
Shape of Convexity: Lateral Edges of Blanks	Parallel: 54% Convergent: 23% Expanding: 20% Ovoid: 13% n=198	Parallel: 39% Convergent: 16% Expanding: 31% Ovoid: 14% n=451	Yes p=.00, G^2=15.456, df=3
Curvature of Convexity: Profile of Blanks	Straight: 64% Curved: 22% Twisted: 14% n=225	Straight: 66% Curved: 15% Twisted: 19% n=465	No p=.06, G^2=5.700, df=2
Lateral Convexity: Cross-Section of Blanks	Triangular: 39% Trapezoidal: 54% Other: 7% n=231	Triangular: 52% Trapezoidal: 27% Other: 20% n=493	Yes p=.00, G^2=56.963, df=2
Vertical Convexity: Width/ Thickness Ratio	mean: 2.83, s.d.: 1.06, n=273	mean: 4.26, s.d.: 1.82 n=542	Yes p=.00, t=-14.123*, df=796.829
Number of Changes/5 Steps			**3/5=0.6**
Total Measure of Difference in Blank Production			**3.27**

*The t-test was calculated with equal variances not being assumed due to a significant F value on Levene's Test for Equality of Variances.

dimensional data for Kůlna layer 7a is affected by its greater degree of retouch intensity is lessened by the results of an independent study. Nejman's (2006) study of these two assemblages according to Kuhn's Mean Geometric Index of Reduction (1990b) and Clarkson's Index of Invasiveness (2002) further supports the comparability of the retouched toolkits within these assemblages:

> It has also been argued in a recent study that the similarities in numerous lithic patterns between the Vedrovice V assemblage and the Kůlna assemblages (layers 6a and 7a) indicate a similar low level of residential mobility, a relatively small home range and a predominant technological strategy of provisioning individuals rather than provisioning of places (Nejman 2006) (Nejman et al. 2011:1062).

Blank Production

The first domain in Table 5.92, "core modification," concerns the establishment and exploitation of core surfaces for debitage. In the case of the core orientation of Kůlna 7a, a bifacial as well as unifacial discoidal form, exploiting equally two surfaces, was evidenced in the majority of cores. In the majority of the Vedrovice V cores, the preferential use of one surface on a longitudinal flake core form is evidenced. For core convexity management, Kůlna 7a flintknappers exploited all margins of the plane of intersection between the two secant core surfaces, thereby constantly refreshing the convexity of each surface. In the assemblage from Vedrovice V, convexities were managed through changes of core orientation from one platform cores to crossed or multiple platform cores, producing lateral core tablets, and refreshing the convexity by using the original platform as a rejuvenating crested flake. These strategies are somewhat related (the Szeletian option is a limited version of the Micoquian option) but are still deviant enough to be significantly different as the Szeletian strategy seems almost opportunistic or expedient compared to the Micoquian strategy.

For platform maintenance options, Kůlna layer 7a and Vedrovice V show similarities only in the frequencies of platform treatment, due to the thinner and steeper platforms in the Vedrovice V debitage. In terms of the direction of core exploitation, Vedrovice V exhibits a distinctly different strategy than at Kůlna layer 7a, as the subcentripetal pattern in early core exploitation in the latter assemblage is replaced by bidirectional blanks in the former. By late core exploitation, crossed blanks remain numerous along with unidirectional blanks in Kůlna, making this portion of the core reduction sequence distinct between the assemblages as well. All but two of the five behavioral categories in the surface convexity domain show significant deviation between the Micoquian and Szeletian assemblages. Behavioral similarities lie in the flake-dominated laminarity ratio and straight profiles.

The final measure of difference between the blank production choices of the Kůlna 7a and Vedrovice V assemblages, weighted by the four knapping domains, produces a value of 3.27 out of a possible maximum difference of 4.0, indicating that the details of blank production in these two assemblages are not at all similar. Significant differences are found in all four knapping domains.

As bifaces were found in Kůlna 7a which were large enough to serve both as cores as well as cutting tools, it is reasonable to inquire to what extent bifacial core reduction should have been considered as a unit of comparison between these two assemblages, for instance in the "core orientation" analytical step of the core modification domain. Both assemblages do indeed exhibit bifacial reduction, although the extant bifaces in the Vedrovice V assemblage are either quite small to

Table 5.93 Comparison of Tool Kit Morphologies for Kolna Cave Layer 7a and Vedrovice V

Tool Kit Morphology Variable	Kolna 7a	Vedrovice V	Significant Difference?
Laminarity	mean: 1.45, s.d.: 0.51, n=162	mean: 1.51, s.d.: 0.51 n=115	No p=.32, t=-1.007, df=275
Width/Thickness	mean: 2.84, s.d.: 1.12, n=162	mean: 3.59, s.d.: 1.56 n=115	Yes p=.00, t=-4.405*, df=194.404
Lateral Edges	Parallel: 55% Convergent: 7% Expanding: 23% Ovoid: 15% n=103	Parallel: 31% Convergent: 12% Expanding: 46% Ovoid: 11% n=61	Yes p=.01, G^2=12.291, df=3
Distal Terminus	Blunt: 74% Pointed: 26% n=65	Blunt: 81% Pointed: 19% n=36	No p=.63, Fisher's Exact
Profile	Straight: 67% Curved: 20% Twisted: 13% n=126	Straight: 64% Curved: 13% Twisted: 23% n=75	No p=.13, G^2=4.019, df=2
Unique Types of Retouch	Flat bifacial retouch	Flat bifacial retouch	No
Tool Types	MP tools dominate	MP tools dominate	No
Number of Changes/7 Steps			**2/7**
Total Measure of Difference in Tool Kit Morphology			**0.29**

*The t-test was calculated with equal variances not being assumed due to a significant F value on Levene's Test for Equality of Variances.

have derived from a nodule or are in fact made on a flake/blade blank. Nevertheless, if "bifacial core reduction" were considered a unit of comparison, both assemblages would be seen as similar in this regard. Such an approach has been used traditionally to argue for genetic relationships based on flat bifacial retouch as a fossile directeur that unites these two industries (Valoch 1990b; Neruda 2000).

The application of bifacial retouch is indeed considered below in terms of the morphology of the tool kit but not discretely for blank production. This is a consequence of the fact that bifacial reduction which produces blanks large enough (i.e., over 2.0 cm in maximum dimension) to be studied as debitage already influences the central tendencies of the debitage's platform maintenance domain, the direction of core exploitation domain, and the dorsal surface convexity system domain. Any products of bifacial reduction are thus already incorporated into the study as is; adding a bifacial reduction characterization to the core modification domain would be redundant and increase the interdependence of the domains. Further, if these products had displayed very similar tendencies between the Micoquian and the Szeletian, the analysis of blank production would have turned out otherwise. In fact, the size of the products removed during bifacial reduction of the small bifaces in

Table 5.94 Measurement of the difference in flintknapping behaviors between Central European assemblages

Comparison of Assemblages Through Time	Measure of Difference in Blank Production (Maximum=4, Minimum=0)	Measure of Difference in Tool Kit Morphology (Maximum=1, Minimum=0)
Kolna Layer 7a vs. Stránská skála IIIa Layer 4	3.13	0.86
Stránská skála IIIa Layer 4 vs. Stránská skála III Layer 5	0.73	0.29
Stránská skála III vs. Vedrovice V	2.67	0.86
Stránská skála IIIa Layer 4 vs. Stránská skála IIIc	0.60	0.14
Stránská skála III Layer 5 vs. Stránská skála IIIc	0.60	0.29
Stránská skála IIIc vs. Vedrovice V	2.47	1.00
Vedrovice V vs. Stránská skála Aurignacian	2.17	0.43
Stránská skála III Layer 5 vs. Stránská skála Aurignacian	2.80	0.57
Kolna Layer 7a vs. Vedrovice V	3.27	0.29

Vedrovice V are likely not usable supports of cutting edges in themselves, unless for the most expedient and light of uses. As Tomka (2001) discusses, while it is possible to butcher an elephant with a tiny flake, the ergonomics of this procedure make it more likely that an alternative cutting edge support would be produced during the balancing between the desired tasks and the practiced organization of technology. Thus, in this context, it is more appropriate to treat bifacial reduction in these assemblages in the tool kit morphology comparison than in the blank production comparison.

Tool Kit Morphology

From the comparison of the morphology of the cutting edge of the tool kits of these two assemblages seen in Table 5.93, only two of the seven variables display significant differences. Despite the concerns that the dimensions of the Kůlna tools would have been reduced to significantly different ratios, only the width/thickness ratio displays a significant difference. The greater application of lateral retouch frequently used in the production of Middle Paleolithic type tools could indeed produce this result. The tool kits are also statistically different in their lateral edges, with higher convergent and expanding frequencies in Vedrovice V than in Kůlna. The remaining two analytical steps within the comparison of tool kit morphology relate to qualitative issues of retouch. The two assemblages are similar in the dominance of Middle Paleolithic tool types. Also, both Kůlna 7a and Vedrovice V possess flat bifacial retouch, applied to small handaxes and bifacial scrapers in the former assemblage, and to bifacial foliate points made on flake blanks in the latter assemblage.

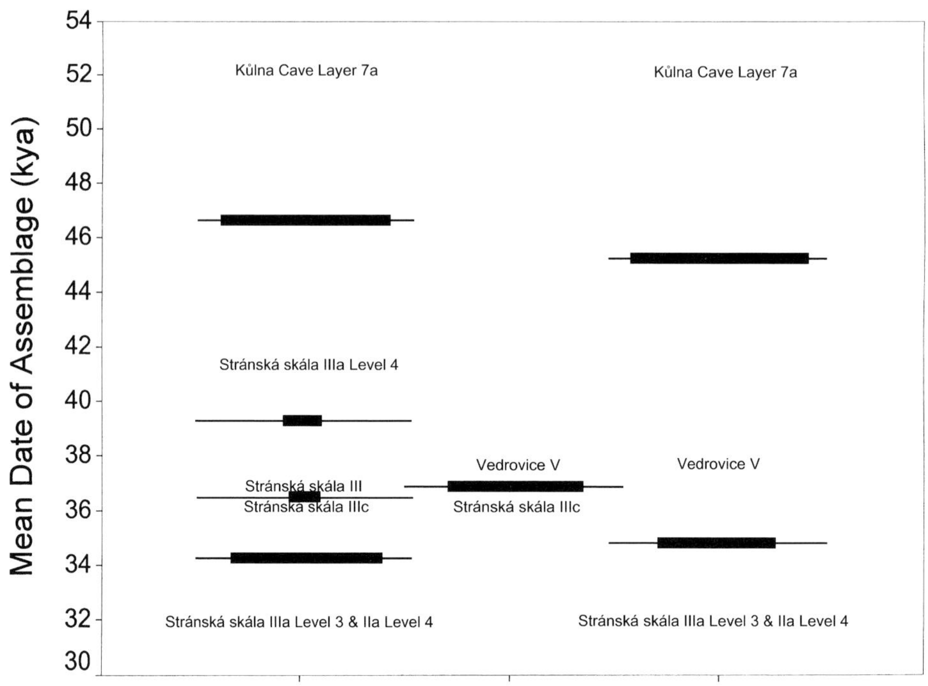

Figure 5.57 Chronological comparison of differences in Blank Production among Central European assemblages. The length of the thick bars represents the difference measure between the two assemblages above and below the bar (the earlier and the later). The length of the thin bars represents the maximum possible difference measure. The three columns represent different sequences for pairing assemblages for comparison.

Considering all 7 analytical steps within the tool kit comparison, these assemblages exhibit a difference measure of two out of seven steps, or 0.29 out of a maximum of 1.0. The tool kits are thus much more similar than the details of the production of the blanks on which the tools were made.

5.8 Significance of the Central European Sequence for Model Expectations 1-4

Now that the Central European sequence of change has been reconstructed, it is possible to assess some of the model expectations for this intra-regional context. It is immediately clear that the details of blank production in the sample assemblages differ to varying degrees. Examining individual variables' continuity between temporally contiguous assemblages (Figures 5.58 and 5.59) demonstrates that all but three technical choices performed at the time Kůlna 7a had changed by the time of Stránská skála IIIa-4. Yet this example does not become a pattern in the region, as the next chronological comparison, between Stránská skála IIIa-4 and III-5 shows the continuity of 9 out of 12 variables. This variance in the enculturating environment that trained the performances we recover as artifacts in each assemblage

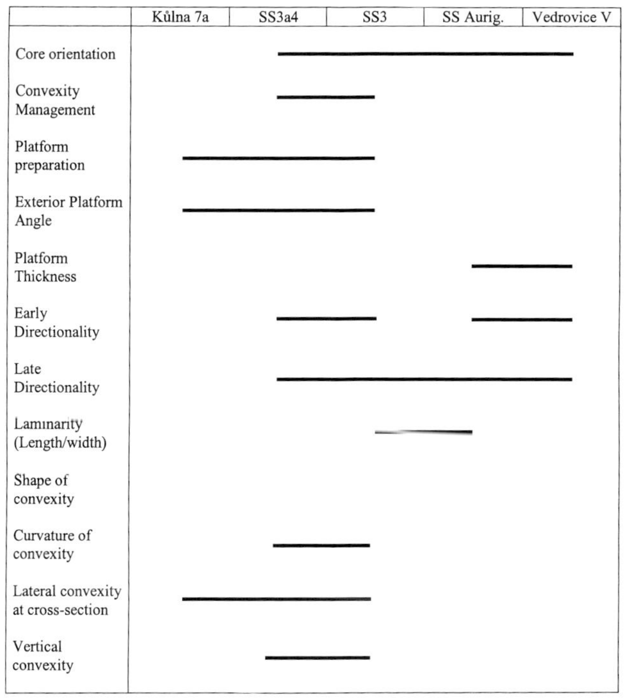

Figure 5.58 Variable by variable view of set one of the pair-wise comparisons within Central Europe between 60 and 30 kya. Bars connecting two assemblages represent the use of statistically similar attribute states between the two assemblages. Similarities are not necessarily implied between assemblages connected by a bar which are not immediately adjacent to each other.

is quantifiable beyond looking at individual technical choice variables, according to Expectation 2 (Number of Changes). Using the measure of difference derived by counting differences in options chosen in knapping steps according to the domain (varying between 0 and 4), it is possible to compress the variability into a unit for easier comparison. By collecting the results of the individual pair-wise comparisons Table 5.94, it is possible to perceive a rough picture of the sequence of behavioral change which occurred in this region between 60 and 30 kya. It must be stressed, however, that this measure of difference compresses all of the variation existing between two operational sequences into a single value, and so may overly simplify the context of any changes.

The first pair-wise comparison, between Kůlna 7a and Stránská skála IIIa level 4, produces a large measure of difference (3.13). The immediately subsequent comparison, between the two Bohunician assemblages of Stránská skála IIIa level 4 and Stránská skála III Layer 5, demonstrates the extremely close similarity (0.73) between these assemblages. The minor

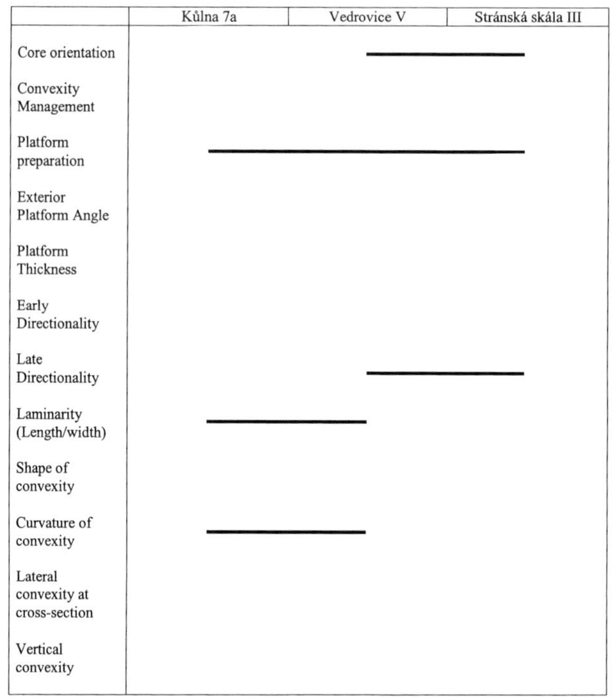

Figure 5.59 Variable by variable view of set two of the pair-wise comparisons within Central Europe between 60 and 30 kya. Bars connecting two assemblages represent the use of statistically similar attribute states between the two assemblages. Similarities are not necessarily implied between assemblages connected by a bar which are not immediately adjacent to each other.

difference between Bohunician assemblages is mainly due to the greater elongation and convergent edges in the latter assemblage. Stránská skála IIIc is even more similar to the other two Bohunician assemblages (0.60). From Expectation Three (Number of Domains), blank production among the Bohunician assemblages shows a difference in only one domain, compared to 4 domains it differs in when compared to the Micoquian.

When we place these measures of difference in their regional and temporal context (Figure 5.57), we see that a marked change in flintknapping behaviors occurred between the activities which created the assemblage of Kůlna layer 7a and those which created the Bohunician assemblages of Stránská skála III, IIIa, and IIIc. The later two Bohunician assemblages differ from the Szeletian assemblage to a lesser degree (2.67 and 2.47) than with the Micoquian site; indeed it is closer in behavioral details to Vedrovice V than Vedrovice V is to the Micoquian (3.27). This turns the traditional view (i.e., Valoch 1990b) of the close relationship between the late Middle Paleolithic of Kůlna Cave and the Central European Szeletian on its

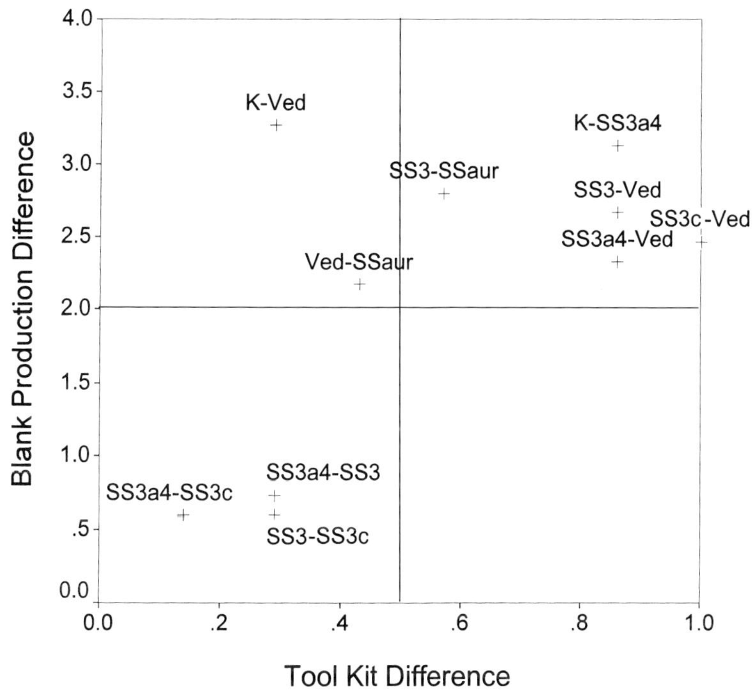

Figure 5.60 Blank Production vs. Tool Kit Differences for pair-wise comparisons for Central European assemblages. Assemblage Key: K: Kůlna Cave Layer 7a; SS3: Stránská skála III; SS3a4: Stránská skála IIIa Level 4; SSaur: Stránská skála IIIa Level 3 and IIa Level 4; SS3c: Stránská skála IIIc; V: Vedrovice V.

head. When retouch index fossils are not given the primary role in determining cultural affiliations, the Szeletian appears more similar to the subsequent Aurignacian (2.17) than it does to the purportedly ancestral Micoquian. A very similar measure of difference is evidenced between the Bohunician and the subsequent Aurignacian (2.80) and between the Szeletian and Aurignacian (2.17) yet the specific options which are shared between the comparisons differ.

This limitation in the utility of a single value measure of difference between assemblages can be removed when one looks at the specific details of each step comparison. This process allows the evaluation of Expectation 1 (Antecedents). It is clear, for instance, that the longitudinal core orientation shared by all later assemblages does not have an antecedent in the earliest assemblage, Kůlna 7a. Nor does this early assemblage possess the antecedents for the particular shift in direction of core exploitation shared by the Bohunician and the later Aurignacian assemblages. It is thus possible to argue based on the presence and absence of antecedent behaviors that Central Europe witnessed a discontinuity, a strong break in flintknapping behaviors between the Micoquian and Bohunician occupation of the landscape. From

Figure 5.57, which depicts thick lines as breaks in the enculturating environment for knapping performance, it is also possible to argue for a discontinuity between the Bohunician and Aurignacian assemblages.

When the differences in blank production and tool kit morphologies are graphed against each other (Figure 5.60), it is possible to evaluate Expectation 4 (the predictions for enculturation vs. stimulus diffusion). Here it becomes apparent that there are more similarities in tool kit morphology between Kůlna 7a and Vedrovice V than among the four domains of blank production. This is also true for the comparison between Vedrovice V and the Aurignacian assemblages. In each case, the tool kit morphologies are just as similar as among the Bohunician assemblages. In the parlance of enculturation, social intimacy, and the functionality of the tool kit, this evidence supports a similar use of tools on the landscape by the makers of Vedrovice V and the makers of Kůlna 7a, but without a commonality to the enculturating environment. If the individuals responsible for Vedrovice V, the chronologically later assemblage, had any cultural transmission with the Neanderthals responsible for Kůlna 7a (requiring the assumption of continued occupation of the region), then the transmission process was stimulus diffusion. With five out of seven tool kit morphologies the same between them, the production of this degree of similarity from adaptive convergence alone seems unlikely although possible. The same conclusion applies to the Aurignacian-Vedrovice V potential contact, as the biplot values for this comparison also fall, if barely, in the upper left corner where tool kits are similar but blank production is different. As the Aurignacian and Vedrovice V share 4 out of 7 tool kit variables, a slightly lower number, adaptive convergence may be more likely than in the Vedrovice V- Kůlna 7a example.

6

Regional Patterns of Behavior: Eastern Europe

The present chapter examines the details of the flintknapping behavior extant between 60,000 BP and 30,000 BP in Eastern Europe, as exemplified by the archaeological sequence in the western Ukraine. Each sample assemblage is described in order of chronological appearance in the region. The technological analysis of the lithic data collected between 1997 and 1999 is presented for each assemblage before the pair-wise comparison of assemblages is conducted. The chapter concludes with a discussion of the significance of the intra-regional, Eastern European sequence of change for the diffusion and independent innovation hypotheses.

6.1 Molodova V, Layer 11

6.1.1 Site Background

Molodova V is an open-air site in the Middle Dnestr region of western Ukraine, excavated by the geologist I. K. Ivanova and archaeologist A. P. Chernysh discontinuously between 1951 and 1964 (Chernysh 1961; 1987; Ivanova and Chernysh 1965; Klein 1973). The site is situated on a headland 50 m above the current Dnestr River level on the western outskirts of the modern village of Molodova. As with the nearby site of Molodova I, the 25 m thick deposits of locality V rest on the 2nd terrace of the Dnestr River, a terrace dated to the Riss/Würm Interglacial by means of correlations with known Black Sea terraces. Ivanova thus judged all of the deposits at both Molodova I and V, which are similar in both artifacts and stratigraphy, to the Last, Würmian Glaciation, OIS 5d-3. With horizontal strata and little solifluction, excellent bone preservation, and covering 935 m^2, Molodova V possesses an excellent sequence of Middle and Upper Paleolithic occupations (see Hoffecker 2002:145, fig. 5.2).

The assemblage of interest for sampling the Late Middle Paleolithic at the site is Layer 11, which produced a minimum radiocarbon date of >40,300 (GrN-4017). This layer is also geologically attributed to the end of the Brörup Interstadial seen in western Europe, witnessing "a cool, then cold climate" (Ivanova and Chernysh 1965:214). Chernysh described this assemblage as a Late Mousterian with abundant Levallois technique, flakes, blades, and Middle Paleolithic tools (Ivanova and Chernysh 1965). This layer is unfortunately separated by 4 m of mostly sterile strata from the next cultural assemblage, in layer 10, a fully Upper Paleolithic industry of Chernysh's second stage in the development of the Upper Paleolithic in the region (1959). Thus, no *in situ* sequence exists at the site between the Middle and Upper Paleolithic occupations. Two very small assemblages of artifacts, layers 10a and 10b, represent occupations of interest between the Middle Paleolithic and fully Upper Paleolithic occupations, but their limited sample size makes them inappropriate for technological analyses. Thus the Korolevo assemblages on the western slopes of the Carpathians are used instead to document the behaviors extant after 40 kya.

The raw material used for making stone tools in the layer 11 occupation is a local, fine-grained chert of good to excellent quality. A high degree of patination and rolling made some technological attributes, such as dorsal scar pattern, difficult to assess.

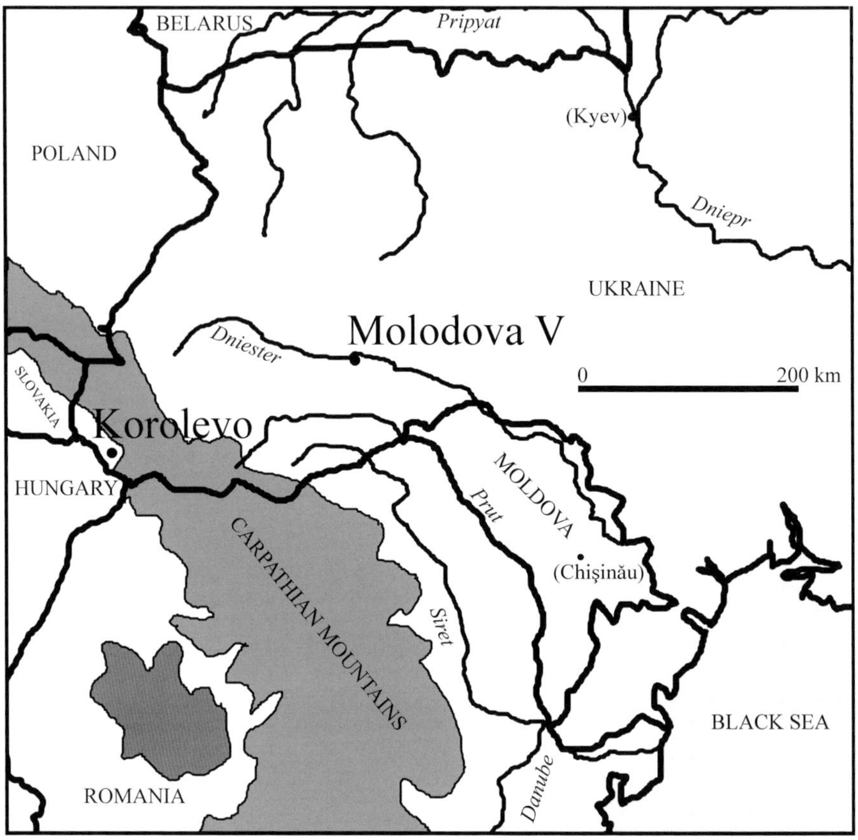

Figure 6.1 Map of Eastern European sites used in the analysis.

6.1.2 Sampling

Layer 11 produced 6047 stone artifacts (Ivanova and Chernysh 1965:201). In an effort to sample a representative portion of these artifacts for a technological and typological study, one column of meter squares was chosen for study. The column labeled with the Roman numeral "X" was studied as it passed through the center of the artifact distribution in this layer (see Ivanova and Chernysh 1965:204, fig. 5). The squares studied thus included XД, XE,XЖ, X3, XИ, XK, XЛ, XM, XO, XЦ, and XX. Only intact debitage was studied. All tools were studied from these squares.

6.1.3 Blank Production

The reconstruction of the options chosen by the varied flintknappers whose cumulative efforts created the palimpsest of the Molodova V Layer 11 assemblage proceeds according to the twelve knapping behaviors or analytical steps within the four domains of the flintknapping process.

Core Modification

The first domain, "core modification," concerns the orientation of the knapping surfaces of the cores. The first analytical step in this domain, "core orientation," is characterized by the distinction

Table 6.1 Molodova V Layer 11, percentage of intact cores (n=20) by number of bulbar negatives in each core sector (A, B, C, D) by surface (1, 2)

Number of bulbar negatives	Surf. 1 Sect. A	Surf. 1 Sect. B	Surf. 1 Sect. C	Surf. 1 Sect. D	Surf. 2 Sect. A	Surf. 2 Sect. B	Surf. 2 Sect. C	Surf. 2 Sect. D
0	5	60	40	50	50	55	55	70
1	25	25	35	30	30	25	20	20
2	55	10	10	20	15	15	15	5
3	10	5	15		0	5	5	5
4	5				5		5	

Table 6.2 Molodova V Layer 11, Core Sector Preparation: Percent of Intact Cores (n=20) by Number of Scars in Sector of Origin (A, B, C, D) by Surface (1, 2)

Number of Scars	Surf. 1 Sect. A	Surf. 1 Sect. B	Surf. 1 Sect. C	Surf. 1 Sect. D	Surf. 2 Sect. A	Surf. 2 Sect. B	Surf. 2 Sect. C	Surf. 2 Sect. D
0	0	60	55	70	20	45	50	45
1	15	15	15	25	5	20	10	20
2	45	15	5	5	0	25	20	20
3	30	10	10		45	10	15	5
4	5		10		5		0	5
5	0		5		10		5	0
6	0				15			5
7	5							

between different core forms, based upon an examination of extant core morphology (Appendix Figures A25-A30) and an evaluation of where platforms were located during most of core exploitation. Table 6.1 presents the cumulative, assemblage-wide tendency for bulbar negatives to appear on given core surfaces ("Surface 1" and "Surface 2") and within given core sectors (A, B, C, and D). In the case of Molodova V layer 11, it is evident that bulbar negatives are slightly more frequent on the broad-faced surface, Surface 1, than on Surface 2. This dominance of Surface 1 is mostly a result of the bulbar negatives in sector A of Surface 1, as the other sectors on both surfaces possess relatively similar frequencies. This shows a moderate preference for one surface over the other but not an exclusive preference, at least as evidenced in the cores at discard. Table 6.2, showing the number of scars within each core sector which originated from the direction of that sector, corroborates the view that sector A on Surface 1 is indeed the main debitage production sector, although Surface 2 sector A also shows significant working according to this measure. The high number of removals from Surface 2 sectors A-D, along with the low number of bulbar negatives on these scars, supports the interpretation of Surface 2 as a circumferential platform for the broad-face of Surface 1.

The second step in this domain, core convexity management, includes the knapping options which allow the knapper to continue reducing the core form to exhaustion through specific modifications of the surface convexities. In the case of Molodova V layer 11, convexity management proceeded by exploitation of the circumference of predominantly one face, in combination with periodic débordant removals along the longer edges. This is adequately evidenced in the core morphologies, the centripetal

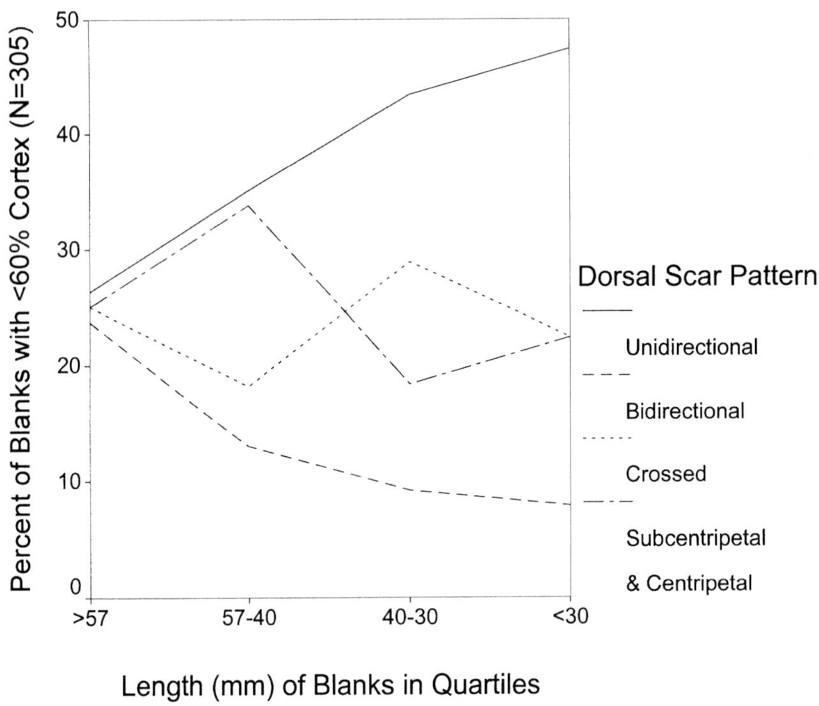

Figure 6.2 Molodova V Layer 11, dorsal scar pattern by blank length (mm) in quartiles cross-tabulation for complete flakes and tools (excluding pieces with a crested scar pattern, an indeterminate scar pattern, greater than 60% cortex, or fewer than three dorsal scars).

pattern of debitage removal, and the presence of crested, twisted profile blanks (serving as débordant removals) in the debitage.

Platform Maintenance

Table 6.3 present the frequencies of platform treatment types in this assemblage. Platform preparation was very frequent, accounting for 61% of the debitage when indeterminates are not counted. Table 6.4 presents the central tendency statistics for the external platform angles and platform thickness, indicating a relatively steep mean EPA of 86 degrees and a relatively thick mean PT of 6 mm.

Direction of Core Exploitation

The next knapping domain, direction of core exploitation, concerns the variability resulting from different options of removing flakes from single to multiple platform cores, creating debitage with particular dorsal scar patterns. The dominant strategy used in the assemblage was determined using the principles of dimensional change of debitage during core reduction (see Chapter 4). Table 6.5 presents the cross-tabulation of debitage and tools according to dorsal scar pattern and blank length (separated into quartiles so that one quarter of the sample population lies within each of the four length

Table 6.3 Molodova V Layer 11, platform types for complete and proximal fragments of flakes and tools

Platform type	Frequency	Percent
Plain	121	28.3
Cortical	14	3.3
Dihedral	29	6.8
Facetted	114	26.7
Chapeau	2	.5
Scarred	78	18.3
Punctiform	2	.5
Linear	7	1.6
Removed	2	.5
Indeterminate	58	13.6
Total	427	100.0

Table 6.4 Molodova V Layer 11, descriptive statistics for continuous platform variables of complete and proximal fragments of flakes and tools

	N	Minimum	Maximum	Mean	Standard Deviation
Exterior Platform Angle (degrees)	301	50.00	135.00	85.60	13.97
Platform Thickness (mm)	355	0.43	26.85	5.62	3.30

Table 6.5 Molodova V Layer 11, dorsal scar pattern by blank length (mm) in quartiles cross-tabulation for complete flakes and tools (excluding flakes with greater than 60% cortex or fewer than 3 dorsal scars)

Dorsal Scar Pattern	>57	57-40	40-30	<30	Total
Unidirectional					
Count	20	27	33	36	116
% within Quartile	22.7%	30.3%	35.5%	33.0%	30.6%
Bidirectional					
Count	18	10	7	6	41
% within Quartile	20.5%	11.2%	7.5%	5.5%	10.8%
Crossed					
Count	19	14	22	17	72
% within Quartile	21.6%	15.7%	23.7%	15.6%	19.0%
Subcentripetal & Centripetal					
Count	19	26	14	17	76
% within Quartile	21.6%	29.2%	15.1%	15.6%	20.1%
Crested					
Count	9	3	3	5	20
% within Quartile	10.2%	3.4%	3.2%	4.6%	5.3%
Indeterminate					
Count	3	9	14	28	54
% within Quartile	3.4%	10.1%	15.1%	25.7%	14.2%
Total					
Count	88	89	93	109	379
% within Cortex	100.0%	100.0%	100.0%	100.0%	100.0%

Table 6.6 Molodova V Layer 11, dorsal scar pattern by blank length (mm) in quartiles cross-tabulation for complete flakes and tools (excluding pieces with a crested scar pattern, an indeterminate scar pattern, greater than 60% cortex, or fewer than three dorsal scars)

Dorsal Scar Pattern	>57	57-40	40-30	<30	Total
Unidirectional					
Count	20	27	33	36	116
% within Quartile	26.3%	35.1%	43.4%	47.4%	38.0%
Bidirectional					
Count	18	10	7	6	41
% within Quartile	23.7%	13.0%	9.2%	7.9%	13.4%
Crossed					
Count	19	14	22	17	72
% within Quartile	25.0%	18.2%	28.9%	22.4%	23.6%
Subcentripetal & Centripetal					
Count	19	26	14	17	76
% within Quartile	25.0%	33.8%	18.4%	22.4%	24.9%
Total					
Count	76	77	76	76	305
% within Cortex	100.0%	100.0%	100.0%	100.0%	100.0%

Table 6.7 Molodova V Layer 11, frequencies of lateral edge, profile, and cross-section types for complete flakes and tools

Type	Frequency	Percent
Lateral edge		
Parallel	139	32.6
Convergent	125	29.3
Expanding	80	18.7
Ovoid	69	16.2
Indeterminate	14	3.3
Total	427	100.0
Profile		
Straight	205	48.0
Curved or Concord	139	32.6
Twisted	81	19.0
Indeterminate	2	0.5
Total	427	100.0
Cross-section		
Triangular	185	43.3
Trapezoidal	223	52.2
Other	13	3.0
Indeterminate	6	1.4
Total	427	100.0

Table 6.8 Molodova V Layer 11, descriptive statistics for blank dimensions for complete flakes and tools

Blank Dimension	N	Range	Mean	Standard Deviation
Length	427	87.45	42.65	16.80
Width	427	61.55	26.29	11.02
Thickness	427	28.37	6.24	3.68
Length/Width Ratio	427	4.97	1.78	0.76
Width/Thickness Ratio	427	19.37	4.94	2.23

ranges). By removing the debitage and tools with dorsal scar patterns noted as indeterminate and crested to clarify the pattern, it is clear (Table 6.6 and Figure 6.2) that all four directions are equally represented in the first quartile. This pattern, highly suggestive of centripetal Levallois reduction (given the surface morphology of the products, below) is thus the signature for early core exploitation. As blank length decreases (and reduction continues), the crossed directions diminish in number, bidirectional and subcentripetal/centripetal patterns oscillate in frequency, and unidirectional patterns continually increase. As unidirectional is the dominant direction by the fourth quartile, this is the signature for late core exploitation.

The dorsal scar patterns are supported by the morphology of the cores at discard. The number of scars originating from the direction of their sector's immediate edge (Table 6.2) shows that sectors A and C, and to a lesser extent B and D, were used frequently to remove flakes, often from both surfaces. The bulbar negative evidence (Table 6.1), however, demonstrates that one sector only tended to be used late in core exploitation: sector A of Surface 1 possesses a disproportionate number of bulbar negatives. This suggests that a continual change in direction of reduction on an assemblage wide basis occurred between the first and fourth length quartiles

Dorsal Surface Convexity

The dorsal surface convexity domain includes the knapping steps which encapsulate the cumulative choice of *nervure-guide* or ridge pattern(s) of the core exteriors for the production of blanks. These steps are quantified by three categorical attributes (lateral edge type, profile type, and cross-section type) and two continuous variables (laminarity or length/width ratio and use of convexity or width/thickness ratio). Table 6.7 presents the counts and frequencies of lateral edge, profile, and cross-section types. Table 6.8 presents the mean, range, and standard deviation for all blank dimensions. Molodova V layer 11 produced only moderately elongated debitage, with a flake-focused length/width mean of 1.78, but very thin blanks (width/thickness mean 4.94) with dispersed ridge patterns (high frequencies of trapezoidal cross-sections).

6.1.4 Tool Kit Morphology

Tool kit morphology refers to the dorsal surface convexity variables which give the form of their cutting edge to the retouched blanks in the assemblage, as well as the diagnostic flakes without retouch traditionally categorized as tools– i.e., Levallois pieces. The modest tool kit evidences flake-based blanks with even thinner cross-sections than found in the debitage. The blanks are also relatively more parallel sided and straight in profile than the debitage.

The last two characterizations of the tool kit morphology refer to two variables describing the application of retouch to blanks: first, the application of distinctive and/or diagnostic forms of retouch and, second, the dominance or equality of Middle Paleolithic (lateral edge) versus Upper Paleolithic (distal and lateral edges) retouched tool types. In this case, Molodova V Layer 11 does not exhibit any unique types of retouch and is dominated by Middle Paleolithic tool types (Ivanova and Chernysh 1965; Chernysh 1961, 1987).

Intensity of Tool and Core Reduction

For the purposes of evaluating the comparability of the Molodova V Layer 11 assemblage with other assemblages in this study, Table 6.11 presents three characterizations of the intensity of use of raw material within the collection. Most of the raw material used in Molodova V layer 11

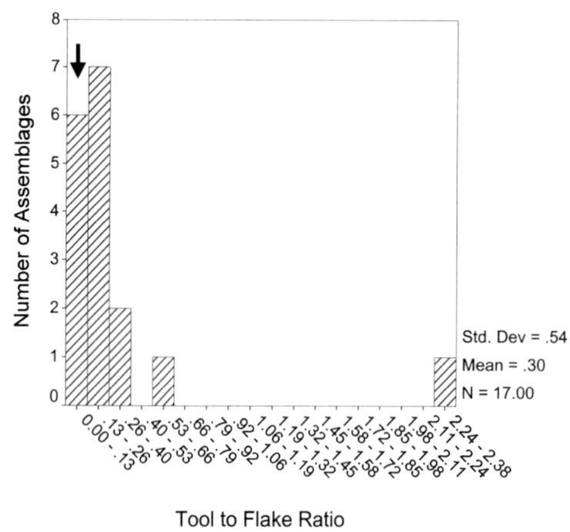

Figure 6.3 Molodova V Layer 11, tool/flake ratio value relative to a histogram of all assemblages.

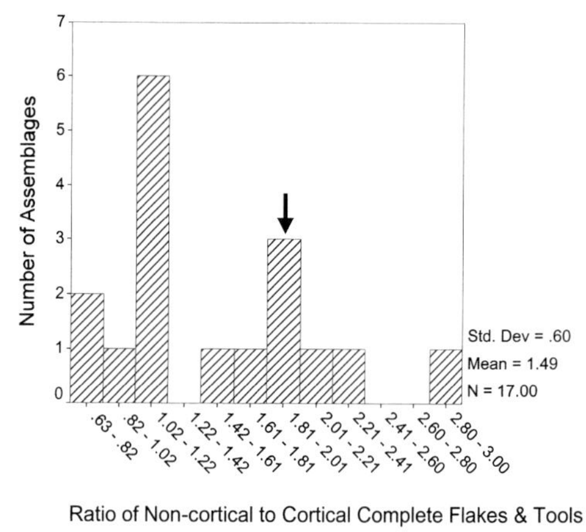

Figure 6.4 Molodova V Layer 11, non-cortical/cortical flake ratio value relative to a histogram of all assemblages.

is a local, although not immediately adjacent, fine-grained chert of good to excellent quality. The abundant access to raw material is evident in the very low tool to flake ratio, one of the lowest of all of the study's assemblages (Figure 6.3), an indication that a sharp cutting edge could be had more easily by production of a new blank than resharpening an existing blank with retouch. The interpretation that the raw material source is not immediately adjacent to the site derives from the assemblage's moderately high non-cortical to cortical blank ratio (Figure 6.4), falling slightly above the mean of all of the assemblages. This ratio indicates that while some reduction of cortical nodules occurred on site, most nodules were decortified into prepared cores before being brought to the site. The relatively high frequency of non-cortical blanks is not likely a result of differentially higher blanks production rates per core in this assemblage as there is very little evidence of intensive core exhaustion. While a pair of cores (two out of 21) possess core facets falling below the 2.0 cm length cut off for the debitage analysis, the dimensional study of the core facets relative to the lengths and widths of the debitage shows that many of the core facets overlap with the upper and middle of the blank distribution. As these cores had significant use-life left, very little core exhaustion is evidenced in this assemblage. The higher non-cortical blank frequencies must, therefore, result from a greater absence of cortex from the site than from greater production of non-cortical artifacts. These data thus indicate that this assemblage captures at least the middle of the core reduction sequence, with a relative absence of the earliest core working and the latest behaviors which would derive from core exhaustion.

6.2 Korolevo II Complex II

6.2.1 Site Background

Korolevo II Complex II is an open-air site located on the western border of Ukraine with Slovakia, on the western side of the Carpathian Mountains. Excavated between 1978 and 1980 by V. N.

Table 6.9 Molodova V Layer 11, descriptive statistics for dimensions of complete tools

Blank Dimension	N	Range	Mean	Standard Deviation
Length/Width Ratio	29	2.34	1.65	0.52
Width/Thickness Ratio	29	7.83	5.40	2.04

Table 6.10 Molodova V Layer 11, frequencies of lateral edge, distal terminus, and profile types for complete tools

Type	Frequency	Percent
Lateral edge		
Parallel	11	37.9
Convergent	7	24.1
Expanding	3	10.3
Ovoid	7	24.1
Indeterminate	1	3.4
Total	29	100.0
Distal Terminus		
Blunt	13	44.8
Pointed	13	44.8
Overshot	1	3.4
Broken	2	6.9
Total	29	100.0
Profile		
Straight	17	58.6
Curved or Concord	8	27.6
Twisted	4	13.8
Total	29	100.0

Table 6.11 Molodova V Layer 11, intensity of core and tool reduction

Variables	Numerator	Denominator	Result
Tools/Flake Ratio	Tools: 29	Flakes: 397	0.07
Non-cortical Debitage/ Cortical Debitage Ratio	Non-cortical Debitage: 279	Cortical Debitage: 147	1.90
Henry's Dimensional Analysis	NA	NA	Core facets overlap with the high to low end of the blank distribution: little to no core exhaustion

Gladilin, the head of the Paleolithic Expedition of the Archaeological Museum of the Zoological Institute of the Academy of Sciences of the Ukrainian SSR, this site contained a single occupation horizon (Demidenko and Usik 1993a; 1993b; Gladilin 1989a; 1989b; Gladilin and Demidenko 1989; 1990; Usik 1989). The stratigraphy of the site was published in terms of a locality-specific stratigraphic profile as well as a generalized section created by means of correlations between the numerous profiles of the other archaeological sites discovered within the Korolevo quarry (Gladilin and Demidenko 1989). As the industrial quarrying operation never stopped, the locality no longer exists and it is not possible to reinvestigate the stratigraphic relations between Korolevo II Complex II and other occupations within the quarry, such as Korolevo I layer 1a. Although little bone was preserved in the loam sedimentary matrix, charcoal

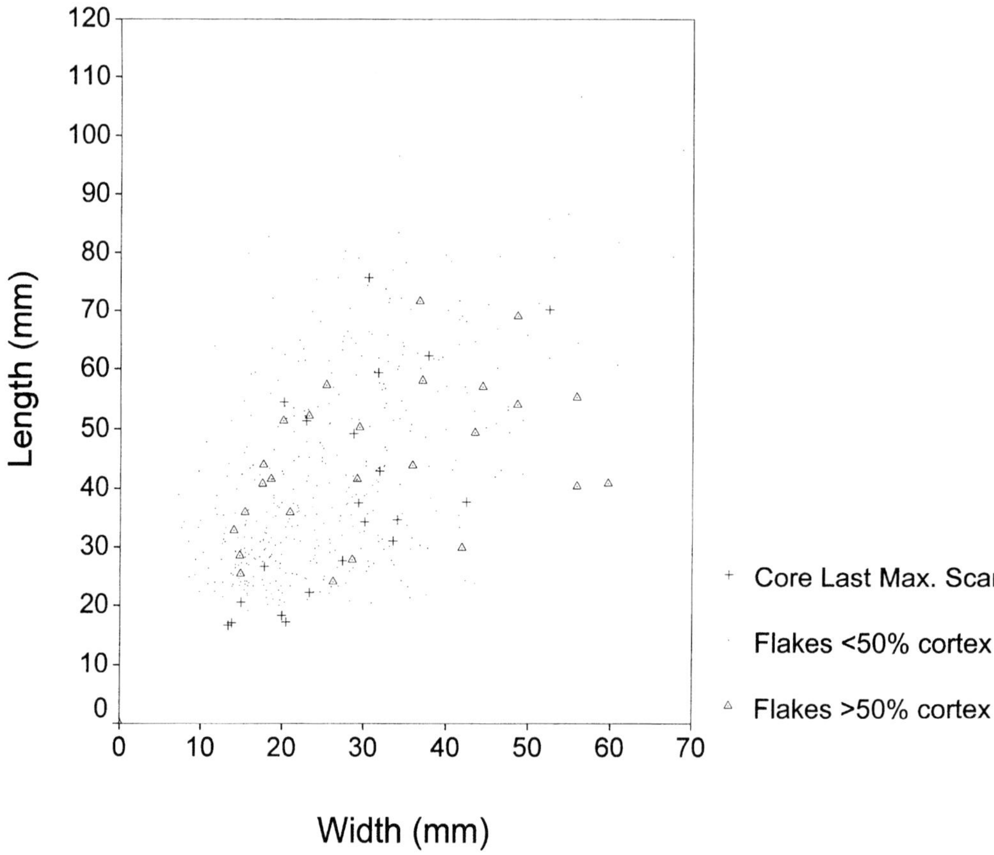

Figure 6.5 Molodova V Layer 11 according to Henry's core reduction intensity measure: length and width of last maximum core facets, highly cortical blanks, and blanks with little cortex.

was present and produced a date of 38,500 ± 1,000 BP (GIN-2774 Moscow). This date disagreed with the geological age of the correlated strata, as Gladilin held that Complex II should date to a pre-Brörup fossil soil. It is more likely that the paleosol in which the material from Korolevo II Complex II was found dates to the last interpleniglacial (Karel Valoch personal communication, 1995). The radiocarbon date would agree with a later pedogenesis.

The raw material represented in this assemblage is almost exclusively a local andesite, 93.8% of the collection according to Gladilin and Demidenko (1989:146). Flint, quartzite, slate, sandstone, and quartz make up the remaining 6.2% of the assemblage. The andesite was "slightly leached, with relatively frequent small cells of leaching" (Gladilin and Demidenko 1989:146). This chemical weathering made the direction of dorsal scar patterns difficult to read at times and also increased the frequency of breakage among the elongated debitage. Together with the higher brittleness (lower elasticity) of this volcanic stone (Usik 1989), the andesite's tendency to weather

Table 6.12 Korolevo II Complex II, percentage of intact cores (n=27) by number of bulbar negatives in each core sector (A, B, C, D) by surface (1, 2)

Number of bulbar Negatives	Surf. 1 Sect. A	Surf. 1 Sect. B	Surf. 1 Sect. C	Surf. 1 Sect. D	Surf. 2 Sect. A	Surf. 2 Sect. B	Surf. 2 Sect. C	Surf. 2 Sect. D
0	0	59	56	44	52	63	67	85
1	22	30	22	48	26	15	15	11
2	41	7	11	4	19	11	19	4
3	30	4	11	4	4	4		
4	7					7		

Table 6.13 Korolevo II Complex II, core sector preparation: percent of intact cores (n=20) by number of scars in sector of origin (A, B, C, D) by surface (1, 2)

Number of Scars	Surf. 1 Sect. A	Surf. 1 Sect. B	Surf. 1 Sect. C	Surf. 1 Sect. D	Surf. 2 Sect. A	Surf. 2 Sect. B	Surf. 2 Sect. C	Surf. 2 Sect. D
0	7	89	48	81	30	63	56	67
1	4	4	15	15	15	4	19	11
2	19	4	19	0	22	11	15	15
3	41	4	7	0	22	4	7	7
4	22		4	4	7	4	0	
5	4		4		0	11	0	
6	4		4		4	0	4	
7						4		

chemically more than other conchoidally fractionable rocks resulted in a much higher frequency of broken artifacts and indeterminate attributes than other assemblages studied in this research. This fact decreased the representation of some attribute analyses. Despite this increased weathering, most technological attributes were still easily read, if they were present, on this material. The cortex on this assemblage takes the form of a rough cratered and pocked surface resulting from intense chemical and physical weathering.

6.2.2 *Sampling*

One two-meter by two-meter square, B4, was chosen as a representative sample of the debitage and tools from this assemblage (see Gladilin and Demidenko 1989:fig. 1 for the site plan). A larger sample would have been useful for inter-regional comparisons but was practically impossible to achieve due to the state of the assemblage's curation. Fortunately, the present attribute analysis is supported by the extensive refittings made by V. Usik (1989; Demidenko and Usik 1991, 1993a,b,c).

6.2.3 *Blank Production*

Core Modification

Core modification in Korolevo II Complex II entails a longitudinal blade/flake core orientation, as evidenced in the identification of the long core face as the debitage surface. This identification is supported by the core morphologies presented in Usik (1989) and Table 6.12, which demonstrates that Surface 1 strongly outweighs Surface 2 in number of bulbar negatives, indicating a preference for exploiting one surface rather than two. The longitudinal core orientation visible in the choice of Surface 1 is also corroborative evidence for the preparation of particular sectors on each surface, as seen in Table 6.13. Sector A on Surface 1 displays the highest frequency of cores with removals from this direction, followed somewhat distantly by sector C, while sectors B and D (the lateral core margins) on Surface 1 show only sporadic use, most likely the result of the preparation for débordant/lateral crested removals. On Surface 2, sectors A and C again show the most working but far less frequently than on Surface 1. Sector B on Surface 2 shows both more bulbar negatives and more scar removals from that sector than all sectors but A on this surface, indicating perhaps the use of sector B removals during late core exploitation.

Core convexity management began with the application of cresting removals to shape the longitudinal convexity of the core surface into a central ridge which was then struck off as a *lame à crête* (Demidenko and Usik 1993b, 1991). This is evidenced in both the presence of crested blades in the assemblage as well as refit sequences which demonstrate the primary role of such removals for the obligatory shaping of the convexity via longitudinal ridges for the subsequent removal of blade blanks (Demidenko and Usik 1993b:figs. 1-4). These "frontal" crests are in fact often lateral to the ultimate debitage surface of the core and so are referred to by Usik as a "crested edge on the longitudinal lateral axis of the precore" (1989:188). Core convexity management was also achieved in this assemblage through the removal of lateral débordant flakes/blades to reposition the dorsal convexity along longitudinal ridges to the middle of debitage surface, when viewed from one of the narrow ends of the core. This is visible both in the debitage and the refittings of V. Usik (1989:195, Figure 14), although it must be pointed out that the voluminous exploitation of the core mass gave these lateral removals a slightly different role and form than seen in strictly Levallois débordant removals (i.e., Meignen and Bar-Yosef 1991). In particular instances, after the flattening of the front face of the core, blades were subsequently removed from the narrow side of the core, a 90 degree rotation to the right or left from the debitage surface when viewed from the platform. This is evidenced in refit sequences such as Usik (1989:181, fig. 1, removals 2, 3, and 4) and Usik (1989:185, fig. 5, removals 4, and 5). This is a pattern also seen in the Bohunician industry and others argued to be related to some degree (see Ginter et al. 1996).

Platform Maintenance

Platform maintenance includes a strong tendency to leave platforms unprepared, accounting for 66% of the debitage when indeterminates are removed (Table 6.14). Measurements of external platform angles and platform thickness (Table 6.15) indicate tendencies around 90 degrees and 8 mm respectively. The very steep exterior platform angle is partly the result of abrasion and beveling behind the platforms in this assemblage.

Direction of Core Exploitation

The attribute analysis of the direction of core exploitation is presented in Table 6.16. The differentially greater chemical weathering of the andesite compared to the raw material in other assemblages resulted in more indeterminate dorsal scar patterns as well as fewer complete blanks eligible for analysis in Korolevo II Complex II. Adding the fact that over 10% of the 7,000 artifacts were refit into unanalyzable reconstructions (Usik 1989:183), a serious lack of eligible complete blanks for analysis faced the attribute analyst of this assemblage. This problem was apparent during the studying of the assemblage but the attribute analysis was conducted as a test nevertheless. The gaps in the representation of several dorsal scar directions in Table 6.16 as well as the extremely small resulting sample size after the removal of

Table 6.14 Korolevo II Complex II, platform types for complete and proximal fragments of flakes and tools

Platform type	Frequency	Percent
Plain	34	30.6
Cortical	6	5.4
Dihedral	2	1.8
Facetted	7	6.3
Chapeau	1	.9
Scarred	12	10.8
Linear	2	1.8
Removed	5	4.5
Indeterminate	42	37.8
Total	111	100.0

Table 6.15 Korolevo II Complex II, descriptive statistics for continuous platform variables of complete and proximal fragments of flakes and tools

	N	Minimum	Maximum	Mean	Standard Deviation
Exterior Platform Angle (degrees)	58	48.00	140.00	90.45	18.18
Platform Thickness (mm)	54	1.77	26.15	7.96	4.48

Table 6.16 Korolevo II Complex II, dorsal scar pattern by blank length (mm) in quartiles cross-tabulation for complete flakes and tools (excluding flakes with greater than 60% cortex or fewer than 3 dorsal scars)

Dorsal Scar Pattern	>78	78-54	54-36	<36	Total
Unidirectional					
Count	2	3	7	2	14
% within Quartile	14.3%	20.0%	58.3%	22.2%	28.0%
Bidirectional					
Count	4	3	2	2	11
% within Quartile	28.6%	20.0%	16.7%	22.2%	22.0%
Crossed					
Count	3	4		4	11
% within Quartile	21.4%	26.7%		44.4%	22.0%
Subcentripetal & Centripetal					
Count	1		1	1	3
% within Quartile	7.1%		8.3%	11.1%	6.0%
Crested					
Count	4	4			8
% within Quartile	28.6%	26.7%			16.0%
Indeterminate					
Count		1	2		3
% within Quartile		6.7%	16.7%		6.0%
Total					
Count	14	15	12	9	50
% within Cortex	100.0%	100.0%	100.0%	100.0%	100.0%

indeterminates indicate, however, that reliance on the relative frequencies in this cross-tabulation would produce a result of dubious accuracy in relation to the known refits and core forms. As a result and fortunately for the present analysis, the determination of the directions of core reduction in early core exploitation and late core exploitation is here based on Usik's reconstructions of refit sequences and not on the dorsal scar pattern versus blank length cross-tabulation. Even though other assemblages in the present study possess as many if not more refit sequences which could be used instead of the attribute analysis to represent the assemblage, the Korolevo II Complex II assemblage, and that of Korolevo I layer 1a below and Boker Tachtit Level 4 (Chapter 7), are the only assemblages in which the refits were used to replace the attribute analyses. This is because only in these three cases was the attribute analysis not representative enough in terms of numbers of complete blanks for the cross-tabulation. A simple disagreement between certain details of the refits and the attribute analysis was deemed insufficient reason to warrant breaking the conventions of comparability for the inter-regional study by throwing out the attribute analysis in favor of the refit data.

Usik's refitted sequences present a clear picture of both early and late core exploitation in terms of the directionality of the removals, based as they are on core forms representing exploitation at discard, core trimming elements such as overshot flakes capturing the entire dorsal surface of the core at various points in exploitation, and the refitted removals which extend from early to late exploitation. As Usik (1989:186) notes:

> In layer II Korolevo II there was a large number of single-platform parallel cores with flat working surface [sic]. After their reconstruction it became evident that while at the beginning of the flaking process the parallel cores were voluminous, towards the end of flaking they were already 'flat' and in the view of numerous prehistorians they were of Mousterian Levallois type. In their prevailing majority these cores were bidirectional with one of the opposite platforms direct (smooth or facetted) and with the second in many cases sharp in profile and heavily beveled on the rear side of the core (Figs. 1:1, 2). In the process of flaking these cores, expecially [sic] in the culminating fase [sic] the flakes taken from the cores covered a great deal of the lower platform, converting them into cores with single flat platform (Figs. 2:3, 3).

This generalized reconstruction made by Usik based on refit sequences, including Usik (1989:194, fig. 13), indicates a strong bidirectional tendency for large blanks and unidirectional pattern for the later and shorter blanks. Further, the frequency of the longest overshot flakes preserving the dorsal surface of bidirectional opposed platform cores and the fact that many of these overshot flakes refitted to cores that were subsequently used for unidirectional exploitation is highly demonstrative of one method of transitioning from bidirectional reduction at the beginning of core exploitation toward unidirectional exploitation at the end of core exploitation. Based on these findings, therefore, the analytical signature for this assemblage for the two steps in this domain are bidirectional in early exploitation and unidirectional in late exploitation.

Dorsal Surface Convexity

The analysis of the attributes serving as the analytical steps within the surface convexity domain are given in Tables 6.17–6.18. The ridge system used at Korolevo II Complex II

Table 6.17 Korolevo II Complex II, frequencies of lateral edge, profile, and cross-section types for complete flakes and tools

Type	Frequency	Percent
Lateral edge		
Parallel	32	28.8
Convergent	19	17.1
Expanding	21	18.9
Ovoid	18	16.2
Indeterminate	21	18.9
Total	111	100.0
Profile		
Straight	51	45.9
Curved or Concord	32	28.8
Twisted	16	14.4
Indeterminate	12	10.8
Total	111	100.0
Cross-section		
Triangular	34	30.6
Trapezoidal	41	36.9
Other	18	16.2
Indeterminate	18	16.2
Total	111	100.0

Table 6.18 Korolevo II Complex II, descriptive statistics for blank dimensions for complete flakes and tools

Blank Dimension	N	Range	Mean	Standard Deviation
Length	111	160.28	63.85	27.80
Width	111	95.86	41.65	18.48
Thickness	111	44.40	12.06	6.98
Length/Width Ratio	111	4.64	1.71	.86
Width/Thickness Ratio	111	9.58	4.10	1.91

Table 6.19 Korolevo II Complex II, descriptive statistics for dimensions of complete tools

Blank Dimension	N	Range	Mean	Standard Deviation
Length/Width Ratio	28	3.25	1.90	0.85
Width/Thickness Ratio	28	6.96	4.20	1.80

shows a propensity for elongated flakes and not true blades (length/width ration of 1.71), thin products (width/thickness mean of 4.10), moderately curved profiles, and dispersed ridges (varied lateral edges and trapezoidal cross-sections). With the increased likelihood of breakage to longer blanks due to weathering in this assemblage, the original length/width ratio may indeed have been higher than currently evidenced in the complete blanks.

6.2.4 *Tool Kit Morphology*

Tool kit morphology refers to the dorsal surface convexity variables which give the form of their cutting edge to the retouched blanks in the assemblage, as well as the diagnostic flakes without retouch traditionally categorized as tools - i.e., Levallois pieces and naturally backed knives. The tool kit in this assemblage, while still in the majority made on the local andesite (85.8%), included a higher proportion of flint, quartzite and slate,

Table 6.20 Korolevo II Complex II, frequencies of lateral edge, distal terminus, and profile types for complete tools

Type	Frequency	Percent
Lateral edge		
Parallel	7	25.0
Convergent	5	17.9
Expanding	4	14.3
Ovoid	7	25.0
Indeterminate	5	17.9
Total	28	100.0
Distal Terminus		
Blunt	10	35.7
Pointed	6	21.4
Overshot	1	3.6
Removed	6	21.4
Broken	2	7.1
Indeterminate	3	10.7
Total	28	100.0
Profile		
Straight	13	46.4
Curved or Concord	4	14.3
Twisted	7	25.0
Indeterminate	4	14.3
Total	28	100.0

Table 6.21 Korolevo II Complex II, intensity of core and tool reduction

Variables	Numerator	Denominator	Result
Tools/Flake Ratio	Tools: 5	Flakes: 53	0.09
Non-cortical Debitage/ Cortical Debitage Ratio	Non-cortical Debitage: 33	Cortical Debitage: 48	0.69
Henry's Dimensional Analysis	NA	NA	Core facets overlap with the middle to low end of the blank distribution: slight core exhaustion

totaling 14.2%, compared to only 6.2% among the debitage (Gladilin and Demidenko 1989:163). Otherwise, the tool kit is similar to the debitage, but with a higher frequency of parallel lateral sides and a higher degree of laminarity, although still not indicative of true blades. The presence of flat bifacial retouch counts for the use of a unique type of retouch. The application of this retouch is actually reconstructed by refitting as well as examination of the bifacial foliate points in the assemblage (Demidenko and Usik 1993c). Upper Paleolithic tool types are more numerous in the tool kit, although many Middle Paleolithic types are also present (Gladilin and Demidenko 1989, 1990).

Intensity of Tool and Core Reduction

From the moderately higher percentage of distant raw materials among the retouched tools, the tool kit may be expected to show some differential reduction of these tools compared to those made on the local andesite. As it is, on the whole,

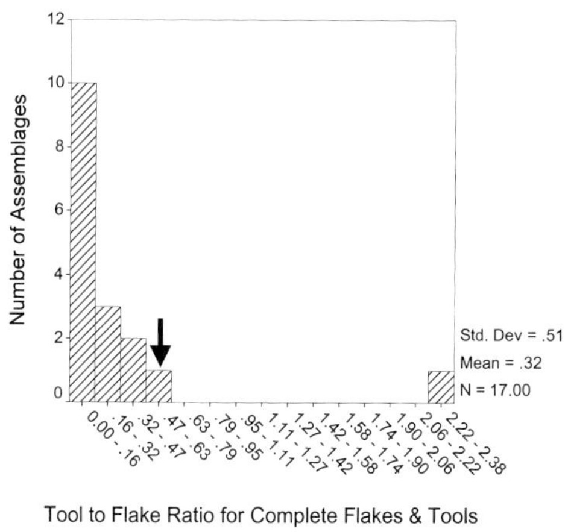

Figure 6.6 Korolevo II Complex II , tool/flake ratio value relative to a histogram of all assemblages.

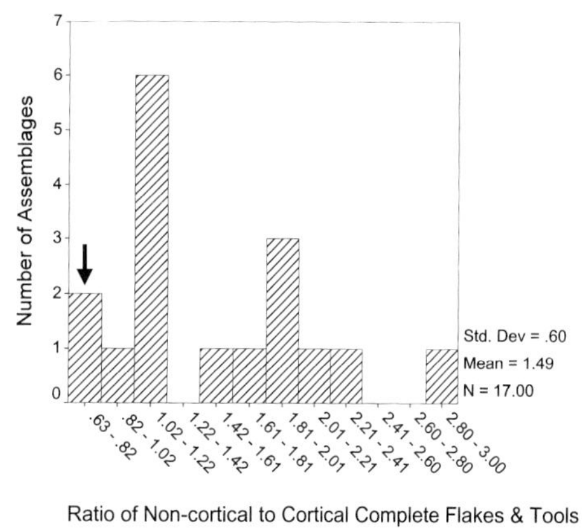

Figure 6.7 Korolevo II Complex II , non-cortical/cortical flake ratio value relative to a histogram of all assemblages.

the tool kit is quite small compared to the debitage, as calculated in the ratio of complete tools to complete flakes from the sample square B4. It is in fact one of the lowest ratios in the study (Table 6.21, Figure 6.6). From the very low ratio of non-cortical blanks to cortical blanks (Figure 6.7), it is clear that cortical nodules of andesite were worked from the very beginning of the core reduction process on site. While this indicates that the early core reduction sequence is represented in this assemblage, the very late portion of the sequence appears to be missing, as there is very little indication of reduction of cores to exhaustion at the site. This is seen partly in the non-cortical to cortical blank ratio but also in the dimensional analysis of core facets relative to the lengths and widths of the blanks. As shown in Figure 6.8, the core facets stretch from the middle to the low end of the blank length distribution, indicating only slight core exhaustion, but the width of the core facets fall mostly below the widths of the blanks. This indicates that the core facets are in fact more elongated than the extant debitage. Given the noted tendency for the longest and therefore the most fragile blanks to break under the weathering conditions in this assemblage, it is likely that the core facets are narrower than the debitage as a result of differential breakage of the more laminar blanks and not because of core exhaustion in the assemblage. This evidence also supports the interpretation that the industry is in fact more elongated if not laminar in blade dimensions than evidenced by the extant debitage. This should be taken into account when comparing the length/width ratio variable between the Korolevo II Complex II and other assemblages.

6.3 Korolevo I Layer 1a

6.3.1 Site Background

Korolevo I Layer 1a is an open-air site located in the same quarry on the western border of Ukraine with Slovakia as Korolevo II Complex II (see above). Begun in 1974, this locality has no direct stratigraphic tie to the profile of Korolevo II Complex II but was correlated based on sediment

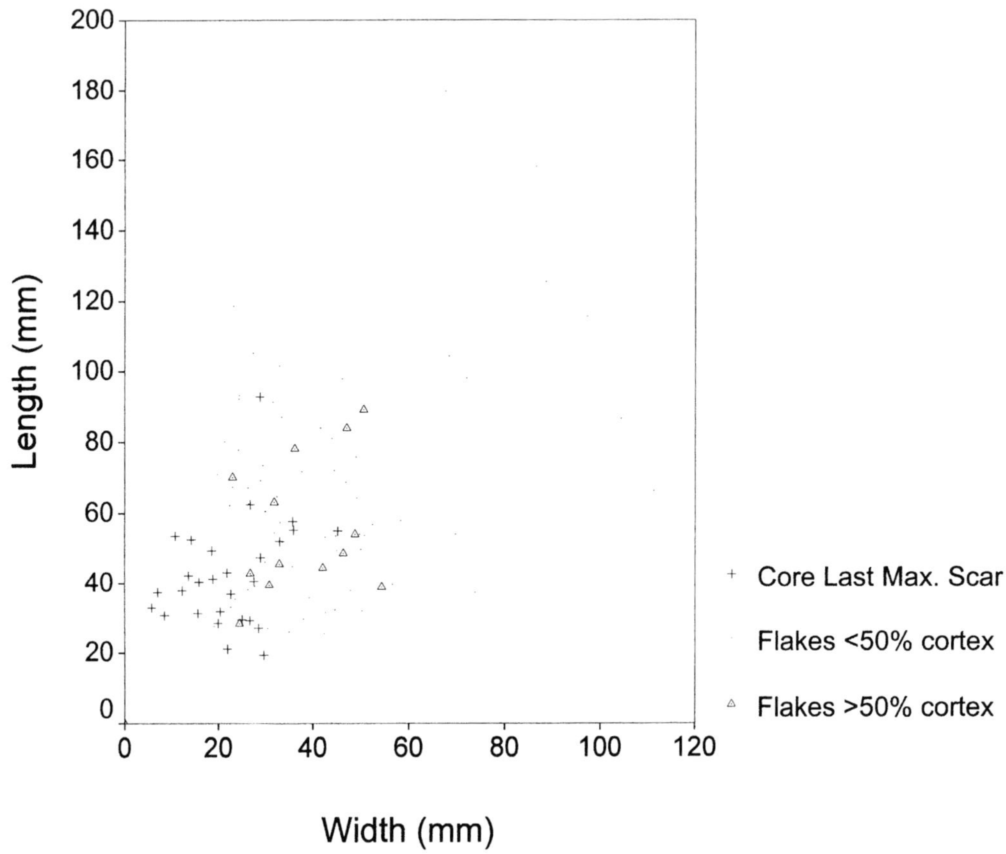

Figure 6.8 Korolevo II Complex II according to Henry's core reduction intensity measure: length and width of last maximum core facets, highly cortical blanks, and blanks with little cortex.

texture, color, and other geological processes to the generalized stratigraphy published by Gladilin (1989a, 1989b; Gladilin and Demidenko 1989, 1990; Usik 1989). This stratigraphic correlation placed Korolevo I layer 1a in a superposed, i.e. later, position compared to Korolevo II. No reliable, uncontroversial radiometric date exists for this assemblage. The one conventional radiocarbon date run by the Geological Institute of the Academy of Sciences of USSR in Moscow (Sulerzhitskiy et al. 1984:168) utilized an aggregate of charcoal particles from the occupation layer and produced a date of 25,700 ± 400 bp (GIN-2773). This date has been rejected by Gladilin and Demidenko (1989:164) as contaminated by carbon from Hallstatt period cremation urn burials in the vicinity. Instead, they provide an interpretation of Layer 1a as dating to the end of the pre-Brörup interval of Würm I. Technologically, Korolevo I layer 1a possesses more Upper Paleolithic features in its debitage and tools as well as less

Table 6.22 Korolevo I-1a, percentage of intact cores (n=28) by number of bulbar negatives in each core sector (A, B, C, D) by surface (1, 2)

Number of bulbar Negatives	Surf. 1 Sect. A	Surf. 1 Sect. B	Surf. 1 Sect. C	Surf. 1 Sect. D	Surf. 2 Sect. A	Surf. 2 Sect. B	Surf. 2 Sect. C	Surf. 2 Sect. D
0	7	46	54	39	86	64	82	79
1	18	21	25	43	11	14	14	11
2	36	25	21	18	0	14	4	7
3	25	7			4	7		4
4	14							

Table 6.23 Korolevo I Layer 1a, core sector preparation: percent of intact cores (n=28) by number of scars in sector of origin (A, B, C, D) by Surface (1, 2)

Number of Scars	Surf. 1 Sect. A	Surf. 1 Sect. B	Surf. 1 Sect. C	Surf. 1 Sect. D	Surf. 2 Sect. A	Surf. 2 Sect. B	Surf. 2 Sect. C	Surf. 2 Sect. D
0	7	57	57	64	32	64	68	71
1	4	11	7	18	39	14	21	14
2	18	25	18	7	11	7	7	11
3	32	7	11	7	0	7	4	4
4	29		7	0	18	4		
5	7			4		4		
6	0							
7	0							
8	4							

weathering of the andesite surfaces; these facts may have contributed to its stratigraphic interpretation. For the present research, Korolevo I layer 1a is treated as post-dating Korolevo II Complex II but pre-dating the 30 kya cut-off for the study.

As with the Korolevo II Complex II assemblage, the raw material represented in Korolevo I layer 1a is almost exclusively the local andesite, representing 97.1% of the assemblage. Radiolarite, chalcedony, sandstone, and quartz account for the remaining 2.9% (Gladilin and Demidenko 1989:164).

6.3.2 *Sampling*

One two-meter by two-meter square, "BV" excavated between 1979 and 1980, was chosen as a representative sample of the debitage from this assemblage (see Gladilin and Demidenko 1989:fig. 16 for the site plan). A random sample of half of all of the tools in this assemblage were studied. The present attribute analysis is supported by the extensive refittings made by V. Usik (1989; Demidenko and Usik 1993a).

6.3.3 *Blank Production*

Core Modification

Core modification in Korolevo I layer 1a entails a longitudinal blade/flake core orientation, as evidenced in the identification of the long core face as the debitage surface. This identification is supported by the core morphologics presented in Usik (1989), Appendix Figures A31-34, and Table 6.22, which demonstrates that Surface 1 strongly outweighs Surface 2 in number of bulbar negatives, indicating a preference for exploiting one surface rather than two.

The longitudinal core orientation visible in the choice of Surface 1 is also corroborated by Table 6.23, from which it is possible to gauge the dominant use of the A sectors on both surfaces

(Surface 2 sector A being a platform surface) with relatively few removals made from the other sectors on either surface. Table 6.23 suggests that the slightly higher number of bulbar negatives (seen in Table 6.22) in sectors B-D of Surface 1 compared to those sectors on Surface 2 is the result of late removals made at the exhaustion of the cores rather than a tendency to use these sectors for centripetal removals.

Core convexity management in Korolevo I layer 1a was achieved through the creation of frontal crests to shape the longitudinal curvature of the debitage surface. These "frontal" crests were in fact formed on what would become the lateral edge of the core's main debitage surface, Surface 1 as defined above, as indicated in Usik's refit sequences (Usik 1989:198). This tendency is partly the result of the frequent use of very large flakes as the supports for core exploitation. The long edge of a large flake was initially crested and, as this edge usually possessed a convexity only on the dorsal surface, only the old ventral surface was struck to remove flakes to shape the edge. These unifacial crests were then removed as reduction progressed from the core-on-flake's edge onto and through the convexity of the old dorsal surface (Usik 1989:figs. 16–19). Depending upon whether or not the unifacial crest was formed on the lateral or distal edge of the original flake, the direction of the old ventral scar would be parallel or crossed to the direction of blank removal during the main sequence of reduction, as Usik (1989:197, fig. 16) demonstrates for the latter situation. Core convexities were also managed in this assemblage through the application of the core tablet technique to repair the loss of proximal convexities at the platform end of the core (Usik 1989:202, fig. 22).

Platform Maintenance

Platform maintenance includes a strong tendency to leave platforms unprepared, accounting for 77% of the debitage when indeterminates are removed from the percentage (Table 6.24). Measurements of external platform angles and platform thickness (Table 6.25) indicate tendencies around 88 degrees and 9 mm respectively.

Direction of Core Exploitation

Over 1,000 artifacts were refitted together by Usik and thus roughly half of the assemblage over 3 cm in size was left unavailable for study (Usik 1989:198) in the Institute of Zoology, Kiev, Ukraine. This still should have left a significant number of intact blanks for attribute analysis. Unfortunately, the sampling strategy applied to Korolevo I layer 1a, like that of Korolevo II Complex II, proved too limited to overcome the higher frequency of broken and indeterminate debitage resulting from the chemical and physical weathering of the andesite assemblage. The resulting cross-tabulation of blank length in quartiles versus dorsal scar pattern is thus biased by its small sample size. The cross-tabulation in Table 6.26 indicates subcentripetal and centripetal removals dominating early core exploitation while unidirectional removals dominate late exploitation. As evidenced by Usik's refit sequences (such as Usik 1989:201, fig. 21) and the morphology of the cores at discard (Tables 6.22-6.23 and Appendix Figures A31-34), the attribute analysis' unidirectional characterization of late core exploitation appears to be correct. The sub/centripetal characterization of early exploitation, however, is apparently a result of sample bias toward secondary crested blanks of large dimensions, such as removal 2 in Usik (1989:197, fig. 16), which are not technically crested as no negative of a bulb of percussion is found on a dorsal ridge but still possess scars originating from the opposite lateral edge(s). Secondary crested blanks usually prove to be infrequent enough that their

Table 6.24 Korolevo I-1a, platform types for complete and proximal fragments of flakes and tools

Platform Type	Frequency	Percent
Plain	69	46.6
Cortical	5	3.4
Dihedral	2	1.4
Facetted	7	4.7
Scarred	15	10.1
Punctiform	1	.7
Linear	6	4.1
Removed	13	8.8
Indeterminate	30	20.3
Total	148	100.0

Table 6.25 Korolevo I-1a, descriptive statistics for continuous platform variables of complete and proximal fragments of flakes and tools

	N	Minimum	Maximum	Mean	Standard Deviation
Exterior Platform Angle (degrees)	87	53.00	135.00	88.32	19.29
Platform Thickness (mm)	86	1.27	40.69	9.24	7.02

Table 6.26 Korolevo I-1a, dorsal scar pattern by blank length (mm) in quartiles cross-tabulation for complete flakes and tools (excluding flakes with greater than 60% cortex or fewer than three dorsal scars)

Dorsal Scar Pattern	>77	77-55	55-40	<40	Total
Unidirectional					
Count	5	8	9	13	35
% within Quartile	20.0%	40.0%	45.0%	61.9%	40.7%
Bidirectional					
Count	4	2	1	1	8
% within Quartile	16.0%	10.0%	5.0%	4.8%	9.3%
Crossed					
Count		3	2	2	7
% within Quartile		15.0%	10.0%	9.5%	8.1%
Subcentripetal & Centripetal					
Count	8	4	4	1	17
% within Quartile	32.0%	20.0%	20.0%	4.8%	19.8%
Crested					
Count	5	3	1	1	10
% within Quartile	20.0%	15.0%	5.0%	4.8%	11.6%
Indeterminate					
Count	3		3	3	9
% within Quartile	12.0%		15.0%	14.3%	10.5%
Total					
Count	25	20	20	21	86
% within Cortex	100.0%	100.0%	100.0%	100.0%	100.0%

subcentripetal or centripetal dorsal scar pattern does not dominate the first length quartile. As this assemblage contains exceedingly large initial removals, however, the signature for this quartile is not representative of the majority of reduction patterns at the beginning of core exploitation, at least as one would conclude from the refit data. Further, the absence of any representation of the crossed dorsal scar pattern in the first quartile in Table 6.22 violates the rules of comparison for the attribute analysis set up at the beginning of the inter-regional study. As a result, the conclusions from Usik's refits, which indicate that the direction of reduction in both early and late core exploitation was wholly unidirectional, are used for the first and second analytical steps in this domain.

Dorsal Surface Convexity

The analysis of the attributes serving as the steps within the surface convexity domain are given in Tables 6.27-6.28. The ridge system used at Korolevo I-1a shows a propensity for elongated flakes (length/width ratio of 1.84), very thin products (width/thickness mean of 4.29), and both straight and curved profiles. The propensity for very elongated products of blade dimensions to break in this assemblage due to the effects of chemical and physical weathering of the andesite material should be kept in mind when considering the length/width ratio. The ratio here should be seen as a minimum value only, as the refits indicate that the removals were very bladey, but broke in several places either at production or as a result of site formation processes.

6.3.4 Tool Kit Morphology

Tool kit morphology refers to the dorsal surface convexity variables which give the form of their cutting edge to the retouched blanks in the assemblage. The tool kit in this assemblage, while still in the vast majority made on the local andesite (78.5%), included a higher proportion of flint, radiolarite, and chalcedony than among the debitage, totaling 21.5 compared to 2.9% (Gladilin and Demidenko 1989:163). Otherwise, the tool kit is roughly similar to the debitage, although slightly thicker and shorter as well as more expansive in lateral edges than the debitage. The tool blanks are also straighter in profile than the debitage. The high frequency of indeterminate distal ends among the tools can be seen as corroboration of the problematic effect of weathering of this raw material on attribute analyses.

The last two characterizations of the tool kit morphology refer to the two variables describing the application of retouch to blanks. Evidence for bifacial retouch in the form of foliate points or bifacial thinning flakes is lacking in this assemblage. A retouched tool type called the "rezchik," however, is defined in this assemblage as a flat burin in which the spall is removed more from the ventral face of the blank than the lateral edge (Gladilin and Demidenko 1989:171). As these pieces are created through the application of a burin blow, are regionally specific as a tool type, and may overlap in some features with truncated-facetted pieces, it was judged that these pieces do not constitute the application of a unique retouch type for the inter-regional study. Other industries contain burins which were not subdivided based on where the burin scar lies. Upper Paleolithic tool types are clearly dominant in the tool kit (Gladilin and Demidenko 1989; 1990).

Intensity of Tool and Core Reduction

From the moderately higher percentage of distant raw materials among the retouched tools compared to the debitage, the tool kit may be expected to show some differential reduction of these tools compared to those made on the local

Table 6.27 Korolevo I-1a, frequencies of lateral edge, profile, and cross-section types for complete flakes and tools

Type	Frequency	Percent
Lateral edge		
Parallel	54	36.5
Convergent	34	23.0
Expanding	24	16.2
Ovoid	14	9.5
Indeterminate	22	14.9
Total	148	100.0
Profile		
Straight	86	58.1
Curved or Concord	33	22.3
Twisted	16	10.8
Indeterminate	13	8.8
Total	148	100.0
Cross-section		
Triangular	52	35.1
Trapezoidal	59	39.9
Other	22	14.9
Indeterminate	15	10.1
Total	148	100.0

Table 6.28 Korolevo I-1a, descriptive statistics for blank dimensions for complete flakes and tools

Blank Dimension	N	Range	Mean	Standard Deviation
Length	148	159.29	62.03	30.64
Width	148	161.78	40.69	26.18
Thickness	148	48.89	11.18	8.17
Length/Width Ratio	148	9.52	1.84	1.10
Width/Thickness Ratio	148	15.17	4.29	2.24

Table 6.29 Korolevo I-1a, descriptive statistics for dimensions of complete tools

Blank Dimension	N	Range	Mean	Standard Deviation
Length/Width Ratio	40	3.98	1.70	0.89
Width/Thickness Ratio	40	15.09	4.06	2.64

andesite. As it is, on the whole, the tool kit as presented in Table 6.31 is quite small compared to the debitage sample (Figure 6.9). Note that only the complete tools and complete flakes from the sample unit of "BV" were used for calculating the tool to flake ratio, even though half of the entire assemblage's tools were studied for the analysis of the tool kit above. From the very high ratio of non-cortical blanks to cortical blanks (Figure 6.10), there are far more non-cortical blanks than cortical blanks in this assemblage than evidenced in all of the other study's assemblages. This ratio can be interpreted in several ways. First, the andesite cores could already have been decortified or worked to some extent before they were brought to the site so that the early, cortical portion of the core reduction sequence is missing from the assemblage. Second, the significant degree of core exhaustion evidenced in the comparison of

Table 6.30 Korolevo I-1a, frequencies of lateral edge, distal terminus, and profile types for complete tools

Platform Type	Frequency	Percent
Lateral edge		
Parallel	12	30.0
Convergent	5	12.5
Expanding	10	25.0
Ovoid	3	7.5
Indeterminate	10	25.0
Total	40	100.0
Distal Terminus		
Blunt	9	22.5
Pointed	3	7.5
Removed	5	12.5
Broken	2	5.0
Indeterminate	21	52.5
Total	40	100.0
Profile		
Straight	23	57.5
Curved or Concord	8	20.0
Twisted	4	10.0
Indeterminate	5	12.5
Total	40	100.0

Table 6.31 Korolevo I-1a, intensity of core and tool reduction

Variables	Numerator	Denominator	Result
Tools/Flake Ratio	Tools: 4	Flakes: 89	0.04
Non-cortical Debitage/ Cortical Debitage Ratio	Non-cortical Debitage: 96	Cortical Debitage: 33	2.91
Henry's Dimensional Analysis	NA	NA	Core facets overlap with the low end of the blank distribution: significant core exhaustion

the dimensions of the last maximum core facets relative to the blank dimensions (Figure 6.11) could indicate that far more blanks were produced per core than in other assemblages. This second interpretation is more likely given the extremely large andesite nodule size evidenced by large primary flakes (serving as cores-on-flake) over 250 mm long and crested blades over 220 mm long (pieces found in Layer 1a outside of unit "BV"). As the non-cortical volume of the core increases by the cube while the cortical surface area increases by the square, a larger initial nodule size would produce more non-cortical pieces per core than a smaller nodule. Thus, as the abundant refit data clearly indicates that the early core reduction sequence is in fact quite well represented in the assemblage, this non-cortical to cortical blank ratio value must be interpreted as corroborative evidence of both longer core reduction, from the earliest rough working of the nodules to the exhaustion of the cores, and the larger size of the initial nodules. Thus for the purpose of comparability with other assemblages, the Korolevo I layer 1a assemblage captures the widest possible representation of the core reduction sequence.

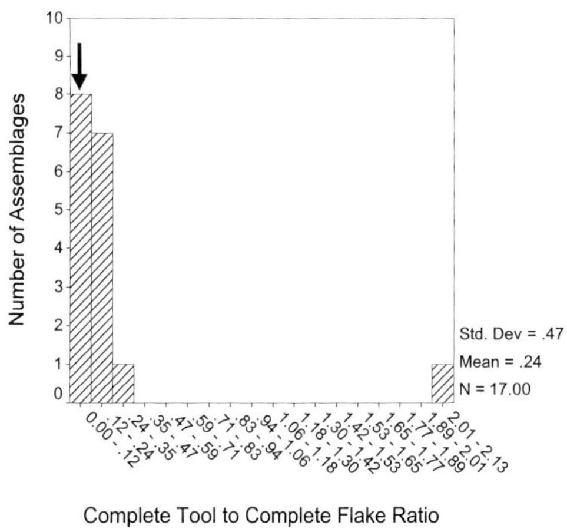

Figure 6.9 Korolevo I Layer I-a, tool/flake ratio value relative to a histogram of all assemblages.

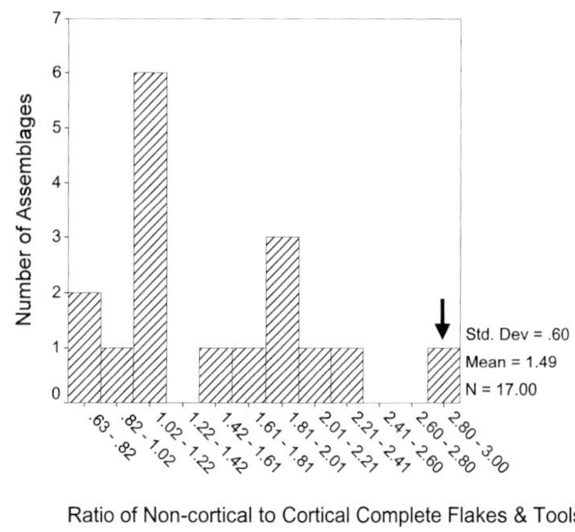

Figure 6.10 Korolevo I Layer I-a, non-cortical/cortical flake ratio value relative to a histogram of all assemblages.

As shown in Figure 6.11, the core facets in this assemblage appear more elongated than the debitage, suggesting, as with Korolevo II Complex II, that differential breakage of blades has biased the extant intact assemblage toward flake dimensions. This should be taken into account when comparing the length/width ratio variable between the Korolevo I layer 1a and other assemblages.

6.4 Comparison of Eastern European Flintknapping Behaviors

The construction of the Eastern European sequence of change between 60 and 30 kya is conducted on the basis of pair-wise comparisons between assemblages as demonstrated in Chapter 5 for the Central European sequence.

6.4.1 Comparison of Molodova V Layer 11 and Korolevo II Complex II

Table 6.32 presents the data for the evaluation of the comparability of the two assemblages of Molodova V layer 11 and Korolevo II Complex II. The application of retouch is extremely similar between the assemblages (Figure 6.12), indicating that resharpening and shaping of cutting edges was of secondary use at both localities, either because of the tasks at hand or the abundance of available raw material. The relative amount of cortical blanks in each assemblage is, however, quite distinct, with Korolevo II Complex II containing far more relative to Molodova V Layer 11 (Figure 6.13). As both assemblages do not appear to be highly reduced, with few cores near exhaustion, it is likely that the two assemblages both possess the middle of the core reduction sequence and lack the exhausted cores of the extreme end of the sequence. Molodova V Layer 11, however, lacks the earliest portion of the sequence as captured at Korolevo. The absence of refitted sequences at Molodova compared to Korolevo is also suggestive of this difference. The two assemblages are thus comparable but with some expected differences due to the wider range of debitage morphologies resulting from initial core working at Korolevo. The close similarity in

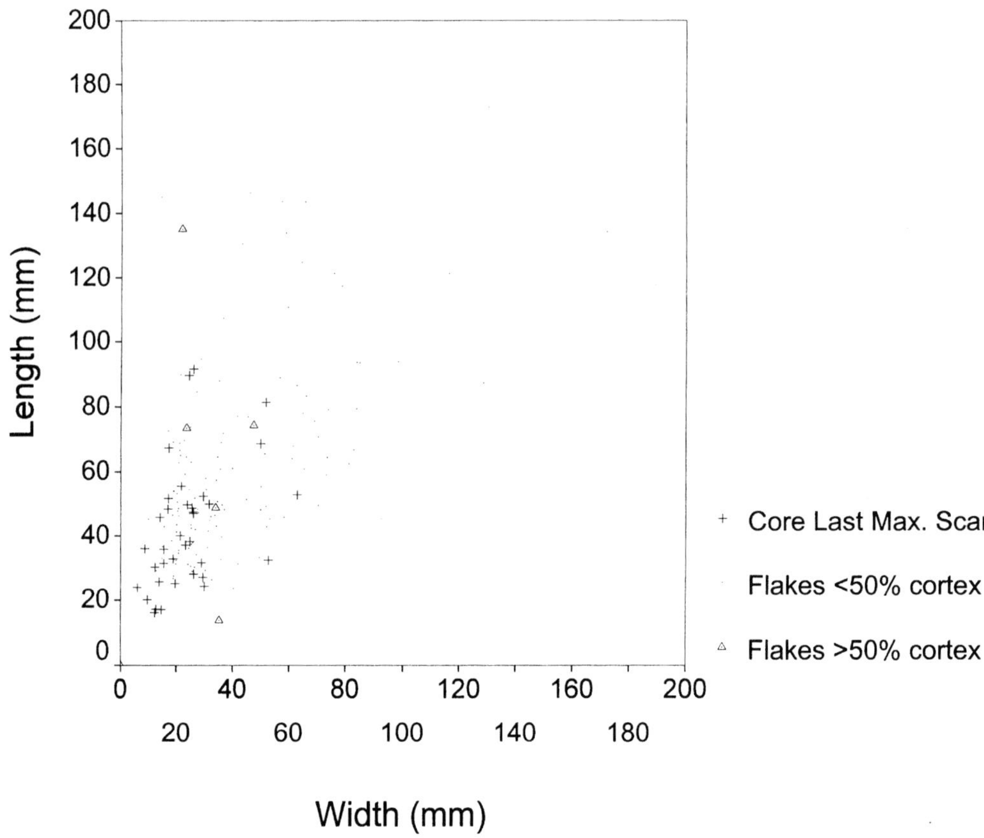

Figure 6.11 Korolevo I Layer I-a according to Henry's core reduction intensity measure: length and width of last maximum core facets, highly cortical blanks, and blanks with little cortex.

the application of retouch, however, indicates that dimensional variables of the tool kits should be quite comparable.

Blank Production

Table 6.33 presents the results of a comparison of the blank production details reconstructed for Molodova V layer 11 and Korolevo II Complex II. The first step in the core modification domain, core orientation, is characterized by the qualitative distinction between methods of exploiting core volumes. In the case of Molodova V layer 11, the majority of cores evidence a broad-faced orientation as well as scars and bulbar negatives indicative of most but not all of the debitage deriving from one surface. In the majority of the Korolevo II-II cores, a quite different, longitudinal narrow core form with preferential use of one surface is evidenced. In the second step in this domain, the choice of core convexity management techniques was distinct between the assemblages, due to the use of frontal crests at Korolevo II-II which were lacking in Molodova V.

Table 6.32 Molodova V Level 11 and Korolevo II Complex II, comparability of the assemblages based on intensity of core and tool reduction

Variables	Molodova V Level 11	Korolevo II-II	Evaluation
Tools/Flake Ratio	0.07	0.09	Similar
Non-cortical Debitage/ Cortical Debitage Ratio	1.90	0.69	Different
Henry's Dimensional Analysis	little to no core exhaustion	slight core exhaustion	Similar

The next knapping domain, platform maintenance, includes the steps whereby the flintknapper modifies the platform surface and platform edge before each removal. In the case of the Molodova V layer 11 and Korolevo II-II comparison, only the mean exterior platform angles were not significantly different. The propensity to prepare platforms in Molodova V and the much thicker platforms of Korolevo II-II, however, made these variables distinctly different.

The next knapping domain, direction of core exploitation, concerns the variability resulting from different options of removing flakes from single to multiple platform cores, creating debitage with particular dorsal scar patterns. In Molodova V layer 11, the balanced frequency of all directions in early core exploitation does not match the bidirectional data from Korolevo II Complex II. The directionalities of late core exploitation, however, are similar, as both assemblages rely on unidirectional reduction. As discussed above, Molodova V Layer 11 lacks part of the early core reduction sequence whereas the Korolevo assemblage does not. Given the fact, however, that nowhere in the refitted core reduction sequences at Korolevo II Complex II do high frequencies of sub/centripetal blanks exit, it is unlikely that the difference reconstructed here for the directionality of early core exploitation is due to the different natures of the two palimpsests.

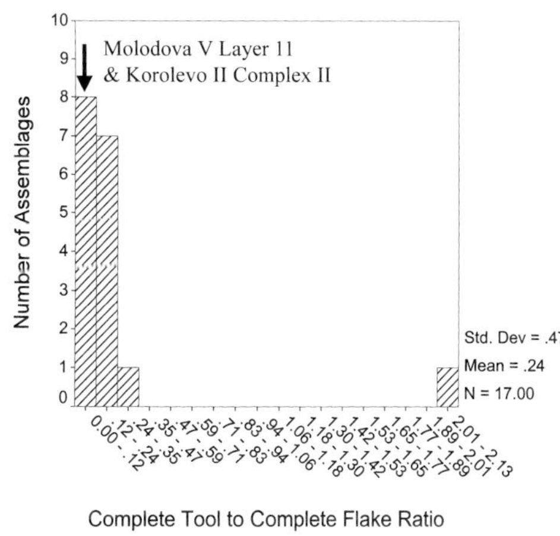

Figure 6.12 Molodova V Level 11 and Korolevo II Complex II compared for tool/flake ratios.

The surface convexity domain includes the knapping steps which encapsulate the cumulative choice of *nervure-guide* or ridge pattern(s) of the core exterior for the production of blanks. Of the five steps within this domain, only the options relating to the choice to follow centralized versus dispersed ridges (i.e., cross-section) and the superficial versus volumetric use of raw material on a flake-by-flake basis (i.e., width/thickness ratio) differed between the two assemblages.

Weighting the final difference measure according to the differences within each of the

Table 6.33 Comparison of Blank Production for Molodova V Level 11 and Korolevo II Complex II

Flintknapping Steps by Domain	Molodova V Level 11	Korolevo II-II	Significant Change?
Core Modification			
Core Orientation	Broad, flat orientation with only partial preferential use of one surface	Preferential use of a Longitudinal Surface	Yes
Core Convexity Management	Débordants & centripetal removals	Débordants & frontal crests	Yes
Number of Changes/2 Steps			**2/2=1**
Platform Maintenance			
Platform Treatment	Unprepared: 39% Prepared: 61% n=367	Unprepared: 66% Prepared: 34% n=64	Yes p=.00, Fisher's Exact
External Platform Angle (degrees)	mean: 85.6, s.d.: 14.0, n=301	mean: 90.4, s.d.: 18.2, n=58	No p=.06, t=-1.926*, df=70.533
Platform Thickness	mean: 5.62, s.d.: 3.30, n=355	mean: 7.96, s.d.: 4.48, n=54	Yes p=.00, t=-3.706*, df=62.036
Number of Changes/3 Steps			**2/3=0.67**
Direction of Core Exploitation			
Direction of Early Exploitation	All four directions equal	Bidirectional	Yes
Direction of Late Exploitation	Unidirectional	Unidirectional	No
Number of Changes/2 Steps			**1/2=0.5**
Dorsal Surface Convexity System			
Laminarity: Length/ Width Ratio	mean: 1.78, s.d.: 0.76, n=427	mean: 1.71, s.d.: 0.86, n=111	No p=.42, t=0.803, df=536
Lateral Edges of Blanks	Parallel: 34% Convergent: 30% Expanding: 19% Ovoid: 17% n=413	Parallel: 36% Convergent: 21% Expanding: 23% Ovoid: 20% n=90	No p=.33, G^2=3.415, df=3
Profile of Blanks	Straight: 48% Curved: 33% Twisted: 19% n=425	Straight: 52% Curved: 32% Twisted: 16% n=99	No p=.76, G^2=0.555, df=2
Cross-Section of Blanks	Triangular: 44% Trapezoidal: 53% Other: 3% n=421	Triangular: 37% Trapezoidal: 44% Other: 19% n=93	Yes p=.00, G^2=26.801, df=2
Convexity: Width/ Thickness Ratio	mean: 4.94, s.d.: 2.23, n=427	mean: 4.10, s.d.: 1.91, n=111	Yes, p=.00, t=3.616, df=536
Number of Changes/5 Steps			**2/5=0.4**
Total Measure of Difference in Blank Production			**2.57**

*The t-test was calculated with equal variances not being assumed due to a significant F value on Levene's Test for Equality of Variances.

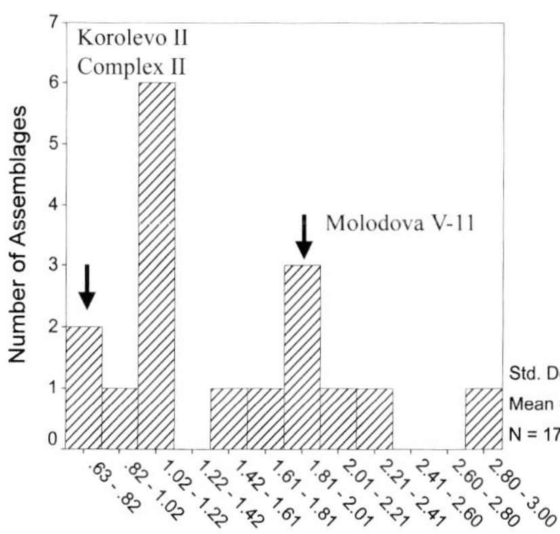

Figure 6.13 Molodova V Level 11 and Korolevo II Complex II compared for non-cortical/cortical flake ratios.

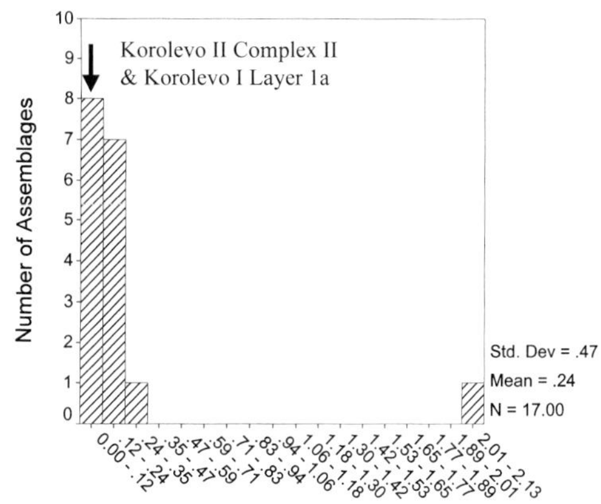

Figure 6.14 Korolevo II Complex II and Korolevo I Layer 1a compared for tool/flake ratios.

four domains, the two assemblages appear more distinct than similar in the details of their blank production, with a value of 2.57 out of a maximum of 4.0. While all four domains are affected, most of the difference is derived from different core modification choices and platform choices.

Tool Kit Morphology

Of the dorsal surface convexity variables which characterize the tool kit morphology of these two assemblages, all but one fail to show a statistically significant difference (Table 6.34). Only the greater surficial exploitation of cores (i.e., wider and thinner blanks) at Molodova V Layer 11 is significantly different between the two tool kits. The assumptions of the G^2 statistical test of the lateral edge type frequencies are violated by two cells with expected counts lower than five but the frequencies themselves are subjectively quite similar. In addition, the higher expanding lateral edge and lower parallel edge frequencies in Korolevo II Complex II are expected, as this assemblage captured more of the early core reduction sequence, during which initial core preforming tends to produce more varied lateral edges. Both assemblages evidence the same unique type of retouch, bifacial retouch in the form of small bifaces at Molodova V and foliate points at Korolevo II, but differ in the dominance of Middle Paleolithic versus Upper Paleolithic tools. Taken together, the final difference measure between the tool kits of 0.29 out of 1.0. This indicates that the tool kits are much more similar than they are different.

6.4.2 Comparison of Korolevo II Complex II and Korolevo I Layer 1a

The next pair-wise comparison in this region between 60 and 30 kya is between Korolevo II Complex II and Korolevo I Layer 1a. Table 6.35 presents the data for the evaluation of the comparability of these assemblages. In terms of the degree to which retouch was applied to blanks (Figure 6.14), both assemblages are quite similar so that no problem exists for the comparison of dimensional variables in the

Table 6.34 Comparison of Tool Kit Morphologies for Molodova V Level 11 and Korolevo II Complex II

Tool Kit Morphology Variable	Molodova V Level 11	Korolevo II-II	Significant Change?
Laminarity	mean: 1.65, s.d.: 0.52, n=29	mean: 1.90, s.d.: 0.85 n=28	No p=.18, t=-1.356*, df=44.388
Width/Thickness	mean: 5.40, s.d.: 2.04, n=29	mean: 4.20, s.d.: 1.80 n=28	Yes p=.02, t=2.334, df=55
Lateral Edges	Parallel: 39% Convergent: 25% Expanding: 11% Ovoid: 25% n=28	Parallel: 30% Convergent: 22% Expanding: 17% Ovoid: 30% n=23	No p=.83, G^2=0.884, df=3†
Distal Terminus	Blunt: 50% Pointed: 50% n=26	Blunt: 63% Pointed: 37% n=16	No p=.53, Fisher's Exact
Profile	Straight: 59% Curved: 27% Twisted: 14% n=29	Straight: 54% Curved: 17% Twisted: 29% n=24	No p=.33, G^2=2.250, df=2‡
Unique Types of Retouch	Bifacial retouch	Flat bifacial retouch	No
Tool Types	MP tools dominate	UP tools dominate	Yes
Number of Changes/7 Steps			**2/7**
Total Measure of Difference in Tool Kit Morphology			**0.29**

*The t-test was calculated with equal variances not being assumed due to a significant F value on Levene's Test for Equality of Variances.

†Two cells (25%) have an expected count less than 5. The minimum expected count is 3.16.

‡One cell (16.7%) has an expected count less than 5. The minimum expected count is 4.98. These results do not invalidate the assumptions of a $\chi2$ test.

Table 6.35 Korolevo II Complex II and Korolevo I layer 1a, comparability of the assemblages based on intensity of core and tool reduction

Variables	Korolevo II-II	Korolevo I layer 1a	Evaluation
Tools/Flake Ratio	0.09	0.04	Very similar
Non-cortical Debitage/ Cortical Debitage Ratio	0.69	2.91	Different but see discussion in text
Henry's Dimensional Analysis	slight core exhaustion	significant core exhaustion	Different

tool kits. Both assemblages are also similar in possessing the very beginning of the core reduction sequence, evidenced by Usik's refit sequences and the dimensions of the crested pieces in layer 1a. The greater degree of core exhaustion in layer 1a, together with the use of very large cores-on-flakes as initial core performs in this assemblage, indicates that the two assemblages differ in the extent to which core reduction continued toward core exhaustion. These two features of the layer 1a assemblage explain the different non-cortical to cortical blank ratios evidenced by the debitage (Figure 6.15). As a whole, these data indicate that the two assemblages are comparable.

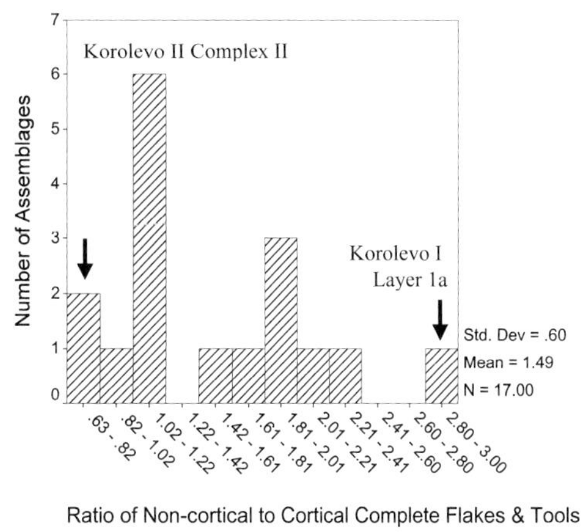

Figure 6.15 Korolevo II Complex II and Korolevo I Layer 1a compared for non-cortical/cortical flake ratios.

Blank Production

Table 6.36 presents the results of a comparison of the blank production details of the assemblages from Korolevo II Complex II and Korolevo I layer 1a. The first step in the core modification domain, core orientation, is characterized by the qualitative distinction between methods of exploiting core volumes. The majority of both the Korolevo II-II and Korolevo I-1a cores evidence the preferential use of one surface on a longitudinal blade/flake core form. In the second step in this domain, the choice of core convexity management techniques was distinct between the assemblages, due to the use of débordant removals but the absence of core tablets in Korolevo II-II, compared to the absence and presence, respectively, of these techniques in Korolevo I-1a.

The next knapping domain, platform maintenance, includes the steps whereby the flintknapper modifies the platform surface and platform edge before each removal. In the case of the Korolevo II-II versus Korolevo I-1a comparison, no significant differences were apparent. In terms of the direction of core exploitation domain, only early core exploitation was significantly different, due to a contrast between bidirectional and unidirectional reduction. Late core exploitation was similar. Of the five steps within the surface convexity domain, no options were judged to be significantly different between the two assemblages.

The blank production details in these two assemblages are quite similar on the whole, measuring 1.00 out of a maximum of 4.00. The differences lie primarily in core convexity management techniques and the directionality of early core exploitation.

Tool Kit Morphology

Of the dorsal surface convexity variables which characterize the cutting edge morphology of the tool kits of these two assemblages, none of the variables are significantly different. The low sample size of the tools produced a problem for the G^2 test of the lateral edge type frequencies as two cells contained expected counts lower than five. The frequencies themselves are subjectively,

Table 6.36 Comparison of Blank Production for Korolevo II Complex II and Korolevo I Level 1a

Flintknapping Steps by Domain	Korolevo II-II	Korolevo I Level 1a	Significant Change?
Core Modification			
Core Orientation	Preferential use of a Longitudinal Surface	Preferential use of a Longitudinal Surface	No
Core Management	Débordants, front crests, & side blade removals	Frontal crests & core tablet	Yes
Number of Changes/2 Steps			**1/2=0.5**
Platform Maintenance			
Platform Treatment	Unprepared: 66% Prepared: 34% n=64	Unprepared: 77% Prepared: 23% n=105	No p=.11, Fisher's Exact
External Platform Angle (degrees)	mean: 90.4, s.d.: 18.2, n=58	mean: 88.3, s.d.: 19.3, n=87	No p=.51, t=0.665, df=143
Platform Thickness	mean: 7.96, s.d.: 4.48, n=54	mean: 9.24, s.d.: 7.02, n=86	No p=.19, t=-1.307*, df=137.956
Number of Changes/3 Steps			**0/3=0**
Direction of Core Exploitation			
Direction of Early Exploitation	Bidirectional	Unidirectional	Yes
Direction of Late Exploitation	Unidirectional	Unidirectional	No
Number of Changes/2 Steps			**1/2=0.5**
Dorsal Surface Convexity System			
Laminarity: Length/Width Ratio	mean: 1.71, s.d.: 0.86, n=111	mean: 1.84, s.d.: 1.10, n=148	No p=.32, t=-0.995, df=257
Lateral Edges of Blanks	Parallel: 36% Convergent: 21% Expanding: 23% Ovoid: 20% n=90	Parallel: 43% Convergent: 27% Expanding: 19% Ovoid: 11% n=126	No p=.20, G^2=4.668, df=3
Profile of Blanks	Straight: 52% Curved: 32% Twisted: 16% n=99	Straight: 64% Curved: 24% Twisted: 12% n=135	No p=.17, G^2=3.496, df=2
Cross-Section of Blanks	Triangular: 37% Trapezoidal: 44% Other: 19% n=93	Triangular: 39% Trapezoidal: 44% Other: 17% n=133	No p=.85, G^2=0.337, df=2
Convexity: Width/Thickness Ratio	mean: 4.10, s.d.: 1.91, n=111	mean: 4.29, s.d.: 2.24, n=148	No p=.47, t=-0.724, df=257
Number of Changes/5 Steps			**0/5=0**
Total Measure of Difference in Blank Production			**1.00**

*The t-test was calculated with equal variances not being assumed due to a significant F value on Levene's Test for Equality of Variances.

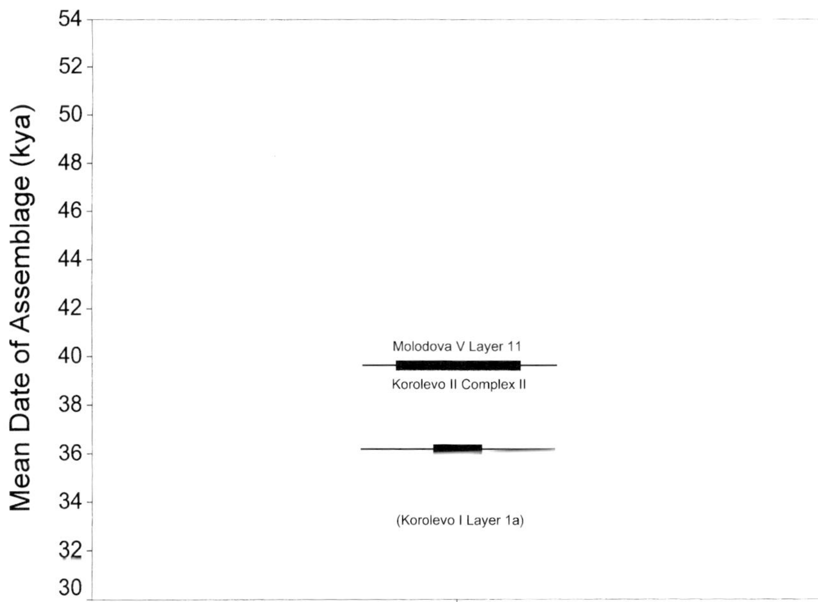

Figure 6.16 Chronological comparison of differences in Blank Production among East European assemblages. The length of the thick bars represents the difference measure between the two assemblages above and below the bar (the earlier and the later). The length of the thin bars represents the maximum possible difference measure.

however, quite different, with Layer 1a displaying more parallel and explanding and less ovoid lateral edges than in Complex II. While it is tempting to indicate a signficant difference for this variable despite the rules of the statistical test, it is possible that such a difference could derive from the relatively longer core reduction sequence, i.e. toward core exhaustion, in Layer 1a. Therefore, the variable is left as non-significant. The assemblages do differ in the presence of unique retouch, as Layer 1a lacks the bifacial reduction of foliate points seen in Complex II. Both tool kits are dominated, however, by Upper Paleolithic tool types, even though Layer 1a possesses far more UP types relative to MP types than Complex II. With only the bifacial retouch at issue in the comparison, the final difference measure between the tool kits is 0.14 out of 1.00. This indicates that the tool kits are extremely similar.

6.5 Significance of the Eastern European Sequence for Model Expectations 1-4

Now that the Eastern European sequence of change has been reconstructed, it is possible to assess some of the model expectations for this intra-regional context. It is immediately clear from Table 6.38 and Figure 6.16 that the Eastern European sequence shows a similar pattern to that of Central Europe: a marked contrast between the last Middle Paleolithic assemblage, Molodova V level 11, and the first non-Middle Paleolithic assemblage, the transitional Korolevo

Table 6.37 Comparison of Tool Kit Morphologies for Korolevo II Complex II and Korolevo I Level 1a

Tool Kit Morphology Variable	Korolevo II-II	Korolevo I-1a	Significant Change?
Laminarity	mean: 1.90, s.d.: 0.85 n=28	mean: 1.70, s.d.: 0.89 n=40	No p=.34, t=0.954, df=66
Width/Thickness	mean: 4.20, s.d.: 1.80 n=28	mean: 4.06, s.d.: 2.64 n=40	No p=.81, t=0.244, df=66
Lateral Edges	Parallel: 30% Convergent: 22% Expanding: 17% Ovoid: 30% n=23	Parallel: 40% Convergent: 17% Expanding: 33% Ovoid: 10% n=30	No p=.20, G^2=4.706, df=3*
Distal Terminus	Blunt: 63% Pointed: 37% n=16	Blunt: 75% Pointed: 25% n=12	No p=.69, Fisher's Exact
Profile	Straight: 54% Curved: 17% Twisted: 29% n=24	Straight: 66% Curved: 23% Twisted: 11% n=35	No p=.23 G^2=2.940, df=2†
Unique Types of Retouch	Flat bifacial retouch	Normal retouch	Yes
Tool Types	UP tools dominate	UP tools dominate	No
Number of Changes/7 Steps			**1/7**
Total Measure of Difference in Tool Kit Morphology			**0.14**

*Two cells (25%) have an expected count less than 5. The minimum expected count is 4.34.

†Two cells (33%) have an expected count less than 5. The minimum expected count is 4.47.

Table 6.38 Measurement of the difference in flintknapping behaviors between East European assemblages

Comparison of of Assemblages Through Time	Measure of Difference in in Blank Production (Maximum=4, Minimum=0)	Measure of Difference in Tool Kit Morphology (Maximum=1, Minimum=0)
Molodova V-11 vs. Korolevo II-II	2.57	0.29
Korolevo II-II vs. Korolevo I-1a	1.00	0.14

II Complex II (2.57). The agreement between Korolevo II Complex II and Korolevo I layer 1a is much closer (1.00). Figure 6.17 indicates that the similarities between Molodova V level 11 and Korolevo II Complex II are found in the dorsal surface convexity domain, whereas almost all domains and steps show continuity between the two Korolevo assemblages.

Examining Figure 6.18, the Molodova V level 11 and the Korolevo II Complex II assemblages are more similar in tool kit morphologies than in blank production. In terms of social intimacy, enculturating environments, and independent innovation, this evidence supports a similar use of tools on the landscape by the makers of Molodova V level 11 and the makers of

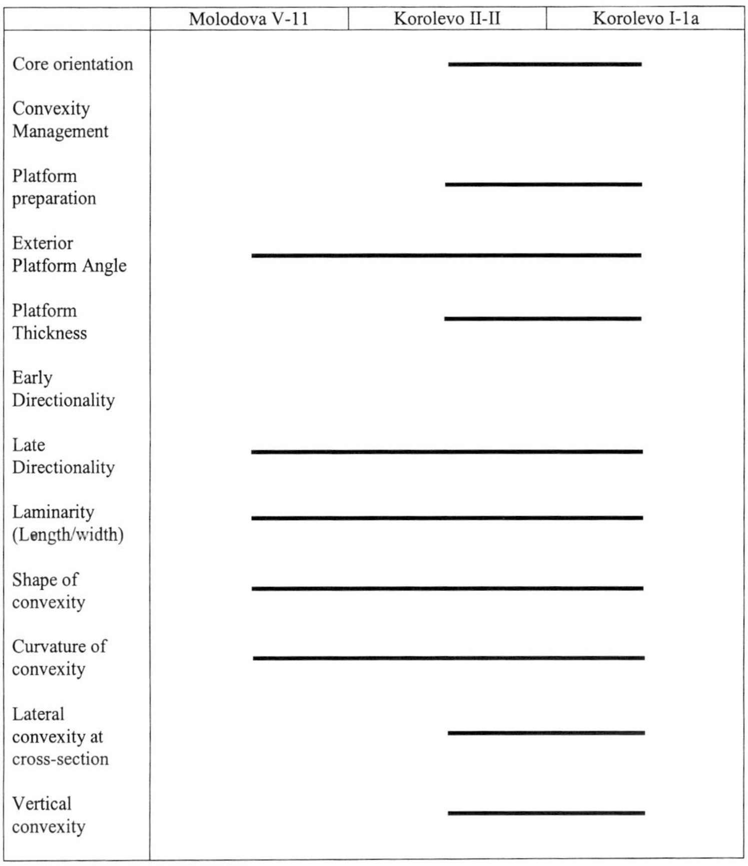

Figure 6.17 Variable by variable view of pair-wise comparisons within Eastern Europe between 60 and 30 kya. Bars connecting two assemblages represent the use of statistically similar attribute states between the two assemblages. Similarities are not necessarily implied between assemblages connected by a bar which are not immediately adjacent to each other.

Korolevo II Complex II, but without a commonality to the enculturating environment. If the individuals responsible for Korolevo II Complex II, the chronologically later assemblage, had any cultural transmission from the individuals (likely Neanderthals) responsible for Molodova V level 11 (requiring the assumption of continued occupation of the region), then the transmission process was stimulus diffusion or adaptive convergence. In contrast, Korolevo II Complex II and Korolevo I-1a share both enculturating environments and tool kit morphologies.

The number of sample assemblages from this region is the smallest in this study. While the representation of the industrial variability in flintknapping could be improved for this region, the problems of finding available collections excavated and curated with adequate care

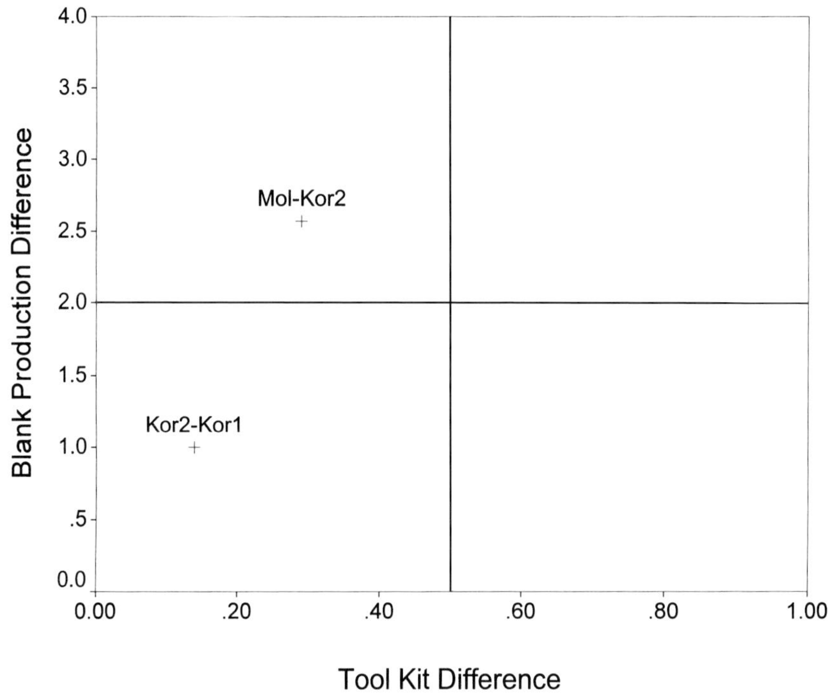

Figure 6.18 Blank Production vs. Tool Kit Differences for pair-wise comparisons for East European assemblages. Assemblage Key: Kor1: Korolevo I Layer 1a; Kor2: Korolevo II Complex II; Mol: Molodova V Layer 11.

remains a problem. Specifically, the problems with the stratigraphic association, and thus the chronological relationship, between the two Korolevo occupations is a case in point. Regardless, the sequence of change in Eastern Europe represented by these three assemblages is the most accurate given the availability of collections during the study seasons of 1996–1999. As pointed out in Meignen et al. (2004), future methodological excavations in this area will be fortuitous for the question of the Middle to Upper Paleolithic transition in this region, as will the full publication of important assemblages such as the lowest layer of Kulychivka (Savich 1975; Demidenko and Usik 1993b; Stepanchuk and Cohen 2000–2001l; Meignen et al. 2004) and Korolevo I layer 2b (Gladilin 1989a, 1989b; Gladilin and Demidenko 1989; 1990; Usik 1989; Demidenko and Usik 1993a).

7

Regional Patterns of Behavior: the Levant

The present chapter examines the details of the flintknapping behavior extant between 60,000 BP and 30,000 BP in the Levant, as exemplified by the archaeological sequence in Israel. Each sample assemblage is described in order of chronological appearance in the region. The technological analysis of the lithic data collected between 1996 and 1999 is presented for each assemblage before the pair-wise comparison of assemblages is conducted. The chapter concludes with a discussion of the significance of the intra-regional, Levantine sequence of change for the diffusion and independent innovation hypotheses.

7.1 Kebara Cave, Unit VI

7.1.1 Site Background

Kebara Cave is situated 65 m above sea level at the western edge of the Mt. Carmel, overlooking the Mediterranean coastal plain. First excavated by Turville-Petre (1932) and later by Stekelis (1951-1965), the cave deposits contain a sequence of cultural horizons stretching from the Middle Paleolithic to the Epipaleolithic. The collections of interest for the present study originate in the excavations of Bar-Yosef, Vandermeersch, Meignen, and Belfer-Cohen between 1982 and 1990 (Valladas et al. 1987; Bar-Yosef et al. 1992; Meignen and Bar-Yosef 1988, 1992; Bar-Yosef et al. 1996; Bar-Yosef and Meignen 2007).

Although the stratigraphic unit which directly underlies the Upper Paleolithic deposits at Kebara is Unit V, Unit VI was chosen to serve as the sample assemblage representative of the Late Middle Paleolithic in this region. This choice was influenced by the difficult stratigraphic boundary noted between the upper Unit IV, possessing Upper Paleolithic artifacts, and the Middle Paleolithic Unit V (Bar-Yosef et al. 1996:301). As technological analyses of debitage are very susceptible to bias introduced by mixed assemblages, Unit VI was considered a safer sample for the Late Levantine Mousterian. Unit V is also stratigraphically a continuation of Unit VI. Unit VI was dated by the AMS method to a minimum date of >48,000 BP (Gif-TAN-90029) and with TL to a mean of 48,300 ± 3,500 (Valladas et al. 1987). Interdisciplinary research at Kebara Cave has provided abundant data on Neanderthal lifeways, although usually focusing on the units earlier than VI: see Goldberg and Bar-Yosef (1998) for the study of site formation processes, Laville and Goldberg (1989) and Goldberg et al. (2007) for sedimentology and stratigraphy, Lev et al. (2005) and Albert et al. (2007) for botanical evidence in the form of carbonized seeds and vegetal materials, Schiegl et al. (1996), Meignen et al. (2007), and Sternberg and Lass (2007) for the study of the hearths, Speth and Tchernov (1998, 2002, 2007) for faunal analysis, Weiner et al. (1993) for bone preservation, Stutz and Lieberman (2007) on seasonality, and Weiner et al. (1995, 2007) for the mineralogy of the cave sediment.

7.1.2 Sampling

This stratigraphic unit in Kebara Cave is relatively rich in debitage and tools. The following meter squares were sufficient for the purposes of

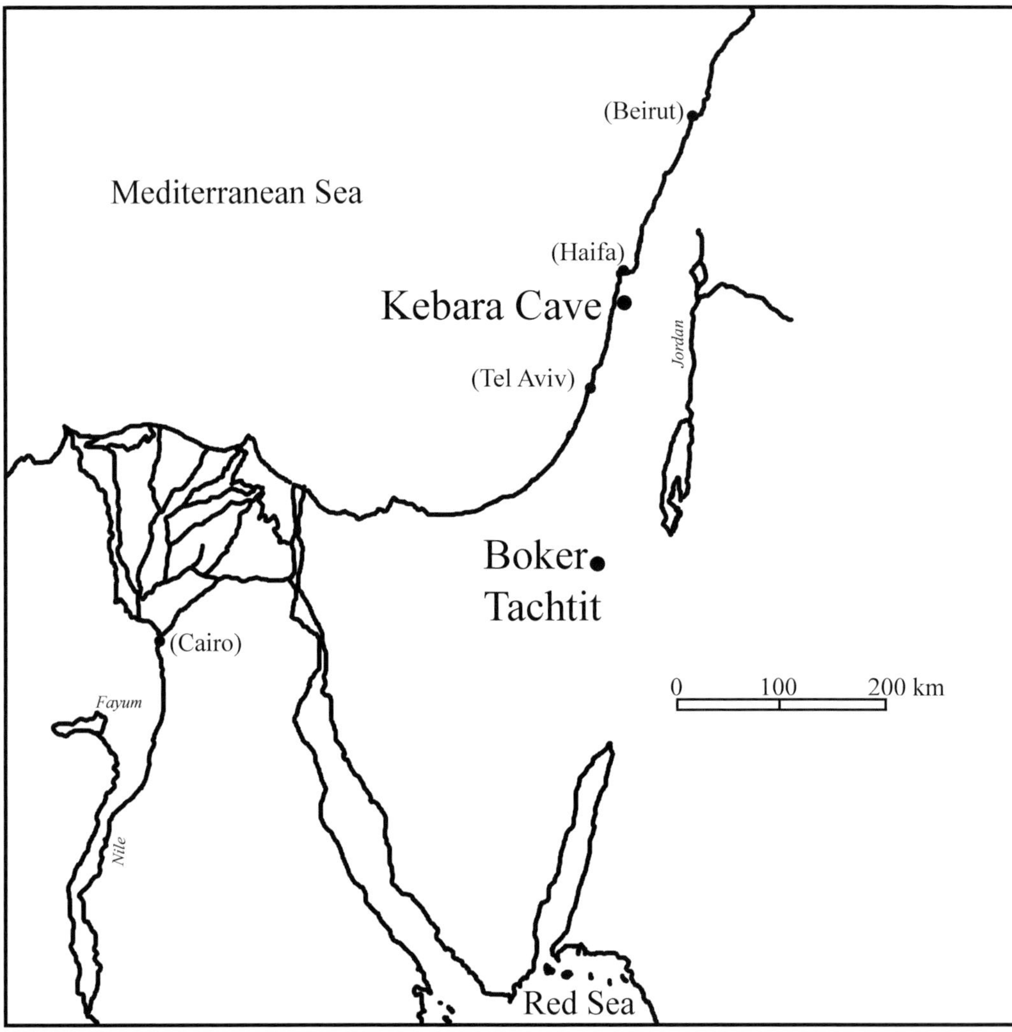

Figure 7.1 Map of Levantine sites used in the analysis.

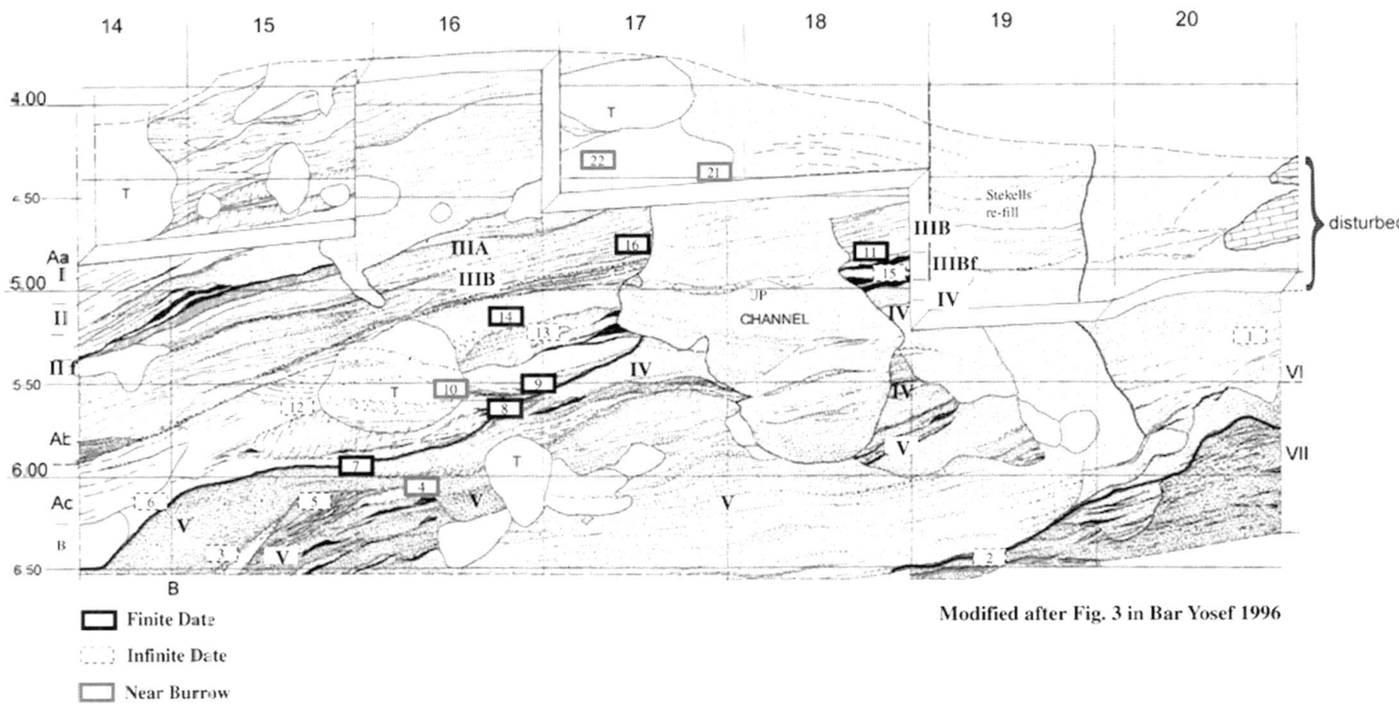

Figure 7.2 Kebara Cave stratigraphy of the south profile after the 1982-1990 excavations, showing Units I-VII. This figure, adapted and reprinted with permission from Rebollo et al. (2011:2427, fig. 2B) subdivides the radiocarbon dates from Bar-Yosef et al. (1996:301, fig. 3) into finite dates, infinite dates, dates near burrows. The profile equates to Square Q in the site grid. The Mousterian Units VI-XII can also be seen in Bar-Yosef et al. (1992: 504, fig. 5).

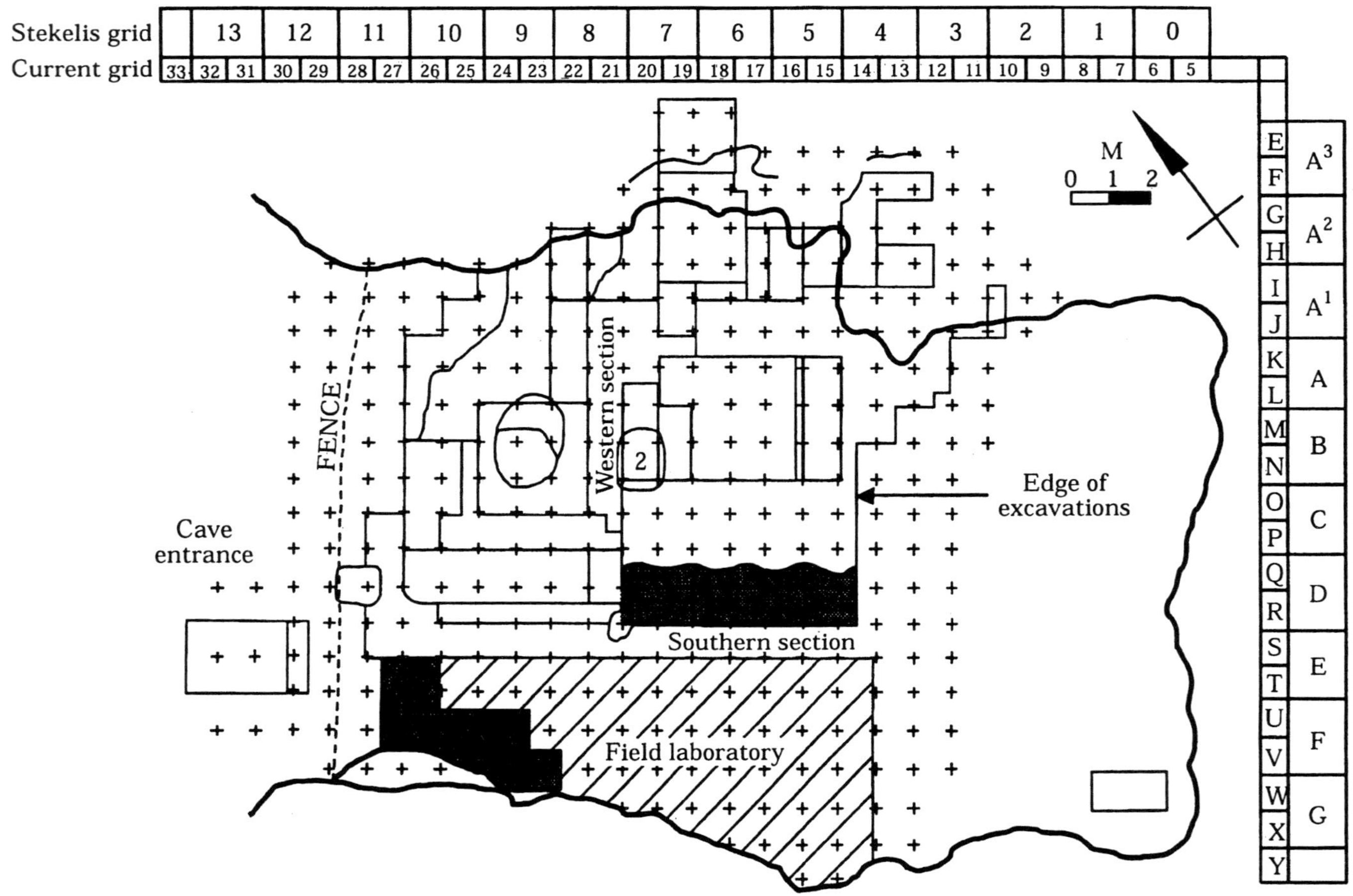

Figure 7.3 Kebara Cave site plan after the 1980s excavations, showing that excavations' 1x1m grid, Stekelis' 2x2m grid, and the locations of the southern and western profile sections from which the Upper Paleolithic and Middle Paleolithic strata, respectively, can be viewed (reprinted with permission from Bar-Yosef et al. 1996: 298, fig. 1).

the technological study: P20, L22, M21, N21, and Q20 (see Figure 7.3). The choice of these squares was influenced by their proximity to the western profile of the 1980s excavations and thus their clearer relation to the geological stratigraphy.

7.1.3 Blank Production

The reconstruction of the options chosen by the varied flintknappers whose cumulative efforts created the palimpsest of the Kebara Cave Unit VI assemblage proceeds according to the twelve knapping behaviors or analytical steps within the four domains of the flintknapping process.

Core Modification

The first domain, "core modification," concerns the orientation of the knapping surfaces of the cores. The first step in this domain, "core orientation," is characterized by the distinction between different methods of exploiting the volume of core forms, based upon an examination of extant core morphology (Appendix Figures A35–A40) and an evaluation of where platforms were located during most of core exploitation. Table 7.1 presents the cumulative, assemblage-wide tendency for bulbar negatives to appear on given core surfaces (Surface 1 and Surface 2) and within given core sectors (A, B, C, and D). In the case of Kebara Unit VI, it is clear that bulbar negatives are more frequent on the flatter, broad-faced surface, Surface 1, than on the convex Surface 2. This preferential exploitation of one surface by striking another surface which is consistently treated as a platform supports a Levallois designation for this assemblage, according to the criteria used by Boëda (1994). Most of the bulbar negatives are located in sector A of Surface 1. This data supports a unidirectional characterization of the core reduction (see below). Table 7.2, showing the number of scars within each core sector which originated from the direction of that sector, corroborates the view that sector A on Surface 1 is indeed the main debitage production sector, although Surface 2 sector A also shows significant working according to this measure. This suggests that sector A of Surface 2 likely served as a platform surface to the broad-face of Surface 1.

The second step in this domain, core convexity management, includes the knapping options which allow the knapper to continue reducing the core volume to exhaustion through specific modifications of the surface convexities relative to the platform(s). In the case of Kebara Unit VI, convexity management proceeded by débordant removals along the lateral edges of the cores, sectors B and D above. This is adequately evidenced in the core morphologies and the identification of these distinctive artifacts among the debitage.

Platform Maintenance

The next knapping domain, platform maintenance, includes the steps whereby the flintknapper modifies the platform surface and platform edge before each removal. These steps include platform treatment options, external platform angle, and platform thickness. Table 7.3 presents the frequencies of platform treatment types in this assemblage. Prepared platforms (dihedral, facetted, scarred, *chapeau de gendarme*) occurred to the same degree as unprepared platforms (plain, cortical, linear, punctiform), accounting for 50% of the debitage after indeterminates and removed pieces are deducted from the percentage. Table 7.4 presents the central tendency statistics for external platform angle and platform thickness, indicating a mean EPA of 87% and mean PT of 5 mm.

Direction of Core Exploitation

The next knapping domain, direction of core exploitation, concerns the variability resulting

Table 7.1 Kebara Cave Unit VI, percentage of intact cores (n=24) by number of bulbar negatives in each core sector (A, B, C, D) by surface (1, 2)

Number of Bulbar Negatives	Surf. 1 Sect. A	Surf. 1 Sect. B	Surf. 1 Sect. C	Surf. 1 Sect. D	Surf. 2 Sect. A	Surf. 2 Sect. B	Surf. 2 Sect. C	Surf. 2 Sect. D
0	8	71	63	79	71	88	79	88
1	63	29	21	4	25	13	13	13
2	29		17	0	4		8	
3				4				

Table 7.2 Kebara Cave Unit VI, core sector preparation: percent of intact cores (n=24) by number of scars in sector of origin (A, B, C, D) by surface (1, 2)

Number of Scars	Surf. 1 Sect. A	Surf. 1 Sect. B	Surf. 1 Sect. C	Surf. 1 Sect. D	Surf. 2 Sect. A	Surf. 2 Sect. B	Surf. 2 Sect. C	Surf. 2 Sect. D
0	4	58	42	71	13	50	46	54
1	38	29	33	8	29	29	25	21
2	38	8	25	13	33	17	8	17
3	17	4		8	17	4	21	8
4	4				8			

Table 7.3 Kebara Cave Unit VI, platform types for complete and proximal fragments of flakes and tools

Platform type	Frequency	Percent
Plain	194	32.2
Cortical	39	6.5
Dihedral	13	2.2
Facetted	94	15.6
Chapeau	32	5.3
Scarred	115	19.1
Punctiform	8	1.3
Linear	9	1.5
Removed	11	1.8
Indeterminate	88	14.6
Total	603	100.0

Table 7.4 Kebara Cave Unit VI, descriptive statistics for continuous platform variables of complete and proximal fragments of flakes and tools

	N	Minimum	Maximum	Mean	Std. Deviation
Exterior Platform Angle (degrees)	474	45.00	150.00	87.37	16.07
Platform Thickness (mm)	475	0.75	15.26	4.78	2.48

from different options of removing flakes from single to multiple platform cores, creating debitage with particular dorsal scar patterns. The dominant strategy used in the assemblage during early exploitation versus late exploitation was determined using the principles of dimensional change during core reduction (see Chapter 4). Table 7.5 presents the cross-tabulation of debitage and tools according to blank length in quartiles and dorsal scar pattern for all categories. By removing the debitage and tools with dorsal scar patterns noted as indeterminate or crested to clarify the pattern as well as blanks with more than 60% cortex or fewer than two dorsal scars, it is evident (Table 7.6 and Figure 7.4) that no clear tendency to shift direction of exploitation during blank production existed. The blanks with unidirectional scar patterns represent the dominant, if not always the majority, of the dorsal pattern types in all length quartiles. There is a relatively consistent decrease in bidirectional blanks and an increase in sub/centripetal blanks through reduction but, together with the oscillation in frequency of the unidirectional and crossed blanks, there is no strong evidence from which to argue a consistent change in reduction strategy through time. Given the dominance of the unidirectional pattern, this pattern is used as the characterization of both the early and late core exploitation.

Given the interest in unidirectional-convergent reduction in the Mousterian levels at Kebara Cave (Meignen and Bar-Yosef 1988b, 1991), it is worth noting that 17.5% of the Unit VI assemblage (as defined for Table 7.5) is composed of unidirectional-convergent blanks while straight unidirectional represents 27.1% of the assemblage; together these pieces account for the 44.6% value for the unidirectional pattern presented in Table 7.5. The frequency of unidirectional-convergent blanks does not consistently change between the length quartiles. Unidirectional-convergent as a dorsal scar pattern was recorded on all suitable artifacts on all assemblages during this research but this type is combined with straight unidirectional in the analyses presented here in order to maximize comparability through the minimization of low cell counts in cross-tabulations with blank length. Please refer to Tostevin (2000c) for data reporting unidirectional-convergent counts and frequencies for these other assemblages.

Dorsal Surface Convexity

The surface convexity domain includes the knapping steps which encapsulate the cumulative choice of *nervure-guide* or ridge pattern(s) of the core exterior for the production of blanks. These steps are quantified by three categorical attributes (lateral edge type, profile type, and cross-section type) and two continuous variables (laminarity or length/width ratio and use of convexity or width/thickness ratio). Table 7.7 presents the counts and frequencies of lateral edge types, profile types, and cross-section types. Table 7.8 provides the mean and standard deviation for all blank dimensions. Kebara Cave Unit VI produced relatively wide and thin flake blanks (length/width mean 1.78; width/thickness mean 5.18) with straight profiles and with diverse ridge patterns (varied lateral edge types).

7.1.4 Tool Kit Morphology

The first five characterizations of the tool kit morphology refer to the dorsal surface convexity variables which give form to the cutting edge to

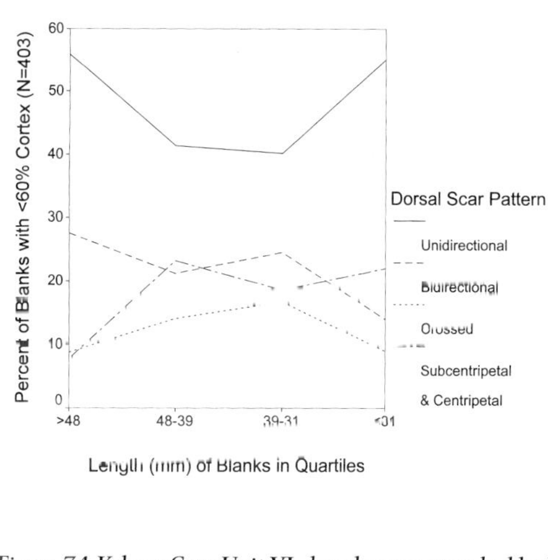

Figure 7.4 Kebara Cave Unit VI, dorsal scar pattern by blank length (mm) in quartiles cross-tabulation for complete flakes and tools (excluding pieces with a crested scar pattern, an indeterminate scar pattern, greater than 60% cortex, or fewer than three dorsal scars).

Table 7.5 Kebara Cave Unit VI, dorsal scar pattern by blank length (mm) in quartiles cross-tabulation for complete flakes and tools (excluding flakes with greater than 60% cortex or fewer than 3 dorsal scars)

Dorsal Scar Pattern	>48	48-39	39-31	<31	Total
Unidirectional					
Count	57	41	41	55	194
% within Quartile	51.8%	38.7%	36.0%	52.4%	44.6%
Bidirectional					
Count	28	21	25	14	88
% within Quartile	25.5%	19.8%	21.9%	13.3%	20.2%
Crossed					
Count	9	14	17	9	49
% within Quartile	8.2%	13.2%	14.9%	8.6%	11.3%
Subcentripetal & Centripetal					
Count	8	23	19	22	72
% within Quartile	7.3%	21.7%	16.7%	21.0%	16.6%
Crested					
Count	5	3	4	2	14
% within Quartile	4.5%	2.8%	3.5%	1.9%	3.2%
Indeterminate					
Count	3	4	8	3	18
% within Quartile	2.7%	3.8%	7.0%	2.9%	4.1%
Total					
Count	110	106	114	105	435
% within Cortex	100.0%	100.0%	100.0%	100.0%	100.0%

Table 7.6 Kebara Cave Unit VI, dorsal scar pattern by blank length (mm) in quartiles cross-tabulation for complete flakes and tools (excluding pieces with a crested scar pattern, an indeterminate scar pattern, greater than 60% cortex, or fewer than 3 dorsal scars)

Dorsal Scar Pattern	>48	48-39	39-31	<31	Total
Unidirectional					
Count	57	41	41	55	194
% within Quartile	55.9%	41.4%	40.2%	55.0%	48.1%
Bidirectional					
Count	28	21	25	14	88
% within Quartile	27.5%	21.2%	24.5%	14.0%	21.8%
Crossed					
Count	9	14	17	9	49
% within Quartile	8.8%	14.1%	16.7%	9.0%	12.2%
Subcentripetal & Centripetal					
Count	8	23	19	22	72
% within Quartile	7.8%	23.2%	18.6%	22.0%	17.9%
Total					
Count	102	99	102	100	403
% within Cortex	100.0%	100.0%	100.0%	100.0%	100.0%

Table 7.7 Kebara Cave Unit VI, frequencies of lateral edge, profile, and cross-section types for complete flakes and tools

Type	Frequency	Percent
Lateral edge		
Parallel	217	36.0
Convergent	170	28.2
Expanding	133	22.1
Ovoid	64	10.6
Indeterminate	19	3.2
Total	603	100.0
Profile		
Straight	407	67.5
Curved or Concord	105	17.4
Twisted	85	14.1
Indeterminate	6	1.0
Total	603	100.0
Cross-section		
Triangular	267	44.3
Trapezoidal	208	34.5
Other	99	16.4
Indeterminate	29	4.8
Total	603	100.0

Table 7.8 Kebara Cave Unit VI, descriptive statistics for blank dimensions for complete flakes and tools

Blank Dimension	N	Range	Mean	Standard Deviation
Length	603	91.95	40.40	13.86
Width	603	57.90	24.83	8.26
Thickness	603	17.71	5.72	2.87
Length/Width Ratio	603	8.27	1.78	0.82
Width/Thickness Ratio	603	21.24	5.18	2.62

Table 7.9 Kebara Cave Unit VI, descriptive statistics for dimensions of complete tools

Blank Dimension	N	Range	Mean	Standard Deviation
Length/Width Ratio	103	2.91	1.84	0.56
Width/Thickness Ratio	103	13.29	5.48	2.21

the retouched blanks in the assemblage, as well as the diagnostic flakes without retouch traditionally categorized as tools, i.e., the typological Levallois pieces frequently found in this assemblage. Possibly due to the inclusion of the typological unretouched Levallois points in this assemblage, the tool kit possesses a higher frequency of convergent lateral edges than the debitage as a whole, 41.9% compared to 29.1% when the indeterminates are removed (Table 7.10). The tools are also less twisted in profile than the debitage, 5.9% compared to 14.2% when indeterminates are removed. Otherwise, the tool kit morphology is similar to that of the debitage.

The last two characterizations of the tool kit morphology refer to two variables describing the application of retouch to blanks: first, the application of distinctive and/or diagnostic forms of retouch and, second, the dominance or equality of Middle Paleolithic (lateral edge) versus Upper

Table 7.10 Kebara Cave Unit VI, frequencies of lateral edge, distal terminus, and profile types for complete tools

Type	Frequency	Percent
Lateral edge		
Parallel	37	37.0
Convergent	39	37.9
Expanding	8	7.8
Ovoid	9	8.7
Indeterminate	10	9.7
Total	103	100.0
Distal Terminus		
Blunt	33	33.0
Pointed	44	42.7
Hinged	7	6.8
Overshot	1	1.0
Broken	5	4.9
Indeterminate	13	12.6
Total	103	100.0
Profile		
Straight	73	70.9
Curved or Concord	23	22.3
Twisted	6	5.8
Indeterminate	1	1.0
Total	103	100.0

Paleolithic (distal and lateral edges) retouched tool types. In this case, Kebara Unit VI does not exhibit any unique types of retouch and is dominated by Middle Paleolithic tool types (Meignen and Bar-Yosef 1988a, 1991).

Intensity of Tool and Core Reduction

Given Kebara Cave's location in the raw material rich Mt. Carmel area of the Levantine coastal plain, it is not surprising that retouch was applied to blanks in Unit VI less frequently than other assemblages, as the tool to flake ratio falls below the mean of all assemblages (Table 7.11, Figure 7.5). While there is more retouch than seen in sites serving as raw material workshops, it is still less frequent than seen in Vedrovice V and much less than in Kůlna Layer 7a. A low frequency of retouch is also a distinguishing character of the Levalloiso-Mousterian. From the ratio of non-cortical to cortical blanks falling below the mean of all assemblages (Figure 7.6),

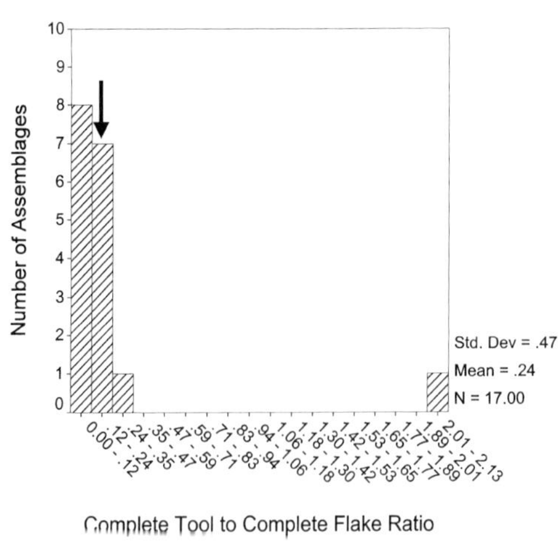

Figure 7.5 Kebara Cave Unit VI, tool/flake ratio value relative to a histogram of all assemblages.

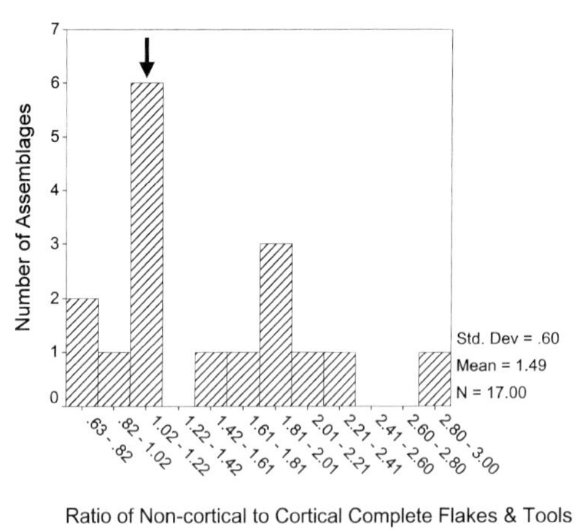

Figure 7.6 Kebara Cave Unit VI, non-cortical/cortical flake ratio value relative to a histogram of all assemblages.

Table 7.11 Kebara Cave Unit VI, intensity of core and tool reduction

Variables	Numerator	Denominator	Result
Tools/Flake Ratio	Tools: 103	Flakes: 495	0.21
Non-cortical Debitage/ Cortical Debitage Ratio	Non-cortical Debitage: 324	Cortical Debitage: 274	1.18
Henry's Dimensional Analysis	NA	NA	Core facets overlap with the low end and middle of the blank distribution: moderate core exhaustion

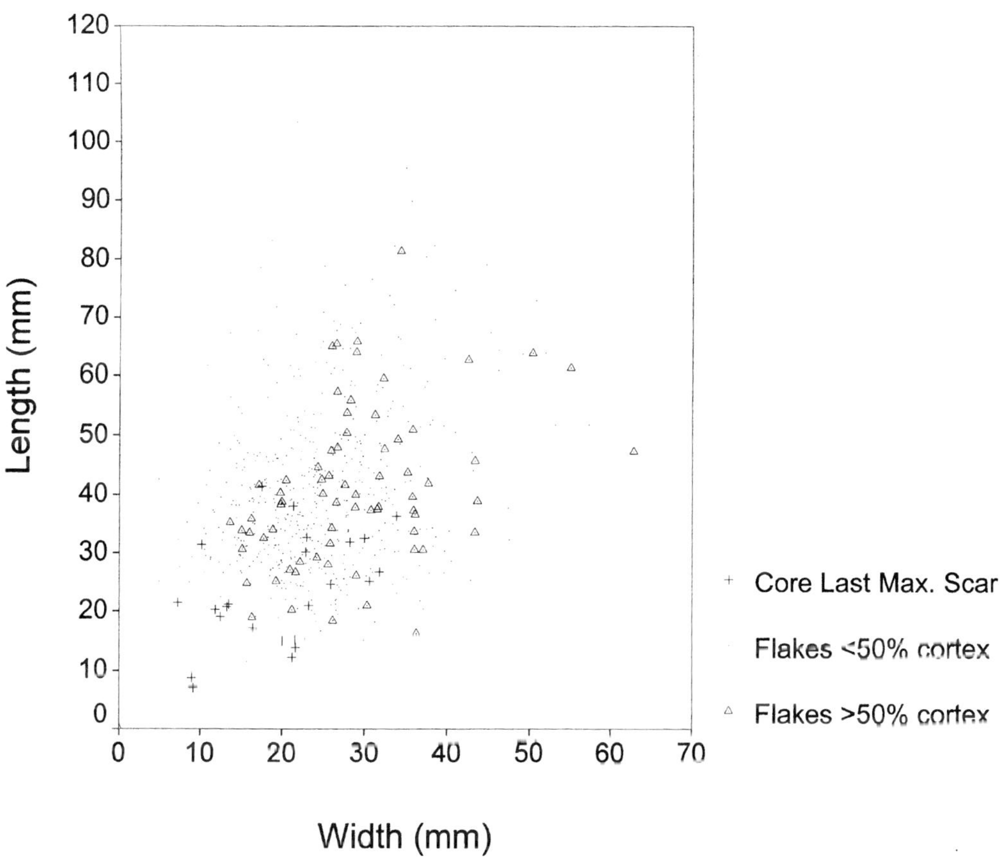

Figure 7.7 Kebara Cave Unit VI according to Henry's core reduction intensity measure: length and width of last maximum core facets, highly cortical blanks, and blanks with little cortex.

it is clear that nodules were worked from a fairly cortical stage on-site. As seen in Figure 7.7, half of the last maximum core facets (13 out of 24) fall at or below the 2.0 cm size cut off for the debitage analysis. With the remaining core facets overlapping only the middle of the length and width distribution of the debitage, this indicates a fair degree of core exhaustion for this assemblage. Taken together, these data suggest that the assemblage captures the beginning but perhaps not the earliest portion of the core reduction sequence as well as the very late behaviors enacted during late core exhaustion. The assemblage may capture more of the early reduction sequence than indicated by the cortical signature, since the relatively high degree of core reduction could elevate the non-cortical blank frequencies relative to less reduced assemblages which captured the same portion of the early sequence. The assemblage is thus highly comparable with other assemblages for both continuous and discontinuous variables.

7.2 Boker Tachtit, Level 1

7.2.1 Site Background

Boker Tachtit is an open-air site in the Central Negev Desert of Israel. Excavated during the Southern Methodist University survey of the Avdat/Aqev area in the late 1970s (Marks 1983; Marks and Volkman 1983), the deposits produced four superimposed cultural horizons, three of which were extremely rich in artifacts. Although no bone was obtained from the site, extensive refitting was conducted on the material, allowing the vast majority of the lithic core reduction sequences to be defined with great accuracy (Volkman 1983, 1989). Long argued to demonstrate an *in situ* development of Upper Paleolithic blade technology from a Middle Paleolithic technology (Marks 1990), the absence of a Middle Paleolithic retouched tool kit (Marks and Kaufman 1983:79; Marks 1992:235) has led to suggestions that the basal level at Boker Tachtit represents a fully Upper Paleolithic occupation (Bar-Yosef 1994; Bar-Yosef et al. 1996). This earliest occupation, Boker Tacthtit level 1, produced several radiocarbon dates, of which 46,930 ± 2,420 (SMU-259) has the smallest standard deviation.

7.2.2 Sampling

The extensive refitting studies conducted by P. Volkman (1983, 1989) allow the reconstruction of almost the whole core reduction sequence for this assemblage. The typological and spatial analysis, as well as technological variables, of the assemblage have also been published in great detail (Marks and Kaufman 1983; Hietala 1983a, 1983b, 1983c). It is fortunate that these studies are published and available since the majority of the artifacts themselves are currently impossible to study due to the state of curation of the debitage. In addition to the fact that an estimated 75% of all debitage has been refitted into glued nodules (Volkman 1989:186–187), the method of storing artifacts in the collection makes it extremely difficult to procure an extensive sample of the remaining unglued artifacts. Within each level, the spatial context of the artifacts is only preserved in the artifacts' storage by square proveniences in the original excavation containers, in this case, flimsy plastic disposable sandwich bags which are now so degraded as to disintegrate upon being touched. With numerous square provenience bags already burst in the museum drawers, the effort to remove bags of artifacts by lifting them from the drawers results in the mixing of artifacts from multiple proveniences. A systematic research project is required to re-curate the whole assemblage, including the refitted nodules which are now

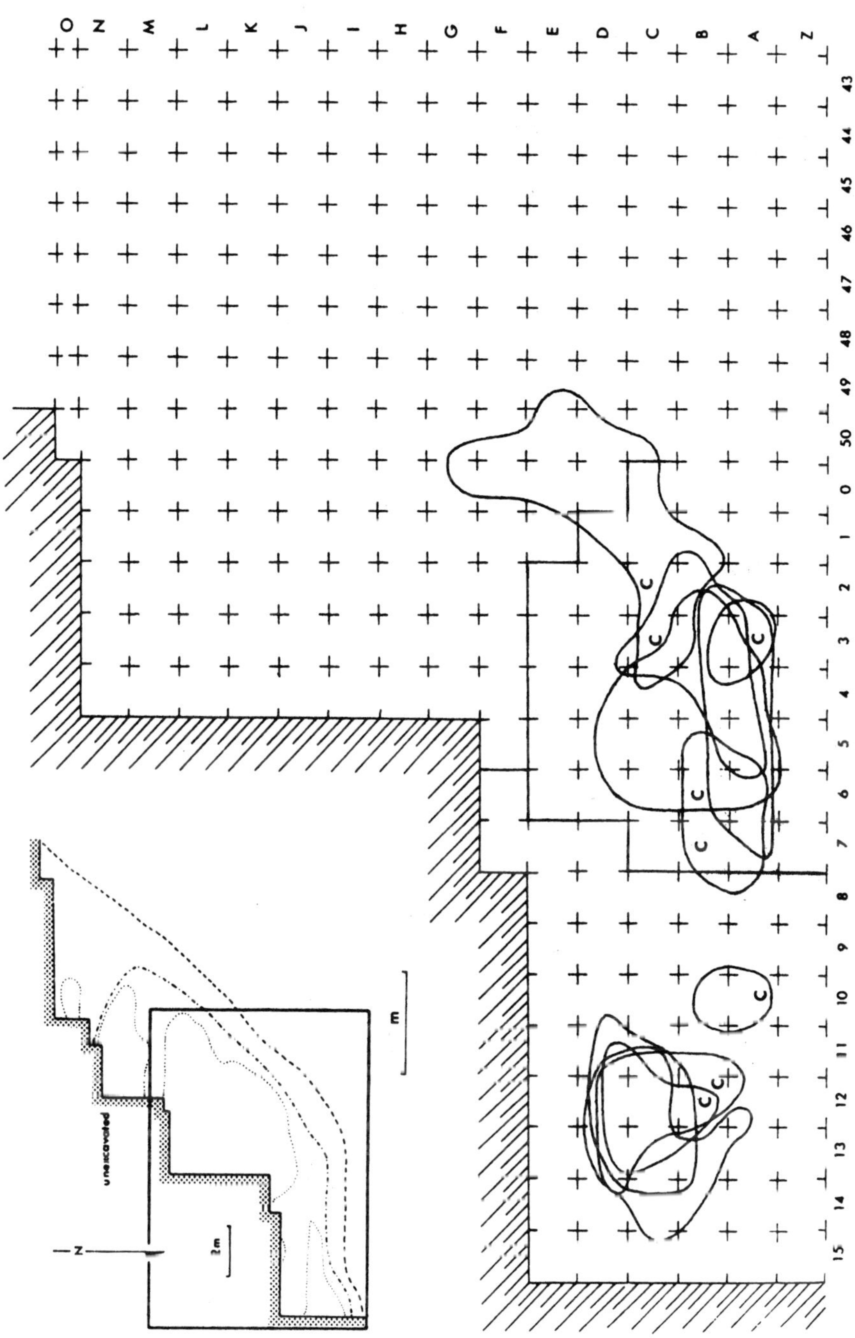

Figure 7.8 Site plan and distribution of core reconstructions for Boker Tachtit Level 1 (from Hietala 1983b: 222, fig. 8-2b, used by permission).

beginning to fall apart as the original glue disintegrates. My request of the Israeli Antiquity Authority for permission to disarticulate a sample of glued, refitted cores for temporary study, with subsequent regluing of the pieces, was denied, upon the reasonable argument that a full scale project to re-curate and digitally record the whole assemblage is warranted. I agree that the incredible resource represented by both the refitted and non-refitted portions of the assemblage is worth a year's curatorial effort in an of itself. Given the scale of the task established at the beginning of the present research, however, correcting the curation of this one collection was beyond the scope of the current research. In the meantime, worthy studies of the extant refits, such as Škrdla (2003a), are currently still possible as they rely on an examination of the morphologies of the refitted nodules, not on the individual attributes of the artifacts within the reconstructions.

The above discussion of the curational problems of the assemblage is not designed to assign blame to either the excavator or the curators for the state of the collection but to point out the context of the assemblage for the attribute analysis presented here. These issues shaped the context of the sampling strategy used to corroborate Volkman's refit data. While the published analyses of the refits serve as the substantive support for the following discussion of the Boker Tachtit level 1 blank production and tool kit morphology, a sample of over 100 complete unretouched artifacts from squares A1, A2, A2/3, A3, A5, and A6 were studied with the attribute analysis advocated by the current research in order to produce data which is comparable with the rest of the studied assemblages. These particular squares were chosen to sample the same or adjacent squares to the locations of major core reconstructions (Figure 7.8).

7.2.3 Blank Production (A Reply to Marks 2003 and Belfer-Cohen and Goring-Morris 2003)

Core Modification

Core orientation in Boker Tachtit Level 1 entails a longitudinal blade core form, as evidenced in the identification of the narrow and long core face as the primary debitage surface among the refitted cores. Modification of the cortical nodules began with a series of blade removals down one side of the core, perpendicular and originating from a natural ridge running the length of the core. This natural ridge, once further isolated by cresting and decortifying removals on either side, formed the basis for the first lame à crête removal, which itself formed the ridges for the production of blade blanks (Volkman 1989:196–202). While there are variations on the subsequent set up of the ridge and opposed platforms, the refits provide a clear indication of the preferential use of one narrow and longitudinally oriented core face.

Core convexity management was achieved through the use of frontal crests as noted above, as well as cresting removals to establish the opposed platforms at either end of the nodules (see Figure 7.9), as Volkman notes (1989:207). Throughout subsequent reduction of the cores, the convexity of the dorsal surface was shaped and isolated toward the center of the surface in order to create ridge patterns conducive to the removal of pointed products. This shaping of the dorsal surface was achieved by means of débordant blade removals from the two opposed platforms (Volkman 1989:195–195), as is clearly seen in the extant morphology of particular point cores such as illustrated by Marks and Kaufman (1983:73, fig. 5-2e).

Platform Maintenance

Platform maintenance includes a slight numerical superiority of unprepared platforms (com-

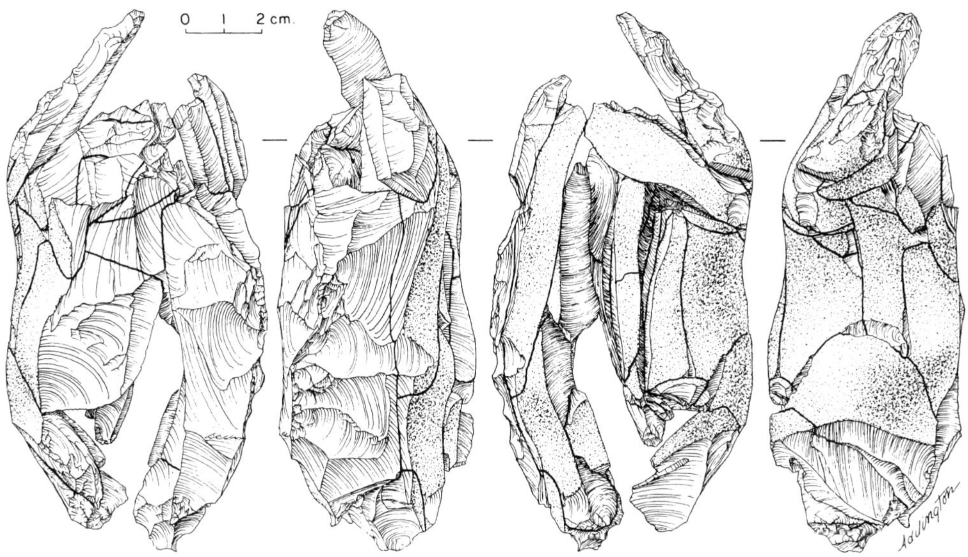

Figure 7.9 Boker Tachtit Level 1. Core reconstruction for Core U, an opposed platform Levallois point core (Volkman 1983: 143, fig. 6-5, used by permission).

bining plain, cortical, punctiform, and linear platforms values in Table 7.12 gives 47.7%), fewer prepared platforms (combining dihedral, facetted, and scarred gives 29.7%), and a fair number of indeterminate platforms (crushed and unknown totaling 21.8%). When removing the indeterminates from the calculation, these values become 60.8% for unprepared and 39.2% for prepared platforms. These latter frequencies (identical to those in Tostevin 2003a:62, tab. 6.2), based on a relatively small sample (n=79), are remarkably similar to the frequencies published by Marks and Kaufman (1983:74, tab. 5-3) on the whole assemblage (see Table 7.13). As Marks and Kaufman cross-tabulate platform types by blank type (blade, flake, primary element, and core trimming element), one must recombine all counts in each blank type into the whole assemblage and then combine the respective prepared vs. unprepared types into the same categories to derive comparable units to those used here. When this is done (see Table 7.13), the present study's data is almost identical to Marks and Kaufman's data. Given the difference in sample size, this is a robust indication of the effectiveness of sampling for technological variables, even with the contingencies of an assemblage with a large portion of the blanks refit into unanalyzable nodules. Given these positive results, it is extremely unfortunate that Belfer-Cohen and Goring-Morris chose to target these frequencies as an example of inter-analyst bias in technological studies (2003:275-276):

> It is quite clear that the different backgrounds and perceptions of the various researchers will influence each and every

Table 7.12 Boker Tachtit Level 1, platform types for complete and proximal fragments of flakes and tools

Platform type	Frequency	Percent
Plain	34	33.7
Cortical	5	5.0
Dihedral	1	1.0
Facetted	10	9.9
Scarred	20	19.8
Punctiform	4	4.0
Linear	5	5.0
Indeterminate	22	21.8
Total	101	100.0

Table 7.13 Boker Tachtit Level 1, platform preparation compared between the present study and Marks and Kaufman 1983, table 5-3

Analytical Platform Type	Frequency calculated from Marks & Kaufman 1983, table 5-3. N=769	Frequency calculated from Table 7.12 N=79
Unprepared	59.8	60.8
Prepared	40.2	39.2
Total	100.0	100.0

Table 7.14 Boker Tachtit Level 1, descriptive statistics for continuous platform variables of complete and proximal fragments of flakes and tools

	N	Minimum	Maximum	Mean	Std. Deviation
Exterior Platform Angle (degrees)	63	58.00	126.00	88.40	13.63
Platform Thickness (mm)	65	1.12	12.65	4.04	2.16

aspect of technological studies. This can even be observed at the most basic level of primary definitions.... When comparing the data presented by Tostevin (herein, Table 6.2) with the original publication of the same data by Marks and Kaufman (1983:74, table 5.3), a major difference in the degree of faceting at Boker Tachtit is observed.

While their goal of identifying methodological inconsistencies between researchers is admirable, accuracy from Belfer-Cohen and Goring-Morris would be equally admirable but is wanting in the present case. First, the values presented here (Table 7.12) and in Tostevin (2003:tab. 6.2) do not use the same "data" as Marks and Kaufman but different samples of artifacts and so might be expected to produce different frequencies. Yet despite these sample size differences, the frequency results are almost identical and yet Belfer-Cohen and Goring-Morris still fail to recognize this. Thus, despite erroneous criticisms to the contrary, this data supports the robust analytical strength of sampling for technological variables as well as the actual miniscule role of inter-analyst bias in the reporting of a variable such as platform preparation.

Inter-analyst bias is a real concern, however, for many other variables, and continuous platform measurements are no exception. For this reason, exterior platform angle and platform thickness are reported not from the literature but from the analyses conducted by Tostevin in order to maintain comparability across regional contexts. Beyond this general reason, however, many

Table 7.15 Boker Tachtit Level 1, dorsal scar pattern by blank length (mm) in quartiles cross-tabulation for complete flakes and tools (excluding flakes with greater than 60% cortex or fewer than three dorsal scars)

Dorsal Scar Pattern	>43	43-33	33-28	<28	Total
Unidirectional					
Count	2	4	6	10	22
% within Quartile	9.5%	16.7%	27.3%	50.0%	25.3%
Bidirectional					
Count	11	6	2	1	20
% within Quartile	52.4%	25.0%	9.1%	5.0%	23.0%
Crossed					
Count	3	6	4	6	19
% within Quartile	14.3%	25.0%	18.2%	30.0%	21.8%
Subcentripetal & Centripetal					
Count	2	3	7	1	13
% within Quartile	9.5%	12.5%	31.8%	5.0%	14.9%
Crested					
Count	3	4	2	1	10
% within Quartile	14.3%	16.7%	9.1%	5.0%	11.5%
Indeterminate					
Count		1	1	1	3
% within Quartile		4.2%	4.5%	5.0%	3.4%
Total					
Count	21	24	22	20	87
% within Cortex	100.0%	100.0%	100.0%	100.0%	100.0%

Table 7.16 Boker Tachtit Level 1, dorsal scar pattern by blank length (mm) in quartiles cross-tabulation for complete flakes and tools (excluding pieces with a crested scar pattern, an indeterminate scar pattern, greater than 60% cortex, or fewer than three dorsal scars)

Dorsal Scar Pattern	>43	43-33	33-28	<28	Total
Unidirectional					
Count	2	4	6	10	22
% within Quartile	11.1%	21.1%	31.6%	55.6%	29.7%
Bidirectional					
Count	11	6	2	1	20
% within Quartile	61.1%	31.6%	10.5%	5.6%	27.0%
Crossed					
Count	3	6	4	6	19
% within Quartile	16.7%	31.6%	21.1%	33.3%	25.7%
Subcentripetal & Centripetal					
Count	2	3	7	1	13
% within Quartile	11.1%	15.8%	36.8%	5.6%	17.6%
Total					
Count	18	19	19	18	74
% within Cortex	100.0%	100.0%	100.0%	100.0%	100.0%

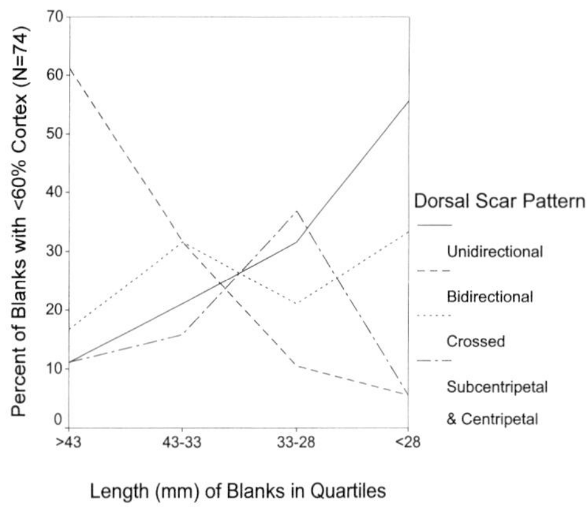

Figure 7.10 Boker Tachtit Level 1, dorsal scar pattern by blank length (mm) in quartiles cross-tabulation for complete flakes and tools (excluding pieces with a crested scar pattern, an indeterminate scar pattern, greater than 60% cortex, or fewer than three dorsal scars).

of the measurements, particularly exterior platform angle, are absent from published literature on debitage studies, since they were not recognized as analytically informative until relatively recently (Dibble and Pelcin 1995; Dibble 1997; but see Dibble and Whittaker 1981 for an earlier discussion). In general, Marks and Kaufman's data presentation is extremely thorough and comparable with many if not most analytical tasks one would want to conduct using literature data. Hietala (1983c) even reports central tendencies for "platform height" in Appendix B but the technique for this measure is not identical to platform thickness as used by the present study, and so the sample acquired by Tostevin is used instead. For this sample from Boker Tachtit Level 1, measurements of external platform angles and platform thickness (Table 7.14) indicate tendencies around 88 degrees and 4 mm respectively.

Direction of Core Exploitation

The analysis of the direction of core exploitation is presented in Tables 7.15–7.16. Figure 7.10 demonstrates that the dominant direction within the longest length quartile, a proxy for early core exploitation, is bidirectional while a unidirectional scar pattern dominates the shortest length quartile, the proxy for the end of core exploitation. The analysis of this domain has received the most criticism to date and so a further elucidation of the data is warranted.

In a summary article for Goring-Morris and Belfer-Cohen's edited monograph of papers on the Upper Paleolithic of the Levant, Marks (2003:264) comments on an earlier version (Tostevin 2003a) of the above reconstruction:

> In the case of Boker Tachtit I (Tostevin herein, Table 6.2), the direction of Blank Removal is described as 'Bidirectional changing to Unidirectional,' while the Direction of Cortex Removal is called Unidirectional. Boker Tachtit Level I, with its extensive core reconstructions (Volkman 1983), is one of those assemblages where it should be impossible to get it wrong. Yet, Tostevin managed to do so. As Tables 6-1 and 6-2 in Volkman (1983:133, 136) clearly show, the discarded cores were almost wholly bidirectional, and after reconstruction there was not a single core that could be classified as unidirectional. There was a single unidirectional reduction, missing the core (Volkman 1983:134, fig. 6-2), but there was no indication that this core, with its crested preparation, was ever bidirectional. The direction of cortex removal is explicitly described by Volkman ((Volkman 1983:137–138) and it is multi-directional, not unidirectional, as claimed by Tostevin. I know Boker Tachtit. How much faith should I have in those characterizations of assemblages I do not know?

Leaving aside the characteristic hyperbole of a Marks criticism (which I am told is a mandatory rite for Levantine prehistorians of the second half of the twentieth century), there are several issues in his comment which require a direct response. First, as previously argued in Chapter 4, I acknowledge that the use of percent cortex on the dorsal surface of debitage is an unreliable proxy for identifying directionalities of decortification relative to the core's ultimate orientation, although it is certainly a credible proxy for the general position of flakes in early versus late core reduction (Dibble 1995a; Baumler1988). The problem rests with the frequent tendency to rotate the core to exploit multiple platforms in decortification, as Volkman notes for Boker Tachtit Level 1 (1989:196-202), so that the dorsal scar patterns of the decortification flakes do not themselves indicate the orientation or rotation of the core itself. The most numerous dorsal scar pattern on these flakes, however, is indeed unidirectional. When examining Volkman's generalized drawings of the two options for performing cores in Level 1 (1983:140-141, fig. 6-3; 1989:198-199, fig. 6-1), few if any of the removals overlap with scars from other directions, leaving unidirectional scars indicative of the cortical debitage, if not of the core rotations during decortification. Thus, there is no inherent error with my basic observations of the cortical artifacts, as fully reported in Tostevin (2000c:255-258) and summarized in (2003a:tab. 6.2). If it were not for the fact that most assemblages lack sufficient cortical pieces to prevent low cell counts in cross-tabulations with dorsal scar pattern types, this form of analysis would still be a justifiable method for characterizing early *debitage from an assemblage, although not core rotations.*

As the characterization of the directionality of debitage during early core reduction using corticality has been replaced in the present study by the identification of the most numerous dorsal scar pattern among the blanks in the longest length quartile, we can move on to Marks' criticism of this aspect of the analysis. From the fact that Marks stresses that almost all cores in Level 1 must be typed "bidirectional," both during refitted reduction and at discard, it is clear that he did not read my article carefully. The analytical description of the direction of blank removal was based on "Correlations of blank length with dorsal scar patterns on debitage and tools" (2003a:tab. 6-1), not on the extant cores or even the core reconstructions. It is surprising to read Marks touting the relevance of discarded cores for, while some argue that the form of discarded cores does inform on the reduction strategy used to produce the debitage (Boëda 1994; Sarel and Ronen 2003; Sarel 2004), Marks has strongly argued the opposite, even in the site report for Boker Tachtit: "...for these assemblages, at least, the final core morphology often gives very little indication of the reduction strategy which produced it, or the nature of the blanks removed during reduction" (Marks and Kaufman 1983:70). The bidirectionality of the discarded cores, however, is not at issue; I thoroughly agree with his characterization of them as bidirectional cores. It is the *debitage* which is not wholly bidirectional, to the extent that the scar pattern in the longer vs. shorter blanks is suggestive of a specific form of bidirectional reduction, one in which an alternating bidirectional strategy is replaced by a sequential bidirectional strategy. Compared to an "alternating bidirectional" scheme in which each removal is followed by a rotation to the opposed platform (producing almost exclusively blanks with a bidirectional scar pattern), a "sequential bidirectional" scheme only rotates to the opposed platform after a series of removals from one platform (producing first bidirectional blanks and

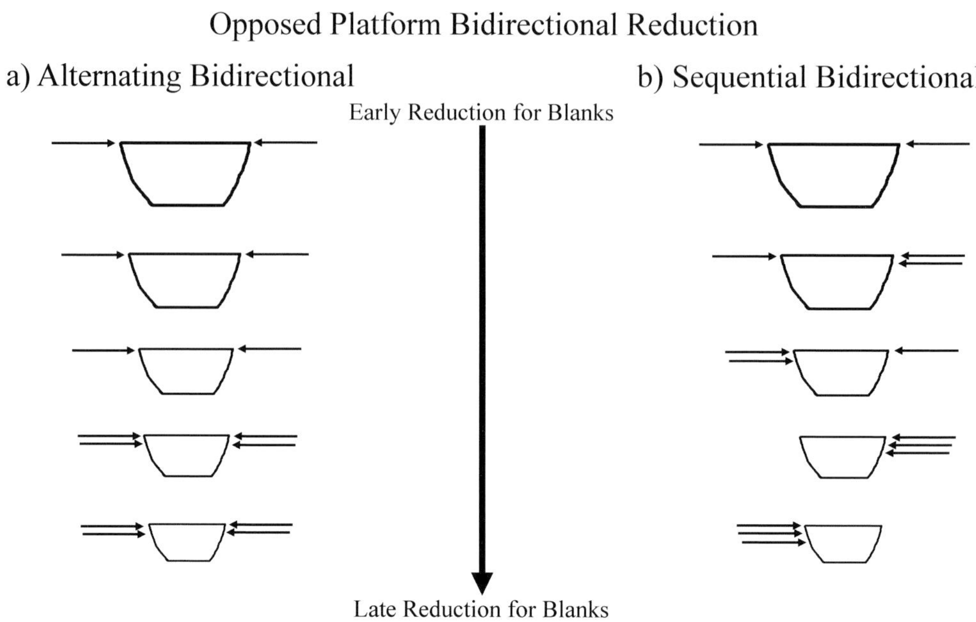

Figure 7.11 Two methods of "bidirectional" reduction: a) alternating bidirectional; b) sequential bidirectional.

subsequently more and more unidirectional blanks as the dorsal surface is cleaned from one platform). Volkman himself recognized how sequential removals can change the scar pattern of the blanks in the Boker Tachtit Level 1 assemblage. In the reconstruction of Core H, within reduction stage I, the first two blades removed after the *lame à crête* show a crossed scar pattern.

> The third blade, as well as the following five blades of the Core H sequence, do not catch the flakes scars produced from the *lame à crête* formation. Instead, their dorsal scar patterns are determined from the *lame à crête* scar and the first two blade removals. Since they were all struck from the same core platform, the dorsal scar patterns of the five are unidirectional along their length axes (Volkman 1989:215).

It was not felt necessary to detail the distinction between alternating bidirectional and sequential bidirectional core reduction in Boker Tachtit Level 1 in Tostevin (2003a), a summary article whose brevity was mandated by the editors of the conference monograph, since the reader was referred to the raw data and the full argumentation for its interpretation in Tostevin (2000c). More importantly, it was felt unnecessary at the time since Marks himself had originally noted this pattern in the site report for Level 1: "Given the dominant core reduction strategy, it is only natural that a high percentage of the blades should exhibit dorsal scars from bidirectional flaking. It is somewhat surprising, therefore, that such a high percentage should show simple unidirectional scars (Table 5-7). This suggests that rigidly alternating flaking of the opposed platforms was not always carried out. In fact, the last two blades, by definition, must have come from the same platform in sequence. The core reconstructions amply confirm the significant number of blades struck in series even in the Levallois point strategy (see Chapter 6)" (Marks and Kaufman 1983:75). The blank length in quartiles cross-tabulated with dorsal scar pattern detailed here thus corroborates Marks' original observations but goes further to point out when in the core reduction sequence this tendency to strike a series of blades from one platform occurs, i.e. late in core reduction. Figure 7.11b illustrates this pattern. Remembering Marks's acidic critique, "I know Boker Tachtit. How much faith should I have in those characterizations of assemblages I do not know?" (2003:264), perhaps one should ask if Marks really remembers Boker Tachtit?

In typing the direction of early core exploitation as bidirectional and the direction of late core exploitation as unidirectional based on Figure 7.10, I am thus using a finer-grained characterization for the assemblage than just bidirectional. If Marks wishes to maintain a more generalized characterization of the assemblage as bidirectional, that is his prerogative and may be appropriate for the scale of resolution of his use of the example of the core reduction sequence of Boker Tachtit Level 1 for interpreting prehistoric change. But his typing the assemblage as bidirectional does not make my analysis incorrect; it just makes mine more specific. Further, failing to distinguish the pattern of sequential bidirectional versus alternating bidirectional reduction in the characterization of an assemblage when comparisons with other assemblages show anthropologically significant patterns is to commit what Sackett calls "the fallacy of the generic" (1990:40–41). This term refers to the inability to interpret correctly a set of archaeological data due to a failure to recognize sufficient details. It would be ironic to stick to a needlessly generalized characterization of the tendencies in this assemblage, given its status as one of the most intricately reconstructed Pleistocene assemblages, when a technique as generally applicable as attribute analysis of individual flakes can produce a more detailed characterization.

Thus, Marks and I do not really differ in how we reconstruct the processes behind the dorsal scar pattern signature of the Boker Tachtit Level 1 debitage, or at least I do not differ with the Marks of 1983. But what of Volkman? There is no need to take Marks' or my data alone for the existence of this pattern in the assemblage. Volkman's graphical presentation of his refitting studies on a core by core basis are uniquely detailed, as they contain information on removals according to multiple variables: which platform was used, when the removal occurred in the reduction sequence relative to platform preparation, and the order of removals relative to each other. The one variable not included is the actual dorsal scar pattern of each removal. While he does comment on the pattern of each core, he only comments on individual removals now and then. Without this direct data now glued inside the reconstructed nodules, we must use another type of data to confirm the tendency to switch through time from alternating bidirectional to sequential bidirectional reduction in this assemblage.

Volkman's graphic representations of individual reduction sequences, such as (1989:230, fig. 6-13), can be used indirectly to test this hypothesis as they contain data on both the progress of reduction from beginning to end and the number of removals made between each rotation of the core for exploitation of the opposed platform. If reduction did proceed as illustrated in Figure 7.11b in a given core reconstruction, the number of removals

Table 7.17 Mean number of removals per rotation event cross-tabulated by the stage of reduction as defined by Volkman (1989) for particular core reconstructions from Boker Tachtit Level 1

Core	Stage 1	Stage 2	Stage 3	Stage 4	Stage 5	Stage 6	Stage 7	Stage 8	Volkman 1989: page, figure
A	2.00	2.00	1.00	3.50	1.00	1.00	3.00		214, 6-7
B	.00	5.00	1.00	2.00	1.00				213, 6-6b
C	.00	2.00	1.00	3.00					222, 6-9b
D	9.00	1.00	1.00	1.00	2.00				229, 6-12a
E	2.00	1.67							229, 6-12b
F	.00	.00	2.00	1.00	2.00	1.50	1.00	1.00	226, 6-11
H	5.00	2.33	1.50	1.00	1.00	2.00	1.00		212, 6-5
I	.00	2.00	1.00	2.00	1.50	2.00	4.00		230, 6-13
K	1.50	2.50	2.00	1.00	1.00				222, 6-9a
L	.00	.00	1.00	2.00	.00	3.00	1.00	2.00	225, 6-10
U	2.00	2.33	2.00	1.75	1.00	1.00			221, 6-8
V	.00	3.00	2.00	7.00	1.00	2.00			213, 6-6a
Mean	1.79	1.99	1.41	2.30	1.28	1.83	2.07	1.50	

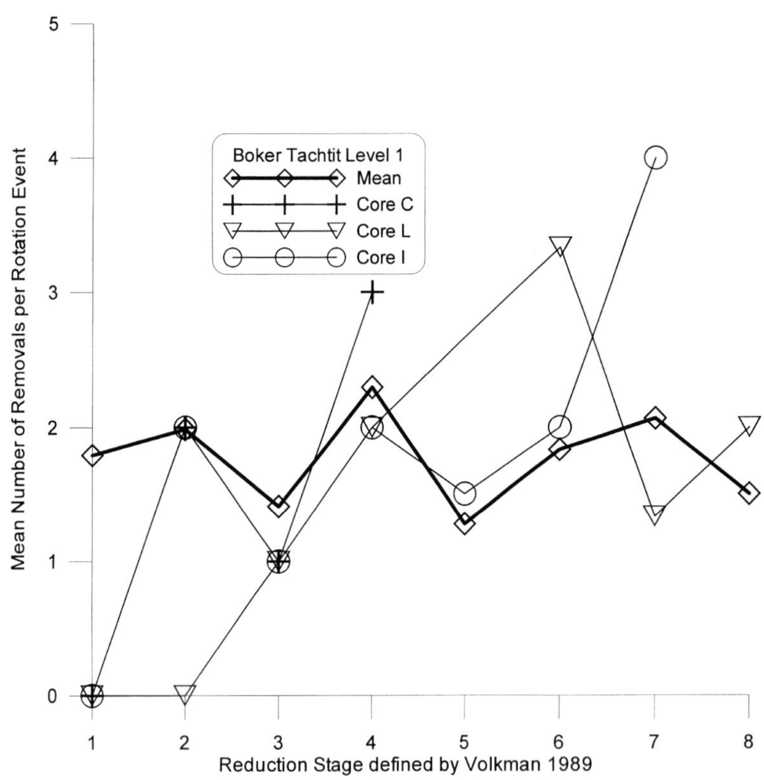

Figure 7.12 Mean number of removals per rotation event by reduction stage for Core reconstructions C, I, L, and the mean of all reconstructions of Boker Tachtit Level 1 (data analyzed from Volkman 1989).

between each rotation of the core should increase through time. Time, or the progress of core reduction, can be measured as periods between platform preparations to get a proxy for blank length, since the length of each removal is primarily dependent upon the length of the debitage surface at the moment of removal, which itself is shortened by each platform preparation. Volkman defined reduction stages based on the interruption of removals from the debitage surface by one of two events: platform preparation or the removal of a Levallois point. The removal of a Levallois point does not actually shorten the debitage surface and so does not really constitute a new reduction stage in the sense implied above. Nevertheless, Volkman's designations of reduction stages (written in Roman numerals in his text) are used here in order to facilitate easier understanding of the present results when compared to Volkman's original publications. Within each reduction stage thus defined by Volkman, the number of arrows drawn at a given platform is counted as the number of removals within one rotation event. For each reduction stage, the average number of removals made during these rotation events is then calculated. The mean number of removals per rotation event can then be compared between earlier and later reduction stages to test whether or not the mean number of removals does in fact increase for individual core sequences and/or the assemblage as a whole.

Table 7.17 presents a cross-tabulation of stage of reduction as defined by Volkman against the mean number of removals per core rotation within each stage for individual core reconstructions, as calculated from his illustrations. The data for the mean of all included core reconstructions per reduction stage is also presented. The rules for the calculation of these values are as follows:

- The counting of removals from each platform begins after the first *lame à crête* removal. If no *lame à crête* removal is made, the count begins after the performing stage as defined by Volkman, with stage I.
- Decortification flakes, cresting blows, and the removal of *lame à crête* blades ("LC" in Volkman's figures) are not counted as rotation events as they do not relate to the bidirectional/unidirectional distinction being pursued here.
- Three core sequences were not included as their exploitation was not comparable to the remaining sample of reconstructions published by Volkman: Core P possessed four platforms and both Core J and Core W were exploited as core burins, with only three reduction stages each.

Examining Table 7.17, is there a change in the mean removals per rotation event as the reduction stages progress? Table 7.17 shows a variety of patterns. Some individual core reconstructions present high mean numbers of removals early in reduction and fewer in later reduction, for example Cores B, D, H, and U. More individual cores show fluctuation without any increase or decrease through reduction, i.e., Cores A, E, F, and K. Still other core reconstructions show a decided increase in the number of removals per event as reduction increases, i.e., Cores C, I, L, and V. The mean across all core reconstructions shows no consistent change in the number of removals per rotation event.

Does the mean of all core reconstructions thus invalidate the blank length vs. dorsal scar pattern cross-tabulation indicating a change from alternating to sequential bidirectional reduction? No, as one must always consider what is being sampled. The excavation squares sampled in the debitage analysis in Tables 7.15-7.16 (squares A1, A2, A2/3, A3, A5, and

A6) are the same squares which produced most of the artifacts for the reconstruction of Core I, which shows a strong tendency to produce more blanks at a time from one platform before rotating the opposed platform. In this and other similar core reconstructions depicted graphically in Figure 7.12, as later stages have a higher mean number of removals per rotation event than the earlier stages, it is more likely that unidirectional dorsal scar patterns are more representative of the shorter products compared to the bidirectional scar patterns of the earlier stages. These core reconstructions can then be said to show a change from "alternating bidirectional" to "sequential bidirectional" core reduction. This can be said to a lesser degree for Cores V and A, both of which produce more blanks per rotation in the middle to end of the reduction than in early reduction. Both of these cores were also refit out of the debitage recovered from the same squares which produced the analysis presented in Figure 7.10. It must be said that Cores U and H, which do not match this pattern, also came from some of these squares. Thus, although the mean of the assemblage shows a consistent alternating pattern, a significant number of core reconstructions show the change from "alternating bidirectional" to "sequential bidirectional" core reduction, enough to give credibility to the characterization of this pattern in this assemblage.

If the mean of all reconstructions is not consistent with the change from alternating to sequential bidirectional reduction, why not characterize late core exploitation in this assemblage as "bidirectional" based on Table 7.17? First of all, the calculation of the mean number of removals per rotation event is an indirect measure of the likelihood to find more unidirectional dorsal scar patterns among shorter than longer blanks. This assemblage mean of an indirect measure should not trump direct observations of dorsal scar patterns, especially when almost a third of the core reconstructions agree with the dorsal scar patterns. More importantly, however, the task at hand is to produce a comparable characterization of the assemblage. Very few assemblages are as extensively refit and none are as well published as Boker Tachtit. Thus, there is no other assemblage in the present study which could offer a correction of the blank length vs. dorsal scar pattern cross-tabulation by means of an analysis of the number of removals per rotation event. One of the central issues raised by the discussion presented in Tostevin (2003a) was the importance of comparable units of analysis for inter-regional studies of lithic core reduction sequences. By characterizing one assemblage with a higher resolution technique (i.e., refitting) that should produce a more accurate result but cannot be applied to all assemblages, there is no guarantee that any comparison between this assemblage and another assemblage which cannot provide this higher resolution will increase the overall accuracy of the comparison. In fact, it is more likely that the low-resolution assemblage will appear different simply due to the difference in resolution. In other words, without comparing data derived from analyses of the same resolution and technique, there is in fact an increased chance of making a type I error, incorrectly rejecting the true null hypothesis that the two assemblages are the same. Thus, in presenting the analytical step of late core exploitation as "unidirectional" here and in Tostevin (2003a, tab. 6.2), I am following the conventions of the inter-assemblage comparisons established at the beginning of the study in order to maximize the likelihood of being accurate at the inter-regional scale, rather than sacrificing this accuracy for the finer resolution offered by one or two assemblages.

Table 7.18 Boker Tachtit Level 1, frequencies of lateral edge, profile, and cross-section types for complete flakes and tools

Type	Frequency	Percent
Lateral edge		
Parallel	72	71.3
Convergent	12	11.9
Expanding	5	5.0
Ovoid	12	11.9
Total	101	100.0
Profile		
Straight	40	39.6
Curved or Concord	40	39.6
Twisted	21	20.8
Total	101	100.0
Cross-section		
Triangular	56	55.4
Trapezoidal	40	39.6
Other	5	5.0
Total	101	100.0

Table 7.19 Boker Tachtit Level 1, descriptive statistics for blank dimensions for complete flakes and tools

Blank Dimension	N	Range	Mean	Standard Deviation
Length	101	89.02	38.98	17.12
Width	101	50.07	20.12	9.21
Thickness	101	26.61	5.40	3.48
Length/Width Ratio	101	6.93	2.25	1.21
Width/Thickness Ratio	101	11.52	4.43	2.33

In all, Marks' reaction to this characterization of the direction of core exploitation based on the analysis of debitage from Level 1 might be understood as a knee-jerk reaction to someone doing something different with artifacts from his site. The fact that he thinks I am saying anything substantially different than what he himself said in 1983 can best be described as forgetful, at worst as essentializing "bidirectional core reduction" in Boker Tachtit Level 1. I tend to think that the former is more likely, as Marks has always been on the fore-front of recognizing equifinality and technological complexity in lithic core reduction sequences. He also deserves much credit for encouraging and fostering the analyses and publications of Philip Volkman which, like those of J. G. D. Clark (1971) at Star Carr, will allow re-analysis based on published data alone for years to come. It is only to be hoped that the actual assemblages from Boker Tachtit get the curatorial attention that Volkman put into the refitting studies.

Dorsal Surface Convexity

The analysis of the attributes serving as the proxies for the surface convexity domain are given in Tables 7.18–7.19. The ridge system used in the Boker Tachtit 1 assemblage shows a propensity for very high blade frequencies (length/width mean of 2.25), very thin products relative to width (width/thickness mean of 4.43), longitudinal ridges (i.e., high frequency of parallel lateral edges) and an equal proportion of straight and curved profiles. The frequencies presented here

Table 7.20 Boker Tachtit Level 1, descriptive statistics for dimensions of complete tools

Blank Dimension	N	Range	Mean	Standard Deviation
Length/Width Ratio	72	3.51	2.06	0.79

Table 7.21 Boker Tachtit Level 1, frequencies of distal terminus for complete tools

Type	Frequency	Percent
Distal Terminus		
Blunt	17	23.6
Pointed	45	62.5
Indeterminate	10	13.9
Total	72	100.0

Table 7.22 Boker Tachtit Level 1, intensity of core and tool reduction

Variables	Numerator	Denominator	Result
Tools/Flake Ratio	Tools: 90	Flakes: 1067	0.08
Non-cortical Debitage/ Cortical Debitage Ratio	Non-cortical Debitage: 523	Cortical Debitage: 481	1.09

are slightly different but of the same order as those presented by Hietala (1983b,c), since the attribute analysis here was conducted across all blank forms together (flakes, blades, core trimming elements, primary flakes). More importantly, the attribute definitions in Chapter 4 are slightly different than Marks' definitions (1976) used by Hietala. Specifically, the profile attribute states defined in Marks (1976:372-373) do not distinguish the degree to which a blank profile can be twisted before it is so typed. Similarly, how far off of the table surface the ventral surface must be, when the extremities are placed on the surface, for the piece to be typed curved is also not clear. These factors, both blank forms included and different definitions, contribute to the higher frequency of straight profiles and lower frequency of curved and twisted profiles in this analysis when compared to Hietala (1983b:243-244). Regardless of these differences, the patterns and trends between Levels 1, 2 and 4 agree with those observed by Hietala

7.2.4 Tool Kit Morphology

The small sample of artifacts available for attribute analysis in the unrefit collections at the Israeli Antiquity Authority was too small to include enough tools to analyze directly. Further, directly comparable data of this type was not published by Marks and Kaufman (1983) or in the otherwise admirable analyses of Hietala (1983b; 1983c), as these attribute studies did not separate the tool kit from the rest of the artifacts. Fortunately, the length and width of the tool kit, both non-retouched points and retouched blanks, were published as scatterplots in Marks and Kaufman (1983:79, fig. 5-5). The length and width of each point representing a tool were measured by caliper off of the publication and transposed into a data file comparable to other assemblages. This allowed the inclusion of the length/width ratio for the Boker Tachtit Level 1 tool kit which shows tools of clearly blade proportions, although no data was published which would allow for the inclusion of

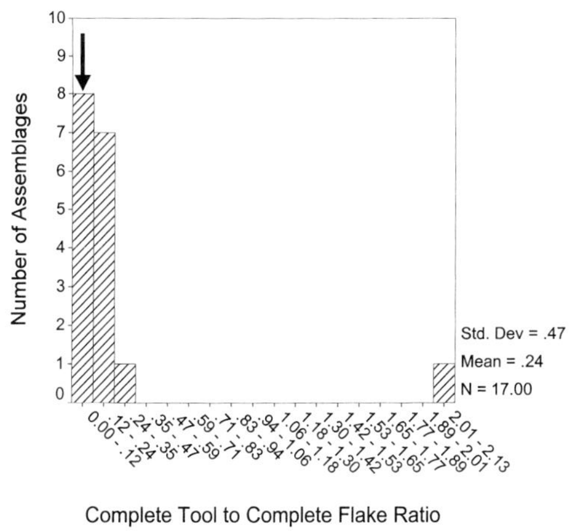

Figure 7.13 Boker Tachtit Level 1, tool/flake ratio value relative to a histogram of all assemblages.

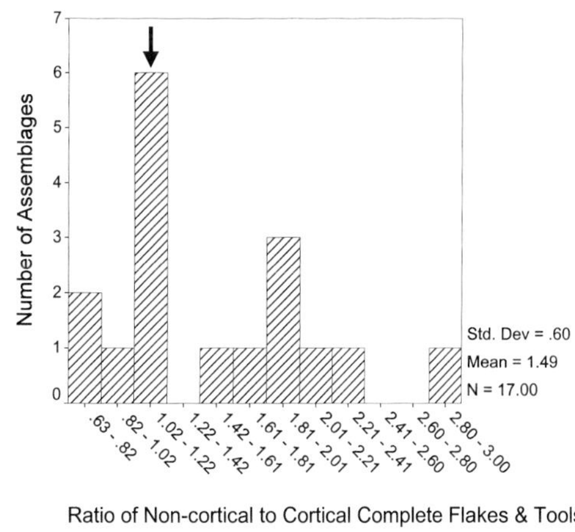

Figure 7.14 Boker Tachtit Level 1, non-cortical/cortical flake ratio value relative to a histogram of all assemblages.

the width/thickness ratio (Table 7.20). Tool type descriptions from this figure were used to ascribe a distal termination type to these pieces, the closest in the Marks and Kaufman publication to descriptions based on complete tools only. While distal terminations (called "tip shape") and profiles were described in Hietala (1983c), tables B-1 through B-4, these statistics are only categorized by blank form (flake, blade, core trimming element, and primary element) and not by application of retouch alone. Only the classification in figure 5.5 of Marks and Kaufman (1983:79) provides an identification of tools that might allow a description of the distal termination apart from the rest of the debitage. Distal terminations were given as follows from the types indicated in this figure: unknown (burins and truncations), blunt (denticulate endscrapers, notches, and scrapers), and pointed (retouched points, unretouched points, Emirah points, point machures, and pointed blades). Table 7.21 shows a clear superiority of pointed distal terminations, as one would expect from the reduction sequences and the inclusion of many unretouched "points." In terms of the presence of unique retouch in Boker Tachtit level 1, fine bifacial retouch of the platform and bulb of percussion, the Emireh-type retouch, represents a unique application of retouch. Only Upper Paleolithic tool types are present in this assemblage.

Intensity of Tool and Core Reduction

Given the abundant amounts of raw material in the form of very large nodules available to residents of the Negev desert, it is not surprising that there was little need to apply extensive retouch to blank cutting edges for the resharpening of tools in this assemblage. The retouch applied appears to shape the blanks to their tasks, rather than resharpen the cutting edges themselves, given the high frequency of burins, endscrapers, and retouched points. The tool vs. debitage ratio, one of the lowest in the study (Table 7.22, Figure 7.13), confirms this view. The value of the tools vs. debitage ratio for Level 1 is calculated from counts of complete tools and debitage taken from Hietala (1983a:199, tab. 7-3), which indicates 1,157 complete pieces of which 90 are tools, leaving 1,067 for the debitage. This assemblage is thus very lightly retouched, so that the dimensional variables of the retouched tool kit should not be biased significantly by reduction.

The non-cortical to cortical blank ratio was taken from counts of the entire assemblage as published by Hietala (1983c:343-346). The counts for non-cortical blanks and cortical blanks were summed from each of the blank types and excavation areas as detailed in Hietala's tables B-1 through B-4. The resulting value of 1.09 falls with the most left grouping of assemblages (Figure 7.14), clearly reflecting the role of the site as a raw material workshop for the working of cortical nodules into blanks. The assemblage possesses relatively more non-cortical blanks than some of other raw material workshops but the surface area to volume relationship of the large nodules used at Boker Tachtit may contribute to this. Otherwise, the refits clearly show the cortical beginning of almost each core reduced on site.

Although no data is available on the dimensions of core facets, since the cores were unavailable for study in the Israeli Antiquity Authority, it is clear from the extant core illustrations that at least a sample of cores from the assemblage were reduced to a fairly exhausted state. The extremely thin cores illustrated in Marks and Kaufman (1983:77, fig. 5-3a-d) reflect a Levallois reduction taken to its limit. The point and non-Levallois cores are not as clearly exhausted, however. Together, these data demonstrate that almost all of the core reduction sequence related to blank production, save perhaps extreme core exhaustion, is represented in the assemblage.

7.3 Boker Tachtit, Level 2

7.3.1 Site Background

The second oldest cultural layer at the open-air site of Boker Tachtit (Marks 1983; Marks and Volkman 1983), Level 2, also possesses numerous refitted cores that allow the accurate identification of the steps within blank production. While no absolute dating technique has been applied to this assemblage, the radiocarbon date from Level 4 of 35,055 ± 4,100 BP provides a *terminus ante quem.*

7.3.2 Sampling

The attribute analysis of the unglued debitage from Boker Tachtit level 2 was conducted using artifacts from squares X43 and X42, in the heart of the refit core sequences.

7.3.3 Blank Production

Core Modification

Core modification in Boker Tachtit level 2 consists of a longitudinal blade core form, as evidenced in the identification of the narrow and long core face as the primary debitage surface among the refitted cores. Core management was achieved first through the use of frontal crests to control the longitudinal curvature of the debitage surface convexity. Lateral convexities were managed by débordant removals to isolate the mass toward the center of the surface in order to create ridge patterns conducive to the removal of pointed products, as evidenced in Volkman's refits. Figure 7.16 illustrates an example of débordant blade removals from the top platform isolating the mass of the debitage surface so that removals from the bottom platform take on a pointed lateral edge type.

Platform Maintenance

Platform maintenance includes a dominance of plain platforms, up to 38% of the debitage (Table 7.23) or 64.5% without indeterminates

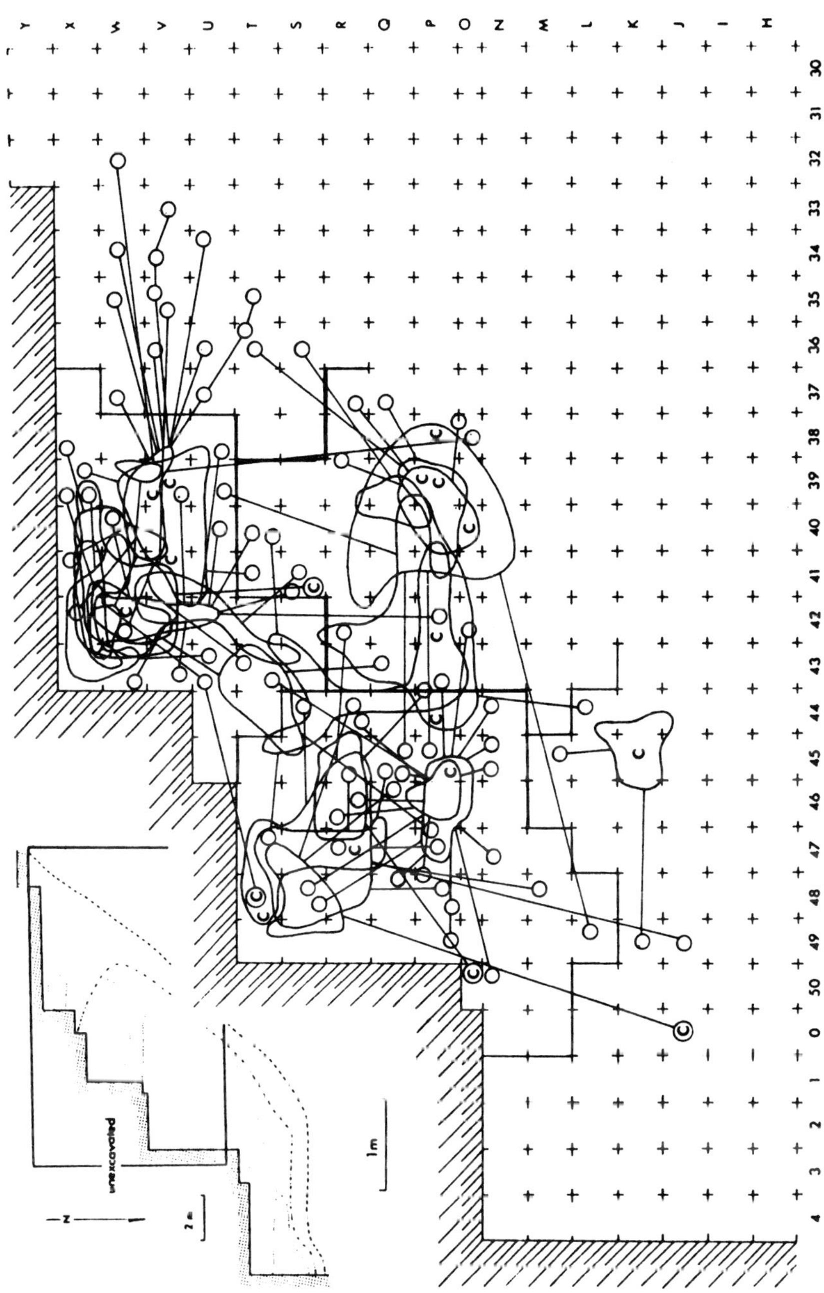

Figure 7.15 Site plan and distribution of core reconstructions for Boker Tachtit Level 2 (from Hietala 1983b: 226, fig. 8-4b, used by permission).

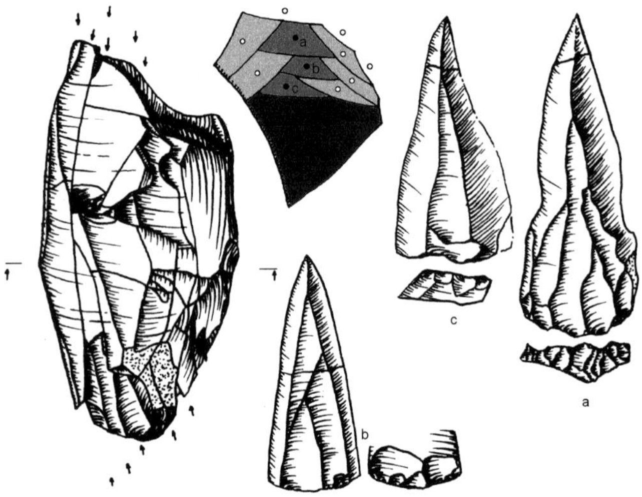

Figure 7.16 Boker Tachtit Level 2, bidirectional Core RRR with three of its refit products. Within the cross-section illustration, filled dots indicate a removal from the lower platform while an empty dot indicates a removal from the top platform. Drawing by P. Škrdla, from Škrdla (2003a: 63, fig. 14), used by permission.

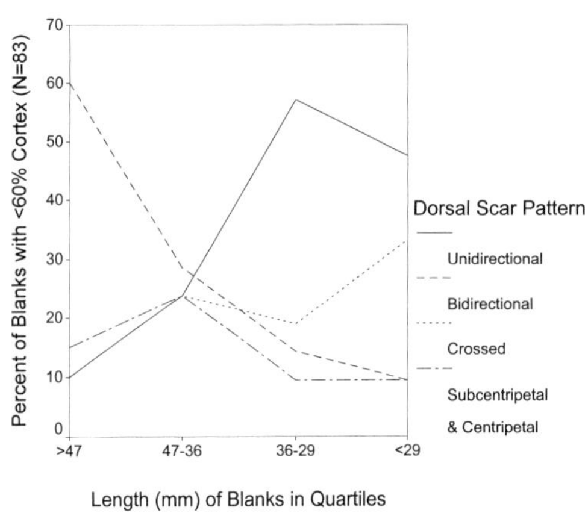

Figure 7.17 Boker Tachtit Level 2, dorsal scar pattern by blank length (mm) in quartiles cross-tabulation for complete flakes and tools (excluding pieces with a crested scar pattern, an indeterminate scar pattern, greater than 60% cortex, or fewer than three dorsal scars).

included. Measurements of external platform angles and platform thickness (Table 7.24) indicate tendencies around 82 degrees and 5 mm respectively.

Direction of Core Exploitation

The analysis of the direction of core exploitation for Boker Tachtit Level 2 is presented in Tables 7.25-7.26. Figure 7.17 demonstrates that the dominant direction for early core exploitation is bidirectional while unidirectional is dominant by late reduction as evidenced in the smallest quartile of blanks.

The debitage sample in this analysis is small, even before crested pieces are removed. Further, as Level 2 has the same issues for the reconstruction of the debitage's directionality changing from bidirectional to unidirectional in early to late core exploitation, an investigation of the pattern of reduction evidenced in Volkman's core

Table 7.23 Boker Tachtit Level 2, platform types for complete and proximal fragments of flakes and tools

Platform type	Frequency	Percent
Plain	38	38.4
Cortical	6	6.1
Dihedral	4	4.0
Facetted	5	5.1
Scarred	18	18.2
Punctiform	1	1.0
Linear	4	4.0
Indeterminate	23	23.2
Total	99	100.0

Table 7.24 Boker Tachtit Level 2, descriptive statistics for continuous platform variables of complete and proximal fragments of flakes and tools

	N	Minimum	Maximum	Mean	Standard Deviation
Exterior Platform Angle (degrees)	58	48.00	112.00	82.07	12.00
Platform Thickness (mm)	62	1.21	10.94	4.54	2.17

Table 7.25 Boker Tachtit Level 2, dorsal scar pattern by blank length (mm) in quartiles cross-tabulation for complete flakes and tools (excluding flakes with greater than 60% cortex or fewer than three dorsal scars)

Dorsal Scar Pattern	>47	47-36	36-29	<29	Total
Unidirectional					
Count	2	5	12	10	29
% within Quartile	8.7%	23.8%	57.1%	47.6%	33.7%
Bidirectional					
Count	12	6	3	2	23
% within Quartile	52.2%	28.6%	14.3%	9.5%	26.7%
Crossed					
Count	3	5	4	7	19
% within Quartile	13.0%	23.8%	19.0%	33.3%	22.1%
Subcentripetal & Centripetal					
Count	3	5	2	2	12
% within Quartile	13.0%	23.8%	9.5%	9.5%	14.0%
Crested					
Count	3				3
% within Quartile	13.0%				3.5%
Total					
Count	23	21	21	21	86
% within Cortex	100.0%	100.0%	100.0%	100.0%	100.0%

reconstructions is warranted for this assemblage. As done with Level 1, Table 7.27 presents the mean number of removals per core rotation event within the reduction stages defined by Volkman.

The removals were tabulated in the same manner as with Level 1, except for situations in which Volkman changed the convention of his abstracted illustrations of the core reconstructions. In particular instances, such as Core TT, stage V, last event in Volkman (1989:287, fig. 6-43a), removals were illustrated from both platforms during the same rotation

Table 7.26 Boker Tachtit Level 2, dorsal scar pattern by blank length (mm) in quartiles cross-tabulation for complete flakes and tools (excluding pieces with a crested scar pattern, an indeterminate scar pattern, greater than 60% cortex, or fewer than three dorsal scars)

Dorsal Scar Pattern	>47	47-36	36-29	<29	Total
Unidirectional					
Count	2	5	12	10	29
% within Quartile	10.0%	23.8%	57.1%	47.6%	34.9%
Bidirectional					
Count	12	6	3	2	23
% within Quartile	60.0%	28.6%	14.3%	9.5%	27.7%
Crossed					
Count	3	5	4	7	19
% within Quartile	15.0%	23.8%	19.0%	33.3%	22.9%
Subcentripetal & Centripetal					
Count	3	5	2	2	12
% within Quartile	15.0%	23.8%	9.5%	9.5%	14.5%
Total					
Count	20	21	21	21	83
% within Cortex	100.0%	100.0%	100.0%	100.0%	100.0%

event. The illustration convention for these examples is not explained by Volkman. It could be taken as indicating either alternating or sequential reduction. It could also be taken as a space saving measure, as the instances of these events are found in illustrations which are limited in the amount of the page the illustration could take up. No core reduction sequence was allowed in Volkman's illustrations to surpass one page and those sequences which possess an example of removals from both platforms within the same rotation event have reasons for limited coverage. Level 2 Core TT has to share the same page with Core UU, as they both derive from the same block, and so each core sequence is only given half a page. The same holds for Core VV sharing the page with XX and WW (Ibid.:289, fig. 6-44). These examples only occur during the last event in the sequence. Thus, as no explicit explication is given by Volkman, the illustration of the use of multiple platforms in the same rotation event is taken as indication of the alternation between platforms from the previous rotation event. This will bias the analysis toward the null hypothesis that no change in number of removals per rotation event occurred as reduction progressed.

The following core reconstructions were not included in the analysis of Table 7.27. Core EE consists of a block which produced five different cores and could not be fit into the comparative analysis. Cores A, P, Z, I I, WW, and XX had too few stages to represent a full reduction. Core M represents an informe core. Cores BB, CC, and DD were too complex for the standardized study. Finally, Cores F, V, and UU all had two opposed platforms but the tilt of the two debitage surfaces was such that reduction was pursued primarily unidirectionally from each platform without impingement upon the other platform's debitage.

Table 7.27 shows divergent patterns among the individual reconstructions but a general trend can be discerned. Some sequences show a decrease in the number of removals as reduction continues, i.e., Cores O, Q, W, OO, PP, and SS. Others show no tendency to change as reduction continues, i.e., Cores C, D, G, J, K, R, T, Y, GG, HH,

Table 7.27 Mean number of removals per rotation event cross-tabulated by the stage of reduction as defined by Volkman (1989) for particular core reconstructions from Boker Tachtit Level 2.

Core	Stage 1	Stage 2	Stage 3	Stage 4	Stage 5	Stage 6	Stage 7	Stage 8	Stage 9	1989: page, figure
B	.00	3.50	2.00	3.00	1.00	10.50				255, 6-25
C	1.00	1.00	1.00							246, 6-17a
D	2.00	1.00	3.50							282, 6-39b
G	.00	1.00	3.00	1.50	4.00	2.00	1.00			277, 6-34
I	2.67	1.00	2.00	1.00	3.00	4.00	3.50			247, 6-18
J	2.00	3.00	1.00	1.00	1.00	3.75				248, 6-19
K	2.67	2.33	2.00	1.00	1.00	3.00	2.00			252, 6-23
N	2.00	1.00	1.00	4.00	1.50	2.33	4.00			284, 6-40
O	3.00	2.20	3.00	1.00	2.00	1.25				279, 6-36
Q	2.00	5.00	4.17	4.00	3.00	3.40				285, 6-41
R	3.00	5.67	1.50							253, 6-24a
S	.00	1.00	2.00	3.00						253, 6-24b
T	.00	3.00	4.50	2.00	1.50					290, 6-45
W	.00	7.00	3.00	7.00	3.00	5.00	4.00	4.00	5.00	280, 6-37
X	.00	1.00	2.00	2.00	1.00	1.00	2.00	1.00	2.50	249, 6-20
Y	.00	1.00	1.50	4.00	2.00	2.00	3.00			267, 6-30a
FF	1.00	1.50	4.00							264, 6-29c
GG	.00	1.00	1.00	1.00						265, 6-29d
HH	.00	1.00	5.00	2.00	1.00	2.00				265, 6-29e
JJ	.00	2.67	3.00							265, 6-29f
LL	.00	1.00	2.00	1.00	1.00	1.00	1.50	2.50		250, 6-21
MM	6.00	4.00	6.50							251, 6-22b
NN	2.00	1.00	2.75							245, 6-16b
OO	1.00	2.00	1.00	1.00	1.67	1.00				273, 6-32
PP	.00	3.00	2.00	2.00						245, 6-16a
QQ	2.00	2.50	3.50	1.00	2.33					251, 6-22a
RR	2.00	3.00	2.00	3.00						246, 6-17b
SS	2.67	2.75	2.00	1.00						274, 6-33
TT	.00	2.50	2.00	2.00	2.50					287, 6-43a
VV	3.33	2.50	9.00							289, 6-44a
YY	.00	.00	3.00	2.00	1.00	2.50	3.00	4.00	4.00	281, 6-38
Mean	2.37	2.34	2.72	2.20	1.86	3.02	2.78	3.00	3.83	

MM, NN, QQ, RR, and TT. Still other reconstructions show a clear increase in the number of removals per rotation of the core, i.e., Cores B, I, N, S, X, JJ, LL, VV, and YY. These latter cores, as illustrated in Figure 7.18, demonstrate a strong tendency to produce more blanks from a given platform at one time later than earlier in reduction, a change from alternating to sequential bidirectional use of platforms. And unlike Level 1, the mean of all reconstructions in each reduction stage shows a clear increase in the number of removals per rotation as reduction continues. Thus, while Level 2 is alike to Level 1 in that a third of the cores show a gradual change from alternating to sequential bidirectional reduction, now these cores outnumber the reconstructions in which sequential removals occur earlier in reduction. This subtle change from Level 1 to 2 in this data is an element of what Volkman saw as the fundamental evolutionary changes within the Boker Tachtit assemblage: a gradual shift in the representation of specific rotational patterns for artifact removals in each level (1989:243, 564). When comparing the dorsal scar pattern by blank length for the two assemblages (Figures 7.10 and 7.17), this gradual change can also be seen in the quartile at which unidirectional removals

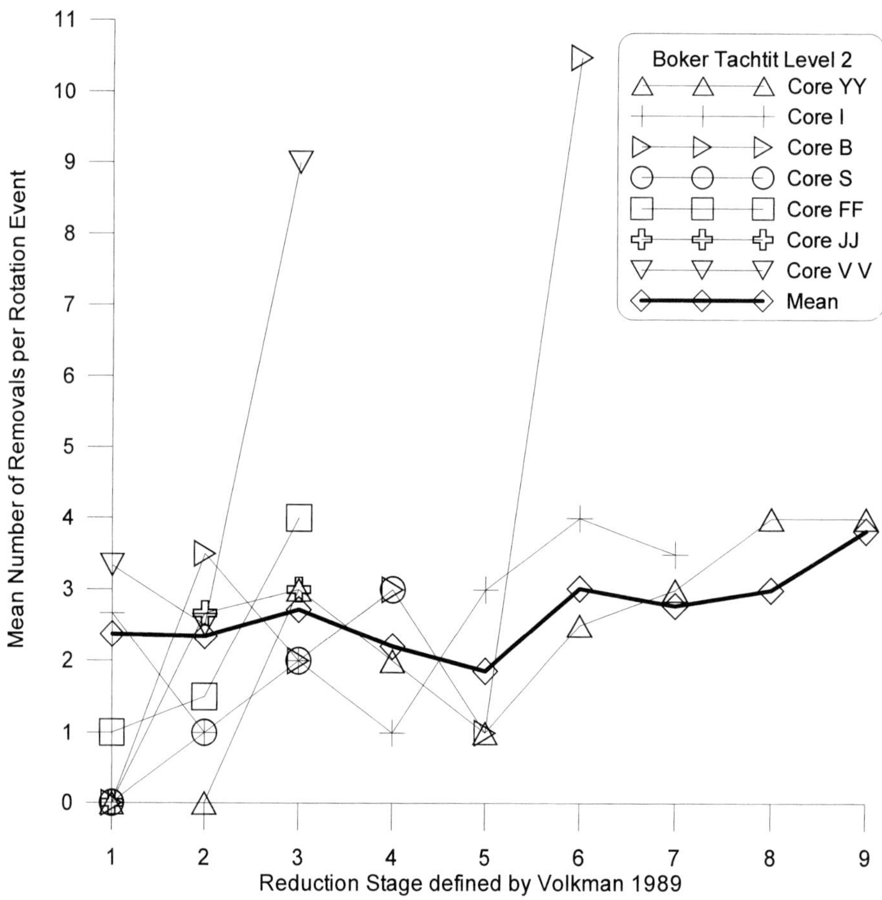

Figure 7.18 Mean number of removals per rotation event by reduction stage for Core reconstructions B, I, S, FF, JJ, VV, YY, and the mean of all reconstructions of Boker Tachtit Level 2 (data analyzed from Volkman 1989).

outnumber the other scar patterns, only by the fourth in Level 1 but by the third in Level 2. It is significant that this subtle distinction in rotational frequency can be picked up by this form of attribute analysis.

Dorsal Surface Convexity

The analysis of the attributes serving as the proxies for the surface convexity domain are given in Tables 7.28-7.29. The ridge system used in the Boker Tachtit Level 2 assemblage shows a propensity for high laminarity (length/width mean of 2.14), very thin products relative to width (width/thickness mean of 4.46), longitudinal ridges (i.e., high frequency of parallel lateral edges) and straighter profiles than in Level 1, another agreement with Hietala (1983b:243-244).

Table 7.28 Boker Tachtit Level 2, frequencies of lateral edge, profile, and cross-section types for complete flakes and tools

Type	Frequency	Percent
Lateral edge		
Parallel	64	64.6
Convergent	18	18.2
Expanding	10	10.1
Ovoid	6	6.1
Indeterminate	1	1.0
Total	99	100.0
Profile		
Straight	50	50.5
Curved or Concord	29	29.3
Twisted	19	19.2
Indeterminate	1	1.0
Total	99	100.0
Cross-section		
Triangular	45	45.5
Trapezoidal	50	50.5
Other	4	4.0
Total	99	100.0

Table 7.29 Boker Tachtit Level 2, descriptive statistics for blank dimensions for complete flakes and tools

Blank Dimension	N	Range	Mean	Standard Deviation
Length	99	72.99	40.80	15.93
Width	99	56.76	21.62	9.12
Thickness	99	16.90	5.71	3.07
Length/Width Ratio	99	6.03	2.14	1.05
Width/Thickness Ratio	99	11.93	4.46	2.35

7.3.4 Tool Kit Morphology

As with Level 1, the small sample of artifacts available for attribute analysis in the unrefit collections at the Israeli Antiquity Authority was too small to include enough tools to analyze directly. From the publications, it is possible to calculate the length and width of the tool kit, both non-retouched points and retouched blanks, as this data could be calculated by measuring off of the scatter plot in Figure 5.13 of Marks and Kaufman (1983:93). While distal terminations (called "tip shape") and profiles were described in Appendix B of Marks (1983), tables B-1 through B-4, these statistics are only categorized by blank form (flake, blade, core trimming element, and primary element) and not by application of retouch alone. Only the classification in figure 5.13 of Marks and Kaufman (1983:93) provides an identification of tools that might allow a description of the distal termination apart from the rest of the debitage for Level 2. Distal terminations were given as follows from the types indicated in this figure: unknown (burins, truncations, and other), blunt (denticulates, burin endscrapers, notches, and scrapers), and pointed (retouched points, unretouched points, Emirah points, point machures, and pointed blades). When compared to the

Table 7.30 Boker Tachtit Level 2, descriptive statistics for dimensions of complete tools

Blank Dimension	N	Range	Mean	Standard Deviation
Length/Width Ratio	210	4.82	1.87	0.87

Table 7.31 Boker Tachtit Level 2, frequencies of distal terminus for complete tools

Type	Frequency	Percent
Distal Terminus		
Blunt	68	32.4
Pointed	91	43.3
Indeterminate	51	24.3
Total	210	100.0

Table 7.32 Boker Tachtit Level 2, intensity of core and tool reduction

Variables	Numerator	Denominator	Result
Tools/Flake Ratio	Tools: 249	Flakes: 3542	0.07
Non-cortical Debitage/ Cortical Debitage Ratio	Non-cortical Debitage: 1991	Cortical Debitage: 2064	.96

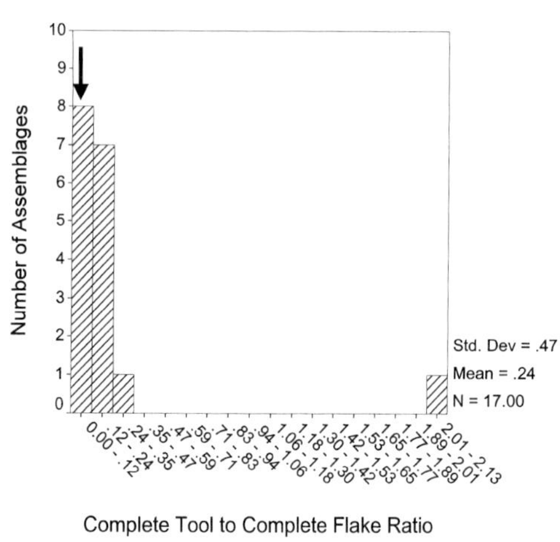

Figure 7.19 Boker Tachtit Level 2, tool/flake ratio value relative to a histogram of all assemblages.

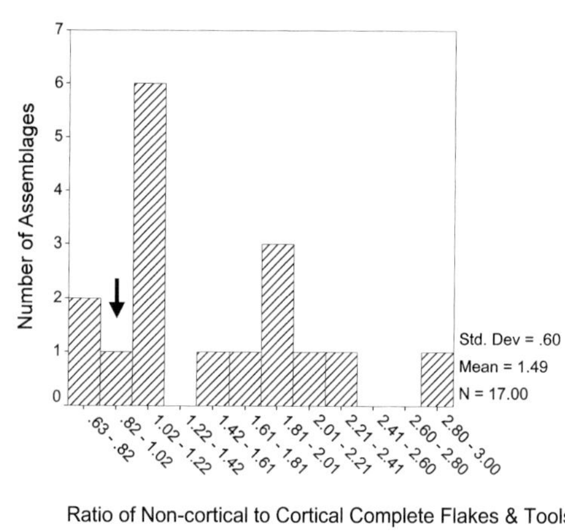

Figure 7.20 Boker Tachtit Level 2, non-cortical/cortical flake ratio value relative to a histogram of all assemblages.

debitage, the tool kit contains fewer blades at discard (Table 7.30) and a high frequency of pointed distal ends (Table 7.31).

For the last two analytical steps in characterizing the tool kit morphology, Boker Tachtit Level 2 exhibits the use of Emireh retouch and a dominance of Upper Paleolithic tool types.

Intensity of Tool and Core Reduction

Given the abundant amounts of raw material in the form of very large nodules available to residents of the Negev desert, it is not surprising that there was little need to apply extensive retouch to blank cutting edges for the resharpening of tools in this assemblage. The retouch applied appears to shape the blanks to their tasks, rather than resharpen the cutting edges themselves, given the high frequency of burins, endscrapers, and retouched points. The tool vs. debitage ratio, one of the lowest in the study (Table 7.32, Figure 7.19), confirms this view. The value of the tools vs. debitage ratio for Level 2 is calculated from counts of complete tools and debitage taken from Hietala (1983a:207, tab. 7-7), which indicates 3,791 complete pieces of which 249 are tools, leaving 3,542 for the debitage. This assemblage is thus very lightly retouched, so that the dimensional variables of the retouched tool kit should not be biased significantly by reduction.

The non-cortical to cortical blank ratio was taken from counts of the entire assemblage as published by Hietala (1983c:343–346). The counts for non-cortical blanks and cortical blanks were summed from each of the blank types and excavation areas as detailed in Hietala's tables B-1 through B-4. The resulting value of 0.96 falls with the most left grouping of assemblages (Figure 7.20), clearly reflecting the role of the site as a raw material workshop for the working of cortical nodules into blanks. The refits corroborate the presence of the earliest portion of the core reduction sequence, as almost each core reduced on site began as a cortical nodule.

Although no data is available on the dimensions of core facets, since the cores were unavailable for study in the Israeli Antiquity Authority, it is clear from the extant core illustrations that at least a few of the cores from the assemblage were reduced to a fairly exhausted state (Marks and Kaufman 1983:90–91, figs. 5-10a and 5-11a-b). The vast amount of blanks with adequate cutting edges, unutilized as far as currently discernable, would suggest, however, that the assemblage does not represent a population of flintknappers under a condition of extreme raw material scarcity requiring conservation. Together, these data demonstrate that almost all of the core reduction sequence related to blank production, save perhaps extreme core exhaustion, is represented in the assemblage.

7.4 Boker Tachtit Level 4

7.4.1 Site Background

The youngest cultural layer at the open-air site of Boker Tachtit (Marks 1983; Marks and Volkman 1983), Level 4, also possesses numerous refitted cores that allow the accurate identification of the steps within the core reduction sequence. Level 3, immediately underlying level 4, was not chosen for the present study due to the paucity of the artifacts and refits available for study. A radiocarbon date of 35,055 ± 4,100 (SMU-579) is associated with the Level 4 assemblage.

7.4.2 Sampling

The attribute analysis of the unglued debitage from Boker Tachtit Level 4 was conducted using artifacts from squares G2, E2, F2, H2, L4, R41, L46, I2, and L47.

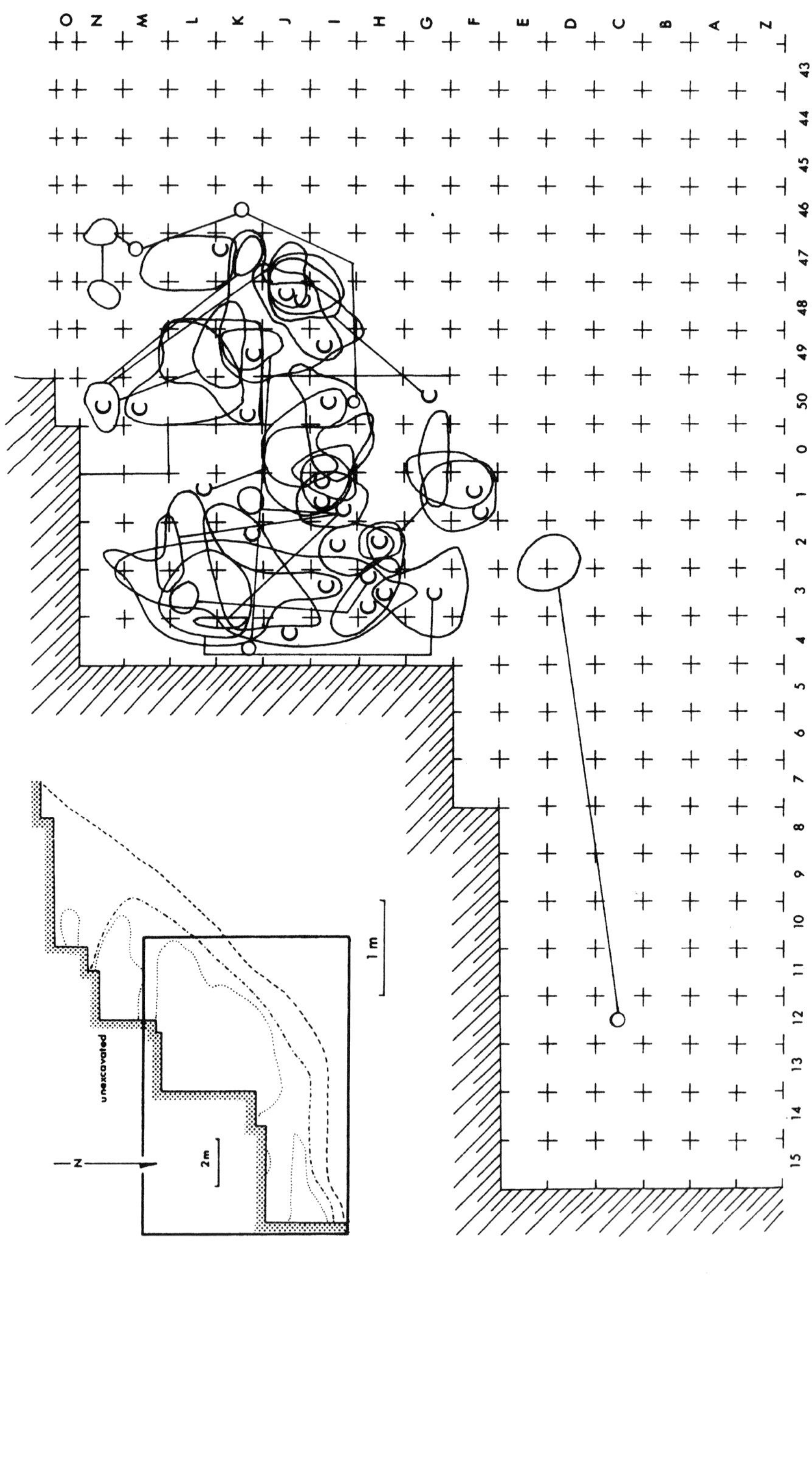

Figure 7.21 Site plan and distribution of core reconstructions for Boker Tachtit Level 4 (from Hietala 1983b: 233, fig. 8-8b), used by permission.

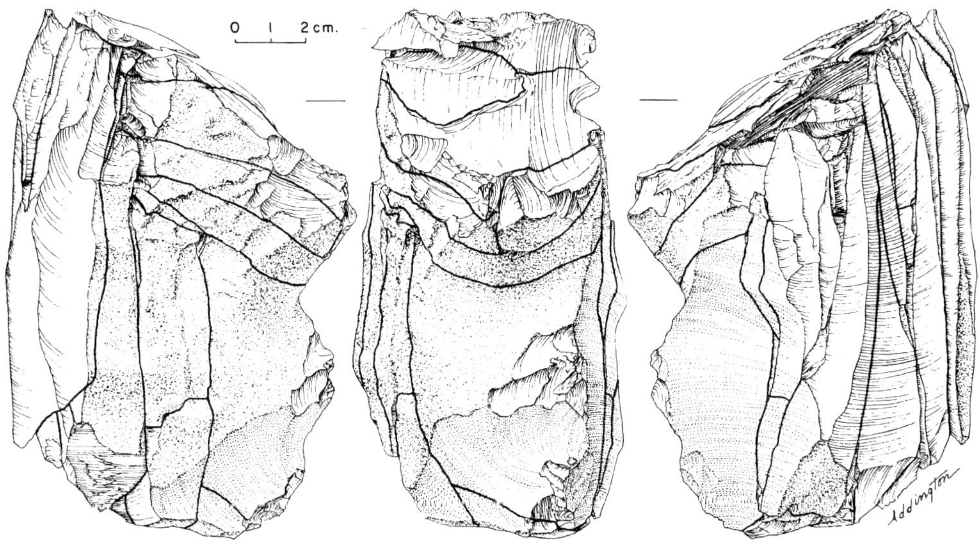

Figure 7.22 (left) Boker Tachtit Level 4. Core reconstruction for Core KK, a single platform blade core with frequent core tablets removed (Volkman 1983:180, fig. 6-35), used by permission.

Table 7.33 Boker Tachtit Level 4, platform types for complete and proximal fragments of flakes and tools

Platform type	Frequency	Percent
Plain	33	33.7
Cortical	6	6.1
Dihedral	2	2.0
Facetted	15	15.3
Scarred	26	26.5
Punctiform	2	2.0
Linear	1	1.0
Indeterminate	13	13.3
Total	98	100.0

Table 7.34 Boker Tachtit Level 4, descriptive statistics for continuous platform variables of complete and proximal fragments of flakes and tools

	N	Minimum	Maximum	Mean	Standard Deviation
Exterior Platform Angle (degrees)	71	56.00	121.00	82.35	12.04
Platform Thickness (mm)	72	1.18	13.86	6.60	3.11

Table 7.35 Boker Tachtit Level 4, dorsal scar pattern by blank length (mm) in quartiles cross-tabulation for complete flakes and tools (excluding flakes with greater than 60% cortex or fewer than three dorsal scars)

Dorsal Scar Pattern	>79	79-51	51-33	<33	Total
Unidirectional					
Count	11	14	11	7	43
% within Quartile	55.0%	73.7%	57.9%	41.2%	57.3%
Bidirectional					
Count	4		2	2	8
% within Quartile	20.0%		10.5%	11.8%	10.7%
Crossed					
Count		3	3	6	12
% within Quartile		15.8%	15.8%	35.3%	16.0%
Subcentripetal & Centripetal					
Count	2		1	1	4
% within Quartile	10.0%		5.3%	5.9%	5.3%
Crested					
Count	2	1		1	4
% within Quartile	10.0%	5.3%		5.9%	5.3%
Indeterminate					
Count	1	1	2		4
% within Quartile	5.0%	5.3%	10.5%		5.3%
Total					
Count	20	19	19	17	75
% within Cortex	100.0%	100.0%	100.0%	100.0%	100.0%

Table 7.36 Boker Tachtit Level 4, frequencies of lateral edge, profile, and cross-section types for complete flakes and tools

Type	Frequency	Percent
Lateral edge		
Parallel	53	54.1
Convergent	29	29.6
Expanding	6	6.1
Ovoid	10	10.2
Total	98	100.0
Profile		
Straight	48	49.0
Curved or Concord	26	26.5
Twisted	24	24.5
Total	98	100.0
Cross-section		
Triangular	44	44.9
Trapezoidal	45	45.9
Other	9	9.2
Total	98	100.0

Table 7.37 Boker Tachtit Level 4, descriptive statistics for blank dimensions for complete flakes and tools

Blank Dimension	N	Range	Mean	Standard Deviation
Length	98	93.57	57.97	25.87
Width	98	61.58	25.57	11.28
Thickness	98	29.32	7.72	4.93
Length/Width Ratio	98	5.05	2.47	1.12
Width/Thickness Ratio	98	9.36	3.86	1.60

7.4.3 Blank Production

Core Modification

Core modification in Boker Tachtit Level 4 consists of a longitudinal blade core form, as evidenced in the identification of the narrow and long core face as the primary debitage surface among the refitted cores. Core management was achieved first through the use of frontal crests to control the longitudinal curvature of the debitage surface convexity. As evidenced in Volkman's refits (Volkman 1983:171-172, figs. 6-26-6-27), lateral convexities were managed by the consistency of the removals from the curved arc of the single platform, with periodic removal of a core tablet to bring the longitudinal convexity back to a straight profile (Figure 7.22). The core tablet technique limited removal errors and hinged distal termini, thereby allowing the straight profile without the assistance of an opposed platform to remove errors.

Platform Maintenance

Platform maintenance includes almost equal proportions of plain and scarred platforms (Table 7.33). This agrees fairly closely with Marks and Kaufman (1983:112, tab. 5-27). Measurements of external platform angles and platform thickness (Table 7.34) indicate tendencies around 82 degrees and 7 mm respectively

Direction of Core Exploitation

The analysis of the direction of core exploitation is presented in Table 7.35. This data indicates that the dominant direction of early exploitation is unidirectional while both unidirectional and crossed patterns are evidenced on the dominant number of blanks in the smallest length quartile. The gaps in the representation of several dorsal scar directions in Table 7.35 as well as the small sample size indicate, however, that reliance on the relative frequencies in this cross-tabulation would produce a result of dubious accuracy in relation to the known refits and core forms. The gaps in the representation of three scar patterns in Table 7.35 also breaks the conventions of comparability for this type of analysis established in Chapter 4. As a result, and fortunately for the present analysis, the determination of the directions of core reduction in early core exploitation and late core exploitation is here based on Volkman's reconstructions of refit sequences and not on the dorsal scar pattern versus blank length cross-tabulation: "Nearly 81% of the reconstructed cores were reduced by strategies which were variations of a single pattern.... Thus, it is not overstated to maintain that the core reduction stragegy of Level 4 was one which predominantly produced blades by hard hammer from single platform cores" (Volkman 1983:185). With single platform exploitation so dominant in the reconstructions, the refits can confirm the characterization of debitage in early core exploitation as unidirectional while correcting the attribute analysis' characterization of debitage in late core exploitation to solely unidirectional.

Table 7.38 Boker Tachtit Level 4, descriptive statistics for dimensions of complete tools

Blank Dimension	N	Range	Mean	Standard Deviation
Length/Width Ratio	157	5.36	2.53	0.89

Table 7.39 Boker Tachtit Level 4, frequencies of distal terminus for complete tools

Type	Frequency	Percent
Distal Terminus		
Blunt	57	36.3
Pointed	84	53.5
Indeterminate	16	10.2
Total	157	100.0

Table 7.40 Boker Tachtit Level 4, intensity of core and tool reduction

Variables	Numerator	Denominator	Result
Tools/Flake Ratio	Tools: 158	Flakes: 1,222	0.13
Non-cortical Debitage/ Cortical Debitage Ratio	Non-cortical Debitage: 892	Cortical Debitage: 784	1.14

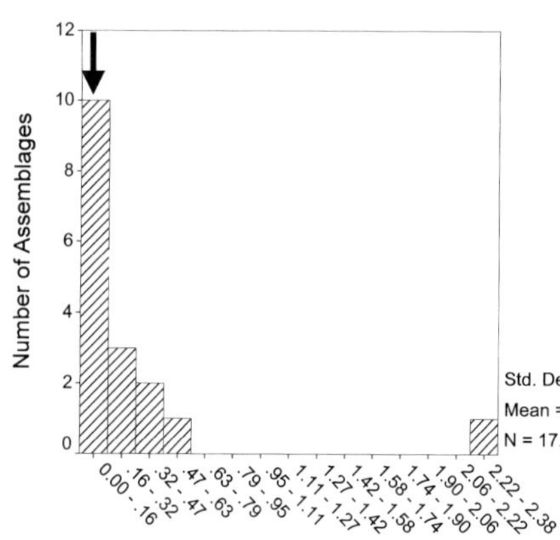

Figure 7.23 Boker Tachtit Level 4, tool/flake ratio value relative to a histogram of all assemblages.

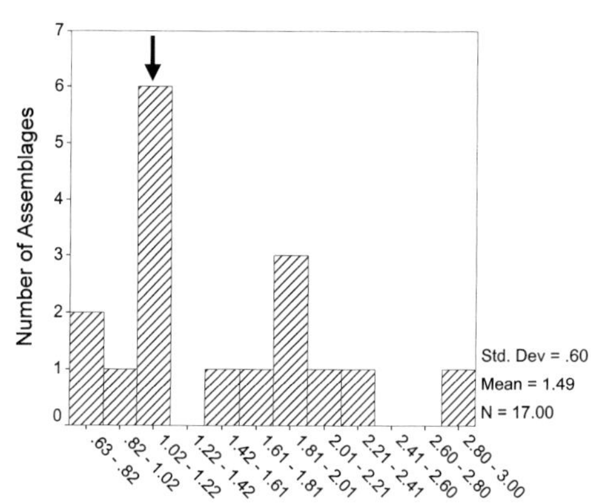

Figure 7.24 Boker Tachtit Level 4, non-cortical/cortical flake ratio value relative to a histogram of all assemblages.

Dorsal Surface Convexity
The analysis of the attributes serving as the proxies for the steps within the blank production domain are given in Tables 7.36-7.37. The ridge system used in the Boker Tachtit 4 assemblage shows a propensity for high laminarity (length/ width mean of 2.47), thicker products than Level 2 but still moderately thin relative to width (width/thickness mean of 3.86), parallel and convergent ridges (i.e., high frequency of parallel and convergent lateral edges) and straight profiles.

7.4.4 Tool Kit Morphology
As with Levels 1 and 2, the small sample of artifacts available for attribute analysis in the unrefit collections at the Israeli Antiquity Authority was too small to include enough tools to analyze directly. From the publications, it is possible to calculate the length and width of the tool kit, both non-retouched points and retouched blanks, as this data could be calculated by measuring off of the scatter plot in figure 5.29 of Marks and Kaufman (1983:119). While distal terminations (called "tip shape") and profiles were described in Appendix B of Marks (1983), tables B-1 through B-4, these statistics are only categorized by blank form (flake, blade, core trimming element, and primary element) and not by application of retouch alone. Only the classification in figure 5.29 of Marks and Kaufman (1983:119) provides an identification of tools that might allow a description of the distal termination apart from the rest of the debitage for Level 4. Distal terminations were given as follows from the types indicated in this figure: unknown (burins and truncations), blunt (denticulates, notches, and scrapers), and pointed (retouched points, unretouched points, point machures, and pointed blades). When compared to the debitage, the tool kit contains relatively the same number of blades at discard (Table 7.38) and a high frequency of pointed distal ends (Table 7.39).

For the last two analytical steps in characterizing the tool kit morphology, Boker Tachtit Level 4 does not exhibit the use of a unique retouch type, as the use of Emireh retouch had stopped by Level 4, but it does possess a dominance of Upper Paleolithic tool types.

Intensity of Tool and Core Reduction
Given the abundant amounts of raw material in the form of very large nodules available to residents of the Negev desert, it is not surprising that there was little need to apply extensive retouch to blank cutting edges for the resharpening of tools in this assemblage. The retouch applied appears to shape the blanks to their tasks, rather than resharpen the cutting edges themselves, given the high frequency of burins, endscrapers, and retouched points. The tool vs. debitage ratio, one of the lowest in the study Table 7.40, Figure 7.23), confirms this view. The value of the tools vs. debitage ratio for Level 4 is calculated from counts of complete tools and debitage taken from Hietala (1983a:213, tab. 7-10), which indicates 1,380 complete pieces of which 158 are tools, leaving 1,222 for the debitage. This assemblage is thus very lightly retouched, so that the dimensional variables of the retouched tool kit should not be biased significantly by reduction.

The non-cortical to cortical blank ratio was taken from counts of the entire assemblage as published by Hietala (1983c:343-346). The counts for non-cortical blanks and cortical blanks were summed from each of the blank types and excavation areas as detailed in Hietala's tables B-1 through B-4. The resulting value of 1.14 falls with the most left grouping of assemblages (Figure 7.24), clearly reflecting the

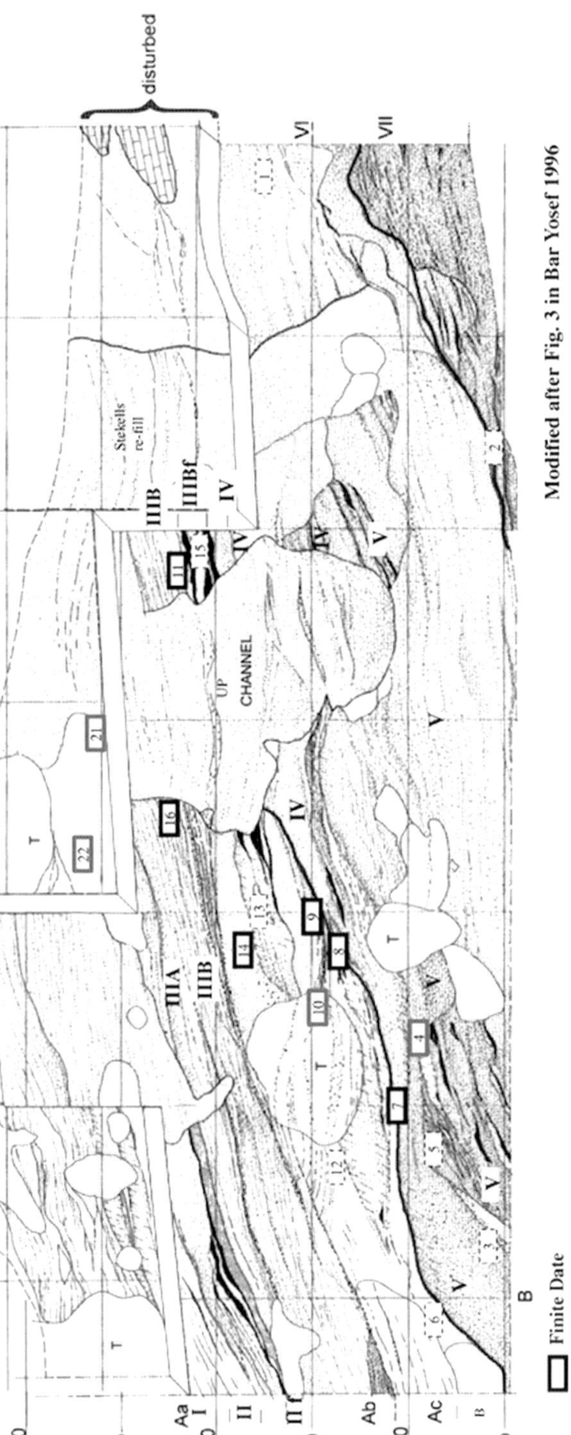

Figure 7.25 Kebara Cave stratigraphy of the south profile after the 1982–1990 excavations, showing Units I-VII. This figure, adapted and reprinted with permission from Rebollo et al. (2011:2427, fig. 2B) subdivides the radiocarbon dates from Bar-Yosef et al. (1996:301, fig. 3) into finite dates, infinite dates, dates near burrows.

Table 7.41 Kebara Cave Unit IV, percentage of intact cores (n=30) by number of bulbar negatives in each core sector (A, B, C, D) by surface (1, 2)

Number of bulbar negatives	Surf. 1 Sect. A	Surf. 1 Sect. B	Surf. 1 Sect. C	Surf. 1 Sect. D	Surf. 2 Sect. A	Surf. 2 Sect. B	Surf. 2 Sect. C	Surf. 2 Sect. D
0	23	60	33	70	70	83	63	73
1	23	40	40	27	23	10	33	20
2	33		20	3	3	7	3	7
3	17		3		3			
4	3		3					

Table 7.42 Kebara Cave Unit IV, core sector preparation: percent of intact cores (n=30) by number of scars in sector of origin (A, B, C, D) by surface (1, 2)

Number of Scars	Surf. 1 Sect. A	Surf. 1 Sect. B	Surf. 1 Sect. C	Surf. 1 Sect. D	Surf. 2 Sect. A	Surf. 2 Sect. B	Surf. 2 Sect. C	Surf. 2 Sect. D
0	13	73	27	83	17	67	13	53
1	10	20	30	13	40	20	43	20
2	23	7	17	3	23	3	23	17
3	13		17		17	7	17	7
4	27		10		3	0	3	3
5	7					3		
6	0							
7	3							
8	3							

role of the site as a raw material workshop for the working of cortical nodules into blanks. The refits corroborate the presence of the earliest portion of the core reduction sequence, as almost each core reduced on site began as a cortical nodule.

Although no data is available on the dimensions of core facets, since the cores were unavailable for study in the Israeli Antiquity Authority, there is no indication from the extant core illustrations that a significant number of the cores from the assemblage were reduced to an exhausted state (Marks and Kaufman 1983:115-117, figs. 5-25 through 5-27). The vast amount of blanks with adequate cutting edges, unutilized as far as currently discernable, would suggest, that the assemblage does not represent a population of flintknappers under a condition of extreme raw material scarcity requiring conservation. Together, these data demonstrate that almost all of the core reduction sequence related to blank production, save perhaps extreme core exhaustion, is represented in the assemblage.

7.5 Kebara Cave, Unit IV

7.5.1 Site Background

Unit IV is the oldest Upper Paleolithic assemblage known from Kebara Cave. Affiliated with the Ahmarian industrial tradition, this unit is dated to 42,500 ± 1,800 (Pta-5002) and 42,100 ± 2,100 (Pta-4987) and represents the northernmost example of this industrial type in the Levant (Bar-Yosef et al. 1996). Please see Rebollo et al. (2011) for the calibrated radiocarbon ages for this assemblage. The Oxford Radiocarbon Acceleration Unit date on charcoal from IVB-Q16b, 28,890 ± 400 (OxA-3978), seems too recent in comparison to the other dates and may indicate movement of charcoal from higher in the stratigraphy. Extensive analyses of the Early

Table 7.43 Kebara Cave Unit IV, platform types for complete and proximal fragments of flakes and tools

Platform type	Frequency	Percent
Plain	149	32.5
Cortical	16	3.5
Dihedral	13	2.8
Facetted	51	11.1
Chapeau	5	1.1
Scarred	84	18.3
Punctiform	18	3.9
Linear	25	5.5
Removed	6	1.3
Indeterminate	91	19.9
Total	458	100.0

Table 7.44 Kebara Cave Unit IV, descriptive statistics for continuous platform variables of complete and proximal fragments of flakes and tools

	N	Minimum	Maximum	Mean	Standard Deviation
Exterior Platform Angle (degrees)	289	47.00	144.00	87.98	14.87
Platform Thickness (mm)	291	0.35	14.49	4.31	2.56

Table 7.45 Kebara Cave Unit IV, dorsal scar pattern by blank length (mm) in quartiles cross-tabulation for complete flakes and tools (excluding flakes with greater than 60% cortex or fewer than three dorsal scars)

Dorsal Scar Pattern	>46	46-35	35-27	<27	Total
Unidirectional					
Count	29	28	31	37	125
% within Quartile	32.6%	33.3%	37.8%	46.8%	37.4%
Bidirectional					
Count	32	25	14	10	81
% within Quartile	36.0%	29.8%	17.1%	12.7%	24.3%
Crossed					
Count	7	7	22	20	56
% within Quartile	7.9%	8.3%	26.8%	25.3%	16.8%
Subcentripetal & Centripetal					
Count	6	13	8	6	33
% within Quartile	6.7%	15.5%	9.8%	7.6%	9.9%
Crested					
Count	6	5	4	3	18
% within Quartile	6.7%	6.0%	4.9%	3.8%	5.4%
Indeterminate					
Count	9	6	3	3	21
% within Quartile	10.1%	7.1%	3.7%	3.8%	6.3%
Total					
Count	89	84	82	79	334
% within Cortex	100.0%	100.0%	100.0%	100.0%	100.0%

Table 7.46 Kebara Cave Unit IV, dorsal scar pattern by blank length (mm) in quartiles cross-tabulation for complete flakes and tools (excluding pieces with a crested scar pattern, an indeterminate scar pattern, greater than 60% cortex, or fewer than 3 dorsal scars)

Dorsal Scar Pattern	>46	46-35	35-27	<27	Total
Unidirectional					
Count	29	28	31	37	125
% within Quartile	39.2%	38.4%	41.3%	50.7%	42.4%
Bidirectional					
Count	32	25	14	10	81
% within Quartile	43.2%	34.2%	18.7%	13.7%	27.5%
Crossed					
Count	7	7	22	20	56
% within Quartile	9.5%	9.6%	29.3%	27.4%	19.0%
Subcentripetal & Centripetal					
Count	6	13	8	6	33
% within Quartile	8.1%	17.8%	10.7%	8.2%	11.2%
Total					
Count	74	73	75	73	295
% within Cortex	100.0%	100.0%	100.0%	100.0%	100.0%

Ahmarian site of Nahal Nizzana XIII, western Negev, Israel, have been published by Davidson and Goring-Morris (2003) and Goring-Morris and Davidson (2006), providing a complementary view from the perspective of refitting analyses for an assemblage ascribed to the same industrial tradition as Kebara Unit IV. The refits from Nahal Nizzana XIII are mentioned here for comparative purposes only, not for use in interpreting Kebara Unit IV.

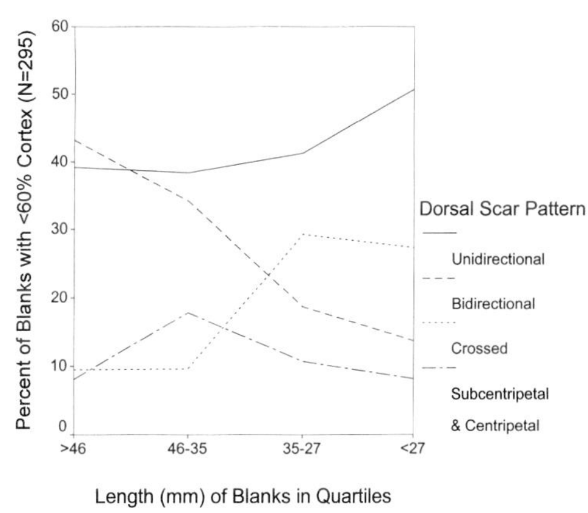

Figure 7.26 Kebara Cave Unit IV, dorsal scar pattern by blank length (mm) in quartiles cross-tabulation for complete flakes and tools (excluding pieces with a crested scar pattern, an indeterminate scar pattern, greater than 60% cortex, or fewer than three dorsal scars).

7.5.2 Sampling

This stratigraphic unit in Kebara Cave is relatively rich in debitage and tools. The following meter squares were included in this study: R18, R19, R20, R21, Q16, Q17, and Q18. These units were chosen for their proximity to the southern section of the 1982-1990 excavations, by which the frequent abnormalities of the cave stratigraphy were avoided on a square by square and unit by unit basis. The presence of the "UP Channel," an erosional gully filled with material believed to be derived from Units III and IV, in the sample squares was avoided in this way.

7.5.3 Blank Production

Core Modification

Core modification in Kebara Unit IV entails a longitudinal blade/flake core form, as evidenced in the identification of the narrow core-face as the debitage surface (Appendix Figures A40–A43). This identification is supported by Table 7.41, which demonstrates that Surface 1 strongly outweighs Surface 2 in number of bulbar negatives, indicating a preference for exploiting one surface rather than two. The longitudinal core orientation visible in the choice of Surface 1 is also corroborated by Table 7.42, which displays the dominance of the working of core sectors A and C on both core surfaces. Lateral débordant removals and frontal crests are used to control the convexity of the debitage surface to perpetuate core reduction as evidenced by the presence of these diagnostic pieces in the debitage sample. The higher frequency of cores with removals in sectors B and D on Surface 2 than Surface 1 is also indicative of the use of these sectors on Surface 2 for the control of lateral convexities.

Platform Maintenance

Platform maintenance includes a good amount of platform preparation, up to 33% of the debitage (Table 7.43), or 42% when indeterminates are removed. The high frequency of linear and punctiform platforms in this assemblage relative to Kebara Unit VI clearly shows the development through time of greater abrasion behind the platform. Measurements of external platform angles and platform thickness (Table 7.44) indicate tendencies around 88 degrees and 4 mm respectively.

Direction of Core Exploitation

The analysis of the direction of core exploitation is presented in Tables 7.45–7.46. Figure 7.26 demonstrates that early core exploitation in Kebara Unit IV is dominated by both bidirectional and unidirectional blanks while late exploitation is dominated by unidirectional blanks. Crossed blanks consistently increase in frequency through the four quartiles but within the smallest quartile crossed blanks do not reach a value within 10% of that displayed by unidirectional blanks. Of the 30 identifiable intact cores studied in this sample of the assemblage, bidirectional cores and unidirectional cores each represent 40% of the assemblage. Given this proportion of core types at discard and the almost equal representation of bidirectional and unidirectional dorsal scar patterns in the first blank length quartile, it is likely that two directionalities of core reduction were pursued independently in this assemblage, at least at the beginning of core exploitation. The gradual replacement of bidirectional scar patterns by unidirectional scar patterns is, however, suggestive of a gradual shift within the bidirectional cores from an alternating bidirectional reduction to a sequential bidirectional reduction. This interpretation is supported by two additional pieces of evidence. First, sectors A and C on Surface 1 both show more scars as evidence of exploitation from that sector than the other sectors of the intact cores (Table 7.42). Second, sector A far outnumbers sector B in possessing bulbar negatives (beyond the definitional requirement to be classified as sector A), indicating that at discard this one sector was being exploited in a sequential fashion.

Dorsal Surface Convexity

The analysis of the attributes serving as the proxies for the surface convexity domain are given in Tables 7.47–7.48. The ridge system used in the

Table 7.47 Kebara Cave Unit IV, frequencies of lateral edge, profile, and cross-section types for complete flakes and tools

Type	Frequency	Percent
Lateral edge		
Parallel	230	52.3
Convergent	96	21.8
Expanding	63	14.3
Ovoid	43	9.8
Indeterminate	8	1.8
Total	440	100.0
Profile		
Straight	253	57.5
Curved or Concord	111	25.2
Twisted	73	16.6
Indeterminate	3	.7
Total	440	100.0
Cross-section		
Triangular	194	44.1
Trapezoidal	203	46.1
Other	37	8.4
Indeterminate	6	1.4
Total	440	100.0

Table 7.48 Kebara Cave Unit IV, descriptive statistics for blank dimensions for complete flakes and tools

Blank Dimension	N	Range	Mean	Standard Deviation
Length	440	78.11	38.01	13.22
Width	440	67.63	20.88	9.92
Thickness	440	19.31	5.49	3.10
Length/Width Ratio	440	6.64	2.17	1.16
Width/Thickness Ratio	440	14.16	4.35	2.11

Table 7.49 Kebara Cave Unit IV, descriptive statistics for dimensions of complete tools

Blank Dimension	N	Range	Mean	Standard Deviation
Length/Width Ratio	52	4.03	2.23	1.12
Width/Thickness Ratio	52	9.46	4.04	1.80

Kebara IV assemblage shows a propensity for high laminarity (length/width mean of 2.17), thin products relative to width (width/thickness mean of 4.35) and longitudinal ridges (i.e., high frequency of parallel lateral edges).

7.5.4 Tool Kit Morphology

The first five characterizations of the tool kit morphology refer to the dorsal surface convexity variables which give form to the cutting edge of the retouched blanks in the assemblage. In the case of Kebara Unit IV, the tool kit is similar to the debitage blanks for all variables. No unique type of retouch is evidenced in the tool types in this assemblage and Upper Paleolithic tool types dominate the tool kit (Bar-Yosef et al. 1996).

Intensity of Tool and Core Reduction

Given Kebara Cave's location in the raw material rich Mt. Carmel area of the Levantine coastal

Table 7.50 Kebara Cave Unit IV, frequencies of lateral edge, distal terminus, and profile types for complete tools

Type	Frequency	Percent
Lateral edge		
Parallel	29	55.8
Convergent	12	23.1
Expanding	3	5.8
Ovoid	7	13.5
Indeterminate	1	1.9
Total	52	100.0
Distal Terminus		
Blunt	16	30.8
Pointed	16	30.8
Hinged	5	9.6
Removed	9	17.3
Broken	4	7.7
Indeterminate	2	3.8
Total	52	100.0
Profile		
Straight	25	48.1
Curved or Concord	12	23.1
Twisted	14	26.9
Indeterminate	1	1.9
Total	52	100.0

Table 7.51 Kebara Cave Unit IV, intensity of core and tool reduction

Variables	Numerator	Denominator	Result
Tools/Flake Ratio	Tools: 52	Flakes: 368	0.14
Non-cortical Debitage/ Cortical Debitage Ratio	Non-cortical Debitage: 290	Cortical Debitage: 130	2.23
Henry's Dimensional Analysis	NA	NA	Core facets overlap with the low end & middle of the blank distribution: moderate core exhaustion

plain, it is not surprising that retouch was applied to blanks in Unit IV less frequently than other assemblages, as the tool to flake ratio falls below the mean of all assemblages (Table 7.51, Figure 7.27). While there is more retouch than seen in sites serving as raw material workshops, it is still less frequent than seen in five other assemblages. This is not surprising given the context of the site as a sheltered cave suitable for longer term use than a raw material workshop. Thus, it is unlikely that the application of retouch has excessively biased the dimensional ratios of the tool kit.

From the ratio of non-cortical to cortical blanks falling clearly above the mean of all assemblages (Figure 7.28), it is clear that nodules were worked and partially decortified before being curated on-site. The size of the raw material available at Kebara Cave is not excessively large and thus a higher frequency of non-cortical blanks to cortical blanks is not likely a result of a higher internal volume of the initial core compared to other sites. As seen in Figure 7.29, only two of the last maximum core facets (out of 30) fall at or below the 2.0 cm size cut off for the debitage analysis. The remaining core facets overlap the middle to lower end of the length and width distribution of the debitage, indicating only a moderate degree of core exhaustion for this

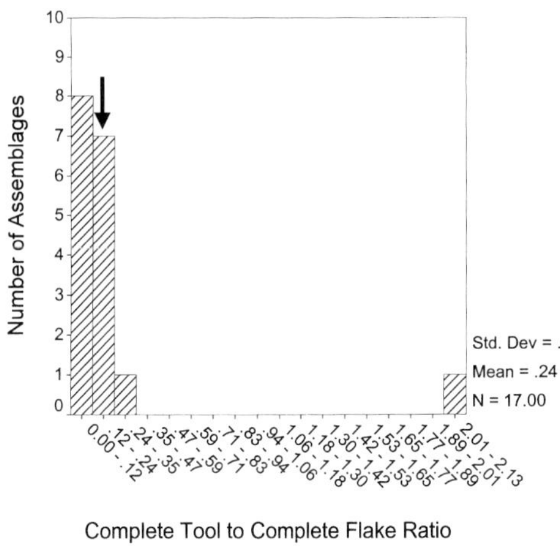

Figure 7.27 Kebara Cave Unit IV, tool/flake ratio value relative to a histogram of all assemblages.

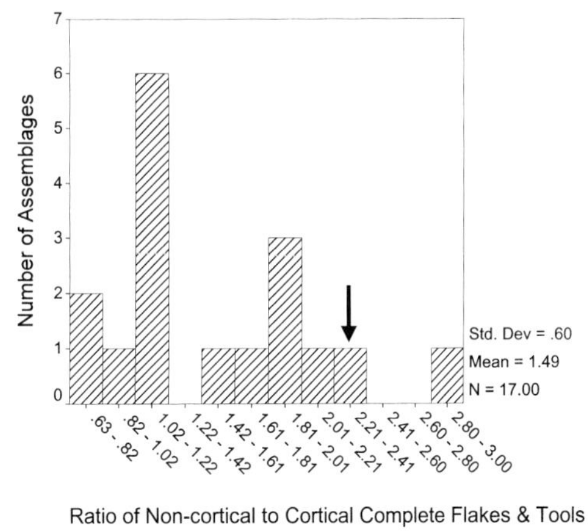

Figure 7.28 Kebara Cave Unit IV, non cortical/cortical flake ratio value relative to a histogram of all assemblages.

assemblage. Taken together, these data suggest that the assemblage captures the middle and end of the core reduction sequence of these hominins, but perhaps not the earliest portion of the sequence in which nodules are decortified. The assemblage is thus fairly comparable with other assemblages for both continuous and discontinuous variables, although raw material workshops may not overlap as much as other types of assemblages.

7.6 Kebara Cave, Unit III

7.6.1 Site Background

Unit III is the second oldest Upper Paleolithic assemblage of the 1980s excavations at Kebara Cave. Also affiliated with the Ahmarian Industrial tradition, this unit is dated between 35,600 + 1,600 (OxA-1567) and 43,500 ± 2,200 (OxA-3976) (Bar-Yosef et al. 1996). Please see Rebollo et al. (2011) for the calbrated radiocarbon ages for this assemblage.

7.6.2 Sampling

This stratigraphic unit in Kebara Cave is very rich in debitage, although less rich in tools. For this reason, only three meter squares were necessary to be chosen for the purposes of the study of debitage. These squares include R16, Q15, and Q18. To enrich the tool sample from this level, the tool collections from the following additional squares were used: Q13, Q14, Q16, Q17, R17, and R18. These units were chosen for their proximity to the southern section of the 1982–1990 excavations, by which the frequent abnormalities of the cave stratigraphy were avoided on a square by square and unit by unit basis. The presence of the "UP Channel," an erosional gully filled with material believed to be derived from Units III and IV, in the sample squares was avoided in this way.

7.6.3 Blank Production

Core Modification

Core modification in Kebara Unit III entails a

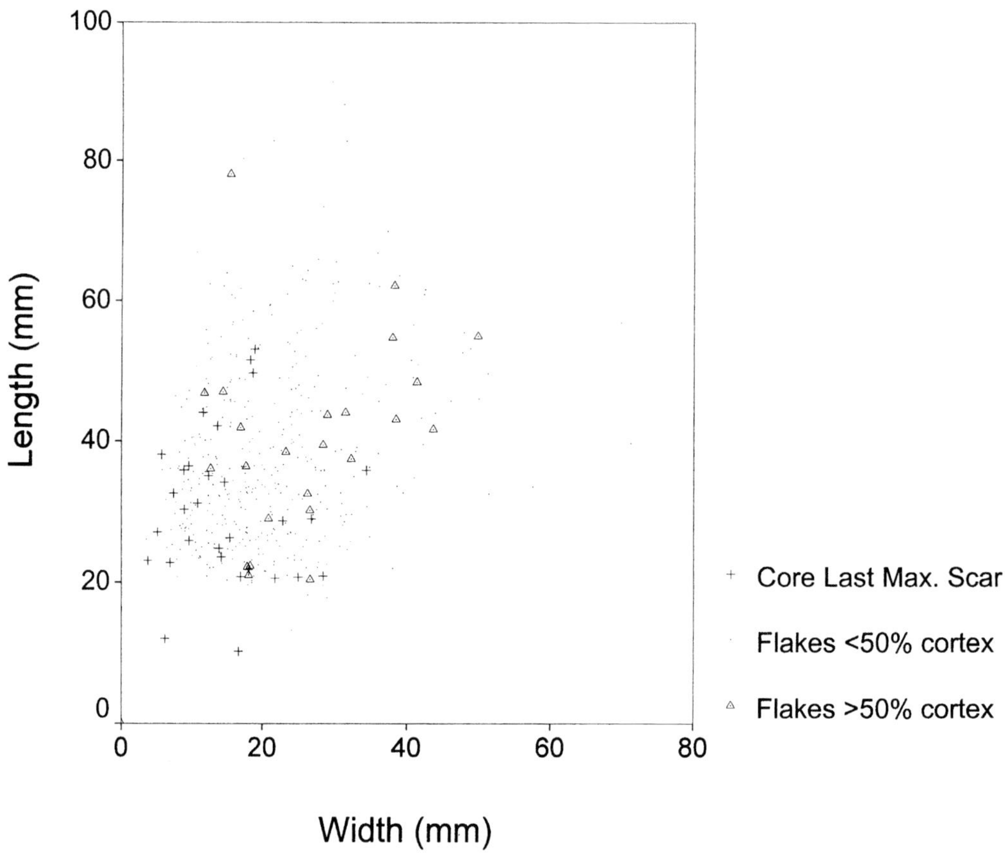

Figure 7.29 Kebara Cave Unit IV according to Henry's core reduction intensity measure: length and width of last maximum core facets, highly cortical blanks, and blanks with little cortex.

Table 7.52 Kebara Cave Unit III, percentage of intact cores (n=64) by number of bulbar negatives in each core sector (A, B, C, D) by surface (1, 2)

Number of bulbar negatives	Surf. 1 Sect. A	Surf. 1 Sect. B	Surf. 1 Sect. C	Surf. 1 Sect. D	Surf. 2 Sect. A	Surf. 2 Sect. B	Surf. 2 Sect. C	Surf. 2 Sect. D
0	2	48	39	58	75	81	77	84
1	28	38	25	28	17	9	16	11
2	44	14	25	9	6	8	5	2
3	23		11	5	2	2	3	2
4	3							0
5								2

Table 7.53 Kebara Cave Unit III, core sector preparation: percent of intact cores (n=64) by number of scars in sector of origin (A, B, C, D) by surface (1, 2)

Number of Scars	Surf. 1 Sect. A	Surf. 1 Sect. B	Surf. 1 Sect. C	Surf. 1 Sect. D	Surf. 2 Sect. A	Surf. 2 Sect. B	Surf. 2 Sect. C	Surf. 2 Sect. D
0	2	94	38	94	17	78	41	78
1	17	2	19	2	38	9	27	6
2	22	5	20	5	28	6	25	11
3	34		16		11	5	6	2
4	17		6		5	2	0	2
5	6		2		2		2	0
6	0							0
7	2							2

Table 7.54 Kebara Cave Unit III, platform types for complete and proximal fragments of flakes and tools

Platform Type	Frequency	Percent
Plain	183	27.8
Cortical	37	5.6
Dihedral	17	2.6
Facetted	42	6.4
Chapeau	4	0.6
Scarred	83	12.6
Punctiform	20	3.0
Linear	53	8.0
Removed	21	3.2
Indeterminate	199	30.2
Total	659	100.0

Table 7.55 Kebara Cave Unit III, descriptive statistics for continuous platform variables of complete and proximal fragments of flakes and tools

	N	Minimum	Maximum	Mean	Standard Deviation
Exterior Platform Angle (degrees)	355	56.00	147.00	91.96	15.72
Platform Thickness (mm)	357	0.61	25.76	3.91	2.57

longitudinal blade core form, as evidenced in the identification of the narrow core-face as the debitage surface (Appendix Figures A44–A45). This identification is supported by Table 7.52, which demonstrates that Surface 1 strongly outweighs Surface 2 in number of bulbar negatives, indicating a preference for exploiting one surface rather than two. The longitudinal core orientation visible in the choice of Surface 1 is also corroborated by Table 7.53, which displays the dominance of the working of core sectors A and C on both core surfaces. The higher frequency of working sectors B and D on Surface 2 compared to Surface 1, as seen in Table 7.53, corresponds to the periodic use of lateral/rear crests to manage lateral core convexities. Frontal crests and core tablets are used to control the convexity of the debitage surface and platform surface to perpetuate core reduction.

Platform Maintenance

Platform maintenance includes a low frequency of platform preparation, only up to 22% of

Table 7.56 Kebara Cave Unit III, dorsal scar pattern by blank length (mm) in quartiles cross-tabulation for complete flakes and tools (excluding flakes with greater than 60% cortex or fewer than three dorsal scars)

Dorsal Scar Pattern	>51	51-41	41-31	<31	Total
Unidirectional					
Count	28	34	37	39	138
% within Quartile	24.6%	28.6%	35.9%	34.5%	30.7%
Bidirectional					
Count	44	36	30	25	135
% within Quartile	38.6%	30.3%	29.1%	22.1%	30.1%
Crossed					
Count	10	12	11	17	50
% within Quartile	8.8%	10.1%	10.7%	15.0%	11.1%
Subcentripetal & Centripetal					
Count	7	9	9	9	34
% within Quartile	6.1%	7.6%	8.7%	8.0%	7.6%
Crested					
Count	13	13	2	4	32
% within Quartile	11.4%	10.9%	1.9%	3.5%	7.1%
Indeterminate					
Count	12	15	14	19	60
% within Quartile	10.5%	12.6%	13.6%	16.8%	13.4%
Total					
Count	114	119	103	113	449
% within Cortex	100.0%	100.0%	100.0%	100.0%	100.0%

the debitage (Table 7.54), or 33% when indeterminates are removed. Measurements of external platform angles and platform thickness (Table 7.55) indicate tendencies around 92 degrees and 4 mm respectively. The mean EPA of 92 degrees reflects the propensity in this assemblage to bevel or abrade off overhangs behind the platform as a means of strengthening platforms and relatively increasing flake length. This tendency is also evident in the frequencies of linear and punctiform platforms, which together represent 16.7% of the identifiable platforms.

Direction of Core Exploitation

The analysis of the direction of core exploitation is presented in Tables 7.56-7.57. Figure 7.30 demonstrates a strong shift from bidirectional scar patterns in the longest length quartile, i.e., at the beginning of core exploitation, to a unidirectional pattern in the smallest length quartile, i.e., at the end of core exploitation. The continuous gradient in the frequency of these two dorsal patterns across the four quartiles, together with the relative stability of the frequencies of crossed and sub/centripetal patterns, is a clear indication of a gradual change in directionality through core exploitation. This is supported by the indications that both sectors A and C on Surface 1 of the extant cores show extensive working from those platform edges (Table 7.53) but that sector A shows significantly more bulbar negatives than sector C at discard (Table 7.52). A gradual transition from the use of both platforms (alternating bidirectional) to more use of one platform at the

Table 7.57 Kebara Cave Unit III, dorsal scar pattern by blank length (mm) in quartiles cross-tabulation for complete flakes and tools (excluding pieces with a crested scar pattern, an indeterminate scar pattern, greater than 60% cortex, or fewer than three dorsal scars)

Dorsal Scar Pattern	>51	51-41	41-31	<31	Total
Unidirectional					
Count	28	34	37	39	138
% within Quartile	31.5%	37.4%	42.5%	43.3%	38.7%
Bidirectional					
Count	44	36	30	25	135
% within Quartile	49.4%	39.6%	34.5%	27.8%	37.8%
Crossed					
Count	10	12	11	17	50
% within Quartile	11.2%	13.2%	12.6%	18.9%	14.0%
Subcentripetal & Centripetal					
Count	7	9	9	9	34
% within Quartile	7.9%	9.9%	10.3%	10.0%	9.5%
Total					
Count	89	91	87	90	357
% within Cortex	100.0%	100.0%	100.0%	100.0%	100.0%

end of reduction (changing to sequential bidirectional reduction) is thus likely in this assemblage. Of the 64 intact cores studied in this sample, 60.9% of them were configured at the time of discard as opposed platform cores and 35.9% of them as unidirectional cores. Without refit data to indicate the number of removals per rotation event, it is impossible to prove that each bidirectional core was transformed into a unidirectionally exploited core during reduction, as either alternating to sequential bidirectional or straight unidirectional reduction. This explanation, however, is more parsimonious than the argument that only small nodules were exploited unidirectionally, especially given the gradual replacement of bidirectional dorsal scars by unidirectional scars from the first to the fourth length quartile.

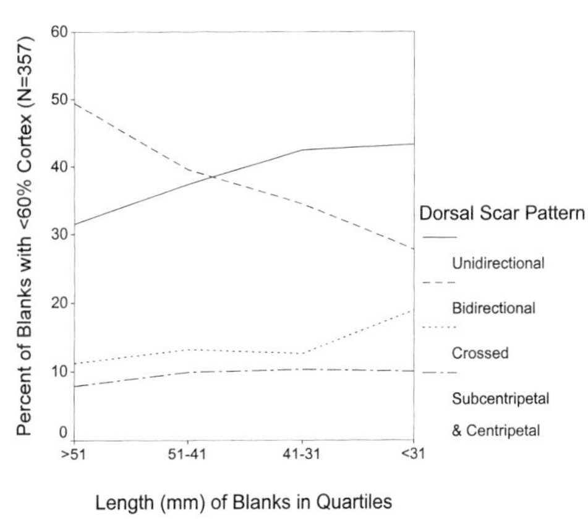

Figure 7.30 Kebara Cave Unit III, dorsal scar pattern by blank length (mm) in quartiles cross-tabulation for complete flakes and tools (excluding pieces with a crested scar pattern, an indeterminate scar pattern, greater than 60% cortex, or fewer than three dorsal scars).

Dorsal Surface Convexity

The analysis of the attributes serving as the proxies for the steps within the blank production domain are given in Tables 7.58-7.59. The

Table 7.58 Kebara Cave Unit III, frequencies of lateral edge, profile, and cross-section types for complete flakes and tools

Type	Frequency	Percent
Lateral edge		
Parallel	365	55.4
Convergent	137	20.8
Expanding	87	13.2
Ovoid	44	6.7
Indeterminate	26	3.9
Total	659	100.0
Profile		
Straight	309	46.9
Curved or Concord	203	30.8
Twisted	132	20.0
Indeterminate	15	2.3
Total	659	100.0
Cross-section		
Triangular	272	41.3
Trapezoidal	299	45.4
Other	71	10.8
Indeterminate	17	2.6
Total	659	100.0

Table 7.59 Kebara Cave Unit III, descriptive statistics for blank dimensions for complete flakes and tools

Blank Dimension	N	Range	Mean	Standard Deviation
Length	659	84.80	41.58	13.74
Width	659	47.49	19.35	8.28
Thickness	659	50.77	5.76	3.83
Length/Width Ratio	659	8.20	2.51	1.24
Width/Thickness Ratio	659	16.51	3.87	1.79

Table 7.60 Kebara Cave Unit III, descriptive statistics for dimensions of complete tools

Blank Dimension	N	Range	Mean	Standard Deviation
Length/Width Ratio	133	7.69	3.09	1.19
Width/Thickness Ratio	133	6.80	3.49	1.29

ridge system used in the Kebara III assemblage shows a propensity for very high laminarity (length/ width mean of 2.51), thick products relative to their width (width/thickness mean of 3.87), a dominance of straight profiles with a substantial minority of curved profiles, and strongly longitudinal ridges (i.e., high frequency of parallel lateral edges).

7.6.4 Tool Kit Morphology

The tool kit of Kebara Unit III is highly bladey and relatively thick in cross-section (width/thickness ratio). The tool kit is also highly pointed, parallel sided, and more curved in profile than the debitage blanks of the assemblage. No unique type of retouch is evidenced in the tool types in this assemblage and Upper Paleolithic tool types dominate the tool kit (Bar-Yosef et al. 1996).

Intensity of Tool and Core Reduction

As only squares R16, Q15, and Q18 were studied for debitage counts of complete unretouched blanks, only those tools found in these three squares were counted for calculating the

Table 7.61 Kebara Cave Unit III, frequencies of lateral edge, distal terminus, and profile types for complete tools

Type	Frequency	Percent
Lateral edge		
Parallel	84	63.2
Convergent	27	20.3
Expanding	6	4.5
Ovoid	7	5.3
Indeterminate	9	6.8
Total	133	100.0
Distal Terminus		
Blunt	21	15.8
Pointed	67	50.4
Hinged	2	1.5
Overshot	2	1.5
Removed	35	26.3
Broken	1	.8
Indeterminate	5	3.8
Total	133	100.0
Profile		
Straight	43	32.3
Curved or Concord	50	37.6
Twisted	36	27.1
Indeterminate	4	3.0
Total	133	100.0

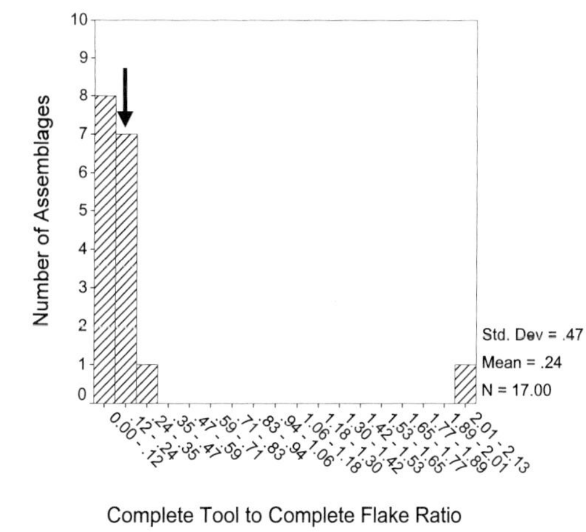

Figure 7.31 Kebara Cave Unit III, tool/flake ratio value relative to a histogram of all assemblages.

Table 7.62 Kebara Cave Unit III, intensity of core and tool reduction

Variables	Numerator	Denominator	Result
Tools/Flake Ratio	Tools: 82	Flakes: 483	0.17
Non-cortical Debitage/ Cortical Debitage Ratio	Non-cortical Debitage: 406	Cortical Debitage: 210	1.93
Henry's Dimensional Analysis	NA	NA	Core facets overlap with the low end & middle of the blank distribution: significant core exhaustion

ratio of complete tools to blanks for Table 7.62. From this data (Figure 7.31), it is apparent that Kebara Unit III is more retouched than many other sites, particularly the raw material workshops, but it possesses far fewer retouched blanks than sites such as Vedrovice V or Kůlna Cave Layer 7a. This is not surprising given the context of the site as a sheltered cave suitable for longer term use than a raw material workshop. Thus, it is unlikely that the application of retouch has excessively biased the dimensional ratios of the tool kit.

From the ratio of non-cortical to cortical blanks falling above the mean and in the middle of the central grouping of assemblages (Figure 7.32), it is clear that nodules were worked and partially decortified before being curated on-site. The size of the raw material

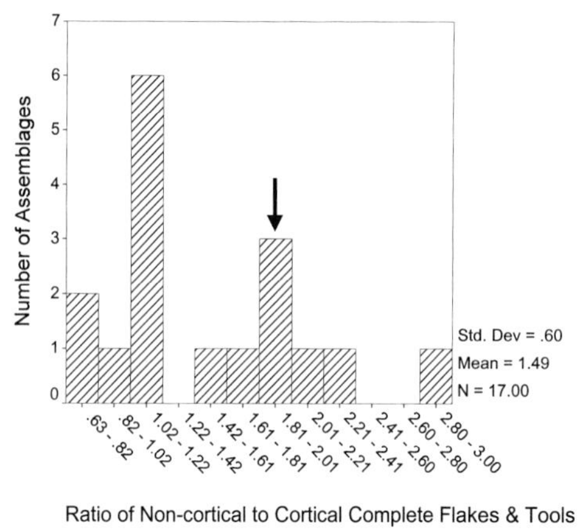

Figure 7.32 Kebara Cave Unit III, non-cortical/cortical flake ratio value relative to a histogram of all assemblages.

available at Kebara Cave is not excessively large and thus a higher frequency of non-cortical blanks to cortical blanks is not likely a result of a higher internal volume of the initial core compared to other sites. As seen in Figure 7.33, as many as ten of the last maximum core facets (out of 67) fall at or below the 2.0 cm size cut off for the debitage analysis. The remaining core facets overlap the middle to lower end of the length and width distribution of the debitage (biased against wide products), indicating a moderate to significant degree of core exhaustion for this assemblage. Taken together, these data suggest that the assemblage captures the middle and end of the core reduction sequence of these hominids, but perhaps not the earliest portion of the sequence in which nodules are decortified. The assemblage is thus fairly comparable with other assemblages for both continuous and discontinuous variables, although raw material workshops may not overlap as much as other types of assemblages.

7.7 Kebara Cave, Unit II

7.7.1 Site Background

Unit II is the first Levantine Aurignacian assemblage at Kebara Cave. With two dates at 36,000 ± 1,600 (OxA-1230) and 34,300 ± 1,100 (Gif-TAN-90028), this assemblage is one of the oldest Aurignacian sites in the Levant (Bar-Yosef et al. 1996).

7.7.2 Sampling

This stratigraphic unit in Kebara Cave is very rich in both debitage and tools. For this reason, only four meter squares were chosen for inclusion in the present study for the analysis of debitage: R16, R17, Q14, and Q15. For the analysis of the tool kit, the retouched tools in squares Q16, Q17, R14, R15, and R18 were also included. These specific units were chosen for their proximity to the southern section of the 1982–1990 excavations, by which the frequent abnormalities of the cave stratigraphy were avoided on a square by square and unit by unit basis. The presence of the "UP Channel," an erosional gully filled with material believed to be derived from Units III and IV, in the sample squares was avoided in this way.

7.7.3 Blank Production

Core Modification

Core modification in Kebara Unit II entails a longitudinal blade core form, as evidenced in the identification of the narrow core-face as the debitage surface. This identification is supported by Table 7.63, which demonstrates that Surface 1 strongly outweighs Surface 2 in number of bulbar negatives, indicating a preference for exploiting one surface rather than two. The longitudinal core orientation visible in the choice of Surface 1 is also corroborated by Table 7.64, which displays the dominance of the working of core sector A on both core surfaces, with

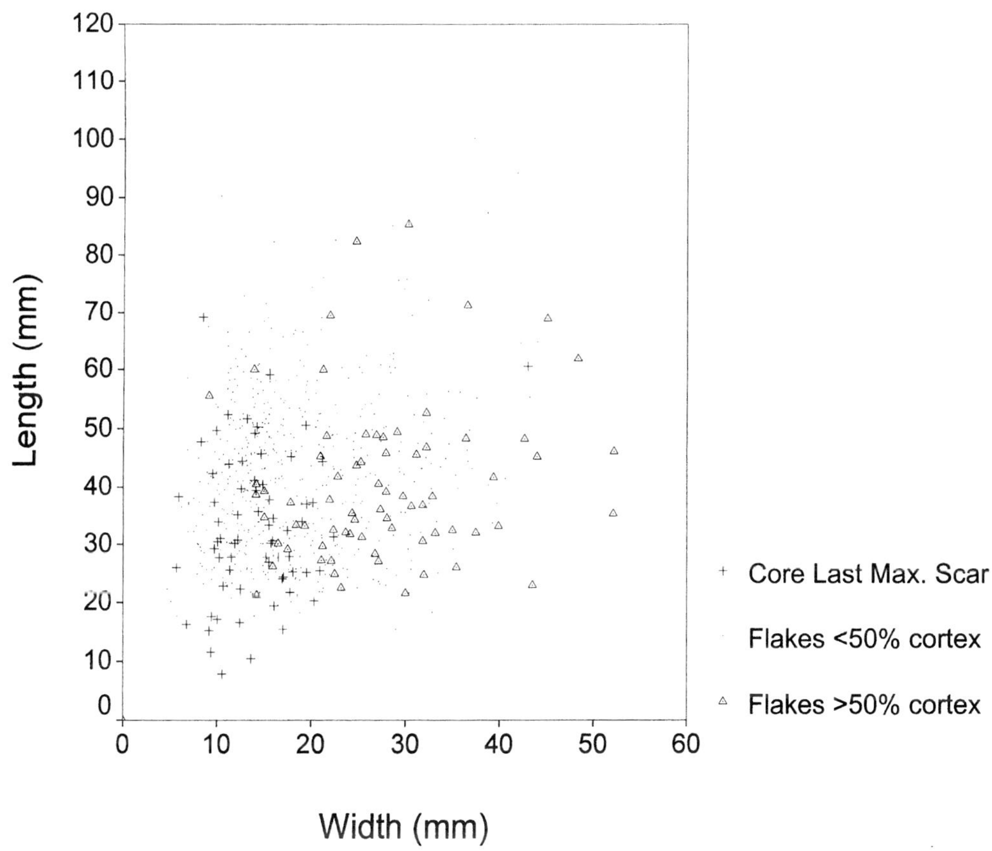

Figure 7.33 Kebara Cave Unit III according to Henry's core reduction intensity measure: length and width of last maximum core facets, highly cortical blanks, and blanks with little cortex.

Table 7.63 Kebara Cave Unit II, percentage of intact cores (n=19) by number of bulbar negatives in each core sector (A, B, C, D) by surface (1, 2)

Number of bulbar negatives	Surf. 1 Sect. A	Surf. 1 Sect. B	Surf. 1 Sect. C	Surf. 1 Sect. D	Surf. 2 Sect. A	Surf. 2 Sect. B	Surf. 2 Sect. C	Surf. 2 Sect. D
0	11	47	68	47	84	84	74	84
1	32	32	21	37	11	0	21	0
2	53	16	5	16	5	11	5	11
3	5	5	5			5		5

Table 7.64 Kebara Cave Unit II, core sector preparation: percent of intact cores (n=19) by number of scars in sector of origin (A, B, C, D) by surface (1, 2)

Number of Scars	Surf. 1 Sect. A	Surf. 1 Sect. B	Surf. 1 Sect. C	Surf. 1 Sect. D	Surf. 2 Sect. A	Surf. 2 Sect. B	Surf. 2 Sect. C	Surf. 2 Sect. D
0	11	89	63	89	42	74	63	63
1	21	5	11	5	37	5	32	26
2	32	5	0	0	5	11	5	5
3	26		21	5	16	5		0
4	11		5			0		0
5						5		

5

Table 7.65 Kebara Cave Unit II, platform types for complete & proximal fragments of flakes & tools

Platform type	Frequency	Percent
Plain	143	32.4
Cortical	22	5.0
Dihedral	9	2.0
Facetted	20	4.5
Chapeau	2	.5
Scarred	53	12.0
Punctiform	20	4.5
Linear	62	14.0
Removed	12	2.7
Indeterminate	99	22.4
Total	442	100.0

Table 7.66 Kebara Cave Unit II, descriptive statistics for continuous platform variables of complete and proximal fragments of flakes and tools

	N	Minimum	Maximum	Mean	Standaard Deviation
Exterior Platform Angle (degrees)	255	49.00	138.00	85.71	14.38
Platform Thickness (mm)	259	0.43	14.05	3.85	2.78

only minor working from sector C. Frontal crests and core tablets, present in the debitage sample, are used to control the convexity of the debitage surface and platform surface to perpetuate core reduction. The moderate frequency of bulbar negatives in sectors B and D on both surfaces indicates the use of lateral crests to control the width of the debitage surface.

Platform Maintenance

Platform maintenance includes a very low frequency of platform preparation on top of the platform (Table 7.65). Preparation behind the platform, in the form of grinding which produces linear and punctiform platforms (together representing 24.8%), is much higher in this assemblage than in Kebara Units IV (11.9%) or III (16.7%) when indeterminates are removed from the calculations. Measurements of external platform angles and platform thickness (Table 7.66) indicate tendencies around 86 degrees, lower than in Kebara Unit III, and 4 mm respectively.

Table 7.67 Kebara Cave Unit II, dorsal scar pattern by blank length (mm) in quartiles cross-tabulation for complete flakes and tools (excluding flakes with greater than 60% cortex or fewer than three dorsal scars)

Dorsal Scar Pattern	>38	38-29	29-24	<24	Total
Unidirectional					
Count	29	28	36	38	131
% within Quartile	37.7%	31.8%	43.9%	52.1%	40.9%
Bidirectional					
Count	17	15	17	12	61
% within Quartile	22.1%	17.0%	20.7%	16.4%	19.1%
Crossed					
Count	13	16	5	12	46
% within Quartile	16.9%	18.2%	6.1%	16.4%	14.4%
Subcentripetal & Centripetal					
Count	10	7	11	5	33
% within Quartile	13.0%	8.0%	13.4%	6.8%	10.3%
Crested					
Count	6	13	3	4	26
% within Quartile	7.8%	14.8%	3.7%	5.5%	8.1%
Indeterminate					
Count	2	9	10	2	23
% within Quartile	2.6%	10.2%	12.2%	2.7%	7.2%
Total					
Count	77	88	82	73	320
% within Cortex	100.0%	100.0%	100.0%	100.0%	100.0%

Direction of Core Exploitation

The analysis of the direction of core exploitation is presented in Tables 7.67-7.68. Figure 7.34 demonstrates that crossed scar patterns fluctuate between quartiles but bidirectional and sub/centripetal patterns do not change much between quartiles. The frequency of unidirectional blanks begins and ends as the dominant reduction pattern, with a slight but continuous increase across all four length quartiles.

Dorsal Surface Convexity

The analysis of the attributes serving as the proxies for the surface convexity domain are given in Tables 7.69-7.70. The ridge system used in the Kebara II assemblage shows a propensity for high laminarity (length/width mean of 2.20), thin products relative to their width (width/thickness mean of 4.00), a strong minority of curved profiles among the dominance of straight profiles, and strong longitudinal ridges (i.e., high frequency of parallel lateral edges).

7.7.4 Tool Kit Morphology

The first five characterizations of the tool kit morphology refer to the dorsal surface convexity variables which give form to the cutting edge of the retouched blanks in the assemblage. The tool kit of Kebara Unit II possesses more elongated but thicker cross-sectioned blades with more curved profiles than the debitage. Statistically, only the differences in laminarity and width/thickness ratios are significant. Carinated tools represent the use of a unique type of retouch in this assemblage.

Table 7.68 Kebara Cave Unit II, dorsal scar pattern by blank length (mm) in quartiles cross-tabulation for complete flakes and tools (excluding pieces with a crested scar pattern, an indeterminate scar pattern, greater than 60% cortex, or fewer than three dorsal scars)

Dorsal Scar Pattern	>38	38-29	29-24	<24	Total
Unidirectional					
Count	29	28	36	38	131
% within Quartile	42.0%	42.4%	52.2%	56.7%	48.3%
Bidirectional					
Count	17	15	17	12	61
% within Quartile	24.6%	22.7%	24.6%	17.9%	22.5%
Crossed					
Count	13	16	5	12	46
% within Quartile	18.8%	24.2%	7.2%	17.9%	17.0%
Subcentripetal & Centripetal					
Count	10	7	11	5	33
% within Quartile	14.5%	10.6%	15.9%	7.5%	12.2%
Total					
Count	69	66	69	67	271
% within Cortex	100.0%	100.0%	100.0%	100.0%	100.0%

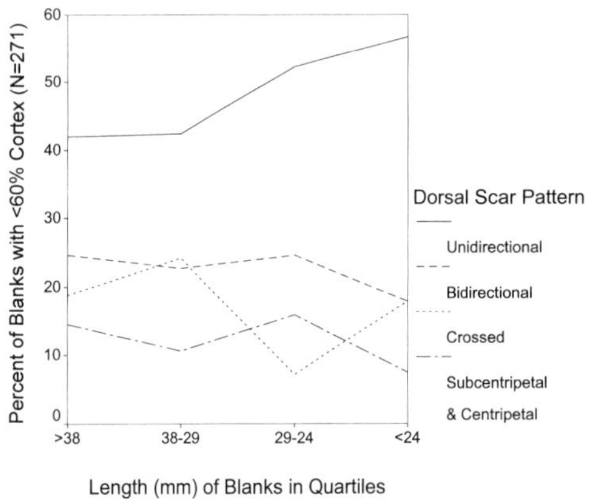

Figure 7.34 Kebara Cave Unit II, dorsal scar pattern by blank length (mm) in quartiles cross-tabulation for complete flakes and tools (excluding pieces with a crested scar pattern, an indeterminate scar pattern, greater than 60% cortex, or fewer than three dorsal scars).

The tool kit is exclusively Upper Paleolithic (Bar-Yosef et al. 1996).

Intensity of Tool and Core Reduction

Only complete retouched tools and complete flakes in squares R16, R17, Q14, and Q15 were included in the calculation of the complete tool to complete flake ratio (Table 7.73). As this value falls with the farthest left group of assemblages, it is clear that relatively little retouch was applied to the blanks in this assemblage (Figure 7.35). While the shelter offered by Kebara Cave gives it a different context than a raw material workshop, the low tool to flake ratio is somewhat surprising given the values for Units III and IV. None of these values, however, are indicative of the application of enough retouch to bias the comparison of the dimensional variables of these tool kits.

From the ratio of non-cortical to cortical blanks falling above the mean and in the middle of the central grouping of assemblages (Figure

Table 7.69 Kebara Cave Unit II, frequencies of lateral edge, profile, and cross-section types for complete flakes and tools

Type	Frequency	Percent
Lateral edge		
Parallel	229	51.8
Convergent	86	19.5
Expanding	60	13.6
Ovoid	40	9.0
Indeterminate	27	6.1
Total	442	100.0
Profile		
Straight	209	47.3
Curved or Concord	136	30.8
Twisted	84	19.0
Indeterminate	13	2.9
Total	442	100.0
Cross-section		
Triangular	181	41.0
Trapezoidal	224	50.7
Other	26	5.9
Indeterminate	11	2.5
Total	442	100.0

Table 7.70 Kebara Cave Unit II, descriptive statistics for blank dimensions for complete flakes and tools

Blank Dimension	N	Range	Mean	Standard Deviation
Length	442	64.32	32.64	10.86
Width	442	58.57	18.40	9.31
Thickness	442	31.22	5.61	4.31
Length/Width Ratio	442	6.45	2.21	1.19
Width/Thickness Ratio	442	11.89	4.00	1.96

Table 7.71 Kebara Cave Unit II, descriptive statistics for dimensions of complete tools

Blank Dimension	N	Range	Mean	Standard Deviation
Length/Width Ratio	47	5.98	2.60	1.58
Width/Thickness Ratio	47	5.25	3.20	1.17

7.36), it is clear that nodules were worked and partially decortified before being curated on-site. The size of the raw material available at Kebara Cave is not excessively large and thus a higher frequency of non-cortical blanks to cortical blanks is not likely a result of a higher internal volume of the initial core compared to other sites. As seen in Figure 7.37, six of the last maximum core facets (out of 23) fall at or below the 2.0 cm size cut off for the debitage analysis. The remaining core facets overlap the low end of the length and width distribution of the debitage, indicating a significant degree of core exhaustion for this assemblage. Taken together, these data suggest that the assemblage captures the middle and end of the core reduction sequence of these hominins, but perhaps not the earliest portion of the sequence in which nodules are decortified. The assemblage is thus fairly comparable with other assemblages for both continuous and discontinuous variables, although raw material workshops may not overlap as much as other types of assemblages.

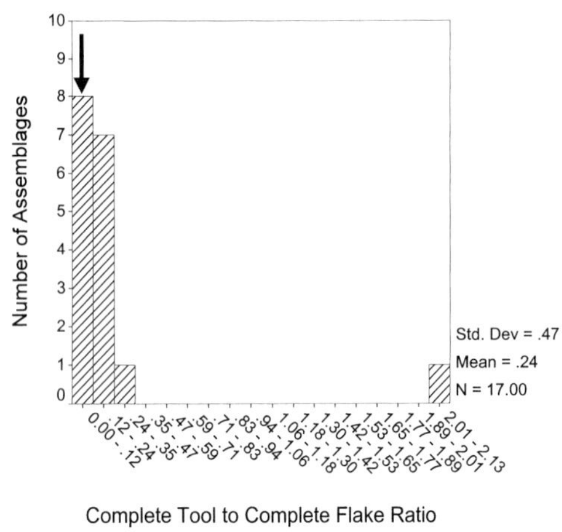

Figure 7.35 Kebara Cave Unit II, tool/flake ratio value relative to a histogram of all assemblages.

Table 7.72 Kebara Cave Unit II, frequencies of lateral edge, distal terminus, and profile types for complete tools

Type	Frequency	Percent
Lateral edge		
Parallel	22	46.8
Convergent	4	8.5
Expanding	3	6.4
Ovoid	4	8.5
Indeterminate	14	29.8
Total	47	100.0
Distal Terminus		
Blunt	5	10.6
Pointed	16	34.0
Hinged	2	4.3
Overshot	3	6.4
Removed	20	42.6
Indeterminate	1	2.1
Total	47	100.0
Profile		
Straight	14	29.8
Curved or Concord	17	36.2
Twisted	13	27.7
Indeterminate	3	6.4
Total	47	100.0

Table 7.73 Kebara Cave Unit II, intensity of core and tool reduction

Variables	Numerator	Denominator	Result
Tools/Flake Ratio	Tools: 27	Flakes: 355	0.08
Non-cortical Debitage/ Cortical Debitage Ratio	Non-cortical Debitage: 264	Cortical Debitage: 138	1.91
Henry's Dimensional Analysis	NA	NA	Core facets overlap with the low end of the blank distribution: significant core exhaustion

7.8 Kebara Cave, Unit I

7.8.1 Site Background

Like Unit II, Unit I is affiliated with the Levantine Aurignacian industrial type. With two dates of 34,510 ± 740 (OxA-3974) and 32,300 ± 630 (Pta-4268) (Bar-Yosef et al. 1996), Unit I is the youngest assemblage in the Levantine sample. The radiocarbon sample taken from the subsurface (22,900 ± 250, Pta-4247) is not considered applicable to the assemblage.

7.8.2 Sampling

This stratigraphic unit in Kebara Cave is very rich in both debitage and tools. For this reason, only two meter squares, R16 and Q14, were chosen for inclusion in the present study for the analysis of debitage. For the analysis of the retouched tool kit, other squares were also included: Q15, Q16, Q17, R14, R15, and R17. Stratigraphic considerations also pointed to the use of these two squares.

7.8.3 Blank Production

Core Modification

Core modification in Kebara Unit I entails a longitudinal blade core form, as evidenced in the identification of the narrow core-face as the

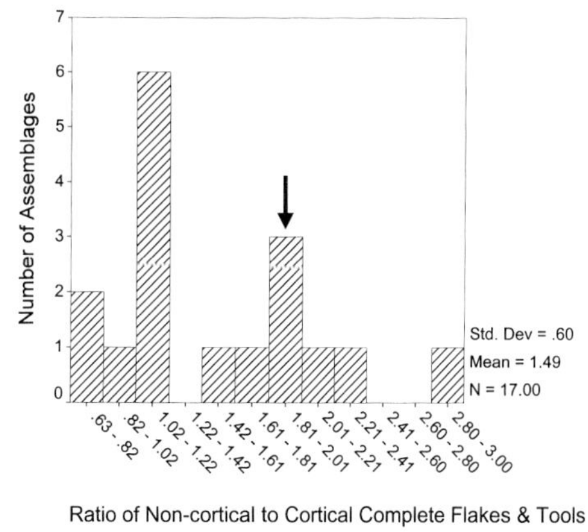

Figure 7.36 Kebara Cave Unit II, non-cortical/cortical flake ratio value relative to a histogram of all assemblages.

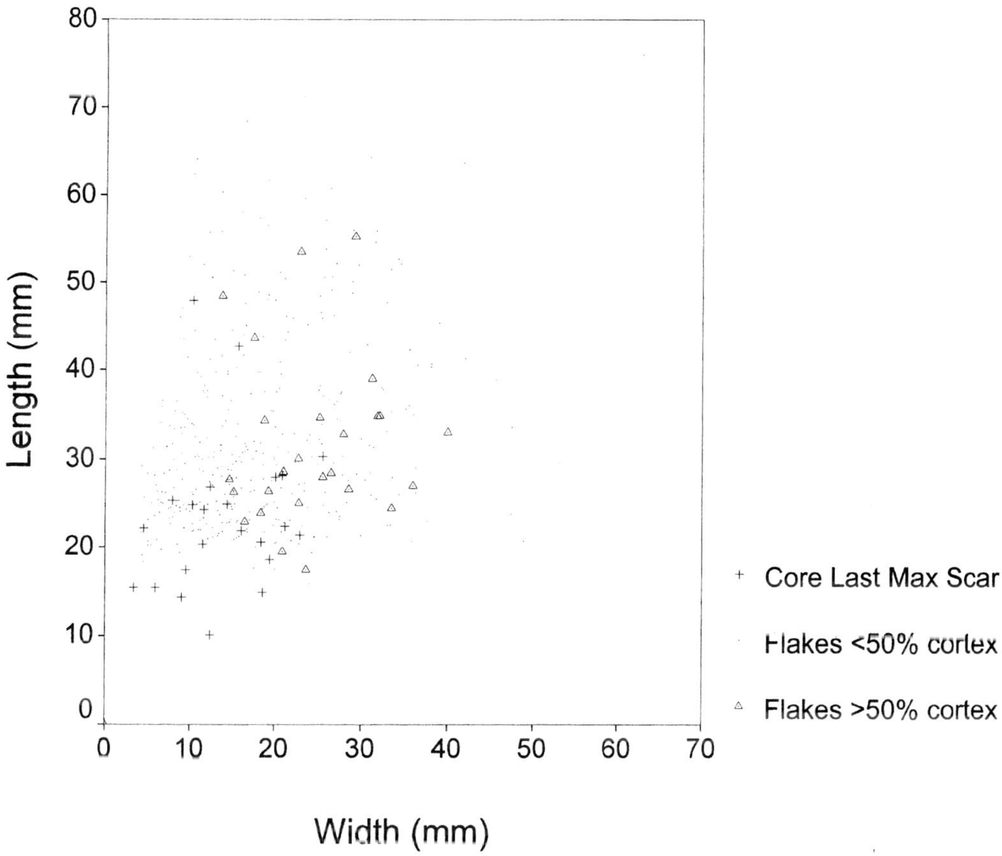

Figure 7.37 Kebara Cave Unit II according to Henry's core reduction intensity measure: length and width of last maximum core facets, highly cortical blanks, and blanks with little cortex.

Table 7.74 Kebara Cave Unit I, percentage of intact cores (n=36) by number of bulbar negatives in each core sector (A, B, C, D) by surface (1, 2)

Number of bulbar negatives	Surf. 1 Sect. A	Surf. 1 Sect. B	Surf. 1 Sect. C	Surf. 1 Sect. D	Surf. 2 Sect. A	Surf. 2 Sect. B	Surf. 2 Sect. C	Surf. 2 Sect. D
0	22	56	69	50	72	75	78	83
1	28	33	19	33	17	11	22	8
2	22	11	8	17	8	11		6
3	28		3		0	3		3
4					3			

Table 7.75 Kebara Cave Unit I, core sector preparation: percent of intact cores (n=36) by number of scars in sector of origin (A, B, C, D) by surface (1, 2)

Number of Scars	Surf. 1 Sect. A	Surf. 1 Sect. B	Surf. 1 Sect. C	Surf. 1 Sect. D	Surf. 2 Sect. A	Surf. 2 Sect. B	Surf. 2 Sect. C	Surf. 2 Sect. D
0	19	94	67	75	42	72	53	72
1	14	6	17	14	28	17	25	8
2	25		8	3	14	8	14	8
3	28		6	8	6	3	3	6
4	14		0		8		6	6
5			3		0			
6					3			

debitage surface. This identification is supported by Table 7.74, which demonstrates that Surface 1 strongly outweighs Surface 2 in number of bulbar negatives, indicating a preference for exploiting one surface rather than two. The longitudinal core orientation visible in the choice of Surface 1 is also corroborated by Table 7.75, which displays the dominance of the working of sector A on both core surfaces, with only minor working from sector C. Frontal crests and core tablets are used to control the convexity of the debitage surface and platform surface to perpetuate core reduction. The moderate frequency of bulbar negatives in Table 7.74 in Surface 1 sectors B and D indicates the use of lateral crests to control the width of the debitage surface. This interpretation is supported by the slightly higher frequency of scars originating in Surface 2 sectors B and D compared to Surface 1.

Platform Maintenance

Platform maintenance includes a very low frequency of platform preparation, only up to 24% of the identifiable platforms (Table 7.76). The frequency of linear and punctiform platforms and the behind-platform abrasion preparation this implies, is equivalent to that seen in Kebara Unit II, 22.3% compared to 24.8% when indeterminates are removed. Measurements of external platform angles and platform thickness (Table 7.77) indicate tendencies around 84 degrees and 4 mm respectively.

Direction of Core Exploitation

The analysis of the direction of core exploitation is presented in Tables 7.78–7.79. Figure 7.38 demonstrates a continuous unidirectional strategy without a shift during reduction. There is a slight increase in the unidirectional frequencies from largest to smallest quartiles, with a concomitant decrease in bidirectional frequencies but this does not change the overwhelming dominance of the unidirectional pattern throughout core exploitation.

Dorsal Surface Convexity

The analysis of the attributes serving as the

Table 7.76 Kebara Cave Unit I, platform types for complete and proximal fragments of flakes and tools

Platform type	Frequency	Percent
Plain	161	30.8
Cortical	29	5.6
Dihedral	3	0.6
Facetted	25	4.8
Chapeau	2	0.4
Scarred	55	10.5
Punctiform	28	5.4
Linear	51	9.8
Removed	22	4.2
Indeterminate	146	28.0
Total	522	100.0

Table 7.77 Kebara Cave Unit I, descriptive statistics for continuous platform variables of complete and proximal fragments of flakes and tools

	N	Minimum	Maximum	Mean	Standard Deviation
Exterior Platform Angle (degrees)	298	47.00	130.00	84.15	13.67
Platform Thickness (mm)	302	0.46	14.87	4.09	2.93

Table 7.78 Kebara Cave Unit I, dorsal scar pattern by blank length (mm) in quartiles cross-tabulation for complete flakes and tools (excluding flakes with greater than 60% cortex or fewer than three dorsal scars)

Dorsal Scar Pattern	>41	41-32	32-26	<26	Total
Unidirectional					
Count	38	40	48	43	169
% within Quartile	45.8%	50.6%	59.3%	59.7%	53.7%
Bidirectional					
Count	14	15	10	6	45
% within Quartile	16.9%	19.0%	12.3%	8.3%	14.3%
Crossed					
Count	12	6	8	11	37
% within Quartile	14.5%	7.6%	9.9%	15.3%	11.7%
Subcentripetal & Centripetal					
Count	5	9	4	9	27
% within Quartile	6.0%	11.4%	4.9%	12.5%	8.6%
Crested					
Count	13	5	9	1	28
% within Quartile	15.7%	6.3%	11.1%	14%	8.9%
Indeterminate					
Count	1	4	2	2	9
% within Quartile	1.2%	5.1%	2.5%	2.8%	2.9%
Total					
Count	83	79	81	72	315
% within Cortex	100.0%	100.0%	100.0%	100.0%	100.0%

Table 7.79 Kebara Cave Unit I, dorsal scar pattern by blank length (mm) in quartiles cross-tabulation for complete flakes and tools (excluding pieces with a crested scar pattern, an indeterminate scar pattern, greater than 60% cortex, or fewer than three dorsal scars)

Dorsal Scar Pattern	>41	41-32	32-26	<26	Total
Unidirectional					
Count	38	40	48	43	169
% within Quartile	55.1%	57.1%	68.6%	62.3%	60.8%
Bidirectional					
Count	14	15	10	6	45
% within Quartile	20.3%	21.4%	14.3%	8.7%	16.2%
Crossed					
Count	12	6	8	11	37
% within Quartile	17.4%	8.6%	11.4%	15.9%	13.3%
Subcentripetal & Centripetal					
Count	5	9	4	9	27
% within Quartile	7.2%	12.9%	5.7%	13.0%	9.7%
Total					
Count	69	70	70	69	278
% within Cortex	100.0%	100.0%	100.0%	100.0%	100.0%

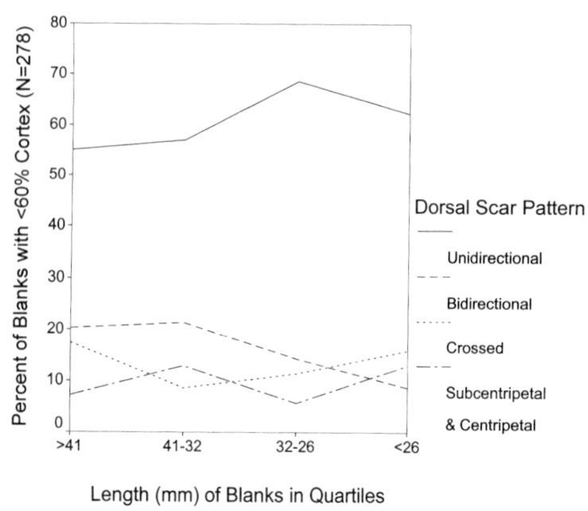

Figure 7.38 Kebara Cave Unit I, dorsal scar pattern by blank length (mm) in quartiles cross-tabulation for complete flakes and tools (excluding pieces with a crested scar pattern, an indeterminate scar pattern, greater than 60% cortex, or fewer than three dorsal scars).

proxies for the surface convexity domain are given in Tables 7.80-7.81. The ridge system used in the Kebara Unit I assemblage shows a propensity for high laminarity (length/width mean of 2.39), thicker products relative to their width (width/thickness mean of 3.76), straight profiles, and longitudinal ridges (i.e., high frequency of parallel lateral edges).

7.8.4 Tool Kit Morphology

The first five characterizations of the tool kit morphology refer to the dorsal surface convexity variables which give form to the cutting edge of the retouched blanks in the assemblage. In this case (Tables 7.82-7.83), the tool kit of Kebara Unit I is almost identical to its debitage in surface convexity variables. Carinated tools represent the use of a unique type of retouch. The tool kit is exclusively Upper Paleolithic (Bar-Yosef et al. 1996).

Intensity of Tool and Core Reduction

Only the tools and debitage from squares R16

Table 7.80 Kebara Cave Unit I, frequencies of lateral edge, profile, and cross-section types for complete flakes and tools

Type	Frequency	Percent
Lateral edge		
Parallel	268	51.3
Convergent	103	19.7
Expanding	62	11.9
Ovoid	39	7.5
Indeterminate	50	9.6
Total	522	100.0
Profile		
Straight	263	50.4
Curved or Concord	142	27.2
Twisted	96	18.4
Indeterminate	21	4.0
Total	522	100.0
Cross-section		
Triangular	206	39.5
Trapezoidal	259	49.6
Other	28	5.4
Indeterminate	29	5.6
Total	522	100.0

Table 7.81 Kebara Cave Unit I, descriptive statistics for blank dimensions for complete flakes and tools

Blank Dimension	N	Range	Mean	Standard Deviation
Length	522	73.52	36.66	12.88
Width	522	60.41	19.54	10.20
Thickness	522	51.93	6.35	5.07
Length/Width Ratio	522	9.54	2.38	1.35
Width/Thickness Ratio	522	18.07	3.76	2.00

Table 7.82 Kebara Cave Unit I, descriptive statistics for dimensions of complete tools

Blank Dimension	N	Range	Mean	Standard Deviation
Length/Width Ratio	101	9.41	2.35	1.50
Width/Thickness Ratio	101	11.80	3.34	1.88

and Q14 were used in calculating the complete tool to complete flake ratio (Table 7.84). As this value falls on the boundary between the farthest and next to farthest left group of assemblages, it is clear that relatively little retouch was applied to the blanks in this assemblage (Figure 7.39). While the shelter offered by Kebara Cave gives it a different context than a raw material workshop, the relatively low tool to flake ratio is surprising when compared to the values for Units III and IV, but it is close to the value for Unit II. None of these values, however, are indicative of the application of enough retouch to bias the comparison of the dimensional variables of these tool kits.

From the ratio of non-cortical to cortical blanks falling above the mean and in the middle of the central grouping of assemblages (Figure 7.40), it is clear that nodules were worked and partially decortified before being curated on-site. The size of the raw material available at Kebara Cave is not excessively large

Table 7.83 Kebara Cave Unit I, frequencies of lateral edge, distal terminus, and profile types for complete tools

Type	Frequency	Percent
Lateral edge		
Parallel	47	46.5
Convergent	17	16.8
Expanding	10	9.9
Ovoid	6	5.9
Indeterminate	21	20.8
Total	101	100.0
Distal Terminus		
Blunt	18	17.8
Pointed	27	26.7
Hinged	11	10.9
Overshot	3	3.0
Removed	35	34.7
Broken	5	5.0
Indeterminate	2	2.0
Total	101	100.0
Profile		
Straight	52	51.5
Curved or Concord	25	24.8
Twisted	16	15.8
Indeterminate	8	7.9
Total	101	100.0

Table 7.84 Kebara Cave Unit I, intensity of core and tool reduction

Variables	Numerator	Denominator	Result
Tools/Flake Ratio	Tools: 41	Flakes: 351	0.12
Non-cortical Debitage/ Cortical Debitage Ratio	Non-cortical Debitage: 290	Cortical Debitage: 162	1.79
Henry's Dimensional Analysis	NA	NA	Core facets overlap the bottom of the blank distribution: significant core exhaustion

and thus a higher frequency of non-cortical blanks to cortical blanks is not likely a result of a higher internal volume of the initial core compared to other sites. As seen in Figure 7.41, almost half of the last maximum core facets (14 out of 35) fall at or below the 2 cm size cut off for the debitage analysis. The remaining core facets clearly overlap the bottom of the length and width distribution of the debitage, indicating a significant degree of core exhaustion for this assemblage. Taken together, these data suggest that the assemblage captures the middle and end of the core reduction sequence of these hominins, but perhaps not the earliest portion of the sequence in which nodules are decortified. The assemblage is thus fairly comparable with other assemblages for both continuous and discontinuous variables, although raw material workshops may not overlap as much as other types of assemblages.

7.9 Comparison of Levantine Flintknapping Behaviors

The construction of the Levantine sequence of

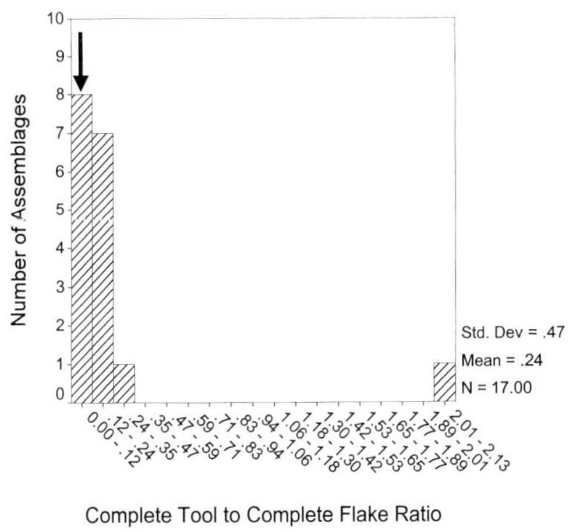

Figure 7.39 Kebara Cave Unit I, tool/flake ratio value relative to a histogram of all assemblages.

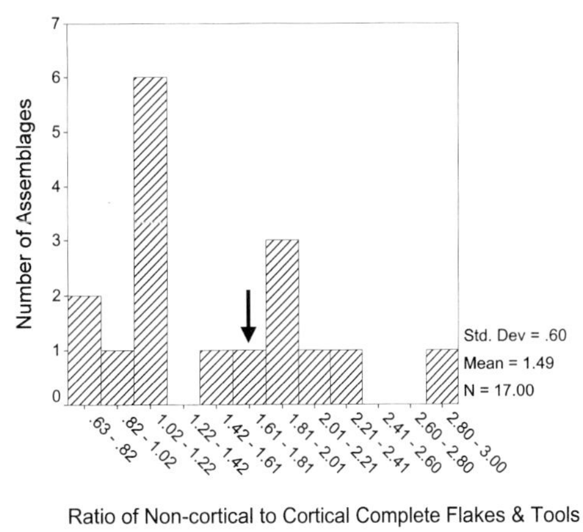

Figure 7.40 Kebara Cave Unit I, non-cortical/cortical flake ratio value relative to a histogram of all assemblages.

change between 60 and 30 kya is conducted on the basis of pair-wise comparisons between assemblages as demonstrated in Chapters 5–6.

7.9.1 Comparison of Kebara Cave Unit VI and Boker Tachtit Level 1

The first pair-wise comparison of the Levantine sample is between the Middle Paleolithic Unit VI of Kebara Cave and the transitional assemblage from Level 1 of Boker Tachtit. The comparability measures of these assemblages are quite similar (Table 7.85). Both the tool to flake ratio (Figure 7.42), in which the values are found in adjacent bars of the histogram, and the non-cortical to cortical blank ratio (Figure 7.43), in which the assemblages fall in the same bar of the histogram, show that these two assemblages overlap in their coverage of their respective core reduction sequences. Kebara VI may not include as much of the initial working of nodules as Boker Tachtit Level 1, based on the latter's refits and the former's higher value in Figure 7.43, but this is unlikely to bias the central tendencies of the assemblages. Both assemblages are lightly retouched enough to argue that no bias should be expected in comparisons of dimensional values between the tool kits.

Blank Production

Table 7.86 presents the results of a comparison of blank production within the assemblages from Kebara Unit VI and Boker Tachtit Level 1. The first step in the core modification domain, core orientation, is characterized by the qualitative distinction between the exploitation of different core morphologies or forms. The Kebara VI cores exhibit a broad-faced orientation, compared to the longitudinal orientation of the Boker Tachtit 1 cores. In the second step in this domain, the choice of core convexity management option was distinct between the assemblages, due to the use of frontal crests in Boker

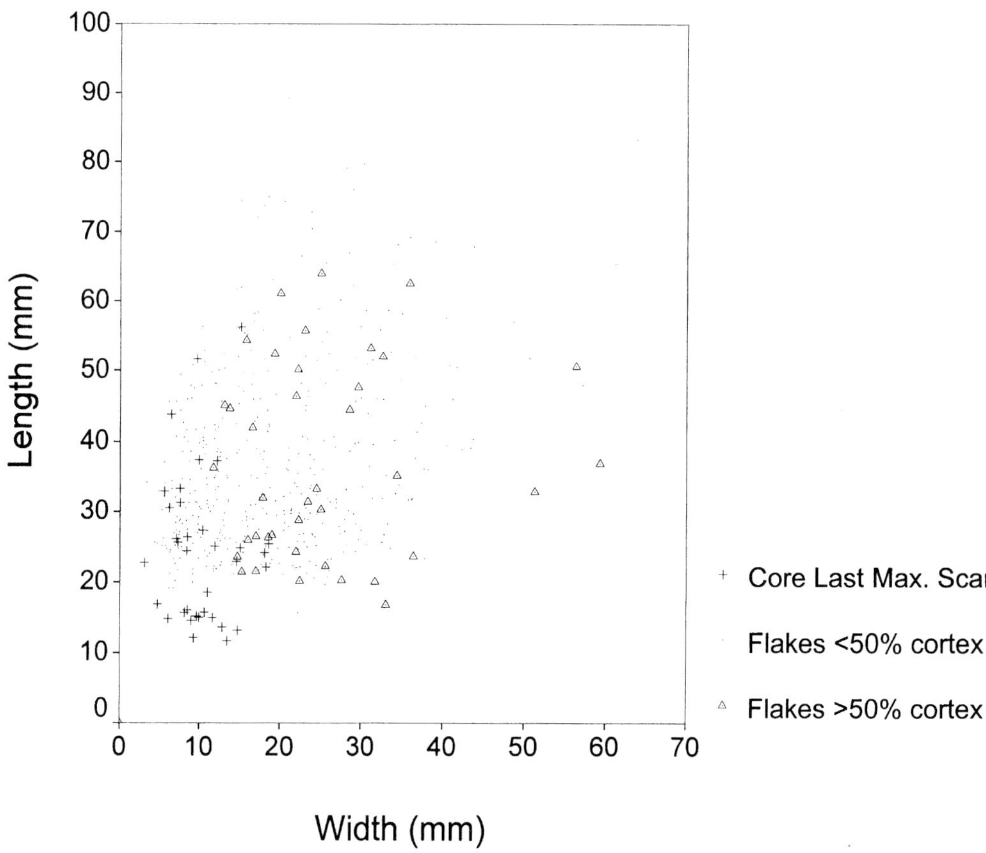

Figure 7.41 Kebara Cave Unit I according to Henry's core reduction intensity measure: length and width of last maximum core facets, highly cortical blanks, and blanks with little cortex.

Tachtit 1. The next knapping domain, platform maintenance, includes the steps whereby the flintknapper modifies the platform surface and platform edge before each removal. In the case of the Kebara VI versus Boker Tachtit 1 comparison, only the choice of platform thickness differed significantly. The direction of early core exploitation shows dissimilar options taken by the flintknappers who created these two assemblages, due to Kebara Unit VI's tendency for unidirectional reduction compared to the bidirectional reduction indicated for Boker Tachtit Level 1. Late core exploitation is considered similar in that unidirectional scar patterns dominate the debitage in the smallest length quartiles in both assemblages. This similarity is known in fact to reflect a sequential bidirectional core rotational pattern at Boker Tachtit compared to a simple continuity of the unidirectional pattern at Kebara VI. This contrast, however, is not visible in the debitage samples themselves, being decipherable only in refits, and so, for the purpose of maintaining the same sampling and analytical resolution across all samples, the similarity rather than contrast is used to

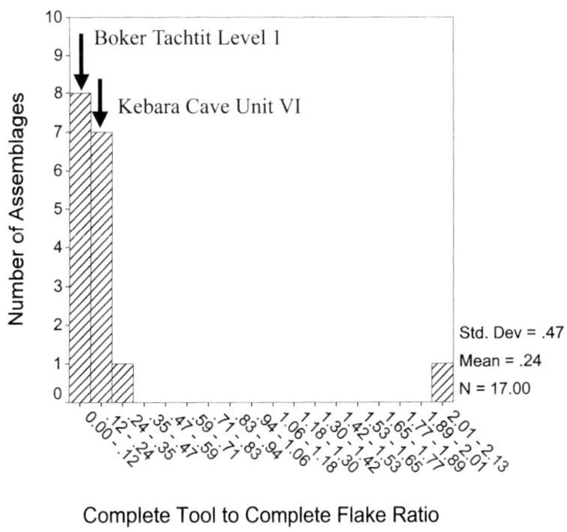

Figure 7.42 Kebara Cave Unit VI and Boker Tachtit Level 1 compared for tool/flake ratio values.

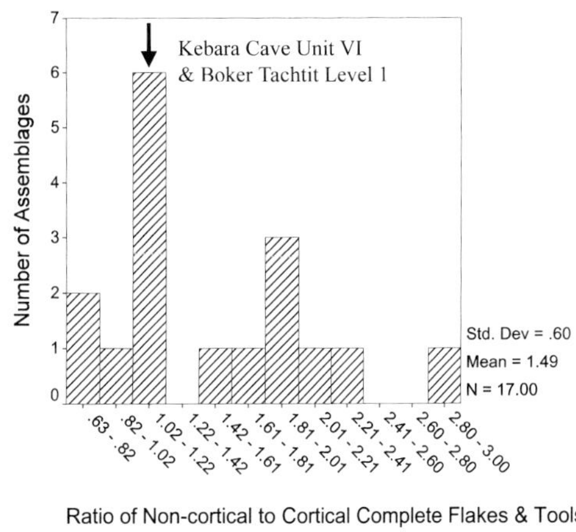

Figure 7.43 Kebara Cave Unit VI and Boker Tachtit Level 1 compared for non-cortical/cortical flake ratio values.

Table 7.85 Kebara Cave Unit VI and Boker Tachtit Level 1, comparability of the assemblages based on intensity of core and tool reduction

Variables	Kebara Cave Unit VI	Boker Tachtit Level 1	Evaluation
Tools/Flake Ratio	0.21	0.08	Moderately Different
Non-cortical Debitage/ Cortical Debitage Ratio	1.18	1.09	Similar
Henry's Dimensional Analysis	Moderate core exhaustion	See text	Similar

characterize the second analytical step in this domain. Of the five steps within the surface convexity domain, all options were judged to be significantly different between the two assemblages. The dominance of the blade/point production techniques at Boker Tachtit Level 1 are responsible for these differences, with their high laminarity values, parallel lateral edges, more curved profiles, and thicker and narrow cross-sections than seen in Kebara Unit VI.

The final measure of difference between the blank production choices within Kebara Cave Unit VI and Boker Tachtit Level 1, weighted by the four knapping domains, produces a value of 2.83 out of a possible maximum difference of 4.0, indicating that the blank production choices are distinctly different.

Tool Kit Morphology

The comparison of the morphology of the cutting edge of the tool kits of these two assemblages utilizes three categorical variables and one continuous variable, due to the sampling problems with the Boker Tachtit Level 1 tool kit. The continuous variable, the length/width ratio, is tested with a t-test for the probability that the samples arose from

Table 7.86 Comparison of Blank Production for Kebara Cave Unit VI and Boker Tachtit Level 1

Flintknapping Steps by Domain	Kebara VI	Boker Tachtit 1	Significant Difference?
Core Modification			
Core Orientation	Broad, flat orientation	Preferential use of a Longitudinal Surface	Yes
Core Management	Débordants	Débordants & frontal crest	Yes
Number of Changes/2 Steps			**2/2=1**
Platform Maintenance			
Platform Treatment	Unprepared: 50% Prepared: 50% n=504	Unprepared: 61% Prepared: 39% n=79	No p=.07, Fisher's Exact
External Platform Angle (degrees)	mean: 87.4, s.d.: 16.1, n=474	mean: 88.4, s.d.: 13.6, n=63	No p=.63, t=-0.483, df=535
Platform Thickness	mean: 4.78, s.d.: 2.48, n=475	mean: 4.04, s.d.: 2.16, n=65	Yes p=.02, t=2.276, df=538
Number of Changes/3 Steps			**1/3=0.33**
Direction of Core Exploitation			
Direction of Early Exploitation	Unidirectional	Bidirectional	Yes
Direction of Late Exploitation	Unidirectional	Unidirectional	No
Number of Changes/2 Steps			**1/2=0.5**
Dorsal Surface Convexity System			
Longitudinal Convexity: Length/Width Ratio	mean: 1.78, s.d.: 0.82, n=603	mean: 2.25, s.d.: 1.21, n=101	Yes p=.00, t=2.908*, df=145.924
Shape of Convexity: Lateral Edges of Blanks	Parallel: 37% Convergent: 29% Expanding: 23% Ovoid: 11% n=584	Parallel: 71% Convergent: 12% Expanding: 5% Ovoid: 12% n=101	Yes p=.00, G^2=50.785, df=3
Curvature of Convexity: Profile of Blanks	Straight: 68% Curved: 18% Twisted: 14% n=597	Straight: 40% Curved: 40% Twisted: 21% n=101	Yes p=.00, G^2=31.368, df=2
Lateral Convexity: Cross-Section of Blanks	Triangular: 47% Trapezoidal: 36% Other: 17% n=574	Triangular: 55% Trapezoidal: 40% Other: 5% n=101	Yes p=.00, G^2=12.616, df=2
Vertical Convexity: Width/Thickness Ratio	mean: 5.18, s.d.: 2.62, n=603	mean: 4.43, s.d.: 2.33, n=101	Yes p=.01, t=2.671, df=702
Number of Changes/5 Steps			**5/5=1**
Total Measure of Difference in Blank Production			**2.83**

*The t-test was calculated with equal variances not being assumed due to a significant F value on Levene's Test for Equality of Variances.

Table 7.87 Comparison of Tool Kit Morphologies for Kebara Cave Unit VI and Boker Tachtit Level 1

Tool Kit Morphology Variable	Kebara VI	Boker Tachtit 1	Significant Difference?
Laminarity	mean: 1.84, s.d.: 0.56 n=103	mean: 2.06, s.d.: 0.79, n=72	Yes p=.05, t=-1.994*, df=118.732
Width/Thickness	mean: 5.48 s.d.: 2.21 n=103	N/A	N/A
Lateral Edges	Parallel: 67% Convergent: 12% Expanding: 9% Ovoid: 12% n=33	N/A	N/A
Distal Terminus	Blunt: 43% Pointed: 57% n=77	Blunt: 27% Pointed: 73% n=62	No p=.08, Fisher's Exact
Profile	Straight: 32% Curved: 39% Twisted: 30% n=44	N/A	N/A
Unique Types of Retouch	Normal retouch	Emireh retouch	Yes
Tool Types	MP tools dominate	UP tools dominate	Yes
Number of Changes/4 Steps			3/4
Total Measure of Difference in Tool Kit Morphology			**0.75**

*The t-test was calculated with equal variances not being assumed due to a significant F value on Levene's Test for Equality of Variances.

the random sampling of the same population. One of the three categorical variables is tested with a G^2 statistic while the other two are judged qualitatively. As no concern for a potential bias by differential tool reduction is warranted by the tool to flake ratio comparison, the dimensional variable is considered valid and shows a statistically significant difference between the assemblages. The distal terminus comparison shows no statistical significance, although the p=0.08 is close to the required value. The remaining two analytical steps within the comparison of tool kit morphology relate to qualitative issues of retouch. Only Boker Tachtit Level 1 possesses a unique retouch type, Emireh retouch. The two assemblages also differ in the dominance of Middle Paleolithic tool types in the earlier assemblage and Upper Paleolithic tool types in the later one. Considering all 4 analytical steps within this tool kit comparison, these assemblages exhibit a difference measure of 3 out of 4 steps, or 0.75 out of a maximum of 1.0.

7.9.2 Comparison of Boker Tachtit Level 1 and Boker Tachtit Level 2

The next pair wise comparison of the Levantine sample is between the two transitional assemblages from Levels 1 and 2 of Boker Tachtit. Given the similarity in contexts, it is not surprising that the comparability measures of these assemblages are almost identical (Table 7.88). Both the tool to

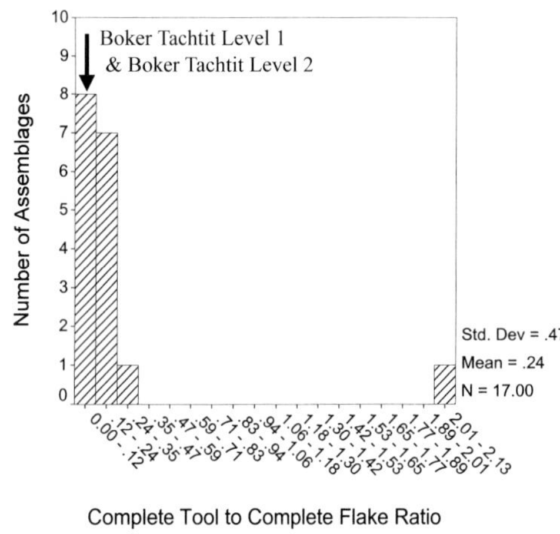

Figure 7.44 Boker Tachtit Levels 1 and 2 compared for tool/flake ratio values.

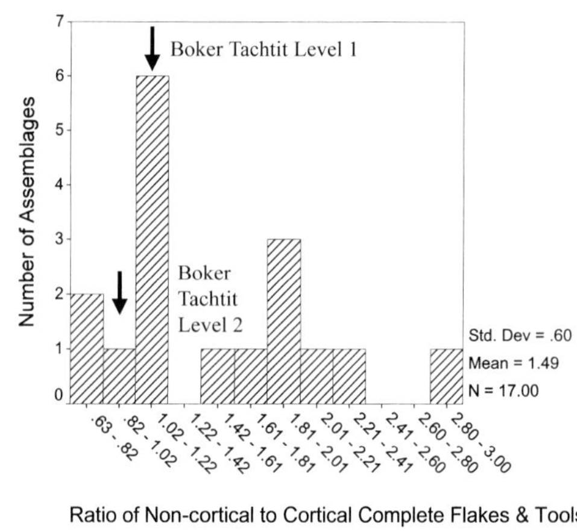

Figure 7.45 Boker Tachtit Levels 1 and 2 compared for non-cortical/cortical flake ratio values

Table 7.88 Boker Tachtit Levels 1 and 2, comparability of the assemblages based on intensity of core and tool reduction

Variables	Boker Tachtit Level 1	Boker Tachtit Level 2	Evaluation
Tools/Flake Ratio	0.08	0.07	Similar
Non-cortical Debitage/ Cortical Debitage Ratio	1.09	.96	Similar
Henry's Dimensional Analysis	See text	See text	Similar

flake ratio (Figure 7.44) and the non-cortical to cortical blank ratio (Figure 7.45) fall in the same grouping of assemblages on their respective histograms, showing that these two assemblages overlap closely in their coverage of their respective core reduction sequences. Both assemblages are lightly retouched enough to argue that no bias should be expected in comparisons of dimensional values between the tool kits.

Blank Production

Table 7.89 presents the results of a comparison of the details of blank production within the assemblages from Boker Tachtit Levels 1 and 2. Core orientation, core modification, platform treatment and platform thickness choices are all similar between these two assemblages. Only external platform angle is significantly different due to the more acute angles of the Boker Tachtit Level 2 debitage. All of the rest of the flintknapping steps are identical between these assemblages, producing the closest measure of difference in the study (0.33). The subtle distinction within the levels between the representation of the change from alternating changing to sequential bidirectional reduction of the debitage is not

Table 7.89 Comparison of Blank Production for Boker Tachtit Levels 1 and 2

Flintknapping Steps by Domain	Boker Tachtit 1	Boker Tachtit 2	Significant Difference?
Core Modification			
Core Orientation	Preferential use of a Longitudinal Surface	Preferential use of a Longitudinal Surface	No
Core Management	Débordants & frontal crest	Débordants & frontal crest	No
Number of Changes/2 Steps			**0/2=0**
Platform Maintenance			
Platform Treatment	Unprepared: 61% Prepared: 39% n=79	Unprepared: 65% Prepared: 36% n=76	No p=.74, Fisher's Exact
External Platform Angle (degrees)	mean: 88.4, s.d.: 13.6, n=63	mean: 82.1, s.d.: 12.0, n=58 Yes	p=.01, p=.01, t=2.701, df=119
Platform Thickness	mean: 4.04, s.d.: 2.16, n=65	mean: 4.54, s.d.: 2.17, n=62	No p=.20, t=-1.292, df=125
Number of Changes/3 Steps			**1/3=0.33**
Direction of Core Exploitation			
Direction of Early Exploitation	Bidirectional	Bidirectional	No
Direction of Late Exploitation	Unidirectional	Unidirectional	No
Number of Changes/2 Steps			**0/2=0**
Dorsal Surface Convexity System			
Longitudinal Convexity: Length/Width Ratio	mean: 2.25, s.d.: 1.21, n=101	mean: 2.14, s.d.: 1.05, mean: 2.14, s.d.: 1.05, n=99	No p=.49, t=0.685, df=198
Shape of Convexity: Lateral Edges of Blanks	Parallel: 71% Convergent: 12% Expanding: 5% Ovoid: 12% n=101	Parallel: 65% Convergent: 18% Expanding: 10% Ovoid: 6% n=98	No p=.15, G^2=5.372, df=3
Curvature of Convexity: Profile of Blanks	Straight: 40% Curved: 40% Twisted: 21% n=101	Straight: 51% Curved: 30% Twisted: 19% n=98	No p=.23, G^2=2.929, df=2
Lateral Convexity: Cross-Section of Blanks	Triangular: 55% Trapezoidal: 40% Other: 5% n=101	Triangular: 46% Trapezoidal: 51% Other: 4% n=99	No p=.30, G^2=2.405, df=2*
Vertical Convexity: Width/Thickness Ratio	mean: 4.43, s.d.: 2.33, n=101	mean: 4.46, s.d.: 2.35, n=99	No p=.95, t=0.061, df=198
Number of Changes/5 Steps			**0/5=0**
Total Measure of Difference in Blank Production			**0.33**

*Two cells (33.3%) have an expected count less than 5. The minimum expected count is 4.46.

Table 7.90 Comparison of Tool Kit Morphologies for Boker Tachtit Levels 1 and 2

Tool Kit Morphology Variable	Boker Tachtit 1	Boker Tachtit 2	Significant Difference?
Laminarity	mean: 2.06, s.d.: 0.79, n=72	mean: 1.87, s.d.: 0.87, n=210	No p=.11, t=1.610, df=280
Width/Thickness	N/A	N/A	N/A
Lateral Edges	N/A	N/A	N/A
Distal Terminus	Blunt: 27% Pointed: 73% n=62	Blunt: 43% Pointed: 57% n=159	Yes p=.05, Fisher's Exact
Profile	N/A	N/A	N/A
Unique Types of Retouch	Emireh retouch	Emireh retouch	No
Tool Types	UP tools dominate	UP tools dominate	No
Number of Changes/4 Steps			**1/4**
Total Measure of Difference in Tool Kit Morphology			**0.25**

Table 7.91 Boker Tachtit Level 2 and Kebara Cave Unit IV, comparability of the assemblages based on intensity of core and tool reduction

Variables	Boker Tachtit Level 2	Kebara Cave Unit IV	Evaluation
Tools/Flake Ratio	0.07	0.14	Slightly Different
Non-cortical Debitage/ Cortical Debitage Ratio	.96	2.23	Different
Henry's Dimensional Analysis	See text	moderate core exhaustion	Similar

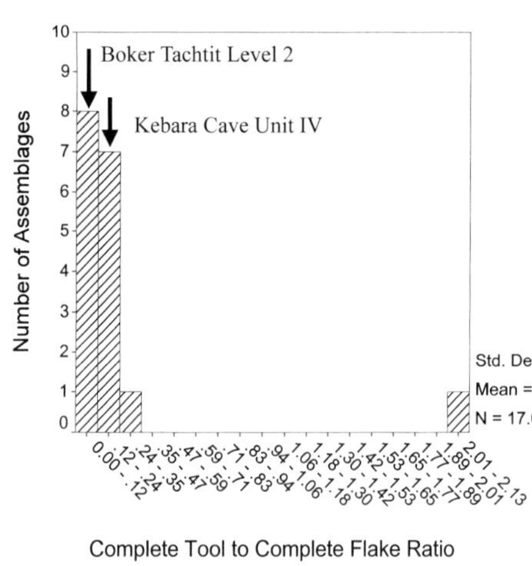

Figure 7.46 Boker Tachtit Level 2 and Kebara Cave Unit IV compared for tool/flake ratio values.

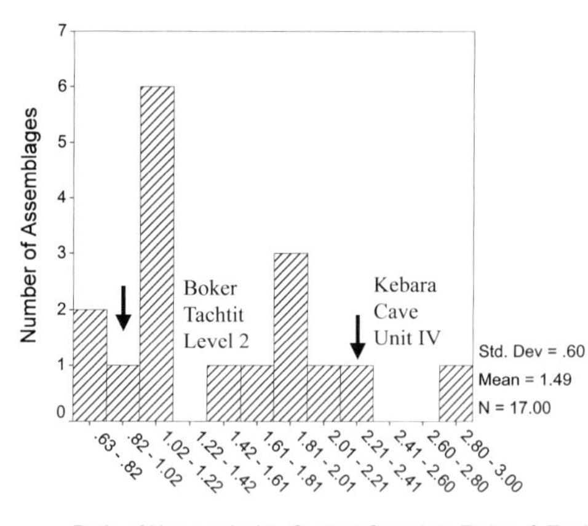

Figure 7.47 Boker Tachtit Level 2 and Kebara Cave Unit IV compared for non-cortical/cortical flake ratio values.

reflected in the characterization of the directionality of core exploitation domain.

Tool Kit Morphology
Of the four variables available for these assemblages for the consideration of the morphology of the tool kit, only one variable resulted in statistically distinct value (tab. 7.90), the higher frequency of pointed distal ends in Level 1. This one variable results in a difference measure of 0.25 out of 1.0.

7.9.3 Comparison of Boker Tachtit Level 2 and Kebara Unit IV
The next pair-wise comparison of the Levantine sample is between the transitional assemblage from Level 2 of Boker Tachtit and the Ahmarian assemblage from Unit IV of Kebara Cave. Given the difference in the site contexts of the two assemblages, one open air and one cave, it is not surprising that the comparability measures of these assemblages are somewhat different (Table 7.91). The tool to flake ratios (Figure 7.46) indicate that Kebara IV has a higher frequency of tools than Boker Tachtit Level 2, and thus may have some tools with dimensions more compromised by reduction than Level 2. The commonality of the low frequency of tools in general in these two sites compared to others such as Vedrovice V and Kůlna Cave Layer 7a makes it more likely that the dimensions of the tool kit are quite comparable. The non-cortical to cortical blank ratios (Figure 7.47) show a greater degree of difference, with Boker Tachtit Level 2 falling with the raw material workshop assemblages far to the left of the histogram and Kebara IV falling in the middle group of assemblages but significantly above the mean. As it appears likely that Kebara IV lacks the earliest portions of the core reduction sequence as captured in Boker Tachtit Level 2, some of the differences noted in the following comparisons may be effected by the different landscape and raw material contexts of the two assemblages.

Blank Production
Table 7.92 presents the results of a comparison of the blank production details of the assemblages from Kebara Unit IV and Boker Tachtit level 2. Both assemblages share the same core modification steps and only differ in the platform maintenance domain in terms of the larger external platform angle in the Kebara assemblage. The assemblages also differ in directionality of early core exploitation, with Kebara IV showing a co-dominance of bidirectional and unidirectional patterns in the longest length quartile while Boker Tachtit Level 2 evidences only one dominant pattern, bidirectional. If indeed Kebara IV lacks some of the earliest portion of the core reduction sequence while Boker Tachtit Level 2 does not, unidirectional blanks may be more represented in the longest quartile in the former site because this quartile represents a stage further on in the reduction of cores than the first quartile in the latter site. In such a case, a bidirectional pattern alone might have been evidenced in this longest quartile if the early sequence were also present. This scenario is not likely, however, as the actual dimensions of the quartiles in the different assemblages are extremely close and if anything, the larger initial nodule size at Boker Tachtit would lend itself to an opposite scenario. For this reason, the characterization remains as read at the first and fourth quartiles (unlike with the comparison of Kebara Units IV and III below). As such, the comparison of early core exploitation is judged to remain different between the two assemblages.

Of the five steps within the blank production domain, no options were judged to be significantly different between the two assemblages. The final measure of difference between the details of the blank production at Kebara Unit IV and Boker

Table 7.92 Comparison of Blank Production for Boker Tachtit Level 2 and Kebara Cave Unit IV

Flintknapping Steps by Domain	Boker Tachtit 2	Kebara IV	Significant Difference?
Core Modification			
Core Orientation	Preferential use of a Longitudinal Surface	Preferential use of a Longitudinal Surface	No
Core Management	Débordants & frontal crest	Débordants & frontal crest	No
Number of Changes/2 Steps			**0/2=0**
Platform Maintenance			
Platform Treatment	Unprepared: 65% Prepared: 36% n=76	Unprepared: 59% Prepared: 41% n=334	No p=.37, Fisher's Exact
External Platform Angle (degrees)	mean: 82.1, s.d.: 12.0, n=58	mean: 88.0, s.d.: 14.9, n=289	Yes p=.01, t=-2.844, df=345
Platform Thickness	mean: 4.54, s.d.: 2.17, n=62	mean: 4.31, s.d.: 2.56, n=291	No p=.52, t=0.648, df=351
Number of Changes/3 Steps			**1/3=0.33**
Direction of Core Exploitation			
Direction of Early Exploitation	Bidirectional	Bidirectional & Unidirectional	Yes
Direction of Late Exploitation	Unidirectional	Unidirectional	No
Number of Changes/2 Steps			**1/2=0.5**
Dorsal Surface Convexity System			
Longitudinal Convexity: Length/Width Ratio	mean: 2.14, s.d.: 1.05, n=99	mean: 2.17, s.d.: 1.16, n=440	No p=.78, t=-0.286, df=537
Shape of Convexity: Lateral Edges of Blanks	Parallel: 65% Convergent: 18% Expanding: 10% Ovoid: 6% n=98	Parallel: 53% Convergent: 22% Expanding: 15% Ovoid: 10% n=432	No p=.16, G^2=5.185, df=3
Curvature of Convexity: Profile of Blanks	Straight: 51% Curved: 30% Twisted: 19% n=98	Straight: 58% Curved: 25% Twisted: 17% n=437	No p=.47, G^2=1.531, df=2
Lateral Convexity: Cross-Section of Blanks	Triangular: 46% Trapezoidal: 51% Other: 4% n=99	Triangular: 45% Trapezoidal: 47% Other: 8% n=434	No p=.26, G^2=2.701, df=2
Vertical Convexity: Width/Thickness Ratio	mean: 4.46, s.d.: 2.35, n=99	mean: 4.35, s.d.: 2.11, n=440	No p=.66, t=0.439, df=537
Number of Changes/5 Steps			**0/5=0**
Total Measure of Difference in Blank Production			**0.83**

Tachtit Level 2, weighted by the four knapping domains, produces a value of 0.83 out of a possible maximum difference of 4.0, indicating that blank production is extremely similar in these two assemblages.

Tool Kit Morphology

Of the four variables available in both of these assemblages for the consideration of the morphology of the tool kit, two show statistically significant differences (Table 7.93). First, the tools in Boker Tachtit Level 2 are more elongated in their length/ width ratio compared to Kebara Unit IV. Second, only the presence of Emireh retouch at Boker Tachtit divides the two assemblages based on retouch itself. Together, these variables reflect a difference measure of 0.50 out of a maximum of 1.0, indicating that the tool kits are moderately different.

7.9.4 Comparison of Boker Tachtit Level 2 and Boker Tachtit Level 4

The next pair-wise comparison of the Levantine sample is between the transitional assemblages from Levels 2 and 4 of Boker Tachtit. Given the similarity in contexts, it is not surprising that the comparability measures of these assemblages are quite similar (Table 7.94). The tool to flake ratios

Table 7.93 Comparison of Tool Kit Morphologies for Boker Tachtit Level 2 and Kebara Cave Unit IV

Tool Kit Morphology Variable	Boker Tachtit 2	Kebara IV	Significant Difference?
Laminarity	mean: 1.87, s.d.: 0.87, n=210	mean: 2.23, s.d.: 1.12 n=52	Yes p=.04, t=-2.138*, df=66.809
Width/Thickness	N/A	mean: 4.04, s.d.: 1.80 n=52	N/A
Lateral Edges	N/A	Parallel: 57% Convergent: 23% Expanding: 6% Ovoid: 14% n=51	N/A
Distal Terminus	Blunt: 43% Pointed: 57% n=159	Blunt: 50% Pointed: 50% n=32 p=.56,	No Fisher's Exact
Profile	N/A	Straight: 49% Curved: 24% Twisted: 27% n=51	N/A
Unique Types of Retouch	Emireh retouch	Normal retouch	Yes
Tool Types	UP tools dominate	UP tools dominate	No
Number of Changes/4 Steps			2/4
Total Measure of Difference in Tool Kit Morphology			0.50

*The t-test was calculated with equal variances not being assumed due to a significant F value on Levene's Test for Equality of Variances.

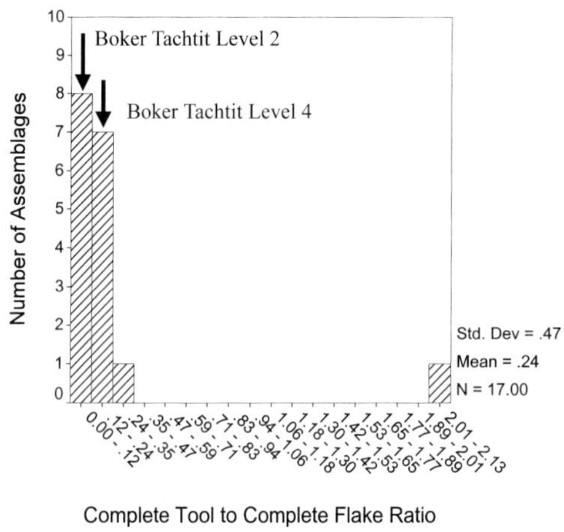

Figure 7.48 Boker Tachtit Levels 2 and 4 compared for tool/flake ratio values.

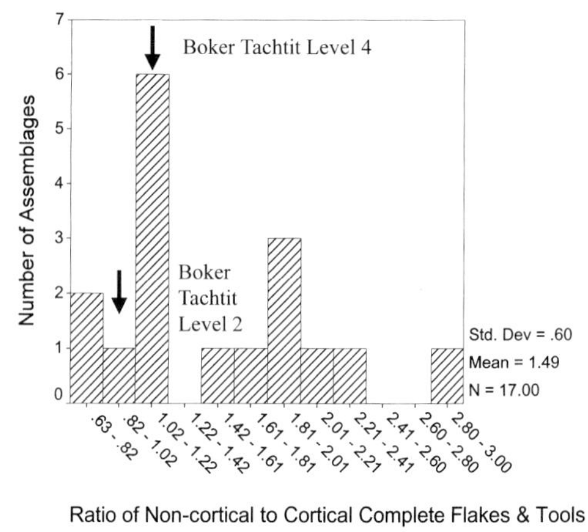

Figure 7.49 Boker Tachtit Levels 2 and 4 compared for non-cortical/cortical flake ratio values.

Table 7.94 Boker Tachtit Levels 2 and 4, comparability of the assemblages based on intensity of core and tool reduction

Variables	Boker Tachtit Level 2	Boker Tachtit Level 4	Evaluation
Tools/Flake Ratio	0.07	0.13	Slightly Different
Non-cortical Debitage/ Cortical Debitage Ratio	.96	1.14	Similar
Henry's Dimensional Analysis	See text	See text	Similar

(Figure 7.48) indicate that Level 4 is slightly more retouched than Level 2 but not to a degree that dimensional variables should be affected by differential reduction. The non-cortical to cortical blank ratios (Figure 7.49) both fall in the same grouping of assemblages, those with the most evidence of initial raw material working on site. In both cases, the refits corroborate the presence of an almost complete core reduction sequence, save for perhaps the activities indicative of extreme raw material conservation and core exhaustion. Together, these measures show that the two assemblages overlap closely in their coverage of their respective core reduction sequences.

Blank Production

Table 7.95 presents the results of a comparison of the blank production details of the assemblages from Boker Tachtit Levels 2 and 4. Both assemblages share the same core orientation but differ in the details of convexity management. Within the platform maintenance domain, only the larger platform thickness choices of Boker Tachtit Level 4 produce a significant difference. The directionality of the debitage of early core exploitation also differs significantly between the two assemblages, although the debitage of late core exploitation shows a similar dominance of unidirectional blanks. While it is clear from the

Table 7.95 Comparison of Blank Production for Boker Tachtit Levels 2 and 4

Flintknapping Steps by Domain	Boker Tachtit 2	Boker Tachtit 4	Significant Difference?
Core Modification			
Core Orientation	Preferential use of a Longitudinal Surface	Preferential use of a Longitudinal Surface	No
Core Management	Débordants & frontal crest	Frontal crest & core tablets	Yes
Number of Changes/2 Steps			**1/2=0.5**
Platform Maintenance			
Platform Treatment	Unprepared: 65% Prepared: 36% n=76	Unprepared: 49% Prepared: 51% n=85	No p=.06, Fisher's Exact
External Platform Angle (degrees)	mean: 82.1, s.d.: 12.0, n=58	mean: 82.4, s.d.: 12.0, n=98	No p=.89, t=-0.133, df=127
Platform Thickness	mean: 4.54, s.d.: 2.17, n=62	mean: 6.60, s.d.: 3.11, n=72	Yes p=.00, t=-4.508*, df=126.771
Number of Changes/3 Steps			**1/3=0.33**
Direction of Core Exploitation			
Direction of Early Exploitation	Bidirectional	Unidirectional	Yes
Direction of Late Exploitation	Unidirectional	Unidirectional	No
Number of Changes/2 Steps			**1/2=0.5**
Dorsal Surface Convexity System			
Longitudinal Convexity: Length/Width Ratio	mean: 2.14, s.d.: 1.05, n=99	mean: 2.47, s.d.: 1.12, n=98	Yes p=.03, t=-2.148, df=195
Shape of Convexity: Lateral Edges of Blanks	Parallel: 65% Convergent: 18% Expanding: 10% Ovoid: 6% n=98	Parallel: 54% Convergent: 30% Expanding: 6% Ovoid: 10% n=98	No p=.13, G^2=5.656, df=3
Curvature of Convexity: Profile of Blanks	Straight: 51% Curved: 30% Twisted: 19% n=98	Straight: 49% Curved: 27% Twisted: 24% n=98	No p=.68, G^2=0.787, df=2
Lateral Convexity: Cross-Section of Blanks	Triangular: 46% Trapezoidal: 51% Other: 4% n=99	Triangular: 45% Trapezoidal: 46% Other: 9% n=98	No p=.33, G^2=2.243, df=2
Vertical Convexity: Width/Thickness Ratio	mean: 4.46, s.d.: 2.35, n=99	mean: 3.86, s.d.: 1.60, n=98	Yes p=.04, t=2.066*, df=173.395
Number of Changes/5 Steps			**2/5=0.4**
Total Measure of Difference in Blank Production			**1.73**

*The t-test was calculated with equal variances not being assumed due to a significant F value on Levene's Test for Equality of Variances.

Table 7.96 Comparison of Tool Kit Morphologies for Boker Tachtit Levels 2 and 4

Tool Kit Morphology Variable	Boker Tachtit 2	Boker Tachtit 4	Significant Difference?
Laminarity	mean: 1.87, s.d.: 0.87, n=210	mean: 2.53, s.d.: 0.89 n=157	Yes p=.00, t=-7.051, df=365
Width/Thickness	N/A	N/A	N/A
Lateral Edges	N/A	N/A	N/A
Distal Terminus	Blunt: 43% Pointed: 57% n=159	Blunt: 40% Pointed: 60% n=141	No p=.73, Fisher's Exact
Profile	N/A	N/A	N/A
Unique Types of Retouch	Emireh retouch	Normal retouch	Yes
Tool Types	UP tools dominate	UP tools dominate	No
Number of Changes/4 Steps			**2/4**
Total Measure of Difference in Tool Kit Morphology			**0.5**

Table 7.97 Kebara Cave Units IV and III, comparability of the assemblages based on intensity of core and tool reduction

Variables	Kebara Cave Unit IV	Kebara Cave Unit III	Evaluation
Tools/Flake Ratio	0.14	0.17	Similar
Non-cortical Debitage/ Cortical Debitage Ratio	2.23	1.93	Similar
Henry's Dimensional Analysis	Moderate core exhaustion	Significant core exhaustion	Moderately different

refit data that Level 4 possesses a straight unidirectional core reduction strategy in contrast to Level 2's shift from an alternating to sequential bidirectional strategy, the debitage patterns in fact are similar in the shortest length quartile. For the purpose of comparability and resolution of data for the inter-regional study, however, these details in how the debitage pattern is produced are ignored. A similarity is thus recorded between the assemblages for the debitage pattern of late core exploitation. Of the five steps within the blank production domain, only the elongation of the debitage and the width/thickness ratios show statistically significant differences. The final measure of difference between the blank production details of Boker Tachtit Levels 2 and 4, weighted by the four knapping domains, produces a value of 1.73 out of a possible maximum difference of 4.0, indicating that the blank production details are marginally different.

Tool Kit Morphology

Of the four variables available in both of these assemblages for the consideration of the morphology of the tool kit, two show statistically significant differences (Table 7.93). First, the tools

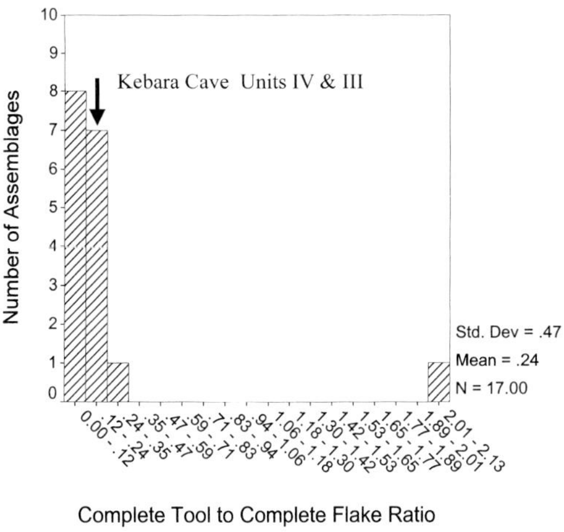

Figure 7.50 Kebara Cave Units IV and III compared for tool/flake ratio values.

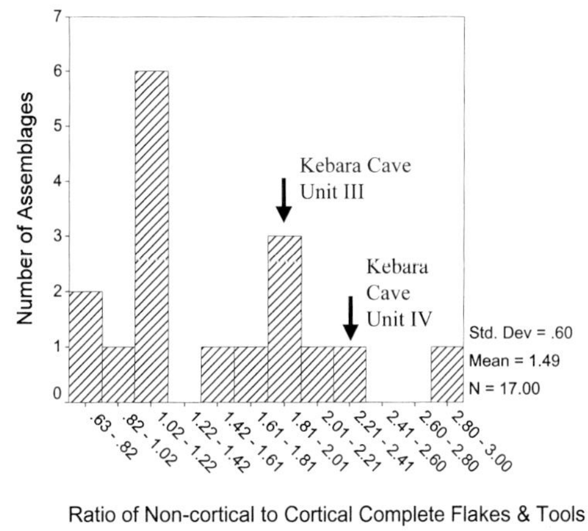

Figure 7.51 Kebara Cave Units IV and III compared for non-cortical/cortical flake ratio values.

in Boker Tachtit Level 2 are less elongated in their length/width ratio compared to Level 4. Second, the absence of Emireh retouch in Level 4 divides the two assemblages. Together, these variables reflect a difference measure of 0.50 out of a maximum of 1.0, indicating that the tool kits are moderately different.

7.9.5 Comparison of Kebara Unit IV and Unit III

The next pair-wise comparison of the Levantine sample is between the Ahmarian assemblages of Units IV and III of Kebara Cave. Given the similarity in contexts, it is not surprising that the comparability measures of these assemblages are quite similar (Table 7.97). The tool to flake ratios (Figure 7.50) indicate that both assemblages are similarly lightly retouched, leaving no problem from the comparability of dimensional data for the tool kits. The non-cortical to cortical blank ratios (Figure 7.51) both fall in the same grouping of assemblages, those with some working of cortical nodules on-site but less than seen in a raw material workshop. The higher degree of core exhaustion evidenced in Unit III than in Unit IV, together with the fact that Unit IV lies on the upper boundary of this group on the non-cortical to cortical ratio histogram, may indicate that Unit IV shares less of the beginning of the core reduction sequence than Unit III does. Even if both assemblages lack the initial working of nodules on site, however, they overlap significantly with each other, if not with raw material workshop assemblages.

Blank Production

Table 7.98 presents the results of a comparison of the blank production details of these assemblages. Both Ahmarian assemblages share the same core orientation but differ in the use of core tablets for convexity management in the latter assemblage. Within the platform maintenance domain, all three steps show significantly

Table 7.98 Comparison of Blank Production for Kebara Cave Unit IV and Unit III

Flintknapping Steps by Domain	Kebara IV	Kebara III	Significant Difference?
Core Modification			
Core Orientation	Use of one longitudinal surface	Use of one longitudinal surface	No
Core Management	Débordants & frontal crest	Frontal crest & core tablets	Yes
Number of Changes/2 Steps			**1/2=0.5**
Platform Maintenance			
Platform Treatment	Unprepared: 59% Prepared: 41% n=334	Unprepared: 67% Prepared: 33% n=439	Yes p=.02, Fisher's Exact
External Platform Angle (degrees)	mean: 88.0, s.d.: 14.9, n=289	mean: 92.0, s.d.: 15.7 n=355	Yes p=.00, t=-3.28,df=642
Platform Thickness	mean: 4.31, s.d.: 2.56, n=291	mean: 3.91, s.d.: 2.57 n=357	Yes p=.05, t=1.98, df=646
Number of Changes/3 Steps			**3/3=1**
Direction of Core Exploitation			
Direction of Early Exploitation	Bidirectional & Unidirectional	Bidirectional & Unidirectional (see text for this exception)	No
Direction of Late Exploitation	Unidirectional	Unidirectional	No
Number of Changes/2 Steps			**0/2=0.0**
Dorsal Surface Convexity System			
Longitudinal Convexity: Length/Width Ratio	mean: 2.17, s.d.: 1.16, n=440	mean: 2.51, s.d.: 1.24 n=659	Yes p=.00, t=-4.55, df=982.4
Shape of Convexity: Lateral Edges of Blanks	Parallel: 53% Convergent: 22% Expanding: 15% Ovoid: 10% n=432	Parallel: 58% Convergent: 22% Expanding: 14% Ovoid: 7% n=633	No p=.28, G^2=3.86, df=3
Curvature of Convexity: Profile of Blanks	Straight: 58% Curved: 25% Twisted: 17% n=437	Straight: 48% Curved: 32% Twisted: 20% n=644	Yes p=.01, G^2=10.28, df=2
Lateral Convexity: Cross-Section of Blanks	Triangular: 45% Trapezoidal: 47% Other: 8% n=434	Triangular: 42% Trapezoidal: 47% Other: 11% n=642	No p=.37, G^2=2.02, df=2
Vertical Convexity: Width/Thickness Ratio	mean: 4.35, s.d.: 2.11, n=440	mean: 3.87, s.d.: 1.79 n=659	Yes p=.00, t=3.89, df=832.2
Number of Changes/5 Steps			**3/5=0.6**
Total Measure of Difference in Blank Production			**2.10**

different use of knapping options. While the directionality of debitage in late core exploitation is similarly unidirectional, early exploitation shows a difference in the co-dominance of bidirectional and unidirectional blanks in Unit IV compared to a clear dominance of only bidirectional patterns in Unit III. The fact that the minimum size value (in mm) for the first quartile of Unit IV lies in the middle of the second quartile for Unit III reinforces the non-cortical to cortical blank ratio differences between the assemblages sufficiently to suggest that landscape and raw material acquisition patterns may be influencing the proportionality of bidirectional and unidirectional blanks in the first quartile between the assemblages. In essence, the pattern graphed for Unit IV for quartiles one through four looks like the pattern graphed for Unit III between quartiles two and four. This is a situation similar to, but more extreme than the comparison of this domain between Stránská skála IIIa Layer 4, Stránská skála III, and Stránská skála IIIc, although there is a greater overlap in the reduction sequences in the Central European case. Thus, unlike the Central European case, the lack of the assemblages' overlap at the beginning of core exploitation warrants a change in the procedure for comparing the behavioral signature of early core exploitation. Instead of characterizing early core exploitation of Unit III for the present comparison with Unit IV at the first quartile, this characterization is made at the second quartile to counterbalance the missing early portion of the reduction sequence at Unit IV. This results in a judgment of "bidirectional and unidirectional" reduction for both assemblages for this analytical step.

Of the five variables within the dorsal surface convexity system, three options were judged to be significantly different between the two assemblages: the length/width ratio, profile, and the width/thickness ratio variables. The final measure of difference between the operational sequences of Kebara Units IV and III, weighted by the four knapping domains, produces a value of 2.10 out of a possible maximum difference of 4.0, indicating that the operational sequences are as dissimilar as they are similar. This is a surprising result given the assemblages' common appellation of "Ahmarian" (Bar-Yosef et al. 1996).

Tool Kit Morphology

Of the variables which characterize the tool kit morphology of these two assemblages, one discontinuous and two continuous variables show statistically significant differences. Kebara Unit III is considerably more bladey than Unit IV as well as thicker relative to width at the cross-section. Unit III is also significantly more pointed at the distal end of its tool blanks than Unit IV. No differences are evidenced in location or type of retouch in the two assemblages. These results provide a difference measure of 0.43 out of a maximum of 1.0, indicating that the tool kits are more similar than dissimilar.

7.9.6 Comparison of Kebara Unit III and Boker Tachtit Level 4

The next pair-wise comparison of the Levantine sample is between the transitional assemblage from Level 4 of Boker Tachtit and the Ahmarian assemblage from Unit III of Kebara Cave. Given the difference in the site contexts of the two assemblages, one open air and one cave, it is not surprising that the comparability measures of these assemblages are somewhat different (Table 7.100). The tool to flake ratios (Figure 7.52) indicate that Kebara Unit III has a slightly higher frequency of tools than Boker Tachtit Level 4, but not enough to be significant. The non-cortical to cortical blank ratios (Figure 7.53) show a greater degree of difference, with Boker Tachtit Level 4

Table 7.99 Comparison of Tool Kit Morphologies for Kebara Cave Unit IV and Unit III

Tool Kit Morphology Variable	Kebara IV	Kebara III	Significant Difference?
Laminarity	mean: 2.23, s.d.: 1.12 n=52	mean: 3.09, s.d.: 1.19, n=133	Yes p=.00, t=-4.473, df=183
Width/Thickness	mean: 4.04, s.d.: 1.80 n=52	mean: 3.49, s.d.: 1.29, n=133	Yes p=.05, t=2.013*, df=72.449
Lateral Edges	Parallel: 57% Convergent: 23% Expanding: 6% Ovoid: 14% n=51	Parallel: 68% Convergent: 22% Expanding: 5% Ovoid: 5% n=124	No p=.32, G^2=3.480, df=3†
Distal Terminus	Blunt: 50% Pointed: 50% n=32	Blunt: 24% Pointed: 76% n=88	Yes p=.01, Fisher's Exact
Profile	Straight: 49% Curved: 24% Twisted: 27% n=51	Straight: 33% Curved: 39% Twisted: 28% n=129	No p=.09, G^2=4.921, df=2
Unique Types of Retouch	Normal retouch	Normal retouch	No
Tool Types	UP tools dominate	UP tools dominate	No
Number of Changes/7 Steps			**3/7**
Total Measure of Difference in Tool Kit Morphology			**0.43**

*The t-test was calculated with equal variances not being assumed due to a significant F value on Levene's Test for Equality of Variances.

†Two cells (25%) have an expected count less than 5. The minimum expected count is 2.62.

falling with the raw material workshop assemblages far to the left of the histogram and Kebara III falling in the middle group of assemblages but significantly above the mean. Kebara Unit III also displays more core exhaustion than evidenced in Boker Tachtit Level 4. As it appears likely that Kebara III lacks somewhat the earliest portions of the core reduction sequence as captured in Boker Tachtit Level 4 while the latter lacks the behavioral output of core exhaustion in the former, some of the differences noted in the following comparisons may be effected by the different landscape and raw material contexts of the two assemblages.

Blank Production

Table 7.101 presents the results of a comparison of the blank production details of the assemblages from Kebara Unit III and Boker Tachtit Level 4. Both assemblages share the same core modification options but differ in all respects to platform maintenance. The direction of core exploitation as evidenced in the dorsal scar pattern of the debitage indicates a similarity at the end of core exploitation but significant difference in the strategy in early exploitation. Of the five steps within the surface convexity system, only the lateral edge types differ significantly. This results in a final measure of difference between the blank

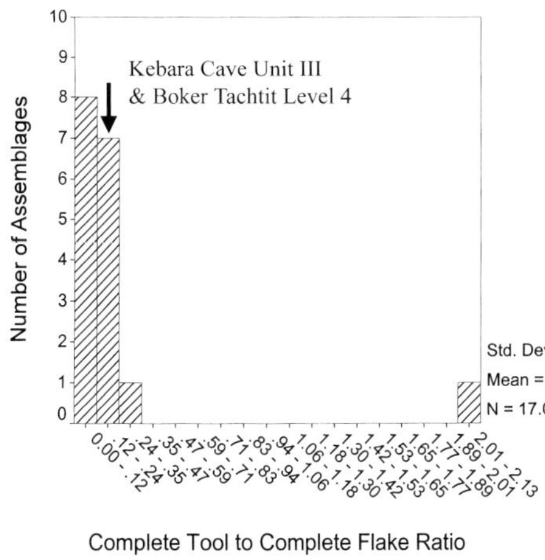

Figure 7.52 Kebara Cave Unit III and Boker Tachtit Level 4 compared for tool/flake ratio values.

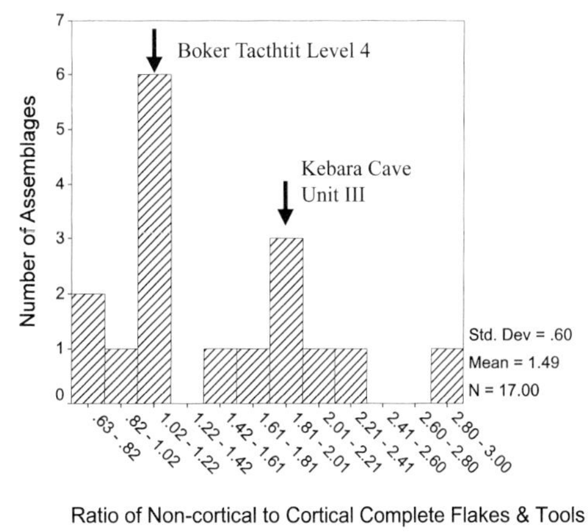

Figure 7.53 Kebara Cave Unit III and Boker Tachtit Level 4 compared for non-cortical/cortical flake ratio values.

Table 7.100 Kebara Cave Unit III and Boker Tachtit Level 4, comparability of the assemblages based on intensity of core and tool reduction

Variables	Kebara Cave Unit III	Boker Tachtit Level 4	Evaluation
Tools/Flake Ratio	0.17	0.13	Similar
Non-cortical Debitage/ Cortical Debitage Ratio	1.93	1.14	Moderately Different
Henry's Dimensional Analysis	Significant core exhaustion	See text	Different

production details of Kebara Unit III and Boker Tachtit Level 4, weighted by the four knapping domains, of 1.70 out of a possible maximum difference of 4.0, indicating that the blank production details are fairly similar.

Tool Kit Morphology

Of the four variables available in both of these assemblages for the consideration of the morphology of the tool kit, two show statistically significant differences (Table 7.102). First, the tools in Boker Tachtit Level 4 are less elongated in their length/width ratio compared to Kebara Unit III. Second, the distal ends of the tool blanks in Boker Tachtit Level 4 are less pointed than those in Kebara Unit III. No differences exist in application or type of retouch. Together, these variables reflect a difference measure of 0.50 out of a maximum of 1.0, indicating that the tool kits are as different as they are alike.

7.9.7 Comparison of Kebara Unit III and Unit II

The next pair-wise comparison of the Levantine sample is between the Ahmarian assemblage of Units III and the Levantine Aurignacian assemblage of Unit II of Kebara

Table 7.101 Comparison of Blank Production for Kebara Cave Unit III and Boker Tachtit 4

Flintknapping Steps by Domain	Kebara III	Boker Tachtit 4	Significant Difference?
Core Modification			
Core Orientation	Preferential use of a Longitudinal Surface	Preferential use of a Longitudinal Surface	No
Core Management	Frontal crest & core tablets	Frontal crest & core tablets	No
Number of Changes/2 Steps			**0/2=0**
Platform Maintenance			
Platform Treatment	Unprepared: 67% Prepared: 33% n=439	Unprepared: 49% Prepared: 51% n=85	Yes p=.00, Fisher's Exact
External Platform Angle (degrees)	mean: 92.0, s.d.: 15.7 n=355	mean: 82.4, s.d.: 12.0, n=71	Yes p=.00, t=5.810*, df=123.042
Platform Thickness	mean: 3.91, s.d.: 2.57 n=357	mean: 6.60, s.d.: 3.11, n=72	Yes p=.00 t=-6.891*, df=91.562
Number of Changes/3 Steps			**3/3=1**
Direction of Core Exploitation			
Direction of Early Exploitation	Bidirectional	Unidirectional	Yes
Direction of Late Exploitation	Unidirectional	Unidirectional	No
Number of Changes/2 Steps			**1/2=0.5**
Dorsal Surface Convexity System			
Longitudinal Convexity: Length/Width Ratio	mean: 2.51, s.d.: 1.24 n=659	mean: 2.47, s.d.: 1.12, n=98	No p=.78, t=0.286, df=755
Shape of Convexity: Lateral Edges of Blanks	Parallel: 58% Convergent: 22% Expanding: 14% Ovoid: 7% n=633	Parallel: 54% Convergent: 30% Expanding: 6% Ovoid: 10% n=98	Yes p=.04, G^2=8.151, df=3
Curvature of Convexity: Profile of Blanks	Straight: 48% Curved: 32% Twisted: 20% n=644	Straight: 49% Curved: 27% Twisted: 24% n=98	No p=.51, G^2=1.351, df=2
Lateral Convexity: Cross-Section of Blanks	Triangular: 42% Trapezoidal: 47% Other: 11% n=642	Triangular: 45% Trapezoidal: 46% Other: 9% n=98	No p=.81, G^2=0.423, df=2
Vertical Convexity: Width/Thickness Ratio	mean: 3.87, s.d.: 1.79 n=659	mean: 3.86, s.d.: 1.60, n=98	No p=.96, t=0.047, df=755
Number of Changes/5 Steps			**1/5=0.2**
Total Measure of Difference in Blank Production			**1.70**

*The t-test was calculated with equal variances not being assumed due to a significant F value on Levene's Test for Equality of Variances.

Table 7.102 Comparison of Tool Kit Morphologies for Kebara Cave Unit III and Boker Tachtit 4

Tool Kit Morphology Variable	Kebara III	Boker Tachtit 4	Significant Difference?
Laminarity	mean: 3.09, s.d.: 1.19, n=133	mean: 2.53, s.d.: 0.89 n=157	Yes p=.00, t=4.460*, df=242.128
Width/Thickness	mean: 3.49, s.d.: 1.29, n=133	N/A	N/A
Lateral Edges	Parallel: 68% Convergent: 22% Expanding: 5% Ovoid: 5% n=124	N/A	N/A
Distal Terminus	Blunt: 24% Pointed: 76% n=88	Blunt: 40% Pointed: 60% n=141	Yes p=.01, Fisher's Exact
Profile	Straight: 33% Curved: 39% Twisted: 28% n=129	N/A	N/A
Unique Types of Retouch	Normal retouch	Normal retouch	No
Tool Types	UP tools dominate	UP tools dominate	No
Number of Changes/4 Steps			**2/4**
Total Measure of Difference in Tool Kit Morphology			**0.50**

*The t-test was calculated with equal variances not being assumed due to a significant F value on Levene's Test for Equality of Variances.

Table 7.103 Kebara Cave Units II and II, comparability of the assemblages based on intensity of core and tool reduction

Variables	Kebara Cave Unit III	Kebara Cave Unit II	Evaluation
Tools/Flake Ratio	0.17	0.08	Moderately Different
Non-cortical Debitage/ Cortical Debitage Ratio	1.93	1.91	Similar
Henry's Dimensional Analysis	Significant core exhaustion	Significant core exhaustion	Similar

Cave. Given the similarity in contexts, it is not surprising that the comparability measures of these assemblages are quite similar (Table 7.103). The tool to flake ratios (Figure 7.54) indicate that both assemblages are lightly retouched when compared to the distribution of all assemblages, but Unit III shows more tools per blank than Unit II. The difference is not great enough to suppose a significant effect on the comparability of dimensional data for the tool kits. The non cortical to cortical blank ratios (Figure 7.55) are almost identical, falling within the same bar of the histogram, in the group of assemblages evidencing some but not much working of cortical nodules on-site. With a relatively equal degree of significant core exhaustion, these two assemblages may be

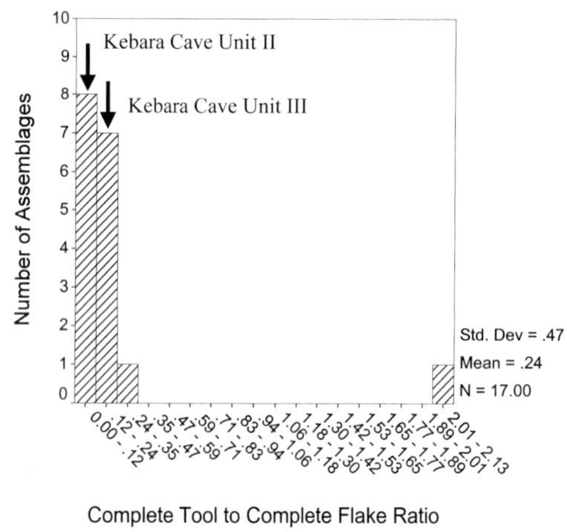

Figure 7.54 Kebara Cave Units III and II compared for tool/flake ratio values.

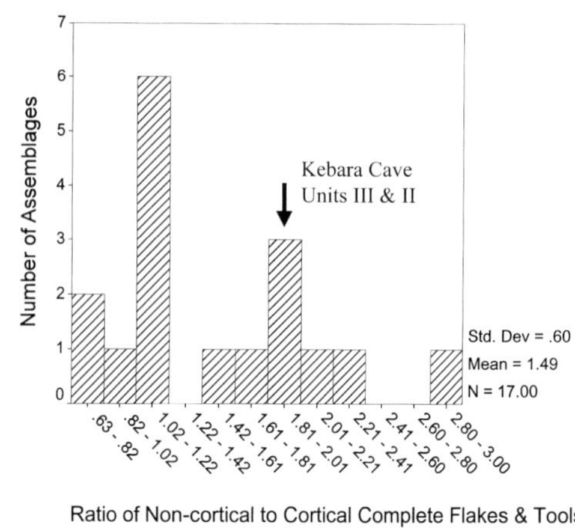

Figure 7.55 Kebara Cave Units III and II compared for non-cortical/cortical flake ratio values.

considered to contain portions of their core reduction sequences which overlap to a very high degree. For this reason, they should be very comparable to each other, but not necessarily to a raw material workshop with little core exhaustion.

Blank Production

Table 7.104 presents the results of a comparison of the blank production details of these assemblages. Both assemblages share the same core modification options but differ in all but one of the options in the platform maintenance domain. The assemblages also differ in the dominant debitage scar pattern in early core exploitation. The small absolute lengths of the quartiles in Unit II compared to Unit III highlights the high frequency of bladelets in the debitage of this assemblage. Of the five steps within the dorsal surface convexity domain, the tendencies toward elongation, width/thickness ratios, and cross-sections also differ significantly. Together, these data result a final difference measure of 1.57, out of a possible maximum difference of 4.0, indicating that some blank production details are dissimilar but most are not.

Tool Kit Morphology

Of the variables which characterize the tool kit morphology of these two assemblages, only one discontinuous variable shows a significant difference. The tools of Kebara Unit III are on the border of significance for their bladey length/width ratios compared to those of Unit II. The contrast between convergent and ovoid lateral edges in the tool kits is also suggestive but not significant, due to the violation of the assumptions of the G^2 statistic as a result of the low sample size of the Unit II tool kit. It is the presence of carination retouch in Unit II and its absence in Unit III which contributes to the final difference measure of 0.14 out of a maximum of 1.0, indicating that the tool kits are extremely similar.

Table 7.104 Comparison of Blank Production for Kebara Cave Unit III and Unit II

Flintknapping Steps by Domain	Kebara III	Kebara II	Significant Difference?
Core Modification			
Core Orientation	Preferential use of a Longitudinal Surface	Preferential use of a Longitudinal Surface	No
Core Management	Frontal crest & core tablets	Frontal crest & core tablets	No
Number of Changes/2 Steps			**0/2=0**
Platform Maintenance			
Platform Treatment	Unprepared: 67% Prepared: 33% n=439	Unprepared: 75% Prepared: 25% n=331	Yes p=.02, Fisher's Exact
External Platform Angle (degrees)	mean: 92.0, s.d.: 15.7 n=355	mean: 85.7, s.d.: 14.4, n=255	Yes p=.00, p=.00, t=5.02, df=608
Platform Thickness	mean: 3.91, s.d.: 2.57 n=357	mean: 3.85, s.d.: 2.78, n=259	No p=.76, t=0.31, df=614
Number of Changes/3 Steps			**2/3=0.67**
Direction of Core Exploitation			
Direction of Early Exploitation	Bidirectional	Unidirectional	Yes
Direction of Late Exploitation	Unidirectional	Unidirectional	No
Number of Changes/2 Steps			**1/2=0.5**
Dorsal Surface Convexity System			
Longitudinal Convexity: Length/Width Ratio	mean: 2.51, s.d.: 1.24 n=659	mean: 2.20, s.d.: 1.19, n=442	Yes p=.00, t=4.03, df=1099
Shape of Convexity: Lateral Edges of Blanks	Parallel: 58% Convergent: 22% Expanding: 14% Ovoid: 7% n=633	Parallel: 55% Convergent: 21% Expanding: 14% Ovoid: 10% n=415	No p=.44, G^2=2.68, df=3
Curvature of Convexity: Profile of Blanks	Straight: 48% Curved: 32% Twisted: 20% n=644	Straight: 49% Curved: 32% Twisted: 20% n=429	No p=.93, G^2=.14, df=2
Lateral Convexity: Cross-Section of Blanks	Triangular: 42% Trapezoidal: 47% Other: 11% n=642	Triangular: 42% Trapezoidal: 52% Other: 6% n=431	Yes p=.01, G^2=9.13, df=2
Vertical Convexity: Width/Thickness Ratio	mean: 3.87, s.d.: 1.79 n=659	mean: 4.00, s.d.: 1.96, n=442	No p=.28, t=-1.09, df=1099
Number of Changes/5 Steps			**2/5=0.4**
Total Measure of Difference in Blank Production			**1.57**

Table 7.105 Comparison of Tool Kit Morphologies for Kebara Cave Unit III and Unit II

Tool Kit Morphology Variable	Kebara III	Kebara II	Significant Difference?
Laminarity	mean: 3.09, s.d.: 1.19, n=133	mean: 2.60, s.d.: 1.58 n=47	No p=.06, t=1.924*, df=65.167
Width/Thickness	mean: 3.49, s.d.: 1.29, n=133	mean: 3.20, s.d.: 1.17 n=47	No p=.17, t=1.395, df=178
Lateral Edges	Parallel: 68% Convergent: 22% Expanding: 5% Ovoid: 5% n=124	Parallel: 67% Convergent: 12% Expanding: 9% Ovoid: 12% n=33	No p=.32, G^2=3.476, df=3†
Distal Terminus	Blunt: 24% Pointed: 76% n=88	Blunt: 24% Pointed: 76% n=21	No p=.99, Fisher's Exact
Profile	Straight: 33% Curved: 39% Twisted: 28% n=129	Blunt: 24% Curved: 39% Twisted: 30% Straight: 32% n=44	No p=.97, G^2=0.054, df=2
Unique Types of Retouch	Normal retouch	Carinated retouch	Yes
Tool Types	UP tools dominate	UP tools dominate	No
Number of Changes/7 Steps			**1/7**
Total Measure of Difference in Tool Kit Morphology			**0.14**

*The t-test was calculated with equal variances not being assumed due to a significant F value on Levene's Test for Equality of Variances.
†Two cells (25%) have an expected count less than 5. The minimum expected count is 1.89.

Table 7.106 Kebara Cave Units II and I, comparability of the assemblages based on intensity of core and tool reduction

Variables	Kebara Cave Unit II	Kebara Cave Unit I	Evaluation
Tools/Flake Ratio	0.08	0.12	Similar
Non-cortical Debitage/ Cortical Debitage Ratio	1.91	1.79	Similar
Henry's Dimensional Analysis	Significant core exhaustion	Significant core exhaustion	Similar

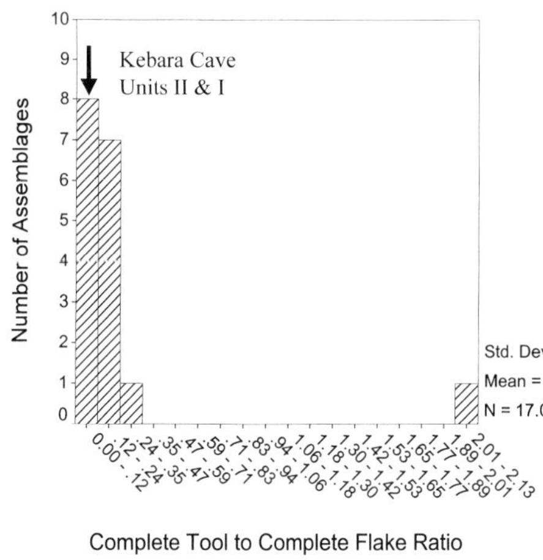

Figure 7.56 Kebara Cave Units II and I compared for tool/flake ratio values.

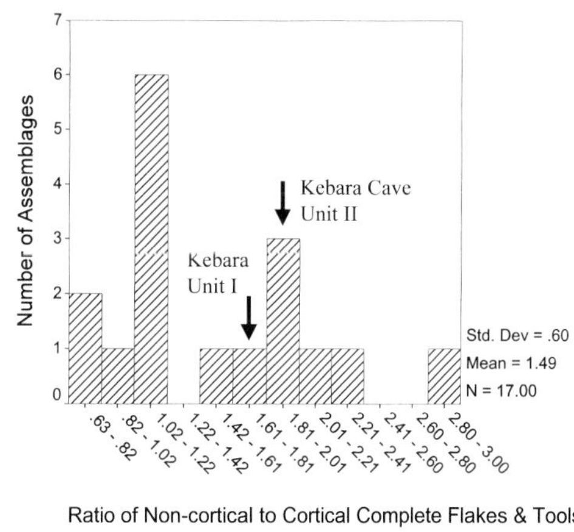

Figure 7.57 Kebara Cave Units II and I compared for non-cortical/cortical flake ratio values.

7.9.8 Comparison of Kebara Unit II and Unit I

The final pair-wise comparison of the Levantine sample is between the two Levantine Aurignacian assemblages of Units II and I of Kebara Cave. Given the similarity in contexts, it is not surprising that the comparability measures of these assemblages are quite similar (Table 7.106). The tool to flake ratios (Figure 7.56) indicate that both assemblages are lightly retouched, when compared to both themselves and the distribution of all assemblages. The non-cortical to cortical blank ratios (Figure 7.57) are similar, falling within adjacent histogram bars within the group of assemblages evidencing some, but not much, working of cortical nodules on-site. With a relatively equal degree of significant core exhaustion, these two assemblages may be considered to contain portions of their core reduction sequences which overlap to a very high degree. For this reason, they should be very comparable to each other, but not necessarily to a raw material workshop with little core exhaustion.

Blank Production

Table 7.107 presents the results of a comparison of the operational sequences of these two Levantine Aurignacian assemblages. Both assemblages share the same core modification choices, platform maintenance choices, and directionality of debitage during core exploitation. Of the five steps within the surface convexity domain, only the length/width tendencies are different, with the later assemblage producing more blades/bladelets. A final difference measure of 0.20, out of a possible maximum difference of 4.0, indicate that the blank production details are almost identical. This result supports the union of these two assemblages under one industrial label.

Tool Kit Morphology

Of the variables which characterize the tool kit morphology of these two assemblages, only one discontinuous variable shows a significant difference. The tools of Kebara Unit II are more curved and twisted in profile compared to the

Table 7.107 Comparison of Blank Production for Kebara Cave Unit II and Unit I

Flintknapping Steps by Domain	Kebara II	Kebara I	Significant Difference?
Core Modification			
Core Orientation	Preferential use of a Longitudinal Surface	Preferential use of a Longitudinal Surface	No
Core Management	Frontal crest & core tablets	Frontal crest & core tablets	No
Number of Changes/2 Steps			**0/2=0**
Platform Maintenance			
Platform Treatment	Unprepared: 75% Prepared: 25% n=331	Unprepared: 76% Prepared: 24% n=354	No p=.72, Fisher's Exact
External Platform Angle (degrees)	mean: 85.7, s.d.: 14.4, n=255	mean: 84.1, s.d.: 13.7, n=298	No p=.19, t=1.308, df=551
Platform Thickness	mean: 3.85, s.d.: 2.78, n=259	mean: 4.09, s.d.: 2.93, n=302	No p=.32, t=-1.004, df=559
Number of Changes/3 Steps			**0/3=0**
Direction of Core Exploitation			
Direction of Early Exploitation	Unidirectional	Unidirectional	No
Direction of Late Exploitation	Unidirectional	Unidirectional	No
Number of Changes/2 Steps			**0/2=0.0**
Dorsal Surface Convexity System			
Longitudinal Convexity: Length/Width Ratio	mean: 2.20, s.d.: 1.19, n=442	mean: 2.39, s.d.: 1.35, n=522	Yes p=.03, t=-2.125*, df=960.403
Shape of Convexity: Lateral Edges of Blanks	Parallel: 55% Convergent: 21% Expanding: 14% Ovoid: 10% n=415	Parallel: 57% Convergent: 22% Expanding: 13% Ovoid: 8% Parallel: 57% n=472	No p=.81, G^2=0.975, df=3
Curvature of Convexity: Profile of Blanks	Straight: 49% Curved: 32% Twisted: 20% n=429	Straight: 53% Curved: 28% Twisted: 19% n=501	No p=.46, G^2=1.542, df=2
Lateral Convexity: Cross-Section of Blanks	Triangular: 42% Trapezoidal: 52% Other: 6% n=431	Triangular: 42% Trapezoidal: 53% Trapezoidal: 53% Other: 6% n=493	No p=.97, G^2=0.065, df=2
Vertical Convexity: Width/Thickness Ratio	mean: 4.00, s.d.: 1.96, n=442	mean: 3.76, s.d.: 2.00, n=522	No p=.06, t=1.881, df=962
Number of Changes/5 Steps			**1/5=0.20**
Total Measure of Difference in Blank Production			**0.20**

*The t-test was calculated with equal variances not being assumed due to a significant F value on Levene's Test for Equality of Variances.

straight profiles dominating the tool kit of Unit I. The profile variable alone contributes to the final difference measure of 0.14 out of a maximum of 1.0, indicating that the tool kits are extremely similar.

7.10 Significance of the Levantine Sequence for Model Expectations 1-4

Now that the Levantine sequence of change has been reconstructed, it is possible to assess some of the model expectations for this intra-regional context. It is immediately clear that the operational sequences of the sample assemblages differ to varying degrees (Table 7.109). In the first chronological comparison in this region, the last Middle Paleolithic assemblage, Kebara Cave Unit VI, and the first non-Middle Paleolithic (or transitional) assemblage of Boker Tachtit level 1 exhibit a relatively large degree of difference, i.e., 2.83. While not as big as the difference between Kůlna 7a and either Stránská skála IIIa-4 or Vedrovice V, it is larger than all of the other comparisons in terms of blank production choices. Despite the arguments for an *in situ* technological transition between the Middle and Upper Paleolithic at this site, Boker Tachtit already differs significantly from the last Middle Paleolithic assemblage in the region as here defined. This indicates that any "transition" was already underway by the creation of the basal layer of Boker Tachtit.

It must be noted that Marks (2003:261) is correct in pointing out that the Mediterranean core region of the Levant may be too restrictive and environmentally inappropriate as a region for looking for antecedent assemblages to Boker Tachtit. While I would very much like to sample more broadly across the Levantine landscape and particularly into the more desertic areas, the sampling methodology used in this research is in no way designed to be a straw man, as Marks implies. Kebara Unit VI is simply the most immediately preceding assemblage in age to Boker Tachtit level 1 in the Levant to which I had practical access. It is necessary to have chronostratigraphic control over all of the assemblages to provide an a priori ordering of the data, otherwise the principle of Expectation 1 (Antecedents) would not provide any discrimination. Significantly, the alternative sites Marks suggests are either an undated Middle Paleolithic site, i.e., Ain Difla (Lindly and Clark 1987; Demidenko and Usik 1993a), or assemblages dated beyond the upper limit of 60 kya for inclusion in the study, i.e., Tor Faraj and Tor Sabiha in Jordan dated to 70 kya (Henry 1992, 1995a, 1995b). Nevertheless, given the similarities in morphology noted by Demidenko and Usik (1993a,b) between the bidirectional, Levallois, blade industries of Jordan and the Negev, it would still be wise to add these three sites to the dataset in the future. Until these assemblages are included, however, in an attribute analysis study which demonstrates their similarities in flintknapping options to the Boker Tachtit lineage, one must look to the last Levantine Mousterian site in the Levant, Kebara Unit VI, for the purposes of understanding how the Boker Tachtit assemblages appeared on the landscape. The alternative of defining ecological zones within the Levant and treating them as separate regions according to the hypotheses would also be possible under the sampling protocols used here. Again, however, the problem of undated or too-widely dated assemblages may make this impractical. There are therefore many avenues to sampling the archaeological record to test a particular question that I would accept and suggestions along this line are useful. What I do not accept is Marks' insinuation that the sampling of this study was rigged simply to annoy him.

Returning to the intra-regional sequence of change, the agreement between the transitional

Table 7.108 Comparison of Tool Kit Morphologies for Kebara Cave Unit II and Unit I

Tool Kit Morphology Variable	Kebara II	Kebara I	Significant Difference?
Laminarity	mean: 2.60, s.d.: 1.58 n=47	mean: 2.35, s.d.: 1.50 n=101	No p=.36, t=0.925, df=146
Width/Thickness	mean: 3.20, s.d.: 1.17 n=47	mean: 3.34, s.d.: 1.88 n=101	No p=.63, t=-0.487, df=146
Lateral Edges	Parallel: 67% Convergent: 12% Expanding: 9% Ovoid: 12% n=33	mean: 3.34, Convergent:21% Parallel: 59% Convergent:21% Expanding: 13% Ovoid: 7% n=80	No p=.54, G^2=2.153, df=3*
Distal Terminus	Blunt: 24% Pointed: 76% n=21	Blunt: 40% Pointed: 60% n=45	No p=.27, Fisher's Exact
Profile	Straight: 32% Curved: 39% Twisted: 30% n=44	Straight: 56% Curved: 27% Twisted: 17% n=93	Yes p=.03, G^2=7.208, df=2
Unique Types of Retouch	Carinated retouch	Carinated retouch	No
Tool Types	UP tools dominate	UP tools dominate	No
Number of Changes/7 Steps			**1/7**
Total Measure of Difference in Tool Kit Morphology			**0.14**

*Two cells (25%) have an expected count less than 5. The minimum expected count is 2.92.

Table 7.109 Measurement of the difference in flintknapping behaviors between Levantine assemblages

Comparison of Assemblages Through Time	Measure of Difference in Blank Production (Maximum=4, Minimum=0)	Measure of Difference in Tool Kit Morphology (Maximum=1, Minimum=0)
Kebara VI versus Boker Tachtit 1	2.83	0.75
Boker Tachtit 1 versus Boker Tachtit 2	0.33	0.25
Boker Tachtit 2 versus Boker Tachtit 4	1.73	0.50
Boker Tachtit 2 versus Kebara IV	0.83	0.50
Kebara IV versus Kebara III	2.10	0.43
Kebara III versus Boker Tachtit 4	1.70	0.50
Kebara III versus Kebara II	1.57	0.14
Kebara II versus Kebara I	0.20	0.14

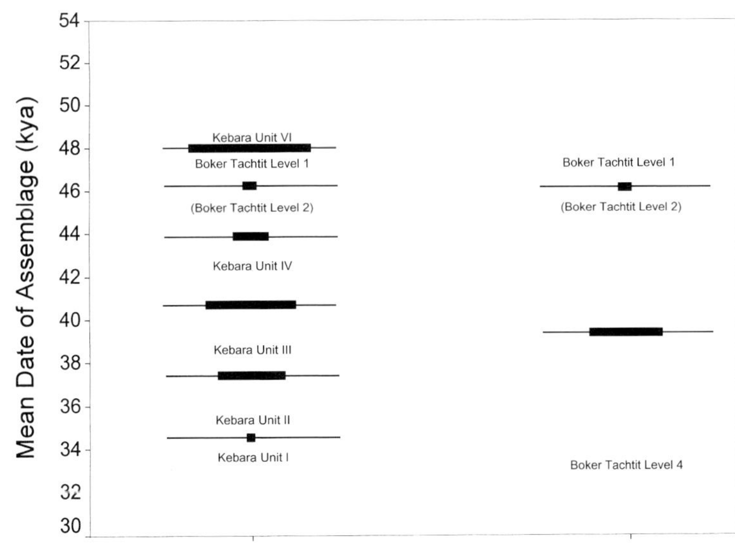

Figure 7.58 Chronological comparison of differences in Blank Production among Levantine assemblages. The length of the thick bars represents the difference measure between the two assemblages above and below the bar (the earlier and the later). The length of the thin bars represents the maximum possible difference measure. The three columns represent different sequences for pairing assemblages for comparison.

levels of Boker Tachtit 1 and 2 is extremely close (0.33), the only closer comparison in the study being the two superimposed Levantine Aurignacian assemblage from Kebara II and I. The two early Boker Tachtit assemblages also show a great deal of similarity to the early Ahmarian of Kebara Unit IV, indicating that the Boker Tachtit lineage likely transmitted some technical choice tendencies to the artisans of the later Upper Paleolithic in this region.

Kebara Unit IV displays a marked difference with its immediate stratigraphic successor, Unit III. The strong deviation between these assemblages (2.10) puts doubt onto the association of these assemblages to a common industrial type. They differ even more than the Ahmarian and Levantine Aurignacian do (Unit III vs. II = 1.57), upon the latter's appearance in the region. This may be the result of biased sampling of the archaeological record from Kebara Unit IV, as Belfer-Cohen and Bar-Yosef (personal communication) have identified a collection of blanks and tools which were missing from the original collection I studied. The possibility of this bias will be evaluated through future research. Within the Levantine Aurignacian itself, flintknapping options change through time but maintain a close similarity of behaviors (0.20).

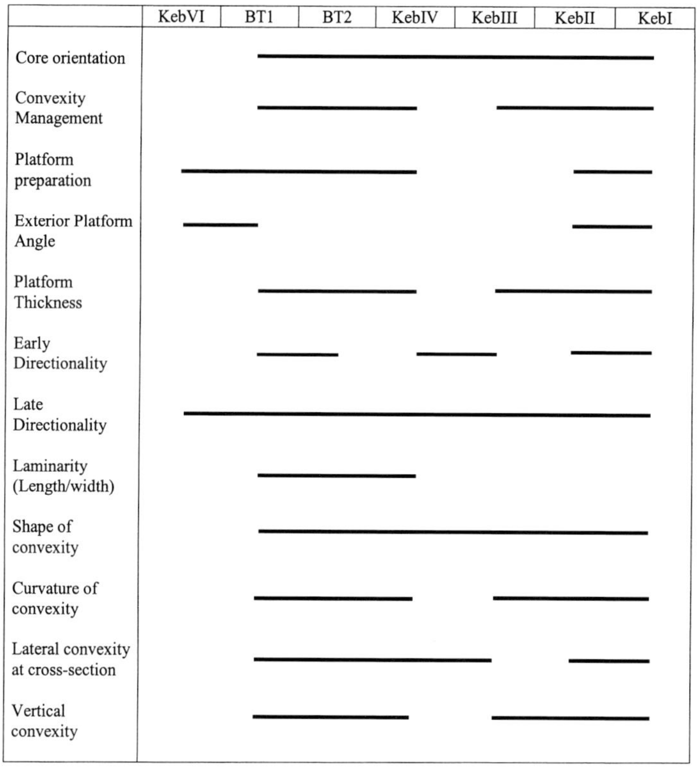

Figure 7.59 Variable by variable view of pair-wise comparisons within the Levant between 60 and 30 kya. Bars connecting two assemblages represent the use of statistically similar attribute states between the two assemblages. Similarities are not necessarily implied between assemblages connected by a bar which are not immediately adjacent to each other.

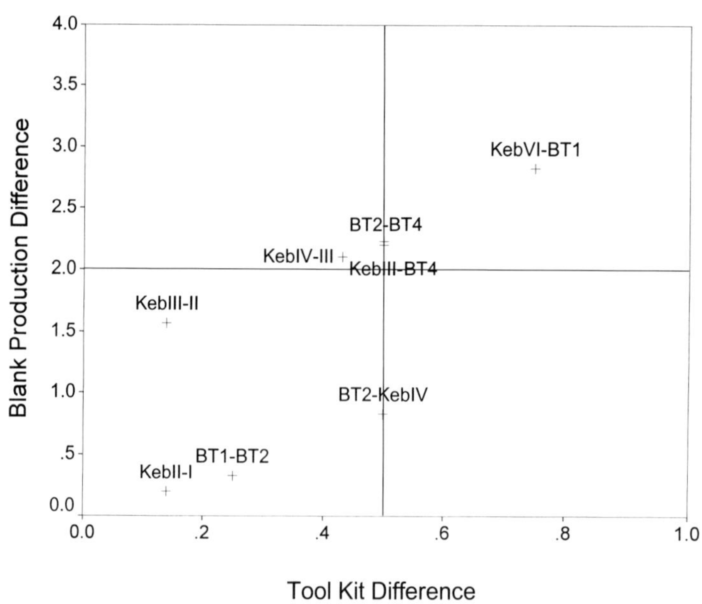

Figure 7.60 Blank Production vs. Tool Kit Differences for pair-wise comparisons for Levantine assemblages. Assemblage Key: BT1: Boker Tachtit Level 1; BT2: Boker Tachtit Level 2; BT4: Boker Tachtit Level 4; KebI: Kebara Cave Unit I; KebII: Kebara Cave Unit II; KebIII: Kebara Cave Unit III; KebIV: Kebara Cave Unit IV; KebVI: Kebara Cave Unit VI.

8

The Inter-Regional Pattern of Change

The intra-regional patterns of change in flintknapping options used by hominins between 60 and 30 kya, described in the last three chapters, illustrate some interesting commonalities. It is apparent after the reconstruction of the Central European, Eastern European, and Levantine sequences that these regional contexts share their most marked difference in flintknapping behavior between the earliest assemblage included in this study and the next subsequent assemblage on the scene, usually a non-Middle Paleolithic assemblage. This commonality between the three regions warrants a direct inter-regional comparison of the assemblages which follow the last Middle Paleolithic assemblage in each region.

8.1 Inter-regional Pair-Wise Comparisons

8.1.1 Comparison of Boker Tachtit Level 1 and Stránská skála IIIa Layer 4

The first pair-wise comparison of the inter-regional study is between the transitional assemblage from Boker Tachtit Level 1 and the Bohunician assemblage from Stránská skála IIIa Layer 4. This represents a systematic assay of the Valoch's (1986) suggestion of similarities between Boker Tachtit Level 1 and the Bohunician, at the time represented by the type collection from Brno-Bohunice. The comparability measures of these assemblages are quite similar (Table 8.1). Both the tool to flake ratios (Figure 8.1), in which the values are found in adjacent bars of the histogram, and the non-cortical to cortical blank ratios (Figure 8.2), in which the assemblages fall in the same bar of the histogram, show that these two assemblages overlap in their coverage of their respective core reduction sequences. Boker Tachtit Level 1 and Stránská skála IIIa Layer 4 both contain the earliest working of cortical nodules at the beginning of the core reduction sequence, indicative of raw material workshops. The refit studies by Volkman and Škrdla, respectively, also highlight the similarity in the nature of the palimpsest formation of these assemblages, catching as they do many sequential knapping operations which could be refit into production sequences. Boker Tachtit Level 1, however, is clearly more completely refit than Stránská skála IIIa Layer 4. The spatial studies of the microdebitage from Boker Tachtit Level 1 (Hietala 1983a) support this functional attribution to the site. While no wet sieving was conducted during the excavation of Stránská skála IIIa Layer 4, such recovery was

Table 8.1 Boker Tachtit Level 1 and Stránská skála IIIa Layer 4, comparability of the assemblages based on intensity of core and tool reduction

Variables	Boker Tachtit Level 1	Stránská skála IIIa-4	Evaluation
Tools/Flake Ratio	0.08	0.16	Slightly Different
Non-cortical Debitage/ Cortical Debitage Ratio	1.09	1.06	Similar
Henry's Dimensional Analysis	Slight core exhaustion; see text	Slight core exhaustion	Similar

Table 8.2 Comparison of Blank Production for Boker Tachtit Level 1 and Stránská skála IIIa Layer 4

Flintknapping Steps by Domain	Boker Tachtit 1	Stránská skála IIIa-4	Significant Difference?
Core Modification			
Core Orientation	Preferential use of a Longitudinal Surface	Preferential use of a Longitudinal Surface	No
Core Management	Débordants & frontal crest	Débordants, front crest, & side blade removals	Yes
Number of Changes/2 Steps			**1/2=0.5**
Platform Maintenance			
Platform Treatment	Unprepared: 61% Prepared: 39% n=79	Unprepared: 58% Prepared: 42% n=448	No p=.71, Fisher's Exact
External Platform Angle (degrees)	mean: 88.4, s.d.: 13.6, n=63	mean: 85.2, s.d.: 15.3, n=425	No p=.12, t=1.567, df=486
Platform Thickness	mean: 4.0, s.d.: 2.16, n=65	mean: 4.8, s.d.: 2.5, n= 433	Yes p=.02, t=-2.434, df=496
Number of Changes/3 Steps			**1/3=0.33**
Direction of Core Exploitation			
Direction of Early Exploitation	Bidirectional	Bidirectional	No
Direction of Late Exploitation	Unidirectional	Unidirectional	No
Number of Changes/2 Steps			**0/2=0**
Dorsal Surface Convexity System			
Longitudinal Convexity: Length/Width Ratio	mean: 2.25, s.d.: 1.21, n=101	mean: 1.71, s.d.: 0.67, n=543	Yes p=.00, t=4.356*, df=111.574
Shape of Convexity: Lateral Edges of Blanks	Parallel: 71% Convergent: 12% Expanding: 5% Ovoid: 12% n=101	Parallel: 49% Convergent: 24% Expanding: 18% Ovoid: 10% n=514	Yes p=.00, G^2=25.123, df=3
Curvature of Convexity: Profile of Blanks	Straight: 40% Curved: 40% Twisted: 21% n=101	Straight: 53% Curved: 29% Twisted: 18% n=522	Yes p=.04, G^2=6.570, df=2
Lateral Convexity: Cross-Section of Blanks	Triangular: 55% Trapezoidal: 40% Other: 5% n=101	Triangular: 44% Trapezoidal: 51% Other: 5% n=527	No p=.09, G^2=4.864, df=2
Vertical Convexity: Width/Thickness Ratio	mean: 4.43, s.d.: 2.33, n=101	mean: 3.99, s.d.: 1.82, n=543	No p=.07, t=1.835*, df=123.695
Number of Changes/5 Steps			**3/5=0.6**
Total Measure of Difference in Blank Production			**1.43**

*The t-test was calculated with equal variances not being assumed due to a significant F value on Levene's Test for Equality of Variances.

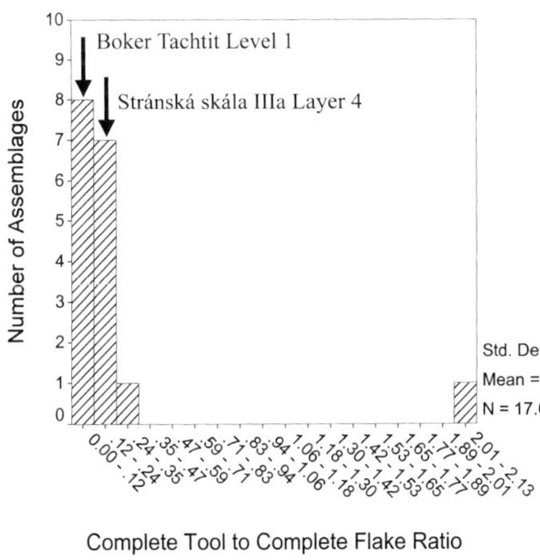

Figure 8.1 Boker Tachtit Level 1 and Stránská skála IIIa Layer 4 compared for tool/flake ratio values.

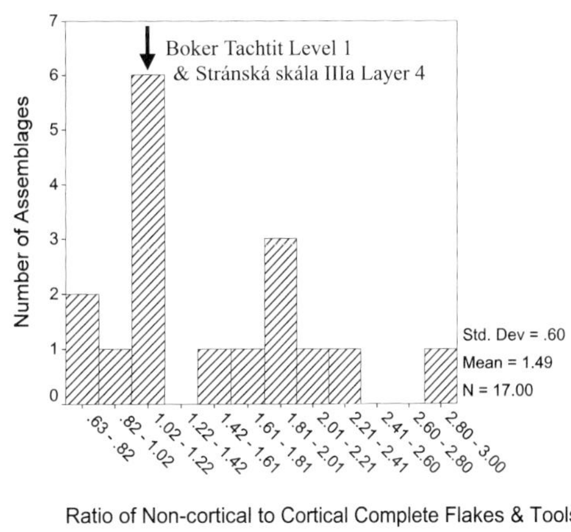

Figure 8.2 Boker Tachtit Level 1 and Stránská skála IIIa Layer 4 compared for non-cortical/cortical flake ratio values.

done during the excavation of the IIIc immediately to the north, producing densities of microdebitage characteristic of primary working of nodules on-site (Monnier 2003; Tostevin 2003b). Both assemblages are lightly retouched enough to argue that no bias should be expected in comparisons of dimensional values between the tool kits.

Blank Production

Table 8.3 presents the results of a comparison of the blank production details of the assemblages from Boker Tachtit Level 1 and Stránská skála IIIa Layer 4. Both assemblages utilize the same core orientation but differ in the details of the convexity management options, namely the lateral blade removal strategy seen in Bohunician assemblages. The refits from Boker Tachtit Level 1 clearly show that almost all core sequences ended with the production or attempted production of a Levallois point from the main debitage surface, not blades from the lateral margins of this surface as in Stránská skála IIIa Layer 4. For platform maintenance, the two assemblages differ only in the tendency for the Bohunician flintknappers to strike deeper into platforms, producing a higher mean platform thickness.

When comparing the directionality of the debitage in early core exploitation and late exploitation between the assemblages, the results of the dorsal scar pattern cross-tabulation with blank length are used to characterize each assemblage. Both assemblages demonstrate a dominance of bidirectional blanks in the longest length quartile and a dominance of unidirectional blanks in the shortest quartile. According to these characterizations, therefore, the assemblages demonstrate similar central tendencies for these technological choices. As argued in Chapter 7 using an analysis of the mean number of removals per rotation event from Volkman's publications, the specific process which resulted in this debitage-length pattern in Boker Tachtit Level 1 is likely a gradual transition from alternating to sequential bidirectional reduction in the cores from which the debitage was sampled for this study. Whether this specific process or another process

Table 8.3 Comparison of Tool Kit Morphologies for Boker Tachtit Level 1 and Stránská skála IIIa Layer 4

Tool Kit Morphology Variable	Boker Tachtit 1	Stránská skála IIIa-4	Significant Difference?
Laminarity	mean: 2.06, s.d.: 0.79, n=72	mean: 1.58, s.d.: 0.50, n=67	Yes p=.00, t=4.334*, df=120.439
Width/Thickness	N/A	mean: 4.28, s.d.: 2.14, n=67	N/A
Lateral Edges	N/A	Parallel: 39% Convergent: 31% Expanding: 16% Ovoid: 14% n=57	N/A
Distal Terminus	Blunt: 27% Pointed: 73% n=62	Blunt: 50% Pointed: 50% n=36	Yes p=.03, Fisher's Exact
Profile	N/A	Straight: 45% Curved: 32% Twisted: 23% n=62	N/A
Unique Types of Retouch	Emireh retouch	Normal retouch	Yes
Tool Types	UP tools dominate	UP tools dominate	No
Number of Changes/4 Steps			**3/4**
Total Measure of Difference in Tool Kit Morphology			**0.75**

*The t-test was calculated with equal variances not being assumed due to a significant F value on Levene's Test for Equality of Variances.

Table 8.4 Boker Tachtit Level 1 and Korolevo II Complex II, comparability of the assemblages based on intensity of core and tool reduction

Variables	Boker Tachtit Level 1	Korolevo II-II	Evaluation
Tools/Flake Ratio	0.08	0.09	Similar
Non-cortical Debitage/ Cortical Debitage Ratio	1.09	0.69	Fairly Similar
Henry's Dimensional Analysis	Slight core exhaustion; see text	Slight core exhaustion	Similar

(such as the gradual transition in individual cores from bidirectional to simple unidirectional exploitation) is responsible for the similar debitage-length pattern seen in Stránská skála IIIa Layer 4 cannot presently be ascertained because the required details are not extant in the published refits. As the specificity of the process responsible for the debitage-length pattern in each assemblage in this study cannot hope to be resolved without the unique data recovery represented by the Boker Tachtit refits, this uncertainty in detail does not affect the similarity in general processes evidenced by the debitage of these two assemblages.

Of the five steps within the blank production domain, the elongation, lateral edge, and profile options were judged to be significantly different between the two assemblages. Specifically, Boker Tachtit Level 1 is more elongated, parallel sided, and curved in profile than the Central European assemblage. The final measure of difference between the operational sequences of Boker Tachtit level 1 and Stránská skála IIIa level 4, weighted by the four knapping domains, thus produces a value of 1.43 out of a possible maximum difference of 4.0, indicating that the blank production details are fairly similar.

Tool Kit Morphology

The comparison of the morphology of the cutting edge of the tool kits of these two assemblages utilizes three categorical variables and one continuous variable, due to the sampling problems with the Boker Tachtit Level 1 tool kit. The continuous variable, the length/width ratio, is tested with a t-test for the probability that the samples arose from the random sampling of the same population. One of the three categorical variables is tested with a G^2 statistic while the other two are judged qualitatively. As no concern for a potential bias by differential tool reduction is warranted by the tool to flake ratio comparison, the dimensional variable is considered valid and shows a statistically significant difference between the assemblages. The distal terminus comparison is statistically significant, with the higher frequency of pointed distal ends distinguishing the Boker Tachtit assemblage. The remaining two analytical steps within the comparison of tool kit morphologies relate to qualitative issues of retouch. Only Boker Tachtit Level 1 possesses a unique retouch type, Emireh retouch. The two assemblages are similar in the dominance of Upper Paleolithic tool types. Considering all 4 analytical steps within this tool kit comparison, these assemblages exhibit a difference measure of 3 out of 4 steps, or 0.75 out of a maximum of 1.0.

8.1.2 Comparison of Boker Tachtit Level 1 and Korolevo II Complex II

The next comparison between regions pairs the first non-Middle Paleolithic assemblage from the Levant with the first such assemblage in the East European sample, Korolevo II Complex II. The comparability measures of these assemblages are quite similar (Table 8.4). The tool to flake ratios (Figure 8.3) possess almost the same values, indicating very little retouch of the blanks in each assemblage. The non-cortical to cortical blank ratios (Figure 8.4) place both assemblages among the raw material workshop sites. Although the ratio for Korolevo II Complex II indicates an assemblage with even more cortical pieces relative to non-cortical pieces than in the Boker Tachtit assemblage, the refit sequences from both sites indicate that the earliest portions of the respective core reduction sequences are similarly present in each assemblage. Little core exhaustion is evidenced in either assemblage, to judge from the extant cores in Boker Tachtit and the dimensional analysis of the core facets from Korolevo. Thus, these two assemblages overlap

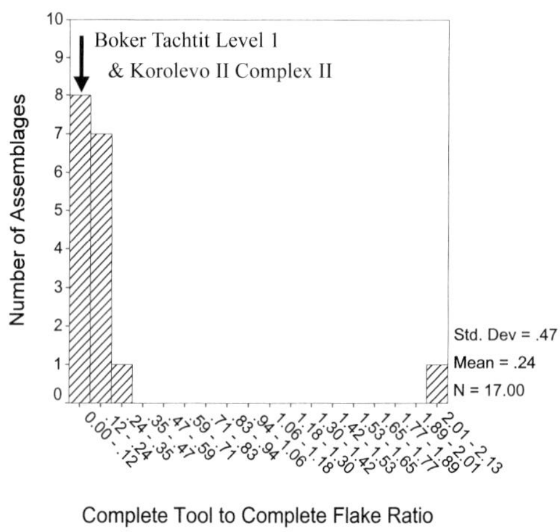

Figure 8.3 Boker Tachtit Level 1 and Korolevo II Complex II compared for tool/flake ratio values.

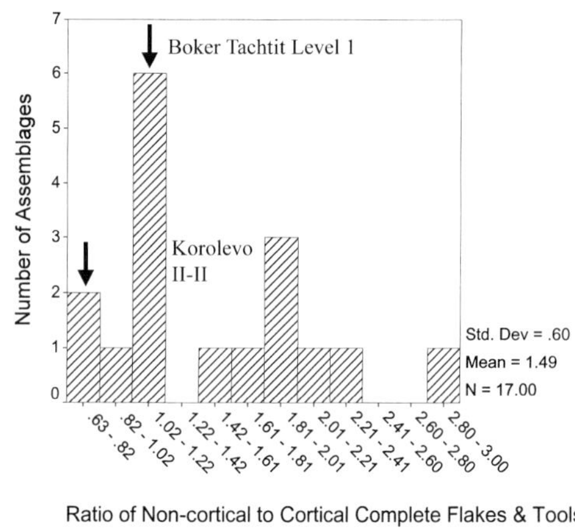

Figure 8.4 Boker Tachtit Level 1 and Korolevo II Complex II compared for non-cortical/cortical flake ratio values.

in their coverage of their respective core reduction sequences to a great degree. The refit studies by Volkman and Usik, respectively, also highlight the similarity in the nature of the palimpsest formation of these assemblages, catching as they do many sequential knapping operations which could be refit into production sequences. Both assemblages are lightly retouched enough to argue that no bias should be expected in comparisons of dimensional values between the tool kits.

Blank Production

Table 8.5 presents the results of a comparison of the blank production details of the assemblages from Boker Tachtit Level 1 and Korolevo II Complex II. Both assemblages utilize the same core orientation and similar convexity management details, save for the practice of removing the lateral side of the Korolevo II cores after the exploitation of the main debitage surface. For platform maintenance, the two assemblages differ only in the tendency for the East European flintknappers to strike deeper into platforms, producing a higher mean platform thickness.

As the debitage sample from Korolevo II Complex II resulted in empty and low cell counts in the dorsal scar pattern cross-tabulation with blank length, Usik's conclusions concerning the switch from a bidirectional to a unidirectional exploitation of individual cores were used to characterize the directionality of the debitage produced during early and late core exploitation in this assemblage (see Chapter 6). This characterization is sufficiently similar in resolution to those of Boker Tachtit Level 1 to warrant a direct comparison. Given these results, no difference exists in the direction of core exploitation between the assemblages. Of the five steps within the dorsal surface convexity domain, all were judged to be significantly different between the two assemblages save for the choice of profiles and vertical convexities (width/thickness ratios). The final measure of difference between the

Table 8.5 Comparison of Blank Production for Boker Tachtit Level 1 and Korolevo II Complex II

Flintknapping Steps by Domain	Boker Tachtit 1	Korolevo II-II	Significant Difference?
Core Modification			
Core Orientation	Preferential use of a Longitudinal Surface	Preferential use of a Longitudinal Surface	No
Core Convexity Management	Débordants & frontal crest	Débordants, front crests, & side blade removals	Yes
Number of Changes/2 Steps			**1/2=0.5**
Platform Maintenance			
Platform Treatment	Unprepared: 61% Prepared: 39% n=79	Unprepared: 66% Prepared: 34% n=64	No p=.60, Fisher's Exact
External Platform Angle (degrees)	mean: 88.4, s.d.: 13.6, n=63	mean: 90.4, s.d.: 18.2, n=58	No, p=.49, t=-0.698*, df=105.326
Platform Thickness	mean: 4.04, s.d.: 2.16, n=65	mean: 7.96, s.d.: 4.48, n=54	Yes p=.00, t=-5.895*, df=73.267
Number of Changes/3 Steps			**1/3=0.33**
Direction of Core Exploitation			
Direction of Early Exploitation	Bidirectional	Bidirectional	No
Direction of Late Exploitation	Unidirectional	Unidirectional	No
Number of Changes/2 Steps			**0/2=0.0**
Dorsal Surface Convexity System			
Longitudinal Convexity: Length/Width Ratio	mean: 2.25, s.d.: 1.21, n=101	mean: 1.71, s.d.: 0.86, n=111	Yes p=.00, t=3.668*, df=179.334
Shape of Convexity: Lateral Edges of Blanks	Parallel: 71% Convergent: 12% Expanding: 5% Ovoid: 12% n=101	Parallel: 36% Convergent: 21% Expanding: 23% Ovoid: 20% n=90	Yes p=.00, G^2=28.544, df=3
Curvature of Convexity: Profile of Blanks	Straight: 40% Curved: 40% Twisted: 21% n=101	Straight: 52% Curved: 32% Twisted: 16% n=99	No p=.24, G^2=2.881, df=2
Lateral Convexity: Cross-Section of Blanks	Triangular: 55% Trapezoidal: 40% Other: 5% n=101	Triangular: 37% Trapezoidal: 44% Other: 19% n=93	Yes p=.00, G^2=12.915, df=2
Vertical Convexity: Width/Thickness Ratio	mean: 4.43, s.d.: 2.33, n=101	mean: 4.10, s.d.: 1.91, n=111	No p=.25, t=1.145, df=210
Number of Changes/5 Steps			**3/5=0.60**
Total Measure of Difference in Blank Production			**1.43**

*The t-test was calculated with equal variances not being assumed due to a significant F value on Levene's Test for Equality of Variances.

blank production details of Boker Tachtit Level 1 and Korolevo II Complex II, weighted by the four knapping domains, thus produces a value of 1.43 out of a possible maximum difference of 4.0, indicating that the operational sequences are fairly similar. This is in fact the same value as seen with the comparison of Boker Tachtit Level 1 and Stránská skála IIIa Layer 4, although different choices of dorsal surface convexity options distinguished the two comparisons.

Tool Kit Morphology

The comparison of the morphology of the cutting edge of the tool kits of these two assemblages utilizes three categorical variables and one continuous variable, due to the sampling problems with the Boker Tachtit Level 1 tool kit. As no concern for a potential bias by differential tool reduction is warranted by the tool to flake ratio comparison, the dimensional variable of length/width is considered valid and shows no statistically significant difference between the assemblages. The distal terminus comparison is statistically significant, with the higher frequency of pointed distal ends distinguishing the Boker Tachtit assemblage. The remaining two analytical steps within the comparison of the tool kit morphologies relate to qualitative issues of retouch. Each assemblage possesses a unique retouch type, although they differ in types employed, with Boker Tachtit Level 1 evidencing Emireh retouch and Korolevo II Complex II evidencing flat bifacial retouch. The two assemblages are similar in the dominance of Upper Paleolithic tool types, although the Korolevo assemblage possesses relatively more Middle Paleolithic retouched types than Boker Tachtit. Considering all 4 analytical steps within this tool kit comparison, these assemblages exhibit a difference measure of 2 out of 4 analytical steps, or 0.50 out of a maximum of 1.0.

8.1.3 Comparison of Stránská skála IIIa Layer 4 and Korolevo II Complex II

The next comparison between regions pairs the first non-Middle Paleolithic assemblage in the Middle Danube region with the first such assemblage in the Carpathian or East European sample, Korolevo II Complex II. The comparability measures of these assemblages are quite similar (Table 8.7). The tool to flake ratios (Figure 8.5) fall in adjacent bars of the histogram, indicating little retouch of the blanks in each assemblage, although more retouch is evidenced in Stránská skála than Korolevo. Both assemblages are lightly retouched enough to argue that no bias should be expected in comparisons of dimensional values between the tool kits. The non-cortical to cortical blank ratios (Figure 8.6) place both assemblages among the raw material workshop sites. Although the ratio for Korolevo II Complex II indicates an assemblage with even more cortical pieces relative to non-cortical pieces than in the Stránská skála assemblage, the refit sequences from both sites indicate that the earliest portions of the respective core reduction sequences are similarly present in each assemblage. Little core exhaustion is evidenced in either assemblage, to judge from the dimensional analysis of the core facets and debitage. Thus, these two assemblages overlap in their coverage of their respective core reduction sequences to a great degree. The refit studies by Škrdla and Usik, respectively, also highlight the similarity in the nature of the palimpsest formation of these assemblages, catching as they do many sequential knapping operations which could be refit into production sequences.

Blank Production

Table 8.8 presents the results of a comparison of the blank production details of the assemblages from Stránská skála IIIa Layer 4 and Korolevo II Complex II. Both assemblages utilize the same

Table 8.6 Comparison of Tool Kit Morphologies for Boker Tachtit Level 1 and Korolevo II Complex II

Tool Kit Morphology Variable	Boker Tachtit 1	Korolevo II-II	Significant Difference?
Laminarity	mean: 2.06, s.d.: 0.79,	mean: 1.90, s.d.: 0.85	No p=.39, df=98
Width/Thickness	N/A	mean: 4.20, s.d.: 1.80 n=28	N/A
Lateral Edges	N/A	Parallel: 30% Convergent: 22% Expanding: 17% Ovoid: 30% n=23	N/A
Distal Terminus	Blunt: 27% Pointed: 73% n=62	Blunt: 63% Pointed: 37% n=16	Yes p=.02, Fisher's Exact
Profile	N/A	Straight: 54% Curved: 17% Twisted: 29% n=24	N/A
Unique Types of Retouch	Emireh retouch	Flat bifacial retouch	Yes
Tool Types	UP tools dominate	UP tools dominate	No
Number of Changes/4 Steps			**2/4**
Total Measure of Difference in Tool Kit Morphology			**0.50**

Table 8.7 Stránská skála IIIa Layer 4 and Korolevo II Complex II, comparability of the assemblages based on intensity of core and tool reduction

Variables	Stránská skála IIIa-4	Korolevo II-II	Evaluation
Tools/Flake Ratio	0.16	0.09	Slightly Different
Non-cortical Debitage/ Cortical Debitage Ratio	1.06	0.69	Fairly Similar
Henry's Dimensional Analysis	Slight core exhaustion	Slight core exhaustion	Similar

core orientation as well as the same details of core convexity management, including the late blade removal strategy from the core lateral edges, to judge from the refit sequences and the discarded cores. In platform maintenance, the two assemblages share only the tendency to use platform preparation to the same degree. No differences in the details of the directionality of debitage in early or late core exploitation are apparent. These conclusions are based on the dorsal scar pattern cross-tabulation with blank length for Stránská skála IIIa level 4 and on the details of Usik's refit sequences for Korolevo II Complex II. Of the five steps within the dorsal surface convexity domain, lateral edge and cross-section options demonstrate significant differences between the two assemblages. To whatever extent the debitage sampled from Korolevo II Complex II derives from bifacial reduction of foliate points rather than from non-bifacial cores in this assemblage,

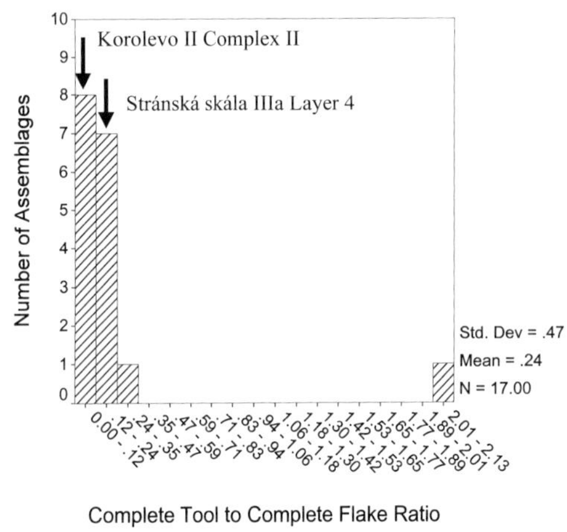

Figure 8.5 Stránská skála IIIa Layer 4 and Korolevo II Complex II compared for tool/flake ratio values.

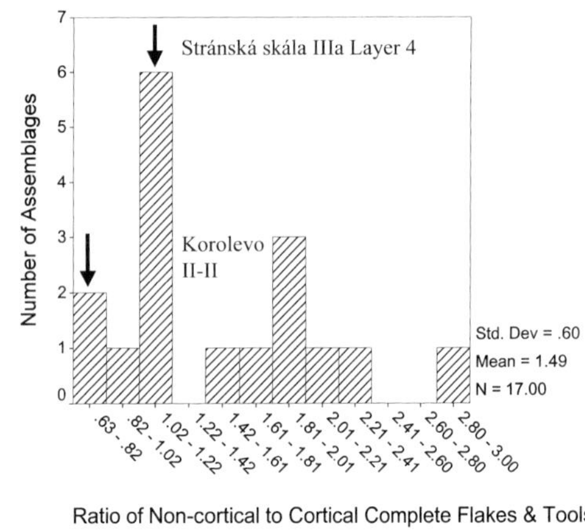

Figure 8.6 Stránská skála IIIa Layer 4 and Korolevo II Complex II compared for non-cortical/cortical flake ratio values.

it is possible that these lateral edge and cross-sectional differences may result from bifacial reduction. The debitage of bifacial reduction tends to possess higher frequencies of ovoid and expanding lateral edges compared to the products known to refit to the blade/flake cores found in the collection. From the illustrations of the bifacial thinning flakes, or flakes of bifacial retouch (Frison 1968:149-150), that refit onto the bifaces and bifacial foliates of the Korolevo II Complex II assemblage (Demidenko and Usik 1990c: figs. 2-4, 9), these reduction products are likely to have been sampled by the present debitage analysis, as most are over the 2.0cm minimum size for inclusion. In fact, within the relatively small sample studied here, seven artifacts were noted as bifacial thinning flake candidates, as they possessed curved profiles, lateral edges with very acute angles, facetted or bifacial platforms, lips between the platform and bulbar surface, and diffuse bulbs of percussion (Andrefsky 1998:118). Despite the fact that a number of pieces match this definition for bifacial thinning flakes, however, it is fairly certain that more of the debitage than this number is likely to have resulted from bifacial reduction, since diagnostically recognizable bifacial thinning flakes frequently underestimate the actual number of flakes produced during bifacial reduction.

The final measure of difference between the blank production details of Stránská skála IIIa Layer 4 and Korolevo II Complex II, weighted by the four knapping domains, thus produces a value of 1.07 out of a possible maximum difference of 4.0, indicating that the blank production details are highly similar.

Tool Kit Morphology

The comparison of the morphology of the cutting edge of the tool kits of these two assemblages utilizes three discontinuous, two categorical, and two continuous variables. As no concern for a potential bias by differential tool reduction is warranted by the tool to flake ratio comparison,

Table 8.8 Comparison of Blank Production for Stránská skála IIIa Layer 4 and Korolevo II Complex II

Flintknapping Steps by Domain	Stránská skála IIIa-4	Korolevo II-II	Significant Difference?
Core Modification			
Core Orientation	Preferential use of a Longitudinal Surface	Preferential use of a Longitudinal Surface	No
Core Management	Débordants, front crests, & side blade removals	Débordants, front crests, & side blade removals	No
Number of Changes/2 Steps			**0/2=0.0**
Platform Maintenance			
Platform Treatment	Unprepared: 58% Prepared: 42% n=448	Unprepared: 66% Prepared: 34% n=64	No p=.28, Fisher's Exact
External Platform Angle (degrees)	mean: 85.2, s.d.: 15.3, n=425	mean: 90.4, s.d.: 18.2, n=58	Yes p=.04, t=-2.096*, df=68.396
Platform Thickness	mean: 4.8, s.d.: 2.5, n= 433	mean: 7.96, s.d.: 4.48, n=54	Yes p=.00, t=-5.062*, df=57.035
Number of Changes/3 Steps			**2/3=0.67**
Direction of Core Exploitation			
Direction of Early Exploitation	Bidirectional	Bidirectional	No
Direction of Late Exploitation	Unidirectional	Unidirectional	No
Number of Changes/2 Steps			**0/2=0.0**
Dorsal Surface Convexity System			
Longitudinal Convexity: Length/Width Ratio	mean: 1.71, s.d.: 0.67, n=543	mean: 1.71, s.d.: 0.86, n=111	No p=.96, t=-0.053*, df=138.008
Shape of Convexity: Lateral Edges of Blanks	Parallel: 49% Convergent: 24% Expanding: 18% Ovoid: 10% n=514	Parallel: 36% Convergent: 21% Expanding: 23% Ovoid: 20% n=90	Yes p=.02, G^2=9.782, df=3
Curvature of Convexity: Profile of Blank	Straight: 53% Curved: 29% Twisted: 18% n=522	Straight: 52% Curved: 32% Twisted: 16% n=99	No p=.81, G^2=0.434, df=2
Lateral Convexity: Cross-Section of Blanks	Triangular: 44% Trapezoidal: 51% Other: 5% n=527	Triangular: 37% Trapezoidal: 44% Other: 19% n=93	Yes p=.00, G^2=17.582, df=2
Vertical Convexity: Width/Thickness Ratio	mean: 3.99, s.d.: 1.82, n=543	mean: 4.10, s.d.: 1.91, n=111	No p=.55, t=-0.601, df=652
Number of Changes/5 Steps			**2/5=0.4**
Total Measure of Difference in Blank Production			**1.07**

*The t-test was calculated with equal variances not being assumed due to a significant F value on Levene's Test for Equality of Variances.

Table 8.9 Comparison of Tool Kit Morphologies for Stránská skála IIIa Layer 4 and Korolevo II Complex II

Tool Kit Morphology Variable	Stránská skála IIIa-4	Korolevo II-II	Significant Difference?
Laminarity	mean: 1.58, s.d.: 0.50, n=67	mean: 1.90, s.d.: 0.85 n=28	No p=.07, t=-1.901*, df=34.915
Width/Thickness	mean: 4.28, s.d.: 2.14, n=67	mean: 4.20, s.d.: 1.80 n=28	No p=.86, t=0.176, df=93 n=57
Lateral Edges	Parallel: 39% Convergent: 31% Expanding: 16% Ovoid: 14% n=57	Parallel: 30% Convergent: 22% Expanding: 17% Ovoid: 30% n=23	No p=.38, G^2=3.068, df=3†
Distal Terminus	Blunt: 50% Pointed: 50% n=36	Blunt: 63% Pointed: 37% n=16	No p=.55, Fisher's Exact
Profile	Straight: 45% Curved: 32% Twisted: 23% n=62	Straight: 54% Curved: 17% Twisted: 29% Straight: 54% n=24	No p=.32 G^2=2.255, df=2
Unique Types of Retouch	Normal retouch	Flat bifacial retouch	Yes
Tool Types	UP tools dominate	UP tools dominate	No
Number of Changes/7 Steps			**1/7**
Total Measure of Difference in Tool Kit Morphology			**0.14**

*The t-test was calculated with equal variances not being assumed due to a significant F value on Levene's Test for Equality of Variances.

†Two cells (25%) have an expected count less than 5. The minimum expected count is 3.74.

the dimensional variables are considered valid and show no statistically significant differences between the assemblages. The dorsal surface convexity variables are also statistically similar. The remaining two analytical steps within the comparison of the tool kit morphologies relate to qualitative issues of retouch. The Korolevo assemblage possesses a unique retouch type, flat bifacial retouch, while the Stránská skála assemblage does not. While other assemblages falling under the Bohunician label have produced *in situ* foliate points in excavated contexts, such as Brno-Bohunice (Tostevin and Škrdla 2003; Škrdla and Tostevin 2005) and Dzierżysław I in southern Poland (Bluszcz et al. 1994; Foltyn and Kozłowski 2003), no such pieces have been found in any Stránská skála Bohunician assemblage. As assemblages and not industrial types are the unit of analysis of this study, the absence of bifacial retouch in the Stránská skála assemblage must be considered a significant deviation from the Korolevo II Complex II assemblage.

Table 8.10 Boker Tachtit Level 2 and Stránská skála IIIa Layer 4, comparability of the assemblages based on intensity of core and tool reduction

Variables	Boker Tachtit Level 2	Stránská skála IIIa-4	Evaluation
Tools/Flake Ratio	0.07	0.16	Slightly Different
Non-cortical Debitage/ Cortical Debitage Ratio	0.96	1.06	Similar
Henry's Dimensional Analysis	Slight core exhaustion; see text	Slight core exhaustion	Similar

The two assemblages are, however, similar in the dominance of Upper Paleolithic tool types. Considering all seven analytical steps within this tool kit comparison, these assemblages exhibit a difference measure of one out of seven analytical steps, or 0.14 out of a maximum of 1.0.

8.1.4 Comparison of Boker Tachtit Level 2 and Stránská skála IIIa Layer 4

As Level 2 of Boker Tachtit is further along than Level 1 in the technological transition from a Middle Paleolithic to Upper Paleolithic core reduction strategy (Marks and Volkman 1983a,b), it is interesting to consider how this slightly later assemblage might compare to the assemblage from Stránská skála IIIa Layer 4, shown above to be quite similar to Level 1. The following comparison evaluates this question. The comparability measures of these assemblages are quite similar (Table 8.10). The tool to flake ratios (Figure 8.7) fall in adjacent bars of the histogram, indicating very little retouch of the blanks in each assemblage, although a slightly higher frequency of tools for the Central European assemblage. Both assemblages are lightly retouched enough when compared to all the study assemblages, however, to argue that no bias should be expected in comparisons of dimensional values between the tool kits. The non-cortical to cortical blank ratios (Figure 8.8) place both assemblages among the raw material workshop sites. Although the ratio for the Stránská skála assemblage indicates even more cortical pieces relative to non-cortical pieces than in the Boker Tachtit assemblage, the refit sequences from both sites indicate that the earliest portions of the respective core reduction sequences are similarly present in each assemblage. The higher frequency of non-cortical blanks in Boker Tachtit is more likely a result of the higher internal volume of the raw material nodules exploited in the assemblage, compared to the nodule size at Stránská skála. Little core exhaustion is evidenced in either assemblage, to judge from the extant cores in Boker Tachtit and the dimensional analysis of the core facets from the Stránská skála assemblage. Thus, these two assemblages overlap in their coverage of their respective core reduction sequences to a great degree. The refit studies by Volkman and Škrdla, respectively, also highlight the similarity in the nature of the palimpsest formation of these assemblages, catching as they do many sequential knapping operations which could be refit into production sequences.

Blank Production

Table 8.11 presents the results of a comparison of the blank production details of the assemblages from Boker Tachtit Level 2 and Stránská skála IIIa Layer 4. Both assemblages utilize the same core orientation but differ in the details of convexity management. Specifically, Level 2 lacks the Bohunician strategy of removing the lateral edge of the core late in reduction, after

Table 8.11 Comparison of Blank Production for Boker Tachtit Level 2 and Stránská skála IIIa Layer 4

Flintknapping Steps by Domain	Boker Tachtit 2	Stránská skála IIIa-4	Significant Difference?
Core Modification			
Core Orientation	Preferential use of a Longitudinal Surface	Preferential use of a Longitudinal Surface	No
Core Management	Débordants & frontal crest	Débordants, front crest, & side blade removals	Yes
Number of Changes/2 Steps			**1/2=0.5**
Platform Maintenance			
Platform Treatment	Unprepared: 65% Prepared: 36% n=76	Unprepared: 58% Prepared: 42% n=448	No p=.32, Fisher's Exact
External Platform Angle (degrees)	mean: 82.1, s.d.: 12.0, n=58	mean: 85.2, s.d.: 15.3, n=425	No p=.08, t=-4.148*, df=128.901
Platform Thickness	mean: 4.54, s.d.: 2.17, n=62	mean: 4.8, s.d.: 2.5, n= 433	No p=.38, t=-3.164, df=640
Number of Changes/3 Steps			**0/3=0**
Direction of Core Exploitation			
Direction of Early Exploitation	Bidirectional	Bidirectional	No
Direction of Late Exploitation	Unidirectional	Unidirectional	No
Number of Changes/2 Steps			**0/2=0**
Dorsal Surface Convexity System			
Longitudinal Convexity: Length/Width Ratio	mean: 2.14, s.d.: 1.05, n=99	mean: 1.71, s.d.: 0.67, n=543	Yes p=.00, t=3.912*, df=112.754
Shape of Convexity: Lateral Edges of Blanks	Parallel: 65% Convergent: 18% Expanding: 10% Ovoid: 6% n=98	Parallel: 49% Convergent: 24% Expanding: 18% Ovoid: 10% n=514	Yes p=.02, G^2=9.907, df=3
Curvature of Convexity: Profile of Blanks	Straight: 51% Curved: 30% Twisted: 19% n=98	Straight: 53% Curved: 29% Twisted: 18% n=522	No p=.89, G^2=0.225, df=2
Lateral Convexity: Cross-Section of Blanks	Triangular: 46% Trapezoidal: 51% Other: 4% n=99	Triangular: 44% Trapezoidal: 51% Other: 5% n=527	No p=.84, G^2=0.347, df=2
Vertical Convexity: Width/Thickness Ratio	mean: 4.46, s.d.: 2.35, n=99	mean: 3.99, s.d.: 1.82, n=543	No p=.06, t=1.891*, df=120.374
Number of Changes/5 Steps			**2/5=0.40**
Total Measure of Difference in Blank Production			**0.90**

*The t-test was calculated with equal variances not being assumed due to a significant F value on Levene's Test for Equality of Variances.

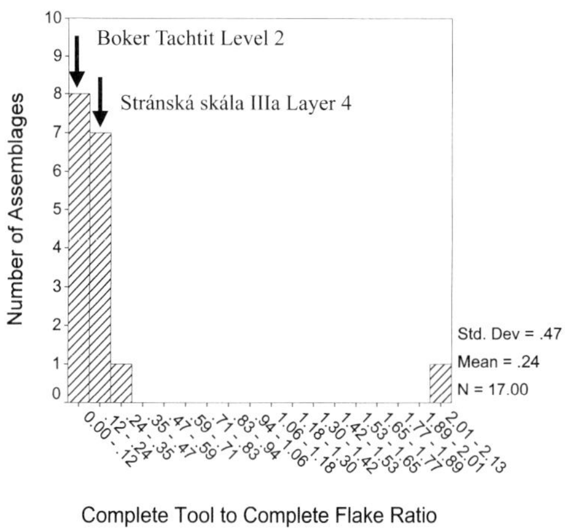

Figure 8.7 Boker Tachtit Level 2 and Stránská skála IIIa Layer 4 compared for tool/flake ratio values.

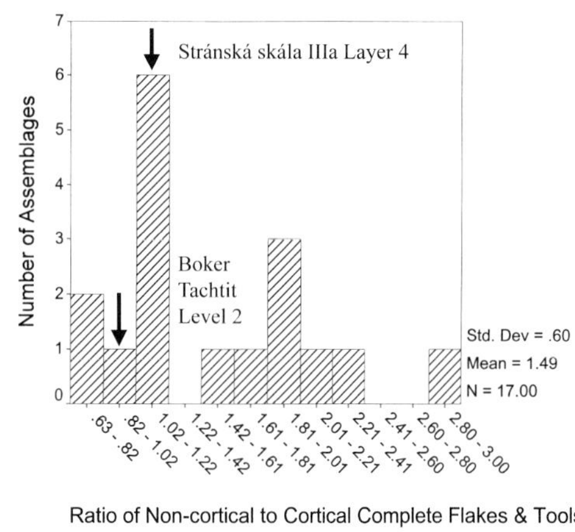

Figure 8.8 Boker Tachtit Level 2 and Stránská skála IIIa Layer 4 compared for non-cortical/cortical flake ratio values.

the flattening of the main debitage surface, through the striking of blades down the side of the core. While this detail of late reduction does not seem to be evidenced in Boker Tachtit Level 2, the refits from this level do clearly show, however, that, unlike Level 1, core reduction most frequently continued with further blade production after the removal of the last Levallois point. This pattern is seen on the discarded cores (Volkman 1989:243) and in the refitted reduction sequences: "Most common in Level 2, however, was a modification in which core reduction was continued after the Levallois point removal. This was not directed toward another point production, but was directed only toward blade blank production" (Volkman 1989:244). With blade production continuing after the flattening of the main debitage surface, this late blade-only production in Level 2 is akin to the laterally-oriented, late blade-only production in Stránská skála IIIa Layer 4. While not explicitly incorporated into the analysis represented by Table 8.11, this technological detail provides added context to the greater degree of similarity evidenced between Stránská skála IIIa Layer 4 and Boker Tachtit Level 2 than Level 1.

The blank production details of these two assemblages are similar in other ways. All three variables relating to platform maintenance are statistically equal and the directionality of the debitage in early and late core exploitation is identical. The latter analytical steps for both assemblages are based on the dorsal scar pattern cross-tabulation with blank length. Of the five steps within the dorsal surface convexity domain, only the laminarity and lateral edge tendencies were judged to be significantly different between the two assemblages. Considering all four knapping domains, the final measure of difference between the blank production details of Boker Tachtit Level 2 and Stránská skála IIIa Layer 4 produces a value of 0.90 out of a possible

Table 8.12 Comparison of Tool Kit Morphologies for Boker Tachtit Level 2 and Stránská skála IIIa Layer 4

Tool Kit Morphology Variable	Boker Tachtit 2	Stránská skála IIIa-4	Significant Difference?
Laminarity	mean: 1.87, s.d.: 0.87, n=210	mean: 1.58, s.d.: 0.50, n=67	Yes p=.00, t=3.485*, df=197.220
Width/Thickness	N/A	mean: 4.28, s.d.: 2.14, n=67	N/A
Lateral Edges	N/A	Parallel: 39% Convergent: 31% Expanding: 16% Ovoid: 14% n=57	N/A
Distal Terminus	Blunt: 43% Pointed: 57% n=159	Blunt: 50% Pointed: 50% n=36	No p=.46, Fisher's Exact
Profile	N/A	Straight: 45% Curved: 32% Twisted: 23% n=62	N/A
Unique Types of Retouch	Emireh retouch	Normal retouch	Yes
Tool Types	UP tools dominate	UP tools dominate	No
Number of Changes/4 Steps			**2/4**
Total Measure of Difference in Tool Kit Morphology			**0.50**

*The t-test was calculated with equal variances not being assumed due to a significant F value on Levene's Test for Equality of Variances.

maximum difference of 4.0, indicating that blank production is very similar in the two assemblages. The differences that do exist are found in the core modification and dorsal surface convexity domains. This comparison is even closer than that of Boker Tachtit Level 1 and Stránská skála IIIa Layer 4, in which a platform maintenance variable is also a point of difference.

Tool Kit Morphology

The comparison of the morphology of the cutting edge of the tool kits of these two assemblages utilizes three categorical variables and one continuous variable, due to the sampling problems with the Boker Tachtit Level 2 tool kit. As no concern for a potential bias by differential tool reduction is warranted by the tool to flake ratio comparison, the dimensional variable of length/width is considered valid and shows a clear statistically significant difference between the assemblages. The distal terminus comparison is not statistically significant. The remaining two analytical steps within the comparison of the tool kit morphologies relate to qualitative issues of retouch. Only Boker Tachtit Level 2 possesses a unique retouch type, as seen on Emireh points. The two assemblages are similar in the dominance of Upper Paleolithic tool types. Considering all 4 analytical steps within this tool kit comparison, these assemblages exhibit a difference measure of two out of four analytical steps, or 0.50 out of a maximum of 1.0.

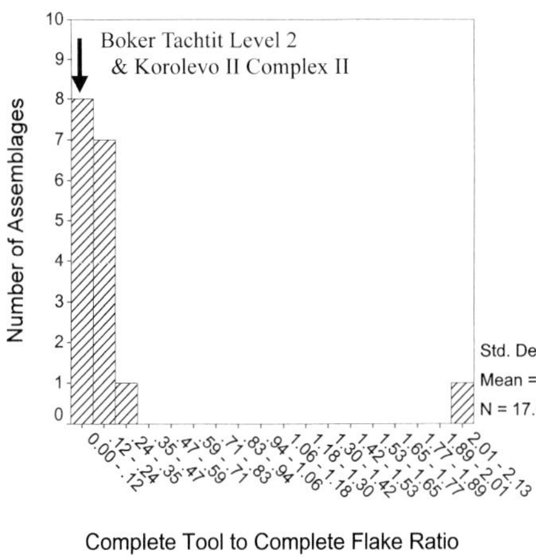

Figure 8.9 Korolevo II Complex II and Stránská skála IIIa Layer 4 compared for tool/flake ratio values.

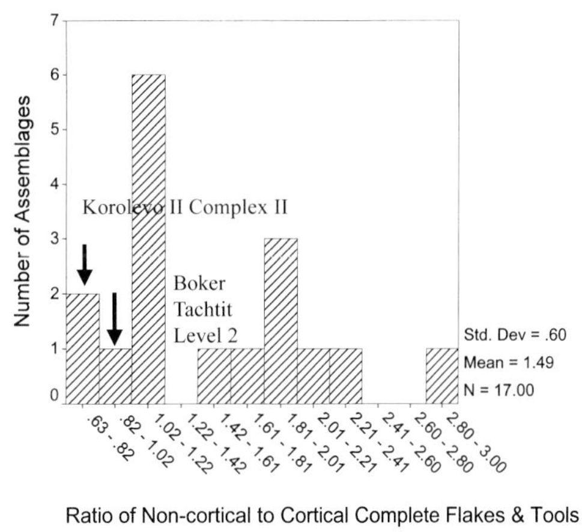

Figure 8.10 Korolevo II Complex II and Stránská skála IIIa Layer 4 compared for non-cortical/cortical flake ratio values.

Table 8.13 Boker Tachtit Level 2 and Korolevo II Complex II, comparability of the assemblages based on intensity of core and tool reduction

Variables	Boker Tachtit Level 2	Korolevo II-II	Evaluation
Tools/Flake Ratio	0.07	0.09	Similar
Non-cortical Debitage/ Cortical Debitage Ratio	0.96	0.69	Similar
Henry's Dimensional Analysis	Slight core exhaustion; see text	Slight core exhaustion	Similar

8.1.5 Comparison of Boker Tachtit Level 2 and Korolevo II Complex II

As with Stránská skála IIIa Layer 4, it is interesting to consider not only the potential similarities between the behavioral signature of Korolevo II Complex II and that of Level 1 of Boker Tachtit but also between the former and Level 2 of Boker Tachtit. The following comparison evaluates this question. The comparability measures of these assemblages are quite similar (Table 8.13). The tool to flake ratios (Figure 8.9) fall in together on the extreme left of the histogram, indicating very little retouch of the blanks in each assemblage. No bias is thus to be expected in comparisons of dimensional values between the tool kits. The non-cortical to cortical blank ratios (Figure 8.10) place both assemblages among the raw material workshop sites. Although the ratio for the Korolevo II assemblage indicates even more cortical pieces relative to non-cortical pieces than in the Boker Tachtit assemblage, the refit sequences from both sites indicate that the earliest portions of the respective core reduction sequences are similarly present in each assemblage. Little core exhaustion is evidenced in either assemblage, to judge from the extant cores in Boker Tachtit and the dimensional analysis of the core facets from the Korolevo assemblage. Thus, these two assemblages overlap in their coverage of their respective core reduction sequences to a great degree. The refit studies by Volkman and Usik, respectively, also highlight the similarity in the nature of the palimpsest formation of these assemblages, catching as they do many sequential knapping operations which could be refit into production sequences.

Blank Production

Table 8.14 presents the results of a comparison of the blank production details of the assemblages from Boker Tachtit Level 2 and Korolevo II Complex II. Both assemblages utilize the same core orientation details but slightly different convexity management details. As with Stránská skála IIIa Layer 4, Korolevo II Complex II evidences the strategy of removing the lateral edge of the core late in reduction, after the flattening of the main debitage surface, through the striking of blades down the side of the core. While Level 2 of Boker Tachtit does not evidence this tactic, it does continue with blade removals after the removal of the last Levallois point. With a shared emphasis on point production in early core exploitation and blade production in late reduction, these details reflect similar choices between the two assemblages, although this fact is not reflected in the analysis presented in Table 8.14. For platform maintenance, the two assemblages differ in the tendency for the East European flintknappers to strike deeper into almost vertical platforms. No difference exists in the directionality of debitage in early vs. late core exploitation, based on a dorsal scar pattern cross-tabulated with blank length for the Levantine sample and Usik's conclusions for the East European sample. Of the five steps within the dorsal surface convexity domain, all were judged to be significantly different between the two assemblages save for the choice of similarly curved longitudinal convexities and relatively thin vertical convexities (width/thickness ratios). Considering all four knapping domains, the final measure of difference between the blank production details of Boker Tachtit Level 2 and Korolevo II Complex II produces a value of 1.77 out of a possible maximum difference of 4.0, indicating that the blank production details are more similar than dissimilar. While Boker Tachtit Level 2 displays closer behavioral similarities to Stránská skála IIIa Layer 4, it deviates from Korolevo II Complex II more than Boker Tachtit Level 1.

Tool Kit Morphology

The comparison of the morphology of the cutting edge of the tool kits of these two assemblages utilizes three categorical variables and one continuous variable, due to the sampling problems with the Boker Tachtit Level 2 tool kit. As no concern for a potential bias by differential tool reduction is warranted by the tool to flake ratio comparison, the dimensional variable of length/width is considered valid and shows no statistically significant difference between the assemblages. The distal terminus comparison is also not statistically significant. The remaining two analytical steps within the comparison of the tool kit morphologies relate to qualitative issues of retouch. Each assemblage possesses a unique retouch type, although they differ in types employed, with Boker Tachtit Level 2 evidencing Emireh retouch and Korolevo II Complex II evidencing flat bifacial retouch. The two assemblages are similar in the dominance of Upper

Table 8.14 Comparison of Blank Production for Boker Tachtit Level 2 and Korolevo II Complex II

Flintknapping Steps by Domain	Boker Tachtit 2	Korolevo II-II	Significant Difference?
Core Modification			
Core Orientation	Preferential use of a Longitudinal Surface	Preferential use of a Longitudinal Surface	No
Core Convexity Management	Débordants & frontal crest	Débordants, front crests & side blade removals	Yes
Number of Changes/2 Steps			**1/2=0.5**
Platform Maintenance			
Platform Treatment	Unprepared: 65% Prepared: 36% n=76	Unprepared: 66% Prepared: 34% n=64	No p=1.0, Fisher's Exact
External Platform Angle (degrees)	mean: 82.1, s.d.: 12.0, n=58	mean: 90.4, s.d.: 18.2, n=58	Yes p=.00, t=-2.930*, df=98.758
Platform Thickness	mean: 4.54, s.d.: 2.17, n=62	mean: 7.96, s.d.: 4.18, n=54	Yes p=.00, t=-5.126*, df=74.187
Number of Changes/3 Steps			**2/3=0.67**
Direction of Core Exploitation			
Direction of Early Exploitation	Bidirectional	Bidirectional	No
Direction of Late Exploitation	Unidirectional	Unidirectional	No
Number of Changes/2 Steps			**0/2=0.0**
Dorsal Surface Convexity System			
Longitudinal Convexity: Length/Width Ratio	mean: 2.14, s.d.: 1.05, n=99	mean: 1.71, s.d.: 0.86, n=111	Yes p=.00, t=3.204, df=208
Shape of Convexity: Lateral Edges of Blanks	Parallel: 65% Convergent: 18% Expanding: 10% Ovoid: 6% n=98	Parallel: 36% Convergent: 21% Expanding: 23% Ovoid: 20% n=90	Yes p=.00, G^2=20.829, df=3
Curvature of Convexity: Profile of Blanks	Straight: 51% Curved: 30% Twisted: 19% n=98	Straight: 52% Curved: 32% Twisted: 16% n=99	No p=.82, G^2=0.410, df=2
Lateral Convexity: Cross-Section of Blanks	Triangular: 46% Trapezoidal: 51% Other: 4% n=99	Triangular: 37% Trapezoidal: 44% Other: 19% n=93	Yes p=.00, G^2=11.877, df=2
Vertical Convexity: Width/Thickness Ratio	mean: 4.46, s.d.: 2.35, n=99	mean: 4.10, s.d.: 1.91, n=111	No p=.23, t=1.207, df=208
Number of Changes/5 Steps			**3/5=0.6**
Total Measure of Difference in Blank Production			**1.77**

*The t-test was calculated with equal variances not being assumed due to a significant F value on Levene's Test for Equality of Variances.

Table 8.15 Comparison of Tool Kit Morphologies for Boker Tachtit Level 2 and Korolevo II Complex II

Tool Kit Morphology Variable	Boker Tachtit 2	Korolevo II-II	Significant Difference?
Laminarity	mean: 1.87, s.d.: 0.87, n=210	mean: 1.90, s.d.: 0.85 n=28	No p=.86, t=-0.177, df=236
Width/Thickness	N/A	mean: 4.20, s.d.: 1.80 n=28	N/A
Lateral Edges	N/A	Parallel: 30% Convergent: 22% Expanding: 17% Ovoid: 30% n=23	N/A
Distal Terminus	Blunt: 43% Pointed: 57% n=159	Blunt: 63% Pointed: 37% n=16	No p=.19, No p=.19, Fisher's Exact
Profile	N/A	Straight: 54% Curved: 17% Twisted: 29% n=24	N/A
Unique Types of Retouch	Emireh retouch	Flat bifacial retouch	Yes
Tool Types	UP tools dominate	UP tools dominate	No
Number of Changes/4 Steps			**1/4**
Total Measure of Difference in Tool Kit Morphology			**0.25**

Paleolithic tool types, although the Korolevo assemblage possesses relatively more Middle Paleolithic retouched types than Boker Tachtit. Considering all four analytical steps within this tool kit comparison, these assemblages exhibit a difference measure of one out of four analytical steps, or 0.25 out of a maximum of 1.0.

8.1.6 Comparison of Kebara Cave Unit I and the Aurignacian of Stránská skála IIa Layer 4 and IIIa Layer 3

The frequent claims for similarity between assemblages designated "Aurignacian" from Western Europe through the Near East warrants a trial of the details of the inter-regional Aurignacian samples in this study. The combined sample of the assemblages from Stránská skála IIa Layer 4 and IIIa Layer 3 serve as the Central European example of this technocomplex while Unit I of Kebara Cave serves as the Levantine example. No Aurignacian assemblage was included in the East European sample, as the Carpathian region does not contain any *sensu stricto* Aurignacian assemblages in the 60–30 kya time frame of the present study (Hoffecker 1988), although the Crimea and the Northwestern Caucasus further to the south would be a prime target for further consideration

Table 8.16 Kebara Cave Unit I and the Aurignacian of Stránská skála IIIa Layer 3 and IIa Layer 4, comparability of the assemblages based on intensity of core and tool reduction

Variables	Kebara Cave Unit I	Stránská skála Aurignacian	Evaluation
Tools/Flake Ratio	0.12	0.16	Similar
Non-cortical Debitage/ Cortical Debitage Ratio	1.79	1.51	Similar
Henry's Dimensional Analysis	Significant core exhaustion	Moderate core exhaustion	Slightly different

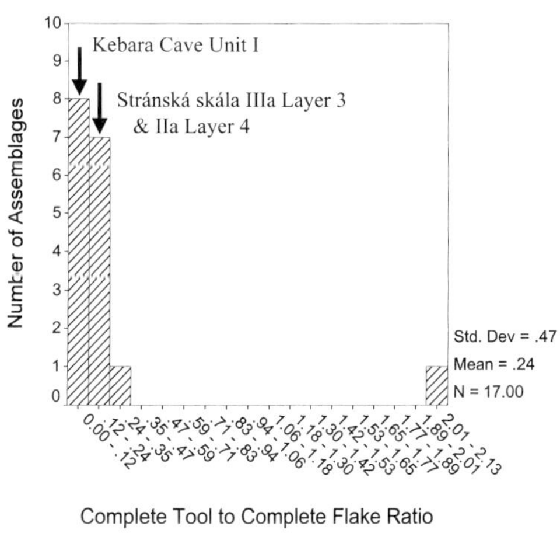

Figure 8.11 Kebara Cave Unit I and Stránská skála IIa Layer 4 and IIIa Layer 3 compared for tool/flake ratio values.

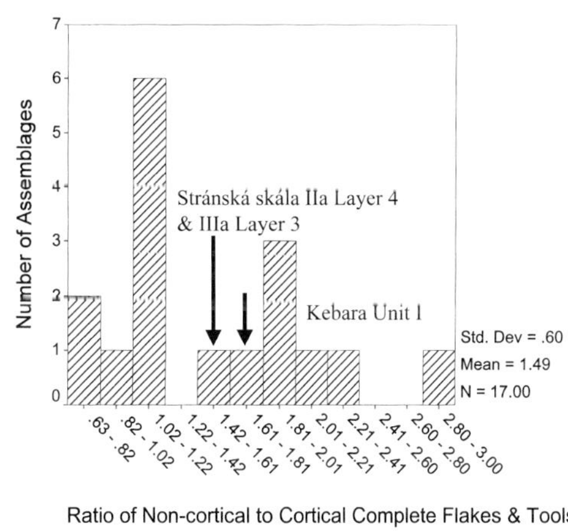

Figure 8.12 Kebara Cave Unit I and Stránská skála IIa Layer 4 and IIIa Layer 3 compared for non-cortical/cortical flake ratio values.

of the role of the Aurignacian in Eastern Europe (Demidenko 2000).

Table 8.16 presents the fairly similar comparability measures for these assemblages. The tool to flake ratios (Figure 8.11) fall in adjacent bars on the left of the histogram, indicating little retouch of the blanks in each assemblage, although more retouch is evidenced in Stránská skála than Kebara. Both assemblages are lightly retouched enough to argue that no bias should be expected in comparisons of dimensional values between the tool kits. The non-cortical to cortical blank ratios (Figure 8.12) place both assemblages among the non-workshop sites in the center of the histogram. The Stránská skála Aurignacian sample falls approximately on the mean of all assemblages while Kebara Unit I falls to the right of it. Both assemblages have more cortical pieces relative to non-cortical than six of the 17 assemblages and so while they possess less of the early portion of the core reduction sequence than a raw material workshop, they are not devoid of early core reduction. Both assemblages evidence moderate to significant

Table 8.17 Comparison of Blank Production for Kebara Cave Unit I and the Aurignacian of Stránská skála IIIa Layer 3 and IIa Layer 4

Flintknapping Steps by Domain	Kebara I	Stránská skála IIIa-3 & IIa-4	Significant Difference?
Core Modification			
Core Orientation	Preferential use of a Longitudinal Surface	Preferential use of a Longitudinal Surface	No
Core Management	Frontal & lateral crests, core tablets	Frontal & lateral crests, core tablets	No
Number of Changes/2 Steps			**0/2=0**
Platform Maintenance			
Platform Treatment	Unprepared: 76% Prepared: 24% n=354	Unprepared: 69% Prepared: 31% n=325	Yes p=.04, Fisher's Exact
External Platform Angle (degrees)	mean: 84.1, s.d.: 13.7, n=298	mean: 82.5, s.d.: 15.1, n=306	No, p=.17, t=1.371, df=602
Platform Thickness	mean: 4.09, s.d.: 2.93, n=302	mean: 4.80, s.d.: 2.52, n=304	Yes p=.00, t=-3.220*, df=589.442
Number of Changes/3 Steps			**2/3=0.67**
Direction of Core Exploitation			
Direction of Early Exploitation	Unidirectional	Bidirectional & Unidirectional	Yes
Direction of Late Exploitation	Unidirectional	Unidirectional	No
Number of Changes/2 Steps			**1/2=0.5**
Dorsal Surface Convexity System			
Longitudinal Convexity: Length/Width Ratio	mean: 2.39, s.d.: 1.35, n=522	mean: 1.73, s.d.: 0.75, n=470	Yes p=.00, t=9.439*, df=834.761
Shape of Convexity: Lateral Edges of Blanks	Parallel: 57% Convergent: 22% Expanding: 13% Ovoid: 8% n=472	Parallel: 57% Convergent: 17% Expanding: 16% Ovoid: 10% n=431	No p=.27, G^2=3.891, df=3
Curvature of Convexity: Profile of Blanks	Straight: 53% Curved: 28% Twisted: 19% n=501	Straight: 58% Curved: 24% Twisted: 18% n=430	No p=.20, G^2=3.204, df=2
Lateral Convexity: Cross-Section of Blanks	Triangular: 42% Trapezoidal: 53% Other: 6% n=493	Triangular: 46% Trapezoidal: 48% Other: 6% n=447	No p=.36, G^2=2.054, df=2
Vertical Convexity: Width/Thickness Ratio	mean: 3.76, s.d.: 2.00, n=522	mean: 3.45, s.d.: 1.51, n=470	Yes p=.01, t=2.732*, df=960.311
Number of Changes/5 Steps			**2/5=0.4**
Total Measure of Difference in Blank Production			**1.57**

*The t-test was calculated with equal variances not being assumed due to a significant F value on Levene's Test for Equality of Variances.

core exhaustion, based on the dimensional analysis of the core facets and debitage. Thus, these two assemblages overlap in their coverage of their respective core reduction sequences to a great degree, from the early/middle portion of the sequence through to the behaviors enacted during the exhaustion of the cores.

Blank Production

Table 8.17 presents the results of a comparison of the blank production details of the Aurignacian assemblages from Kebara Cave Unit I and Stránská skála IIIa Layer 3 and IIa Layer 4. Both assemblages utilize the same core modification options for both core orientation and convexity management details. Platform treatment options are barely statistically significant while the Stránská skála Aurignacian evidences significantly thicker platforms. The directionality of debitage in early core exploitation is also significantly different, with a near equivalence of bidirectional and unidirectional blanks present in the first quartile of the Central European sample and only unidirectional blanks dominating the first quartile in the Levantine sample. Late core exploitation is considered similar, however. Of the five steps within the dorsal surface convexity domain, only the laminarity and vertical convexity (width/thickness ratio) tendencies were judged to be significantly different between the two assemblages. The final measure of difference between the Aurignacian blank production details of Kebara Unit I and those of Stránská skála IIIa Layer 3 and IIa Layer 4, weighted by the four knapping domains, thus produces a value of 1.57 out of a possible maximum difference of 4.0, indicating that blank production is quite similar

Tool Kit Morphology

The comparison of the morphology of the cutting edge of the tool kits of these two assemblages utilizes three discontinuous, two categorical, and two continuous variables. As no concern for a potential bias by differential tool reduction is warranted by the tool to flake ratio comparison, the dimensional variables are considered valid. The length/width ratios of the tool kits are statistically different while the width/thickness ratios are not. The dorsal surface convexity variables are also statistically similar. The remaining two analytical steps within the comparison of the tool kit morphologies relate to qualitative issues of retouch. Both assemblages possess a unique retouch type, the carinated retouch which produces curved and twisted bladelets The two assemblages are also similar in the dominance of Upper Paleolithic tool types. Considering all seven analytical steps within this tool kit comparison, these assemblages exhibit a difference measure of one out of seven analytical steps, or 0.14 out of a maximum of 1.0. To the extent that the industrial homogeneity of the European Aurignacian with the Levantine Aurignacian may be taken as a measure of internal difference in a recognized industrial type or technocomplex, these technological signatures of blank production and tool kit morphology can be used to evaluate whether or not the deviation between two assemblages might be recognized as larger or smaller than that known from recognized industrial types. This is an alternative use of the approach advocated here and in Tostevin and Škrdla (2006).

8.2 An Evaluation of the Inter-regional Sequence of Change

The continuity of blank production behaviors between the inter-regional comparisons is depicted in Figure 8.13. The consolidation of this variability into the measure of difference in blank production is depicted chronostratigraphically for the inter-regional comparisons in Figure 8.14. The measures of difference in blank production across

Table 8.18 Comparison of Tool Kit Morphologies for Kebara Cave Unit I and the Aurignacian of Stránská skála IIIa Layer 3 and IIa Layer 4

Tool Kit Morphology Variable	Kebara I	Stránská skála IIIa-3 & IIa-4	Significant Difference?
Laminarity	mean: 2.35, s.d.: 1.50 n=101	mean: 1.73, s.d.: 0.78 n=52	Yes p=.00, t=3.364*, df=150.992
Width/Thickness	mean: 3.34, s.d.: 1.88 n=101	mean: 3.33, s.d.: 1.48 n=52	No p=.97, t=0.035, df=151
Lateral Edges	Parallel: 59% Convergent:21% Expanding: 13% Ovoid: 7% n=80	Parallel: 48% Convergent: 13% Expanding: 17% Ovoid: 22% n=46	No p=.08, G^2=6.694, df=3
Distal Terminus	Blunt: 40% Pointed: 60% n=45	Blunt: 63% Pointed: 37% n=19	No p=.11, Fisher's Exact
Profile	Straight: 56% Curved: 27% Twisted: 17% n=93	Straight: 63% Curved: 25% Twisted: 12% n=40	No p=.72, G^2=0.655, df=2
Unique Types of Retouch	Carinated retouch	Carinated retouch	No
Tool Types	UP tools dominate	UP tools dominate	No
Number of Changes/7 Steps			**1/7**
Total Measure of Difference in Tool Kit Morphology			**0.14**

*The t-test was calculated with equal variances not being assumed due to a significant F value on Levene's Test for Equality of Variances.

all regions is depicted in Figure 8.15. The biplot contrast of blank production versus tool kit morphology differences for the inter-regional comparisons is provided in Figure 8.16. Finally, the biplot contrast is provided for all pair-wise comparisons in this study in Figure 8.17.

Quantitatively, the pair-wise comparisons between Boker Tachtit Level 1 and the first Bohunician assemblage in Central Europe (producing a blank production difference value of 1.43) and between Boker Tachtit Level 1 and the first non-Middle Paleolithic assemblage in Eastern Europe, Korolevo II Complex II (producing a blank production difference value of 1.43), are extremely surprising given their geographical separation (Table 8.19, Figures 8.13-8.14). The first comparison is less than twice as different as the value between Stránská skála IIIa-4 and III (0.73), assemblages universally grouped together under the Bohunician label. Surprisingly, it is also less than the blank production difference value between the European Aurignacian of Stránská skála IIIa-3 and IIa-4 and the Levantine Aurignacian of Kebara Unit I (blank production difference value of 1.57).

The comparison of Boker Tachtit level 2 to Stránská skála IIIa level 4 produces a value (0.90) which is even closer, indicating more similarity between these assemblages than with Boker Tachtit level 1. This conclusion is corroborated by the independent data and evaluation of Škrdla (2003a), who uses a purely refit approach to reconstructing lithic technology. Boker Tachtit 2 shows fewer similarities with Korolevo II Complex II (1.77), however, than it does with Stránská skála IIIa Level 4. These comparisons

Table 8.19 Difference Measures between Assemblages for all Regions

Comparison of Assemblages Through Time	Measure of Difference in Blank Production (Maximum=4, Minimum=0)	Measure of Difference in Tool Kit Morphology (Maximum=1, Minimum=0)
Central Europe		
Kůlna Layer 7a vs. Stránská skála IIIa-4	3.13	0.86
Stránská skála IIIa-4 vs. Stránská skála III	0.73	0.29
Stránská skála III vs. Vedrovice V	2.67	0.86
Stránská skála IIIa-4 vs. Stránská skála IIIc	0.60	0.14
Stránská skála IIIc vs. Stránská skála III	0.60	0.29
Stránská skála IIIc vs. Vedrovice V	2.47	1.00
Vedrovice V vs. Stránská skála IIIa-3 & IIa-4	2.17	0.43
Stránská skála III vs. Stránská skála IIIa-3 & IIa-4	2.80	0.57
Kůlna 7a vs. Vedrovice V	3.27	0.29
Eastern Europe		
Molodova V-11 vs. Korolevo II-II	2.57	0.29
Korolevo II-II vs. Korolevo I-1a	1.00	0.14
Levant		
Kebara VI vs. Boker Tachtit 1	2.83	0.75
Boker Tachtit 1 vs. Boker Tachtit 2	0.33	0.25
Boker Tachtit 2 vs. Boker Tachtit 4	1.73	0.50
Boker Tachtit 2 vs. Kebara IV	0.83	0.50
Kebara IV vs. Kebara III	2.10	0.43
Kebara III vs. Boker Tachtit 4	1.70	0.50
Kebara III vs. Kebara II	1.57	0.14
Kebara II vs. Kebara I	0.20	0.14
Inter-Regional Sequence		
Boker Tachtit 1 vs. Stránská skála IIIa-4	1.43	0.75
Boker Tachtit 1 vs. Korolevo II-II	1.43	0.50
Stránská skála IIIa-4 vs. Korolevo II-II	1.07	0.14
Boker Tachtit 2 vs. Stránská skála IIIa-4	0.90	0.50
Boker Tachtit 2 vs. Korolevo II-II	1.77	0.25
Kebara I vs. Stránská skála IIIa-3 & IIa-4	1.57	0.14

point to a common behavioral phenomenon appearing after the last Middle Paleolithic assemblage in each region.

It must be stressed that this measure of difference compresses all of the variation existing between two operational sequences into a single value, simplifying the differences. It is useful, however, for viewing the rough intra-regional and inter-regional patterns. Also, while quantitative in nature, the evaluation of this difference figure between two pair-wise comparisons is clearly a continuum between similarity and dissimilarity with clear boundaries missing between differences modest enough to fit a cultural transmission

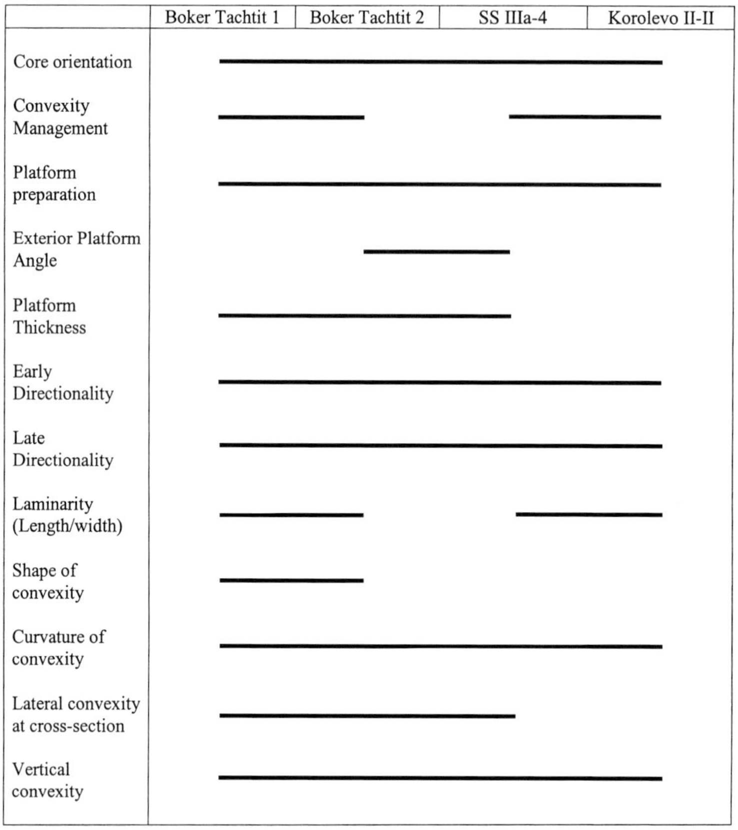

Figure 8.13 Variable by variable view of inter-regional pair-wise comparisons. Bars connecting two assemblages represent the use of statistically similar attribute states between the two assemblages. Similarities are not necessarily implied between assemblages connected by a bar which are not immediately adjacent to each other.

scenario and differences significant enough to point to an innovation scenario.

When one increases the analytical resolution, however, by examining the actual behavioral options employed in each of the assemblages' operational sequences to determine which knapping options appeared and disappeared through time within and between adjacent regions (compare Figures 5.58–5.59, 6.17, 7.59, and 8.13), one sees that the three terminal Middle Paleolithic operational sequences differ greatly from each other (Figure 8.15). One also sees that the behavioral options employed in the different steps of the operational sequences of Boker Tachtit level 1, Stránská skála IIIa level 4, and Korolevo II Complex II (the first non-Middle Paleolithic assemblages in each regions) are predominantly the same (Figure 8.13, Table 8.20), even when compared to extremely similar assemblages such as Boker Tachtit levels 1 and 2. The similarities are found in a very specific combination of knapping options across all four domains of blank production (Table 8.20).

Examining the presence and absence of antecedent elements for the flintknapping options employed in each assemblage (compare Figures 5.58–5.59, 6.17, 7.59, and 8.13), it is apparent that few antecedents can be found for the behaviors used in the operational sequences of these three post-Middle

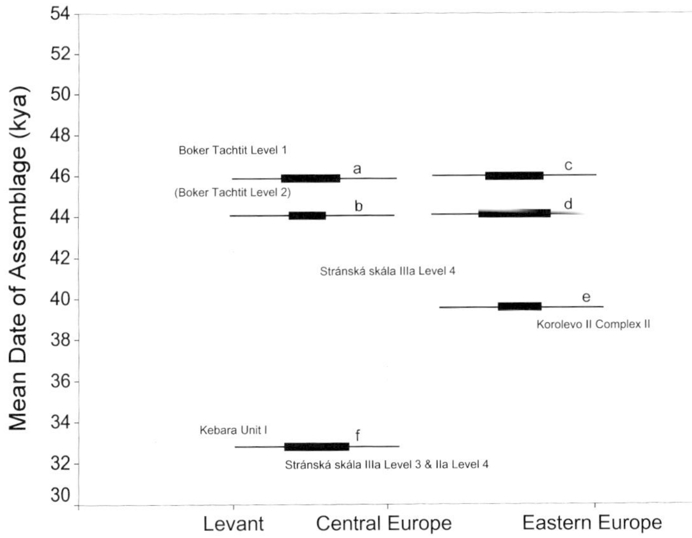

Figure 8.14 Chronological comparison of differences in Blank Production among specific assemblages from different regions. The length of the thick bars represents the difference measure between the two assemblages above and below the bar (the earlier and the later). The length of the thin bars represents the maximum possible difference measure. The three columns represent different sequences for pairing assemblages for comparison. Key: a: Boker Tachtit Level 1 vs. Stránská skála IIIa Layer 4; b: Boker Tachtit Level 2 vs. Stránská skála IIIa Layer 4; c: Boker Tachtit Level 1 vs. Korolevo II Complex II; d: Boker Tachtit Level 2 vs. Korolevo II Complex II; e: Stránská skála IIIa Layer 4 vs. Korolevo II Complex II; f: Kebara Unit I vs. Stránská skála IIa-4 and IIIa-3.

Figure 8.15 Chronological comparison of differences in Blank Production among assemblages in all three regions. The length of the thick bars represents the difference measure between the two assemblages above and below the bar (the earlier and the later). The length of the thin bars represents the maximum possible difference measure. The three columns represent the regional sequences for pairing assemblages for comparison.

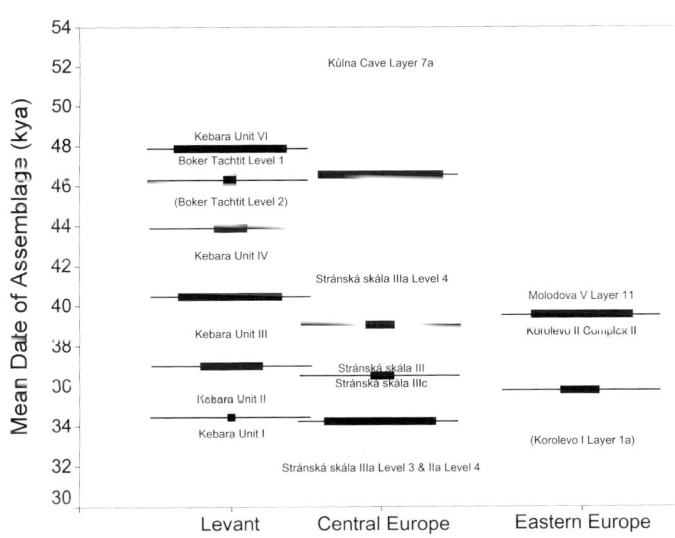

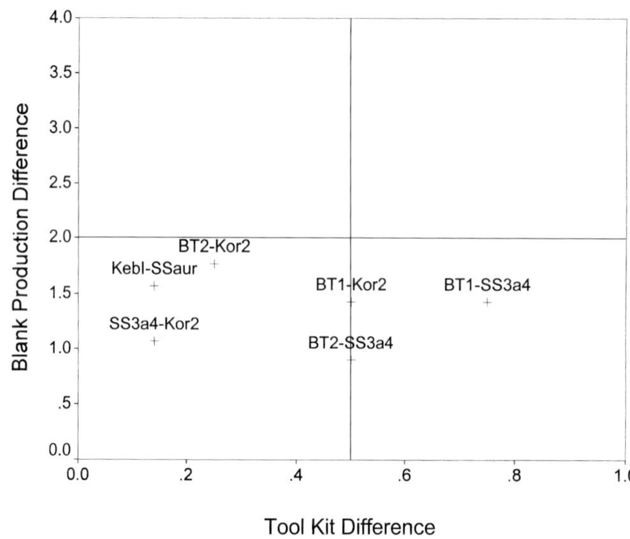

Figure 8.16 Blank Production vs. Tool Kit Differences for pair-wise inter-regional comparisons. Assemblage Key: BT1: Boker Tachtit Level 1; BT2: Boker Tachtit Level 2; KebI: Kebara Cave Unit I; Kor2: Korolevo II Complex II; SS3a4: Stránská skála IIIa Level 4; SSaur: Stránská skála IIIa Level 3 and IIa Level 4.

Table 8.20 Central tendencies by domain common to assemblages defining the Bohunician Behavioral Package

Core Modification:
Longitudinal Orientation; Débordant & Crested Blade Core Management

Platform Maintenance:
~40% Facetted Platforms, ~86 degree External Platform Angle, ~4 mm PT

Direction of Core Exploitation:
Bidirectional changing to Unidirectional debitage exploitation

Dorsal Surface Convexity
Length/Width Ratio of 1.80, Width/Thickness Ratio of 4.25

Tool Kit Morphology:
Levallois point & distally retouched tool kit

Assemblages:
Boker Tachtit 1 & 2, Stránská skála IIIa, III, IIIc, Korolevo II-II, Brno-Bohunice (see Tostevin & Škrdla 2006; Škrdla & Tostevin 2005)

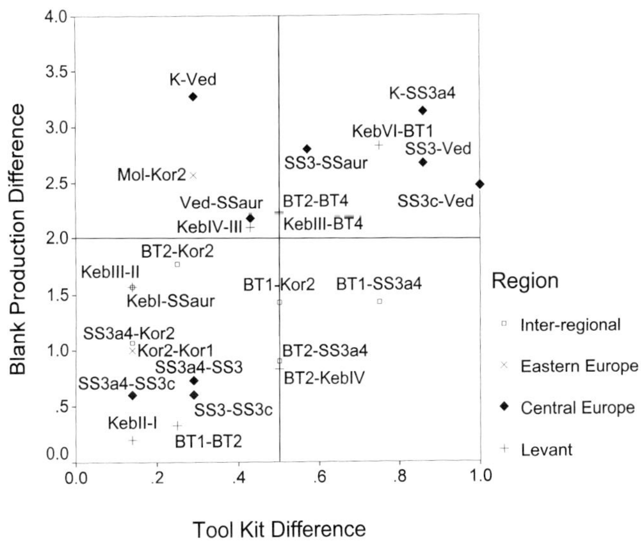

Figure 8.17 Blank Production vs. Tool Kit Differences for all pair-wise comparisons. Assemblage Key: BT1: Boker Tachtit Level 1; BT2: Boker Tachtit Level 2; BT4: Boker Tachtit Level 4; KebI: Kebara Cave Unit I; KebII: Kebara Cave Unit II; KebIII: Kebara Cave Unit III; KebIV: Kebara Cave Unit IV; KebVI: Kebara Cave Unit VI; Kor1: Korolevo I Layer 1a; Kor2: Korolevo II Complex II; K: Kůlna Cave Layer 7a; Mol: Molodova V Layer 11; SS3: Stránská skála III; SS3a4: Stránská skála IIIa Level 4; SSaur: Stránská skála IIIa Level 3 and IIa Level 4; SS3c: Stránská skála IIIc; V: Vedrovice V.

Table 8.21 Central tendencies by domain common to assemblages defining the Aurignacian Behavioral Package

Core Modification:
Longitudinal Orientation; Front & Rear Crest Core Management

Platform Maintenance:
~70% Plain & Linear Platforms, ~85 degree External Platform Angle

Direction of Core Exploitation:
Unidirectional debitage exploitation through reduction with some independent Bidirectional reduction on separate cores.

Dorsal Surface Convexity:
Parallel edged blanks, Length/Width Ratio of ca. 2.10, Width/Thickness of ca. 3.75

Tool Kit Morphology:
Carination & distally retouched tool kit

Assemblages:
Kebara Units II & I, Stránská skála IIa-4 & IIIa-3

Paleolithic assemblages within the last Middle Paleolithic assemblage of each region. For instance, the shift from a bidirectional to a unidirectional method of core exploitation, as seen in Boker Tachtit level 1, Stránská skála IIIa level 4, and Korolevo II Complex II, is not present in any of the preceding Middle Paleolithic assemblages in any of the three regions. Similarly, no evidence of the recurrence or clustering of the specific knapping options seen in these three later assemblages (Table 8.20) is seen in the earlier assemblages.

Due to the lack of antecedents in each region (Expectation 1) and the high number of similar knapping options (Expectation 2) used in all four independent domains of the operational sequence (Expectation 3) evidenced in Boker Tachtit level 1, Stránská skála IIIa level 4, and Korolevo II Complex II, parsimony favors the conclusion that all of these assemblages possess the same behavioral package as part of their technological signature. Evaluating these expectations against the possible processes which could have caused this pattern (see Figure 3.7), these assemblages' signatures and their relationships within the intra- and inter-regional sequences of change best fit the model expectations of a cultural transmission event. Independent innovation through adaptive convergence is unlikely to have produced such similarities, without antecedents, given the fact that the more functionally-sensitive tool kit morphologies deviate more within these three assemblages than do the blank production behaviors, i.e., they vary more along the x-axis than the y-axis in Figure 8.16. If all three regions were subjected to environmental change or other common selective force unrelated to cultural transmission so that each regional population produced the same functionality of cutting edges, one would expect the actual functional retouched tool kits to be similar. As they are much less similar than are the blank production choices which affect cutting edge morphologies far less, adaptive convergence is far from the most parsimonious argument. Independent innovation through cultural drift (Figure 3.7) is even more unlikely, given the high number of similarities in all domains and the consistent chronological progression of the geographical appearances of these regions (see below). Thus, according to the expectations of cultural transmission theory vs. independent innovation established in Chapter 3, these assemblages represent cultural transmission.

While the entire technological signature is not exactly the same in each assemblage, one would not expect them to be, given the tendency for diffused behaviors to deviate through time and space via cultural drift from errors of transmission, a process Deetz and Dethlefsen called the Doppler Effect (1965). Regardless of this expected deviation, which is greatest in the most distant pair of assemblages (Boker Tachtit 2 and Korolevo II-II), these assemblages are still startlingly alike.

Furthermore, the temporal and geographic progression of the appearances of these particular assemblages is indeed historically logical (Expectation 5). From the regions studied during this project, this cluster of knapping options appears first in the southern Levant at 47/46 kya, next at Stránská skála IIIa level 4 in Central Europe at 41 kya, and finally in Eastern Europe at Korolevo II-II by 38 kya. The time difference between the appearances in Central and Eastern Europe, with Stránská skála being roughly 3,500 km and Korolevo II-II being roughly 3,300 km from Boker Tachtit in the Negev (using coastal routes), can be understood as either differences in the speed of demic diffusion by natal dispersion, or random sampling of the archaeological record, error ranges in the chronometric dating methods, or all three. There is no requirement within Expectation 5 that end points of a cultural transmission event must evidence exactly similar diffusion rates.

This cultural transmission event may be called the "Bohunician Behavioral Package." Tostevin (2000c, 2003a,b) used an earlier and significantly different analytical structure that produced a moderately different set of behavioral choices as representing the transmitted package with this name (compare Table 6.5 in Tostevin 2003a with Table 8.20 above). The conclusions from the two analytical methods are comparable, however, and using the same name, the "Bohunician Behavioral Package," for the analytical construct still has utility as long as the newer details for the behavioral constituents of the behavioral packages identified in this research are used (Tables 8.19–8.21) instead of the older versions. In naming the behavioral package as an analytical generalization above the level of the individual assemblage, the aim is of course to isolate trends in the data for how stone tools were made across space and time such that the data supporting or rejecting a cultural transmission hypothesis can be evaluated according to appropriate data (i.e., the technological signatures of individual assemblages) rather than the essentialism of technocomplex or industrial-type labels. The "behavioral package" concept has significant methodological differences from the industrial type concept, as I take pains to point out elsewhere:

> The term "behavioral package" thus serves as a noun to describe the physical evidence of a particular cultural transmission event. The definition of a specific behavioral package consists of the grouping or association of flintknapping behaviors that *link one assemblage to the next in the order of the package's chronological appearance*... Unlike an industrial type, a behavioral package may *gain or lose constituents as it spreads through time and space.* Industrial types are often depicted as evolving through time, say from an "Early Aurignacian" to an "Evolved Aurignacian," but this is frequently presented (e.g., Otte and Kozłowski 2003) as a transformative process akin to Whitean-Spencerian change (*sensu* Tschauner 1994:82), contradictory to evolutionary theory as understood since the Modern Synthesis (Mayr 1982:Chap. 11). The behavioral package concept, however, does not have this problem as it is composed of different elements that can vary between assemblages; it is the consistency of the package from one assemblage to the next chronogeographically that serves as an indicator of the transmission strength of the package, rather than an absolute similarity between end points (Tostevin 2009:170, emphasis original).

In naming the package after the Central European industrial-type, despite its first appearance in the Levant at Boker Tachtit, is merely an homage to the Central and Eastern European prehistorians (Valoch 1990a; Kozłowski 1992; Demidenko and Usik 1993a; Škrdla 1996; Ginter et al. 1996) who first noticed the morphological similarities among these disparate assemblages. It does not imply that any assemblage ascribed to the Bohunician Industrial-type should automatically be expected to include technological elements of the Bohunician Behavioral Package; that would be a hypothesis for future testing. This should also not be considered a slight to Levantine prehistorians, as there is currently evidence against the interpretation that the package originated there, nor to western European prehistorians, as there is a paucity of evidence that the package ever appeared in this region.

In evaluating Expectation 5 with assemblages studied according to the procedures dictated by Chapters 3-4, one may also include the Brno-Bohunice assemblage excavated in 2002 by Škrdla and Tostevin (2003, 2005). Tostevin and Škrdla (2006) evaluate this assemblage according to the method and theory used here, with the result that the Brno-Bohunice 2002 assemblage is directly comparable to the present data set. The Brno-Bohunice 2002 assemblage has a blank production difference value of 1.57 and a toolkit morphology difference value of 0.29 when compared to Stránská skála IIIa level 4, blank production difference value of 1.57 and a toolkit morphology difference value of 0.14 when compared to Stránská skála IIIc, and a blank production difference value of 1.57 and a toolkit morphology difference value of 0.43 when compared to Stránská skála III (see Tostevin and Škrdla 2006:tables 2-3).[1] These results indicate that the Brno-Bohunice 2002 assemblage has a mean blank production difference value of 1.64 when compared to assemblages already concluded to contain the Bohunician Behavioral Package. The mean toolkit morphology difference value is 0.29. These values place the Brno-Bohunice 2002 assemblage well within the range of blank production difference values of the Bohunician Behavioral Package assemblages identified here, and closer in values for the toolkit morphology comparison to the other Central European assemblages with this package (compare with Figure 8.16).[2] The details of the specific technological choices in each domain of blank production (Tostevin and Škrdla 2006:Table 2) in the Brno-Bohunice 2002 assemblage also support the interpretation that the Bohunician Behavioral Package is represented in this assemblage.

As the Brno-Bohunice 2002 assemblage was clearly made with technological choices matching those of the Bohunician Behavioral Package, the dating of this assemblage has implications for the logic of the evaluation of Expectation 5. Richter et al.'s (2008) weighted mean of 48.2 ±1.9 ka BP_{TL} for 11 heated flint samples from Brno-Bohunice pushes the first appearance date of the Bohunician Behavioral Package back to a calendrical age of 48.2 ±1.9 ka BP_{TL}. This leads Richter et al. (2008) to conclude:

> Had the TL results placed Brno-Bohunice earlier (i.e., outside the 2 σ range) than the 47/46 ka ^{14}C BP age (~51/50 ka cal BP) for Boker Tachtit Level 1, the Bohunician Behavioral Package hypothesis would have been disproved (Richter et al. 2008:883).

Beyond this additional testing of Expectation 5 for this research, the TL, OSL, and calibrated AMS dating program reported in Richter et al. (2009) provides other important data on regional occupational continuity for interpreting the goodness of fit with Expectation 5, as it:

> ... draws into question a 10 ka hiatus observed in the chronometric data for the hominin occupation of Moravia between 53 and 43 ka cal BP (Zilhão 2006:189; Valladas et al. 2003). With the Neanderthal occupation of Kůlna Cave layer 7a dated by ESR (Rink et al., 1996) to 50±5 ka for a linear uptake (LU) model (53±6 ka recent uptake (RU) model) and the Bohunician at Brno-Bohunice 2002 dated to 48.2±1.9 ka (Richter et al., 2008), there is overlap already at the 1-σ probability level which indicates a more or less continuous occupation and makes it difficult to argue for a hiatus, as is clearly observed elsewhere, for instance in the stratigraphies of the Swabian Alb (Conard et al., 2006) (Richter et al. 2009:718).

Under Expectation 5, the assemblages containing the behaviors listed as constituting through time the Bohunician Behavioral

Package have a continuous temporal and geographical appearance and so cultural transmission is a parsimonious argument in the present case. If the same similarity data were produced for comparison of these assemblages with a geographically distant assemblage without geographic/temporal intermediates, say from Central Asia, Expectation 5 would strongly suggest a combination of independent innovation by adaptive convergence and cultural drift. Brantingham et al. (2000) have even provided a good example whereby raw material quality limitations would encourage the independent innovation, through guided variation, of Levallois-like surficial reduction strategies in a context where they did not exist previously. More models such as this will aid the evaluation of adaptive convergence from cultural transmission. As such, however, the assemblages showing great similarity in this study are geographically and temporally logical in their appearance and in no case is raw material quality poor enough to encourage convergence of the type Brantingham et al. (2000) suggested for the Initial Upper Paleolithic in North China.

As with the material culture evidence for the cultural transmission of the Bohunician Behavioral Package, similar reasoning can be used to recognize the appearance of a second transmission event, spreading what may be called the "Aurignacian Behavioral Package." In the Central European sequence, one sees a break in flintknapping traditions with the appearance of Stránská skála IIIa-3 and IIa-4 (2.80 when compared to Stránská skála III; 2.17 when compared to Vedrovice V). Neither the Szeletian nor the Bohunician which occupied Central Europe before the Aurignacian's appearance are likely to have been its progenitor in this region, at least as reflected in the behaviors extant in these assemblages. This is also indicated by the earlier dates for the Aurignacian in the Upper Danube (Conard and Bolus 2003). In the Near Eastern context, the first appearance of the Levantine Aurignacian assemblage of Kebara Cave Unit II brings very moderately different flintknapping options to the region when compared to the earlier Kebara Unit III (1.57). Alone, this measure of difference would not be enough to indicate an external origin for the Levantine Aurignacian, as these differences are not as great as seen between the preceding (and likely continuous) Ahmarian assemblages of Kebara Units IV and III, which produce a measure of difference of 1.70. Indeed one could argue for an *in situ* appearance of the Aurignacian in the Levant but for two factors. First and more importantly, earlier dates for the Aurignacian exist in Western and Central Europe (Conard and Bolus 2003, 2008; Conard et al. 2006; Nigst 2009), even if the date of >43,000 BP (GrN-7545) from Bacho Kiro Layer 11 Levels I-IV (Kozlowski and Ginter 1982: 168) can no longer be argued to be antecedent to the Aurignacian, as Teyssandier has reasoned (2006, 2008). Further, when one compares the Levantine Aurignacian to the Aurignacian assemblages of Central Europe, the measure of difference is too small and the clustering of behavioral options is too consistent to ignore the likelihood of inter-regional relationships between these assemblages. Eastern Europe received the "Aurignacian Behavioral Package" only after the temporal limit of this study (Anikovich 1992), and so is not represented here.

Thus, both of these transmission events brought about the performance of unprecedented technological behaviors to each region and set the stage for the florescence of the Upper Paleolithic. While other regions are likely to have witnessed quite different processes of change, for instance the Crimea (Chabai et al. 2004) and Western Europe where the Bohunician Behavioral Package never reached, the intriguing result of this study is the fact that the Middle to Upper Paleolithic transition in these three regions appears to result from the transmission of TWO behavioral packages or at least two discontinuities in technical learning sets. It is also significant that the first process, the Bohunician Behavioral Package, does not appear to have contributed its particular technological style to the subsequent Upper Paleolithic industries of Central and Eastern Europe (although there is evidence that it contributed to the later Levantine Upper Paleolithic in the form of the Ahmarian industry). The role played by the Bohunician Behavioral Package in these regions was thus not a herald of the "whole" Upper Paleolithic phenomenon but a first step within a historical process that led to the Upper Paleolithic.

8.2.1 The Relevance of Regions Not Sampled in this Research

The discussion above of the significance of the dating of Brno-Bohunice for the lack of an occupational hiatus within Moravia, as seen from neighboring regions, leads directly to a needed discussion of the role of assemblages and regions unexamined through this research protocol in supporting or weakening the evaluations of the cultural transmission hypotheses tested here. Richter et al.'s (2009) statement above is in direct response to a critique presented by Zilhão (2006) of the interpretation of cultural connections between assemblages of the Bohunician Industrial-type in Central Europe with assemblages such as those at Boker Tachtit in the Levant. Strangely, Zilhão does not cite in his critique Tostevin (2003b) nor Škrdla (2003c) but only the editors of the monograph in which these chapters are published (Svoboda and Bar-Yosef 2003). Zilhão also does not cite Škrdla's (2003a) systematic comparison of Volkman's refits at Boker Tachtit with Škrdla's refits at Stránská skála. He makes no reference to the "Bohunician Behavioral Package" discussed in Tostevin (2003b) but suggests that parsimony would argue for a closer autochthonous or *in situ* origin for the Bohunician phenomenon in southern Poland rather than looking to the Levant. He points to the sites of Piekary IIa (Sitlivy et al.1999b; Valladas et al. 2003; Sitlivy and Zięba 2006; Zięba et al. 2008) and Księcia Józefa (Sitlivy et al.1999a; Sitlivy and Zięba 2006; Zięba et al. 2008) near Krakow, Poland, as documenting:

> the *in-situ* development of volumetric Upper Paleolithic methods of blade debitage out of Levallois flake-based technologies during the time interval of the Moravian hiatus (ca. 53–43 kyr cal BP)[60] (Fig. 5). Parsimony dictates that there is no need to look into the Middle East for the source of the Bohunician if a better local alternative is available (Zilhão 2006:187, 189).

In many ways, this suggestion is exactly what I hope the present research agenda will elicit as a response: suggestions for further sites and regions to examine in testing the hypotheses that were unavailable when the initial data collection was undertaken. But these responses have to be considered as statements of future hypotheses, not data that by its essential nature can be assumed to refute the conclusions of the present research. While laudable for themselves as contributing to our knowledge of prehistory, the *chaîne opératoire* and lithic economy approaches of studies such as Sitlivy et al. and Zięba (2006), and Zięba et al. (2008) were not designed from first principles to be sensitive to cultural transmission processes (i.e., knapper's choices that are learnable, efficacious, and etically-observable to the archaeologist). "Levallois" is not an analytical category that is congruent with a single behavior that is learnable and performable but instead an analytical abstraction that confounds many variables of knapper choice. It is also problematic when viewed etically by archaeologists; this is why the participants of the Levallois Conference of May 1993 considered removing Levallois elements from Bordes' type-list due to the large variance in the definition of Levallois among researchers. Similarly, providing an illustration of artifacts from an assemblage does not constitute data, especially when the illustrations do not conform to geoarchaeologically-valid assemblages: Zilhão's figure 5 showing artifacts from Level 7b of Piekary IIa mixes artifacts recovered from different decades' excavations and even levels within the site. Sitlivy and Zięba (2006, fig. 19-19) indicate that Zilhão's figure 5 artifact "i" is in fact from W. Morawski's 1960s and 1980s excavations from level 7c, not level 7b, and at a locality on Piekary hill that is over 20 m from the locality of the excavations discussed in Sitlivy et al. (1991b). Valladas et al. (2003:fig. 7) indicate that Zilhão's figure 5 artifacts "c" and "g" also come from level 7c, not 7b, and derive from Morawski's trench XIII whereas Valladas et al. (2003:figure 8) indicate Zilhão's figure 5 artifact "h" comes from the 2000s excavation 20 meters away in trench XXII. From my own examination of the surface conditions of the artifacts from Morawski's excavations and those of the more recent excavations of Sobczyk, Sitlivy, and Kozłowski, the artifacts' life histories are significantly different between the assemblages, with striking contrasts in patination, weathering, rolling, and edge damage. Beyond questioning how collections of illustrations are assembled as published figures or "data", or how the geoarchaeological discreteness of different assemblages are handled, I would point out that Zilhão's suggestions to look to the late Middle Paleolithic of southern Poland for possible antecedents for the Bohuncian Behavioral Package is of course a good idea (he could be correct). Such a study is currently underway. Yet we cannot pretend that such a suggestion is credible archaeological data for drawing conclusions.

But what can we learn now from widening the evaluation of Expectation 5 to include interpretations derived from published literature in regions not sampled in the present project? Using published data gathered under analytical protocols different from the one espoused here can be problematic but, again, represents further hypotheses for future testing. For instance, one could hypothesize that the following assemblages may also contain large numbers of behavioral choices that are also found within the Bohunician Behavioral Package. Going northward, after the first appearance of the Bohunician Behavioral Package in the southern Levant at 47/46 kya, one could hypothesize its next appearance in the northern Levant at Üçagizli Cave layers G-I at least by 39 kya (Kuhn et al. 1999, 2004), possibly next in the Balkans if the preliminary results from Temnata Cave layer VI, sector TD-II (Ginter et al. 1996; Drobniewicz et al. 2000) are confirmed, next in Central Europe at 41 kya, and finally in Eastern Europe by 38 kya. The geographic trajectory is even more complete in Eastern Europe if the highly suggestive reports on the lowest complex of Kulychivka (Savich 1975; Demidenko and Usik 1993b; Stepanchuk and Cohen 2000-2001; Meignen et al. 2004) and Korolevo I layer 2b (Gladilin 1989a, 1989b; Gladilin and Demidenko 1989, 1990; Usik 1989; Demidenko and Usik 1993a) are confirmed by rigorous attribute analysis coupled with geoarchaeological investigation of the integrity of the artifact associations. The Early Upper Paleolithic assemblage from Sokirnitsa IA level 3 in western Ukraine (Monigal et al. 2004) also warrants future testing using the protocol presented here.

The archaeological record of Bulgaria as a region of great importance for the hypothesized cultural transmission of the Bohunician Behavioral Package needs to be addressed more specifically. The original conclusions drawn by Tostevin (2000c, 2003a,b) were informed by the literature coming from the initial publication of the excavations at Temnata Cave layer VI, sector TD-II, which was not available for my direct study at the time of the data collection period for the project. The analysis of this assemblage by Drobniewicz et al. (2000) and the conclusions of Kozłowski (2004) were highly suggestive of fitting the knapping patterns of the Bohunician Behavioral Package, leading to its inclusion as part of the argument for the goodness of fit between the archaeological record and Expectation 5. Subsequent reanalysis of the collections from Temnata Cave layer VI, sector TD-II by T. Tsanova (2006, 2008) has, however, thrown doubt onto the geoarchaeological integrity of this "transitional" assemblage. Using the method of lithic taphonomy championed by J.-G. Bordes (2003) in French Early Upper Paleolithic sites, Tsanova, with the guidance of J.-G. Bordes, restudied the collections in an effort to evaluate the spatial distribution of the artifacts according to their technological and temporal association. She evaluates the spatial patterning of discoidal cores and Levallois blade cores as representing Middle Paleolithic period artifacts, and semi-tournant blade cores as representing Upper Paleolithic period artifacts. She concludes that the Levallois elements are found more spatially associated within layer VI with the semi-tournant Upper Paleolithic cores and debitage further down the talus slope of this area of the site, while the Levallois and Middle Paleolithic elements are more isolated from the UP elements in the sediment of layer VI found further up the slope. She concludes, therefore, that there existed two strata within the cave or near the entrance, originally superimposed and temporally and technologically distinct, that subsequently were eroded and post-depositionally mixed on the talus slope and excavated as a "transitional" assemblage within an *in situ* stratum, layer VI.

Given the uncertainties of geological processes on cave talus slopes, Tsanova's conclusions may indeed be correct. But this is more of a question for a geoarchaeologist or micromorphologist to evaluate than for a lithic analyst. The lithic taphonomy approach that J.-G. Bordes and T. Tsanova brought to the Temnata assemblage derives from a geographical context, southwestern France, where the prehistoric record is the richest and most complete in the world, where there are many reasons to assume that "we know what we know." Yet, in Southeastern Europe, is this still true? Each of the elements mapped by Tsanova as being temporally distinct as Middle Paleolithic or Upper Paleolithic can in fact be found within the same refit sequence from Stránská skála IIIa level 4 (for instance, refitted core No. S01/84, Škrdla 2003c:figs. 9.1–9.2), in a region far closer to Temnata than is Roc de Combe in France. The circularity of the lithic taphonomy argument, in the face of extant refitted sequences of demonstrated "transitional" technologies closer to the site, makes her conclusions less certain. She still could be correct, given the geological issue, but it is by no means absolutely demonstrated. As a result, it is worth the effort to test the blank production and toolkit morphology difference values between Bohunician Behavioral Package assemblages and Temnata Cave layer VI, sector TD-II (as is currently being pursued), as well as calling for more explicit geoarchaeological testing of the postdepositional mixing hypothesis.

9

Conclusions

This book reports on a research project designed to fill a gap in archaeological method and theory for understanding how different cultural transmission processes create patterning in the material culture of foragers at the resolution of Paleolithic palimpsests. The research project attempts to bridge this gap with a middle-range theory connecting cultural transmission and dual inheritance theory with the archaeological study of flintknappers' flake by flake choices in the production of lithic assemblages. The project thus combines a new middle-range theory as well as a new approach to characterizing Paleolithic assemblages for systematic comparison of units of analysis appropriate to distinguishing forces of change in cultural evolution. This approach is applied to the Middle to Upper Paleolithic transition in three regions of western Eurasia in order to demonstrate its utility in making quantitative distinctions between continuity and discontinuity in cultural learning sets. This is the first step in addressing the debate over the extent to which the Middle to Upper Paleolithic transition and the dispersal of anatomically modern humans from Africa into Eurasia at the end of the Pleistocene were the result of the same process, related processes, or unrelated but coincident processes.

Eighteen archaeological assemblages dating between 60 and 30 kya in three regions of western Eurasia were analyzed in order to identify which flintknapping behaviors changed during the Middle to Upper Paleolithic transition and to compare the sequences of these diachronic behavioral changes within and between each region to determine how the transition occurred. Specifically, this research tested three hypotheses concerning change in flintknapping behaviors between 60 and 30 kya in Central Europe, Eastern Europe, and the Levant:

- Hypothesis 1 (Innovation): flintknapping behaviors changed due to *in situ* behavioral change by means of independent innovation in each region. In this case, any similarities in blank production and/or tool kit morphology are judged to be the result of adaptive convergence and/or cultural drift.
- Hypothesis 2 (Cultural Transmission): flintknapping behaviors in a region changed due to the spread of behaviors from one region to another as a result of cultural transmission across space. This hypothesis is further subdivided according to the distinction between symmetric and asymmetric transmission of learned behaviors from cultural transmission theory (Boyd and Richerson 1985).
- Hypothesis 3 (Combination): flintknapping behaviors changed due to independent innovation in one or more regions with the subsequent cultural transmission of the behaviors into adjacent regions.

These hypotheses were tested by separating the research program into two parts, asking first a culture historical question (what changed in the transition?), and using the answer to that question to address its processual counterpart (how did this change occur?). By learning from the mistakes of earlier approaches, from the Kulturkreislehre to the culture historians of the Classificatory-Historical Period of American archaeology (Willey and Sabloff 1993), a unit of analysis appropriate for culture historical questions was created. As Chapter 3 describes, technical choices made by artisans during the artifact production sequence, rather than merely the shape of the final artifact, are measures sensitive to variation between material culture traditions (Hodder 1979; Longacre 1981; Braithwaite 1982; Wiessner 1990; Childs 1991; Lemonnier 1992; Aronson et al. 1994; Stark 1995; Carr 1995b). Using the differences in the taskscape visibility of these behaviors during the flintknapping process, a middle-range theory was proposed to connect specific portions of the lithic archaeological record with behaviors of enculturation which result from transmission of knapping behaviors by observation within a socially intimate group. The middle-range theory further developed the direct correlation between social intimacy in modern hunter-gatherers and the dominance of vertical cultural transmission from parents and older relatives (Shennan and Steele 1999). As such contexts of intimacy with vertical transmission tie cultural and biological inheritance together within symmetric transmission (Boyd and Richerson 1985), the archaeological proxies for knapping choices made within such contexts of social intimacy should distinguish the action of specific culture historical processes, such as symmetric transmission and asymmetric transmission, from independent innovation of the details within lithic operational sequences during the Middle to Upper Paleolithic transition.

The three hypotheses described above were tested by assessing the goodness of fit between the archaeological record and a series of model expectations derived from a combination of archaeological (Andrefsky 1987; Willey et al. 1956; Lechtman 1977; Stark 1998; Carr 1995a,b; Eerkens and Lipo 2007:240), history of science (Gille 1978; Basalla 1988), and social anthropological thought (Steward 1929; Kroeber 1931, 1940; Kluckhohn 1936; Barnett 1953; Rands and Riley 1958; Lemonnier 1992) concerning the problem of distinguishing cultural transmission from independent innovation. The first step in this procedure was the description of the technological signature for each assemblage according to the behaviors of flintknapping that are learnable, efficacious in producing artifactual variation, and etically-observable by archaeologists (Chapter 4). The second step was the comparison of the technological signatures for assemblages that succeed each other through time within the same region. Determining which flintknapping behaviors in each step of the operational sequence appeared for the first time in a particular region or disappeared in later assemblages in that region produced an intra-regional sequence of behavioral change (Chapters 5-7). The third step was accomplished by charting which behavioral options appeared and disappeared through time between adjacent regions, i.e., comparing the intra-regional sequences to produce an inter-regional sequence of change (Chapter 8). The final step in testing the hypotheses was the evaluation of the inter-regional sequence against the five model expectations, each of which was based upon a theoretical principle which predicts how the intra-regional sequences and inter-regional sequence should appear given Hypothesis 1 (Innovation), Hypothesis 2 (Cultural Transmission), or Hypothesis 3 (Combination).

9.1 Results

The final analysis of the goodness of fit between the archaeological pattern of behavioral change between 60 and 30 kya in these three regions and the five model expectations supports the cultural transmission hypothesis (Hypothesis 2) rather than the independent innovation hypothesis (Hypothesis 1). Surprisingly, the archaeological record reconstructed by this research demonstrates the existence of not one—as traditionally proposed—but TWO discontinuities in learned behaviors in these regions, one between 46-41,000 BP which spread by demic diffusion via natal dispersion first to the Levant and then to Central and Eastern Europe, and another between 36-32,000 BP. Without antecedents in any of the three regions, the first diffusion event brought an extremely homogeneous set of flintknapping behaviors to each region. This cluster, labeled the "Bohunician Behavioral Package" after the Czech industry which marks its westernmost distribution, is distinguished not only by the appearance of specific behaviors but also by their consistent reoccurrence in different regional settings despite the functional independence of the domains in which the similarities are found. The second cluster of behaviors, the "Aurignacian Behavioral Package," appears intrusively in the Levant and later in Central Europe. It appears after the chronological limit of this study in Eastern Europe (Anikovich 1992).

9.2 Implications for the Origins of Anatomically Modern Humans

The implications of these results for the debate over the origins of anatomically modern humans must be evaluated at several levels of resolution. At the largest level, the two cultural transmission events evidenced between 60 and 30 kya in these three regions disprove the archaeological expectations of behavioral continuity argued by Clark and Lindly (1989a; 1989b) in favor of the Multi-Regional Hypothesis. This research thus provides support for the Out of Africa Hypothesis in terms of the expectations for cultural continuity during this period. At a finer resolution of scale, however, the extent to which the spread of the Bohunician Behavioral Package evidences the gene flow of modern humans through these regions requires a more complex evaluation.

9.2.1 *Symmetric or Asymmetric Transmission*

The predictions for the likelihood of gene flow being responsible for the spread of the Bohunician Behavioral Package arise from the fact that modern hunter-gatherers structure their social environment through mating networks and systems for the exchange of social obligations (Lee and DeVore 1976; Kelly 1995, 161-202, 261-92). The different degrees of gene flow during culture contact events known from ethnohistoric and ethnographic contexts clearly reflect levels of social intimacy (Van Kirk 1983; Merrell 1999; White 1991; Fix 1999; Cook 1998). Thus, a case of high social intimacy producing cultural transmission of the knowledge of how to make lithic artifacts a certain way most likely involves individuals mixing their residential and biological lives in order for them to learn the other's behaviors of low taskscape visibility. Conversely, learning only behaviors of high taskscape visibility, as in a case of stimulus diffusion, would on average require little social intimacy that would result in frequent mating opportunities. These are probability statements, however, and do not rule out the odd contrary example. All things being equal, these probabilities make it more likely that the Bohunician Behavior Package spread by symmetric transmission involving gene flow rather than the asymmetric transmission of cultural traits without gene flow.

All things are not equal in Paleolithic archaeology, however. While one might be able to distinguish symmetric and asymmetric transmission in societies which possess multiple lines of material culture evidence of their social intimacy, Paleolithic archaeologists possess only stone tools for most periods, making Expectation 3 (number of domains of material culture) weaker than it should be. Other time periods thus have the ability to improve the power of their tests through multiple lines of evidence, resulting in the "consilience of induction" (Whewell 1858: 70, 85-96; Gould 1989:282). Further, our present control over the chronology and resolution of Paleolithic assemblages is far from perfect, so that the actual process of innovation and transmission for some behaviors may be invisible in the archaeological record (Zilhão and d'Errico 2003). Nevertheless, there are details in the Pleistocene record which make one mode of cultural transmission more likely than the other.

9.2.2 Why did the Bohunician Behavioral Package spread between regions?

The blank production choices characteristic of the Bohunician Behavioral Package could have been moved between the Negev and Central Europe without the wholesale displacement of resident populations by an intrusive population of artisans. Specifically, the mating networks with which hunter-gatherers structure their social environment are more likely to be responsible for the diffusion of the Bohunician Behavioral Package behaviors. The behaviors are learnable only at levels of close social intimacy, since the individual behaviors are restricted to the taskscape visibilities of blank production. In modern foragers, this level of social intimacy would involve individuals mixing their residential lives, via mating networks and the exchange of social obligations. Thus, assuming that the resident population into which the dispersing population moved was not beyond the Paterson (1985) point of speciation by mate recognition and assuming that there were hominins resident in the area (an empty landscape has been argued for the arrival of Aurignacian hominins in the Upper Danube - Conard and Bolus 2003), it is likely that gene flow would be one result of this cultural transmission event from the Levant to Central Europe. The interpretation of the results of this research thus depends to some degree on whether you assume Neanderthals were not there on the landscape to engage in gene flow or were not there simply as mates. But this statement does not explain why the transmission occurred.

Mating networks could be responsible for the Bohunician Behavioral Package cultural transmission event even if there were no advantage for greater population dispersal associated with the lithic package itself. Such an advantage need not derive from the lithics at all, since selectively neutral (*sensu* O'Brien 1996 as well as *sensu* Sackett 1982, 1990) flintknapping choices could have been carried along a mating network experiencing another adaptive advantage, such as the more secure pooling of subsistence risk through long distance reciprocal altruistic relationships (Wiessner 1982) evidenced by raw material exchange (Roebroeks et al 1988; Adler et al. 2006). The behaviors recognized as the transmitted package would have been preferentially learned by individuals through the effect of either indirect or frequency-dependent biased transmission (see Chapter 3 and Boyd and Richerson 1985). In indirect biased transmission, an individual viewed as successful in society is taken by the learner as the best role model from whom to learn many traits, including traits that in truth are not responsible for the role model's success. The Bohunician Behavioral Package behaviors could easily fall into this category and were simply transmitted on the "coat tails" of other successful adaptations. Frequency-dependent biased transmission, in which the learner is influenced to adopt the most frequent or at times most rare trait in the population, could also aid the transmission of the package behaviors if they just happened to be the more numerous in the society as a result of other factors, such as the presence of more fit adaptations of those who happen to use the package behaviors.

Conversely, it is also possible that the tool kit itself was more advantageous than alternative technologies so that it spread preferentially through populations by means of direct biased transmission, in which the properties of the chosen trait itself make it more likely to be learned than an alternative variant (cf. Bettinger and Eerkens 1999). The elongated Levallois-like point characteristic of the blanks and tool kit of the Bohunician Behavioral Package could possibly have offered an advantage as an armature for use with a true projectile technology such as the spear thrower (*sensu* Hughes 1998; Shea 2003:368–373, 2006). Shea's (2006) evaluation of the Emireh points from Boker Tachtit levels 1–3 as projectile armatures showed that they were unlikely to have served as tips for spear thrower darts. More likely according to the tip cross-sectional area measurement, these were thrusting spear points. Sisk and Shea (2009), however, have demonstrated how Levallois points could serve as efficient projectile armatures. Only later retouched point types such as the el-Wad point and Ksar 'Akil points of Üçagizli Cave (Kuhn et al. 1999, 2004), however, fall clearly into the dart tip variability of the tip cross-sectional area measurement. While tip cross-sectional area data for the Levallois points of the Bohunician is not currently available, it is likely that these pieces will, like those of Boker Tachtit, not show dart tip values. This is certainly true for the foliate points in the Szeletian and Bohunician, which Shea (2006) judges to be knives or points for throwing spears. Thus it is currently unlikely that the Bohunician Behavioral Package itself represented a true dart tip projectile technology that could have offered its learners-carriers an adaptive advantage. It is possible, however, that the points produced in these assemblages could represent the initial development of the dart projectile system but with armature tips not yet honed by cultural and natural selection to the refined tip cross-sectional areas of points serving this function among the recent Holocene hunter-gatherers from whose armatures this measurement was deduced by Hughes (1998) and used by Shea (2006). In this way, it may be proposed that the pointed products of the Bohunician Behavioral Package may themselves have offered an adaptive advantage that would have favored their transmission by both symmetric and asymmetric means.

Thus, there is no easy answer to why the package spread. Nor is there a reason for why it failed to spread beyond its last westernmost extant example, Hradsko in Bohemia if Neruda and Nerudová (2000) are correct in ascribing this assemblage to the Bohunician. In the process of recognizing where the package had an adaptive advantage in spreading through space and time versus where it did not, we are significantly further along in recognizing the local context (*sensu* Svoboda 2004) which disrupts the adaptive advantage. This is one direction for which the present approach offers entirely new avenues of research, since having identified a cultural transmission event it is now possible to devote research effort into this particular adaptive context which otherwise might have been ignored as "yet another late Middle Paleolithic technocomplex."

9.2.3 *Who made the Bohunician Behavioral Package?*

If gene flow was involved in the spread of these flintknapping behaviors, then the hominins responsible for making the Bohunician Behavioral Package include all of the hominin morphospecies at either end of the transmission across space. In other words, if it spread by gene flow, starting with modern humans at one end of the transmission and including Neanderthals at the other, then both morphotypes must be acknowledged as producers of the package. But can we point to any specific hominins associated with the package?

No direct association of a hominin exits with any of the assemblages containing the package studied within this sample. Tantalizing if inconclusive possibilities exist, however, for both Neanderthals and modern humans being involved in this cultural transmission event. Specifically, a Neanderthal tooth is associated with an Initial Upper Paleolithic industry from Lakonis Cave, Greece (Harvati et al. 2003; Panagopoulou et al. 2004, fig. 8; Elefanti et al. 2008). Situated in the path of the transmission event, this Initial Upper Palaeolithic assemblage is chronologically appropriate (ca. 40 kya) to evidence the diffused behavioral package. Despite the highly suggestive laminar Levallois component, the current published evidence is unfortunately too preliminary to determine the relationship between this assemblage's blank production behaviors and those of the Bohunician Behavioral Package.

For the modern human connection, a modern human child burial is associated with a Middle Palaeolithic assemblage from Taramsa 1, Egypt (Vermeersch et al. 1998; Van Peer and Vermeersch 2007), directly ancestral to the transitional Levallois blade reduction sequences Van Peer (1998) has pointed to as likely antecedents to technological performances of level 1 of Boker Tachtit. As level 1 of Boker Tachtit is the earliest example of an assemblage containing the Bohunician Behavioral Package in the regions studied in this research project, Van Peer's reconstructions from Taramsa 1 make it more likely that the specifics of the behavioral package arose in the Nile Valley rather than in the Levant (see also Van Peer et al. 2010). Even Marks (2003) admits that a late Middle Paleolithic antecedent to the Emiran is unknown within the Levant. Except for the difficulties with the OSL dating of the burial (Van Peer and Vermeersch 2007), it would be possible to add "modern human" to the Egyptian origins of the Bohunician Behavioral Package if the Taramsa assemblage proved to be similar in its blank production behaviors, as analyzed here, to level 1 of Boker Tachtit. Mellars (2005) has also argued that an anatomically modern human, "Egbert," is associated with an early Ahmarian assemblage (Level XVII of Ewing's excavations) from Ksar 'Akil situated between dated levels of 32,000 ± 1500 BP (Monaco dates, MC-1192, in Ewing's level X) and 43,750 ± 1500 BP (Ewing's level XXVI) (Bergman and Stringer 1989; Mellars and Tixier 1989). This is highly suggestive of anatomically modern human involvement in the transmission of the Package but as Egbert is more likely to date at the upper end of the expected range above, the package would already have reached Central Europe long before the involvement of Egbert in the production of his assemblage. Also, as this argument rests upon the hypothesis, rather than demonstration, of symmetric transmission between the period during which the package would have been extant in Lebanon and the period in which Egbert lived at Ksar 'Akil, the role of anatomically modern humans in the initial spread of the package remains equivocal, despite the certainty that anatomically moderns were the producers of assemblages which inherited many cultural traits from that time period, as the Initial Upper Paleolithic does show strong continuity with the later Upper Paleolithic of the Levant (Belfer-Cohen and Goring-Morris 2007; Bar-Yosef 1994). Nevertheless, this argument is further encouragement to include Ksar 'Akil in analyses of the type presented here, despite the issues with the stratigraphic agreement between the Ewing and Murphy excavations (Copeland 1987; Bergman 1987).

Altogether, given the OSL difficulties with the Taramsa 1 burial, the preliminary nature of the report on Lakonis Cave, and the indirect evidence of Egbert at Ksar 'Akil, no hominin association with the Bohunician Behavioral Package is very secure. The timing of the arrival of the Package in Central Europe in relation to the appearance of the Aurignacian, which still cannot be separated from anatomically modern humans on the extant evidence (Churchill and Smith 2000; Bailey et al. 2009), is, however, highly suggestive.

9.2.4 *The Role of the Bohunician Behavioral Package in the Danube Corridor Model*

Conard and Bolus (2003; Bolus and Conard 2001) have presented two models to explain the involvement of anatomically modern humans in the origins of the Aurignacian in Europe. The Danube Corridor Model proposes that the earliest anatomically modern humans dispersed into Europe from the Near East via the Danube River Valley. The related hypothesis, the Kulturpumpe Model, proposes that modern humans invented the material culture archaeologists recognize as the Aurignacian once they arrived in the Upper Danube basin, as a culture contact response to the Neanderthal occupation in neighboring regions. The Kulturpumpe Model also argues for the origin of the Gravettian in the Upper Danube but this aspect of the model is not relevant here as it is beyond the chronological limit of the study. Together the Danube Corridor and Kulturpumpe Models try to account for the early AMS and TL dates for the Aurignacian at Geiβenklösterle (Conard and Bolus 2003; Richter et al. 2000; Higham et al. 2012) by placing the *in situ* development of the art and lithics of the Aurignacian in the Upper Danube at the hands of newly arrived modern humans. The chronological arguments of Conard and Bolus (2003; Conard et al. 2003) have not been received without significant debate (Zilhão and d'Errico 1999, 2003) but those well argued critiques are not the present concern. Rather, the Danube Corridor and Kulturpumpe Models raise interesting possibilities for the intersection of the enculturating environments of archaics and moderns in the Middle Danube, as this region lies astride the route to the proposed Kulturpumpe of southern Germany. Yet the complexity of the archaeological record in the Middle Danube region, with its numerous technocomplexes, has been glossed over in the presentation of the models. Specifically, the Danube Corridor Model fails to explain what archaeological signature was left along the path of dispersal up the Danube by the anatomically modern humans before they invented the Aurignacian in the Upper Danube. It is known that Neanderthals created the Central European Micoquian in the Middle Danube, as seen in Kůlna Cave Layer 7a at 50 kya (Valoch 1988; Rink et al. 1996). The archaeological signature of the moderns must therefore be one of the two archaeological technocomplexes which postdate the Micoquian but predate the Aurignacian in the Middle Danube, namely the Bohunician or the Szeletian. The fact that no identifiable occupational hiatus, given our current dating resolution, existed between the Neanderthals at Kůlna Cave Layer 7a and makers of the 2002 assemblage from Brno-Bohunice (Richter et al. 2008, 2009) makes the Middle Danube even more significant for the Danube Corridor Model.

As discussed above, it would be presumptuous to make an absolute claim for either the Bohunician or the Szeletian as the handiwork of anatomically modern humans, as no fossil specimens have been recovered from either

Bohunician or Szeletian contexts. Nevertheless, the Szeletian is unlikely to be the archaeological signature of modern humans, given the arguments made for the Szeletian's having evolved *in situ* from the Neanderthal-made Micoquian (Valoch 1990a; Neruda 2000; Allsworth-Jones 1986, 1989). The conclusions of this study's comparison of Kůlna Cave Layer 7a and Vedrovice V, and that of the three-way acculturation approach using Stránská skála IIIa Layer 4 as well (Tostevin 2007), do point to continuity but only in the toolkit morphology, not in the blank production sequence sensitive to enculturation and symmetric transmission. The Bohunician, on the other hand, is a far better candidate for being the handiwork of modern humans. This book has presented arguments for an intrusive origin of the technological choices which produced the assemblages archaeologists recognize as Bohunician in Central Europe. A specific suite of blank production choices unites particular assemblages between the Levant and Central and Eastern Europe as well as distinguishing them from their predecessors in each region, as recognized by other scholars (Bar-Yosef 2002; Škrdla 2003a; Kozłowski 2004). The geographic and chronological context of this pattern of learned behavior suggests that if the Danube Corridor Model is correct in identifying the route for modern human dispersal into the Aurignacian Kulturpumpe of southern Germany, the Bohunician Behavioral Package is likely involved. In fact, it is the only epistemologically and methodological demonstrated cultural transmission with both the correct geographic vector and temporal context to serve as the archaeological proxy for the arrival of anatomically modern humans in Europe.

I therefore recommend that the Bohunician Behavioral Package transmission event must be integrated into Conard and Bolus's (2003) Danube Corridor Model as a related hypothesis for the gene flow of anatomically modern humans into Europe.

9.3 Methodological Contributions of this Research

It must be emphatically stressed that the appellation "behavioral package" is not a synonym for "industrial type" (Tostevin 2009). The approach espoused by this research uses behavioral details as the units of analysis for answering questions of culture historical process. Instead of mapping the peregrinations of industrial types (*sensu* Clark and Lindly 1991), this research maps the appearance, disappearance, and associations between individual flintknapping options through time and space to answer questions of inter-regional scale related to the existence of social intimacy through time which would allow the learning versus the necessity of independent innovation of these behavioral options.

As the recent Neanderthal acculturation debate has shown (d'Errico et al. 1998; d'Errico 2003; Mellars et al. 1999; Mellars 2005), the current arguments using industrial types and technocomplexes as the units of analysis have proven unproductive and inappropriate for use outside of the regions in which the industries were first defined. Behavioral steps within the lithic operational sequence, however, do not require inclusive and exclusive principles of definition. Through the recognition of technical choices of artisans according to the social intimacy and taskscape visibility concepts, the behavioral approach taken here avoids the ubiquitous problem of defining acceptable limits to industrial variability and does not constrain assemblage variability into industrial types when we specifically want to understand change in hominin material culture behavior. The term "behavioral package" thus serves only as a noun to describe a particular grouping or association of flintknapping behaviors in a particular context. Unlike an industrial type, a behavioral package may gain or lose constituents as it spreads through time and space. Theoretically, industrial types also possess spatial variability but are rarely used in this way in practice. The behavioral package concept, however, does not have this problem. For instance, a behavioral package in two regions adjacent to the source region might include 12 similar knapping steps within their operational sequences. A third region, even further from the region of origin, might only bear witness to 8 similar steps within the transmitted behavioral package. The most distant region receiving the diffused package may exhibit only a few of the specific knapping options, although the transmission interpretation becomes less and less likely as fewer behaviors are shared between regions.

The approach exemplified by this research has another methodological advantage. While many have tried to understand the Middle to Upper Paleolithic transition by comparing the Middle Paleolithic record as a whole to the Upper Paleolithic record as a whole, the broad distinctions thus found cannot be as readily recognized when comparing the earliest Upper Paleolithic assemblages to their immediate Middle Paleolithic predecessors (Straus 1983). This generalized approach has resulted in our failure to understand the specific mechanism by which the Upper Paleolithic replaced the Middle Paleolithic on the Pleistocene landscape (Gilman 1983; Harrold 1991). The research approach advocated here, however, solves this problem by first determining what features of human behavior changed, and then determining how those features changed. This structure is possible because the archaeological record was defined by abstract chronological limits, 60,000 BP to 30,000 BP, rather than by *a priori* analytical units such as "Middle Paleolithic," Transitional," and "Upper Paleolithic."

The structure created for this research would also serve as a new theoretical and methodological approach for asking questions of culture historical process in other periods as well. In fact, the five model expectations would allow for even greater resolution in a potential diffusion or innovation scenario. Specifically, Expectation 3 asserts that a transmission event causes many more types of material culture to change at one time than an innovation event (Steward 1929; Linton 1936:372; Lechtman 1977). This is due to the fact that although an instance of innovation in one material culture domain, ceramics for instance, may affect another material culture domain, such as metallurgy, it is unlikely to affect several independent domains simultaneously. Thus, one can use the number of independent material culture domains that show changes at the same time as an indicator of a diffusion event. As later archaeological periods possess much richer and more numerous independent domains of material culture, archaeologists addressing these later periods may find an added utility in the present research that is not available to Paleolithic archaeologists.

9.3.1 Paths Not Taken: Alternative Structures of Analysis

The general principles outlined in Chapter 3 for a behavioral approach to the structuring of technological signatures, and their evaluation against defined culture historical processes, can take one of many forms. This book has outlined one version of such a structure, but others are possible. In particular, different structures would be necessary in different contexts of lithic variability, as for instance, in the biface-dominated New World. The taskscape visibility concept of core exploitation for blanks at residential sites may have to be altered in a setting in which bifaces were both mobile tools and cores (Kelly 1988) moving between residential sites and the pathways of the landscape in a way distinct from that modeled here for the heavier prepared core traditions of Pleistocene western Eurasia.

Other methodologies might be preferred in contexts in which assemblages are not absolutely dated or are dated but with error ranges that overlap too much for a clear chronological superposition of assemblages. In such a context, the chronological ordering of pair-wise assemblage comparisons would be impossible, leading to the likely use of a similarity matrix of comparisons between all assemblages and not just a subset as demonstrated with this research. An example of the use of such a similarity matrix in a context with multiple contemporaneous assemblages is presented by Tostevin and Škrdla (2006) for a new Bohunician assemblage from the type-site of Brno-Bohunice.

Cases in which it is logical to produce a complete matrix of variation of all assemblages against each other would also allow a different statistical representation of the behaviors documented in each case. Staller and Thompson (2002) have used squared chord distance analysis as well as detrended correspondence analysis, techniques frequently used by palynologists (e.g., Gavin et al. 2003), for the analysis of assemblage differences in phytolith morphologies for the documentation of the dispersal of varieties of maize in the past. As the structure of Staller and Thompson's assemblage data is quite similar to the lithic assemblage data used here, it will be possible in the future to use such multivariate clustering techniques to present the behavioral approach to lithics and cultural transmission theory in new ways.

Other multivariate techniques for the presentation of technological data at the assemblage level are also available. Both Wurz et al. (2003) and Hovers and Raveh (2000) have presented graphically-advanced multivariate methods of presenting assemblage data. While these methods are likely the future of the field, I prefer the methods developed here, primarily because they are low powered and therefore less threatening to the traditionalist heart of most Paleolithic archaeologists; the use of cultural transmission theory is perhaps threatening enough for most of my readers. More importantly, the behavioral-choice by behavioral-choice structure to the analysis, with individually visible univariate tests, keeps the methodology close to the core idea of cultural transmission theory and the model predictions for distinguishing independent innovation from other cultural transmission processes. The methodology is thus more transparent in how the variability affects the results of the analysis than most multivariate methods.

That being said, the great potential of cladistic approaches (*sensu* Hennig 1966) made available by the use of multivariate analyses, as demonstrated by Shennan and Collard (2000), Collard and Shennan (2000), O'Brien et al. (2001), Tehrani and Collard (2002), O'Brien and Lyman (2003), Mace et al. (2005), Jordan and Shennan (2005), Lycett et al. (2006), Lycett and von Cramon-Taubadel (2008), and Lycett (2010) make it likely that my approach to cultural transmission theory will converge with the similar approaches taken by these scholars. The first caveat against this convergence, however, is the assumption in cladistic modeling of bifurcation of variants at an innovation event, a feature of biological inheritance mechanisms which is not a given in cultural inheritance (Boyd and Richerson 1985; Morris 1994; Barnet and Eldredge 2004). This is fortunately, however, an empirical issue that can be tested (Collard and Shennan 2000; Collard et al. 2006) for each material culture type and societal context, rather than an *a priori* assumption.

The second worry with the application of cladistic approaches to lithic technology, even with the extremely laudable examples where cultural transmission theory is explicitly tested using artifact variability rather than norms (Lycett and von Cramon-Taubadel 2008; Lycett 2010), is the unproblematized assumption that artifact form, as represented by a 3D geometric morphometrics analysis of the shape of artifacts at discard, is what we should be examining. This issue goes back at least to the classic paper of Jelinek (1976) on the Frison Effect, if not earlier to Lowie's comment that "The comparison of form can never do more than establish the identity of forms; that such identity is to be explained by genetic relationship is an hypothesis..." (1912:28). In analyzing biface and core forms discarded into the archaeological record in diverse localities, with potential differences in initial raw material package size and shape as well as availability, it is unwise to assume that artifact shape—particularly with a core-tool form—equates to a mental template when in fact it could equate more to variation in the degree of core exhaustion at discard. McPherron (1994, 2000) and Iovita and McPherron (2011) have shown that biface shape can be systematically affected by relative degrees of resharpening, a potentially unique aspect of each assemblage's raw material context. Since different geological sources for tool stone often produce a different initial nodule size, one cannot simply correct for differential core-tool exhaustion between assemblages by measuring the strength of size in the multivariate analyses. It is really a question of size relative to the unworked nodule. So, to what extent does the variation in artifact life histories within an assemblage affect the test of intra-assemblage variation for an iterative founder effect model? More importantly, however, given the conceptual modeling of the observational learning of the flintknapping process, the pure shape of discarded tools is unlikely to represent the units of cultural transmission in flintknapping, i.e., what was actually observed, learned, and then retransmitted through time.

A behavioral approach such as the present project is more theoretically justifiable from the point of view of cultural transmission theory than a purely quantitative method for evaluating variables as useful units of transmission (e.g., Pocklington 2006) without a theoretical understanding of how they are transmitted and replicated. What we have learned from cultural transmission theory so far makes it clear that how a trait is transmitted, i.e., its context of learning, alteration during transmission, and then replication are as important to the study of cultural evolution as mating and genetic replication and recombination are to biological evolution.

Thus I believe that the future of the multivariate cladistic approaches to the study of cultural transmission of lithic technology must entail a 3D geometric morphometrics approach that uses landmarks that are behaviorally-relevant to what units were learnable in the past and etically-observable in the present (Tostevin 2008). Yet the essence of the 3D geometric morphometrics approach is about variation and as such it is still leaps and bounds in advance of the traditionalist culture historical approaches to Paleolithic technologies. As the best examples of the application of cultural transmission theory to lithic technology have shown (in my view, Bettinger and Eerkens 1999 and Eerkens and Lipo 2007), there is great analytical power in using changes in artifactual variability through time as a way to study modes of transmission. But there is also great promise in combining this with a behavioral approach to how the specific technology is learned and performed (Bamforth and Finlay 2008).

Another approach recommended by one of the anonymous reviewers of this book was to introduce an intermediate level of generalization, between that of the assemblage as the population of knappers and that of the individual flake contributed by one knapper into the assemblage. The suggested addition involved analyzing subassemblages by raw material types. I cannot agree more, particularly if it were done in a way similar to the Minimum Analytical Nodule Analysis (MANA) approach introduced by Larson and Kornfeld (1997). By treating each identifiable raw material type separately, and even distinguishing individual nodules within each raw material type (if the variability is sufficient), a finer resolution of knapping behavior is discernable, as Baumler and Davis (2004) and Sellet (2004) have shown with their applications of MANA to North American Paleoindian assemblages. While the analytical power of the approach taken by the present research would have been expanded by a MANA-based protocol, the practicalities of the original data collection for this research, i.e., the investigation of 18 assemblages over three regions within the period of a doctoral degree, precluded my training in the recognition of the raw material variability within each assemblage and region. However, while I was unable to take this approach, there is no reason not to apply it in future contexts when practicalities allow.

Other scholars have already begun to use and improve upon the behavioral approach to Pleistocene lithic technology advocated by this research. The approach has been adjusted and improved by Nigst (2009, 2010, 2012) in studying the Early Upper Paleolithic sequence at Willendorf II, Austria. Using the earlier version of analytical protocol in Tostevin (2000c) and the social intimacy concept from Tostevin (2007), Nigst increased the importance of the reduction sequence analysis within the strategic decision nodes and used the modified approach to analyze previously unpublished assemblages from Willendorf II. He was able to recognize knapping behaviors within layer AH-2 that, when compared to the data from Vedrovice V in Tostevin (2000c), placed this previously unaffiliated assemblage within the Szeletian Industrial-type. Further to the west, in the context of the archaeological record of the French Upper Paleolithic open air site of Solvieux (Sackett 1999), Grimm and Koetje (2008) have also applied the approach, using an early draft of this book. While using the domain system explained in Chapter 4, Grimm and Koetje added a behavioral step related to the application of platform abrasion to the platform maintenance domain to more accurately address the variability in their context. Even more recently, Roussel (2011) has used the social intimacy and taskscape visibility concepts from Tostevin (2007), although not the analytical characterization of the lithic assemblages, to compare the Châtelperronian assemblages at la Grande-Roche-de-la-Plématrie à Quinçay, France, with the *chaîne opératoire* generalized by Bon (2002) for the Proto-Aurignacian. Roussel's conclusion that the Châtelperronian at Quinçay evidences stimulus diffusion from the Proto-Aurignacian shows the interesting results that derive from using more anthropological theory in studying Pleistocene cultural historical processes as well as highlights the need to increase the application of quantitative, comparable analyses in western Europe to better connect the archaeological record with cultural transmission theory. As I hope will soon become evident, the behavioral approaches to comparisons of technological variability through time and space are likely to take many forms before an optimal consensus appears. These are all welcome additions.

9.4 Predictions for Further Research

As none of the three regions in this study possess antecedent behaviors to indicate a local origin for the Bohunician or Aurignacian Behavioral Package, Hypothesis 3 for the Middle to Upper Paleolithic transition in these regions, a combination of independent innovation in one or more regions and subsequent diffusion in other regions, is currently unlikely. One may suggest the possibility that the Bohunician Behavioral Package originated in a region adjacent to the Levant, perhaps the Nile Valley given Van Peer's research (1998; Van Peer et al. 2010), but this idea will remain tentative until such regional sequences can be studied using a similar analytical structure, as argued by Vermeersch (2001). Other regions, such as southern Poland, should also be investigated further with this methodology, to ascertain any possible relationship between enculturating environments of the assemblages of Stránská skála IIIa-4, Korolevo II Complex II, and the Middle Paleolithic layers 7a to 7c of Piekary II near Krakow (Valladas et al. 2003; see also Sitlivy et al. 1999, 2001). These newly excavated assemblages contain bladey artifacts, dated between 60 and 35 kya, which are candidates for antecedents, descendants or convergences to the Bohunician Behavioral Package. The data provided by Conard and Bolus (2003) also make it clear that an approach similar to the one advocated here may be productive in the Upper Danube as well.

Thus future research programs are needed to replicate the present approach in regions adjacent to the three studied here. Just as important, however, is the necessity to expand each region's sample of assemblages to bolster the confidence of this study's results. The East European regional sample in particular needs bolstering with analysis, as pointed out by Sitlivy and Zięba (2006), but more importantly with more field excavations using modern artifact retrieval and dating techniques. Meignen et al.'s (2004) mention of new work along this line at Molodova 5 will be particularly useful. Further work of this type at Kulychivka and Molodova 1, would also go a long way to improving our ability to evaluate Meignen et al.'s arguments for technological similarities between the Bohunician assemblages, Kulychivka, and the Middle Paleolithic Levalloisian assemblages of Molodova V levels 11 and 12 and Molodova I layers 4 and 5. From the data presented in Chapter 6, Molodova V level 11 is noticeably distinct in blank production details from Korolevo II Complex II, but this is the smallest of the differences between the last Middle Paleolithic and the next assemblage on the scene of all three regions (Figure 8.19). Also, the tool kit morphologies are the closest across this divide in this region compared to the others. These points suggest more research is needed, for instance in Molodova I, to elucidate the role of cultural transmission in this region.

Not only does the Eastern European sample warrant augmentation but each regional sequence of change would benefit from more representation of the variation in assemblages and knapping behaviors between 60 and 30 kya, particularly across environmental zones within each region. This is especially true for the Levant, as the views of Marks (2003) and the research of Sarel (2004) make clear. Such research will hopefully avoid the frequent pitfalls identified, if not solved, by the present study in its search for an ultimate explanation of the meaning of the Middle to Upper Paleolithic transition for human evolution.

Given the limitations of the sampling of assemblages included in this study, it is probable that the hypothesized cultural transmission events supported by this research will be modified, if not discarded as new data are acquired in the next decade. This is only to be expected: a collection of eighteen out of thousands of Paleolithic assemblages is a poor sample. Not only new assemblages but also new dates could disarticulate the assemblages within two proposed cultural transmission events, or for instance simply push the whole Bohunician Behavioral Package back in time such that it becomes a purely Neanderthal-associated event in OIS 4 between the Levant and Central and Eastern Europe (although this would still be intriguing). Similarly, much work still needs to be done through controlled flake fracture studies, such as those begun by Dibble and Rezek (2009) and Rezek et al. (2011), to test further the causal association of knapper-controlled variables as reduction of the core continues. Such work is needed to strengthen the separation of knapping variables into the four blank production domains identified here. Given all of this, is it not fair to label the new bridging arguments presented for the middle-range theory that stands at the core of this research project as "A Bridge Too Far"?[1]

I would disagree for two important reasons. First, even though neither the archaeological record nor this study's sample is perfect, waiting to explore cultural evolutionary questions of this detail until the record is complete is not really an option. Paleolithic archaeology needs to maintain its partnership within Paleoanthropology or it will be left behind as the other branches of the discipline, i.e., biological anthropologists studying fossils and molecular anthropologists studying the structure of hominin genomes, make our conclusions irrelevant. A solid middle-range theory that connects artifacts with the evolutionary theory of dual inheritance modeling is critical for investigating culture historical processes as part of human biological and cultural evolution in the Pleistocene. Second, Paleolithic archaeology is in desperate need for an approach to lithic technological variability that offers the possibility of meta-analysis, the incorporation of data from independent studies conducted with compatible methods that allows larger-scale questions to be analyzed. Consider what Bordes' (1961a) typology did for the comparability of the quantitative meta-analysis of assemblages of retouched tools (e.g., Dibble and Roland 1992; Dibble 1995b; Hiscock and Clarkson 2009b). Nigst (2009, 2012) and Grimm and Koetje (2008) have already shown the potential for constructive meta-analysis using the middle-range theory presented here. In an effort to increase the potential for such meta-analysis, the data set representing all 18 assemblages examined in this project is available on-line. So while I expect that the specific culture historical reconstruction presented here will be overturned in the due course of new field research, dating techniques, and analytical methods, I hope that the middle-range theory at the heart of this book will at least serve as an example for how future Paleolithic archaeologists model and study the connections between cultural transmission processes, cultural adaptation, and the production of material culture variability through evolutionary time.

Appendix

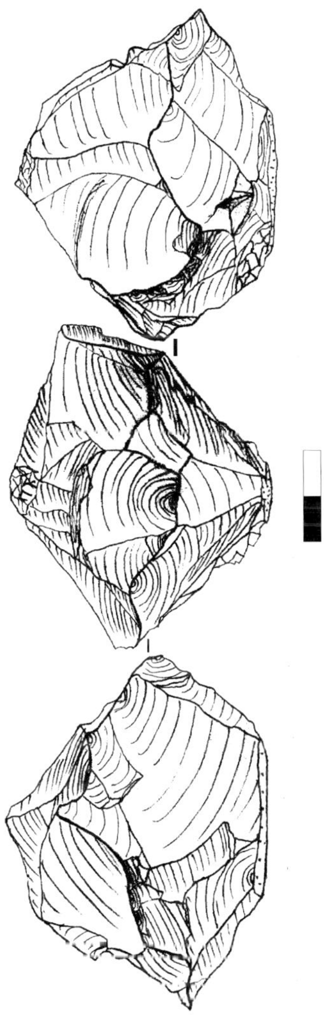

Figure A1 Kůlna Layer 7a core. Scale 2 cm.

a)

b)

c)

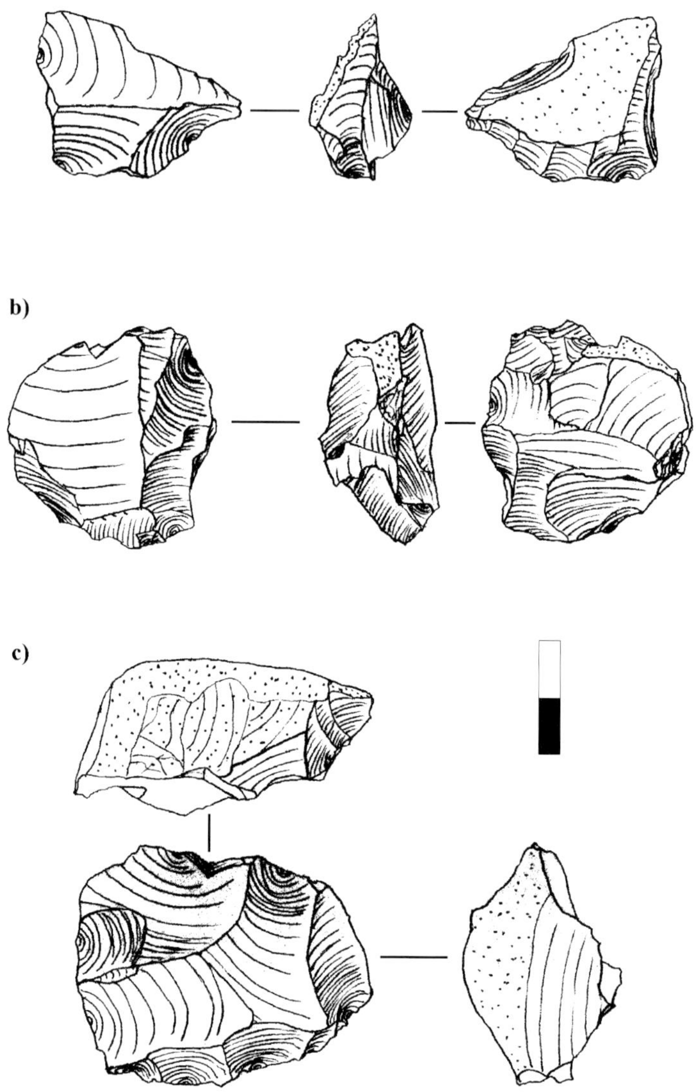

Figure A2 Kůlna Layer 7a cores. Scale 2 cm.

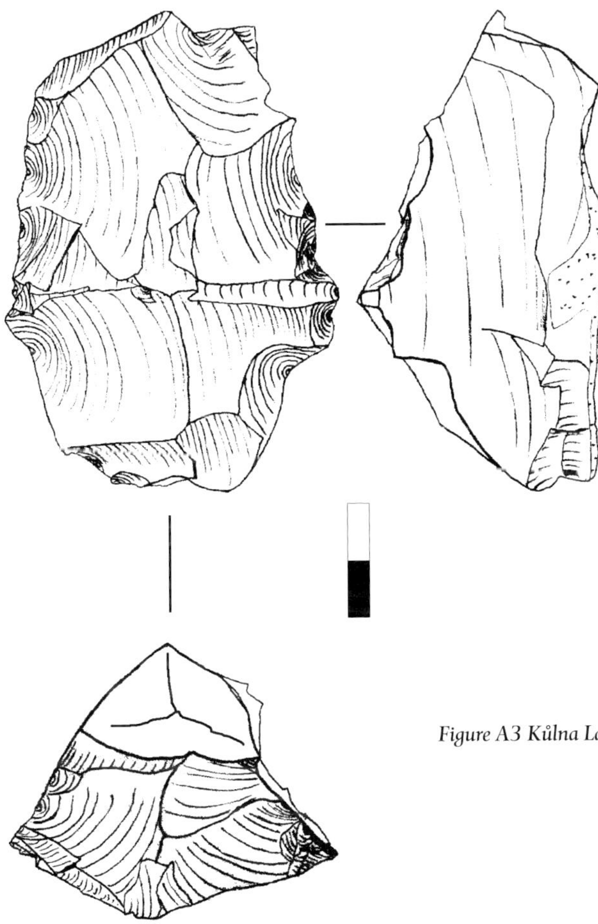

Figure A3 Kůlna Layer 7a core. Scale 2 cm.

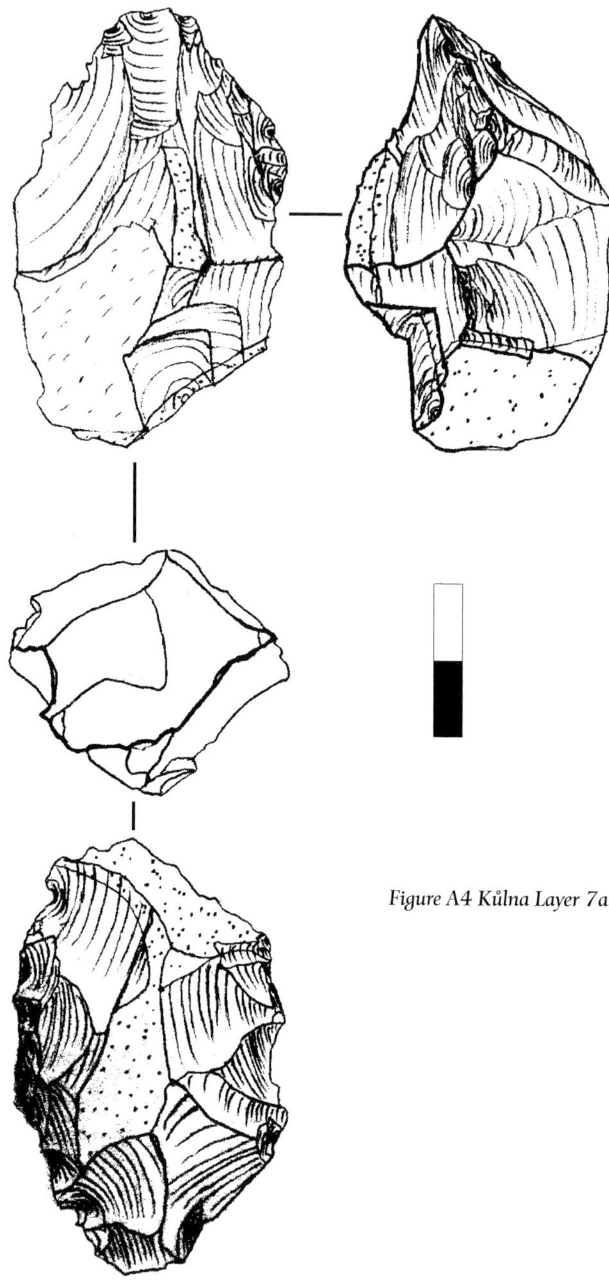

Figure A4 Kůlna Layer 7a core. Scale 2 cm.

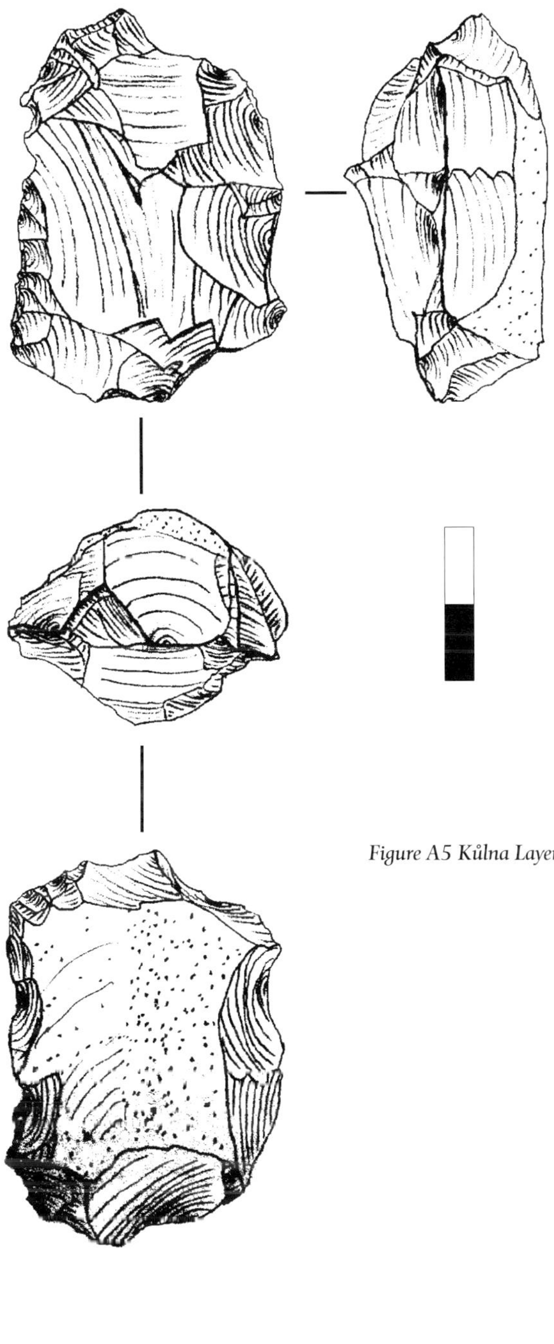

Figure A5 Kůlna Layer 7a core. Scale 2 cm.

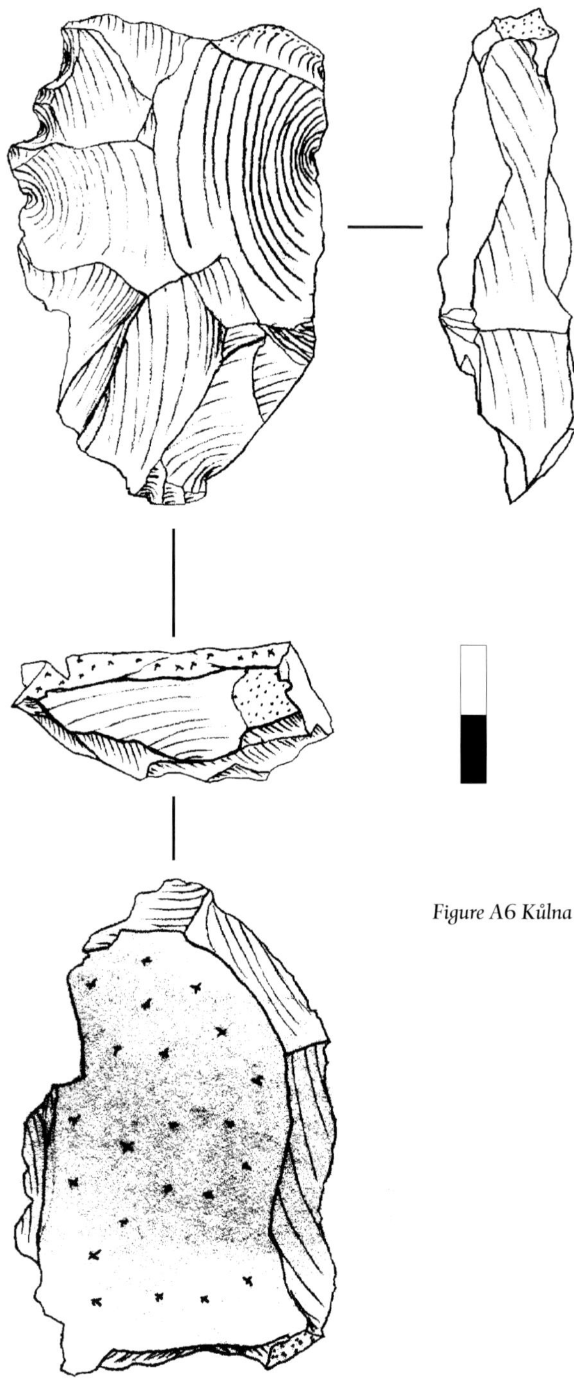

Figure A6 Kůlna Layer 7a core. Scale 2 cm.

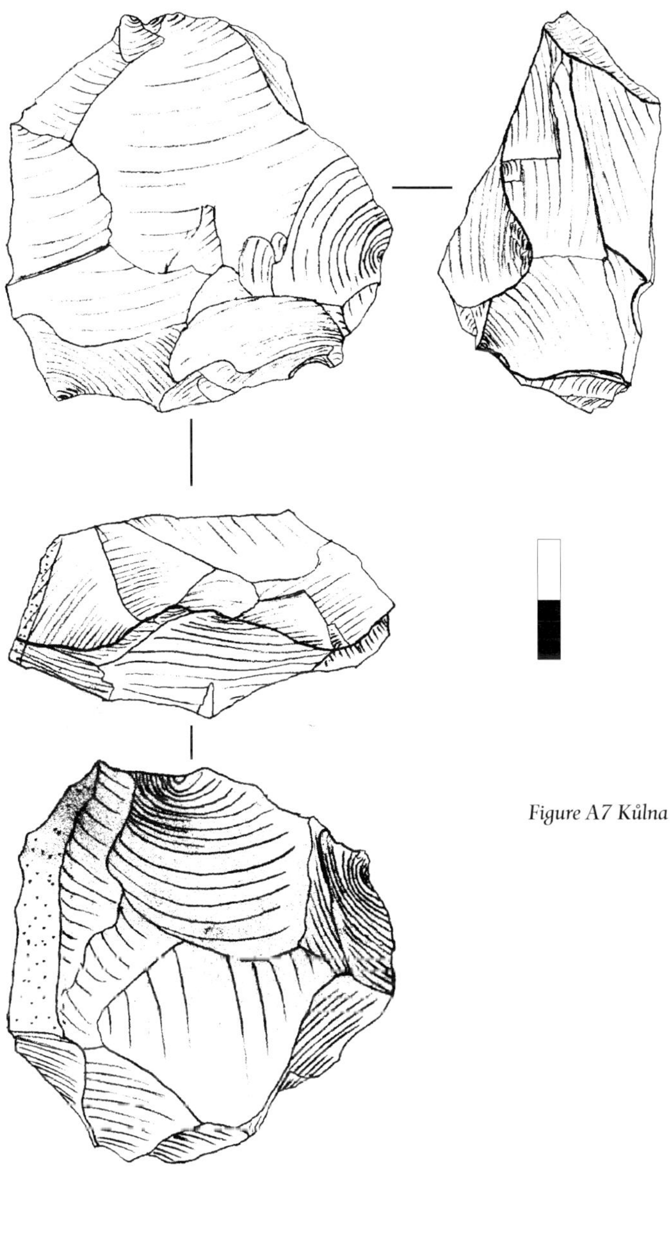

Figure A7 Kůlna Layer 7a core. Scale 2 cm.

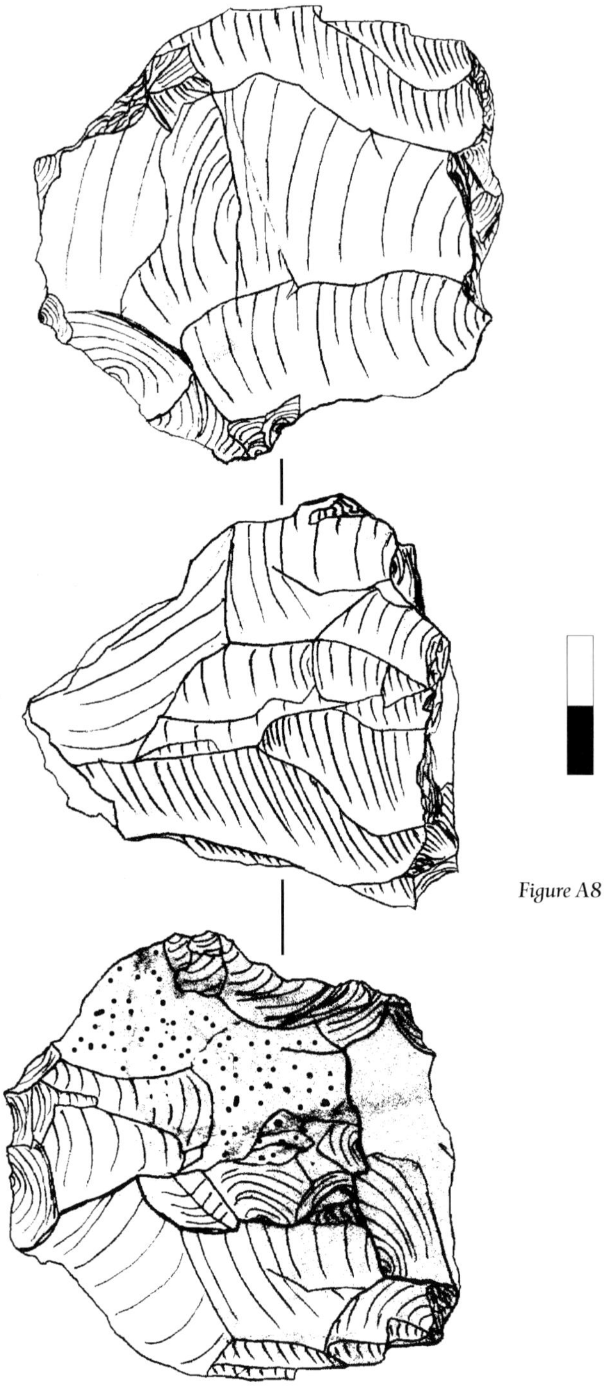

Figure A8 Stránská skála IIIa Layer 4 core. Scale 2 cm.

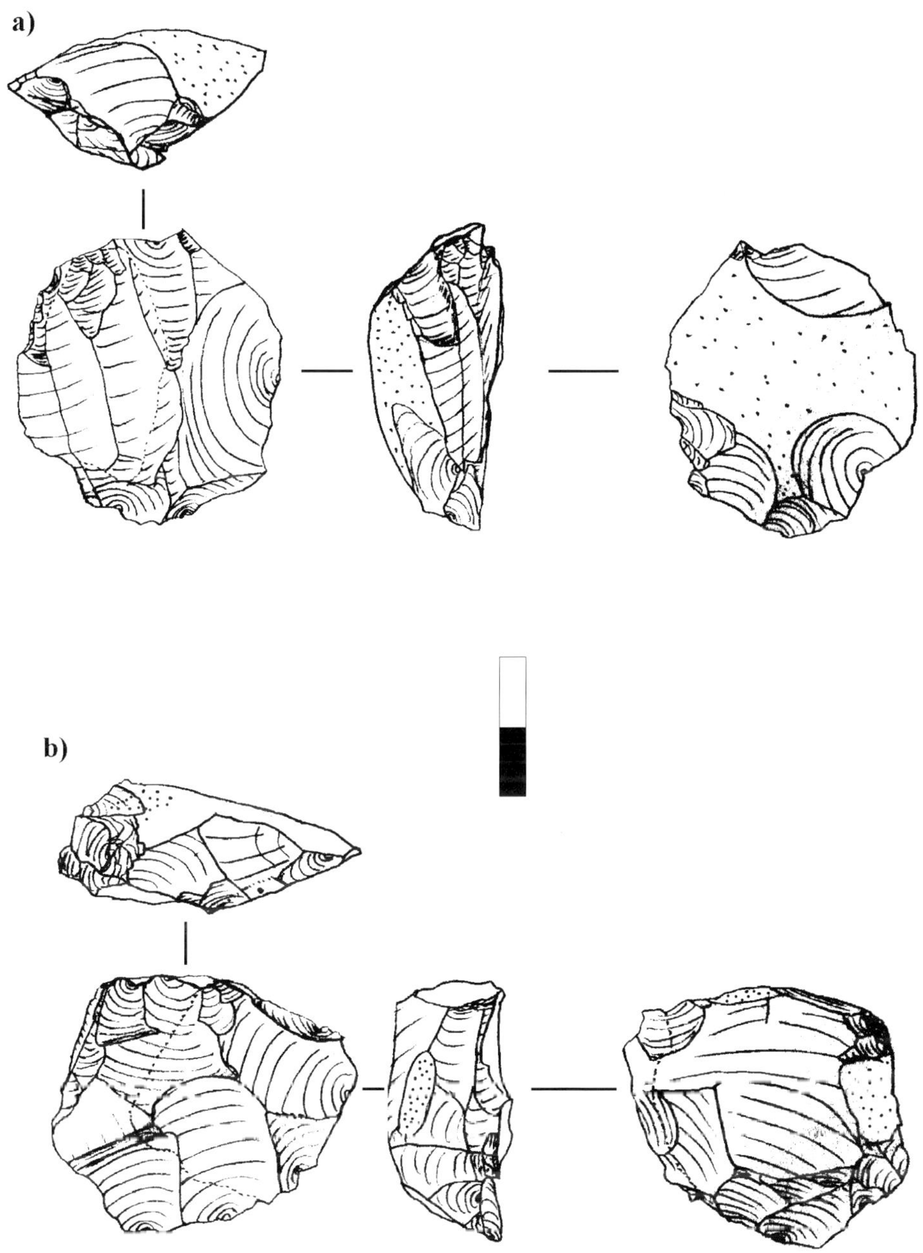

Figure A9 Stránská skála IIIa Layer 4 cores. Scale 2 cm.

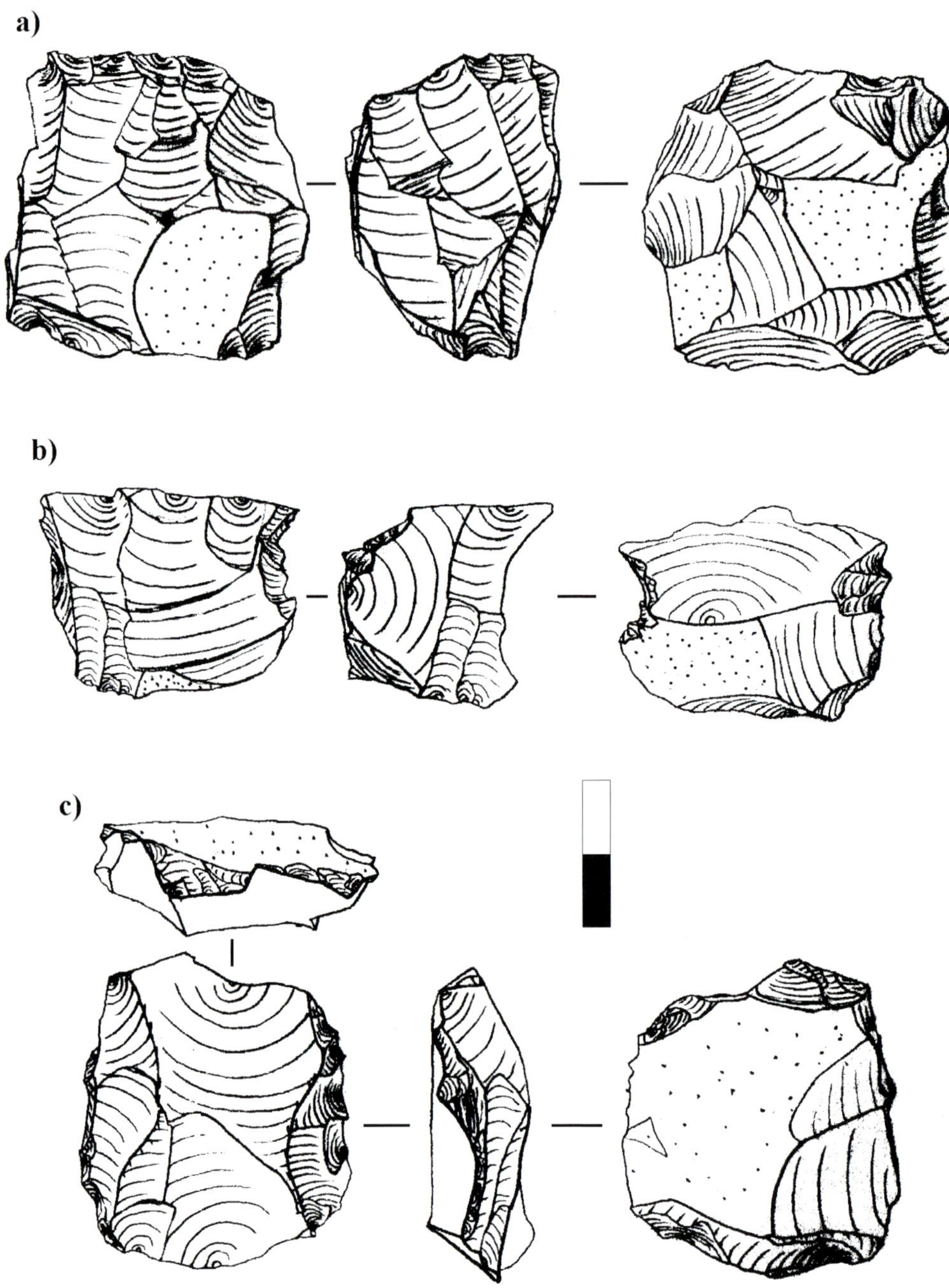

Figure A10 Stránská skála IIIa Layer 4 cores. Scale 2 cm.

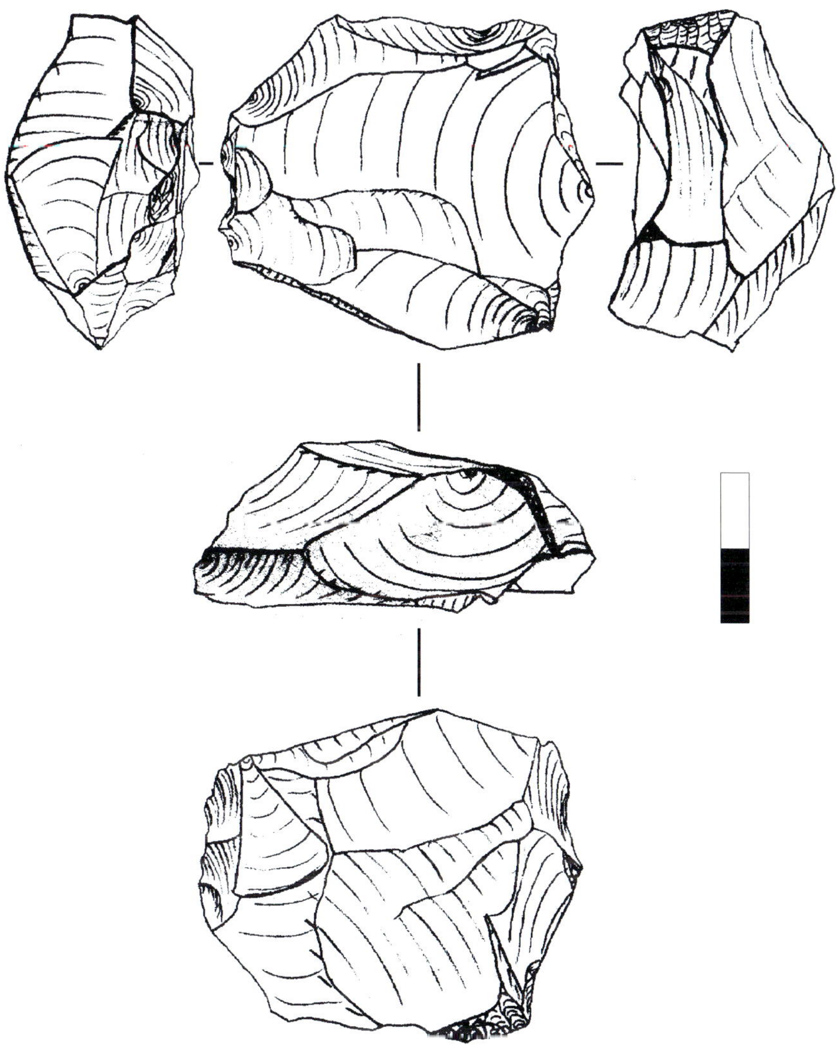

Figure A11 Stránská skála IIIa Layer 4 core. Scale 2 cm.

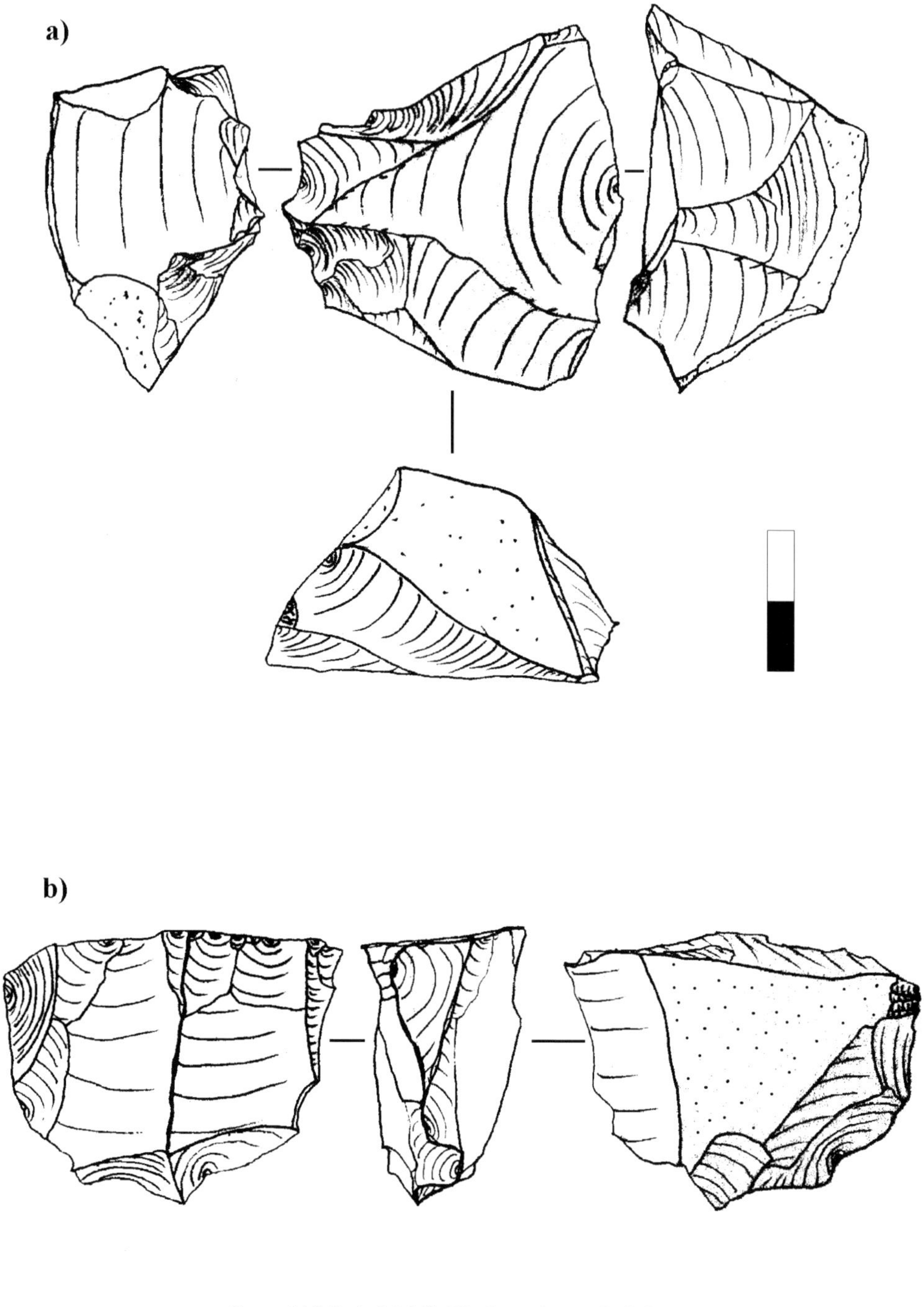

Figure A12 Stránská skála IIIa Layer 4 cores. Scale 2 cm.

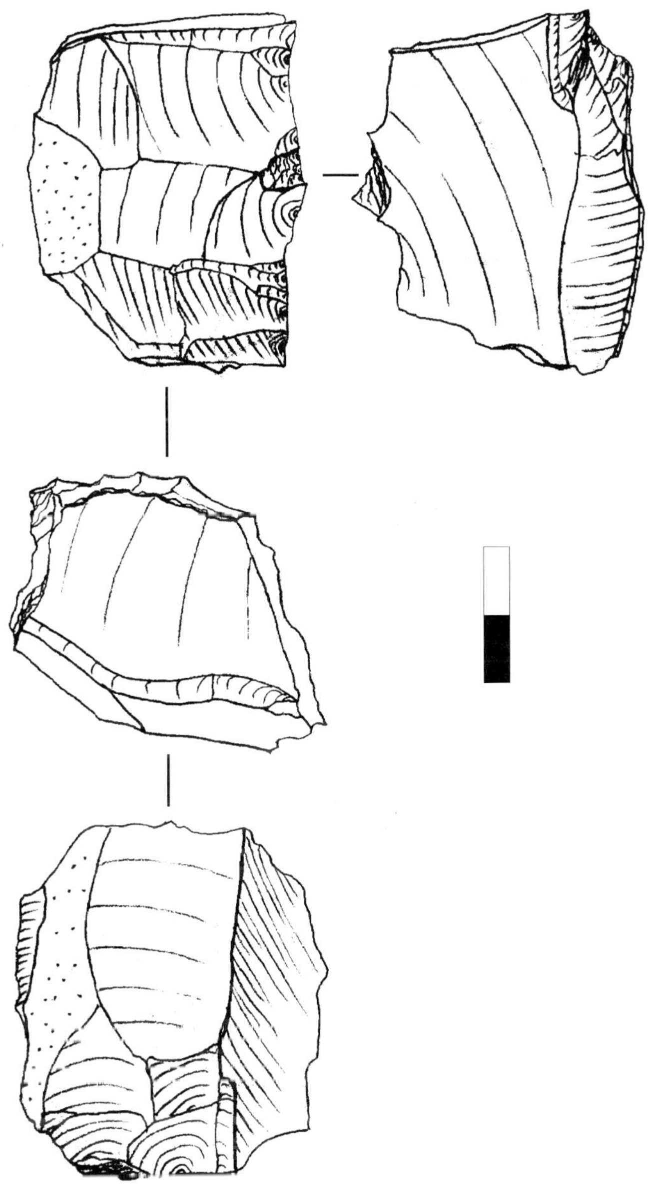

Figure A13 Stránská skála IIIa Layer 4 core. Scale 2 cm.

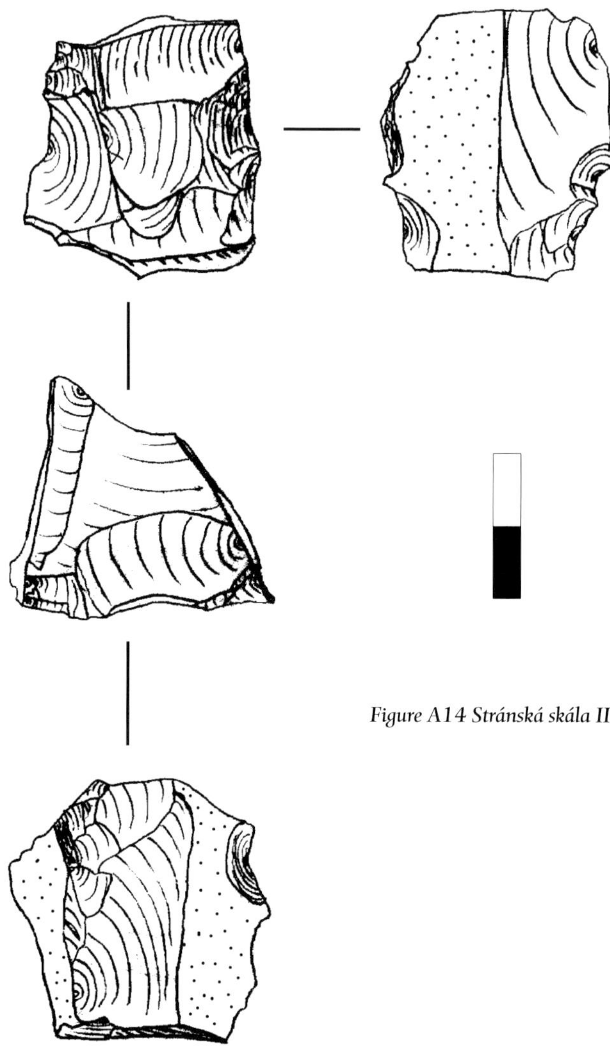

Figure A14 Stránská skála IIIa Layer 4 core. Scale 2 cm.

a)

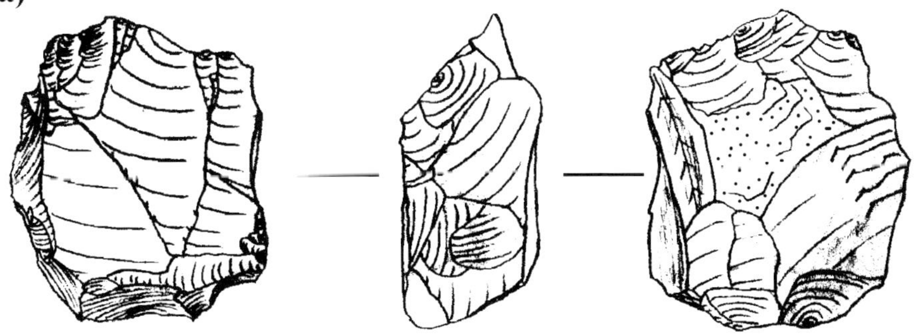

b)

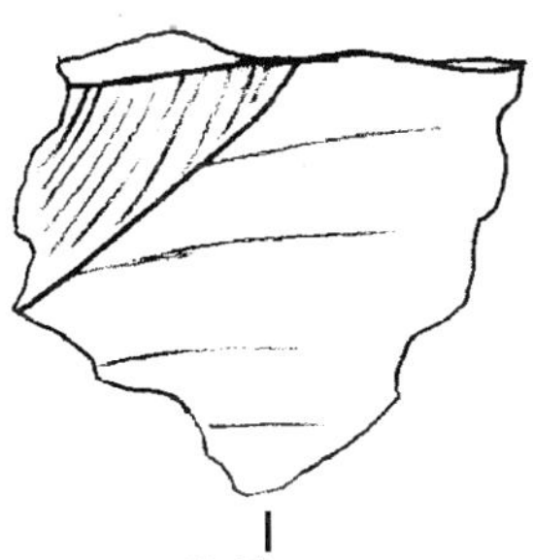

Figure A15 *Stránská skála IIIa Layer 4 cores. Scale 2 cm.*

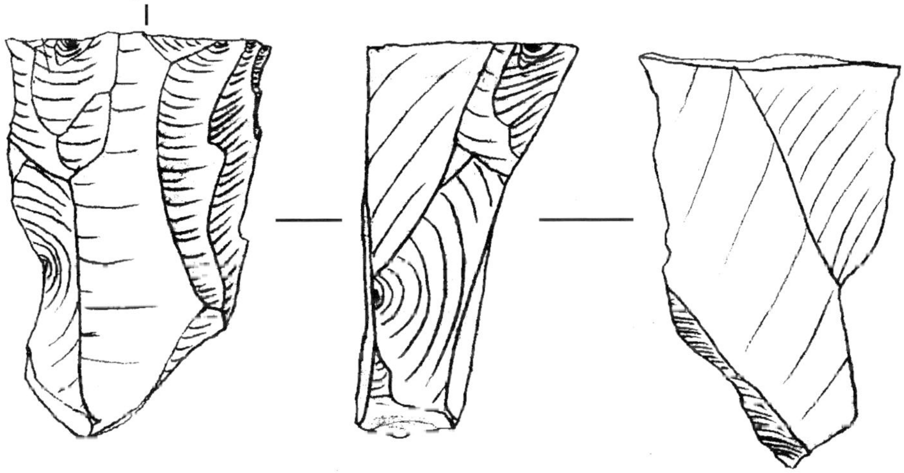

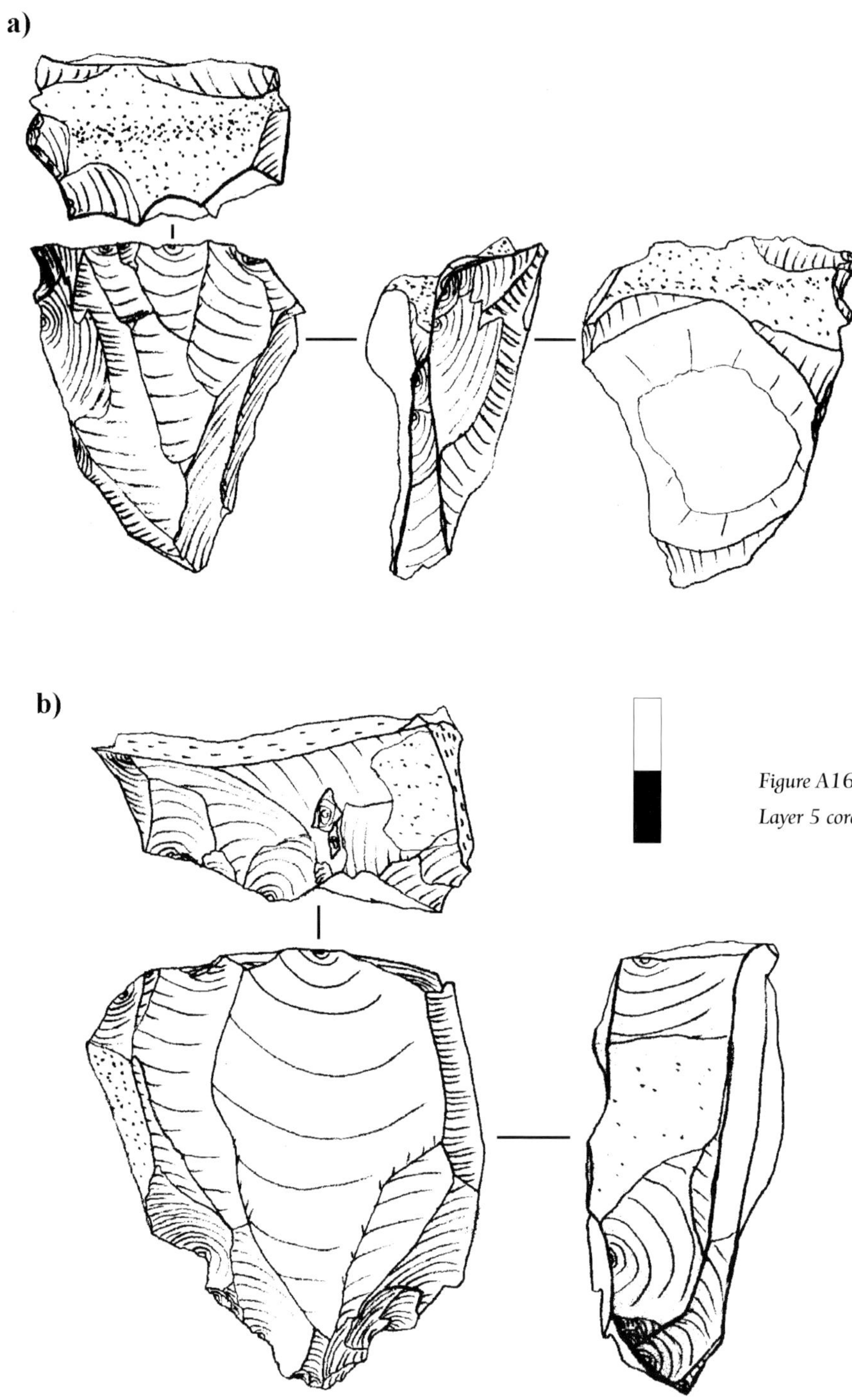

Figure A16 Stránská skála III Layer 5 cores. Scale 2 cm.

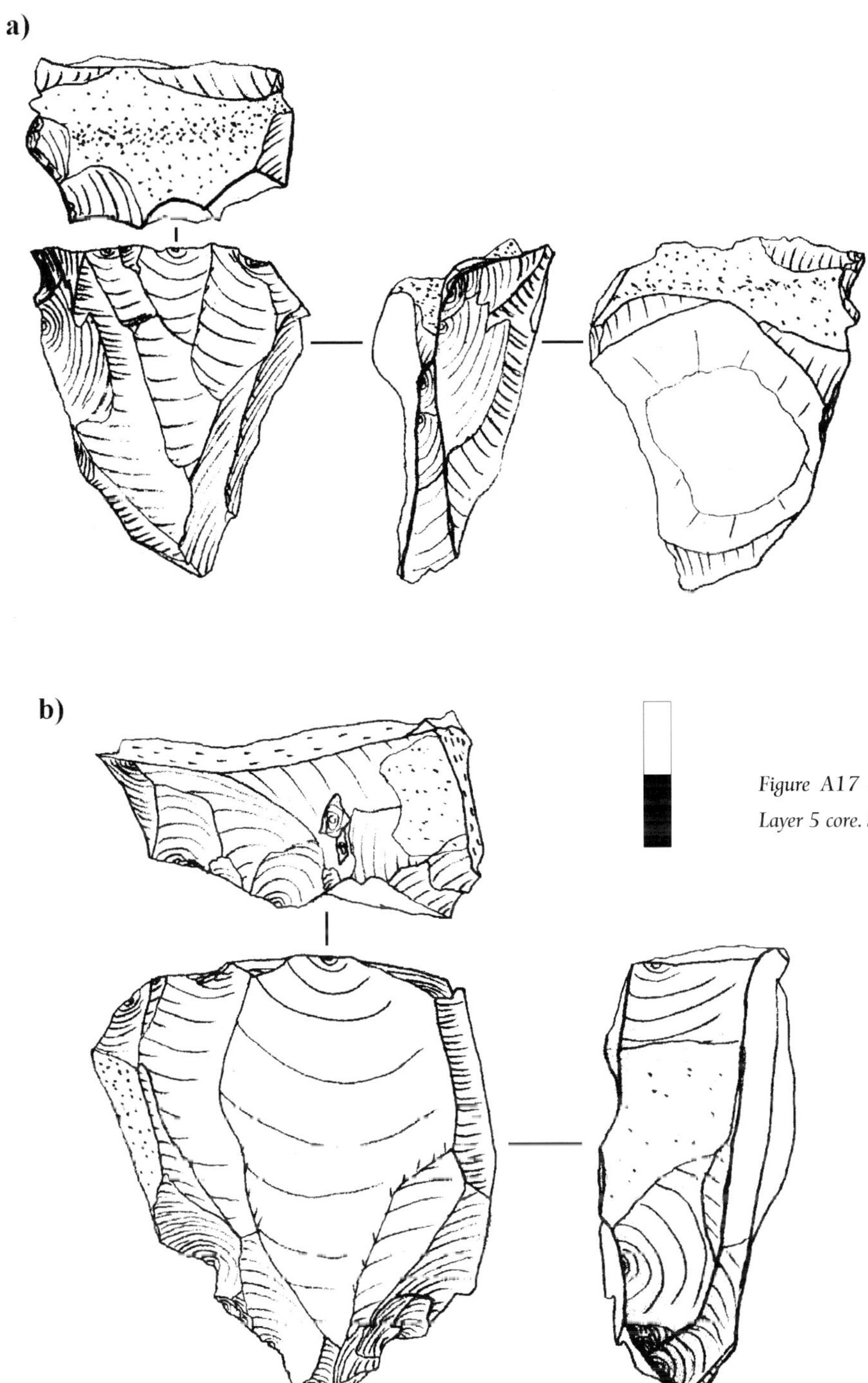

Figure A17 Stránská skála III Layer 5 core. Scale 2 cm.

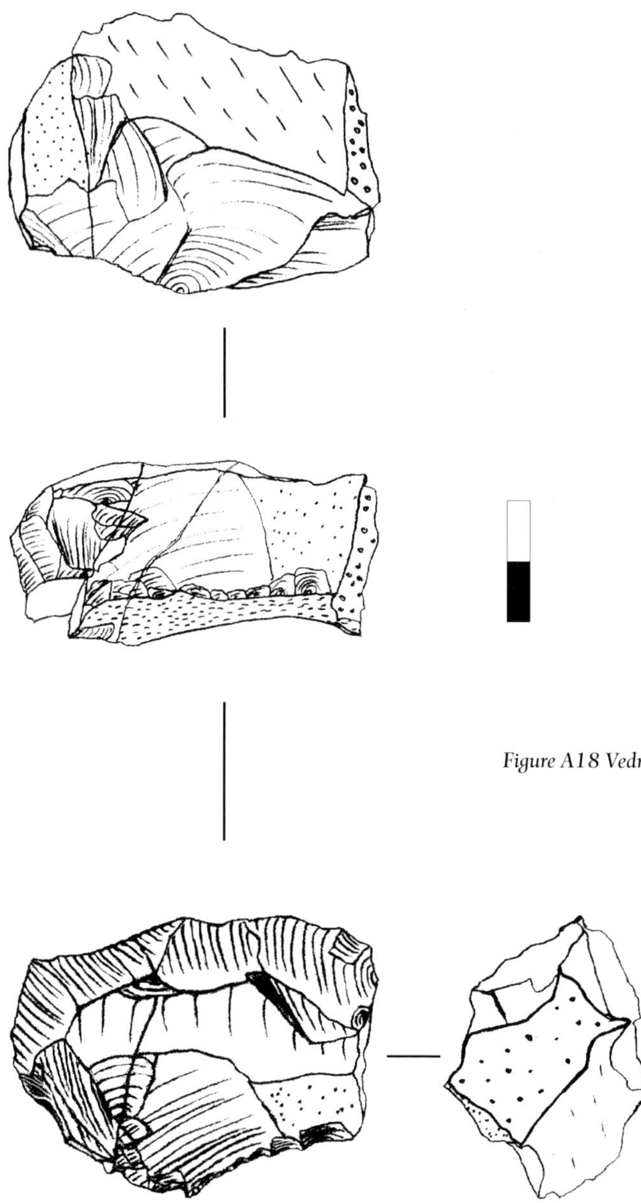

Figure A18 Vedrovice V core. Scale 2 cm.

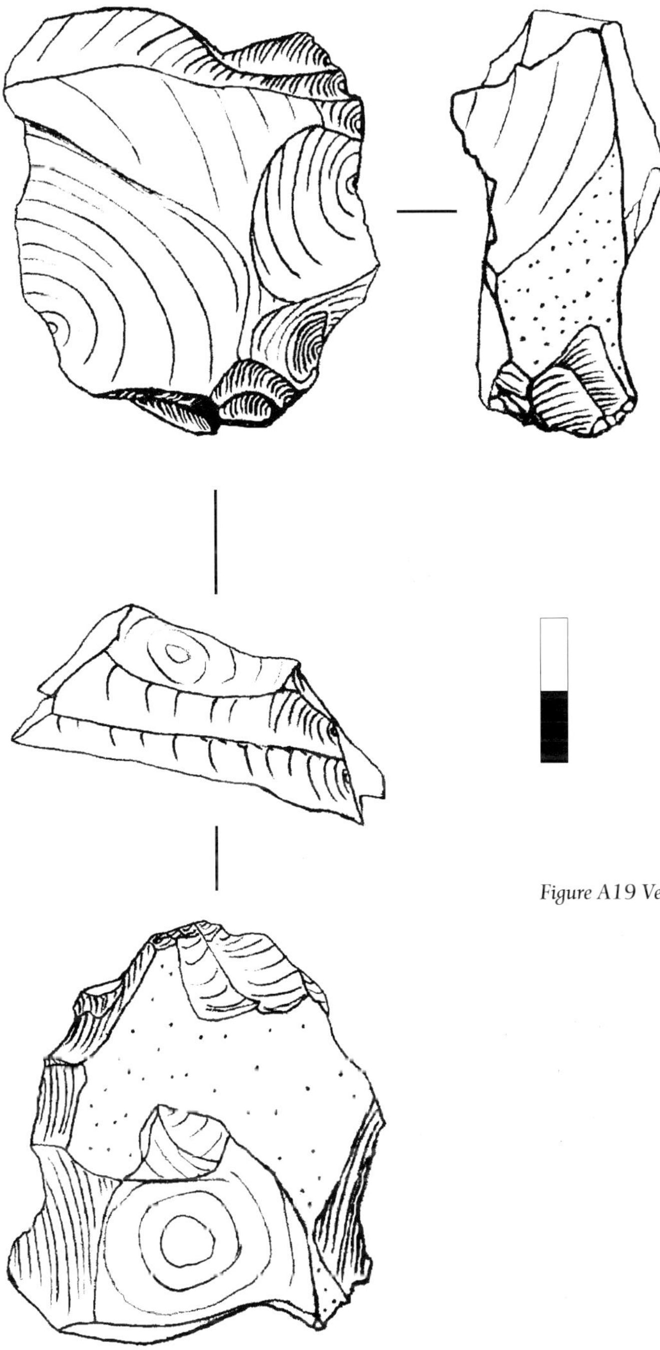

Figure A19 Vedrovice V core. Scale 2 cm.

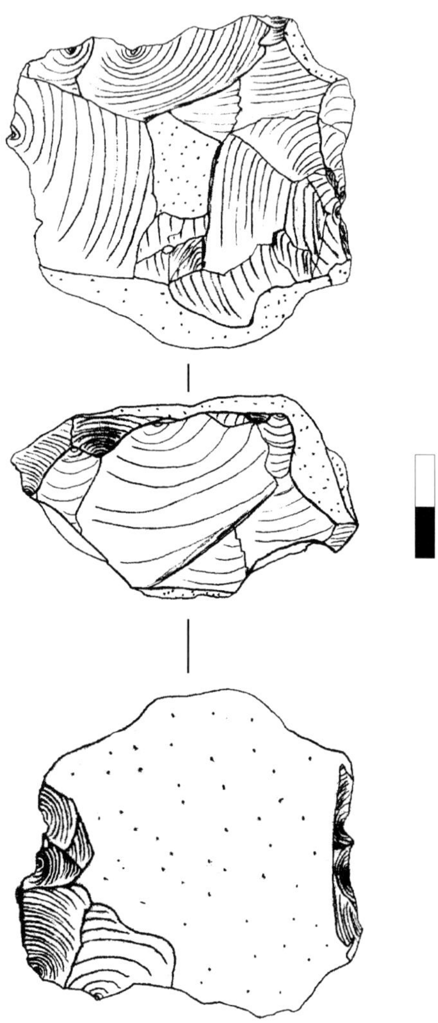

Figure A20 Vedrovice V core. Scale 2 cm.

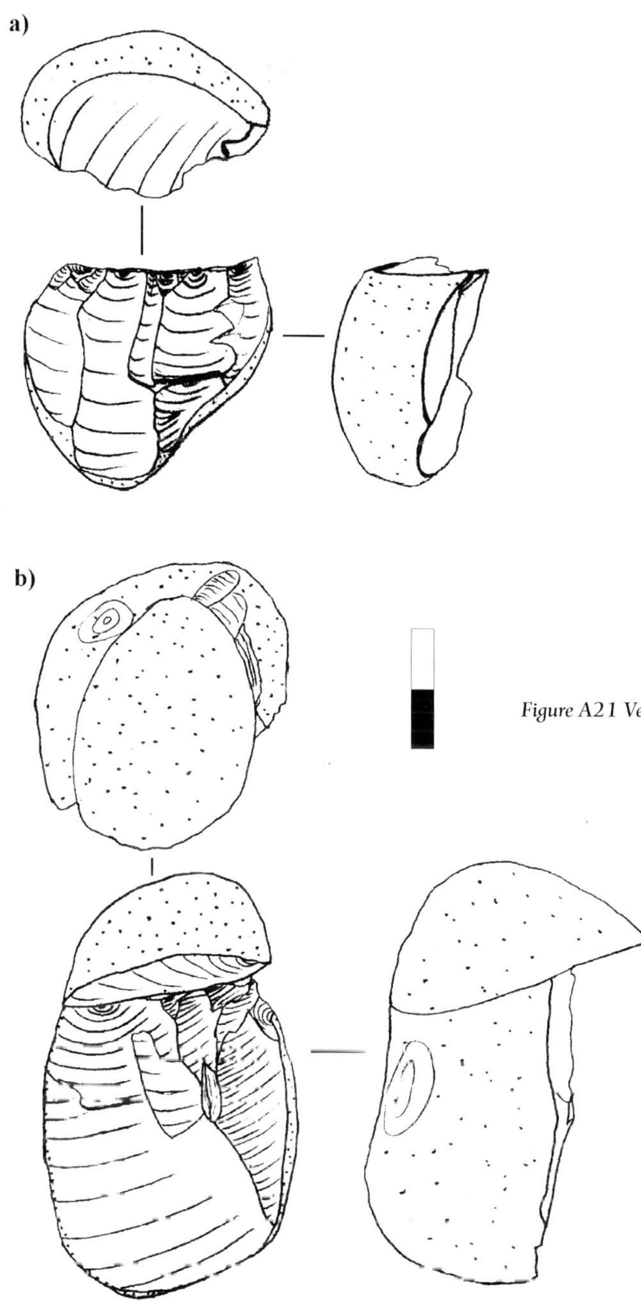

Figure A21 Vedrovice V cores. Scale 2 cm.

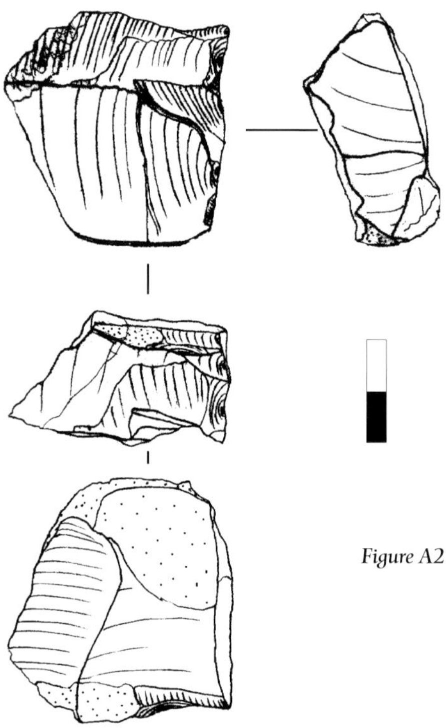

Figure A22 Vedrovice V core. Scale 2 cm.

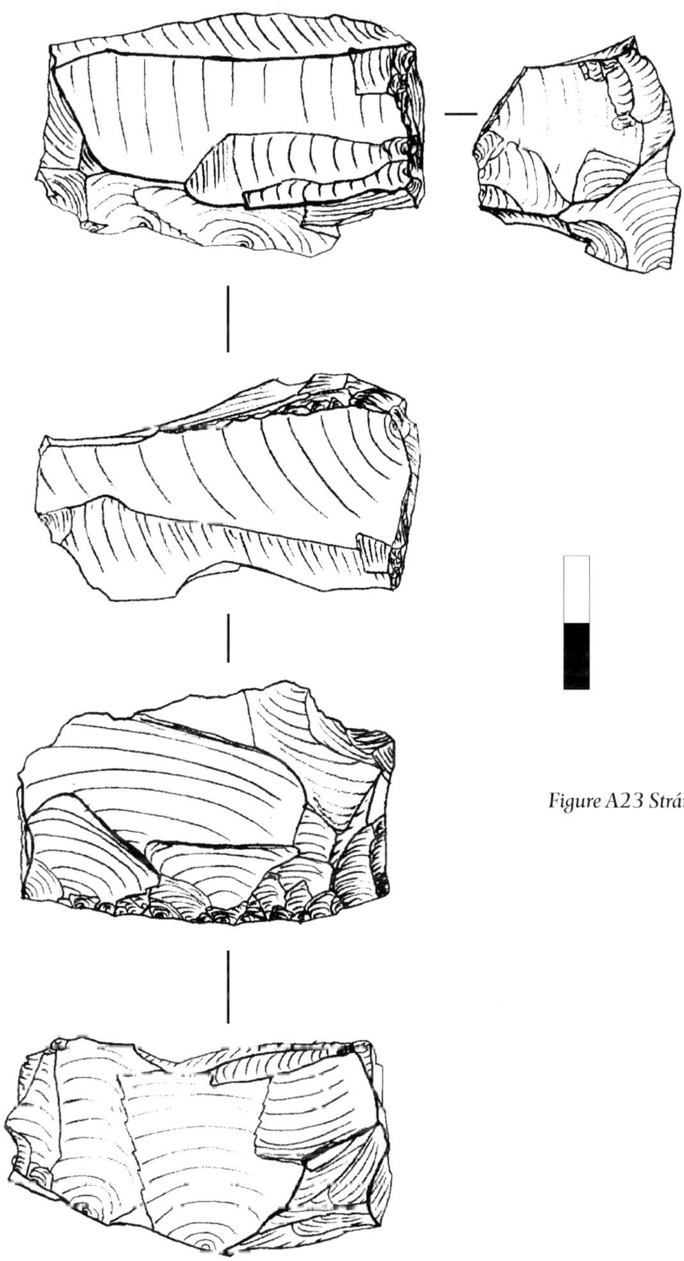

Figure A23 Stránská skála IIIa Layer 3 core. Scale 2 cm.

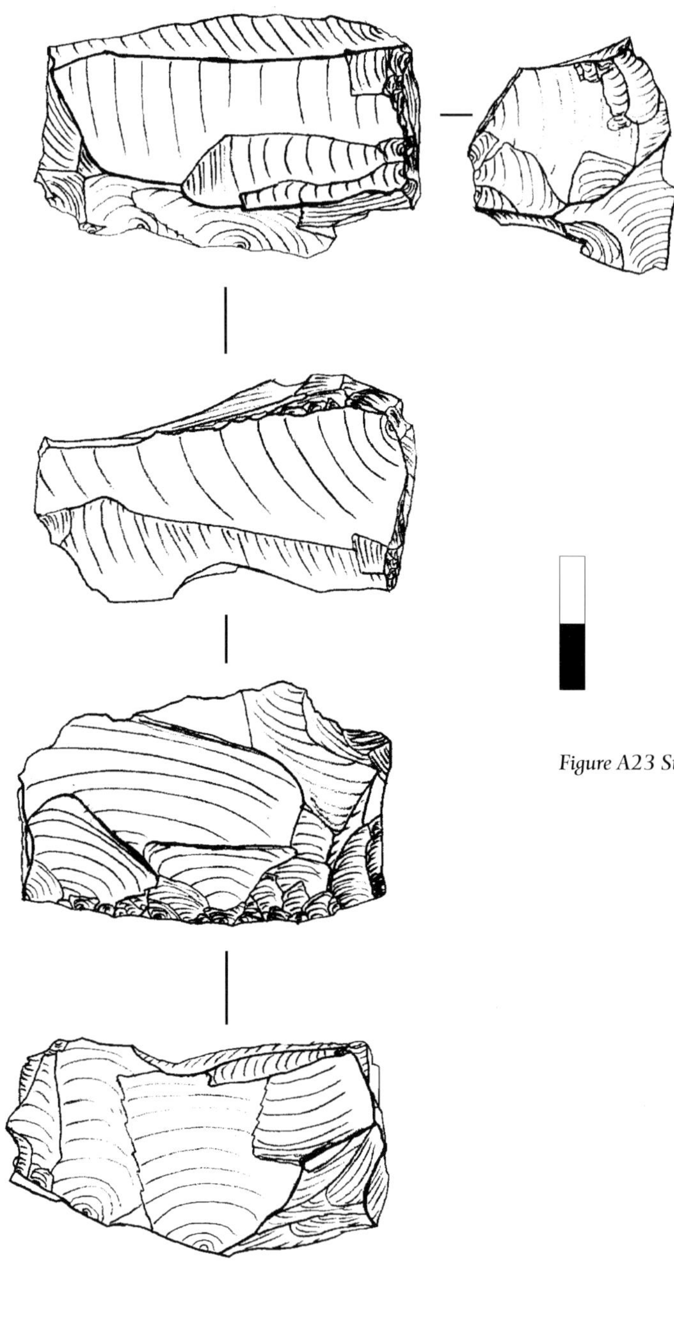

Figure A23 Stránská skála IIIa Layer 3 core. Scale 2 cm.

a)

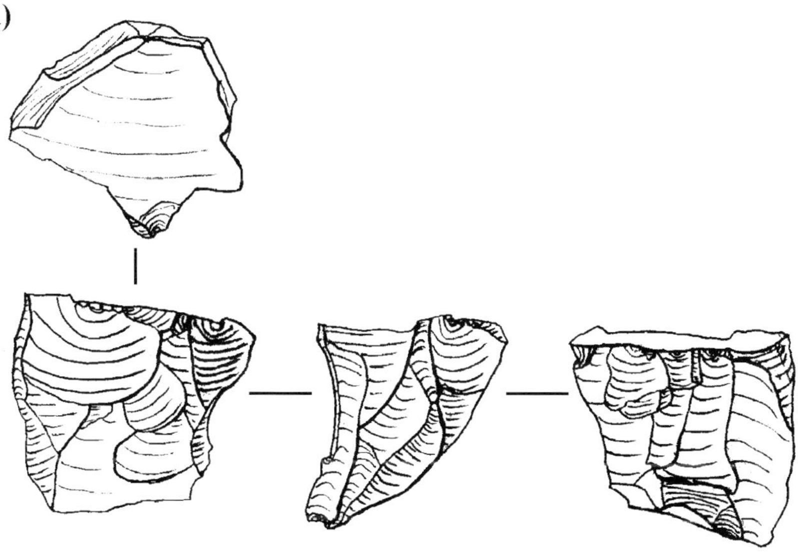

Figure A24 Stránská skála IIIa Layer 3 cores. Scale 2 cm.

b)

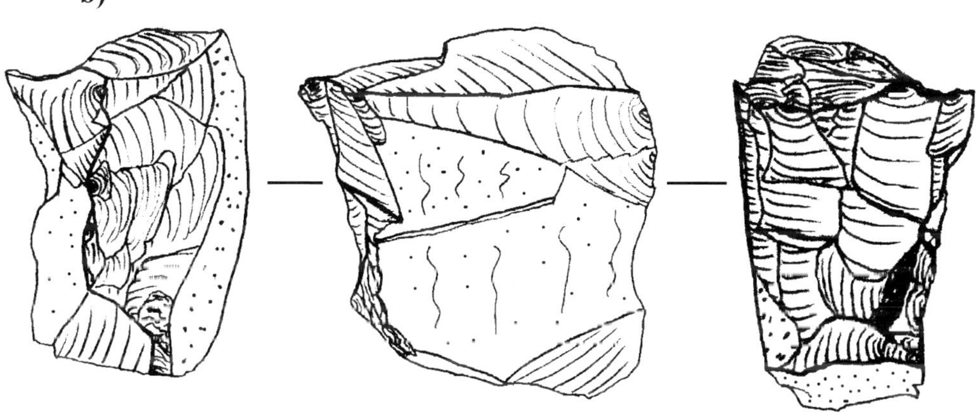

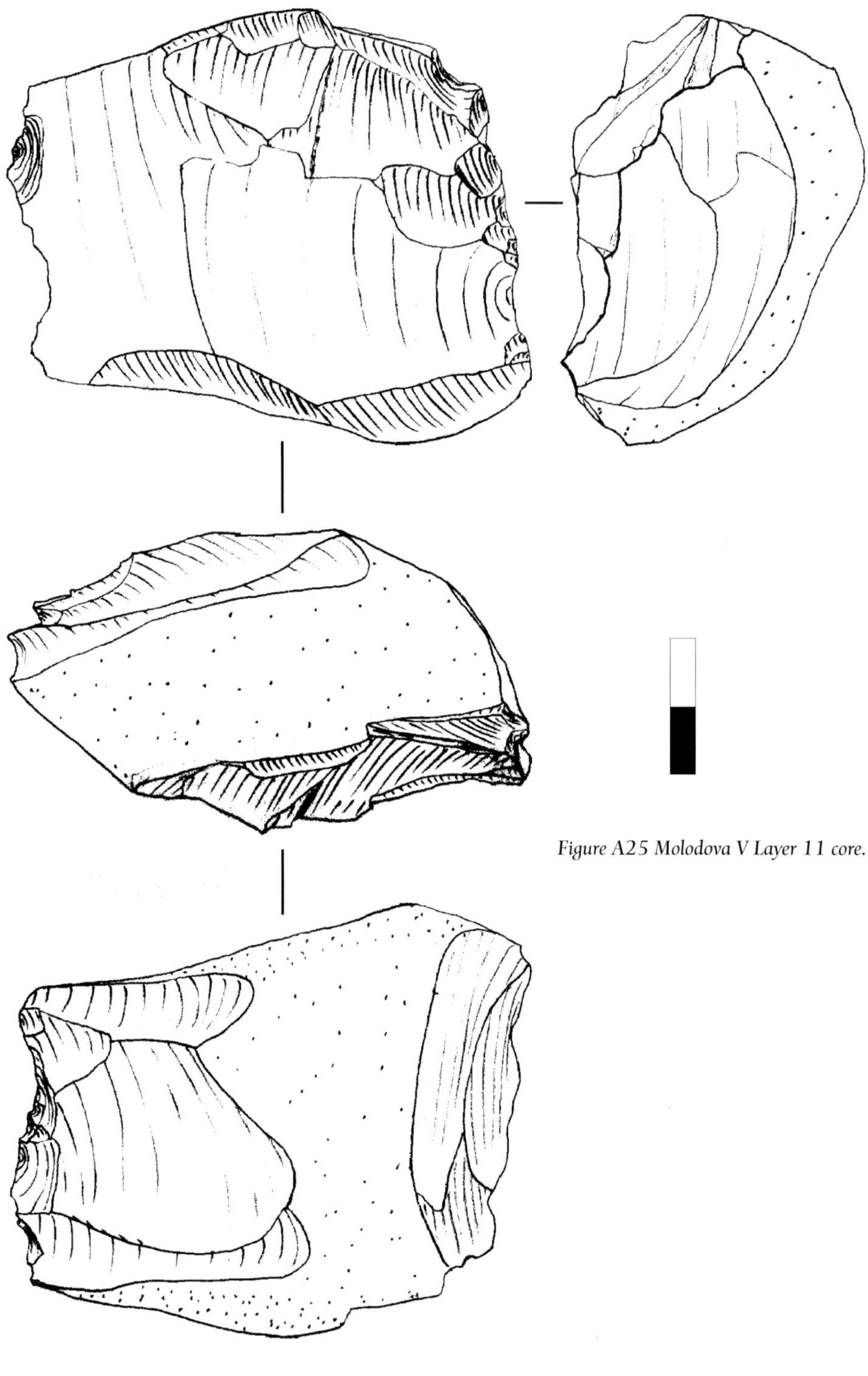

Figure A25 Molodova V Layer 11 core. Scale 2 cm.

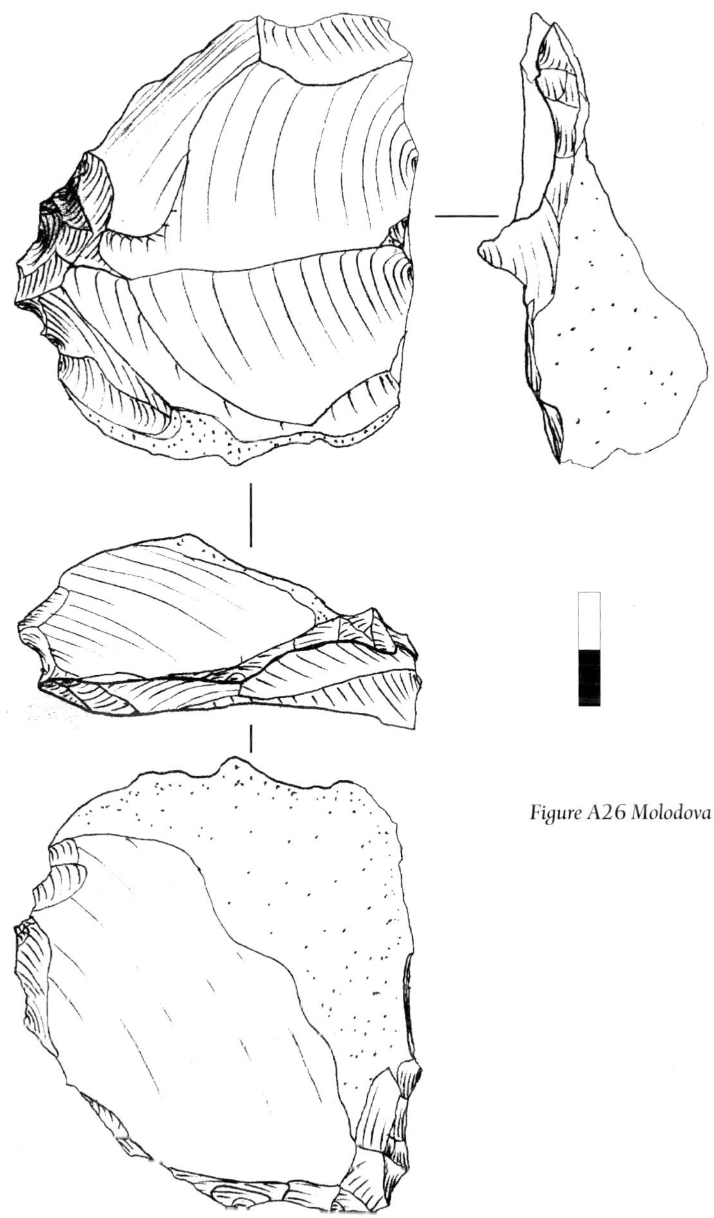

Figure A26 Molodova V Layer 11 core. Scale 2 cm.

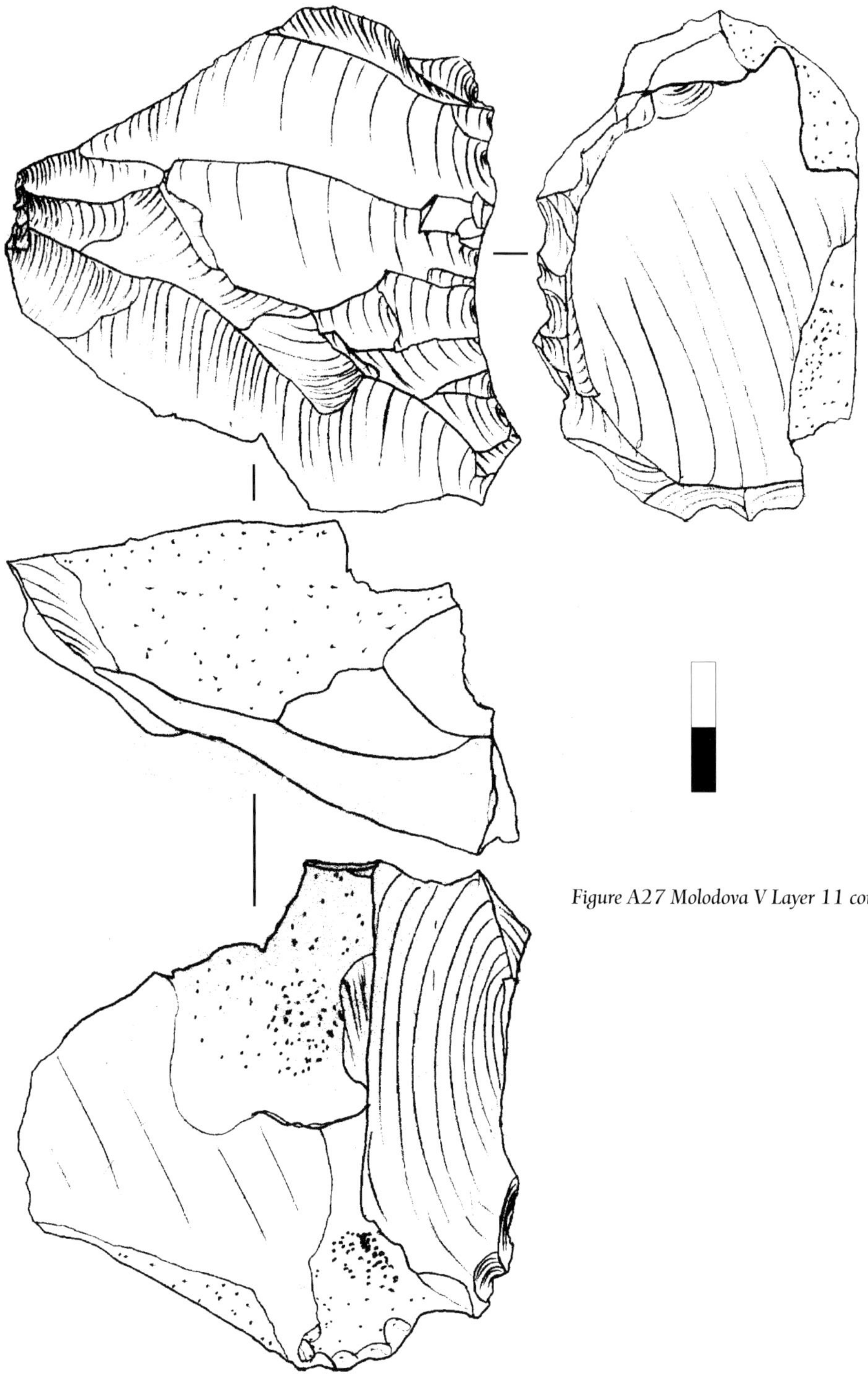

Figure A27 Molodova V Layer 11 core. Scale 2cm.

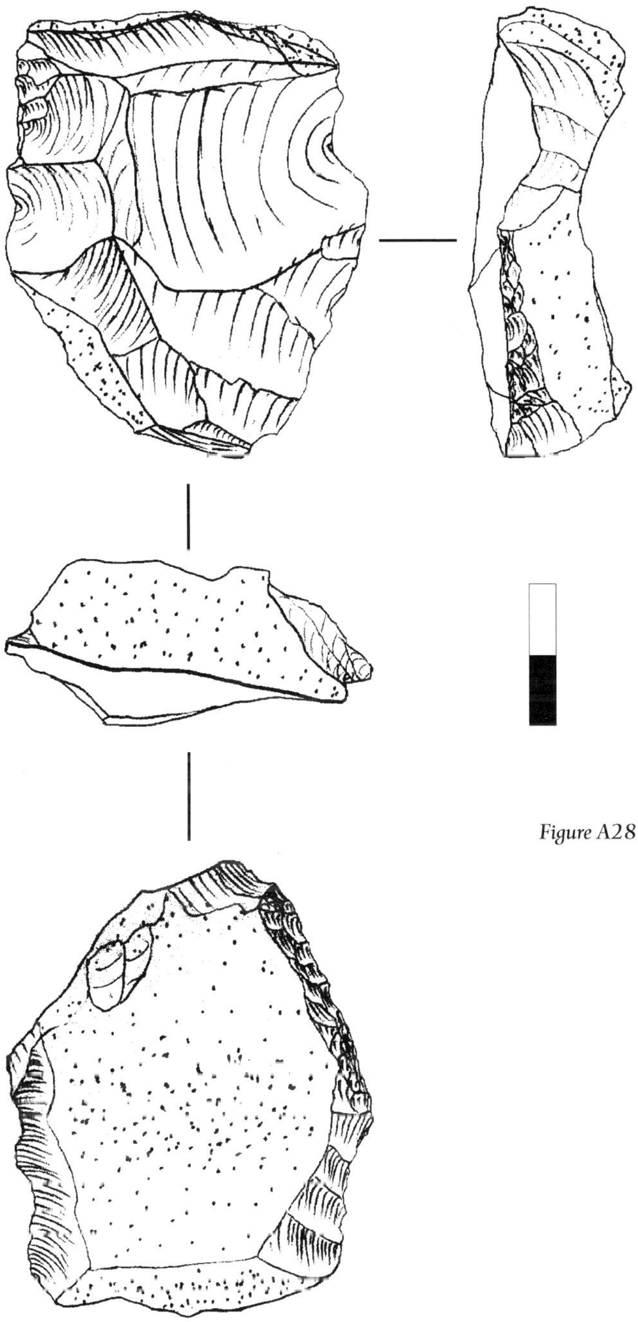

Figure A28 Molodova V Layer 11 core. Scale 2 cm.

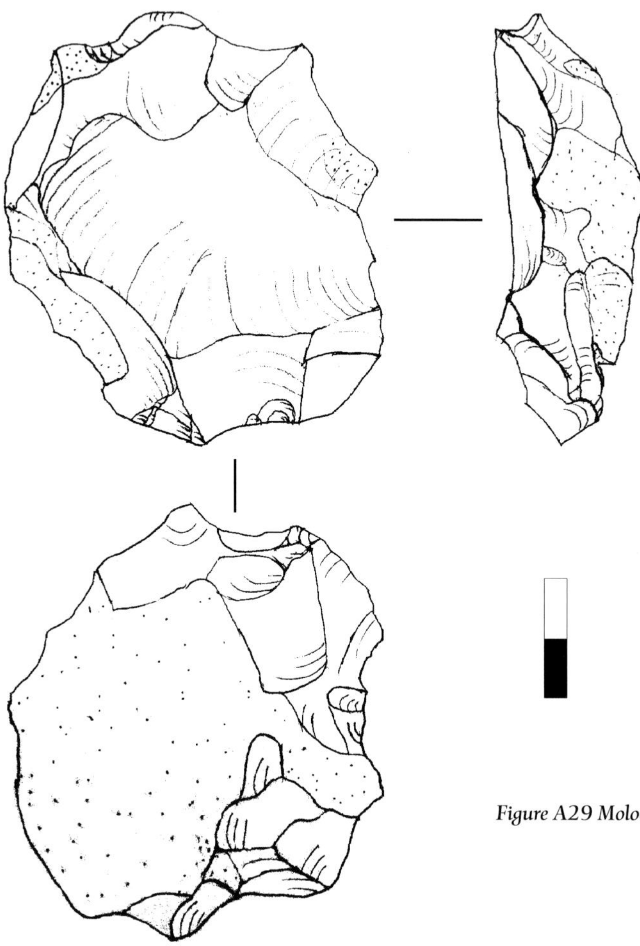

Figure A29 Molodova V Layer 11 core. Scale 2 cm.

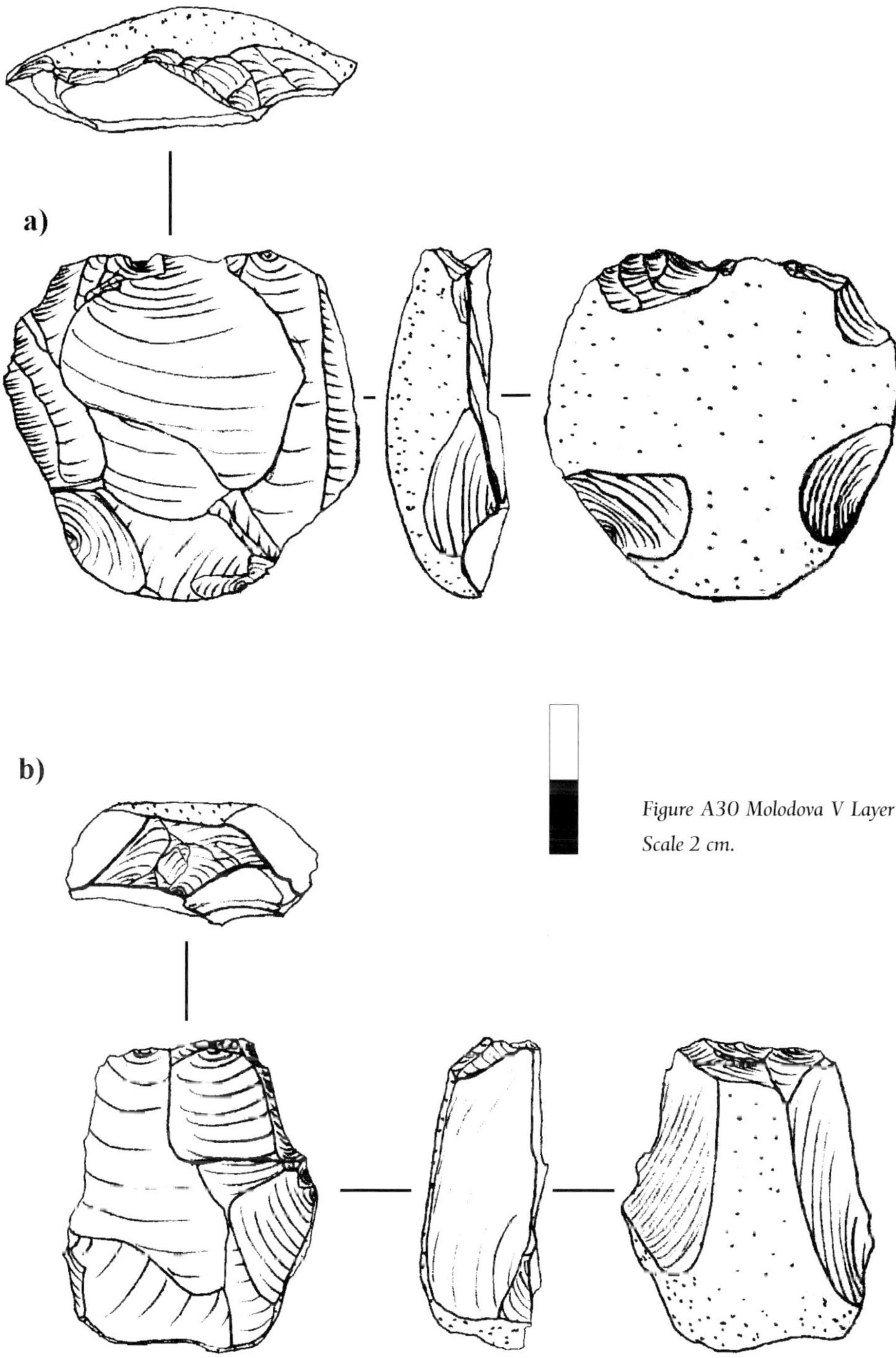

Figure A30 Molodova V Layer 11 cores. Scale 2 cm.

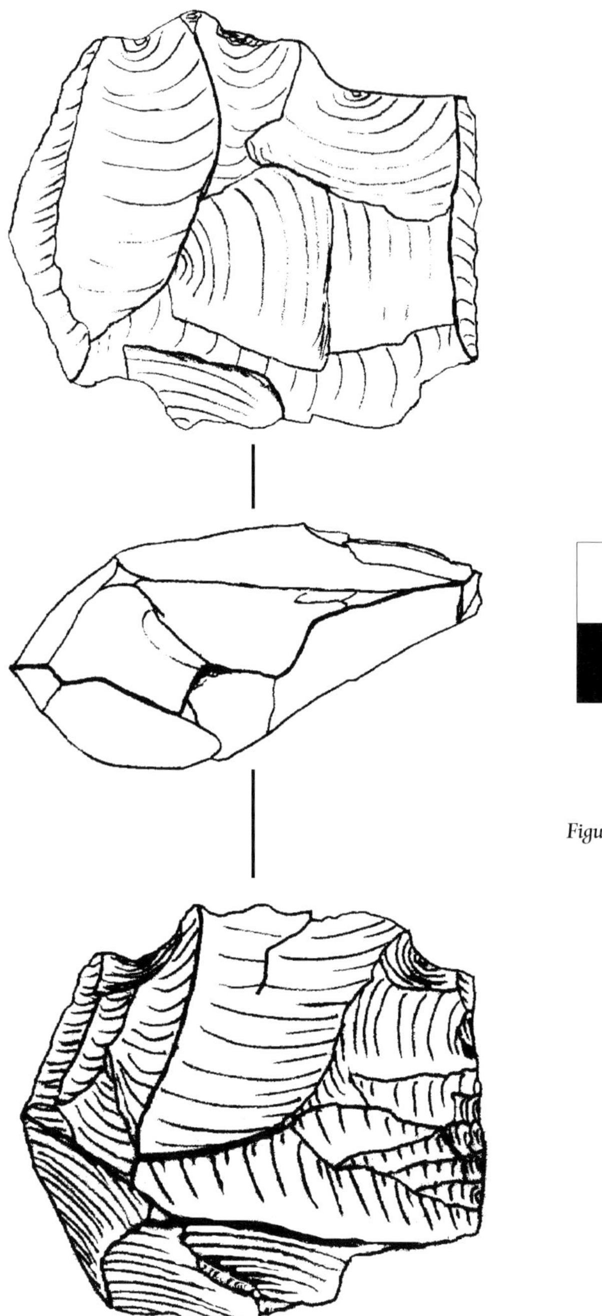

Figure A31 Korolevo I Layer 1a core. Scale 2 cm.

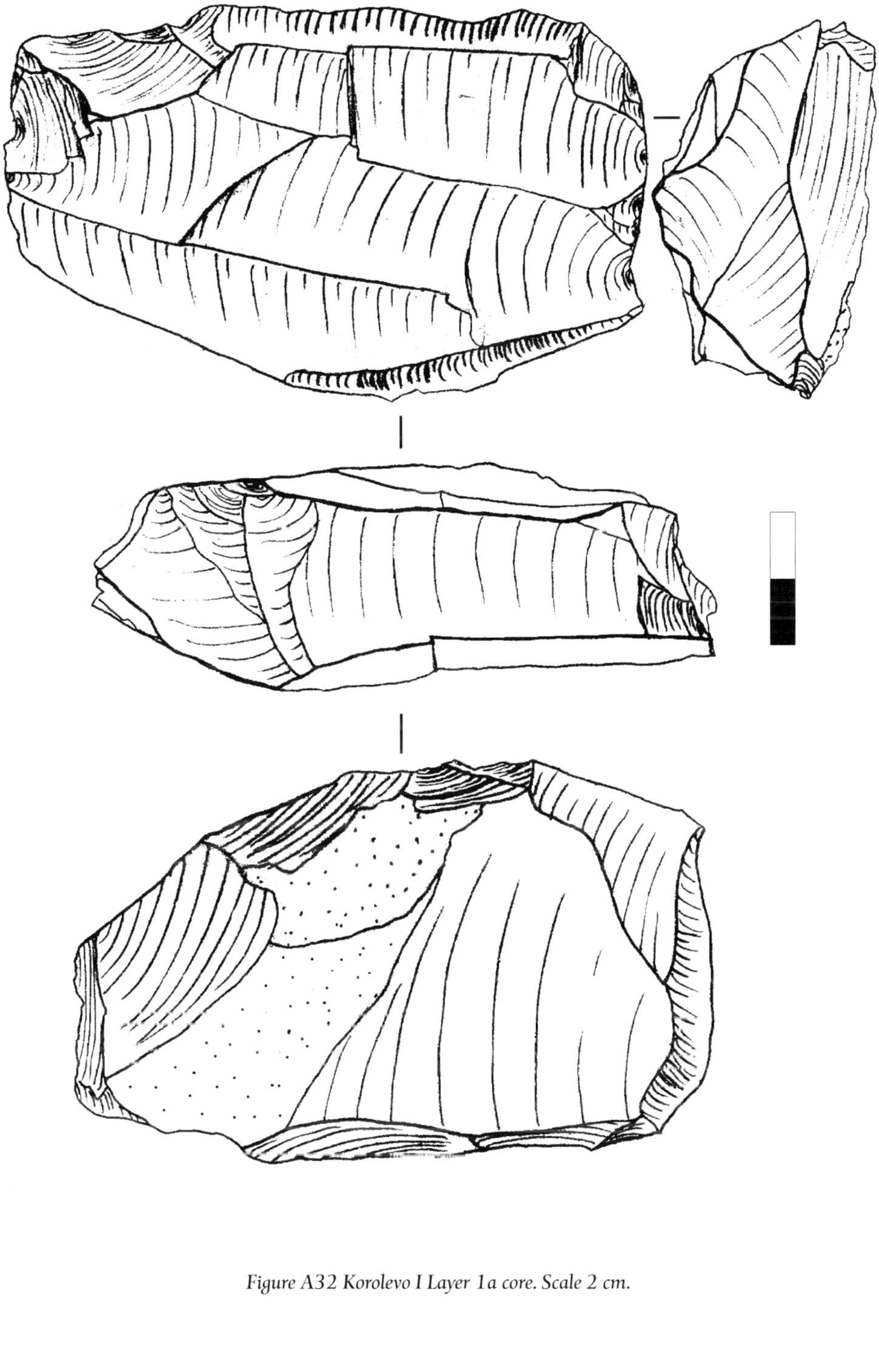

Figure A32 Korolevo I Layer 1a core. Scale 2 cm.

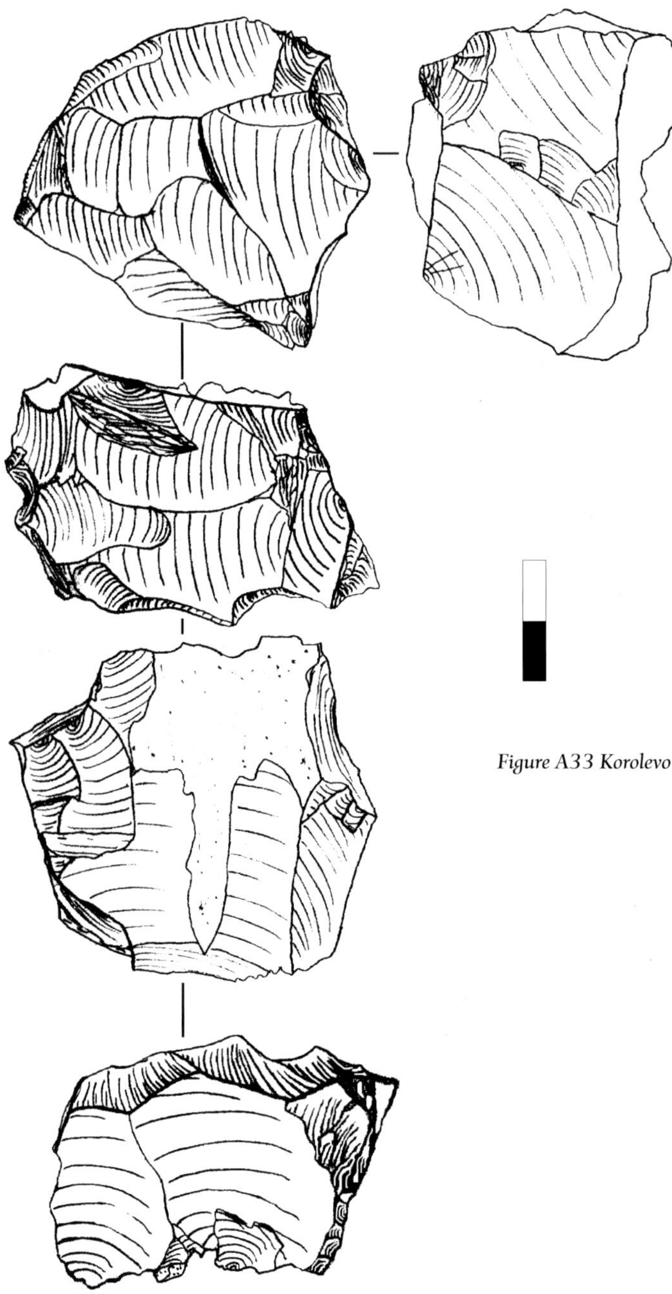

Figure A33 Korolevo I Layer 1a core. Scale 2 cm.

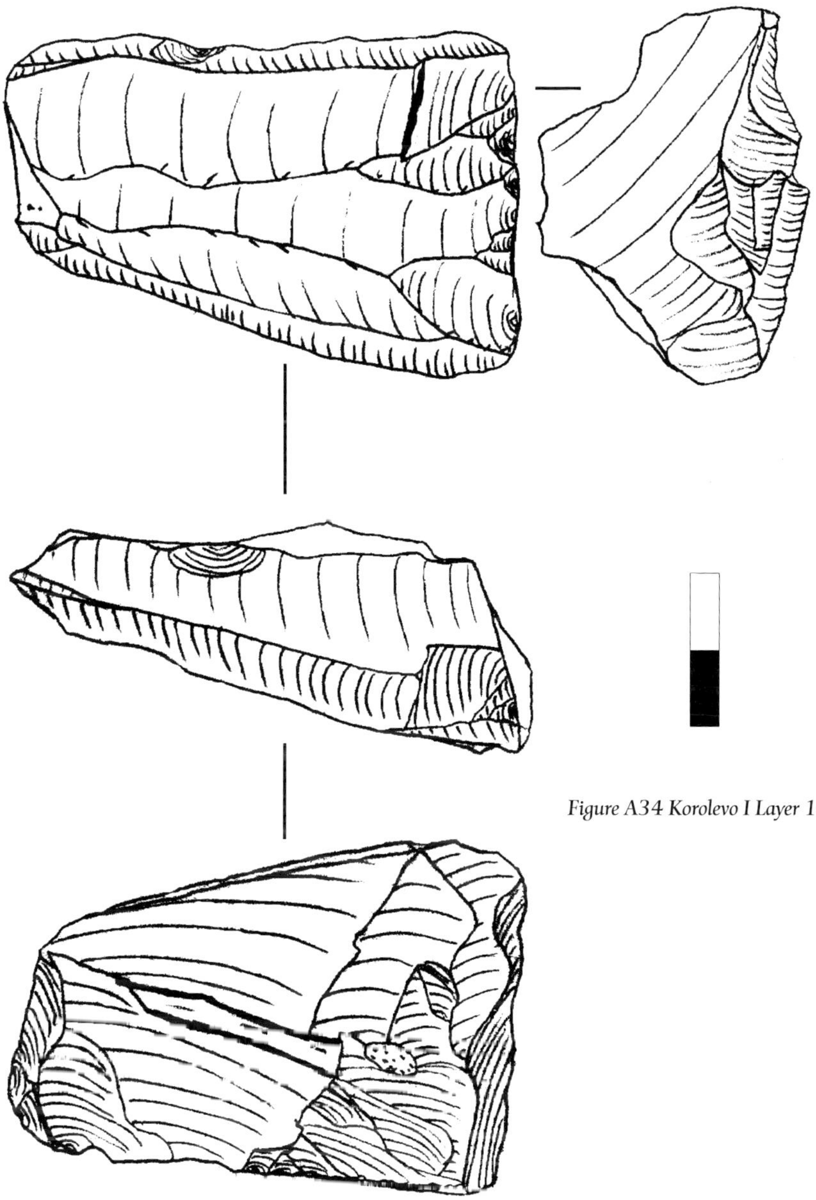

Figure A34 Korolevo I Layer 1a core. Scale 2 cm.

a)

b)

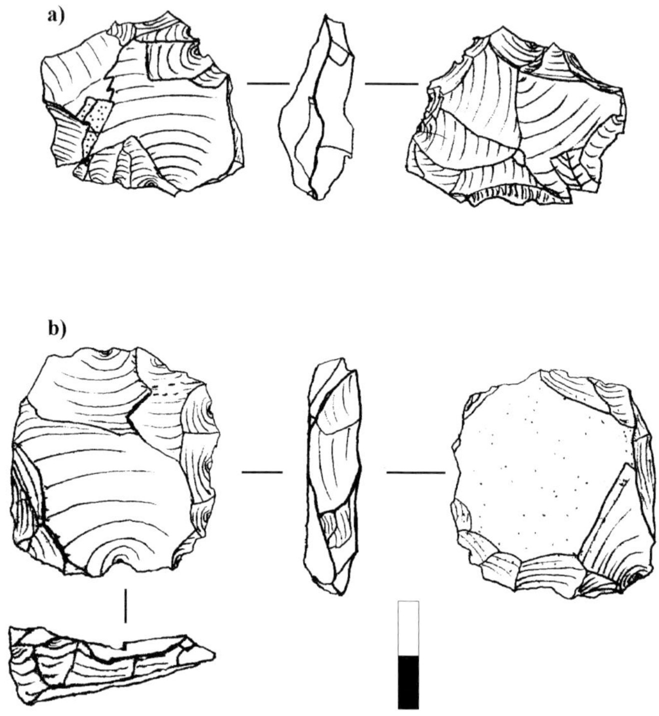

Figure A35 Kebara Cave Unit VI cores. Scale 2 cm.

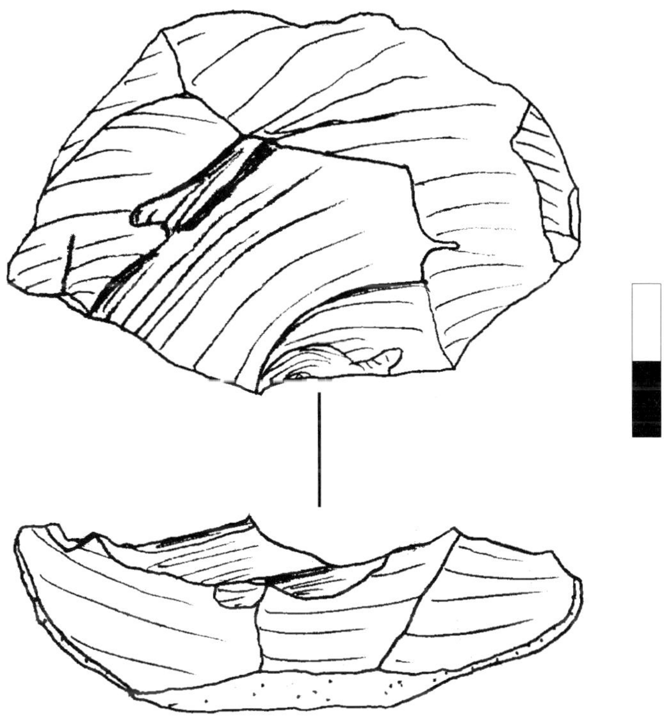

Figure A36 Kebara Cave Unit VI core. Scale 2 cm.

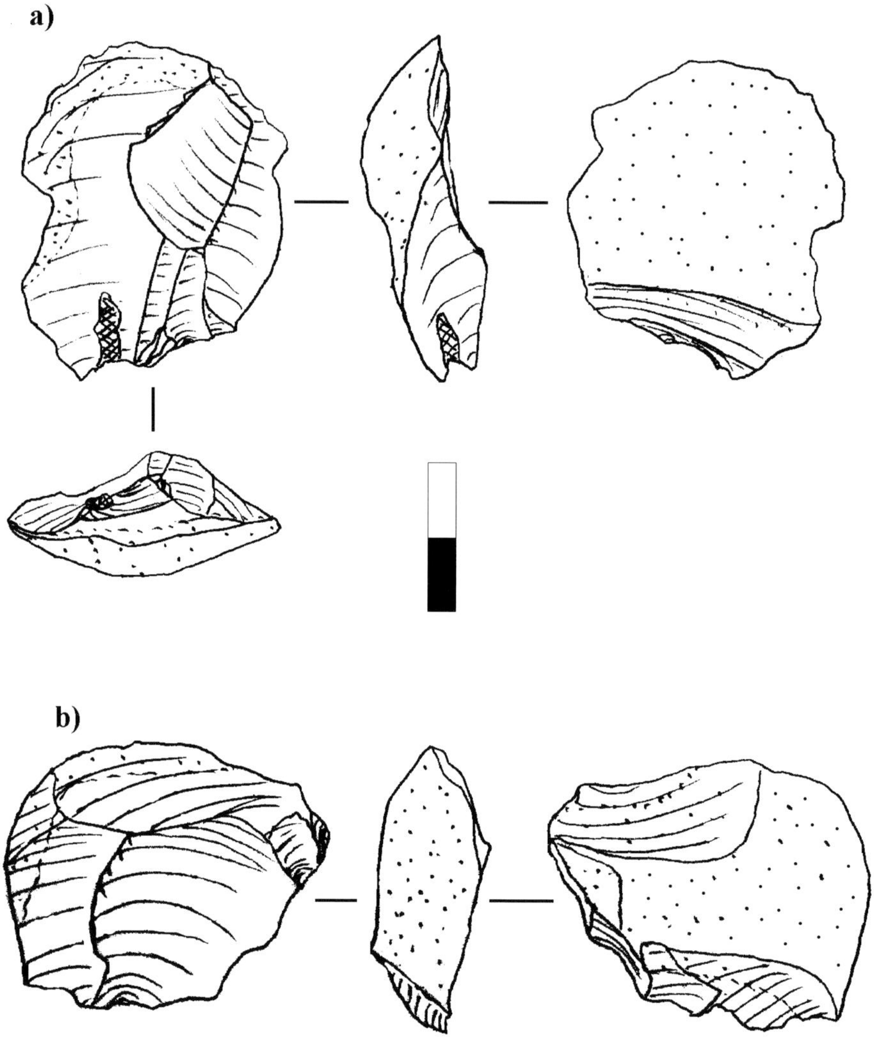

Figure A37 Kebara Cave Unit VI cores. Scale 2 cm.

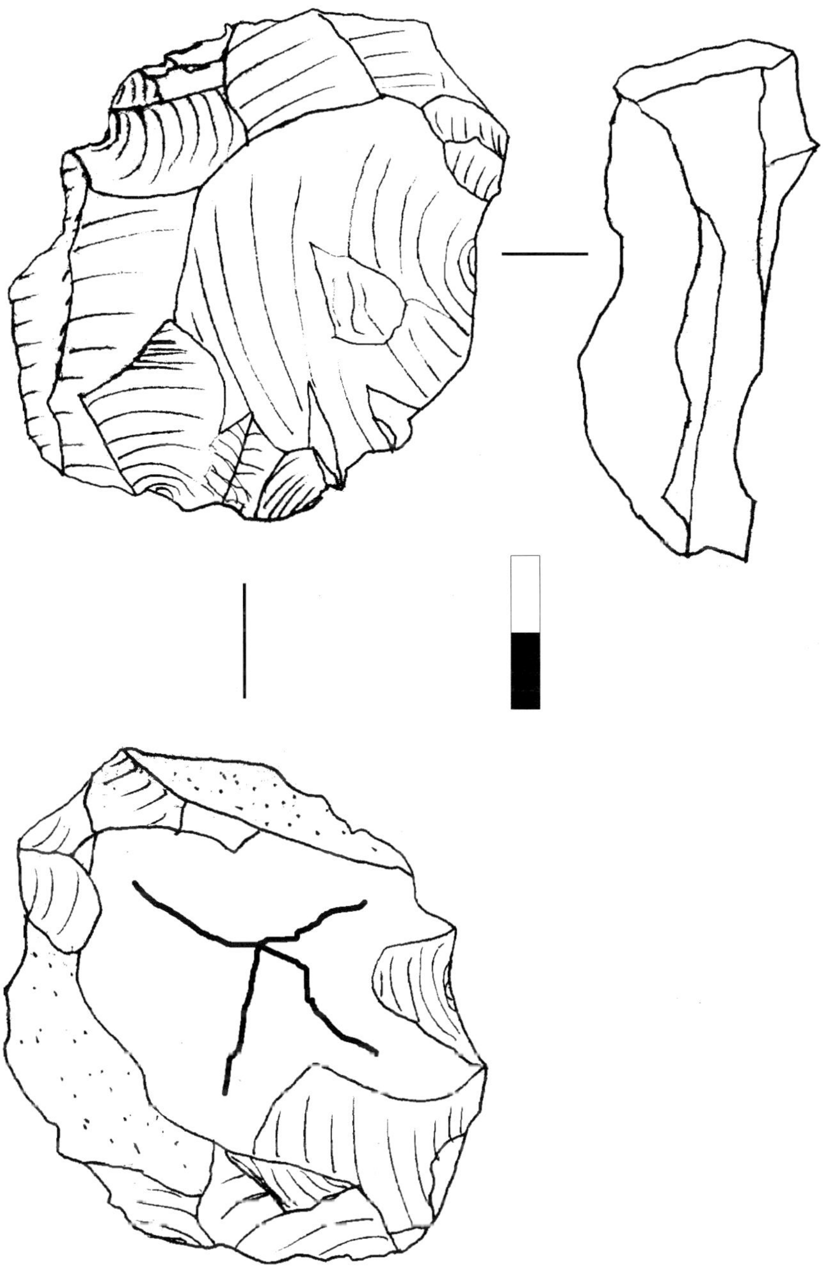

Figure A38 Kebara Cave Unit VI core. Scale 2 cm.

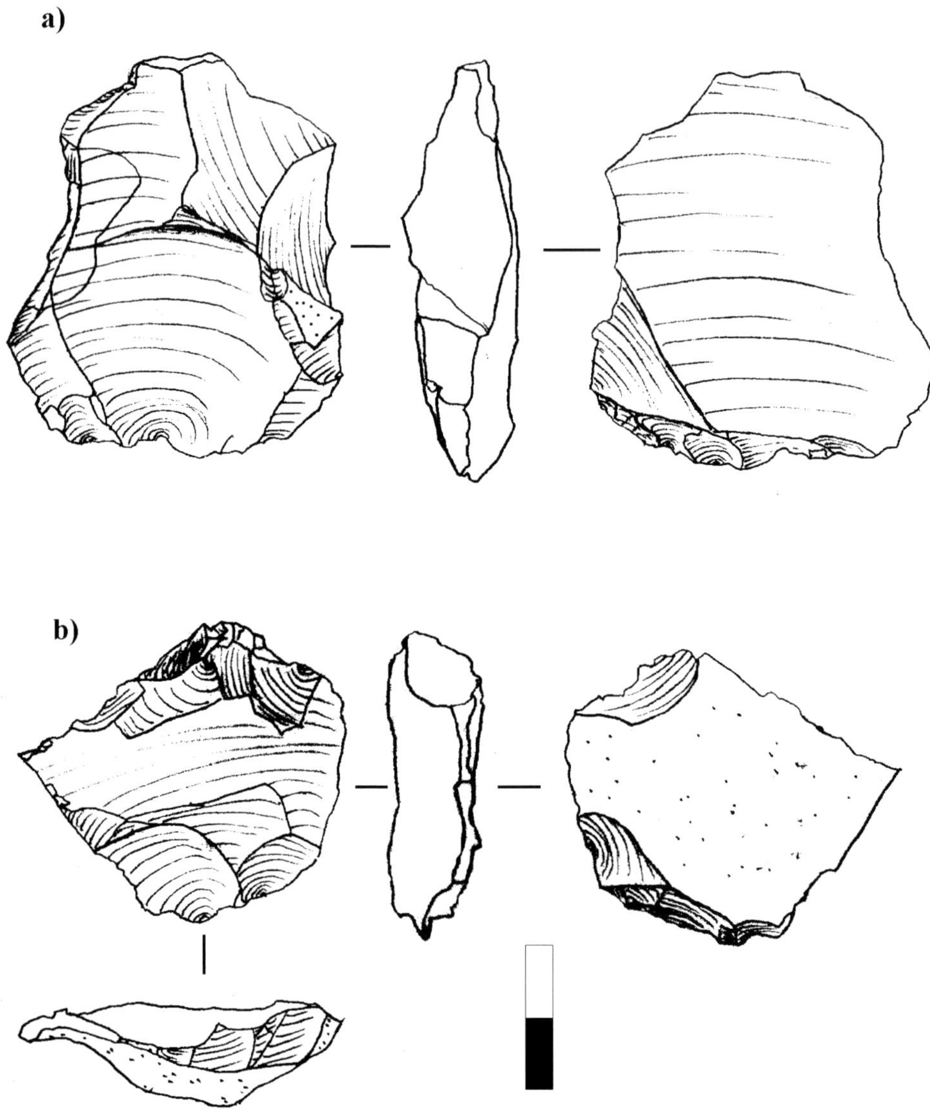

Figure A39 Kebara Cave Unit VI cores. Scale 2 cm.

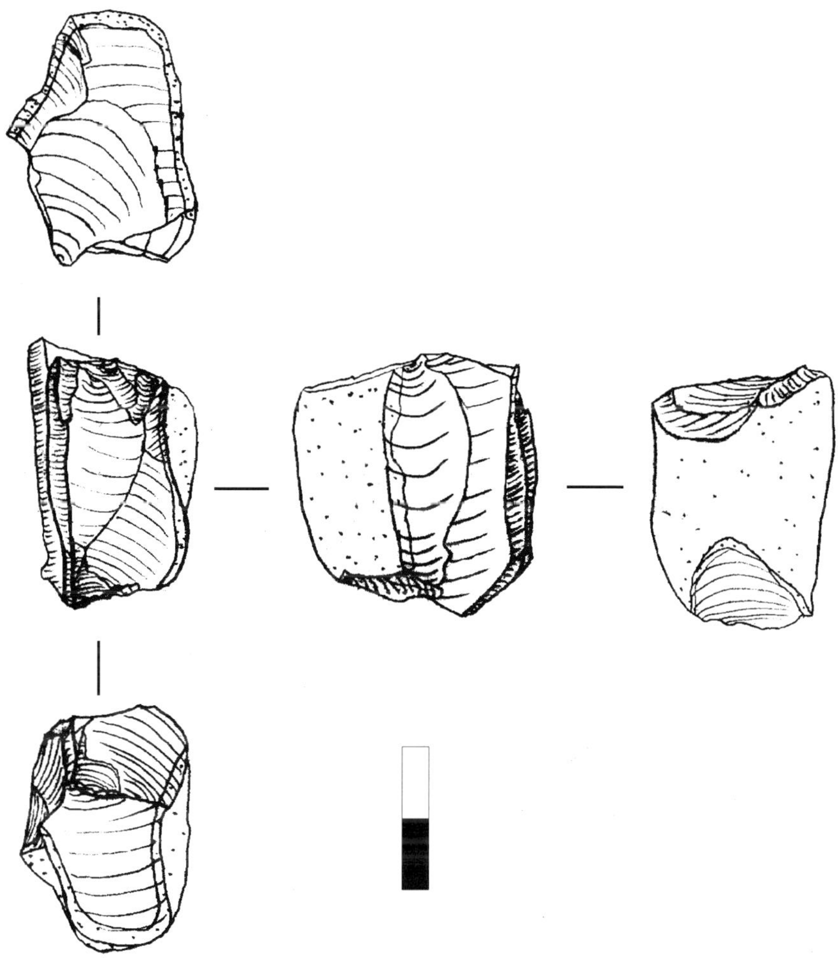

Figure A40 Kebara Cave Unit IV core. Scale 2 cm.

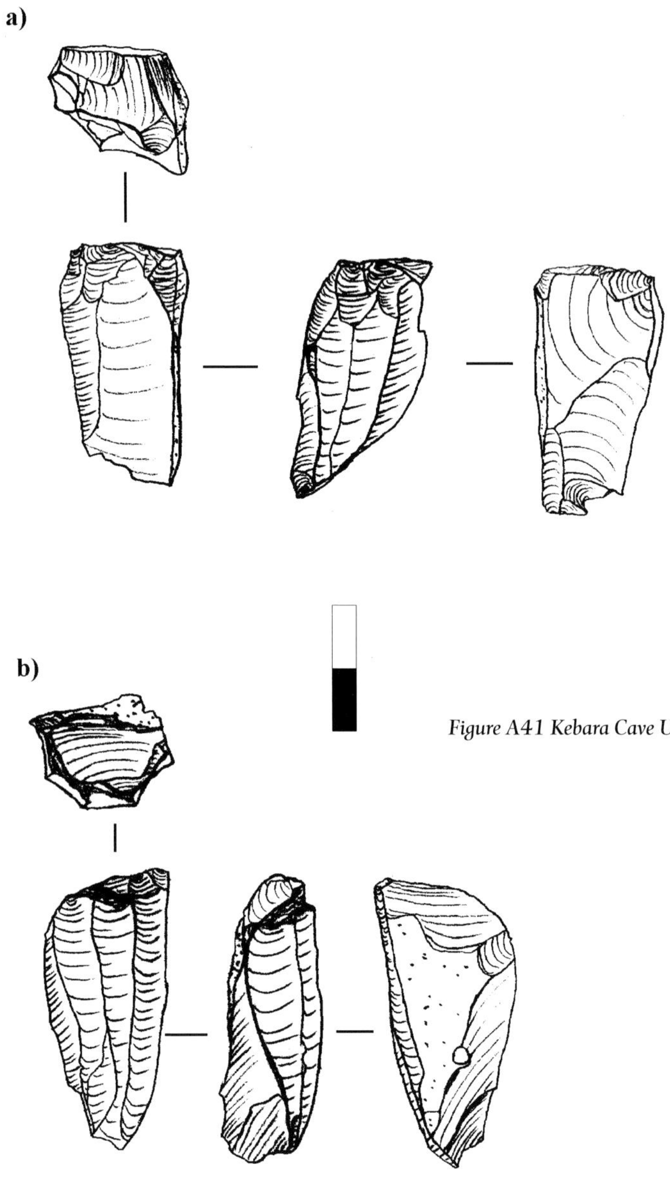

Figure A41 Kebara Cave Unit IV cores. Scale 2 cm.

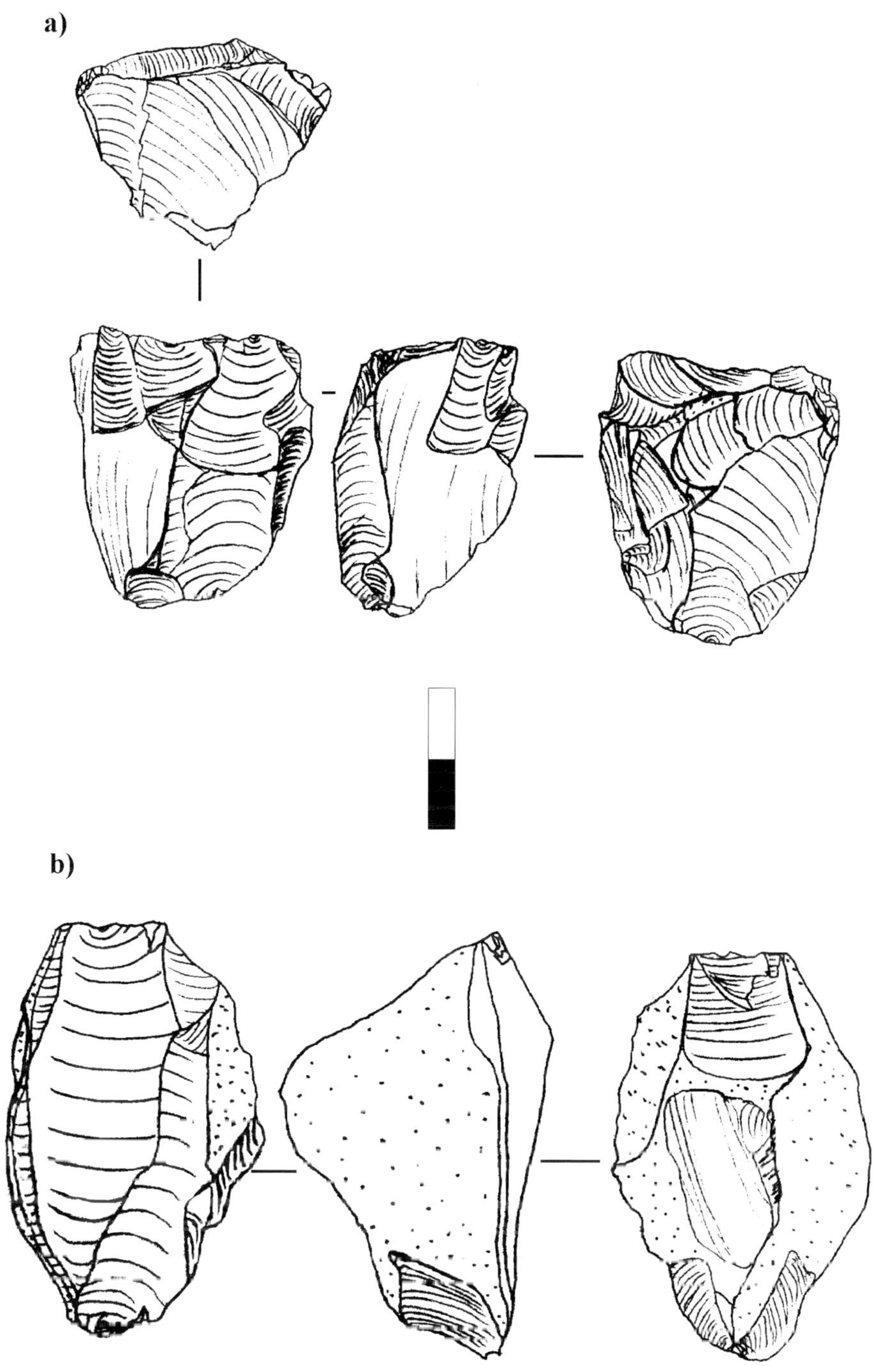

Figure A42 Kebara Cave Unit IV cores. Scale 2 cm.

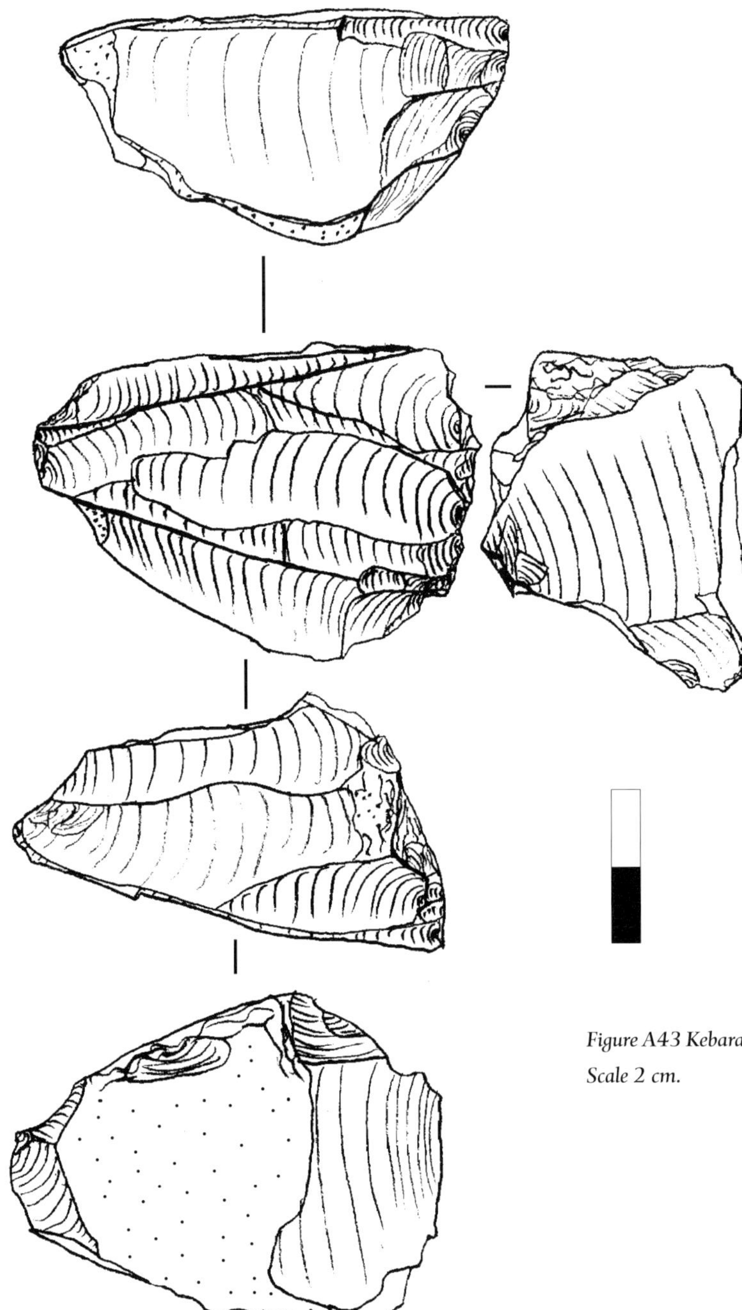

Figure A43 Kebara Cave Unit IV core. Scale 2 cm.

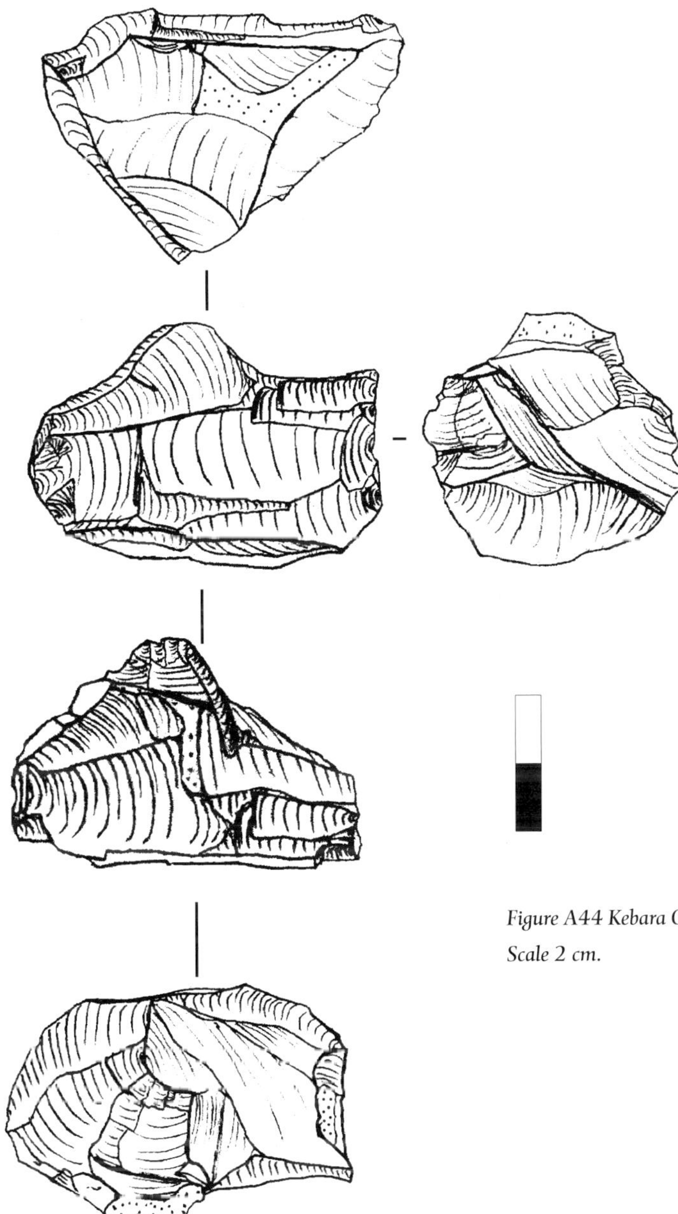

Figure A44 Kebara Cave Unit III core. Scale 2 cm.

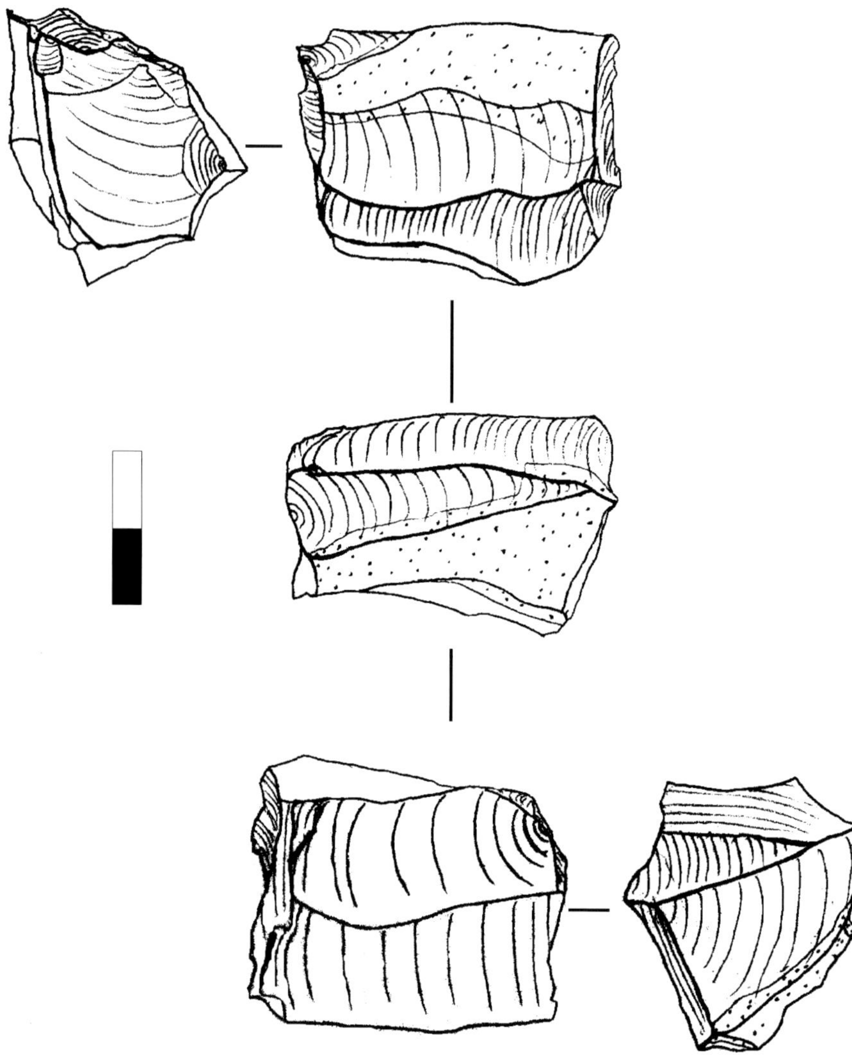

Figure A45 Kebara Cave Unit III core. Scale 2 cm.

Endnotes

Chapter 3

1 A shorter and abridged version of this section, presenting the middle range theory concept of social intimacy and taskscape visibility, is presented in Tostevin (2007).

Chapter 4

1 From, *A Treatise on Painting*, translated in 1802 by J. F. Rigaud.

2 Van Peer et al.'s (2010:43–45) distinction between plane fracture and curvilinear fracture types, using Wurz et al. (2005), has implications for differences in the relationships between size variables (flake length, width, thickness), platform variables, and core surface morphology. Although promising, the distinction requires further study through controlled experiments to demonstrate the distinction as well as elucidate any contextual changes in these relationships.

3 These model expectations are built off of earlier and less sophisticated versions presented in Tostevin (2000a,b,c). In addition to incorporating cultural transmission theory into each model, the expectation "Enculturation vs. Stimulus Diffusion" in the present research replaces the expectation "Technological Style" in these earlier works as the primary middle-range theory for the test between the hypotheses. The "Technological Style" concepts of Tostevin (2000a,b,c), drawing mainly from Sackett (1990), were incorporated in the present work within the development of the social intimacy and taskscape visibility middle-range theory.

Chapter 8

1 Please note that the values in Tostevin and Škrdla (2006) tables 2–3 are correct and that the value of "1.77" for the blank production difference value between Brno-Bohunice Lower Assemblage and Stránská skála IIIc in table 4 is a typographical error and should read "1.57".

2 The direct pair-wise comparison of the Brno-Bohunice 2002 assemblage with the other Bohunician Behavioral Package assemblages outside of Central Europe, i.e., Korolevo II-II and Boker Tachtit Level 1 and 2, are pursued in a separate, future publication.

Chapter 9

1 For those readers who did not grow up watching World War II movies, I am referring to the 1977 film based on the 1974 book of the same name by Cornelius Ryan, which depicted the failure of the Allied attempt to break through German lines and seize several bridges in the occupied Netherlands during Operation Market Garden.

References

Adams, B.

1998 *The Middle to Upper Paleolithic Transition in Central Europe: The Record from the Bükk Mountain Region.* BAR International Series 693. Archeopress, Oxford.

Adams, B., and A. Ringer

2004 New ^{14}C Dates for the Hungarian Early Upper Palaeolithic. *Current Anthropology* 45(4):541–551.

Adams, E. C.

1996 The Pueblo III–Pueblo IV Transition in the Hopi Area, Arizona. In *The Prehistoric Pueblo World, A.D. 1150–1350*, edited by Michael A. Adler, pp. 48–58. University of Arizona Press, Tucson.

Adams, W. Y., D. P. Van Gerven, and R. S. Levy

1978 The Retreat from Migrationism. *Annual Review of Anthropology* 7:483–532.

Adams, W. Y., and E. W. Adams

1991 *Archaeological Typology and Practical Reality: A Dialectical Approach to Artifact Classification and Sorting.* Cambridge University Press, Cambridge.

Adler, D. S., G. Bar-Oz, A. Belfer-Cohen, and O. Bar-Yosef

2006 Ahead of the Game: Middle and Upper Palaeolithic Hunting Behaviors in the Southern Caucasus. *Current Anthropology* 47(1):89–188.

Ahler, S.

1989a Experimental Knapping with KRF and Mid-Continent Cherts: Overview and Applications. In *Experiments in Lithic Technology*, edited by D. Amick and R. Mauldin, pp. 199–234. British Archaeological Reports International Series 528, Oxford.

1989b Mass Analysis of Flaking Debris: Studying the Forest Rather than the Tree. In *Alternative Approaches to Lithic Analysis*, edited by D. Henry and G. H. Odell, pp. 85–118. Archeological Papers of the American Anthropological Association Number 1, Washington, D.C.

Aitken, M. J., C. B. Stringer, and P. A. Mellars (Editors)

1993 *The Origin of Modern Humans and the Impact of Chronometric Dating.* Princeton University Press, Princeton.

Akazawa, T., K. Aoki, and T. Kimura (Editors)

1992 *The Evolution and Dispersal of Modern Humans in Asia.* Hokusen-sha, Tokyo.

Akazawa, T., K. Aoki, and O. Bar-Yosef (Editors)

1998 *Neanderthals and Modern Humans in Western Asia.* Plenum Press, New York.

Albert, R. M., O. Bar-Yosef, and S. Weiner

2007 Use of Plant Materials in Kebara Cave: Phytoliths and Mineralogical Analyses. In *Kebara Cave, Mt. Carmel, Israel: The Middle and Upper Paleolithic Archaeology, Part I.* American School of Prehistoric Research Bulletin 49, edited by O. Bar-Yosef and L. Meignen, pp. 147–164. Peabody Museum of Archaeology and Ethnology, Harvard University, Cambridge.

Alcock, J.

1993 *Animal Behavior: An Evolutionary Approach, 5th Edition.* Sinauer Associates, Sunderland.

Alcock, J., and P. Sherman

1994 The Utility of the Proximate-Ultimate Dichotomy in Ethology. *Ethology* 96:58–62.

Allsworth-Jones, P.

1986 *The Szeletian and the Transition from Middle to Upper Palaeolithic in Central Europe.* Clarendon Press, Oxford.

1989 The Szeletian and the Stratigraphic Succession in Central Europe and Adjacent Areas: Main Trends, Recent Results, and Problems for Solution. In *The Emergence of Modern Humans: An Archaeological Perspective*, edited by P. Mellars, pp. 60-242. Edinburgh University Press, Edinburgh.

Alvarez, R. R., Jr.

1987 *Familia: Migration and Adaptation in Baja and Alta California, 1800-1975.* University of California Press, Berkeley.

Amick, D., and R. Mauldin (Editors)

1989 *Experiments in Lithic Technology.* British Archaeological Reports International Series 528, Oxford.

Ammerman, A. J., and L. L. Cavalli-Sforza

1984 *The Neolithic Transition and the Genetics of Populations in Europe.* Princeton University Press, Princeton.

Andrefsky, W., Jr.

1986 A Consideration of Blade and Flake Curvature. *Lithic Technology* 15:48-54.

1987 Diffusion and Innovation from the Perspective of Wedge Shaped Cores in Alaska and Japan. In *The Organization of Core Technology*, edited by J. K. Johnson and C. A. Morrow, pp. 13-44. Westview Press, Boulder.

1994 Raw-Material Availability and the Organization of Technology. *American Antiquity* 59(1):21-34.

2005 *Lithics: Macroscopic Approaches to Analysis. Second Edition.* Cambridge Manuals in Archaeology. Cambridge University Press, Cambridge.

Andrefsky, W., Jr. (Editor)

2008 *Lithic Technology: Measures of Production, Use, and Curation.* Cambridge University Press, Cambridge.

Andriolo, K. R.

1979 Kulturkreislehre and the Austrian Mind. *Man* 14:133-144.

Anikovich, M.

1992 Early Upper Paleolithic Industries of Eastern Europe. *Journal of World Prehistory* 6(2):205-245.

Anthony, D. W.

1990 Migration in Archaeology: The Baby and the Bathwater. *American Anthropologist* 92:895-914.

1997 Prehistoric Migrations as Social Process. In *Migrations and Invasions in Archaeological Explanation*, edited by J. Chapman and H. Hamerow, pp. 21-32. BAR International Series 664, Oxford.

Appadurai, A.

1986 Toward an Anthropology of Things, Introduction: Commodities and the Politics of Value. In *The Social Life of Things: Commodities in Cultural Perspective*, edited by A. Appadurai, pp. 3-63. Cambridge University Press, Cambridge.

Argue, D., M. J. Morwood, T. Sutikna, Jatmiko, E. W. Saptomo

2009 *Homo floresiensis*: A Cladistic Analysis. *Journal of Human Evolution* 57:623-639.

Ariew, A.

2003 Ernst Mayr's 'Ultimate /Proximate' Distinction Reconsidered and Reconstructed. *Biology and Philosophy* 18:553-565.

Arnold, J. E., A. Preziosi, and P. Shattuck

2001 Flakes Stone Craft Production and Exchange in Island Chumash Territory. In *The Origins of a Pacific Coast Chiefdom: The Chumash of the Channel Islands*, edited by Arnold, pp. 113-132. University of Utah Press, Salt Lake City.

Aronson, M., J. M. Skibo, and M. T. Stark

1994 Production and Use Technologies in Kalinga Pottery. In *Kalinga Ethnoarchaeology: Expanding Archaeological Method and Theory*, edited by W. A. Longacre and J. Skibo, pp. 83-111. Smithsonian Institution Press, Washington, D.C.

Audouze, F.

1994 Verberie. In *Environments et Habitats Magdaléniens dans le Centre du Bassin Parisien*, edited Y. Taborin, pp. 167–172. Documents d'Archéologie Française No. 43. Editions de la Maison des Sciences de L'Homme, Paris.

1999 New Advances in French Prehistory. *Antiquity* 73:167–175.

Audouze, F., and D. Cahen

1984 L'Occupation Magdalénienne de Verberie et sa Chronologie. *Jungpaläolitische Siedlung Strukturen in Europa* 8:143–159.

Azoury, I.

1986 *Ksar 'Akil, Lebanon: a Technological and Typological Analysis of the Transitional and Early Upper Palaeolithic Levels of Ksar Akil and Abu Halka*. British Archaeological Reports International Series 289, Oxford.

Baab, K., and K. McNulty

2009 Size, Shape, and Asymmetry in Fossil Hominins: The Status of the LB1 Cranium Based on 3D Morphometric Analyses. *Journal of Human Evolution* 57:608–622.

Bailey, S. E., and J.-J. Hublin

2006 Dental Remains from the Grotte du Renne at Arcy-sur-Cure (Yonne). *Journal of Human Evolution* 50:485–508.

Bailey, S. E., T. D. Weaver, J.-J. Hublin

2009 Who Made the Aurignacian and Other Early Upper Paleolithic Industries? *Journal of Human Evolution* 57:11–26.

Bamforth, D.

1986 Technological Efficiency and Tool Curation. *American Antiquity* 51(1):38–50.

2002 Evidence and Metaphor in Evolutionary Archaeology. *American Antiquity* 67(3):435–452.

Bamforth, Douglas B., and Nyree Finlay

2008 Introduction: Archaeological Approaches to Lithic Production Skill and Craft Learning. *Journal of Archaeological Method and Theory* 15:1–27.

Bar-Yosef, O.

1991 The Search for Lithic Variability among Levantine Epi-Paleolithic Industries. In *25 Ans d'Études Technologiques en Préhistoire: XI Recontres Internationales d'Archéologie et d'Histoire d'Antibes*, pp. 319–336. Éditions APDCA, Juan-les-Pins.

1993 The Role of Western Asia in Modern Human Origins. In *The Origin of Modern Humans and the Impact of Chronometric Dating*, edited by M. J. Aitken, C. B. Stringer, and P. A. Mellars, pp. 132–147. Princeton University Press, Princeton.

1994 The Contributions of Southwest Asia to the Study of the Origin of Modern Humans. In *Origins of Anatomically Modern Humans*, edited by M. H. Nitecki and D. V. Nitecki, pp. 23–66. Plenum Press New York.

1998 On the Nature of Transitions: the Middle to Upper Palaeolithic and the Neolithic Revolution. *Cambridge Archaeological Journal* 8(2):141–163.

2002 The Upper Paleolithic Revolution. *Annual Review of Anthropology* 31:363–93.

2003 Reflections on the Selected Issues of the Upper Paleolithic. In *More than Meets the Eye: Studies on Upper Palaeolithic Diversity in the Near East*, edited by A. N. Goring-Morris and A. Belfer-Cohen, pp. 265–273. Oxbow Press, Oxford.

Bar-Yosef, O., and A. Belfer-Cohen

1988 The Early Upper Paleolithic in Levantine Caves. In *The Early Upper Paleolithic: Evidence from Europe and the Near East*, edited by J. H. Hoffecker and C. A. Wolf, pp. 23–42. British Archaeological Reports International Series 437, Oxford.

Bar-Yosef, O., and J.-G. Bordes

2010 Who Were the Makers of the Châtelperronian Culture? *Journal of Human Evolution* 59(5):586–593.

Bar-Yosef, O., and S. L. Kuhn

1999 The Big Deal about Blades: Laminar Technologies and Human Evolution. *American Anthropologist* 101(2):322–338.

Bar-Yosef, O., and L. Meignen (editors)

2007 *Kebara Cave, Mt. Carmel, Israel: The Middle and Upper Paleolithic Archaeology, Part I.* American School of Prehistoric Research Bulletin 49. Peabody Museum of Archaeology and Ethnology, Harvard University, Cambridge.

Bar-Yosef, O., and P. Van Peer

2009 The *Chaîne Opératoire* Approach in Middle Paleolithic Archaeology. *Current Anthropology* 50(1):103-131.

Bar-Yosef, O., B. Vandermeersch, B. Arensburg, A. Belfer-Cohen, P. Goldberg, H. Laville, L. Meignen, Y. Rak. J. D. Speth, E. Tchernov, A-M. Tillier, and S. Weiner

1992 The Excavations in Kebara Cave, Mt. Carmel. *Current Anthropology* 33(5):497-550.

Bar-Yosef, O., M. Arnold, N. Mercier, A. Belfer-Cohen, P. Goldberg, R. Housley, H. Laville, L. Meignen, J. C. Vogel, B. Vandermeersch

1996 The Dating of the Upper Paleolithic Layers in Kebara Cave, Mt. Carmel. *Journal of Archaeological Science* 23:297-306.

Bar-Yosef, O., A. Belfer-Cohen, and D. Adler

2006 The Implications of the Middle-Upper Paleolithic Chronological Boundary in the Caucasus to Eurasian Prehistory. *Anthropologie* (Brno) XLIV(1):49-60.

Barkow, J., L. Cosmides, and J. Toby (Editors)

1992 *The Adapted Mind, Evolutionary Psychology and the Generation of Culture.* Oxford University Press, New York.

Barnet, B., and N. Eldredge

2004 Material Cultural Evolution: An Interview with Niles Eldredge. *Fibreculture the Journal* 2004, Issue 3 (ISSN 1449-1443). Link accessed 10/10/11: http://journal.fibreculture.org/issue3/issue3_barnet.html.

Barnett, H. G.

1953 *Innovation: The Basis of Cultural Change.* McGraw-Hill, New York.

Barth, F.

1969a Introduction. In *Ethnic Groups and Boundaries: The Social Organization of Culture Difference*, edited by F. Barth, pp. 9-38. Little and Brown Series in Anthropology. George Allen and Unwin, London.

1969b Pathan Identity and its Maintenance. In *Ethnic Groups and Boundaries: The Social Organization of Culture Difference*, edited by F. Barth, pp. 117-134. Little and Brown Series in Anthropology. George Allen and Unwin, London.

Basalla, G.

1988 *The Evolution of Technology.* Cambridge History of Science Series. Cambridge University Press, Cambridge.

Baumler, M. F.

1988 Core Reduction, Flake Production, and the Middle Paleolithic Industry of Zobiste (Yugoslavia). In *Upper Pleistocene Prehistory of Western Eurasia*, edited by H. L. Dibble and A. Montet-White, pp. 255-274. University Museum Monograph 54, University Museum Symposium Series 1. University Museum, University of Pennsylvania, Philadelphia.

Baumler, M. F., and L. B. Davis

2004 The Role of Small-Sized Debitage in Aggregate Lithic Analysis. In *Aggregate Analysis in Chipped Stone*, edited by C. Hall and M. L. Larson, pp. 45-64. University of Utah Press, Salt Lake City.

Beatty, J.

1994 The Proximate/ Ultimate Distinction in the Multiple Careers of Ernst Mayr. *Biology and Philosophy* 9:333-356.

de Beaune, S.

2000 *Pour une Archéologie du Geste: Broyer, Moudre, Piler, des Premiers Chasseurs aux Premiers Agriculteurs.* CNRS Éditions, Paris.

2002 Comment on D. Stout's 'Skill and Cognition in Stone Tool Production: An Ethnographic Case Study from Irian Jaya. *Current Anthropology* 43(5):716.

de Beaune, S. (Editor)

2007 *Chasseurs-Cueilleurs: Comment Vivaient nos Ancêtres du Paléolithique Supérieur. Méthodes d'Analyse et d'Interprétation en Préhistoire.* CNRS Éditions, Paris.

Beeman, William O.

2008 The Performance Hypothesis: Practicing Emotions in Protected Frames. In *The Emotions: A Cultural Reader*, edited by Helena Wulff, pp. 273-298. Berg Publishers, London.

Belfer-Cohen, A., and O. Bar-Yosef

1981 The Aurignacian of Hayonim Cave. *Paléorient* 7:19-42.

2006 The Geography and Chronology of Carinated Core *Chaîne Opératoire*, a paper published on-line in the electronic symposium entitled, *Core Reduction Sequence, Chaîne Opératoire, and Other Methods: the Epistemologies of Different Approaches to Lithic Analysis*, organized by G. Tostevin at the 71st Annual Meeting of the Society for American Archaeology, San Juan, Puerto Rico.

Belfer-Cohen, A., and A. N. Goring-Morris

2003 Final Remarks and Epilogue. In *More than Meets the Eye: Studies on Upper Palaeolithic Diversity in the Near East*, edited by A. N. Goring-Morris and A. Belfer-Cohen, pp. 274-280. Oxbow Press, Oxford.

2007 From the Beginning: Levantine Upper Palaeolithic Cultural Continuity. In *Rethinking the Human Revolution: New Behavioural and Biological Perspectives on the Origins and Dispersal of Modern Humans*, edited by P. Mellars, K. Boyle, O. Bar-Yosef, and C. Stringer, pp. 199-206. MacDonald Institute Research Monograph, Cambridge.

Bentley, G. C.

1987 Ethnicity and Practice. *Society for the Comparative Study of Society and History* 29:24-55.

Bergman, C.

1987 *Ksar Akil, Lebanon: A Technological and Typological Analysis of the Later Upper Palaeolithic Levels.* British Archaeological Reports International Series 329, Oxford.

Bergman, C., and C. B. Stringer

1989 Fifty years after: Egbert, an Early Upper Paleolithic Juvenile from Ksar 'Akil, Lebanon. *Paléorient* 15(2):99-111.

Bertouille, H.

1989 Théories physiques et mathématiques de la taille des outils préhistoriques. *Cahiers du Quaternaire* 15. Éditions du CNRS, Paris.

Bettinger, R. L., and M. A. Baumhoff

1982 The Numic Spread: Great Basin Cultures in Competition. *American Antiquity* 47(3):485-503.

Bettinger, R. L., and J. Eerkens

1999 Point Typologies, Cultural Transmission, and the Spread of Bow and Arrow Technology in the Prehistoric Great Basin. *American Antiquity* 64:231-242.

Bettinger, R. L., and P. J. Richerson

1996 The State of Evolutionary Archaeology: Evolutionary Correctness, or the Search for the Common Ground. In *Darwinian Archaeologies*, edited by H. D. G. Maschner, pp. 221-232. Plenum, New York.

Bettinger, R. L., R. Boyd, and P. J. Richerson

1996 Style, Function, and Cultural Evolutionary Processes. In *Darwinian Archaeologies*, edited by H. D. G. Maschner, 133-164. Plenum, New York.

Beyries, S.

1988 Functional Variability of Lithic Sets in the Middle Paleolithic. In *Upper Pleistocene Prehistory of Western Eurasia*, edited by H. Dibble and A. Montet-White, pp. 213-224. University Museum Monographs No. 4. University of Pennsylvania, Philadelphia.

Bijker, W. E.

1987 The Social Construction of Bakelite: Toward a Theory of Invention. In *The Social Construction of Technological Systems: New Directions in the Sociology and History of Technology*, edited by W. E. Bijker, T. P. Hughes, and T. F. Pinch, pp. 159-187. MIT Press, Cambridge.

Binford, L. R.

1962 Archeology as Anthropology. *American Antiquity* 28:217-225.

1965 Archaeological Systematics and the Study of Culture Process. *American Antiquity* 31(2):203-210.

1968 Archaeological Perspectives. In *New Perspectives in Archaeology*, edited by S. R. Binford and L. R. Binford, pp. 5-32. Aldine Publishing Company, Chicago.

1973 Interassemblage Variability-the Mousterian and the 'Functional' Argument. In The *Explanation of Culture Change: Models in Prehistory*, edited by C. Renfrew, pp. 227-253. Duckworth, London.

1977 General Introduction. In *For Theory Building in Archaeology: Essays on Faunal Remains, Aquatic Resources, Spatial Analysis and Systematic Modeling*, edited by L. Binford, pp. 1-10. Academic Press, New York.

1978 *Nunamiut Ethnoarchaeology*. Academic Press, New York.

1979 Organization and Formation Processes: Looking at Curated Technologies. *Journal of Anthropological Research* 35(3):255-273.

1980 Willow Smoke and Dog's Tails: Hunter-Gatherer Settlement Systems and Archaeological Site Formation. *American Antiquity* 45:4-20.

1982 The Archaeology of Place. *Journal of Anthropological Archaeology* 1:5-31.

1983 *Working at Archaeology*. Academic Press, New York.

Binford, L. R., and S. R. Binford

1966 A Preliminary Analysis of Functional Variability in the Mousterian of Levallois Facies. *American Antiquity* 68(2):238-295.

Bischoff, J. L., N. Soler, J. Maroto, and R. Julia

1989 Abrupt Mousterian/Aurignacian Boundary at c. 40 kya bp: Accelerator ^{14}C Dates from L'Abreda Cave (Catalunya, Spain). *Journal of Archaeological Science* 16:563-576.

Bisson, M.

2000 Nineteenth Century Tools for Twenty-First Century Archaeology? Why the Middle Paleolithic Typology of François Bordes Must Be Replaced. *Journal of Archaeological Method and Theory* 7(1):1-48.

2001 Interview with a Neanderthal: An Experimental Approach for Reconstructing Scraper Production Rules, and their Implications for Imposed Form in Middle Palaeolithic Tools. *Cambridge Archaeological Journal* 11(2):165-84.

Blackmore, Susan

1999 *The Meme Machine*. Oxford University Press, Oxford.

Blackwell, W.

2004 *Flintknapping: The Art of the Ancient*. McLean Design/Opposable Thumb Productions, Digital Video.

Blades, Brooke S.

2000 *Aurignacian Lithic Economy: Ecological Perspectives from Southwestern France*. Kluwer Academic/Plenum Press, New York.

Bleed, P.

1986 The Optimal Design of Hunting Weapons: Maintainability or Reliability. *American Antiquity* 51:547-562.

2001 Trees or Chains, Links or Branches: Conceptual Alternatives for Consideration of Stone Tool Production and Other Sequential Activities. *Journal of Archaeological Method and Theory* 8(1):101-127.

2011 Loosening our *Chaînes*: Cognitive Insights for the Archaeological Application of Sequence Models. *PaleoAnthropology* 2011:297-304.

Bluszcz, A. J. Kozłowski, and E. Foltyn

1994 New Sequence of EUP Leaf Point Industries in Southern Poland. *Préhistoire Européenne* 6:197-222.

Bocquet-Appel, J.-P., and P. Yves Demars

2000 Neanderthal Contraction and Modern Human Colonization of Europe. *Antiquity* 74:544-552.

Boëda, E.

1986 Approche Technologique du Concept Levallois et Evaluation de son Champ d'Application: Etude de Trois Gisements Saaliens et Weichséliens de la France Septentrionale. Unpublished Ph.D. Dissertation, Université de Paris X.

1988a Analyse Technologique du Débitage du Niveau IIA. In *Le Gisement Paleolithique Moyen de Biache-Saint-Vaast (Pas-de-Calais)*, Vol. 1, edited by A. Tuffreau and J. Sommé, pp. 185-214. Mémoires de la Société Préhistorique Française, Paris.

1988b Le Concept Laminaire: Rupture et Filiation ave le Concept Levallois. In *L'Homme de Néandertal, La Mutation*, Volume 8, edited by J. Kozłowski, pp. 41-59. ERAUL, Liège.

1993 Le débitage discoïde et le débitage levallois récurrent centripète. *Bulletin de la Société Préhistorique Française* 90(6):392-404.

1994 *Le Concept Levallois: Variabilité des Méthodes*. Monographie du CRA 9. Centre National de la Recherche Scientifique, Paris.

1995 Levallois: A Volumetric Construction, Methods, a Technique. In *The Definition and Interpretation of Levallois Technology*, edited by H. L. Dibble and O. Bar-Yosef, pp. 41-68. Prehistory Press, Madison.

Boëda, E., J.-M. Geneste, and L. Meignen

1990 Identification de *Chaînes Opératoires* Lithiques du Paléolithique Ancien et Moyen. *Paléo* 2:43-80.

Boesch, C.

1994 Cooperative Hunting in Wild Chimpanzees. *Animal Behaviour* 48:653-667.

Boesch, C., and M. Tomasello

1998 Chimpanzee and Human Cultures. *Current Anthropology* 39(5):591-614.

Bodu, P.

1990 L'Application de la Method des Remontages à l'Etude du Materiel Lithique des Premiers Niveaux Châtelperroniens d'Archy-sur-Cure. In *Paléolithique moyen récent et Paléolithique supérieur ancien en Europe, Mémoires du Musée de Préhistoire d'Ile-de-France* 3, edited by C. Farizy, pp. 309-312. Édition de l'Association pour la Promotion de la Recherche Archéologique en Ile de France, Nemours.

Bolus, M., and N. J. Conard

2001 The Late Middle Paleolithic and Earliest Upper Paleolithic in Central Europe and their Relevance for the Out of Africa Hypothesis. *Quaternary International* 75:29-40.

Bon, F.

2002 *L'Aurignacien Entre Mer et Océan. Réflexion sur l'Unité des Phases Anciennes de l'Aurignacien dane le sud de la France*. Société Préhistorique Française Mémoire 29, Paris.

2005 Little Big Tool. Enquête Autour due Success de la Lamelle. In *Productions Lamelleaires attribuées à l'Aurignacien: Chaînes Opératoires et Perspectives Technoculturelles*, edited by F. Lebrun-Ricalens, pp. 479-484. ArchéoLogiques No. 1, Luxemburg.

Bon, F., and P. Bodu

2002 Analyse Technologique du Débitage Aurignacien. In *L'Aurignacien de la Grotte du Renne. Les fouilles d'André Leroi-Gourhan à Arcy-sur-Cure (Yonne)*, edited by B. Schmider, pp. 115-133. Supplement XXXIV of Gallia Préhistoire, CNRS, Paris.

Bonnichsen, R.

1977 *Models for Deriving Cultural Information from Stone Tools*. National Museum of Man Mercury Series, Archaeological Survey of Canada Paper 60. Ottawa.

Boone, J. J., and E. A. Smith

1998 Is It Evolution Yet? A Critique of Evolutionary Archaeology. *Current Anthropology* 39(Supplement):S141-S173.

Bordes, F.

1950 L'Evolution Buissonnante des Industries en Europe Occidentale: Considérations Théoriques sur le Paléolithique Ancien et Moyen. *L'Anthropologie* 54:393-420.

1958 Le passage du Paléolithique moyen au Paléolithique supérieur. In *Hundert Jahre Heandethaler*, edited by G. H. R. von Koenigswald, pp. 175-181. Kemink en Zoon, Utrecht.

1961a *Typologie du Paléolithique Ancien et Moyen*. Publication de l'Institut de Préhistoire de l'Université de Bordeaux, Bordeaux.

1961b Mousterian Cultures in France. *Science* 134:803-810.

1965 A propos de typologie. *L'Anthropologie* (Paris) 69(3-4):369-377.

1968a La Question Périgordienne. In *La Préhistoire: Problèmes et Tendances*, edited by J. Piveteau, pp. 59-70. Éditions du Centre National de la Recherche Scientifique, Paris.

1968b *The Old Stone Age*. McGraw-Hill Book Company, New York.

1972 Du Paléolithique Moyen au Paléolithique Supérieur: Continuité ou Discontinuité? In *The Origin of Homo Sapiens*, edited by F. Bordes, pp. 211-218. Centre National de la Recherche Scientifique, Paris.

Bordes, F., and D. de Sonneville-Bordes

1970 The Significance of Variability in Palaeolithic Assemblages. *World Archaeology* 2(1):61-73.

Bordes, F., and J. Labrot

1967 La Stratigraphie d Roc de Combe (Lot) et ses Implications. *Bulletin de la Société Préhistorique Française* 64:15-28.

Bordes, J.-G.

2003 Lithic Taphonomy of the Châtelperronian/Aurignacian Interstratifications in Roc de Combe and Le Piage (Lot, France). In *The Chronology of the Aurignacian and of the Transitional Technocomplexes: Dating, Stratigraphies, Cultural Implications*, edited by J. Zilhão and F. d'Errico, pp. 223-244. Trabalhos de Arquelogia 33. Instituto Português de Arqueologia, Lisboa.

Bourdieu, P.

1977 *Outline of a Theory of Practice*. Cambridge University Press, Cambridge.

1980 *Le sens Pratique*. Les Éditions de Minuit, Paris.

Boyd, R., and P. J. Richerson

1985 *Culture and the Evolutionary Process*. University of Chicago Press, Chicago.

1987 Evolution of Ethnic Markers. *Cultural Anthropology* 2(1):65-79.

1996 Why Culture is Common but Cultural Evolution is Rare. *Proceedings of the British Academy* 88:77-93.

2000 Meme Theory Oversimplifies Cultural Change. *Scientific American* October: 54-55.

2005 *The Origin and Evolution of Cultures*. Oxford University Press, New York.

Boyd, R., and J. Silk

2000 *How Humans Evolved, Second Edition*. W. W. Norton and Company, New York.

Boydston, R.

1989 A Cost-Benefit Study of Functionally Similar Tools. In *Time, Energy, and Stone Tools*, edited by R. Torrence, pp. 67-77. Cambridge University Press, Cambridge.

Bradbury, A., and P. Carr

1999 Examining Stage and Continuum Models of Flake Debris Analysis: An Experimental Approach. *Journal of Archaeological Science* 26:105-116.

Bradley, B.

1975 Lithic Reduction Sequences: a Glossary and Discussion. In *Lithic Technology: Making and Using Stone Tools*, edited by E. Swanson, pp. 5-14. Mouton, The Hague.

1989 Flintknapping with Bruce Bradley, Ph.D. Video documentary. INTERpark and Prim Teck Enterprises, Cortez, CO.

Bradley, B., and D. Stanford

2004 The North Atlantic Ice-Edge Corridor: A Possible Palaeolithic Route to the New World. *World Archaeology* 36(4):459-478.

Braithwaite, M.

1982 Decoration as Ritual Symbol: A Theoretical Proposal and an Ethnographic Study in Southern Sudan. In *Symbolic and Structural Archaeology*, edited by I. Hodder, pp. 80-88. Cambridge University Press, Cambridge.

Brandt, S. A., K. J. Weedman, and G. Hundie

1996 Gurage Hide Working Stone Tool Use and Social Identity: An Ethnoarchaeological Perspective. In *Essay on Gurage Language and Culture: Dedicated to Wolf Leslau on the Occasion of his 90th Birthday November 14, 1996*, edited by G. Hudson, pp. 35-51. Harrassowitz, Verlag.

Brandtner, F.

1996 Zur Geostratigraphischen und Kulturellen Zuordnung de Paläolithstation Grubgraben bei Kammern, NÖ. In *Paleolithic in the Middle Danube Region: Anniversary Volume to Bohuslave Klima*, edited by J. Svoboda, pp. 121-145. Archeologický ústav AV CR, Brno.

Brantingham, P. J., J. W. Olsen, J. A. Rech, and A. I. Krivoshapkin

2000 Raw Material Quality and Prepared Core Technologies in Northeast Asia. *Journal of Archaeological Science* 27:255-271.

Brantingham, P. J., S. L. Kuhn, and K. W. Kerry (Editors)

2004 *The Early Upper Paleolithic beyond Western Europe*. University of California Press, Berkeley.

Braukämper, U.

1979 The Enigma of the Austrian Mind. *Man* 14:560-561.

Braun, D. P.

1983 Pots as Tools. In *Archaeological Hammers and Theories*, edited by A. Keene and J. Moore, pp. 107-134. Academic Press, New York.

1995 Style, Selection, and Historicity, In *Style, Society, and Person: Archaeological and Ethnological Perspectives*, edited C. Carr and J. Neitzel, pp. 123-141. Plenum Press, New York.

Brettell, C. B., and J. F. Hollifield (Editors)

2000 *Migration Theory: Talking Across Disciplines*. Routledge, New York.

Breuil, H.

1906 *Essai Startigraphique des Dépôts de l'Âge du Renne*. Actes du Congrès Préhistorique de France. Première Session, 1905, pp. 74-83.

1907a Les Gisements Présolutréens du Type d'Aurignac. Coup d'Oil Sure le Plus Ancien Âge du Renne. *Comptes Rendus de la XIIIe Congrès Internationale d'Anthropologie et d'Archéologie Préhistorique, Monaco, 1906, Tôme 1*, pp. 323-346.

1907b La Question Aurignacienne. Étude Critique de Stratigraphie Comparée. *Revue Préhistorique* 2:171-219.

1909a La Transition du Mousterien vers l'Aurignacien à l'Abri Audi (Dordogne) et au Moustier. *Revue de l'École d'Anthropologie* 19:320-340.

1909b L'Aurignacien Présolutréen. Epilogue d'une controverse. *La Revue Préhistorique* 4:229-248, 265-286.

1913 Les Subdivisions du Paléolithique Supérieur et leur Signification. In *Congrès International d'Anthropologie et d'Archéologie Préhistorique*, pp. 165-238. Geneva, 1912.

Bright, J., A. Ugan, and L. Hunsaker

2002 The Effect of Handling Time on Subsistence Technology. *World Archaeology* 34(1):164-181.

Broglio, A., and G. Laplace

1966 Études de Typologie Analytique des Complexes Leptolithiques de l'Europe Centrale. I. Les Complexes Aurignacoïdes de la Basse Autriche. *Rivista di Scienze Peristoriche* (Firenze) 21(1):61-121.

Brooks, A. S., J. E. Yellen, L. Nevell, G. Hartman

2005 Projectile Technologies of the African MSA: Implications for Modern Human Origins. In *Transitions before the Transition*, edited by E. Hovers and S. L. Kuhn, pp. 233-256. Plenum/Kluwer, New York.

Brown, B. B .

1983 The Impact of Male Labour Migration on Women in Botswana. *African Affairs* 82:367-88.

Brown, P., T. Sutikna, M. J. Morwood, R. P. Soejono, Jatmiko, E. Wayhu Saptomo, and R.Awe Due

2004 A New Small-Bodied Hominin from the Late Pleistocene of Flores, Indonesia. *Nature* 431:1055-1061.

Brun-Ricalens, F., Le

1993 Reflexions Preliminaires sur le Comportement Litho-Technologique et l'Occupation du Territoire du Pays des Serres a l'Aurignacien: Le Gisement de 'Toulousete' à Beauville (Lot-et-Garonne), une Occupation Mousterienne et Aurignacienne de Plein Air. *Paléo* 5:127-153.

Bulliet, R.

1994 Determinism and Pre-Industrial Technology. In *Does Technology Drive History? The Dilemma of Technological Determinism*, edited by Merritt Roe Smith and Leo Marx, pp. 200-216. MIT Press, Cambridge.

Burmeister, S.

2000 Archaeology and Migration: Approaches to an Archaeological Proof of Migration. *Current Anthropology* 41(4):539-567.

Burton, J.

1984 Quarrying in a Tribal Society. *World Archaeology* 16:234-247.

Butzer, K. W.

1982 *Archaeology as Human Ecology*. Cambridge University Press, Cambridge.

Byers, A. M.

1994 Symboling and the Middle-Upper Palaeolithic Transition: A Theoretical and Methodological Critique. *Current Anthropology* 35(4):369-400.

Callahan, Erret

1979 The Basics of Biface Knapping in the Eastern Fluted Point Tradition: a Manual for Flint Knappers and Lithic Analysis. *Archaeology of Eastern North America* 7:1-180.

Cameron, C.

1995 Migration and the Movement of Southwestern Peoples. *Journal of Anthropological Archaeology* 14:104-124.

Camilli, E.

1989 The Occupational History of Sites and the Interpretation of Prehistoric Technological Systems: An Example from Cedar Mesa, Utah. In *Time, Energy, and Stone Tools*, edited by R. Torrence, pp. 17-26. Cambridge University Press, Cambridge.

Camps, M., and P. Chauhan (Editors)

2009 *Sourcebook of Paleolithic Transitions: Methods, Theories, and Interpretations.* Springer, New York.

Cann, R.

2001 Genetic Clues to Dispersal in Human Populations: Retracing the Past to the Present. *Science* 291:1742-1748.

Cann, R., M. Stoneking, and A. Wilson

1987 Mitochondrial DNA and human evolution. *Nature* 325:31-36.

Carr, C.

1995a Buiding a Unified Theory of Artifact Design. In *Style, Society, and Person: Archaeological and Ethnological Perspectives*, edited by C. Carr and J. Neitzel, pp. 151-170. Plenum Press, New York.

1995b A Unified Middle-Range Theory of Artifact Design. In *Style, Society, and Person: Archaeological and Ethnological Perspectives*, edited by C. Carr and J. Neitzel, 171-258. Plenum Press, New York.

Carr, C., and R. F. Maslowski

1995 Cordage and Fabrics: Relating Form, Technology, and Social Processes. In *Style, Society, and Person: Archaeological and Ethnological Perspectives*, edited by C. Carr and J. Neitzel, pp. 297-344. Plenum Press, New York.

Carr, C., and J. E. Neitzel (Editors)

1995 *Style, Society, and Person: Archaeological and Ethnological Perspectives.* Plenum Press, New York.

Carr, P. J. (Editor)

1994 *The Organization of North American Prehistoric Chipped Stone Tool Technologies.* International Monograpns in Prehistory, Ann Arbor, Michigan.

Carr, P. J., and A. P. Bradbury

2001 Flake Debris Analysis, Levels of Production, and the Organization of Technology. In *Lithic Debitage: Context, Form, Meaning*, edited by William Andrefsky, Jr., pp. 126-146. University of Utah Press, Salt Lake City.

Castles, S., and M. Miller

1993 *The Age of Migration: International Population Movements in the Modern World.* Macmillan Press, Basingstoke.

Cavalli-Sforza, L. L., and M. W. Feldman.

1981 *Cultural Transmission and Evolution: A Quantitative Approach.* Princeton University Press, Princeton.

Cavalli-Sforza, L. L., P. Menozzi, and A. Piazza

1993 Demic Expansions and Human Evolution. *Science* 259:639-646.

Cervinka, J. L.

1927 Pravek zemi Ceských. *Pravek* 1-2:33-80.

Chabai, V., and V. Sitlivyi

1994 The Blade Component in the Middle Palaeolithic of Ukraine: Origin and Evolution. In *Les Industries Laminaires au Paléolithique Moyen*, edited by S. Révillion and A. Tuffreau, pp. 161-178. Dossier de Documentation Archéologique No. 18. CRA. Éditions du Centre National de la Recherche Scientifique, Paris.

Chabai, V., K. Monigal, and A. E. Marks (Editors)

2004 *The Middle Paleolithic and Early Upper Paleolithic of Eastern Crimea.* ERAUL, Liège.

Chabai, V., J. Richter, and T. Uthmeier (Editors)

2006 *Kabazi II: The 70 000 Years Since the Last Interglacial.* Shlyakh, Simferopol, Cologne.

Champagne, F., and R. Espitalié

1981 *Le Piage, Site Préhistorique du Lot.* Mémoires de la Société Préhistorique Française 15, Paris.

Chang, K. C.

1978 Some Theoretical Issues in the Archaeological Study of Historical Reality. In *Archaeological Essays in Honor of Irving B. Rouse,* edited by R. Dunnell and E. Hall, pp. 13-26. Mouton Publishers, The Hague.

Chapman, J., and H. Hamerow

1997 On the Move Again—Migrations and Invasions in Archaeological Explanation. In *Migrations and Invasions in Archaeological Explanation,* edited by J. Chapman and Helena Hamerow, pp. 1-10. BAR International Series 664, Oxford.

Chase, P. G.

1987 Spécialisation de la Chasse et Transition vers le Paléolithique Supérieur. *L'Anthropologie* 91:175-188.

1989 How Different was Middle Palaeolithic Subsistence? A Zooarchaeological Perspective on the Middle to Upper Palaeolithic Transition. In *The Human Revolution: Behavioural and Biological Perspectives in the Origins of Modern Humans,* edited by P. Mellars and C. Stringer, pp. 321-337. Edinburgh University Press, Edinburgh.

1991 Symbols and Paleolithic Artifacts: Style, Standardization, and the Imposition of Arbitrary Form. *Journal of Anthropological Archaeology* 10:193-214.

Chase, P. G., and H. L. Dibble

1987 Middle Paleolithic Symbolism: A Review of Current Evidence and Interpretations. *Journal of Anthropological Archaeology* 6:263-293.

Cheney, D. L., and R. M. Seyfarth

1987 The influence of intergroup competition on the survival and reproduction of female vervet monkeys. *Behavioral Ecology and Sociobiology* 21:375-386.

Chernysh, A. P.

1961 *Palaeolitiginia Stoanka Molodava V.* An. Ukr. SSR, Kiev.

1982 Mnogoslojnaya paleoliticheskaya stoyanka Molodova I. In *Molodova I. Unikal'noe must'erskoe poselenie na Srednem Dnestre,* edited by G. I. Goretskij and I. K. Ivanova, pp. 6-102. Nauka, Moscow.

1987 Etalonnaya Mnogoslojnaya Stoyanka Molodova V. Arkheologia. In *Mnogoslojnaya Paleoliticheskaya Stoyanka Molodova V. Lyudi Kammennogo Veka: Okruzhayushchaya Sreda,* edited by I. K. Ivanova and S. M. Tsejtlin, pp. 7-93. Nauka, Moscow.

Childe, V. G.

1929 *The Danube in Prehistory.* Oxford University Press, Oxford.

1951 *Social Evolution.* Watts and Company, London.

1956 *Piecing Together the Past: The Interpretation of Archaeological Data.* Routledge and Kegan Paul, London.

Childs, S. T.

1991 Style, Technology, and Iron Smelting Furnaces in Bantu-Speaking Africa. *Journal of Anthropological Archaeology* 10:332-359.

Churchill, S. E., and F. Smith

2000 Makers of the Early Aurignacian of Europe. *Yearbook of Physical Anthropology* 43:61-115.

Ciolek-Torrello, R. S.

1997 Prehistoric Settlement and Demography in the Lower Verde River Region. In *Vanishing River: Landscapes and Lives of the Lower Verde River Valley,* edited by S. M. Whittlesey, R. Ciolek-Torrello, and J. Altschul, pp. 531-595. SRI Press, Tucson.

Clark, G. A.

1993 Paradigms in Science and Archaeology. *Journal of Archaeological Research* 1:203-234.

1994 Migration as an Explanatory Concept in Paleolithic Archaeology. *Journal of Archaeological Method and Theory* 1(4):305-343.

2000 Deconstructing the North Atlantic Connection. *Current Research in the Pleistocene* 17:11-14.

2005 Modern Approaches to Paleolithic Archaeology in Europe - A Sampler of Research Traditions. *American Antiquity* 70(2):376-384.

Clark, G. A., and J. M. Lindly

1989a Modern Human Origins in the Levant and Western Asia: The Fossil and Archaeological Evidence. *American Anthropologist* 91:962-985.

1989b The Case for Continuity: Observations on the Biocultural Transition in Europe and Western Asia. In The *Human Revolution: Behavioural and Biological Perspectives in the Origins of Modern Humans*, edited by P. Mellars and C. Stringer, pp. 626-676. University of Edinburgh Press, Edinburgh.

1991 On Paradigmatic Biases and Paleolithic Research Traditions. *Current Anthropology* 32(5):577-587.

Clark, G. A., and C. M. Willermet

1997 *Conceptual Issues in Modern Human Origins Research*. Aldine de Gruyter, New York.

Clark G. A., and J. Riel-Salvatore

2003 What's in a Name? An Empirical Review of the Compositional Integrity of the Eurasian Aurignacian. Paper presented at the 2003 annual meetings of the Paleoanthropology Society, Tempe, AZ.

Clark, J. Desmond, Y. Beyene, G. WoldeGabriel, W. K. Hart, P. R. Renne, H. Gilbert, A. Defleur, G. Suwa, S. Katoh, K. R. Ludwig, J.-R. Boisserie, B. Asfawkk, and T. D. White

2003 Stratigraphic, Chronological and Behavioural Contexts of Pleistocene *Homo sapiens* from Middle Awash, Ethiopia. *Nature* 423(12):74-752.

Clark, J. E.

1987 Politics, Prismatic Blades, and Mesoamerican Civilizations. In *The Organization of Core Technology*, edited by J. K. Johnson and C. A. Morrow, pp. 259-284. Westview Press, Boulder, CO.

1991 Flintknapping and Debitage Disposal among the Lacandon Maya of Chiapas, Mexico. In *The Ethnoarchaeology of Refuse Disposal*, edited by E. L. Staski and L. D. Sutro, pp. 63-78. Arizona State University, Tempe.

Clark, J. G. D.

1968 *World Prehistory: A New Outline. Second Edition.* Cambridge University Press, Cambridge.

1971 *Excavations at Star Carr: An Early Mesolithic Site at Seamer Near Scarborough, Yorkshire.* Cambridge University Press, Cambridge.

Clark, J. J.

2001 *Tracking Prehistoric Migrations: Pueblo Settlers among Tonto Basin Hohokam.* Anthropological Papers of the University of Arizona No. 65. University of Arizona Press, Tucson.

Clarkson, C.

2002 An Index of Invasiveness for the Measurement of Unifacial and Bifacial Retouch: A Theoretical, Experimental and Archaeological Verification. *Journal of Archaeological Science* 29:65-75.

2010 Regional Diversity Within the Core Technology of the Howiesons Poort Techno-Complex. In *New Perspectives on Old Stones: Analytical Approaches to Paleolithic Technologies*, edited by S. J. Lycett and P. R. Chauhan, pp. 43-59. Springer Science+Business Media, New York.

Clarkson, C., and P. Hiscock

2008 Tapping into the Past: Exploring the Extent of Palaeolithic Retouching through Experimentation. *Lithic Technology* 33:5-16.

2011 Estimating Original Flake Mass from 3D Scans of Platform Area. *Journal of Archaeological Science* 38:1062-1068.

Clarkson, C., L. Vinicius, and M. M. Lahr

2006 Quantifying Flake Scar Patterning on Cores Using 3D Recording Techniques. *Journal of Archaeological Science* 33:132-142.

Cleland, C. E.

1976 The Focal-Diffuse Model: An Evolutionary Perspective on the Prehistoric Cultural Adaptations of the Eastern United States. *Mid-Continental Journal of Archaeology* 1(1):59-76.

Close, A. E.

1989 Identifying Style in Stone Artefacts: a Case Study from the Nile Valley. In *Alternative Approaches to Lithic Analysis*, edited by D. Henry and G. H. Odell, pp. 3-26. Archeological Papers of the American Anthropological Association Number 1, Washington D.C.

2000 Reconstructing Movement in Prehistory. *Journal of Archaeological Method and Theory* 7(1):49-77.

Cohen, R.

1978 Ethnicity: Problem and Focus in Anthropology. *Annual Review of Anthropolology* 7:379-403

Collard, M., and S. Shennan

2000 Processes of Culture Change in Prehistory: A Case Study from the European Neolithic. In *Molecular Genetics and Early Europe: Working Papers in Population Prehistory*, edited by C. Renfrew and P. Forster, pp. 89-97. McDonald Institute for Archaeological Research, Cambridge.

Collard, M., and J. Tehrani

2005 Phylogenesis versus Ethnogenesis in Turkmen Cultural Evolution. In *The Evolution of Cultural Diversity: A Phylogenetic Approach*, edited by R. Mace, C. Holden, and S. Shennan, pp. 109-131. UCL Press, London.

Collard, M., S. J. Shennan, and J. Tehrani

2006 Branching, Blending, and the Evolution of Cultural Similarities and Differences among Human Populations. *Evolution and Human Behavior* 27:169-184.

Collins, D.

1970 Stone Artefact Analysis and the Recognition of Culture Traditions. *World Archaeology* 2:17-27.

Collins, M.

1975 Lithic Technology as a Means of Processual Inference. In *Lithic Technology: Making and Using Stone Tools*, edited by E. H. Swanson, pp. 15-34. Mouton, The Hague.

Comaroff, J. L.

1987 Of Totemism and Ethnicity: Consciousness, Practice and Signs of Inequality. *Ethnos* 52:301-323.

Comaroff, J., and J. Comaroff

1992 *Ethnography and the Historical Imagination*. Westview Press, Boulder, CO.

Conard, Nicholas

1990 Laminar Lithic Assemblages from the Last Interglacial Complex in Northwestern Europe. *Journal of Anthropological Research* 46(3):243-262.

2002 The Timing of Cultural Innovations and the Dispersal of Modern Humans in Europe. *Terra Nostra* 6:82-94.

Conard, N. J., and D. S. Adler

1997 Lithic Reduction and Hominid Behavior in the Middle Paleolithic of the Rhineland. *Journal of Anthropological Research* 53:147-175.

Conard, N. J., and M. Bolus

2003 Radiocarbon Dating the Appearance of Modern Humans and Timing of Cultural Innovations in Europe: New Results and New Challenges. *Journal of Human Evolution* 44:331-371.

2008 Radiocarbon Dating the Late Middle Paleolithic and the Aurignacian of the Swabian Jura. *Journal of Human Evolution* 55:886-897.

Conard, N. J., M. Bolus, and T. Schneidermeier

1999 Current Middle and Early Upper Paleolithic Research in southwestern Germany. In *Krapina 1899-1999, International Conference, The Krapina Neanderthals and Human Evolution in Central Europe; Zabreb-Krapina, August 23-26, 1999*, pp. 17-18. Book of Abstracts.

Conard, N. J., M. Bolus, P. Goldberg, S. Munzel

2006 The Last Neanderthals and First Modern Humans in the Swabian Jura. In *When Neanderthals and Modern Human Met. Tübingen Publications in Prehistory*, edited by N. Conard, pp. 305-342. Kerns Verlag, Tübingen.

Cook, N. D.

1998 *Born to Die: Disease and New World Conquest, 1492-1650*. Cambridge University Press, Cambridge.

Coolidge, F. L., and T. Wynn

2005 Working Memory, its Executive Functions, and the Emergence of Modern Thinking. *Cambridge Archaeological Journal* 15(1):5-26.

Copeland, L.

1975 The Middle and Upper Paleolithic of Lebanon and Syria in the Light of Rcent Research. In *Problems in Prehistory: North Africa and the Levant*, edited by F. Wendorf and A. Marks, pp. 317-350. Southern Methodist University Press, Dallas.

1987 Preface to Volume II. In *Ksar Akil, Lebanon: A Technological and Typological Analysis of the Later Paleolithic Levels of Ksar Akil, Volume II: Levels XIII-VI*, edited by C. Bergman, pp. vi-x. British Archaeological Reports International Series 329.

Cosmides, L., and J. Tooby

1992 Cognitive Adaptations for Social Exchange. In *The Adapted Mind: Evolutionary Psychology and the Generation of Culture*, edited by J. Barkow, L. Cosmides, and J. Tooby, pp. 163-228. Oxford University Press, New York.

Cotterell, B. and J. Kamminga

1987 The Formation of Flakes. *American Antiquity* 52(4):675-708.

Cotterell, B., J. Kamminga, and F. D. Dickson

1985 The Essential Mechanics of Concoidal Flaking. *International Journal of Fracture* 20:205-221.

Courty, M. A., P. Goldberg, and R. Macphail

1989 *Soils and Micromorphology in Archaeology*. Cambridge Manuals in Archaeology. Cambridge University Press, Cambridge.

Cowan, R. S.

1985 How the Refrigerator Got its Hum. In *The Social Shaping of Technology: How the Refrigerator Got its Hum*, edited by D. MacKenzie and J. Wajcman, pp. 202-218. Open University Press, Milton Keynes and Philadelphia.

Crabtree, D.

1967 Notes on Experiments in Flintknapping: 4-Tools Used for Making Flaked Stone Artefacts. *Tebiwa* 10:60-73

1972 *An Introduction to Flintworking*. Occasional Papers of the Idaho State Museum Number 28.

Crew, H. L.

1975 An Examination of the Variability of the Levalloisian Method: Its Implications for the Internal and External Relationships of the Levantine Mousterian. University Microfilms International, Ph.D. Dissertation, University of California at Davis.

Croes, D. R.

1989 Prehistoric Ethnicity on the Northwest Coast of North America: An Evaluation of Style in Basketry and Lithics. *Journal of Anthropological Archaeology* 8:101-130.

Czudek, T., L. Smolíková, and J. Svoboda

1991 Profil IIIa na Stránské skále v Brně. *Anthropozoikum, N. S.* 20:203–223.

Daly, M., and M. Wilson

1983 *Sex, Evolution, and Behavior*. Second Edition. Wadsworth Publishing Co., Belmont.

Damblon, F., P. Haesaerts, and J. van der Plicht

1996 New Datings and Considerations on the Chronology of Upper Palaeolithic Sites in the Great Eurasiatic Plain. *Préhistoire Européenne* 9:177–231.

Daniel, G.

1975 *A Hundred and Fifty Years of Archaeology*. Duckworth, London.

Darwent, J., and M. J. O'Brien

2006 Using Cladistics to Construct Lineages of Projectile Points from Northeastern Missouri. In *Mapping Our Ancestors: Phylogenetic Approaches in Anthropology and Prehistory*, edited by C. Lipo, M. J. O'Brien, M. Collard, and S. J. Shennan, pp. 185–208. Aldine Transaction Publishers, New Brunswick and London.

Darwin, C.

1964 *On the Origin of Species. Facsimile of the First Edition, 1859*, edited by E. Mayr. Harvard University Press, Cambridge.

Davidson, I., and W. Noble

1993 Tools and Language in Human Evolution. In *Tools, Language and Cognition in Human Evolution*, edited by K. Gibson and T. Ingold, pp. 363–388. Cambridge University Press, Cambridge.

Davidson, A., and N. Goring-Morris

2003 Sealed in Stone: the Upper Palaeolithic Early Ahmarian Knapping Method in the Light of Refitting Studies at Nahal Nizzana XIII, Western Negev, Israel. *Journal of the Israel Prehistoric Society* 33:75–205.

Davis, D.

1983 Investigating the Diffusion of Stylistic Innovations. *Advances in Archaeological Method and Theory* 6:53–89.

Davis, Z. J., and J. J. Shea

1998 Quantifying Lithic Curation: An Experimental Test of Dibble and Pelcin's Original Flake-Tool Mass Predictor. *Journal of Archaeological Science* 25(7):603–610.

Dawkins, R.

1976 *The Selfish Gene*. Oxford University Press, Oxford.

Dean, J.

1978 Independent Dating in Archaeological Analysis. *Advances in Archaeological Method and Theory* 1:223–255.

DeBoer, W. R.

1990 Interaction, Imitation, and Communication as Expressed in Style: The Ucayali Experience. In *The Uses of Style in Archaeology*, edited by Margaret Conkey and Christine Hastorf, pp. 82–104. Cambridge University Press, Cambridge.

Deetz, J.

1965 *The Dynamics of Stylistic Change in Arikara Ceramics*. Illinois Studies in Anthropology No. 4. University of Illinois Press, Urbana.

1977 *In Small Things Forgotten: The Archaeology of Early American Life*. Anchor, Garden City, New Jersey.

Deetz, J., and E. Dethlefsen

1965 The Doppler Effect and Archaeology: A Consideration of the Spatial Aspects of Seriation. *Southwestern Journal of Anthropology* 21(3):196–206.

Demars, P.-Y., and J. J. Hublin

1989 LaTransition Néandertaliens/Hommes de Type Modern en Europe Occidentale: Aspects Paléontologiques et Culturels. In *L'Homme de Néandertal, Vol. 7, L'Extinction*, edited by M. Otte, pp. Études et Recherches Archéologiques de l'Université de Liège 34, Liège.

Demidenko, Y., and V. Usik

1991 On the Levallois Technique in the Upper Palaeolithic. In *Aurignacien en Europe et en Proche Orient. Résumés des Communications*, edited by L. Bánes and J.K. Kozłowski, pp. 239-242. Actes du XIIe Congrés International des Sciences Préhistoriques et Protohistoriques, Bratislava.

1993a The Problem of Changes in Levallois Technique during the Technological Transition from the Middle to Upper Palaeolithic. *Paléorient* 19(2):5-15.

1993b On the Lame à Crête Technique in the Palaeolithic. *Préhistoire Européene* 4:33-48.

1993c Leaf Points of the Upper Palaeolithic Industry from the 2nd Complex of Korolevo II and Certain Morphological Problems in Description and Interpretation of the Category of Palaeolithic Tools. *Préhistoire Européene* 4:49-62.

Demidenko, Y.

2000 The European Early Aurignacian of Krems-Dufour Type Industries: A View from Eastern Europe. *Préhistoire Européenne* 16-17:147-162.

Dennell, R.

1985 The Hunter-Gatherer/Agricultural Frontier in Prehistoric Temperate Europe. In *The Archaeology of Frontiers and Boundaries*, edited by S. W. Green and S. M. Perlman, pp. 113-139. Academic Press, New York.

Dennett, D. C.

1996 *Darwin's Dangerous Idea*. W. W. Norton, New York.

d'Errico, F.

2003 The Invisible Frontier. A Multiple Species Model for the Origin of Behavioral Modernity. *Evolutionary Anthropology* 12:188-202.

d'Errico, F., J. Zilhão, M. Julien, D. Baffier, and J. Pelegrin

1998 Neanderthal Acculturation in Western Europe? A Critical Review of the Evidence and Its Interpretation. *Current Anthropology* 39 (Supplement):S1-S44.

d'Errico, F., C. Henshilwood, G. Lawson, M. Vanhaeren, A.-M. Tillier, M. Soressi, F. Bresson , B. Maureille, A. Nowell, J. Lakarra, L. Backwell, and M. Julien

2003 Archaeological Evidence for the Emergence of Language, Symbolism, and Music - An Alternative Multidisciplinary Perspective. *Journal of World Prehistory* 17(1):1-70.

d'Errico, F., J.-G. Bordes, A. Lenoble, and J. Zilhão

2006 No Neandertal/Early Modern Human Interstratification at the Châtelperronian Type-Site. Paper presented at the Paleoanthropology Society Meetings, April, 2006.

Dethlefsen, E., and J. Deetz

1966 Death's Heads, Cherubs, and Willow Trees: Experimental Archaeology in Colonial Cemeteries. *American Antiquity* 31:502-510.

Diamond, J.

1997 *Guns, Germs, and Steel: The Fates of Human Societies*. First Edition. W. W. Norton, New York.

Dibble, H. L.

1987a The Interpretation of Middle Paleolithic Scraper Morphology. *American Antiquity* 52:109-117.

1987b Reduction Sequences in the Manufacture of Mousterian Implements of France. In *The Pleistocene Old World: Regional Perspectives*, edited by O. Soffer, pp. 33-45. Plenum Press, New York.

1988 Typological Aspects of Reduction and Intensity of Utilization of Lithic Resources in the French Mousterian. In *Upper Pleistocene Prehistory of Western Eurasia*, edited by H. L. Dibble and A. Montet-White, pp. 181-187. University Museum Monograph 54, University Museum Symposium Series 1. University Museum, University of Pennsylvania, Philadelphia.

1995a Biache Saint-Vaast, Level IIA: A Comparison of Analytical Approaches. In *The Definition and Interpretation of Levallois Variability*, edited by H. L. Dibble and O. Bar-Yosef, pp. 93-116. Prehistory Press, Madison.

1995b Middle Paleolithic Scraper Reduction: Background, Clarification, and Review of the Evidence to Date. *Journal of Archaeological Method and Theory* 2(4):299-368.

1997 Platform Variability and Flake Morphology: A Comparison of Experimental and Archaeological Data and Implications for the Interpreting Prehistoric Lithic Technological Strategies. *Lithic Technology* 22(2):150-170.

Dibble, H. L., and M. Bernard

1980 A Comparison of Basic Edge Angle Measurement Techniques. *American Antiquity* 45:857-865.

Dibble, H. L., and P. G. Chase

1990 Current Anthropology Reply to Lindly and Clark's "Symbolism and Modern Human Origins". *Current Anthropology* 31:233-261.

Dibble, H. L., and A. Pelcin

1995 The Effect of Hammer Mass and Velocity on Flake Mass. *Journal of Archaeological Science* 22:429-439.

Dibble, H. L., and Z. Rezek

2009 Introducing a New Experimental Design for Controlled Studies of Flake Formation: Results for Exterior Platform Angle, Platform Depth, Angle of Blow, Velocity, and Force. *Journal of Archaeological Science* 36:1945-1954.

Dibble, H. L., and N. Rolland

1992 On Assemblage Variability in the Middle Paleolithic of Western Europe: History, Perspectives, and a New Synthesis. In *The Middle Paleolithic: Adaptation, Behavior, and Variability*, edited by H. Dibble and P. Mellars, pp. 1-28. University Museum, University of Pennsylvania, Philadelphia.

Dibble, H. L. and J. Whittaker

1981 New Experimental Evidence on the Relation between Percussion Flaking Flake Variation. *Journal of Archaeological Science* 6:283-296.

Dibble, H. L., S. P. McPherron, and B. J. Roth

2003 *Virtual Dig: A Simulated Archaeological Excavation of a Middle Paleolithic Site in France.* Second Edition. McGraw-Hill/Mayfield, New York.

Dibble, H. L., U. Schurmans, R. Iovita, and M. McLaughlin

2005 The Measurement and Interpretation of Cortex in Lithic Assemblages. *American Antiquity* 70(3):545-560.

Dietler, M., and I. Herbich

1989 Tich Matek: The Technology of Luo Pottery Production and the Definition of Ceramic Style. *World Archaeology* 21:148-164.

1998 Habitus, Techniques, Style: An Integrated Approach to the Social Understanding of Material Culture and Boundaries. In *The Archaeology of Social Boundaries*, edited by M. T. Stark, pp. 232-263. Smithsonian Institution Press, Washington, D.C.

Dobres, M.-A., and C. R. Hoffman

1994 Social Agency and the Dynamics of Prehistoric Technology. *Journal of Archaeological Method and Theory* 1(3):211-258.

Dobyns, H. F.

1993 Disease Transfer at Contact. *Annual Review of Anthropology* 22:273–291.

Donald, M.

1991 *Origins of the Human Mind: Three Stages in the Evolution of Culture and Cognition*. Harvard University Press, Cambridge.

1998 Hominid Enculturation and Cognitive Evolution. In *Cognition and Material Culture: the Archaeology of Symbolic Storage*, edited by Colin Renfrew and Chris Scarre, pp. 7–18. McDonald Institute Monographs. McDonald Institute for Archaeological Research, Cambridge.

Drobniewicz, B., B. Ginter, B. Kazior, and J. K. Kozłowski

2000 'Transitional' Industry from Layer VI, Trench TD-II. In *Temnata Cave: Excavations in Karlukovo Karst Area, Bulgaria 2.1*, edited by B. Ginter, J. Kozłowski, J-L. Guadelli, and H. Laville, pp. 243–316. Jagellonian University, Krakow.

Drout, M. D. C.

2006 *How Tradition Works: A Meme-Based Cultural Poetics of the Anglo-Saxon Tenth Century*. Arizona Center for Medieval and Renaissance Studies, Tempe, AZ.

Duff, A. I. L.

1998 The Process of Migration in the Late Prehistoric Southwest. In *Migration and Reorganization: The Pueblo IV Period in the American Southwest*, edited by K. A. Spielmann, pp. 31–52. Anthropological Research Papers No. 51. Arizona State University, Tempe.

Duke, P. G., J. Ebert, G. Langeman, and A. P. Buchner (Editors)

1978 *Diffusion and Migration: Their Roles in Cultural Development. Proceedings of the Tenth Annual Chacmool Conference*. Archaeological Association of the University of Calgary, Calgary.

Dunbar, R. I. M.

2003 The Social Brain: Mind, Language, and Society in Evolutionary Prospective. *Annual Review of Anthropology* 32:163–81.

Dunnell, R. C.

1978 Style and Function: A Fundamental Dichotomy. *American Antiquity* 43:192–202.

1980 Evolutionary Theory and Archaeology. In *Advances in Archaeological Method and Theory, Vol. 3*, edited by M. B. Schiffer, pp. 35–99. Academic Press, New York.

1989 Aspects of the Application of Evolutionary Theory in Archaeology. In *Archaeological Thought in America*, edited by C. C. Lamberg-Karlovsky, pp. 35–49. Cambridge University Press, Cambridge.

1990 The Role of the Southeast in American Arcaheology. *Southeastern Archaeology* 9(1):11–22.

Durham, W. H

1990 Advances in Evolutionary Culture Theory. *Annual Review of Anthropology* 19:187–210.

1991 *Coevolution: Genes, Culture, and Human Diversity*. Stanford University Press, Stanford.

1992 Applications of Evolutionary Culture Theory. *Annual Review of Anthropology* 21:331–355.

Dyson-Hudson, R., and E. A. Smith

1978 Human Territoriality: An Ecological Reassessment. *American Anthropologist* 80:21–41.

Dzambazov, N.

1959 Razkopki v Pesterata Samuilitsa II. *Arkheologiya* 1:47–53.

1967 Les Pointes Bifaciales dans les Grottes Samouilitsa I et II et les Autres Atations du Paléolithique Récent en Bulgarie. *Quartär* 18:195–199.

Eerkens, J. W.

1999 Common Pool Resources, Buffer Zones, and Jointly Owned Territories: Hunter-Gatherer Land and Resource Tenure in Fort Irwin, Southeastern California. *Human Ecology* 27(2):297–318.

Eerkens, J. W., and R. L. Bettinger

2001 Techniques for Assessing Standardization in Artifact Assemblages: Can We Scale Material Variability? *American Antiquity* 66:493–504.

Eerkens, J. W., and C. P. Lipo

2005 Cultural Transmission, Copying Errors, and the Generation of Variation in Material Culture in the Archaeological Record. *Journal of Anthropological Archaeology* 24:316–334.

2007 Cultural Transmission Theory and the Archaeological Record: Context to Understanding Variation and Temporal Changes in Material Culture. *Journal of Archaeological Research* 15:239–274.

Elefanti, P., E. Panagopoulou, and P. Karkanas

2008 The Transition from the Middle to the Upper Paleolithic n the Southern Balkans: The Evidence from the Lakonis Cave, Greece. *Eurasian Prehistory* 5(2):85–95.

Emberling, G.

1997 Ethnicity in Complex Societies: Archaeological Perspectives. *Journal of Archaeological Research* 5:295–345.

Eren, M. I., M. Dominguez-Rodrigo, S. L. Kuhn, D. S. Adler, I. Le, O. Bar-Yosef

2005 Defining and Measuring Reduction in Unifacial Stone Tools. *Journal of Archaeological Science* 32:1190–1201.

Eren, M. I., and M. E. Prendergast

2008 Comparing and Synthesizing Unifacial Stone Tool Reduction Indices. In *Lithic Technology: Measures of Production, Use, and Curation*, edited by W. Andrefsky, Jr., pp. 49–85. Cambridge University Press, Cambridge.

Eren, M., and C. G. Sampson

2009 Kuhn's Geometric Index of Unifacial Stone Tool Reduction (GIUR): Does it Measure Missing Flake Mass? *Journal of Archaeological Science* 36:1243–1247

Fagan, B. M.

1998 *Clash of Cultures. Second Edition.* Altamira Press, Walnut Creek.

Fallows, J.

1985 The American Army and the M-16 Rifle. In *The Social Shaping of Technology: How the Refigerator Got its Hum*, edited by D. MacKenzie and J. Wajcman, pp. 239–251. Open University Press, Milton Keynes and Philadelphia.

Farizy, C.

1988 Arcy-sur-Cure: Un Site Clef pour la Transition du Paléolithique Moyen au Paléolithique Supérieur. In *De Néandertal à Cro Magnon*, edited by J-B. Roy and A-S. Leclerc pp. 77–84. Musée de Préhistoire d'Ile-de-France, Nemours.

1990 The Transition from Middle to Upper Palaeolithic at Arcy-sur-Cure (Yonne, France): Technological, Economic, and Social Aspects. In *The Emergence of Modern Humans: An Archaeological Perspective*, edited by P. Mellars, pp. 303–326. Edinburgh University Press, Edinburgh.

Farizy, C., and F. David

1992 Subsistence and Behavioral Patterns of Some Middle Paleolithic Local Groups. In *The Middle Paleolithic: Adaptation, Behavior, and Variability*, edited by H. L. Dibble and A. Montet-White, pp. 87–96. University Museum Monograph 72. University Museum of the University of Pennsylvania, Philadelphia.

Farrand, W. R.

1994 Confrontation of Geological Stratigraphy and Radiometric Dates from Upper Pleistocene Sites in the Levant. In *Late Quaternary Chronology and Paleoclimates of the Eastern Mediterranean*, edited by O. Bar-Yosef and R. S. Kra, pp. 33–54. Radiocarbon, Tucson.

Féblot-Augustins, J.

1993 Mobility Strategies in the Late Middle Palaeolithic of Central Europe and Western Europe: Elements of Stability and Variability. *Journal of Anthropological Archaeology* 12(3):211–65.

1994 La Circulation des Matières Premières Lithiques au Paléolithique. Synthèse des Données, Perspectives Comportementales. Unpublished Ph.D. Dissertation, University of Paris X.

1997 *La Circulation des Matières Premières au Paléolithique*. ERAUL 75, Liège.

1999 Raw Material Transport Patterns and Settlement Systems in the European Lower and Middle Palaeolithic: Continuity, Change and Variability. In *The Middle Paleolithic Occupation of Europe*, edited by W. Roebroeks and C. Gamble, pp. 193–214. University of Leiden Press, Leiden.

Ferguson, Jeffrey R.

2008 The When, Where, and How of Novices in Craft Production. *Journal of Archaeological Method and Theory* 15:51–67.

Fix, A. G.

1999 *Migration and Colonization in Human Microevolution*. Cambridge University Press, Cambridge.

Flenniken, J. J.

1985 Stone Tool Reduction Technologies as Cultural Markers. In *Stone Tool Analysis: Essays in Honor of Don E. Crabtree*, edited by M. G. Plew, J. C. Woods, and M. G. Pavesic, pp. 265–276. University of New Mexico Press, Albuquerque.

Foley, J. M.

1988 *The Theory of Oral Composition: History and Methodology*. Indiana University Press, Bloomington, IN.

Foley, R.

1984 Putting People into Perspective: An Introduction to Community Evolution and Ecology. In *Hominid Evolution and Community Ecology: Prehistoric Human Adaptation in Biological Perspective*, edited by R. Foley, pp. 1–24. Academic Press, London.

1985 Optimality Theory in Anthropology. *Man* 20:222–242.

1987 *Another Unique Species: Patterns in Human Evolutionary Ecology*. Longman Scientific and Technical, Essex.

Foley, R., and C. Gamble

2009 The Ecology of Social Transitions in Human Evolution. *Philosophical Transactions of the Royal Society* B 364:3267–3279.

Foley, R., and M. Lahr

1997 Mode 3 Technologies and the evolution of Modern Humans. *Cambridge Archaeological Journal* 7:3–36.

2003 On Stony Ground: Lithic Technology, Human Evolution, and the Emergence of Culture. *Evolutionary Anthropology* 12:109–122.

2005 The Origins of Modern Humans: Insights into Speciation. A paper presented at the conference, Rethinking the Human Revolution: New Behavioural and Biological Perspectives on the Origins and Dispersal of Modern Humans, organized by P. Mellars, K. Boyle, O. Bar Yosef, and C. Stringer, at the MacDonald Institute for Archaeological Research, Cambridge.

Foltyn, E., and J. K. Kozłowski

2003 The Lower Level of the Site of Dzierżysław I, Opole Voivodship (Silesia, Poland) and the Problem of the Bohunician. *Eurasian Prehistory* 1(2):79-116.

Ford, J. A.

1952 Measurements of Some Prehistoric Design Developments in the Southeastern States. *American Museum of National History, Anthropological Papers* 44(3):313-384.

Forster, P.

2004 Ice Ages and the Mitochondrial DNA Chronology of Human Dispersals: A Review. Philosophical Transactions of the Royal Society of London B 359:255-264.

Friedrich, M. H.

1970 Design Structure and Social Interaction. Archaeological Implications of an Ethnographic Analysis. *American Antiquity* 35(3):332-343.

Frison, G.

1968 A Functional Analysis of Certain Chipped Stone Tools. *American Antiquity* 33:149-155.

Fritz, J. M., and F. T. Plog

1970 The Nature of Archaeological Explanation. *American Antiquity* 35(4):405-412.

Gallagher, J.

1977 Contemporary Stone Tools in Ethiopia: Implications for Archaeology. *Journal of Field Archaeology* 4:407-414.

Gamble, C.

1986 *The Palaeolithic Settlement of Europe.* Cambridge University Press, Cambridge.

1991 The Social context for European Palaeolithic Art. *Proceedings of the Prehistoric Society* 57(1):3-15.

1992 Ancestors and Agendas. In *Archaeological Theory–Who Sets the Agenda?*, edited by N. Yoffee and A. Sherratt, pp. 39-52. Cambridge University Press, Cambridge.

1993 People on the Move: Interpretations of Regional Variation in Palaeolithic Europe. In *Cultural Transformations and Interactions in Eastern Europe*, edited by J. Chapman and P. Dolukhanov, pp. 37-55. Aldershot, Avebury.

1999 *The Palaeolithic Societies of Europe.* Cambridge University Press, Cambridge.

Gamble, C., and M. Porr (Editors)

2005 *The Hominid Individual in Context: Archaeological Investigations of Lower and Middle Palaeolithic Landscapes, Locales and Artefacts.* Routledge, London.

Garrod, D. A. and D. Bates

1937 *The Stone Age of Mount Carmel. Excavations at the Wadi el Mughara.* Clarendon Press, Oxford.

Garrod, D. A., B. Howe, and J. H. Gaul

1939 Excavations in the Cave of Bacho Kiro, North-East Bulgaria. *American School of Prehistoric Research Bulletin* 15:46-126.

Gauthreaux, S. A., Jr.

1980 The Influences of Long-Term and Short-Term Climatic Changes on the Dispersal and Migration of Organisms. In *Animal Migration, Orientation, and Navigation*, edited by S. A. Gauthreaux, Jr., pp. 103-174. Academic Press, New York.

Gavin, D. G., W. W. Oswald, E. R. Wahl, and J. W. Williams

2003 A Statistical Approach to Evaluating Distance Metrics and Analog Assignments for Pollen Records. *Quaternary Research* 60(2003):356-367.

Geneste, J.-M.

1985 Analyse Lithique d'Industries Moustériennes du Périgord: Une Approche Technologique du Comportement des Groupes Humains au Paléolithique Moyen. Thèse a l'Université de Bordeaux I, Bordeaux.

1988 Systèmes d'Approvisionnement en Matières Premières au Paléolithique Moyen et au Paléolithique Supérieur en Aquitaine. In *L'Homme de Neandertal*, Vol 8, La Mutation, J. Kozlowski, pp. 61-70. Études et Recherches Archéologiques de l'Université de Liège 35, Liège.

Geneste, J.-M., and H. Plisson

1990 Technologie Fonctionelle des Pointes a Cran Solutreennes: L'apport des Nouvelles Donnees de la Grotte de Combe Sauniere (Dordogne). In *Feuilles de Pierre: Les Industries a pointes foliacees de Paleolithique Superieur Europeen*, edited by J. Kozlowski, pp. 293-320. Etudes et Recherches Archeologiques de l'Universite de Liege, No. 42. Universite de Liege, Liege.

Gero, J.

1989 Assessing Social Information in Material Objects: How Well Do Lithics Measure Up? In *Time, Energy, and Stone Tools*, edited by R. Torrence, pp. 92-105. Cambridge University Press, Cambridge.

Gibson, J. J.

1979 *The Ecological Approach to Visual Perception*. Erlbaum Associates, Hillsdale.

Giddens, A.

1979 *Central Problems in Social Theory*. Cambridge University Press, Cambridge.

1984 *The Constitution of Society: Outline of the Theory of Structuration*. University of California Press, Berkeley.

Gilby, I. C.

2006 Meat Sharing among the Gombe Chimpanzees: Harassment and Reciprocal Exchange. *Animal Behaviour* 71:953-963.

Gilby, I. C., L. E. Eberly, and A. E. Pusey

2006 Ecological and Social Influences on the Hunting Behaviour of Wild Chimpanzees (*Pan troglodytes schweinfurthii*). *Animal Behaviour* 71:953-963.

Gilby, I. C., L. E. Eberly, and R. W. Wrangham

2008 Economic Profitability of Social Predation among Wild Chimpanzees: Individual Variation Promotes Cooperation. *Animal Behaviour* 75:351-360.

Gilead, I.

1991 The Upper Paleolithic Period in the Levant. *Journal of World Prehistory* 5(2):105-145.

Gille, B.

1978 *Histoire des Techniques: Technique et Civilisation, Technique et Sciences*. Gallimand, Paris.

Ginter, B., J. Kozlowski, H. Laville, N. Sirakov, and R. E. M. Hedges

1996 Transition in the Balkans: News from the Temnata Cave, Bulgaria. In *The Last Neandertals, the First Anatomically Modern Humans: A Tale about the Human Diversity. Cultural Changes and Human Evolution: The Crisis at 40 ka BP*, edited by E. Carbonell and M. Vaquero, pp. 169-200. Universitat Rovira i Virgili, Barcelona.

Ginter, B., J. Kozlowski, J.-L. Guadelli, H. Laville, and N. Sirakov

2000 *Temnata Cave: Excavations in Karlukovo Karst Area, Bulgaria 2.1*. Jagellonian University, Institute of Archaeology, Kraków.

Gladilin, V. N.

1989a Problemy ranego paleolita. Doctorate of Science, Dissertation.

1989b The Korolevo Palaeolithic site: Research Methods, Stratigraphy. *L'Anthropologie* (Brno) XXVII/2-3:93-103.

Gladilin, V. N., and Y. E. Demidenko

1989 Upper Palaeolithic Stone Tool Complexes from Korolevo. *L'Anthropologie* (Brno) XXVII/2-3:143-178.

1990 On the Origins of Early Upper Paleolithic Industries with Leaf Points in the Carpatho-Balkan Region. In *Feuilles de Pierre: Les Industries à Pointes Foliacées Paléolithique Supérieur Européen*, edited by J. K. Kozłowski, pp. 115-124. Études et Recherches archéologiques de l'Université de Liège No. 42, Liège.

Glassie, H.

1975 *Folk Housing in Middle Virginia: A Structural Analysis of Historic Artefacts.* University of Tennessee Press, Knoxville.

Gmelch, G.

1980 Return Migration. *Annual Review of Anthropology* 9:135–159.

Goldberg, P.

1981 Late Quaternary Stratigraphy of Israel: An Eclectic View. In *Préhistoire du Levant*, edited by J. Cauvin and P. Sanlaville, pp. 55–66. Editions du Centre National de la Recherche Scientifique, Paris.

Goldberg, P., and B. Brimer

1983 Late Pleistocene Geomorphic Surfaces and Environmental History of Avdat/Havarim Area, Nahal Zin. In *Prehistory and Paleoenvironments in the Central Negev, Israel, Volume III, The Avdat/Aqev Area, Part 3*, edited by A. E. Marks, pp. 1–14. Southern Methodist University Press, Dallas.

Goldberg, P., and O. Bar-Yosef

1998 Site Formation Processes in Kebara and Hayonim Caves and their Significance in Levantine Prehistoric Caves. In *Neandertals and Modern Humans in Western Asia*, edited by T. Akazawa, K. Aoki and O. Bar-Yosef, pp. 107–125. Plenum Press, New York.

Goldberg, P., H. Laville, L. Meignen, and O. Bar-Yosef

2007 Stratigraphy and Geoarchaeological History of Kebara Cave, Mount Carmel. In *Kebara Cave, Mt. Carmel, Israel: The Middle and Upper Paleolithic Archaeology, Part I*, edited by O. Bar-Yosef and L. Meignen, pp. 49–90. American School of Prehistoric Research Bulletin 49. Peabody Museum of Archaeology and Ethnology, Harvard University, Cambridge.

Golledge, R. G., and R. J. Stimson

1997 Chapter 12: The Causes and Nature of Migration. In *Spatial Behavior: A Geographic Perspective*, pp. 424–458. Guilford Press, New York.

González Echegaray, J. (Editor)

1980 El Yacimiento de la Cueva de "El Pendo" (Excavaciones 1953–1957). *Bibliotheca Praehistorica Hispana* 17.

Goodall, J.

1970 Tool-Using in Primates and Other Vertebrates. In *Advances in the Study of Behaviour, Vol. 3*, edited by D. S. Lehrman, R. A. Hinde, and E. Shaw, pp. 195–249. Academic Press, New York.

1973 Cultural Elements in a Chimpanzee Community. In *Precultural Primate Behaviour, Vol. 1.*, edited E. W. Menzel. Fourth IPA Symposia Proceedings, Karger.

1986 *The Chimpanzees of Gombe: Patterns of Behaviour*. The Belknap Press of Harvard University Press, Cambridge.

Goodenough, W.

1971 *Culture, Language, and Society*. McCaleb Module in Anthropology. Addison-Wesley, Reading.

Goren-Inbar, N.

1990 *Quneitra: A Mousterian Site on the Golan Heights*. Qedem, Monographs of the Institute of Archaeology, Hebrew University, Jerusalem.

Goren-Inbar, N., and J. D. Speth (Editors)

2004 *Human Paleoecology in the Levantine Corridor*. Oxbow Books, Oxford.

Goring-Morris, N., and A. Davidson

2006 Straight to the Point: Upper Paleolithic Ahmarian Lithic Technology in the Levant. *Anthropologie (Brno)* XLIV(1):93–111.

Goring-Morris, N., O. Marder, A. Davidson, and F. Ibrahim

1998 Putting Humpty Together Again: Preliminary Observations on Refitting Studies in the Eastern Mediterranean. In *The Organization of Lithic Technology in Late Glacial Early Postglacial Europe*, edited by S. Milliken, pp. 149-182. British Archaeological Reports International Series 700. Archaeopress, Oxford.

Gosselain, O. P.

1992 Technology and Style: Potters and Pottery among Bafia of Cameroon. *Man* 27(3):559-586.

1998 Social and Technical Identity in a Clay Crystal Ball. In *The Archaeology of Social Boundaries*, edited by M. T. Stark, pp. 78-106. Smithsonian Institution Press, Washington, D. C.

Gould, R. A.

1990 *Recovering the Past*. University of New Mexico Press, Albuquerque.

Gould, R. A., and S. Saggers

1985 Lithic Procurement in Central Australia: A Closer Look at Binford's Idea of Embeddedness in Archaeology. *American Antiquity* 50:117-135.

Gould, R. A., D. A. Koster, and A. H. Sontz

1971 The Lithic Assemblage of the Western Desert Aborigines of Australia. *American Antiquity* 36(2):149-169.

Gould, S. J.

1978 Sociobiology: The Art of Storytelling. *New Scientist* Vol. 80.

1989 *Wonderful Life: The Burgess Shale and the Nature of History*. W. W. Norton and Company, New York.

1990 Down on the Farm. Book review of *From Foraging to Agriculture: The Levant at the End of the Ice Age* (D. O. Henry). *New York Review of Books* Vol. 36, January 18, 1990, pp. 26-27.

2002 *The Structure of Evolutionary Theory*. The Belknap Press of Harvard University Press, Cambridge.

Graebner, F.

1911 *Methode der Ethnologie*. Universitätsbuchhandlung, Heidelberg.

Gramly, R.

1980 Raw Material Source Areas and 'Curated' Tool Assemblages. *American Antiquity* 45(4):823-833.

Gravina, B., P. Mellars, C. Bronk Ramsey

2005 Radiocarbon Dating of Interstratified Neanderthal and Early Modern Human Occupations at the Châtelperronian Type-Site. *Nature* 438:51-56.

Grayson, D. K.

1986 Eoliths, Archaeological Ambiguity, and the Generation of 'Middle-Range' Research. In *American Archaeology Past and Future*, edited by D. J. Meltzer, D. D. Fowler, and J. A. Sabloff, pp. 135-162. Smithsonian Institution, Washington, D.C.

Green, R. E., J. Krause, A. W. Briggs, T. Maricic, U. Stenzel, M. Kircher, N. Patterson, H. Li, W. Zhai, M. H. Fritz, N. F. Hansen, E. Y. Durand, A.-S. Malaspinas, J. D. Jensen, T. Marques Bonet, C. Alkan, K. Prüfer, M. Meyer, H. A. Burbano, J. M. Good, R. Schultz, A. Aximu-Petri, A. Butthof, B. Höber, B. Höffner, M. Siegemund, A. Weihmann, C. Nusbaum, E. S. Lander, C. Russ, N. Novod, J. Affourtit, M. Egholm, C. Verna, P. Rudan, D. Brajkovic, Ž. Kucan, I. Gušic, V. B. Doronichev, L. V. Golovanova, C. Lalueza-Fox, M. de la Rasilla, J. Fortea, A. Rosas, R. W. Schmitz, P. L. F. Johnson, E. E. Eichler, D. Falush, E. Birney, J. C. Mullikin, M. Slatkin, R. Nielsen, J. Kelso, M. Lachmann, D. Reich, and S. Paabo

2010 A Draft Sequence of the Neandertal Genome. *Science* 328:710-722.

Green, Stanton W., and S. M. Perlman

1985 Frontiers, Boundaries, and Open Social Systems. In *The Archaeology of Frontiers and Boundaries*, edited by S. Green and S. Perlman, pp. 3-13. Academic Press, San Diego.

Greenfield, P. M.

1984 A theory of the teacher in the learning activities of everyday life. In *Everyday Cognition: Its Development in Social Context*, edited by B. Rogoff and J. Lave, pp. 117-138. Harvard University Press, Cambridge.

Grimm, L., and T. Koetje

2008 Patterns of Technological Organization in the Beauronnian at Solvieux. Paper presented at the 73rd Annual Meeting of the Society for American Archaeology, Vancouver, British Columbia, Canada, March, 2008.

Groenen, M.

1994 *Pour une Histoire de la Préhistoire: Le Paléolithique*. Jérôme Millon, Grenoble.

Grootes, P. M., M. Stuiver, J. W. C. White, S. Johnsen, and J. Jouzel

1993 Comparison of Oxygen Isotope Records from the GISP2 and GRIP Greenland Ice Cores. *Nature* 366:552-554.

Groves, R. H., and F. Di Castri

1991 *Biogeography of Mediterranean Invasions*. Cambridge University Press, Cambridge.

Guichard, J., and G. Guichard

1965 The Early and Middle Palaeolithic of Nubia: A Preliminary Report. In *Contributions to the Prehistory of Nubia*, edited by F. Wendorf, pp. 57-116. Southern Methodist University Press, Dallas.

Guilbaud, M.

1993 Debitage from the Upper Castelperronian Level at Saint-Césaire: Methodological Approach and Implications for the Transition from Middle to Upper Paleolithic. In *Context of a Late Neandertal: Implications of Multidisciplinary Research for the Transition to Upper Paleolithic Adaptations at Saint-Césaire, Charente-Maritime, France*, edited by F. Lévêque, A. M. Backer, and M. Guilbaud, pp. 37-58. Monographs in World Prehistory 16. Prehistory Press, Madison.

1996 Psychotechnic Analysis and Culture Change: Origins of the Upper Paleolithic as Seen through the Example of Saint-Césaire. In *The Last Neandertals The First Anatomically Modern Humans: Cultural Change and Human Evolution: The Crisis at 40 KA BP*, edited by E. Carbonell and M. Vaquero, pp. 337-354. Universitat Rovira i Virgili, Barcelona.

Haesaerts, Paul

1990a Stratigraphy of the Grubgraben Loess Sequence. In *The Epigravettian Site of Grubgraben, Lower Austria: The 1986 and 1987 Excavations*, edited by A. Montet-White, pp. 15-35. Études et Recherches Archéologiques de l'Université de Liège 40, Liège.

1990b Nouvelles recherches au gisement de Willendorf (Basse Autriche). *Bulletin de l'Institut royal des Sciences de la Terre* 60:203-218.

1990c Evolution de l'Environnement et du Climat au Cours de l'Interpléniglaciaire en Basse Autriche et en Moravie. In *Feuilles de Pierre: Les industries à pointes foliacées du Paléolithique Supérieur Européen*, edited by J. Kozłowski, pp. 523-538. Etudes et Recherches Archéologiques de l'Université de Liège 42, Liège.

Haesaerts, P., F. Damblon, M. Bachner, G. Trnka

1996 Revised Stratigraphy and Chronology of the Willendorf II Sequence, Lower Austria. *Arch. Austriaca* 80:24-29.

Haesaerts, P., and N. Teyssandier

2003 The Early Upper Paleolithic Occupations of Willendorf II (Lower Austria): A Contribution to the Chronostratigraphic and Cultural Context of the Beginning of the Upper Paleolithic in Central Europe. In *The Chronology of the Aurignacian and of the Transitional Technocomplexes: Dating, Stratigraphies, Cultural Implications*, edited by J. Zilhão and F. d'Errico, pp. 133-151. Proceedings of the Symposium 6.1 of the XIVth Congress of the UISPP. Trabalhos de Arquelogia 33. Instituto Português de Artquelogia, Lisbon.

Hahn, J.

1972 Aurignacian Signs, Pendants and Art Objects in Central and Eastern Europe. *World Archaeology* 3:233-244.

Hahn, J., and L. Owen

1985 Blade Technology in the Aurignacian and Gravettian of Geissenklösterle Cave, Southwest Germany. *World Archaeology* 17(1):61–75.

Hallos, J.

2004 Artefact Dynamics in the Middle Pleistocene: Implications for Hominid Behavior. In *Lithics in Action: Papers form the Conference Lithic Studies in the Year 2000*, edited by E. A. Walker, F. Wenban-Smith, and F. Healy, pp. 26–37. Oxbow Books, London.

2005 '15 Minutes of Fame': Exploring the Temporal Dimension of Middle Pleistocene Lithic Technology. *Journal of Human Evolution* 49:155–179.

Hamilton, W. D.

1964 The Genetical Evolution of Social Behaviour. I and II. *Journal of Theoretical Biology* 7:1–52.

Hardesty, D.

1980 The Use of General Ecological Principles in Archaeology. *Advances in Archaeological Method and Theory* 3:157–187.

Harlan, J. R.

1971 Agricultural Origins: Centers and Non-centers. *Science* 174:468–474.

Harris, D. R. (Editor)

1996 *The Origins and Spread of Agriculture and Pastoralism in Eurasia*. Smithsonian Institution Press, Washington, D.C.

Harris, M.

1968 *The Rise of Anthropological Theory*. Crowell, New York.

Harrold, F.

1983 The Châtelperronian and the Middle-Upper Paleolithic Transition. In *The Mousterian Legacy: Human Biocultural Change in the Upper Pleistocene*, edited by E. Trinkaus, pp. 123–140. British Archaeological Reports International Series 164, Oxford.

1989 Mousterian, Châtelperronian, and Early Aurignacian in Western Europe: Continuity or Discontinuity? In *The Human Revolution: Behavioural and Biological Perspectives in the Origins of Modern Humans*, edited by P. Mellars and C. Stringer, pp. 677–713. Edinburgh University Press, Edinburgh.

Harvati., K., E. Panagopoulou, and P. Karkanas

2003 First Neanderthal Remains from Greece: The Evidence from Lakonis. *Journal of Human Evolution* 45:465–473.

Harvati, K., S. R. Frost, and K. P. McNulty

2004 Neanderthal Taxonomy Reconsidered: Implications of 3D Primate Models of Intra- and Interspecific Differences. *Proceedings of the National Academy of Sciences* 101(5):1147–1152.

Haudricourt, Andrés-Georges

1987 *La Technologie Science Humaine: Recherches d'Histoire et d'Ethnologie des Techniques*. Éditions de la Maison des Sciences de l'Homme, Paris.

Haury, Emil W.

1958 Evidence at Point of Pines for a Prehistoric Migration from Northern Arizona. In *Migrations in New World Culture History*, edited by Raymond H. Thompson, pp. 1–6. University of Arizona Bulletin 29(2). Social Science Bulletin 27, University of Arizona, Tucson.

Hauser, M.

1997 *The Evolution of Communication*. The MIT Press, Cambridge.

Hawkes, C.

1954 Archeological Theory and Method: Some Suggestions from the Old World. *American Anthropologist* 56:155–168.

Hawkes, K.

1991 Showing Off: Tests of an Hypothesis about Men's Foraging Goals. *Ethology and Sociobiology* 12:29-54.

Hawkes, K., and R. Bird

2002 Showing Off, Handicap Signaling, and the Evolution of Men's Work. *Evolutionary Anthropology* 11(2):58-67.

Hayden, B.

1979 *Paleolithic Reflections: Lithic Technology and Ethnographic Excavation Among the Australian Aborigines.* Humanities Press, New Jersey.

1989 From Chopper to Celt: The Evolution of Resharpening Techniques. In *Time, Energy, and Stone Tools*, edited by R. Torrence, pp. 7-16. Cambridge University Press, Cambridge.

Hayden, B. and M. Nelson

1981 The Use of Chipped Lithic Material in the Contemporary Maya Highlands. *American Antiquity* 45:885-98.

Hays, M., and P. T. Thacker (Editors)

2001 *Questioning the Answers: Re-solving Fundamental Problems of the Early Upper Paleolithic.* British Archaeological Reports 1005, Oxford.

Hegmon, M.

1992 Archaeological Research on Style. *Annual Review of Anthropology* 21:517-536.

1998 Technology, Style, and Social Practices: Archaeological Approaches. In *The Archaeology of Social Boundaries*, edited by M. T. Stark, pp. 264-279. Smithsonian Institution Press, Washington, D.C.

2000 *The Archaeology of Regional Interaction: Religion, Warfare, and Exchange across the American Southwest and Beyond*, edited by M. Hegmon. University Press of Colorado, Boulder.

Heine-Geldern, R.

1964 One Hundred Years of Ethnological Theory in the German-Speaking Countries: Some Milestones. *Current Anthropology* 5(5):407-418.

Hennig, W.

1966 *Phylogenetic Systematics.* University of Illinois Press, Urbana.

Henry, D. O.

1982 The Prehistory of Southern Jordan and Relationships with the Levant. *Journal of Field Archaeology* 9:417-444.

1989 Correlations between Reduction Strategies and Settlement Patterns. In *Alternative Approaches to Lithic Analysis*, edited by D. Henry and G. H. Odell, pp. 139-155. Archeological Papers of the American Anthropological Association Number 1, Washington D.C.

1992 Transhumance during the Late Levantine Mousterian. In *The Middle Paleolithic: Adaptation, Behavior, and Variability*, edited by H. L. Dibble and P. Mellars, pp. 143-162. The University Museum Press, Philadelphia.

1995a *Prehistoric Cultural Ecology and Evolution: Insights From Southern Jordan.* Plenum, New York.

1995b The Influence of Mobility Levels on Levallois Point Production, Late Levantine Mousterian, Southern Jordan. In *The Definition and Interpretation of Levallois Variability*, edited by H. L. Dibble and O. Bar-Yosef, pp. 185-200. Prehistory Press, Madison.

2003 *Neanderthals in the Levant: Behavioral Organization and the Beginnings of Human Modernity*, edited by D. O. Henry. Continuum, London.

Henry, D. O., and G. H. Odell (Editors)

1989 *Alternative Approaches to Lithic Analysis.* Archeological Papers of the American Anthropological Association Number 1, Washington D.C.

Hensilwood, C. S., and C. Marean

2002 The Origin of Modern Human Behaviour: A Review and Critique of Models and Test Implications. *Current Anthropology* 44:627-651.

Heskovits, M. J.

1948 *Man and His Works: The Science of Cultural Anthropology*. Alfred A. Knopf, New York.

Hewlett, B. S., and L. L. Cavalli-Sforza

1986 Cultural Transmission among Aka Pygmies. *American Anthropologist* 88:922-934.

Hietala, H. J.

1983a Boker Tachtit: Spatial Distributions. In *Prehistory and Paleoenvironments in the Central Negev, Israel, Volume III The Avdat/Aqev Area, Part 3*, edited by A. E. Marks, pp. 191-216. Southern Methodist University Press, Dallas.

1983b Boker Tachtit: Intralevel and Interlevel Spatial Analysis. In *Prehistory and Paleoenvironments in the Central Negev, Israel, Volume III The Avdat/Aqev Area, Part 3*, edited by A. E. Marks, pp. 217-282. Southern Methodist University Press, Dallas.

1983c Boker Tachtit: Summary of Technological Data. In *Prehistory and Paleoenvironments in the Central Negev, Israel, Volume III The Avdat/Aqev Area, Part 3*, edited by A. E. Marks, pp. 343-349. Southern Methodist University Press, Dallas.

Higham, T., R. Jacobi, M. Julien, F. David, L. Basell, R. Wood, W. Davies, and C. B. Ramsey

2010 Chronology of the Grotte du Renne (France) and Implications for the Context of Ornaments and Human Remains within the Châtelperronian. *Proceedings of the National Academy of Sciences* 107(47): 20234-20239.

Higham, Thomas, Laura Basell, Roger Jacobi, Rachel Wood, Christopher Bronk Ramsey, and Nicholas J. Conard

2012 Testing models for the beginnings of the Aurignacian and the advent of figurative art and music: The radiocarbon chronology of Geißenklösterle. *Journal of Human Evolution* 62:664-676.

Hill, J. N.

1985 Style: A Conceptual Evolutionary Framework. In *Decoding Prehistoric Ceramics*, edited by B. A. Nelson, pp. 362-385. Southern Illinois University Press, Carbondale.

Hill, K., and K. Hawkes

1983 Neotropical Hunting Among the Aché of Eastern Paraguay. In *Adaptive Responses of Native Amazonians*, edited by R. B. Hames and W. T. Vicker, pp. 139-188. Academic Press, New York.

Hill, R. A., and R. I. M. Dunbar

2003 Social Network Size in Humans. *Human Nature* 14(1):53-72.

Hiscock, P., and C. Clarkson

2005 Experimental Evaluation of Kuhn's Geometric Index of Reduction and the Flat-Flake Problem. *Journal of Archaeological Science* 32:1015-1022.

2007 Retouched Notches at Combe Grenal (France) and the Reduction Hypothesis. *American Antiquity* 72:176-190.

2009a The Reality of Reduction Experiments and the GIUR: Reply to Eren and Sampson. *Journal of Archaeological Science* 36:1576-1581.

2009b The Construction of Morphological Diversity: A Study of Mousterian Implement Retouching at Combe Grenal. In *Lithic Technology: Measures of Production, Use, and Curation*, edited by W. Andrefsky, Jr., pp. 106-135. Cambridge University Press, Cambridge.

Hodder, I.

1979a Pottery Distributions: Service and Tribal Areas. In *Pottery and the Archaeologist*, edited by M. Millett, pp. 7-24. Institute of Archaeology Occasional Publication No. 4, London.

1979b Economic and Social Stress and Material Culture Patterning. *American Antiquity* 44:446-454.

1982 *Symbols in Action: Ethnoarchaeological Studies of Material Culture.* Cambridge University Press, Cambridge.

1985 Boundaries as Strategies: An Ethnoarchaeological Study. In *The Archaeology of Frontiers and Boundaries*, edited by S. W. Green and S. M. Perlman, pp. 141-159. Academic Press, New York.

1990 Style as Historical Quality. In *The Uses of Style in Archaeology*, edited by M. Conkey and C. Hastorf, pp. 44-51. Cambridge University Press, Cambridge.

Hoffecker, J. F.

1988 Early Upper Paleolithic Sites of the European USSR. In *The Early Upper Paleolithic*, edited by J. F. Hoffecker and C. A. Wolf, pp. 237-272. BAR International Series 437, Oxford.

2002 *Desolate Landscapes: Ice-Age Settlement in Eastern Europe.* Rutgers University Press, New Brunswick.

Hoffman, C. M.

1989 Projectile Point Maintenance and Typology. In *For Concordance in Archaeological Analysis*, edited by C. Carr, pp. 566-612. Waveland Press, Prospect Heights, IL.

Holliday, V., and P. Goldberg

1993 *Soils in Archaeology*. Smithsonian Institution Press, Washington, D.C.

Holmes, W. H.

1894 Natural History of Flaked Stone Implements. In *Mémoirs of the International Conference on Anthropology*, edited by C. S. Wake, pp. 120-139. Schulte, Chicago.

1897 Stone Implements of the Potomac-Chesapeake Tidewater Province. *Bureau of American Ethnography Annual Report* 15:13-152.

1919 Handbook of Aboriginal American Antiquities, Part I: Introduction, the Lithic Industries. *Bulletin of American Ethnology* 60. Bureau of American Ethnology (1894), Washington, D. C.

Horowitz, A.

1976 Late Quaternary Paleoenvironments of Prehistoric Settlements. In *Prehistory and Paleoenvironments in the Central Negev, Israel, Volume 1, The Avdat/Aqev Area, Part 1*, edited by A. E. Marks, pp. 57-68. Southern Methodist University Press, Dallas.

Hours, F.

1974 Remarques sur l'Utilization de Listes-Types pour l'Étude du Paléolithique Supérieur et de l'Epipaléolithique du Levant. *Paléorient* 2(1):3-18.

Hovers, E.

1998 The Lithic Assemblages of Amud Cave: Implications for Understanding the End of the Mousterian in the Levant. In *Neandertals and Modern Humans in Western Asia*, edited by T. Akazawa, K. Aoki, O. Bar-Yosef, pp. 143-164. Plenum Press, New York.

2001 Territorial Behavior in the Middle Paleolithic of the Southern Levant. In *Settlement Dynamics of the Middle Paleolithic and Middle Stone Age*, edited by N. J. Conard, pp. 123-152. Kerns Verlag, Tübingen.

Hovers, E., and A. Raveh

2000 The Use of a Multivariate Graphic Display Technique as an Exploratory Tool in the Analysis of Inter-Assemblage Lithic Variability: a Case Study from Qafzeh Cave, Israel. *Journal of Archaeological Science* 27:1023-1038.

Hublin, J.-J.

1994 The Zafarraya Mousterian Site: New Evidence on the Contemporaneity of Modern Humans and Neanderthals in South-Western Europe. Paper presented at the Paleoanthropology Society Meetings, Anaheim, April 19, 1994.

Hublin, J.-J., F. Spoor, M. Braun, F. Zonneveld, and S. Condemi

1996 A Late Neanderthal Associated with Upper Palaeolithic Artefacts. *Nature* 381(6579):224–226.

Hudson, J. C.

1977 Theory and Methodology in Comparative Frontier Studies. In *The Frontier: Comparative Studies*, edited by David H. Miller and Jerome O. Steffen, pp. 11–32. Norman: University of Oklahoma Press.

Hughes, S.

1998 Getting to the Point: Evolutionary Change in Prehistoric Weaponry. *Journal of Archaeological Method and Theory* 5(4):345–408.

Hughes, T. P.

1987 The Evolution of Large Technological Systems. In *The Social Construction of Technological Systems: New Directions in the Sociology and History of Technology*, edited by W. E. Bijker, T. P. Hughes, and T. F. Pinch, pp. 51–82. MIT Press, Cambridge.

Ingold, T.

1993 The Temporality of the Landscape. *World Archaeology* 25(2):152–174.

Inizan, M.-L., H. Roche, and J. Tixier

1992 Technology of Knapped Stone. Préhistoire de la Pierre Taillée Vol. 3. Centre National de la Recherche Scientifique, Meudon.

Iovita, R., and S. McPherron

2011 The Handaxe Reloaded: A Morphometric Reassessment of Acheulian and Middle Paleolithic Handaxes. *Journal of Human Evolution* 61:61–74.

Isbell, L. A., D. L. Cheney, and R. M. Seyfarth

1991 Costs and Benefits of Home Range Shifts Among Vervet Monkeys (*Cercopithecus aethiops*) in Amboseli National Park, Kenya. *Behavioral Ecology and Sociobiology* 27:351–358.

Ivanova, I. K., and A. P. Chernysh

1965 The Paleolithic Site of Molodova V on the Middle Dnestr (USSR). *Quaternaria* 7:197–217.

Ivanova, I. K.

1966 The Stratigraphy of the Upper Pleistocene of Central and Eastern Europe According to Data from the Study of Loesses. In *Verkhnij Plejstotsen*, pp. 32–66. (In Russian.)

1987 Paleogeografiya I Paleoekologiya Sredy Obitaniya Lyudei Kamennogo Veka na Srednem Dnestre. Stoyanka Molodova V. In *Mnogosloinaya Paleoliticheskaya Stoyanka Molodova V: Lyudi Kamennogo Veka I Okruzhayushchaya Sreda*, edited by I .K. Ivanova and S. M Tseitlin, pp. 94–123. Nauka, Moscow.

Jacobsen, G.

1988 Female Migration and the Late Medieval Town. In *Migration in der Feudalgesellschaft*, edited by G. Jaritz and A. Müller, pp. 43–55. Campus-Verlag, Frankfurt/Main.

Jaubert, J.

1993 Le Gisement Paléolithique Moyen de Mauran (Haute-Garonne): Techno-économie des Industries Lithiques. *Bulletin de la Société Préhistorique Française* 90(5):328–335.

Jaubert, J., M. Lorblanchet, and H. Laville

1990 *Les Chasseurs d'Aurochs de la Borde. Un Site du Paléolithique Moyen (Livernon, Lot)*. Documents Archéologiques Françaises. Éditions de la Maison des Sciences de l'Homme, Paris.

Jelinek, A.

1976 Form, Function and Style in Lithic Analysis. In *Cultural Change and Continuity: Essays in Honor of James Bennett Griffin*, edited by C. Cleland, pp. 19-33. Academic Press, New York.

1977 The Lower Paleolithic: Current Evidence and Interpretations. *Annual Review of Anthropology* 6:11-32.

1982 The Middle Palaeolithic in the Southern Levant, with Comments on the Appearance of Modern *Homo Sapiens*. In *The Transition from Lower to Middle Palaeolithic and the Origin of Modern Man*, edited by A. Ronen, pp. 57-101. British Archaeological Reports International Series 151, Oxford.

1989 The Amudian in the Context of the Mugharan Tradition at the Tabun Cave (Mount Carmel), Israel. In *The Emergence of Modern Humans: An Archaeological Perspective*, edited by P. Mellars, pp. 81-90. Edinburgh University Press, Edinburgh.

Jelinek, J.

1988 Anthropologische Funde aus der Kůlna-Hölhe. In *Die Erforschung der Kůlna-Hölhe 1961-1976*, edited by Karel Valoch, pp. 261-284. Anthropos, Brno.

Jeske, R.

1989 Economies in Raw Material Use by Prehistoric Hunter-Gatherers. In *Time, Energy, and Stone Tools*, edited by R. Torrence, pp. 34-45. Cambridge University Press, Cambridge.

Jochim, M.

1989 Optimization and Stone Tool Studies: Problems and Potentials. In *Time, Energy, and Stone Tools*, edited by R. Torrence, pp. 106-111. Cambridge University Press, Cambridge.

Johnson, Jay K., and C. A. Morrow (Editors)

1987 *The Organization of Core Technology*. Westview Press, Boulder.

Jones, S.

1996 Discourses of Identity in the Interpretation of the Past. In *Cultural Identity in Archaeology*, edited by P. Graves-Brown, S. Jones, and C. Gable, pp. 62-80. Routledge, London.

1997 *The Archaeology of Ethnicity, Constructing Identities in the Past and Present*. Routledge, London.

Jones, T. L., and K. A. Klar

20005 Diffusionism Reconsidered: Linguistic and Archaeological Evidence for Prehistoric Polynesian Contact with Southern California. *American Antiquity* 70(3):457-484.

Jordan, P., and S. Shennan

2005 Cultural Transmission in Indigenous California. In *The Evolution of Cultural Diversity: A Phylogenetic Approach*, edited by R. Mace, C. Holden, and S. Shennan, pp. 165-196. UCL Press, London.

Jöris, O., E. Álvarez Fernández, and B. Weninger

2003 Radiocarbon Evidence of the Middle to Upper Palaeolithic Transition in Southwestern Europe. *Trabajos de Prehistoria* 60(2):15-38.

Kadic, O.

1915 A Szeleta-Barlang Kutatásának Eredményei (Ergenbnisse der Erforschung der Szeletahölhe). *Jahresbericht der Geologischen Gesellschaft* 23:148-296.

Kaplan, B. A.

1988 Migration and Disease. In *Biological Aspects of Human Migration*, edited by C. G. N. Mascie-Taylor and G. W. Lasker, pp. 216-245. Cambridge University Press, Cambridge.

Kaplan, H., and K. Hill

1985 Food-Sharing among Aché Foragers: Tests of Explanatory Hypotheses. *Current Anthropology* 26:223-246.

Karlin, C., S. Ploux, P. Bodu, and N. Pigeot

1993 Some Socio-Economic Aspects of the Knapping Process among Groups of Hunter-Gatherers in the Paris Basin Area. In *The Use of Tools by Human and Non-Human Primates*, edited by A. Berthelet and J. Chavaillon, pp. 318-340. Clarendon Press, Oxford.

Kelly, R. L.

1988 The Three Sides of a Biface. *American Antiquity* 53:717-734.

1995 *The Foraging Spectrum: Diversity in Hunter-Gatherer Lifeways*. Smithsonian Institution Press, Washington, D.C.

Kerr, R. A.

1993 The Whole World had a Case of the Ice Age Shivers. *Science* 262:1972-1973.

Kershaw, A. C.

1978 Diffusion and Migration Studies in Geography. In *Diffusion and Migration: Their Roles in Cultural Development*, edited by P. G. Duke, J. Ebert, G. Langeman, and A. P. Buchner, pp. 6-13. Proceedings of the Tenth Annual Chacmool Conference. Archaeological Association of the University of Calgary, Calgary.

Klein, R.

1973 *Ice-Age Hunters of the Ukraine. Prehistoric Archaeology and Ecology Series*. University of Chicago Press, Chicago.

1995 Anatomy, Behaviour, and Modern Human Origins. *Journal of World Prehistory* 9:167-198.

Klingelhofer, E.

1987 Aspects of Early Afro-American Material Culture: Artifacts from the Slave Quarters at Garrison Plantation, Maryland. *Historical Archaeology* 21(2):112-119.

Kluckhohn, C.

1936 Some Reflections on the Method and Theory of Kulterkreislehre. *American Anthropologist* 38:157-196.

Knecht, H.

1991 The Role of Innovation in Changing Early Upper Paleolithic Organic Projectile Technologies. *Techniques et Culture* 17-18:115-144.

1993 Early Upper Paleolithic Approaches to Bone and Antler Projectile Technology. In *Hunting and Animal Exploitation in the Later Palaeolithic and Mesolithic of Eurasia*, edited by G. Larsen Peterkin, H. Bricker, and P. Mellars, pp. 33-48. Archaeological Papers of the American Anthropological Association, Vol. 4, Washington D. C.

Knecht, H., A. Pike-Tay, and R. White (Editors)

1993 *Before Lascaux: The Complex Record of the Early Upper Paleolithic*. CRC Press, Boca Raton.

Koenig, W. D.

1989 Sex-biased Dispersal in the Contemporary United States. *Ethology and Sociobiology* 10:263-278.

Koppers, W.

1959 Grundsätzliches und Geschichtliches zur ethnologischen Kulturkreislehre. In *Beiträge Österreichs zur Erforschung der Vergangenheit und Kulturgeschichte des Menschheit. Bericht über das erste österreicheische Symposion auf Burg Wartenstein bei Gloggnitz, Sept. 8-12, 1958*, pp. 110-126. Ferdinand Berger, Horn.

Kowalski, K.

1982 Animal Remains: General Remarks. In *Excavation in the Bacho Kiro Cave (Bulgaria). Final Report*, J. K. Kozłowski and B. Ginter, pp. 66-74. Panstwowe Wydawnictwo Naukowe, Warsaw.

Kozłowski, J. K.

1972 The Origin of Lithic Raw Materials Used in the Palaeolithic of the Carpathian Countries. *Acta Archaeologica Carpathica* 13:5-19.

1979 Le Bachokirien - La Plus Ancienne Industrie du Paléolithique Supérieur en Europe. In *Middle and Early Upper Palaeolithic in Balkans*, edited by J. K. Kozlowski. *Prace Archeologiczne* 28:77-99.

1988 L'apparition du Paléolithique Supérieur. In *L'Homme de Néandertal, Vol. 8, La Mutation*, edited by J. Kozlowski, pp. 11-21. Études et Recherches Archéologiques de ll'Université de Liège 35, Liège.

1989 A Multiaspectual Approach to the Origins of the Upper Palaeolithic in Europe. In *The Emergence of Modern Humans: An Archaeological Perspective*, edited by P. Mellars, pp. 419-437. Edinburgh University Press, Edinburgh.

1990 Certains Aspects Techno-Morphologiques des Pointes Foliacées de la Fin du Paléolithique Moyen au Début du Paléolithique Supérieur en Europe Centrale. In *Paléolithique Moyen Récent et Paléolithique Supérieur Ancien en Europe*, edited by C. Farizy, pp. 125-134. Mémoires du Musée de Préhistoire d'Ile de France. Centre National de la Recherche Scientifique, Nemours.

2000 The Problem of Cultural Continuity between the Middle and Upper Paleolithic in Central and Eastern Europe. In *The Geography of Neandertals and Modern Humans in Europe and the Greater Mediterranean*, edited by O. Bar-Yosef and D. Pilbeam, pp. 77-105. Peabody Museum, Harvard University, Cambridge.

2004 Early Upper Paleolithic Levallois-Derived Industries in the Balkans and in the Middle Danube Basin. *Anthropologie* (Brno) XLII (3):263-280.

Kozłowski, J. K., and M. Pawlikowski

1989 Investigations into the Northern Lithic Raw Materials in Upper Silesia (Poland). In *"Northern" (Erratic and Jurassic) Flint of South Polish Origin in the Upper Palaeolithic of Central Europe*, edited by J. K. Kozłowski, pp. 17-46. University of Kansas/ Jagellonian University, Lawrence/Kraków.

Kozłowski, J. K., and B. Ginter

1982 *Excavation in the Bacho Kiro Cave (Bulgaria): Final Report.* Panstwowe Wydawnictwo Naukowe, Warsaw.

Kozłowski, J. K., H. Laville, and N. Sirakov

1989 Une Nouvelle Séquence Géologique et Archéologique dans les Balkans: La Grotte Temnata à Karloukovo (Bulgarie Nord). *L'Anthropologie* (Paris) 93:159-172.

Krantz, G. S.

1976 On the Nonmigration of Hunting Peoples. *Northwest Anthropological Research Notes* 10(2):209-216.

Krause, J., Q. Fu, J. M. Good, B. Viola, M. V. Shunkov, A. P. Derevianko, and S. Pääbo

2010 The Complete Mitochondrial DNA Genome of an Unknown Hominin from Southern Siberia. *Nature* 464:894-897.

Krebs, J. R.

1978 Optimal Foraging: Decision Rules for Predators. In *Behavioural Ecology: An evolutionary Approach*, edited by J. R. Krebs and N. B. Davies, pp. 23-63. Blackwell, Oxford.

Krebs, J. R., and N. B. Davies

1997 *Behavioral Ecology: An Evolutionary Approach. 4th Edition.* Blackwell, Oxford.

Krings, M., A. Stone, R. W. Schmitz, H. Krainitzki, M. Stoneking, and Svante Pääbo

1997 Neandertal DNA Sequences and the Origin of Modern Humans. *Cell* 90:19-30.

Kristiansen, K.

1989 Prehistoric Migrations - the Case of the Single Grave and Corded Ware Cultures. *Journal of Danish Archaeology* 8:211-225.

Kroeber, A. L.

1931 Historical Reconstruction of Culture Growths and Organic Evolution. *American Anthropologist* 33(2):149-156.

1940 Stimulus Diffusion. *American Anthropologist* 42(1):1-20.

1948 *Anthropology*. Harcourt, Brace and Co., New York.

Kroeber, A. L., and C. Kluckhohn

1952 *Culture, a Critical Review of Concepts and Definitions*. Papers of the Peabody Museum of American Archaeology and Ethnology XLVII (1). Peabody Museum, Harvard University, Cambridge.

Kuhn, S. L.

1989 Hunter-Gatherer Foraging Organization and Strategies of Artifact Replacement and Discard. In *Experiments in Lithic Technology*, edited by D. Amick and R. Mauldin,pp. 33-47. British Archaeological Reports International Series 528, Oxford.

1990a Diversity within Uniformity: Tool Manufacture and Use in the Pontinian Mousterian of Latium (Italy). Ph.D. Dissertation, University of New Mexico.

1990b A Geometric Index of Reduction for Unifacial Stone Tools. *Journal of Archaeological Science* 17:583-593.

1991 Unpacking Reduction: Lithic Raw Material Economy in the Mousterian of West-Central Italy. *Journal of Anthropological Archaeology* 10:76-106.

1992 On Planning and Curated Technologies in the Middle Paleolithic. *Journal of Anthropological Research* 48(3):185-214.

1994 A Formal Approach to the Design and Assembly of Mobile Toolkits. *American Antiquity* 59(3):426-442.

1995 *Mousterian Lithic Technology: An Ecological Perspective*. Princeton University Press, Princeton.

1996 The Trouble with Ham Steaks: A Reply to Morrow. *American Antiquity* 61(3):591-595.

2003 In What Sense is the Levantine Initial Upper Paleolithic a "Transitional" Industry"? In *The Chronology of the Aurignacian and of the Transitional Technocomplexes: Dating, Stratigraphies, Cultural Implications*, edited by J. Zilhão and F. d'Errico, pp. 61-69. Proceedings of the Symposium 6.1 of the XIVth Congress of the UISPP. Trabalhos de Arquelogia 33. Instituto Português de Artquelogia, Lisbon.

2004 Upper Paleolithic Raw Material Economies at Üçağizli Cave, Turkey. *Journal of Anthropological Archaeology* 23:431-448.

Kuhn, S. L., and M. C. Stiner

2006 What's a Mother to Do? The Division of Labor among Neandertals and Modern Humans in Eurasia. *Current Anthropology* 47(6):953-980.

2007 Body Ornamentation as Information Technology: Towards an Understanding of the Significance of Beads. In *Rethinking the Human Revolution: New Behavioural and Biological Perspectives on the Origins and Dispersal of Modern Humans*, edited by P. Mellars, K. Boyle, O. Bar-Yosef, and C. Stringer, pp. 45-54. MacDonald Institute Research Monograph, Cambridge.

Kuhn, S. L., M. C. Stiner, E.Güleç

1999 Initial Upper Palaeolithic in South-central Turkey and its Regional Context: A Preliminary Report. *Antiquity* 73:505-517.

2004 New Perspectives on the Initial Upper Paleolithic: the View from Üçagizli Cave, Turkey. In *The Early Upper Paleolithic Beyond Western Europe*, edited by P. J. Brantingham, S. L. Kuhn, K. W. Kerry, pp. 113-128. University of California Press, Berkeley.

Kuhn, Steven L., Mary C. Stiner, Erksin Güleç, Ismail Özer, Hakan Yılmaz, Ismail Baykara, Ayşen Açıkkol, Paul Goldberg, Kenneth Martínez Molina, Engin Ünay, Fadime Suata-Alpaslan

2009 The early Upper Paleolithic occupations at Üçağizli cave, (Hatay, Turkey). *Journal of Human Evolution* 56:87-113.

Kuijt, I., W. C. Prentiss, and D. L. Pokotylo

1995 Bipolar Reduction: An Experimental Study of Debitage Variability. *Lithic Technology* 20(2):116-127.

Laden, G. T.

1992 Ethnoarchaeology and Land Use Ecology of the Efe (Pygmies) of the Ituri Rain Forest, Zaire. Ph.D. Dissertation, Department of Anthropology, Harvard University, Cambridge.

Lahr, M. M., and R. A. Foley

1998 Towards a Theory of Modern Human Origins: Geography, Demography, and Diversity in Recent Human Evolution. *Yearbook of Physcial Anthropology* 41:137-176.

2003 Demography, Dispersal and Human Evolution in the Last Glacial Period. In *Neanderthals and Modern Humans in the European Landscape during the Last Glaciation: Archaeological Results of the Stage 3 Project*, edited by T. H. van Andel and W. Davies, pp. 241-256. McDonald Institute for Archaeological Research, Cambridge.

2004 Human Evolution Writ Small. *Nature* 431:1043-1044.

Larick, R.

1987 Men of Iron and Social Boundaries in Northern Kenya. In *Ethnicity and Culture*, edited by Réginald Auger, Margaret F. Glass, Scott MacEachern, and Peter H. McCartney, pp. 67-76. University of Calgary Press, Calgary.

Larson, M. L., and M. Kornfeld

1997 Chipped Stone Nodules: Theory, Method, and Examples. *Lithic Technology* 22(1):4-18.

Larson, S. G. W. L. Jungers, M. J. Morwood, T. Sutikna, Jatmiko, E. W. Saptomo, R. A. Due, and T. Djubiantono

2007 *Homo floresiensis* and the Evolution of the Hominin Shoulder. *Journal of Human Evolution* 53:718-731.

Lartet, E., and H. Christy

1865-75 *Reliquiae Aquitanicae: Being Contributions to the Archaeology and Palaeontology of Perigord and the Adjoining Provinces of Southern France*. T. Rupert Jones, London and Edinburgh.

Lasker, G. W., and C. G. N. Mascie-Taylor

1988 The Framework of Migration Studies. In *Biological Aspects of Human Migration*, C. G. N. Mascie-Taylor and G. W. Lasker, eds., pp. 1-13. Cambridge University Press, Cambridge.

Laughlin, John P., and Robert L. Kelly

2010 Experimental analysis of the practical limits of lithic refitting. *Journal of Archaeological Science* 37:427-433.

Laville, H.

1976 Deposits in Calcareous Rock Shelters: Analytical Methods and Climatic Interpretation. In *Geoarchaeology*, edited by D. A. Davidson and M. L. Shackley, pp. 137-155. Duckworth, London.

Laville, H., and P. Goldberg

1989 The Collapse of the Mousterian Sedimentary Regime and the Beginning of the Upper Palaeolithic at Kebara Cave, Mount Carmel. In *Investigations in South Levantine Prehistory (Préhistoire du Sud-Levant)*, edited by O. Bar-Yosef and B. Vandermeersch, pp. 75-95. BAR International Series 497. British Archaeological Reports, Oxford.

Laville, Henri, J-P. Rigaud, and J. R. Sackett

1980 *Rock Shelters of the Perigord: Geological Stratigraphy and Archaeological Succession*. Academic Press, Orlando.

Lebrun-Ricalens, F.

1993 Réflexions Préliminaires sur le Comportement Lithotechnologique et l'Occupation de Territoire du pays des Serre à l'Aurignacien: Le Gisement de Toulousète à Beauville (Lot-et-Garonne), une Occupation Moustérienne et l'Aurignacienne de Pleine Air. *Paléo* 5:127-53.

Lebrun-Ricalens, F. (Editor)

2005 *Productions Lamellaires Attribuées à l'Aurignacien: Chaînes Opératoires et Perspectives Techno-Culturelles*. Musée National d'Histoire et d'Art, Luxembourg.

Lebrun-Ricalens, F., J.-G. Bordes, L. Eizenberg

2009 A Crossed-Glance between Southern European and Middle-Near Eastern Early Upper Palaeolithic Lithic Complexes. Existing Models, New Perspectives. In *The Mediterrranean from 50 000 to 25 000 BP. Turning Points and New Directions*, edited by M. Camps et C. Szmidt, pp. 11-34. Oxbow Books, Oxford.

Lechtman, H.

1977 Style in Technology Some Early Thoughts. In *Material Culture: Styles, Organization, and Dynamics of Technology*, edited by H. Lechtman and R. Merrill, pp. 3-20. West Publishing Co., St. Paul.

Lechtman, H., and R. S. Merrill (Editors)

1977 *Material Culture: Styles, Organization, and Dynamics of Technology. 1975 Proceedings of the American Ethnological Society*. West Publishers, New York.

Lee, E.

1966 A Theory of Migration. *Demography* 3(1):47-57.

Lee, R., and I. DeVore (Editors)

1976 *Kalahari Hunter-Gatherers: Regional Studies of the !Kung-San and Their Neighbors*. Harvard University Press, Cambridge.

Lefferts, H. L., Jr.

1977 Frontier Demography: An Introduction. In *The Frontier: Comparative Studies*, edited by D. H. Miller and J. O. Steffen, pp. 33-56. University of Oklahoma Press, Norman.

Lekson, S. H., and C. M. Cameron

1995 The Abandonment of Chaco Canyon, the Mesa Verde Migrations, and the Reorganization of the Pueblo World. *Journal of Anthropological Archaeology* 14:184-202.

LeMoine, G.

2002 Dorset-Thule Contact: A Proposal for Research. In *The Archaeology of Contact: Processes and Consequences*, K. Lesick, B. Kulle, C. Cluney, and M. Peuramaki-Brown, pp. 424-430. Proceedings of the Twenty-Fifth Annual Conference of the Archaeological Association of the University of Calgary. Archaeological Association of the University of Calgary, Calgary.

Lemonnier, P.

1986 The Study of Material Culture Today: Towards an Anthropology of Technology. *Journal of Anthropological Archaeology* 5:147-186.

1989 Bark Capes, Arrowheads and Concorde: On Social Representations of Technology. In *The Meaning of Things. Material Culture and Symbolic Expression*, edited by I. Hodder, pp. 156-171. Unwin Hyman, London.

1992 *Elements for an Anthropology of Technology*. Anthropological Papers, Museum of Anthropology, University of Michigan No. 88. Regents of the University of Michigan, Ann Arbor.

Lemonnier, P. (Editor)

1993 *Technological Choices. Transformation in Material Cultures Since the Neolithic*. Routledge, London.

Leonard, R. D., and G. T. Jones

1987 Elements of an Inclusive Evolutionary Model for Archaeology. *Journal of Anthropological Archaeology* 6:199-219.

Leroi-Gourhan, André

1943 *Evolution et Techniques: L'Homme et la Matière*. A. Michel, Paris.

1945 *Milieu et Techniques*. A. Michel, Paris.

1964 *Le Geste et la Parole I and II. Gesture and Speech*, translated by Anna Bostock Berger. MIT Press, Cambridge, 1993.

1976 Les Structures d'Habitat au Paléolithique Supérieur. In *La Préhistoire Françaises, Vol. 1, Les civilisations Paléolithiques et Mésolithiques*, edited by H. de Lumley, pp. 656-663. Centre National de la Recherche Scientifique, Paris.

Leroi-Gourhan, A., and M. Brézillon

1983 *Fouilles de Pincevent: Essai d'Analyse Ethnographique d'un Habitat Magdalénien.* VIIe Supplément à Gallia Préhistoire. Centre National de la Recherche Scientifique, Paris.

Leroi-Gourhan, A., and A. Leroi-Gourhan

1965 Chronologie des Grottes d'Arcy-sur-Cure (Yonne). *Gallia-Préhistoire* 7:1-64.

Leroi-Gourhan, Arlette

1984 La Place du Néandertalien de St.-Césaire dans la Chronologie Würmienne. *Bulletin de la Société Préhistorique Française* 81:196-198.

Leroyer, C.

1987 Les Gisements Castelperroniens de Quinçay et de Saint-Césaire: Quelques Comparaisons Préliminaires des Etudes Palynologiques. In *Préhistoire de Poitou-Charentes: Problèmes Actuels*, edited by B. Vandermeersch, pp. 125-134. Éditions du Comité des Travaux Historiques et Scientifiques, Paris.

Lesick, K., B. Kulle, C. Cluney, and M. Peuramaki-Brown (Editors)

2002 *The Archaeology of Contact: Processes and Consequences. Proceedings of the Twenty-Fifth Annual Conference of the Archaeological Association of the University of Calgary*. Archaeological Association of the University of Calgary, Calgary.

Leslie, G., and A. Richardson

1961 Life-cycle, Career Pattern, and the Decision to Move. *American Sociological Review* 26:894-902.

Lev, E., M. E. Kislev, and O. Bar-Yosef

2005 Mousterian Vegetal Food in Kebara Cave, Mt. Carmel. *Journal of Archaeological Science* 32:475-484.

Lévêque, F.

1987 Les Gisements Castelperroniens de Quinçay et de Saint-Césaire: Comparaisons Préliminaires, Stratigraphie et Industries. In *Préhistoire de Poitou-Charentes: Problèmes Actuels*, edited by B. Vandermeersch, pp. 91-98. Éditions du Comité des Travaux Historiques et Scientifiques, Paris.

1993 *The Castelperronian Industry of Saint-Césaire: The Upper Level.* In *Context of a late Neandertal. Implications of Multidisciplinary Research for the Transition to Upper Paleolithic adaptations at Saint-Césaire, Charente-Maritime, France*, edited by F. Lévêque, A. Backer and M. Guilbaud, pp. 25-35. Prehistory Press, Monographs in World Archaeology No.16, Madison.

Lévêque, F., and B. Vandermeersch

1980 Découverte de Restes Humains dans un Niveau Castelperronien à Saint-Césaire (Charente-Maritime). *Comptes Rendus de l'Académie des Sciences de Paris, Series* 2, 291:187-189.

Lévêque, F., A. Mary Backer, and M. Guilbaud (Editors)

1993 *Context of a Late Neandertal: Implications of Multidisciplinary Research for the Transition to Upper Paleolithic Adaptations at Saint-Césaire, Charente-Maritime, France.* Monographs in World Archaeology No. 16. Prehistory Press, Madison.

Lewontin, R. C.

1970 The Units of Selection. *Annual Review of Ecology and Systematics* 1:1-14.

1974 *The Genetic Basis of Evolutionary Change.* Columbia University Press, New York.

Lieberman, D. E., and J. J. Shea

1994 Behavioral Differences between Archaic and Modern Humans in the Levantine Mousterian. *American Anthropologist* 96(2):300-332.

Lieberman, D. E., R. C. McCarthy, K. M. Hiiemae, and J. B. Palmer

2001 Ontogeny of Postnatal Hyoid and Larynx Descent in Humans. *Archives of Oral Biology* 46:117-128.

Lightfoot, K. G., and A. Martinez

1995 Frontiers and Boundaries in Archaeological Perspective. *Annual Review of Anthropology* 24:471–492.

Lightfoot, K. G., A. Martinez, and A. M. Schiff

1998 Daily Practice and Material Culture in Pluralistic Social Settings: An Archaeological Study of Culture Change and Persistence from Fort Ross, California. *American Antiquity* 63:199–222.

Lindly, J. and G. Clark

1987 A Preliminary Lithic Analysis of the Mousterian Site of Ain Difla, WHS Site 634, in the Wadi Ali, West-Central Jordan. *Proceedings of the Prehistoric Society* 53:279–292.

1990 Symbolism and Modern Human Origins. *Current Anthropology* 31:233–261.

Linton, R.

1936 *The Study of Man.* Appleton-Century-Crofts, New York.

Lipo, C.

2006 The Resolution of Cultural Phylogenies Using Graphs. In *Mapping Our Ancestors: Phylogenetic Approaches in Anthropology and Prehistory*, edited by C. Lipo, M. J. O'Brien, M. Collard, and S. J. Shennan, pp. 89–107. Aldine Transaction Publishers, New Brunswick and London.

Lipo, C., M. J. O'Brien, M. Collard, and S. J. Shennan (Editors)

2006 *Mapping Our Ancestors: Phylogenetic Approaches in Anthropology and Prehistory*. Aldine Transaction Publishers, New Brunswick and London.

Longacre, W. A.

1964 Archaeology as Anthropology: A Case Study. *Science* 144:1454–1455.

1981 Kalinga Pottery: An Ethnoarchaeological Study. In *Pattern of the Past: Studies in Honor of David Clarke*, edited by I. Hodder, G. Isaac, and N. Hammond, pp. 49–66. Cambridge University Press, Cambridge.

Lord, A. B.

1960 The Singer of Tales. Harvard Studies in Comparative Literature, 24. Harvard University Press, Cambridge, MA.

2000 *The Singer of Tales, 2nd Edition*, edited by S. Mitchell and G. Nagy. Harvard Studies in Comparative Literature, 24. Harvard University Press, Cambridge, MA.

Lowie, R. H.

1912 The Principle of Convergence in Ethnology. *Journal of American Folk-Lore* 25(45):24–42.

1937 *The History of Ethnological Theory*. Farrar and Rinehart, New York.

Lurie, R.

1989 Lithic Technology and Mobility Strategies: The Koster Site Middle Archaic. In *Time, Energy, and Stone Tools*, edited by R. Torrence, pp. 46–56. Cambridge University Press, Cambridge.

Lycett, S. J.

2007 Is the Soanian Techno-Complex a Mode 1 or Mode 3 Phenomenon? A Morphometric Assessment. *Journal of Archaeological Science* 34:1434–1440.

2010 Cultural Transmission, Genetic Models and Palaeolithic Variability: Integrative Analytical Approaches. In *New Perspectives on Old Stones: Analytical Approaches to Paleolithic Technologies*, edited by S. J. Lycett and P. R. Chauhan, pp. 207–234. Springer Science+Business Media, New York.

Lycett, S. J., and N. von Cramon-Taubadel

2008 Acheulean Variability and Hominin Dispersals: A Model-Bound Approach. *Journal of Archaeological Science* 35:553–562.

Lycett, S. J., N. von Cramon-Taubadel, and R. A. Foley

2006 A Crossbeam Co-ordinate Caliper for the Morphometric Analysis of Lithic Nuclei: A Description, Test and Empirical Examples of Application. *Journal of Archaeological Science* 33(6):847-861.

Lycett, S. J., N. von Cramon-Taubadel, and J. A. J. Gowlett

2010 A Comparative 3D Geometric Morphometric Analysis of Victoria West Cores: Implications for the Origins of Levallois Technology. *Journal of Archaeological Science* 37:1110-1117.

Lyman, R. L., M.J. O'Brien, and R. C. Dunnell

1997 *The Rise and Fall of Culture History*. Plenum Press, New York.

Lynch, C.

1966 The 'Lower Perigordian' in French Archaeology. *Proceedings of the Prehistoric Society* 32:156-198.

Lyons, P.

1987 Language and Style in the Peruvian Montaña. In *Ethnicity and Culture*, edited by Réginald Auger, Margaret F. Glass, Scott MacEachern, and Peter H. McCartney, pp. 101-114. University of Calgary Press, Calgary.

Lyons, P. D.

2003 *Ancestral Hopi Migrations. Archaeological Papers of the University of Arizona No.* 68. University of Arizona Press, Tucson.

Mace, R., C. Holden, and S. Shennan (Editors)

2005 *The Evolution of Cultural Diversity: A Phylogenetic Approach*. UCL Press, London.

MacDonald, L. J., and J. MacDonald

1964 Chain Migration, Ethnic Neighborhood Formation, and Social Networks. *Milbank Memorial Fund Quarterly* 42:82-97.

MacKenzie, D.

1987 Missile Accuracy: A Case Study in the Social Processes of Technological Change. In *The Social Construction of Technological Systems: New Directions in the Sociology and History of Technology*, edited by W. E. Bijker, T. P. Hughes, and T. F. Pinch, pp. 195-222. MIT Press, Cambridge.

Madsen, B.

2003 *The Cutting Edge: A Million Years of Stone Technology*. Video documentary, Institute of Archaeology, Hebrew University, Jerusalem.

Mahias, M.-C.

1993 Pottery Techniques in India: Technical Variants and Social Choice. In *Technological Choices: Transformation in Material Cultures since the Neolithic*, edited by P. Lemonnier, pp. 157-180. Routledge, London.

Marean, C.

2007 Heading North: an Africanist Perspective on the Replacement of Neanderthals by Modern Humans. In *Rethinking the Human Revolution: New Behavioural and Biological Perspectives on the Origins and Dispersal of Modern Humans*, edited by P. Mellars, K. Boyle, O. Bar-Yosef, and C. Stringer, pp. 367-379. MacDonald Institute Research Monograph, Cambridge.

2010 Pinnacle Point Cave 13B (Western Cape Province, South Africa) in Context: The Cape Floral Kingdom, Shellfish, and Modern Human Origins. *Journal of Human Evolution* 59:425-443.

Marks, A. E.

1976 Glossary. In *Prehistory and Paleoenvironments in the Central Negev, Israel, Volume I, The Avdat/Aqev Area, Part 1*, edited by A. E. Marks, pp. 371-383. Southern Methodist University, Dallas.

1983a The sites of Boker Tachtit and Boker: A Brief Introduction. In *Prehistory and Palaeoenvironments in the Central Negev, Israel, Vol. III, The Avdat/Aqev Area, Part 3*, edited by A. E. Marks, pp. 15-37. Southern Methodist University Press, Dallas.

1983b The Middle to Upper Palaeolithic Transition in the Levant. *Advances in World Archaeology* 2:51-98.

1988 The Middle to Upper Paleolithic Transition in the Southern Levant: Technological Change as an Adaptation to Increasing Mobility. In *L'Homme de Néandertal, Vol. 8, La Mutation*, edited by M. Otte, pp. 109-124. Études et Recherches Archéologiques de l'Université de Liège 35, Liège.

1990 The Upper and Middle Palaeolithic of the Near East and the Nile Valley: The Problem of Cultural Transformations. In *The Emergence of Modern Humans: An Archaeological Perspective*, edited by P. Mellars, pp. 56-80. Edinburgh University Press, Edinburgh.

2003 Reflections on Levantine Upper Palaeolithic Studies: Past and Present. In *More than Meets the Eye: Studies on Upper Palaeolithic Diversity in the Near East*, edited by A. N. Goring-Morris and A. Belfer-Cohen, pp. 249-264. Oxbow Press, Oxford.

Marks, A. E., and R. Ferring

1988 The Early Upper Paleolithic of the Levant. In *The Early Upper Paleolithic: Evidence from Europe and the Near East*, edited by J. F. Hoffecker and C. A. Wolf, pp. 43-72. British Archaeological Reports International Series 437, Oxford.

Marks, A. E., and D. Kaufman

1983 Boker Tachtit: The Artifacts. In *Prehistory and Palaeoenvironments in the Central Negev, Israel, Vol. III, The Avdat/Aqev Area, Part 3*, edited by A. E. Marks, pp. 69-126. Southern Methodist University Press, Dallas.

Marks, A. E., and P. Volkman

1983a Changing Core Reduction Strategies: A Technological Shift from the Middle to the Upper Paleolithic in the Southern Levant. In *The Mousterian Legacy: Human Biocultural Change in the Upper Pleistocene*, edited by E. Trinkaus, pp. 13-33. British Archaeological Reports International Series 164, Oxford.

1983b Technological Variability and Change Seen through Core Reconstruction. In *The Human Uses of Flint and Chert*, edited by G. de G. Sieveking and M. H. Newcomer, pp. 11-20. Proceedings of the Fourth International Flint Symposium Held at Brighton Polytechnic 1983. Cambridge University Press, Cambridge.

1986 The Mousterian of Ksar Akil: Levels XXVIA through XXVIIIB. *Paléorient* 12(1):5-20.

Marks, A. E., H. J. Hietala, and J. K. Williams

2001 Tool Standardization in the Middle and Upper Palaeolithic: A Closer Look. *Cambridge Archaeological Journal* 11(1):17-44

Marshack, A.

1972 *The Roots of Civilization*. Weidenfeld and Nicolson, London.

Marshall, F.

1998 Early Food Production in Africa. *The Review of Archaeology* 19(2):47-58.

Mauldin, R., and D. Amick

1989 Investigating Patterning in Debitage from Experimental Bifacial Core Reduction. In *Experiments in Lithic Technology*, edited by D. Amick and R. Mauldin, pp. 67-88. British Archaeological Reports International Series 528, Oxford.

Mauss, M.

1935 Les Techniques du Corps. *Journal de Psychologie* 32(3-4):271-293.

Mayr, E.

1961 Cause and Effect in Biology. *Science* 134(3489):1501-1506.

1982 *The Growth of Biological Thought: Diversity, Evolution, and Inheritance*. The Belknap Press of Harvard University Press, Cambridge.

1993 Proximate and Ultimate Causation. *Biology and Philosophy* 8:95-98.

1994 Response to John Beatty. *Biology and Philosophy* 9:359-371.

McBrearty, S., and A. S. Brooks

2000 The Revolution that Wasn't: A New Interpretation of the Origin of Modern Human Behavior. *Journal of Human Evolution* 39:453-563.

McElreath, R., R. Boyd, and P. J. Richerson

2003 Shared Norms and the Evolution of Ethnic Markers. *Current Anthropology* 44(1):122-129.

McGovern, T. H.

1981 The Economies of Extinction in Norse Greenland. In *Climate and History: Studies in Past Climates and their Impact on Man*, edited by. T. M. L. Wigley, M. J. Ingram, and G. Farmer, pp. 404-433. Cambridge University Press, Cambridge.

1994 Management for Extinction in Norse Greenland. In *Historical Ecology: Cultural Knowledge and Changing Landscapes*, edited by C. Crumley, pp. 127-154. School of American Research Press, Santa Fe.

McGrew, W. C.

1992 *Chimpanzee Material Culture: Implications for Human Evolution*. Cambridge University Press, Cambridge.

1994 *Tools Compared: The Material of Culture*. In *Chimpanzee Cultures*, edited by R. Wrangham, W. C. McGrew, F. de Waal, and P. Heltne, pp. 25-41. Harvard University Press, Cambridge.

1998 Culture in Nonhuman Primates? *Annual Review of Anthropology* 27:301-28.

McGrew, W. C., C. E. G. Tutin, and P. J. Baldwin

1979 Chimpanzees, Tools, and Termites: Cross-Cultural Comparisons of Senegal, Tanzania, and Río Muni. *Man* 14:185-214.

McGuire, K., and W. Hildebrandt

2005 Re-Thinking Great Basin Foragers: Prestige Hunting and Costly Signaling During the Middle Archaic Period. *American Antiquity* 70(4):695-712.

McGuire, R., and M. B. Schiffer

1983 A Theory of Architectural Design. *Journal of Anthropological Archaeology* 2:277-303.

McPherron, S. P.

1994 A Reduction Model for Variability in Acheulian Biface Morphology. Ph.D. Thesis. University of Pennsylvania.

2000 Handaxes as a Measure of the Mental Capabilities of Early Hominids. *Journal of Archaeological Science* 27:655-663.

Meggers, B. J., C. Evans, and E. Estrada

1965 *Early Formative Period of Coastal Ecuador: The Valdivia and Machalilla Phases*. Smithsonian Contributions to Anthropology Vol. 1. Smithsonian Institution Press, Washington, D.C.

Meignen, L.

1994 Le Paléolithique Moyen au Proche-Orient: Le Phénomène Laminaire. In *Les Industries Laminaires au Paléolithique Moyen*, edited by S. Révillion and A. Tuffreau, pp. 125-160. Dossier de Documentation Archéologique No. 18. Éditions du Centre National de la Recherche Scientifique, Paris.

1995 The Mousterian assemblages of Hayonim Cave in their Levantine Context. Paper Presented at the Paleoanthropology Society Meetings, New Orleans, April 10, 1995.

2006 From the Late Middle Paleolithic to the Early Upper Paleolithic, Between the Adriatic and the Caspian Sea: Continuity or Discontinuity? An Introduction. *Anthropologie* (Brno) XLIV(1):1-7.

Meignen, L., and O. Bar-Yosef

1988a Kebara et le Paléolithique moyen du Mont Carmel. *Paléorient* 14:123-130.

1988b Variabilité Technologique au Proche Orient: L'Example de Kebara. In *L'Homme de Néandertal, Vol. 4, La Technique*, edited by M. Otte, pp. 81-95. Études et Recherches Archéologiques de l'Université de Liège 35, Liège.

1988c Un Example de Comportement Technologique Différentiel selon les Matières Premières: Marillac, Couches 9 et 10. In *L'Homme de Néandertal, Vol. 4, La Technique*, edited by M. Otte, pp. 71–80. Études et Recherches Archéologiques de l'Université de Liège 35, Liège.

1991 Les Industries Mousterienne de Kebara. In *Une Squelette Mousteriénne à Kebara, Mt. Carmel Israel*, edited by O. Bar-Yosef and B. Vandermeersch. Éditions du Centre National de la Recherche Scientifique, Paris.

Meignen, L., J.-M. Geneste, L. Koulakovskaia, and A. Sytnik

2004 Koulichivka and Its Place in the Middle-Upper Paleolithic Transition in Eastern Europe. In *The Early Upper Paleolithic beyond Western Europe*, edited by P. J. Brantingham, S. L. Kuhn, and K. W. Kerry, pp. 50–79. University of California Press, Berkeley.

Meignen, L., P. Goldberg, and O. Bar-Yosef

2007 The Hearths at Kebara Cave and Their Role in Site Formation Processes. In *Kebara Cave, Mt. Carmel, Israel: The Middle and Upper Paleolithic Archaeology, Part I*, edited by O. Bar-Yosef and L. Meignen, pp. 91–122. American School of Prehistoric Research Bulletin 49, Peabody Museum of Archaeology and Ethnology, Harvard University, Cambridge.

Mellars, P.

1970 Some Comments on the Notion of 'Functional Variability' in Stone-Tool Assemblages. *World Archaeology* 2:74–90.

1989 Major Issues in the Emergence of Modern Humans. *Current Anthropology* 30(3):34–385.

1992 Technological Change in the Mousterian of Southwest France. In *The Middle Paleolithic: Adaptation, Behavior, and Variability*, edited by H. L. Dibble and P. Mellars, pp. 29–44. University Museum, University of Pennsylvania, Philadelphia.

1993 Archaeology and the Population-Dispersal Hypothesis of Modern Human Origins in Europe. In *The Origin of Modern Humans and the Impact of Chronometric Dating*, edited by M. J. Aitken, C. B. Stringer, and P. A. Mellars, pp. 196–216. Princeton University Press, Princeton.

1996 *The Neanderthal Legacy: An Archaeological Perspective from Western Europe*. Princeton University Press, Princeton.

2005 The Impossible Coincidence. A Single-Species Model for the Origins of Modern Human Behavior in Europe. *Evolutionary Anthropology* 14:12–27.

2006a A New Radiocarbon Revolution and the Dispersal of Modern Humans in Eurasia. *Nature* 439:931–935.

2006b Archeology and the Dispersal of Modern Humans in Europe: Deconstructing the Aurignacian. *Evolutionary Anthropology* 15:167–182.

2010 Neanderthal Symbolism and Ornament Manufacture: The Bursting of a Bubble? *Proceedings of the National Academy of Sciences* 107(47):20147–20148.

Mellars, P. (Editor)

1990 *The Emergence of Modern Humans: An Archaeological Perspective*. Edinburgh University Press, Edinburgh.

Mellars, P., and J. C. French

2011 Tenfold Population Increase in Western Europe at the Neandertal-to-Modern Human Transition. *Science* 333:623–627

Mellars, P., and B. Gravina

2008 Châtelperron: Theoretical Agendas, Archaeological Facts, and Diversionary Smoke-Screens. *PaleoAnthropology* 2008:1–42.

Mellars, P., and C. Stringer (Editors)

1989 *The Human Revolution: Behavioural and Biological Perspectives in the Origins of Modern Humans*. Edinburgh University Press, Edinburgh.

Mellars, P., and J. Tixier

1989 Radiocarbon-Accelerator Dating of Ksar 'Akil (Lebanon) and the Chronology of the Upper Palaeolithic Sequence in the Middle East. *Antiquity* 63:761–768.

Mellars, P., M. Otte, L. Straus, J. Zilhão, and F. D'Errico

1999 The Neanderthal Problem Continued. CA Forum on Theory in Anthropology. *Current Anthropology* 40(3):341-364.

Mellars, P., K. Boyle, O. Bar-Yosef, and C. Stringer (Editors)

2007 *Rethinking the Human Revolution: New Behavioural and Biological Perspectives on the Origins and Dispersal of Modern Humans.* MacDonald Institute Research Monograph, Cambridge.

Merrell, J. H.

1999 *Into the American Woods: Negotiators on the Pennsylvania Frontier.* W. W. Norton, New York.

Milardo, R. M.

1988 *Families and Social Networks: An Overview of Theory and Methodology.* In *Families and Social Networks*, edited by R. M. Milardo, pp. 13-47. Sage, Newbury Park.

Miller, D.

1985 *Artefacts as Categories: A Study of Ceramic Variability in Central India.* Cambridge University Press, Cambridge.

Miller, D. H., and J. O. Steffen (Editors)

1977 *The Frontier: Comparative Studies.* University of Oklahoma Press, Norman.

Misa, T. J.

2000 Controversy and Closure in Technological Change: Constructing Steel. In *Shaping Technology/Building Society: Studies in Sociotechnical Change*, edited by W. Bijker and J. Law, pp. 109-139. MIT Press, Cambridge.

Mitani, J., and D. Watts

1999 Demographic Influences on the Hunting Behavior of Chimpanzees. *American Journal of Physical Anthropology* 109:439-454.

Moch, L.

1992 *Moving Europeans: Migration in Western Europe since 1650.* Indiana University Press, Bloomington.

Monigal, K., V. Usik, L. Koulakovskaya, N. P. Gerasimenko

2006 The Beginning of the Upper Paleolithic in Transcarpathia, Ukraine. *Anthropologie* (Brno) XLIV(1):61-74.

Monnier, G. F.

2003 The Spatial Distribution of Lithic Artifacts at Stránská skála IIIc and IIId: 1997-1999 Excavations. In *Stránská skála: Origins of the Upper Paleolithic in the Brno Basin*, edited by J. Svoboda and O. Bar-Yosef, pp. 45-57. American School of Prehistoric Research Bulletin 47. Dolní Věstonice Studies 10. Peabody Museum of Archaeology and Ethnology, Harvard University, Cambridge.

2006 The Lower/Middle Paleolithic Periodization in Western Europe: An Evaluation. *Current Anthropology* 47(5):709-744.

2007 Middle Palaeolithic Scraper Morphology, Flaking Mechanics and Imposed Form: Revisiting Bisson's 'Interview with a Neanderthal'. *Cambridge Archaeological Journal* 17:(3)341-50.

Moore, Frank W. (Editor)

1961 *Readings in Cross-Cultural Methodology.* Human Relations Area Files Press, New Haven.

Moore., J. H.

1994 Putting Anthropology Back Together Again: The Ethnogenetic Critique of Cadistic Theory. *American Anthropologist* 96(4):925-948.

Morrill, R. L., and D. Manninen

1975 Critical Parameters of Spatial Diffusion Processes. *Economic Geography* 51:269.

Morrow, C. A.

1989 Trade or Embedded Procurement?: A Test Case from Southern Illinois. In *Time, Energy, and Stone Tools*, edited by R. Torrence, pp. 27-33. Cambridge University Press, Cambridge.

Morrow, T. A.

1996 Bigger is Better: Comments on Kuhn's Formal Approach to Mobile Tool Kits. *American Antiquity* 61(3):581-590.

Mortillet, G. de

1867 Promenades préhistoriques à l'Exposition Universelle. *Matériaux pour l'Histoire Positive et Philosophique de l'Homme* 3:181-368.

1873 Classification des diverses périodes de l'Age de la Pierre. In *Comptes Rendus de la VIe Congrès Internationale d'Anthropologie et d'Archéologie Préhistorique*, pp. 432-456. Brussels, 1872.

1883 *Le Préhistorique. Antiquité de l'homme.* Reinwald, Paris.

Movius, H., N. David, H. Bricker, and B. Clay

1968 *The Analysis of Certain Major Classes of Upper Palaeolithic Tools.* American School of Prehistoric Research Bulletin No. 26. Peabody Museum, Harvard University, Cambridge.

Mulvaney, D. J.

1976 The Chain of Connection. In *Tribes and Boundaries in Australia*, edited by N. Peterson, pp 72-94. Australian Institute of Aboriginal Studies, Canberra.

Murdock, G. P.

1967 *World Ethnographic Sample.* University of Pittsburgh Press, Pittsburgh.

Murdock, J. P.

1960 How Culture Changes. In *Man, Culture and Society*, edited by H. S. Shapiro, pp. 247-60. Oxford University Press, Oxford.

Musil, R.

1988 Ökostratigraphie der Sedimente in der Kůlna-Höhle. *Anthropologie* (Brno) 24:215-255.

Musil, R., and K. Valoch

1968 Stránská skála: Its Meaning for Pleistocene Studies. *Current Anthropology* 9:534-539.

Myers, A.

1989 Reliable and Maintainable Technological Strategies in the Mesolithic of Mainland Britain. In *Time, Energy, and Stone Tools*, edited by R. Torrence, pp. 78-91. Cambridge University Press, Cambridge.

Needham, J.

1954-84 *Science and Civilization in China. Six Volumes.* Cambridge University Press, Cambridge.

Neitzel, J. E.

1995 Elite Styles in Hierarchically Organized Societies: The Chacoan Regional System. In *Style, Society, and Person: Archaeological and Ethnological Perspectives*, edited by C. Carr and J. Neitzel, pp. 393-417. Plenum Press, New York.

Nejman, L.

2006 Lithic Patterning and Land-Use During the Middle-Upper Palaeolithic Transition in Moravia (Czech Republic). Doctoral Dissertation, Australian National University.

2009 A Re-Interpretation of Early Upper Palaeolithic Assemblages from Stránská skála: The Differences in Lithic Economy Between the Aurignacian and the Bohunician Assemblages. *Přehled výzkumů* 49:23-45.

Nejman, L., and C. Clarkson

2008 Testing Models of Intra-Assemblage Variability in Late Middle and Early Upper Palaeolithic Lithic Assemblages in Central Europe. *Lithic Technology* 32(1):17-30.

Nejman, L., E. Rhodes, P. Škrdla, G. Tostevin, P. Neruda, Z. Nerudová, K. Valoch, M. Oliva, L. Kaminská, J. A. Svoboda, and R. Grün

2011 New Chronological Evidence for the Middle to Upper Palaeolithic Transition in the Czech Republic and Slovakia: New Optically Stimulated Luminescence Dating Results. *Archaeometry* 53(5):1044-1066.

Nelson, M.

1991 The Study of Technological Organization. In *Archaeological Method and Theory, Vol.* 3, edited by M. Schiffer, pp. 57-100. University of Arizona Press, Tucson.

Neruda, P.

2000 The Cultural Significance of Bifacial Retouch. The Transition from the Middle to Upper Paleolithic Age in Moravia. In *Neanderthals and Modern Humans: Discussing the Transition. Central and Eastern Europe from 50,000-30,000 BP*, edited by G.-C. Weniger and J. Orschiedt, pp. 151-158. Neanderthal Museum, Düsseldorf.

Neruda, P., and Z. Nerudová

2000 The Upper Palaeolithic Levallois Industry from Hradsko (Mělnik District, Czech Republic). *Anthropologie* (Brno) 38(3):271-281.

Netting,R. McC.

1993 *Smallholders, Householders: Farm Families and the Ecology of Intensive, Sustainable Agriculture.* Stanford University Press, Stanford.

Newcomer, M. H.

1971 Some Quantitative Experiments in Hand-Axe Manufacture. *World Archaeology* 3:85-94.

Niebel, B. W.

1988 *Motion and Time Study, Eighth Edition.* IRWIN, Homewood, IL.

Nigst, P. R.

2009 The Early Upper Palaeolithic in the Middle Danube Region: A Regional Study using the Evidence of Willendorf II, Stratzing 94, Vedrovice V, and Stránká skála IIIc. Doctoral Dissertation, University of Leipzig, Germany.

2010 The Early Upper Paleolithic of the Middle Danube Region: Modern Human Dispersal or Local Evolution? Paper presented at the Paleoanthropology Society Meetings, St. Louis, April, 2010.

2012 *The Early Upper Paleolithic of the Middle Danube Region.* Studies in Human Evolution Series. Leiden University Press, Leiden.

Nishida T. , T. Hasegawa, H. Hayaki, Y. Takahata, and S. Uehara

1992 Meat-sharing as a Coalition Strategy by an Alpha Male Chimpanzee? In *Topics in Primatology*, Vol. I, Human origins, edited by T. Nishida T, W. McGrew, P. Marler, M. Pickford, F. deWaal, pp. 159-174. University of Tokyo Press, Tokyo.

Nishiaki, Y.

1989 Early Blade Industries in the Levant: The Placement of Douara IV Industry in the Context of the Levantine Early Middle Paleolithic. *Paléorient* 15(1):215-229.

Nitecki, M. H., and D. V. Nitecki (Editors)

1994 *Origins of Anatomically Modern Humans.* Plenum Press, New York.

Nowak, E.

1975 The Range Expansion of Animals and Its Causes (as demonstrated by 28 presently spreading species from Europe). Translated from Polish. *Zeszyty Naukowe* 3:1-255. Published for the Smithsonian Institution and the National Science Foundation. Institute of Ecology, Polish Academy of Sciences.

O'Brien, M. J. (Editor)

1996 *Evolutionary Archaeology: Theory and Application.* University of Utah Press, Salt Lake City.

O'Brien, M. J., and R. L. Lyman

1999 *Seriation, Stratigraphy, and Index Fossils: The Backbone of Archaeological Dating*. Kluwer Academic/Plenum, New York.

2000 *Applying Evolutionary Archaeology: A Systematic Approach*. Kluwer Press, New York.

2003 *Cladistics and Archaeology*. University of Utah Press, Salt Lake City.

O'Brien, M. J., and T. D. Holland

1995 The Nature and Premise of a Selection-Based Archaeology. In *Evolutionary Archaeology: Methodological Issues*, edited by Patrice A. Teltser, pp. 175-200. University of Arizona Press, Tucson.

O'Brien, M. J., T. D. Holland, R. J. Hoard, and G. L. Fox

1994 Evolutionary Implications of Design and Performance Characteristics of Prehistoric Pottery. *Journal of Archaeological Method and Theory* 1:259-304.

O'Brien, M. J., J. Darwent, and R. L. Lyman

2001 Cladistics Is Useful for Reconstructing Archaeological Phylogenies: Palaeoindian Points from the Southeastern United States. *Journal of Archaeological Science* 28:1115-1136.

O'Connell, J.

1995 Ethnoarchaeology Needs a General Theory of Behavior. *Journal of Archaeological Research* 3:205-55.

Odell, G. H.

1996 *Stone Tools and Mobility in the Illinois Valley: From Hunter-Gatherer Camps to Agricultural Villages*. International Monographs in Prehistory, Ann Arbor.

2004 *Lithic Analysis*. Kluwer Academic/Plenum Publishers, New York.

Odell, G. H., and D. O. Henry

1989 Summary of Discussions. In *Alternative Approaches to Lithic Analysis*, edited by D. Henry and G. H. Odell, pp. 237-256. Archeological Papers of the American Anthropological Association Number 1, Washington D.C.

Ohnuma, K.

1986 A Technological Study of the Upper Palaeolithic Material from Levels XXV-XIV from Ksar Akil. Unpublished Ph.D. Dissertation, University of London.

Ohnuma, K., and C. Bergman

1990 A Technological Analysis of the Upper Palaeolithic Levels (XXV-VI) of Ksar 'Akil, Lebanon. In *The Emergence of Modern Humans: An Archaeological Perspective*, edited by P. Mellars, pp. 91-138. Edinburgh University Press, Edinburgh.

Ohnuma, K., K.Aoki, and T. Akazawa

1997 Transmission of Tool-making through Verbal and Non-verbal Communication: Preliminary Experiments in Levallois flake Production. *Anthropological Science* 105:159-168.

Oliva, M.

1981 Die Bohunicien-Station bei Podolí (Bez. Brno-land) und ihre Stellung im beginnen Jungpaläolithikum. *Časopis Moravského muzea Sc. soc.* 66:7-45.

1984 Le Bohunicien, un nouveau groupe culturel en Moravie. *L'Anthropologie* (Paris) 88:209-220.

1989 Excavations in the Palaeolithic Site of Milovice I (Southern Moravia) in the Year 1988. *Anthropologie* (Brno) 26:105-112.

Opravil, E.

1988 Ergebnisse der Holzkohlenanalyse aus der Kůlna- Höhle. *Anthropos* (Brno) 24:211-214.

Otte, M.

1995 The Nature of Levallois. In *The Definition and Interpretation of Levallois Variability*, edited by H. L. Dibble and O. Bar-Yosef, pp. 117-124. Prehistory Press, Madison.

Otte, M., and V. Chirica

1993 Atelier Aurignacien à Mitoc Malul Galben (Moldavie roumaine). *Préhistoire Européenne* 3:55-66.

Otte, Marcel, and L. Keeley

1990 The Impact of Regionalisation on Palaeolithic Studies. *Current Anthropology* 31:577-582.

Otte, M., and J. K. Kozłowski

2003 Constitution of the Aurignacian through Eurasia. In *The Chronology of the Aurignacian and of the Transitional Technocomplexes: Dating, Stratigraphies, Cultural Implications*, edited by J. Zilhão and F. d'Errico, pp. 19-28. Trabalhos de Arquelogia 33. Instituto Português de Arqueologia, Lisboa.

Otte, M., E. Boëda, and P. Haesaerts

1990 Rocourt: Industrie Laminaire Archaïque. *Hélinium*, XXIX/1:3-13.

Pääbo, S., H. Poinar, D. Serre, V. Jaenicke-Després, J. Hebler, N. Rohland, M. Kuch, J. Krause, L. Vigilant, and M. Hofreiter

2004 Genetic Analyses from Ancient DNA. *Annual Review of Genetics* 38:645-79.

Panagopoulou, E., P. Karkanas, G. Tsartsidou, E. Kotjabopoulou, K. Harvati, and M. Ntinou

2004 Late Pleistocene Archaeological and Fossil Human Evidence from Lakonis Cave, Southern Greece. *Journal of Field Archaeology* 29(3-4):323-349.

Parry, W. J., and R. Kelly

1987 Expedient Core Technology and Sedentism. In *The Organization of Core Technology*, edited by J. K. Johnson and C. A. Morrow, pp. 285-304. Westview Press, Boulder.

Paterson, H. E. H.

1985 The Species Recognition Concept. In *Species and Speciation*, edited by E. S. Vrba, pp. 21-29. Transvaal Museum Monograph 4, Pretoria.

Pelcin, A.

1996 Controlled Experiments in the Production of Flake Attributes. Ph.D. Dissertation, University of Pennsylvania, Philadelphia.

1997a The Effect of Indentor Type on Flake Attributes: Evidence from a Controlled Experiment. *Journal of Archaeological Science* 24:613-621.

1997b The Effect of Core Surface Morphology on Flake Attributes: Evidence from a Controlled Experiment. *Journal of Archaeological Science* 24:749-756.

1997c The Formation of Flakes: The Role of Platform Thickness and Exterior Platform Angle in the Production of Flake Initiations and Terminations. *Journal of Archaeological Science* 24:1107-1113.

1998 The Threshold Effect of Platform Width: A Reply to Davis and Shea. *Journal of Archaeological Science* 25:615-620.

Pelegrin, J.

1990a Observations Technologiques sur Quelques Séries du Châtelperronien et du MTA B du Sub-Ouest de la France: Une Hypothèse d'Evolution. In *Paléolithique Moyen Récent et Paléolithique Supérieur Ancien en Europe*, edited by C. Farizy, pp. 195-202. Mémoires du Musée de Préhistoire d'Ile-de-France 3. Édition de l'Association pour la Promotion de la Recherche Archéologique en Ile de France, Nemours.

1990b Prehistoric Lithic Technology: Some Aspects of Research. *Archaeological Review from Cambridge* 9(1):116-125.

1993 A Framework for Analyzing Prehistoric Stone Tools Manufacture and a Tentative Application to Some Early Stone Industries. In *The Use of Tools by Human and Non-human Primates*, edited by A. Berthelet and J. Chavaillon, pp. 302-314. Symposiua of the Fyssen Foundation 3. Clarendon Press, Oxford.

1995 *Technologie Lithique: Le Châtelperronien de Roc-de-Combe (Lot) et de La Côte (Dordogne).* Cahiers du Quaternaire No. 20. Centre National de la Recherche Scientifique, Paris.

Perdue, P. C.

1994 Technological Determinism in Agrarian Societies. In *Does Technology Drive History? The Dilemma of Technological Determinism*, edited by M. R. Smith and L. Marx, pp. 169-200. MIT Press, Cambridge.

Perlès, C.

1990 L'Outillage de Pierre Taillée Néolithique en Grèce: Approvisionnement et Exploitation des Matières Premières. *Bulletin de Correspondance Hellénique* 1:1-42.

1991 Économie des matières premières et économie du débitage: deux conceptions opposées? In *25 Ans d'Études Technologiques en Préhistoire: XI Recontres Internationales d'Archéologie et d'Histoire d'Antibes*, pp. 35-45. Éditions APDCA, Juan-les-Pins.

Peterson, N.

1986 Australian Territorial Organization. *Oceania Monograph* 30, Sydney.

Pétrequin, P.

1993 North Wind, South Wind: Neolithic Technical Choices in the Jura Mountains, 3700-2400 BC. In *Technological Choices: Transformation in Material Cultures since the Neolithic*, edited by P. Lemonnier, pp. 36-76. Routledge, London.

Pétrequin, P., and A.-M. Pétrequin

1994 *Écologie d'un Outil: La Hache de Pierre en Irian Jaya (Indonésie).* Monographie du CRA 12. Éditions du Centre National de la Recherche Scientifique, Paris.

Peyrony, D.

1922 Nouvelles Observations sur le Moustérian final et l'Aurignacien Inférieur. Association Française pour l'Avancement de la *Science* 46:511-514.

Pfaffenberger, B.

1992 Social Anthropology of Technology. *Annual Review of Anthropology* 21:491-516.

Pianka, E.

1978 *Evolutionary Ecology, Second Edition.* Harper and Row, New York.

Pigeot, Nicole

1987 *Magdaléniens d'Étiolles: Économie de Débitage et Organization Sociale.* XXVe Supplément à Gallia Préhistoire. Éditions du Centre National de la Recherche Scientifique, Paris.

1990 Techniques and Social Actors: Flintknapping Specialists and Apprentices at Magdalenian Etiolles. *Archaeological Review from Cambridge* 9(1):126-141.

Pinch, T. J., and W. E. Bijker

1987 The Social Construction of Facts and Artifacts: Or How the Sociology of Science and the Sociology of Technology Might Benefit Each Other. In *The Social Construction of Technological Systems: New Directions in the Sociology and History of Technology*, edited by W. E. Bijker, T. P. Hughes, and T. F. Pinch, pp. 17-50. MIT Press, Cambridge.

Pitts, M., and M. Roberts

1997 *Fairweather Eden: Life in Britain Half a Million Years Ago as Revealed by the Excavations at Boxgrove.* Century Books, London.

Pletka, S.

2001 Bifaces and the Institutionalization of Exchange Relationships in the Chumash Sphere. In *The Origins of a Pacific Coast Chiefdom: The Chumash of the Channel Islands*, edited by J. Arnold, pp. 133-150. University of Utah Press, Salt Lake City.

Plog, S.

1980 *Stylistic Variation in Prehistoric Ceramics: Design Analysis in the American Southwest.* Cambridge University Press, Cambridge.

Ploux, S., and C. Karlin

1993 Fait Technique et Degré de Sens dans l'Analyse d'un Processus de Débitage Magdalénien. *Techniques et Culture* 21:61-78.

Pocklington, R.

2006 What is a Culturally Transmitted Unit, and How Do We Find One? In *Mapping Our Ancestors: Phylogenetic Approaches in Anthropology and Prehistory*, edited by C. Lipo, M. J. O'Brien, M. Collard, and S. J. Shennan, pp. 19-31. Aldine Transaction Publishers, New Brunswick and London.

Pollack, S.

1983 Style and Information: An Analysis of Susiana Ceramics. *Journal of Anthropological Archaeology* 2:354-390.

Pond, A. W.

1930 Primitive Methods of Working Stone, Based on Experiments of Halvor L. Skavlem. *Logan Museum, Beloit College Bulletin* 2(1).

Pope, M., and M. Roberts

2005 Observations on the Relationship between Palaeolithic Individuals and Artefact Scatters at the Middle Pleistocene Site of Boxgrove, UK. In *The Hominid Individual in Context: Archaeological Investigations of Lower and Middle Palaeolithic Landscapes, Locales and Artefacts*, edited by C. Gamble and M. Porr, pp. 81-97. Routledge, London.

Porraz, G., P. Simon, A. Pasquini, and G. Onoratini

2010 Identité Technique et Comportements Economiques des Groupes Protoaurignaciens à la Grotte de l'Observatoire (Principauté de Monaco). *Gallia Préhistoire* 52:33-59.

Potts, R.

1998 Variability Selection in Hominid Evolution. *Evolutionary Anthropology* 7:81-96.

Premo, L. S., and Jean-Jacques Hublin

2009 Culture, population structure, and low genetic diversity in Pleistocene hominins. *Proceedings of the National Academy of Sciences of the United States* 106(1):33-37.

Premo, L. S., and Steven L. Kuhn

2010 Modeling Effects of Local Extinctions on Culture Change and Diversity in the Paleolithic. *PLoS One* 5(12):e15582.

Preziosi, A.

2001 Standardization and Specialization: The Island Chumash Microdrill Industry. In *The Origins of a Pacific Coast Chiefdom: The Chumash of the Channel Islands*, edited by J. Arnold, pp. 151-164. University of Utah Press, Salt Lake City.

Price, T. D., and A. B. Gebauer

1992 The Final Frontier: Foragers to Farmers in Southern Scandinavia. In *Transitions to Agriculture in Prehistory*, edited by A. B. Gebauer and T. D. Price, pp. 97-116. Monographs in World Archaeology No. 4. Prehistory Press, Madison.

Prošek, F.

1953 Szeletien na Slovensku. *Slovenská Archeológia* 1:133-194.

Pryor, J., and C. Carr

1995 Basketry of Northern California Indians: Interpreting Style Hierarchies. In *Style, Society, and Person: Archaeological and Ethnological Perspectives*, edited by C. Carr and J. Neitzel, pp. 259-296. Plenum Press, New York.

Pusey, A. E.

1992 The Primate Perspective on Dispersal. In *Animal Dispersal: Small Mammals as a Model*, edited by N. C. Stenseth and W. Z. Lidicker, pp. 243-259. Chapman and Hall, London.

Quinn, G. D.

2007 *Fractography of Ceramics and Glasses*. NIST Recommended Practice Guide, Special Publication 960-16. National Institute of Standards and Technology (NIST), U.S. Department of Commerce, Washington D.C.

Ramenofsky, A. F.

1995 Evolutionary Theory and Native American Artifact Change in the Postcontact Period. In *Evolutionary Archaeology: Methodological Issues*, P. Teltser, pp. 129–47. University of Arizona Press, Tucson.

Rands, R. L., and C. L. Riley

1958 Diffusion and Discontinuous Distribution. *American Anthropologist* 60:274–297.

Rasic, J., and W. Andrefsky, Jr.

2001 Alaskan Blade Cores as Specialized Components of Mobile Toolkits: Assessing Design Parameters and Toolkit Organization through Debitage Analysis. In *Lithic Debitage: Context, Form, Meaning*, edited by W. Andrefsky, Jr., pp. 6–79. University of Utah Press, Salt Lake City.

Rebollo, N. R., S. Weiner, F. Brock, L. Meignen, P. Goldberg, A. Belfer-Cohen, O. Bar-Yosef, and E. Boaretto

2011 New radiocarbon dating of the transition from the Middle to the Upper Paleolithic in Kebara Cave, Israel. *Journal of Archaeological Science* 38:2424-2433.

Reich, D., R. E. Green, M. Kircher, J. Krause, N. Patterson, E. Y. Durand, B. Viola, A. W. Briggs, U. Stenzel, P. L. F. Johnson, T. Maricic, J. M. Good, T. Marques-Bonet, C. Alkan, Q. Fu, S. Mallick, H. Li, M. Meyer, E. E. Eichler, M. Stoneking, M. Richards, S. Talamo, M. V. Shunkov, A. P. Derevianko, J.-J. Hublin, J. Kelso, M. Slatkin, and S. Pääbo

2010 Genetic History of an Archaic Hominin Group from Denisova Cave in Siberia. *Nature* 468:1053–1060.

Reid, J. J.

1997 Return to Migration, Population Movement, and Ethnic Identity in the American Southwest. In *Vanishing River: Landscapes and Lives of the Lower Verde River Valley*, edited by S. M. Whittlesey, R. Ciolek-Torrello, and J. Altschul, pp. 629–638. SRI Press, Tucson.

Reid, J. J., M. B. Schiffer, and W. L. Rathje

1975 Behavioral Archaeology: Four Strategies. *American Anthropologist* 77:864–869.

Relethford, J. H.

1998 Genetics of Modern Human Origins and Diversity. *Annual Review of Anthropology* 27:1–23.

Renfrew, C.

1978 The Anatomy of Innovation. In *Social Organization and Settlement: Contributions from Anthropology, Archaeology and Geography, Part I*, edited by D. Green, C. Haselgrove, and M. Spriggs, pp. 89–117. BAR International Series (Supplementary) 47(i), Oxford.

Renfrew, C., and P. Bahn

2000 *Archaeology: Theories, Methods, and Practice. Third Edition.* Thames and Hudson, London.

Révillion, S.

1989 Le Débitage du Gisement Paléolithique Moyen de Seclin (Nord). In *Paléolithique et Mésolithique du Nord de la France: Nouvelles Recherches*, pp. 79–89. Publication of CERP 1. Université des Sciences et Techniques de Lille, Flandres-Artois, Lille.

Révillion, S., and A. Tuffreau (Editors)

1994 *Les Industries Laminaires au Paléolithique Moyen.* Dossier de Documentation Archéologique No. 18. CRA. Éditions du Centre National de la Recherche Scientifique, Paris.

Rezek, Z., S. Lin, R. Iovita, and H. L. Dibble

2011 The Relative Effects of Core Surface Morphology on Flake Shape and Other Attributes. *Journal of Archaeological Science* 38:1346–1359.

Richter, D., J. Waiblinger, W. J., Rink, G.A. Wagner

2000 Thermoluminescence, Electron Spin Resonance and ^{14}C dating of the Late Middle and Early Upper Palaeolithic Site of Geißenklösterle Cave in southern Germany. *Journal of Archaeological Science* 27:71-89.

Riel-Salvatore, J., and C. M. Barton

2004 Late Pleistocene Technology, Economic Behavior, and Land-Use Dynamics in Southern Italy. *American Antiquity* 69(2):257-274.

Riel-Salvatore, J., A. E. Miller, and G. A. Clark

2008 An Empirical Evaluation of the Case for a Châtelperronian-Aurignacian Interstratification at Grotte des Fées de Châtelperron. *World Archaeology* 40(4):480-492.

Riel-Salvatore, J., G. Popsecu, and C. M. Barton

2008 Standing at the Gates of Europe: Human Behavior and Biogeography in the Southern Carpathians during the Late Pleistocene. *Journal of Anthropological Archaeology* 27:399-417.

Rigaud, J. F. (Translator)

1802 *A Treatise on Painting by Leonardo Da Vinci*. Taylor, London.

Rindos, D. R.

1989 Diversity, Variation, and Selection. In *Quantifying Diversity in Archaeology*, edited by R. D. Leonard and G. T. Jones, pp. 13-23. Cambridge University Press, Cambridge.

Rink, W. J., H. P. Schwarcz, H. K. Lee, V. Cabrera Valdés, F. Bernaldo de Quirós, and M. Hoyos

1996 ESR Dating of Tooth Enamel: Comparison with AMS ^{14}C at El Castillo Cave, Spain. *Journal of Archaeological Science* 23:945-951.

Richerson, P. J., and R. Boyd

1978 A Dual Inheritance Model of the Human Evolutionary Process: I. Basic Postulates and a Simple Model. *Journal of Social and Biological Structures* 1:127-154.

2001 Institutional Evolution in the Holocene: The Rise of Complex Societies. In *The Origin of Human Social Institutions*, edited by W. G. Runciman, pp. 197-234. Published for the British Academy by the Oxford University Press, Vol. 110. Oxford.

2002 Culture is Part of Human Biology: Why the Superorganic Concept Serves the Human Sciences Badly. In *Probing Human Origins*, edited by P. J. Richerson, pp. 59-85. American Academy of Arts and Sciences, Cambridge.

2005 *Not by Genes Alone: How Culture Transformed Human Evolution*. University of Chicago Press, Chicago.

Rink, W. J., H. P. Schwarcz, K. Valoch, L. Seitl, and C. B. Stringer

1996b ESR Dating of the Micoquian Industry and Neanderthal Remains at Kulna Cave, Czech Republic. *Journal of Archaeological Science* 23:889-901.

Roberts, F.

1935 A Folsom Complex: Preliminary Report on Investigators at the Lindemeier Site in Northern Colorado. *Smithsonian Miscellaneous Collections* 94(4).

Roe, P. G.

1995 Style, Society, Myth, and Structure. In *Style, Society, and Person: Archaeological and Ethnological Perspectives*, edited by C. Carr and J. Neitzel, pp. 27-76. Plenum Press, New York.

Roebroeks, W.

1989 *From Find Scatters to Early Hominid Behaviour: A Study of Middle Palaeolithic Riverside Settlements at Maastricht-Belvédère (The Netherlands)*. Analecta Praehistorica Leidensia 21. University of Leiden, Leiden.

Rogers, E. M.

1995 *Diffusion of Innovations. Fourth Edition.* The Free Press, New York.

Rolland, N.

1981 The Interpretation of Middle Paleolithic Variability. *Man* 16:15-42.

Rolland, N., and H. L. Dibble

1990 A New Synthesis of Middle Paleolithic Variability. *American Antiquity* 55(3):480-499.

Ronen, A., and B. Vandermeersch

1972 The Upper Paleolithic Sequence in the Cave of Qafza (Israel). *Quaternaria* XVI:189-202.

Rose, L. M.

1999 Book Review: Chimpanzee and Red Colobus and *The Hunting Apes* by Craig Stanford. *International Journal of Primatology* 20(6):1029-1031.

Rouse, I. B.

1958 The Inference of Migration from Anthropological Evidence. In *Migrations in New World Culture History*, edited by R. H. Thompson, pp. 63-68. University of Arizona Social Science Bulletin 27. University of Arizona, Tucson.

1986 *Migrations in Prehistory: Inferring Population Movement from Cultural Remains.* Yale University Press, New Haven.

Roussel, M.

2011 Normes et Variations de la Production Lithique Durant le Châtelperronien: La Séquence de la Grande-Roche-de-la-Plématrie à Quinçay. Doctoral Thesis, Université Paris Ouest Nanterre-La Défense.

Roux, V., B. Bril, and G. Dietrich

1995 Skills and Learning Difficulties Involved in Stone Knapping: The Case of Stone-Bead Knapping in Khambhat, India. *World Archaeology* 27(1):63-87.

Roux, V., and B. Bril (Editors)

2005 *Stone Knapping: The Necessary Conditions for a Uniquely Hominin Behaviour.* McDonald Institute for Archaeological Research, Cambridge.

Rust. A.

1933 Beitrag zur Erkenntnis der Abwicklung der vorgeschitctlichen Kulturperioden in Syrien. *Praehistorische Zeitschrift* 24:205-218.

1950 *Die Höhlenfunde von Yabrud (Syrien).* Karl Wachholtz, Neumunster.

Sackett, J. R.

1966 Quantitative Analysis of Upper Paleolithic Stone Tools. *American Anthropologist* 68:356-394.

1973 Style, Function and Artifact Variability in Palaeolithic Assemblages. In *The Explanation of Culture Change: Models in Prehistory*, edited by C. Renfrew, pp. 317-325. Duckworth, London.

1982 Approaches to Style in Lithic Analysis. *Journal of Anthropological Archaeology* 1:59-112.

1985 Style and Ethnicity in the Kalahari: A Reply to Wiessner. *American Antiquity* 50(1):154-159.

1986a Isochrestism and Style: A Clarification. *Journal of Anthropological Archaeology* 5:266-277.

1986b Style, Function and Artifact Variability. A Reply to Binford. *American Antiquity* 51(3):628-634.

1989 Statistics, Attributes, and the Dynamics of Burin Typology. In *Alternative Approaches to Lithic Analysis*, edited by D. O. Henry and G. Odell, pp. 51-82. Archaeolgoical Papers of the American Anthropological Association 1, Washington, D.C.

1990 Style and Ethnicity in Archaeology: The Case for Isochrestism. In *The Uses of Style in Archaeology*, edited by M. Conkey and C. Hastorf, pp. 32-43. Cambridge University Press, Cambridge.

1991 Straight Archaeology French Style: The Phylogenetic Paradigm in Historic Perspective. In *Perspectives on the Past: Theoretical Biases in Mediterranean Hunter-Gatherer Research*, edited by G. A. Clark, pp. 109–140. University of Pennsylvania Press, Philadelphia.

Sackett, J. R. (Editor)

1999 *The Archaeology of Solvieux: An Upper Paleolithic Open Air Site in France.* Monumenta Archaeologica 19. Institute of Archaeology, University of California, Los Angeles.

Sancetta, C., J. Imbrie, and N. G. Kipp

1973 Climatic Record of the Past 130,000 Years in North Atlantic Deep-Sea Core V23-82: Correlation with theTerrestrial Record. *Quaternary Research* 3:110–116.

Sandgathe, D. M.

2004 Alternative Interpretation of the Levallois Reduction Technique. *Lithic Technology* 29(2):147–159.

Sapir, Edward

1916 *Time Perspective in Aboriginal American Culture: A Study in Method.* Canada Department of Mines, Geological Survey, Memoir 90, Anthropological Series, No. 13, Ottawa.

Sarel, J.

2004 *The Middle-Upper Paleolithic Transition in Israel: Technological Analysis.* British Archaeological Reports International Series 1229, Oxford.

Sarel, J., and A. Ronen

2003 The Middle/Upper Palaeolithic Transition in Northern and Southern Israel: A Technological Comparison. In *More than Meets the Eye: Studies on Upper Palaeolithic Diversity in the Near East*, edited by A. N. Goring-Morris and A. Belfer-Cohen, pp. 68–79. Oxbow Press, Oxford.

Savage, W. W., Jr., and S. I. Thopmson (Editors)

1979 *The Frontier: Comparative Studies, Volume Two.* University of Oklahoma Press, Norman.

Savich, V. P.

1975 *Pizn'opaleolithischne nasalennya Pivdenno-Zakhidnoj Volyni.* Naukova dumka, Kiev.

Schiffer, M. B.

1975 Behavioral Chain Analysis: Activities, Organization, and the Use of Space. *Chapters in the Prehistory of Eastern Arizona IV, Fieldiana: Anthropology* Vol. 65, pp. 103–119. Field Museum of Natural History, Chicago.

1976 *Behavioral Archaeology*. Academic Press, New York.

1987 *Formation Processes of the Archaeological Record.* University of New Mexico Press, Albuquerque.

1991 *The PorRadio in American Life.* University Arizona of Press, Tucson.

1993 Cultural Imperatives and Product Development: The Case of the Shirt-Pocket Radio. *Technology and Culture* 34:98–113.

1995 Social Theory and History in Behavioral Archaeology. In *Expanding Archaeology*, edited by J. M. Skibo, W. H. Walker, and A. E. Nielson, pp. 22–35. University of Utah Press, Salt Lake City.

1996 Some Relationships Between Behavioral and Evolutionary Archaeologies. *American Antiquity* 61:643–662.

1999 *The Material Life of Human Beings: Artifacts, Behavior, and Communication.* Routledge, New York.

2005 The Devil is in the Details: The Cascade Model of Invention Processes. *American Antiquity* 70(3):485–502.

Schiffer, M. B., and J. Skibo

1987 Theory and Experiment in the Study of Technological Change. *Current Anthropology* 28:595–622.

1997 The Explanation of Artifact Variability. *American Antiquity* 62:27–50.

Schiffer, M. B., T. C. Butts, and K. K. Grimm

1994 *Taking Charge: The Electric Car in America*. Smithsonian Institution Press, Washington, D.C.

Schlanger, N.

1990 Technique as Human Action: Two perspectives. *Archaeological Review from Cambridge* 9:18-26.

1996 Understanding Levallois: Lithic Technology and Cognitive Archaeology. *Cambridge Archaeological Journal* 6(2):231-254.

Schmider, B., and M. Perpère

1995 Production et Utilization des Lamelles dans l'Aurignacien de la Grotte du Renne à Arcy-sur-Cure. In *Paléolithique Supérieur et Epipaléolithique dans le Nord-Est de la France*, edited by Y. Pautrat, pp. 4-10. Direction Régionale des Affaires Culturelles, Dijon.

Schmidt, Wilhelm

1939 *The Cultural Historical Method of Ethnology*, translated by S. A. Sieber. Fortuny's, New York.

Schmidt, W., and W. Koppers

1924 *Völker und Kulturen: Gesellschaft und Wirtschaft der Völker*. Verlag Josef Habbel, Regensberg.

Schurmans, U.

2007 Refitting in the Old and New Worlds. In *Fitting Rocks: Lithic Refitting Examined*, edited by U. Schurmans and M. De Bie, pp. 7-24. British Archaeological Reports International Series 1596. Archaeopress, Oxford.

Schwarcz, H. P., W. M. Buhay, R. Grun, H. Valladas, E. Tchernov, O. Bar-Yosef, and B. Vandermmeersch

1989 ESR Dating of the Neanderthal Site, Kebara Cave, Israel. *Journal of Archaeological Science* 16:653-659.

Schiegl, S., P. Goldberg, O. Bar-Yosef, and S. Weiner

1996 Ash Deposits in Hayonim and Kebara Caves, Israel: Macroscopic, Microscopic and Mineralogical Observations, and their Archaeological Implications. *Journal of Archaeological Science* 23:763-781.

Secakuku, A. H.

1995 *Following the Sun and Moon: Hopi Kachina Tradition*. Northland Publishing and the Heard Museum, Flagstaff.

Sellet, F.

2004 Beyond the Point: Projectile Manufacture and Behavioral Inference. *Journal of Archaeological Science* 31:1553-1566.

Serre, D., A. Langaney, M. Chech, M. Teschler-Nicola, M. Paunovic, P. Mennecier, M. Hofreiter, G. Possnert, S. Pääbo

2004 No Evidence of Neandertal mtDNA Contribution to Early Modern Humans. *PLoS Biology* 2(3):313-317.

Shackleton, N. J.

1987 Oxygen Isotopes, Ice Volume, and Sea Level. *Quaternary Science Reviews* 6:183-190.

Shea, J. J.

1998 Neandertal and Early Modern Human Behavioral Variability: A Regional-Scale Approach to Lithic Evidence for Hunting in the Levanting Mousterian. *Current Anthropology* 39(2):S45-S78.

2003 The Middle Paleolithic of the East Mediterranean Levant. *Journal of World Prehistory* 17(4):313-94

2006 The Origins of Lithic Projectile Point Technology: Evidence from Africa, the Levant, and Europe. *Journal of Archaeological Science* 33:823-846.

2007 Behavioral Differences between Middle and Upper Paleolithic *Homo sapiens* in the East Mediterranean Levant: The Roles of Intraspecific Competition and Dispersal from Africa. *Journal of Anthropological Research* 63(4):449-488.

2008a The Middle Stone Age Archaeology of the Lower Omo Valley Kibish Formation: Excavations, Lithic Assemblages, and Inferred Patterns of Early *Homo sapiens* Behavior. *Journal of Human Evolution* 55:418-485.

2008b The Lower and Middle Paleolithic in the Middle East and Neighboring Regions. *Evolutionary Anthropology* 17:205-207.

2011a Stone Tool Analysis and Human Origins Research: Some Advice from Uncle Screwtape. *Evolutionary Anthropology* 20:48-53.

2011b *Homo sapiens* Is as *Homo sapiens* Was. *Current Anthropology* 52(1):1-35.

Shea, J. J., and O. Bar-Yosef

2005 Who Were The Skhul/Qafzeh People? An Archaeological Perspective on Eurasia's Oldest Modern Humans. *Journal of the Israel Prehistoric Society* 35:451-468.

Shea, J. J., Z. Davis, and K. Brown

2001 Experimental Tests of Middle Palaeolithic Spear Points Using a Calibrated Crossbow. *Journal of Archaeological Science* 28:807-816.

Shennan, S. J.

1989a Cultural Transmission and Cultural Change. In *What's New: A Closer Look at the Process of Innovation*, edited by S. E. van der Leeuw and R. Torrence, pp. 330-346. Unwin Hyman, London.

1989b Introduction. In *Archaeological Approahes to Cultural Identity*, edited by S. J. Shennan, pp. 1-32. Unwin and Hyman, London.

1991 Tradition, Rationality, and Cultural Transmission. In *Processual and Postprocessual Archaeologies: Multiple Ways of Knowing the Past*, edited by R. W. Preucel, pp. 197-208. Center for Archaeoligcal Investigations Occassional Paper 10. Southern Illinois University, Carbondale.

1997 *Quantifying Archaeology, Second Edition.* Edinburgh University Press, Edinburgh.

2000 Population, Culture History, and the Dynamics of Culture Change. *Current Anthropology* 41(5):811-835.

2002 Archaeology and Evolutionary Ecology. *World Archaeology* 34(1):1-5.

2003 *Genes, Memes and Human History: Darwinian Archaeology and Cultural Evolution.* Thames and Hudson, London.

Shennan, S. J., and J. Steele

1999 Cultural Learning in Hominids: A Behavioural Ecological Approach. In *Mammalian Social Learning: Comparative and Ecological Perspectives*, edited by H. O. Box and K. R. Gibson, pp. 367-388. Symposia of the Zoological Society of London 72. Cambridge University Press, Cambridge.

Shennan, S. J., and M. Collard

2000 The Evolutionary Analysis of Cultural Behaviour. *Archaeology International* 4:21-23.

2005 Investigating Processes of Cultural Evolution on the North Coast of New Guinea with Multivariate and Cladistic Analyses. In *The Evolution of Cultural Diversity: A Phylogenetic Approach*, edited by R. Mace, C. Holden, and S. Shennan, pp. 133-164. UCL Press, London.

Sherman, P. W.

1977 Nepotism and the Evolution of Alarm Calls. *Science* 197:1246-1253.

Shimelmitz, R., R. Barkai, and A. Gopher

2011 Systematic Blade Production at Late Lower Paleolithic (400–200 kyr) Qesem Cave, Israel. *Journal of Human Evolution* 61:458–479.

Shott, M.

1986 Technological Organization and Settlement Mobility: An Ethnographic Examination. *Journal of Anthropological Research* 42(1):15-51.

1989 On Tool-Class Use Lives and the Formation of Archaeological Assemblages. *American Antiquity* 54(1):9-30.

1996 Stage Versus Continuum in the Debris Assemblage from Production of a Fluted Biface. *Lithic Technology* 21:6-22.

2003 *Chaîne Opératoire* and Reduction Sequence. *Lithic Technology* 28(2):95-105.

Shott, M. J., and P. Sillitoe

2004 Modeling Use-Life Distributions in Archaeology Using New Guinea Wola Ethnographic Data. *American Antiquity* 69(2):339-355.

Shott, M. J., A. P. Bradbury, P. J. Carr, and G. H. Odell

2000 Flake Size from Platform Attributes: Predictive and Empirical Approaches. *Journal of Archaeological Science* 27:877-894.

Shott, M. J., J. M. Lindly, and G.A. Clark

2011 Continuous Modeling of Core Reduction: Lessons from Refitting Cores from WHS 623x, An Upper Paleolithic Site in Jordan. *PaleoAnthropology* 2011:320-333.

Sillitoe, P., and K. Hardy

2003 Living Lithics: Ethnoarchaeology in Highland New Guinea. *Antiquity* 77:555-566.

Sinclair, A. R. E.

1983 The Functions of Distance Movements in Vertebrates. In *The Ecology of Animal Movement*, edited by I. R. Swingland and P. J. Greenwood, pp. 240-258. Clarendon Press, Oxford.

Sinopoli, C.

1991 Style in Arrows: A Study of an Ethnographic Collection from the Western United States. In *Foragers in Context: Long-Term, Regional and Historical Perspectives in Hunter-Gatherer Studies*, edited by P. Miracle, L. Fisher, and J. Brown, pp. 63-87. Michigan Discussions in Anthropology 10. University of Michigan, Ann Arbor.

Sirakov, N.

1979 Analytical Study on Blanks from the Cave Samuilitsa II (An Example of Application of Attribute Correlation Method). In *Middle and Early Upper Palaeolithic in Balkans*, edited by J. K. Kozlowski, *Prace Archeologiczne* 28:65-76.

1983 Reconstruction of the Middle Palaeolithic Flint Assemblages from the Cave Samuilitsa II (Northern Bulgaria) and their Taxonomical Position seen against the Palaeolithic of South-Eastern Europe. *Folia Quaternaria* 55.

Sisk, M. L., and J. J. Shea

2009 Experimental Use and Quantitative Performance Analysis of Triangular Flakes (Levallois Points) Used as Arrowheads. *Journal of Archaeological Science* 36:2039-2047.

Sitlivy, V., and A. Zięba

2006 Eastern and Central Europe before 30 kyr BP: Mousterian, Levallois and Blade Industries. In *Kabazi II: The 70 000 Years Since the Last Interglacial*, edited by V. Chabai, J. Richter, and T. Uthmeier, pp. 361-419. Shlyakh, Simferopol - Cologne.

Sitlivy, V., K. Sobczyk, T. Kalicki, C. Escutenaire, A. Zięba, K. Kaczor

1999a The New Palaeolithic Site of Księcia Józefa (Kraków, Poland) with Blade and Flake Productions. *Préhistoire Européenne* 15:87-111.

Sitlivy, V., K Sobczyk, W. Morawski, A. Zięba, and C. Escutenaire

1999b Piekary IIa Palaeolithic Industries: Preliminary Results of New Multidisciplinary Investigations. *Préhistoire Européenne* 15:45-61.

Skibo, J. M, W. H. Walker, and A. E. Nielson (Editors)

1995 *Expanding Archaeology*. University of Utah Press, Salt Lake City.

Škrdla, P.

1994 Rekonstrukce Paleolitických Technologií na Stránské skále. *Pravěk* 4:5-15.

1996 The Bohunician Reduction Strategy. *Quaternaria Nova* VI:93-107.

2003a Comparison of Boker Tachtit and Stránská skála MP/UP Transitional Industries. *Journal of the Israel Prehistoric Society* 33:33-69.

2003b Bohunician and Aurignacian Technologies: Morphological Description. In *Stránská skála: Origins of the Upper Paleolithic in the Brno Basin*, edited by J. Svoboda and O. Bar-Yosef, pp. 65-76. American School of Prehistoric Research Bulletin 47. Dolní Věstonice Studies 10. Peabody Museum of Archaeology and Ethnology, Harvard University, Cambridge.

2003c Bohunician Technology: A Refitting Approach. In *Stránská skála: Origins of the Upper Paleolithic in the Brno Basin*, edited by J. Svoboda and O. Bar-Yosef, pp. 119–151. American School of Prehistoric Research Bulletin 47. Dolní Věstonice Studies 10. Peabody Museum of Archaeology and Ethnology, Harvard University, Cambridge.

Škrdla, P., and G. Tostevin

2003 Brno (k.ú. Bohunice, okr. Brno-město). *Přehled výzkumů* 44:188–192. In Czech and English.

Škrdla, P., and G. Tostevin

2005 Brno-Bohunice, Analýza Materiálu z Výzkumu v Roce 2002. Brno-Bohunice, Analysis of the Material from the 2002 Excavation. *Přehled výzkumů* 46:35–61. In Czech and English.

Skutil, J.

1938 *Praveké nálezy v Mladci u Litovle na Moravé.* Litovel.

Smith, E. A.

1983 Anthropological Applications of Optimal Foraging Theory: A Critical Review. *Current Anthropology* 24(5):625–651.

Smith, E. A., and B. Winterhalder

1992 Natural Selection and Decision Making: Some Fundamental Principles. In *Evolutionary Ecology and Human Behavior*, edited by E. A. Smith and B. Winterhalder, pp. 25–60. Aldine de Gruyter, New York.

Smith, F. H., I. Janković, and I. Karavanic

2005 The Assimilation Model, Modern Human Origins in Europe, and the Extinction of Neandertals. *Quaternary International* 137:7–19.

Smith, M. R.

1994 Technological Determinism in American Culture. In *Does Technology Drive History? The Dilemma of Technological Determinism*, edited by M. R. Smith and L. Marx, pp. 1–35. MIT Press, Cambridge.

Smith, M. L.

1994 Recourse of Empire: Landscapes of Progress in Technological America. In *Does Technology Drive History? The Dilemma of Technological Determinism*, edited by M. R. Smith and L. Marx, pp. 37–52. MIT Press, Cambridge.

Snow, D. R.

1995 Migration in Prehistory: The Northern Iroquoian Case. *American Antiquity* 60(1):59–79.

1996 More on Migration in Prehistory: Accommodating New Evidence in the Northern Iroquoian Case. *American Antiquity* 61(4):791–796.

Sokal, R. R., and F. J. Rohlf

1995 *Biometry*. W. H. Freeman, New York.

Soltis, J., R. Boyd, and P. J. Richerson

1995 Can Group-Functional Behaviors Evolve by Cultural Group Selection? *Current Anthropology* 36(3):473–494.

Sonneville-Bordes, D. de

1960 *Le Paléolithique Supérieur en Périgord.* Delmas, Bordeaux.

Sonneville-Bordes, D. de and J. Perrot

1954-56 Lexique Typologique du Paléolithique Supérieur. *Bulletin de la Société Préhistorique Française* 51:327–335; 52:76–79; 53:408-412, 547–559.

Sorel, J.

2004 *The Middle-Upper Paleolithic Transition in Israel: Technological Analysis.* BAR International Series 1229, Oxford.

Sorel, J., and A. Ronen

2003 The Middle/Upper Palaeolithic Transition in Northern and Southern Israel: A Technological Comparison. In *More than Meets the Eye: Studies on Upper Palaeolithic Diversity in the Near East*, edited by A. N. Goring-Morris and A. Belfer-Cohen, pp. 68-79. Oxbow Press, Oxford.

Soressi, M.

2002 Le Moustérien de Tradition Acheuléenne du Sud-ouest de la France. Discussion sur la Signification du Faciès à Partir de L'étude Comparée de Quatre Sites: Pech-de-l'Azé I, Le Moustier, La Rochette et la Grotte XVI. Thèse de l'Université Bordeaux I.

Soressi, M., and J.-M. Geneste

2011 The History and Efficacy of the *Chaîne Opératoire* Approach to Lithic Analysis: Studying Techniques to Reveal Past Societies in an Evolutionary Perspective. *PaleoAnthropology* 2011:334-350.

Soressi, M., W. Rendu, J.-P. Texier, É. Claud, L. Daulny, F. D'Errico, V. Laroulandie, B. Maureille, M. Niclot, S. Schwortz, and A.-M. Tillier

2008 Pech-de-l'Azé I (Dordogne, France). Nouveau Regard sur un Gisement Moustérien de Tradition Acheuléene Connu depuis le XIXe Siècle. In Les Societies du Paléolithique dans un Grand Sud-Ouest de la France: Nouveaux Gisements, Nouveaux Resultants, Nouvelles Méthodes, *Société Préhistorique Français Mémoire* 47:95-132. Paris.

Spaulding, A. C.

1953 Statistical Techniques for the Discovery of Artifact Types. *American Antiquity* 18:305-313.

Sperber, D.

1996 *Explaining Culture: A Naturalistic Approach.* Blackwell, Oxford.

Speth, J. D.

1972 Mechanical Basis of Percussion Flaking. *American Antiquity* 37:34-60.

1974 Experimental Investigations of Hard-Hammer Percussion Flaking. *Tebiwa* 17:7-36.

1975 Miscellaneous Studies in Hard-Hammer Percussion Flaking: The Effects of Oblique Impact. *American Antiquity* 40:203-207.

1981 The Role of Platform Angle and Core Size in Hard Hammer Percussion Flaking. *Lithic Technology* 10:16-21.

Speth, J. D., and E. Tchernov

1998 The Role of Hunting and Scavenging in Neandertal Procurement Strategies: New Evidence from Kebara Cave (Israel). In *Neandertals and Modern Humans in Western Asia*, edited by T. Akazawa, K. Aoki and O. Bar-Yosef, pp. 223-239. Plenum Press, New York.

2002 Middle Paleolithic Tortoise Use at Kebara Cave (Israel). *Journal of Archaeological Science* 29:471-483.

2007 The Middle Paleolithic Occupations at Kebara Cave. A Faunal Perspective. In *Kebara Cave, Mt. Carmel, Israel: The Middle and Upper Paleolithic Archaeology, Part I*, edited by O. Bar-Yosef and L. Meignen, pp. 165-260. American School of Prehistoric Research Bulletin 49. Peabody Museum of Archaeology and Ethnology, Harvard University, Cambridge.

Spielmann, K. A. (Editor)

1998 *Migration and Reorganization: The Pueblo IV Period in the American Southwest.* Anthropological Research Papers No. 51, Arizona State University, Tempe.

Spratt, D. A.

1982 The Analysis of Innovation Processes. *Journal of Archaeological Science* 9:79-94.

Stahle, D. W., and J. E. Dunn

1982 An Analysis and Application of the Size Distribution of Waste Flakes from the Manufacture of Bifacial Stone Tools. *World Archaeology* 14:84-97.

Staller, J. E., and R. G. Thompson

2002 A Multidisciplinary Approach to Understanding the Initial Introduction of Maize into Coastal Ecuador. *Journal of Archaeological Science* 29(1):33-50.

Stanford, C. B.

1998 *Chimpanzee and Red Colobus: The Ecology of Predator and Prey*. Harvard University Press, Cambridge.

1999 *The Hunting Apes: Meat Eating and the Origins of Human Behavior*. Princeton University Press, Princeton.

2001 A Comparison of Social Meat-Foraging by Chimpanzees and Human Foragers. In *Meat-Eating and Human Evolution*, edited by C. B. Stanford and H. T. Bunn, pp. 122-140. Oxford University Press, Oxford.

Stanford, C. B., J. Wallis, E. Mpongo, and J. Goodall

1994 Hunting Decisions in Wild Chimpanzees. *Behaviour* 131:1-20.

Stanford, Dennis J., and Bruce A. Bradley

2012 *Across Atlantic Ice: The Origin of America's Clovis Culture*. University of California Press, Berkeley.

Stanner, W. E. H.

1965 Aboriginal Territorial Organization: Estate, Range, Domain and Regime. *Oceania* 36:1-26.

Stark, M. T.

1995 Economic Intensification and Ceramic Specialization in the Philippines: A View from Kalinga. *Research in Economic Anthropology* 16:179-226.

1998 Technical Choices and Social Boundaries in Material Culture Patterning: An Introduction. In *The Archaeology of Social Boundaries*, edited by M. T. Stark, pp. 1-11. Smithsonian Institution Press, Washington, D.C.

Stark, M. T., J. J. Clark, and M. D. Elson

1995 Social Boundaries and Cultural Identity in the Tonto Basin. In *The Roosevelt Community Development Study: New Perspectives on Tonto Basin Prehistory*, edited by M. D. Elson, M. T. Stark, and D. A. Gregory, pp. 343-368. Anthropological Papers 15. Center for Desert Archaeology, Tucson.

Stark, M. T., M. D. Elson, and J. J. Clark

1998 Social Boundaries and Technical Choices in Tonto Basin Prehistory. In *The Archaeology of Social Boundaries*, edited by M. T. Stark, pp. 208-231. Smithsonian Institution Press, Washington, D.C.

Stearns, S. C.

1992 *The Evolution of Life Histories*. Oxford University Press, Oxford.

Steffen, A., E. J. Skinner, and P. W. Ainsworth

1998 A View to the Core: Technological Units and Debitage Analysis. In *Unit Issues in Archaeology: Measuring Time, Space, and Material*, edited by A. F. Ramenofsky and A. Steffen, pp. 131-146. University of Utah Press, Salt Lake City.

Stenseth, N. C., and W. Z. Lidicker, Jr.

1992 The Study of Dispersal: A Conceptual Guide. In *Animal Dispersal: Small Mammals as a Model*, edited by N. C. Stenseth and W. Z. Lidicker, pp. 5-20. Chapman and Hall, London.

Stepanchuk, V., and V. Cohen

2000-2001 The Kremenician, a Middle to Upper Palaeolithic Transitional Industry in the Western Ukraine. *Préhistoire Européenne* 16-17:75-110.

Sternberg, R., and E. H. E. Lass

2007 An Archaeomagnetic Study of Two Hearths from Kebara Cave, Israel. In *Kebara Cave, Mt. Carmel, Israel: The Middle and Upper Paleolithic Archaeology, Part I*, edited by O. Bar-Yosef and L. Meignen, pp. 123-130. American School of Prehistoric Research Bulletin 49. Peabody Museum of Archaeology and Ethnology, Harvard University, Cambridge.

Steward, J. H.

1929 Diffusion and Independent Invention: A Critique of Logic. *American Anthropologist* 31:491-495.

1955 *Theory of Culture Change*. University of Illinois Press, Urbana.

Stiner, M.

1995 *Honor among Thieves: A Zooarchaeological Perspective on Neandertal Ecology*. Princeton University Press, Princeton.

Stiner, M., and S. Kuhn

1992 Subsistence, Technology, and Adaptive Variation in Middle Paleolithic Italy. *American Anthropologist* 94(2):306-339.

Stojanowski, C. M.

2004 Population History of Native Groups in Pre- and Postcontact Spanish Florida: Aggregation, Gene Flow, and Genetic Drift on the Southeastern U.S. Atlantic Coast. *American Journal of Physical Anthropology* 123:316-332.

Stone, T.

2003 Social Identity and Ethnic Interaction in the Western Pueblos of the American Southwest. *Journal of Archaeological Method and Theory* 10(1):31 67.

Stout, D.

2002 Skill and Cognition in Stone Tool Production: An Ethnographic Case Study from Irian Jaya. *Current Anthropology* 43(5):693-722.

2005 The Social and Cultural Context of Stone-knapping Skill Acquisition. In *Stone Knapping: The Necessary Conditions for a Uniquely Hominin Behaviour*, edited by V. Roux and B. Bril, pp. 331-340. McDonald Institute for Archaeological Research, Cambridge.

Straus, L. G.

1983 *From Mousterian to Magdalenian: Cultural Evolution Viewed from Vasco-Cantabrian Spain Pyrenean France*. BAR International Series 164, Oxford.

1989 Age of the Modern Europeans. *Nature* 342:476-477.

1994 Upper Paleolithic Origins and Radiocarbon Calibration: More New Evidence from Spain. *Evolutionary Anthropology* 2:195-198.

2003 The Aurignacian? Some Thoughts. In *The Chronology of the Aurignacian and of the Transitional Technocomplexes: Dating, Stratigraphies, Cultural Implications*, edited by J. Zilhão and F. d'Errico, pp. 11-18. Proceedings of the Symposium 6.1 of the XIVth Congress of the UISPP. Trabalhos de Arquelogia 33. Instituto Português de Arqueologia, Lisbon.

Straus, L. G., and G. Clark

1986 La Riera Archaeological Remains—Level Contents and Characteristics. In *La Riera Cave: Stone Age Hunter-Gatherer Adaptations in Northern Spain*, edited by L. Straus and G. Clark, pp. 75-187. Arizona State University Anthropological Papers No. 36, Tempe.

Stringer, C. B., and P. Andrews

1988 Genetic and Fossil Evidence for the Origin of Modern Humans. *Science* 239:1263-1268

Stutz, A., and G. Estabrook

2004 Computationally Intensive Multivariate Statistics and Relative Frequency Distributions in Archaeology (with an Application to the Early Epipaleolithic of the Levant). *Journal of Archaeological Science* 31:1643-1658.

Stutz, A., and D. Lieberman

2007 Cementum Increment Analysis of Ungulate Teeth from the Kebara Cave Faunal Assemblages: Discriminating Seasonal Signals from Their Diagenetic Mimics. In *Kebara Cave, Mt. Carmel, Israel: The Middle and Upper Paleolithic Archaeology, Part I*, edited by O. Bar-Yosef and L. Meignen, pp. 261-278. American School of Prehistoric Research Bulletin 49. Peabody Museum of Archaeology and Ethnology, Harvard University, Cambridge.

Sulerzhickiy, L. D.

1984 *Radiouglerodyne daty GIN AN SSSR*. Soobshchenie XL BKIChP No. 54, p. 168. Moscow.

Sullivan, A., and K. Rosen

1985 Debitage Analysis and Archaeological Interpretation. *American Antiquity* 50(4):755-779.

Svoboda, J.

1980 Kîcmencova Industrie z Ondratic. K Problému Pocátku Mladého Paleolitu. *Studie AU CSAV v Brne* IX/I:1-109. Academia, Prague.

1983 Raw Material Sources in Early Upper Paleolithic Moravia. The Concept of Lithic Exploitation Areas. *Anthropologie* (Brno) XXI/2:147-158.

1987 *Stránská skála. Bohunický Typ v Brnenské Kotline*. Academia, Prague.

1991 Stránská skála. Výsledky Výzkumu v Letech 1985-87. *Památky Archeologické (PA)* 82:5-47.

2000 Brno-Stránská skála (k.o. Slatina, okr. Brno-město. Přehled výzkumů 41:76-80

2003a The Bohunician and the Aurignacian. In *The Chronology of the Aurignacian and of the Transitional Technocomplexes: Dating, Stratigraphies, Cultural Implications*, edited by J. Zilhão and F. d'Errico, pp. 123-131. Proceedings of the Symposium 6.1 of the XIVth Congress of the UISPP. Trabalhos de Arquelogia 33. Instituto Português de Artquelogia, Lisbon.

2003b Bohunician and Aurignacian Typology at Stránská skála. In *Stránská skála: Origins of the Upper Paleolithic in the Brno Basin*, edited by J. Svoboda and O. Bar-Yosef, pp. 153-165. American School of Prehistoric Research Bulletin 47. Dolní Věstonice Studies 10. Peabody Museum of Archaeology and Ethnology, Harvard University, Cambridge.

2003c Chronostratigraphic Background, Environment, and Formation of the Archaeological Layers. In *Stránská skála: Origins of the Upper Paleolithic in the Brno Basin*, edited by J. Svoboda and O. Bar-Yosef, pp. 15-26. American School of Prehistoric Research Bulletin 47. Dolní Věstonice Studies 10. Peabody Museum of Archaeology and Ethnology, Harvard University, Cambridge.

2004 From the Nile to the Danube: An Introduction. *Anthropologie* (Brno) XLII(3):199-201.

Svoboda, J., and O. Bar-Yosef (Editors)

2003 *Stránská skála: Origins of the Upper Paleolithic in the Brno Basin*. American School of Prehistoric Research Bulletin 47. Dolní Věstonice Studies 10. Peabody Museum of Archaeology and Ethnology, Harvard University, Cambridge.

Svoboda, J., and K. Simán

1989 The Middle-Upper Paleolithic Transition in Southeastern Central Europe (Czechoslovakia and Hungary). *Journal of World Prehistory* 3(3):283-322.

Svoboda, J., and P. Škrdla

1995 Bohunician Technology, In *The Definition and Interpretation of Levallois Technology*, edited by H. L. Dibble and O. Bar-Yosef, pp. 429-438. Prehistory Press, Madison.

Svoboda, J., and H. Svobodá

1985 Les Industries de Type Bohunice dans Leur Cadre Stratigraphique et Écologique. *L'Anthropologie* (Paris) 89(4):505-514.

Svoboda, J., and K. Valoch

2003 History and Strategy of Research at Stránská skála. In *Stránská skála: Origins of the Upper Paleolithic in the Brno Basin*, edited by J. Svoboda and O. Bar-Yosef, pp. 1–14. American School of Prehistoric Research Bulletin 47. Dolní Věstonice Studies 10. Peabody Museum of Archaeology and Ethnology, Harvard University, Cambridge.

Svoboda, J., V. Ložek, and E. Vlček

1996 *Hunters Between East and West: The Paleolithic of Moravia*. Plenum Press, New York.

Svobodová, H.

1987 Přirodní Prostředí. In *Stránská skála, Bohunický typ v brněnské kotlině*, edited by J. Svoboda, pp. 18–21. Academia, Prague.

Szombathy, J.

1925 Die diluvialen Menschenreste aus der Fürst Johanns Höhle bei Lautsch in Märhen. *Die Eiszeit* 2, pp. 1–34, 73–95. Vienna.

Taborin, Y.

1977 Quelques Objets de Parure. Etude Technologique: Les Percements d'Incisives de Bovinés et des Canines de Renards. In *Methodologie Appliquée à l'Industrie de l'Os Préhistorique*, edited by H. Caps-Fabrer, pp. 303–310. Centre National de la Recherche Scientifique, Paris.

1985 Les Origines des Coquillages Paléolithiques en France. In *La Signification Culturelle des Industries Lithiques*, edited by M. Otte, pp. 278–301. British Archaeological Reports International Series S239, Oxford.

1994 *La Parure en Coquillage au Paleolithique*. Supplément à Gallia Préhistoire 29. Centre National de la Recherche Scientifique, Paris.

Tattersall, I.

1998 Neanderthal Genes: What do they mean? *Evolutionary Anthropology* 6(5):157–158.

Taylor, W. A.

1948 *A Study of Archaeology*. American Anthropological Association Memoir 69. American Anthropological Association, Washington, D.C.

Tchernov, E.

1988 The Paleobiogeographical History of the Southern Levant. In *The Zoogeography of Israel*, edited by Y. Yom-Tov and E. Tchernov, pp. 159–250. W. Junk, The Hague.

1994 New Comments on the Biostratigraphy of the Middle and Upper Pleistocene of the Southern Levant. In *Late Quaternary Chronology and Paleoclimates of the Eastern Mediterranean*, edited by O. Bar-Yosef and R. S. Kra, pp. 333–350. Radiocarbon, Tucson.

Tehrani, J., and M. Collard

2002 Investigating Cultural Evolution through Biological Phylogenetic Analyses of Turkmen Textiles. *Journal of Anthropological Archaeology* 21:443–463.

Teltser, P. A.

1995a The Methodological Challenge of Evolutionary Theory in Archaeology. In *Evolutionary Archaeology: Methodological Issues*, edited by P. A. Teltser, pp. 1–11. University of Arizona Press, Tucson.

1995b *Evolutionary Archaeology: Methodological Issues*, edited by P. A. Teltser. University of Arizona Press, Tucson.

Templeton, A.

2002 Out of Africa Again and Again. *Nature* 416:45–51.

Teyssandier, N.

2006 Questioning the First Aurignacian: Mono or Multi Cultural Phenomenon during the Formation of the Upper Paleolithic in Central Europe and the Balkans. *Anthropologie* (Brno) XLIV/1:9–29.

2008 Revolution or Evolution: The Emergence of the Upper Paleolithic in Europe. *World Archaeology* 40(4):493–519.

Teyssandier, N., and D. Liolios

2003 Defining the Earliest Aurignacian in the Swabian Alp: The Relevance of the Technological Study of the Geissenklösterle (Baden-Württemberg, Germany) Lithic and Organic Productions. In *The Chronology of the Aurignacian and of the Transitional Technocomplexes: Dating, Stratigraphies, Cultural Implications*, edited by J. Zilhão and F. d'Errico, pp. 179–198. Trabalhos de Arquelogia 33. Instituto Português de Arqueologia, Lisboa.

Teyssandier, N., F. Bon, and J.-G. Bordes

2010 Within Projectile Range. Some thoughts on the Appearance of the Aurignacian in Europe. *Journal of Anthropological Research* 66(2):209–229.

Thomas, C.

1894 *Report of the Mound Explorations of the Bureau of Ethnology*. American Bureau of Ethnology, Washington, D.C.

Thomas, D. H.

1986 *Refiguring Anthropology: First Principles of Probability and Statistics*. Waveland Press, Inc., Prospect Heights, Illinois.

1998 *Archaeology*. Third Edition. Harcourt College Publishers, Fort Worth.

Thorne, A. G. and M. H. Wolpoff

1992 The Multiregional Evolution of Humans. *Scientific American* 266:76–83.

Thorpe, W.

1956 *Learning and Instinct in Animals*. Methuen, London.

Tilly, C.

1978 Migration in Modern European History. In *Human Migration: Patterns and Policies*, edited by W. McNeill and R. Adams, pp. 48–74. Indiana University Press, Bloomington.

Tinbergen, N.

1963 On Aims and Methods of Ethology. *Zeitschrift für Tierpsychologie* 20:410–433.

Tixier, J.

1984 Lames. In *Préhistoire de la Pierre Taillée 2: Economie du Débitage Laminaire*, edited by J. Tixier, pp. 13–19. Cercle de Recherches et d'Études Préhistoriques, Paris.

Tomasello, M.

1994 The Question of Chimpanzee Culture. In *Chimpanzee Cultures*, edited by R. Wrangham, W. C. McGrew, F. B. M. de Waal, and P. G. Heltne, pp. 301–318. Harvard University Press, Cambridge.

1999 The Human Adaptation for Culture. *Annual Review of Anthropology* 28:509–29.

Tomasello, M., A. Cale Kruger, and H. Horn Ratner

1993 Cultural Learning. *Behavioral and Brain Sciences* 16:495–552.

Tomka, S. A.

2001 The Effect of Processing Requirements on Reduction Strategies and Tool Form: A New Perspective. In *Lithic Debitage: Context, Form, Meaning*, edited by William Andrefsky, Jr., pp. 207–223. University of Utah Press, Salt Lake City.

Torrence, R.

1983 Time Budgeting and Hunter-Gatherer Technology. In *Hunter-Gatherer Economy in Prehistory*, edited by G. Bailey, pp. 11–22. Cambridge University Press, Cambridge.

1989a Tools as Optimal Solutions. In *Time, Energy, and Stone Tools*, edited by R. Torrence, pp. 1–6. Cambridge University Press, Cambridge.

1989b Re-Tooling: Towards a Behavioral Theory of Stone Tools. In *Time, Energy, and Stone Tools*, edited by R. Torrence, pp. 57–66. Cambridge University Press, Cambridge.

Tostevin, G. B.

2000a The Middle to Upper Paleolithic Transition from the Levant to Central Europe: *In Situ* Development or Diffusion? In *Neanderthals and Modern Humans: Discussing the Transition. Central and Eastern Europe from 50,000–30,000 BP*, edited by G.-C. Weniger and J. Orschiedt, pp. 90–109. Neanderthal Museum, Düsseldorf.

2000b It Happened Twice: Inter-regional Diffusion between 60 and 30 kya in Central Europe, Eastern Europe, and the Levant. In *Vremya Poslednih Neandertaltsev (Time of the Last Neanderthals). Special Edition of Stratum Plus*, edited and translated by L. Vishnyatsky, pp. 216–235. High Anthropological School, St. Petersburg. (In Russian.)

2000c Behavioral Change and Regional Variation across the Middle to Upper Paleolithic Transition in Central Europe, Eastern Europe, and the Levant. Ph.D. Dissertation, Department of Anthropology, Harvard University, Cambridge.

2003a A Quest for Antecedents: A Comparison of the Terminal Middle Paleolithic and Early Upper Paleolithic of the Levant. In *More than Meets the Eye: Studies on Upper Palaeolithic Diversity in the Near East*, edited by A. N. Goring-Morris and A. Belfer-Cohen, pp. 54–67. Oxbow Press, Oxford.

2003b Attribute Analysis of the Lithic Technologies of Stránská skála II–III in their Regional and Inter-regional Context. In *Stránská skála: Origins of the Upper Paleolithic in the Brno Basin*. edited by J. Svoboda and O. Bar-Yosef, pp. 77–118. American School of Prehistoric Research Bulletin 47. Dolní Věstonice Studies 10. Peabody Museum of Archaeology and Ethnology, Harvard University, Cambridge.

2005 The Use and Abuse of the *Chaîne Opératoire* vs. Reduction Sequence Debate in Paleolithic Archaeology. Paper presented at the Paleoanthropology Society Meetings, Milwaukee, WI, April 7, 2005. Abstract available at http://www.paleoanthro.org/abst2005.htm.

2006 Levels of Theory and Social Practice in the Reduction Sequence and *Chaîne Opératoire* Methods of Lithic Analysis. In the electronic symposium *Core Reduction, Chaîne Opératoire, and Other Methods: The Epistemologies of Different Approaches to Lithic Analysis*, organized by G. Tostevin, at the 71st Annual Meeting of the Society for American Archaeology, San Juan, Puerto Rico.

2007 Social Intimacy, Artefact Visibility, and Acculturation Models of Neanderthal-Modern Human Interaction. In *Rethinking the Human Revolution: New Behavioural and Biological Perspectives on the Origins and Dispersal of Modern Humans*, edited by P. Mellars, K. Boyle, O. Bar-Yosef, and C. Stringer, pp. 341–357. MacDonald Institute Research Monograph, Cambridge.

2008 Flakes in the Round: Combining a Behavioral Approach to Attribute Analysis with 3D geometric Morphometrics. Paper presented in the symposium Analytical approaches to Palaeolithic technologies, S. Lycett and P. Chauhan organizers, at the 73rd Annual Meeting of the Society for American Archaeology, Vancouver, British Columbia, Canada, March, 2008. Presentation pdf available on request.

2009 The Importance of Process and Historical Event in the Study of the Middle to Upper Paleolithic Transition. In *Transitions in Prehistory: Papers in Honor of Ofer Bar Yosef*, edited by J. Shea and D. Lieberman, pp. 166–183. American School of Prehistoric Research Monograph. Oxbow Books, Oxford.

2011 Levels of Theory and Social Practice in the Reduction Sequence and *Chaîne Opératoire* Methods of Lithic Analysis: An Introduction to the Epistemologies of Different Approaches to Lithic Analysis. *PaleoAnthropology* 2011:351–375.

Tostevin, G. B., and P. Škrdla

2003 New Research on the Early Upper Paleolithic of Central Europe: The Re-excavation of Brno-Bohunice. Paper presented at the Paleoanthropology Society 2003 Meetings, Tempe, AZ. Abstract at http://www.paleoanthro.org/abst2003.htm.

Tostevin, G. B., and P. Škrdla

2006 New Excavations at Bohunice and the Question of the Uniqueness of the Type-site for the Bohunician Industrial Type. *Anthropologie* (Brno) XLIV(1).

Traphagan, J. W.

2000 The Liminal Family: Return Migration and Intergenerational Conflict in Japan. *Journal of Anthropological Research* 56(3):365-85.

Trigger, B. G.

1968 *Beyond History: the Methods of Prehistory*. Holt, Rinehart, and Winston, New York.

1985 The Past as Power: Anthropology and the North American Indian. In *Who Owns the Past*, edited by I. McBryde, pp. 11-40. Oxford University Press, Oxford.

1989 *A History of Archaeological Thought*. Cambridge University Press, Cambridge.

Trinkaus, E., O. Moldovan, Ş. Milota, A. Bîlgăr, Laurenţiu Sarcina, S. Athreya, S. E. Bailey, R. Rodrigo, G. Mircea, T. Higham, C. Bronk Ramsey, and J. van der Plicht

2003a An Early Modern Human from the Peştera cu Oase, Romania. *Proceedings of the National Academy of Sciences* 100(20):11231-11236

Trinkaus, E., O. Moldovan, Ş. Milota, R. Rodrigo, G. Mircea, and O. Moldovan

2003b Early Modern Human Cranial Remains from the Peştera cu Oase, Romania. *Journal of Human Evolution* 45:245-253.

Trinkaus, E., and J. Zilhão

2002 Chapter 32. Phylogenetic Implications. In *Portrait of the Artist as a Child: The Gravettian Human Skeleton from Abrigo do Lagar Velho and its Archaeological Context*, edited by J. Zilhão and E. Trinkaus, pp. 497-518. Trabalhos de Arqueologia 22, IPA. Instituto Português de Arqueologia, Lisbon.

Trivers, R. L.

1971 The Evolution of Reciprocal Altruism. *Quarterly Review of Biology* 46:35-57.

1985 *Social Evolution*. Benjamin/Cummings Publishing Co., Menlo Park.

Tsanova, T.

2006 Les Débuts du Paléolithique Supérieur dans l'Est des Balkans. Réflexion à Partir de l'étude Taphonomique et Techno-Economique des Ensembles Lithiques des Sites de Bacho Kiro (couche 11), Temnata (couches VI et 4) et Kozarnika (niveau VII). Doctoral Thesis, Bordeaux 1.

2008 *Les Débuts du Paléolithique Supérieur dans l'Est des Balkans. Réflexion à partir des études taphonomique et techno-économique des ensembles lithiques des Grottes Bacho Kiro (couche 11), Temnata (couche VI et 4) et Kozarnika (niveau VII)*. BAR International Series 1752. Oxbow, Oxford.

Tschauner, H.

1994 Archaeological Systematics and Cultural Evolution: Retrieving the Honour of Culture History. *Man* 29(1):77-93.

Tuffreau, A., S. Révillion, J. Sommé, M. J. Aitken, J. Huxtable, and A. Leroi-Gourhan

1985 Le Gisement Paléolithique Moyen de Seclin (Nord, France). *Archäologisches Korrespondenzblatt* 15:131-138.

Turner, A.

1984 Hominids and Fellow Travellers: Human Migration into High Latitudes as Part of a Large Mammal Community. In *Hominid Evolution and Community Ecology: Prehistoric Human Adaptation in Biological Perspective*, edited by R. Foley, pp. 193-217. Academic Press, London.

1991 Origine des Grand Mannifères Plio-Pléistocènes d'Europe et Migrations Humaines. In *Les Premiers Européens*, edited by E. Bonifay and B. Vandermeersch, Actes du 114e Congrès National des Sociétés Savantes. Éditions du C. T. H. S., Paris.

1992 Large Carnivores and Earliest European Hominids: Changing Determinants of Resource Availability during the Lower and Middle Pleistocene. *Journal of Human Evolution* 22:109-126.

Turq, A.

1992 Raw Material and Technological Studies of the Quina Mousterian in Perigord. In *The Middle Paleolithic: Adaptation, Behavior, and Variability*, edited by H. L. Dibble and P. Mellars, pp. 75-85. University Museum, University of Pennsylvania, Philadelphia.

Turville-Petre, F.

1932 Excavations in the Mugharet el-Keberah. *Journal of the Royal Anthropological Institute* 62:271-276.

Tylor, E. B.

1896 On American Lot-Games, as Evidence of Asiatic Intercourse before the Time of Columbus. *Internationales Archiv für Ethnographie*, Vol. 9, supplement: 55-67. Leiden.

Upton, D.

1996 Ethnicity, Authenticity and Invented Traditions. *Historical Archaeolology* 30(2):1-7.

Usik, V. I.

1989 Korolevo: Transition from Lower to Upper Palaeolithic According to Reconstitution Data. *Anthropologie* (Brno) XXVII/2-3:179-212.

Valladas, H., J. L. Joron, G. Valladas, B. Arensburg, O. Bar-Yosef, A. Belfer-Cohen, P. Goldberg, H. Laville, L. Meignen, Y. Rak, E. Tchernov, A.-M. Tillier, and B. Vandermeersch

1987 Thermoluminescence Dates for the Neanderthal Burial Site at Kebara in Israel. *Nature* 330(6144):159-160.

Valladas, Helene, Norbert Mercier, Catherine Escutenaire, Tomacz Kalicki, Janusz Kozlowski, Valery Sitlivy, Krzysztaf Sobczyk, Aleksandra Zięba, and Brigitte Van Vliet-Lanoë

2003 The Late Middle Paleolithic Blade Technologies and the Transition to the Upper Paleolithic in Southern Poland: TL Dating Contribution. *Eurasian Prehistory* 1(1):57-82.

Valoch, Karel

1967 Paleolitické osidleni jaskynê Kulny na Sloupu v Moravskem Krasu. *Archeologicke Rozhledy* 19:566-575.

1976 *Die Altsteinzeitliche Fundstelle in Brno-Bohunice*. Studie AU CSAV IV/I. Academia, Prague.

1982 Neue Paläolithische Funde von Brno-Bohunice. *Časopis Moravského muzea Sc. soc.* 67:31-48.

1984 Výzkum Paleolitu ve Vedrovicích V (okr. Znojmo). *Časopis Moravského muzea* 69:5-22.

1986 Stone Industries of the Middle/Upper Palaeolithic Transition. in *The Pleistocene Perspective, Vol. 1*, edited by M. Day and R. Foley, pp. 1-15. Proceedings of the World Archaeological Congress. Allen and Unwin, London.

198 The Early Palaeolithic Site Stranska skala I Near Brno (Czechoslovakia). *Anthropologie* (Brno) 25:125-42.

1988 Die Erforschung der Kůlna Höhle 1961-1976. *Anthropos* 24. Moravské Muzeum, Brno.

1990a La Moravie il y a 40,000 Ans. In *Paléolithique Moyen Récent et Paléolithique Supérieur Ancien en Europe*, edited by C. Farizy, pp. 115-124. Mémoires du Musée de Préhistoire d'Ile de France No. 3. Édition de l'Association pour la Promotion de la Recherche Archéologique en Ile de France, Nemours.

1990b Le Szeletien en Moravie. In *Feuilles de Pierre. Les Industries à Pointes Foliacées du Paléolithique supérieur Européen*, edited by J. Kozłowski, pp. 213-221. Études de Recherches archéologiques de l'Université de Liège 42, Liège.

1993 Vedrovice V, eine Siedlung des Szeletien in Südmähren. *Quartär* 43/44:7-93.

2003 The Archaeology of Stránská skála III-1. In *Stránská skála: Origins of the Upper Paleolithic in the Brno Basin*, edited by J. Svoboda and O. Bar-Yosef, pp. 27-35. American School of Prehistoric Research Bulletin 47. Dolní Věstonice Studies 10. Peabody Museum of Archaeology and Ethnology, Harvard University, Cambridge.

Valoch, K., Z. Nerudova, and P. Neruda

2000 Stránská skála III: Ateliers des Bohunicien. *Památky Archaeologické* 91:5–113.

van Andel, J. H., and W. Davies

2003 *Neanderthals and Modern Humans in the European Landscape during the Last Glaciation: Archaeological Results of the Stage 3 Project*. McDonald Institute Research Monographs, Cambridge.

Van Bulck, V.

1932 *Beiträge zur Methodik der Völkerkunde*. Vienna.

van den Berghe, P. L.

1979 *Human Family Systems: An Evolutionary View*. Elsevier, New York.

van der Leeuw, S. E., and R. Torrence

1989 *What's New: A Closer Look at the Process of Innovation*. Unwin Hyman, London.

Van Kirk, S.

1983 *Many Tender Ties: Women in Fur-Trade Society, 1670–1870*. University of Oklahoma Press, Norman.

Van Peer, P.

1992 *The Levallois Reduction Strategy*. Monographs in World Prehistory No. 13. Prehistory Press, Madison.

1998 The Nile Corridor and the Out-of-Africa Model. *Current Anthropology* 39 (Supplement):S115–S140.

2007 Refitting of Reduction Sequences, Formal Classification Systems and Middle Palaeolithic Individuals at Work. In *Fitting Rocks: Lithic Refitting Examined*, edited by U. Schurmans and M. De Bie, pp. 91–104. British Archaeological Reports International Series 1596. Archaeopress, Oxford.

Van Peer, P., and P. M. Vermeersch

2007 An Interpretation of Behavioural Change in North-East Africa during the Late Middle and Early Upper Pleistocene with a Specific Focus on the Taramsa 1 Burial. In *Rethinking the Human Revolution: New Behavioural and Biological Perspectives on the Origins and Dispersal of Modern Humans*, edited by C. Stringer and P. Mellars, pp.187–198. MacDonald Institute Research Monograph, Cambridge.

Van Peer, P., and S. Wurz

2006 The *Chaîne Opératoire* Concept in Middle Palaeolithic Lithic Studies: Some Problems and Solutions. In *Core Reduction, Chaîne Opératoire, and Other Methods: The Epistemologies of Different Approaches to Lithic Analysis*, edited by G. Tostevin, monograph of the electronic symposium at the 71st Annual Meeting of the Society for American Archaeology, San Juan, Puerto Rico. Publication in preparation.

Van Peer, P., P. M. Vermeersch, and E. Pailissen

2010 *Chert Quarrying, Lithic Technology and a Modern Human Burial at the Palaeolithic Site of Taramsa 1, Upper Egypt*. Leuven University Press, Leuven.

van Schaik, C. P., R. Deaner, and M. Merrill

1999 The Conditions for Tool use in Primates: Implications for the Evolution of Material Culture. *Journal of Human Evolution* 36:719–741.

Vermeersch, P. M.

2001 'Out of Africa' from an Egyptian Point of View. *Quaternary International* 75:103–112.

Vermeersch, P. M., E. Paulissen, S. Stokes, C. Charlier, P. Van Peer, C. Stringer, and W. Lindsay

1998 A Middle Palaeolithic Burial of a Modern Human at Taramsa Hill, Egypt. *Antiquity* 72:475–484.

Vértes, L.

1955 Neuere Ausgrabungen und Paläolithische Funde in der Höhle von Istalloskö. *Acta Archaeologica Acad. Sc. Hung.* Vol. 5:111-131

1961 Das Verhältniss des Aurignacien zum Szeletien in der Istállósköen Höhle. *Germania* 39:295-298.

1964 *Tata-Eine Mittelpalaolithische Travertinsiedlung in Ungarin.* Akademiai Kiado, Budapest.

Vishnyatsky, L.

2006 Book Review of *Stránská Skála: Origins of the Upper Paleolithic in the Brno Basin, Moravia, Czech Republic*, edited by J. A. Svoboda and O. Bar-Yosef. *PaleoAnthropology* 2006(1):27–28.

Volkman, P. W.

1983 Boker Tachtit: Core Reconstructions. In *Prehistory and Paleoenvironments in the Central Negev, Israel, Volume III The Avdat/Aqev Area, Part 3*, edited by A. E. Marks, pp. 127-190. Southern Methodist University Press, Dallas.

1989 Boker Tachtit: The Technological Shift from the Middle to the Upper Paleolithic in the Central Negev, Israel. University Microfilm International, Ph.D. Dissertation, Southern Methodist University, Dallas.

Volkman, P. W., and D. Kaufman

1983 A Reassessment of the Emireh Point as a Possible Type Fossil for the Technological Shift from the Middle to the Upper Paleolithic in the Levan. In *The Early Upper Paleolithic: Evidence from Europe and the Near East*, edited by J. F. Hoffecker and C. A. Wolf, pp. 35-51, British Archaeological Reports International Series 437, Oxford.

Waldorf, D. C.

2006 *The Art of Flintknapping, 5th Edition.* Mound Builder Arts and Trading Co., Branson, MO.

Wallace, I., and J. J. Shea

2006 Mobility Patterns and Core Technologies in the Middle Paleolithic of the Levant. *Journal of Archaeological Science.*

Wason, P.

1966 Reasoning. In *New Horizons in Psychology*, edited by B. M. Foss. Penguin, Harmondsworth.

Waters, M. R.

1992 *Principles of Geoarchaeology: A North American Perspective.* University of Arizona Press, Tucson.

Watson, V. D.

1995 Simple and Significant: Stone Tool Production in highland New Guinea. *Lithic Technology* 20(2):89-99.

Weedman, K. J.

2000 An Ethnoarchaeological Study of Stone Scrapers among the Gamo People of Southern Ethiopia. Ph.D. Dissertation, Anthropology, University of Florida.

2002a On the Spur of the Moment: Effect of Age and Experience on Hafted Stone Scraper Morphology. *American Antiquity* 67(4):731-44.

2002b White Washing Cultural Stains. An Ethnoarchaeological Study of Ritual-Political Membership, Kinship, and Stone Tools Among Gamo of Southern Ethiopia. Poster presented March 21, 2002, at the 67th Annual Meeting of the Society for American Archaeology, Denver.

Weiner, S., P. Goldberg, and O. Bar-Yosef

1993 Bone Preservation in Kebara Cave, Israel Using On-site Fourier Transform Infrared Spectrometry. *Journal of Archaeological Science* 20:613-627.

Weiner, S., S. Schiegl, P. Goldberg, and O. Bar-Yosef

1995 Mineral Assemblages in Kebara and Hayonim, Israel: Excavation Strategies, Bone Preservation and Wood Ash Remnants. *Israel Journal of Chemistry* 35:143-154.

Weiner, S., F. Berna, I. Cohen-Ofri, R. Shahack-Gross, R. M. Albert, P. Karkanas, L. Meignen, and O. Bar-Yosef

2007 Mineral Distributions in Kebara Cave: Diagenesis and Its Affect on the Archaeological Record. In *Kebara Cave, Mt. Carmel, Israel: The Middle and Upper Paleolithic Archaeology, Part I*, edited by O. Bar-Yosef and L. Meignen, pp. 131-146. American School of Prehistoric Research Bulletin 49. Peabody Museum of Archaeology and Ethnology, Harvard University, Cambridge.

Weiss, K. M.

1988 In Search of Times Past: Gene Flow and Invasion in the Generation of Human Diversity. In *Biological Aspects of Human Migration*, edited by C. G. N. Mascie-Taylor and G. W. Lasker, pp. 130-166. Cambridge University Press, Cambridge.

Wells, P.

1999 *The Barbarians Speak: How the Conquered Peoples Shaped Roman Europe*. Princeton University Press, Princeton.

Weninger, B., O. Jöris, and U. Danzeglocke

2007 CalPal-2007. Cologne Radiocarbon Calibration and Palaeoclimate Research Package. http://www.calpal.de/.

Weninger, B., and O. Jöris

2008 A ^{14}C age Calibration Curve for the Last 60 ka: The Greenland-Hulu U/Th Time-Scale and its Impact on Understanding the Middle to Upper Paleolithic Transition in Western Eurasia. *Journal of Human Evolution* 55:772-781.

Wenzel, K. E., and P. H. Shelley

2001 What Put the Small in the Arctic Small Tool Tradition: Raw Material Constraints on Lithic Technology at the Mosquito Lake Site, Alaska. In *Lithic Debitage: Context, Form, Meaning*, edited by W. Andrefsky, Jr., pp. 106-123. University of Utah Press, Salt Lake City.

West, D.

1996 Horse Hunting, Processing, and Transport in the Middle Danube. In *Paleolithic in the Middle Danube Region*, edited by J. Svoboda, pp. 209-246. Institute of Archaeology, Brno.

Whallon, Robert

2006 Social networks and information: Non-"utilitarian" mobility among hunter-gatherers. *Journal of Anthropological Archaeology* 25:259-270.

Whewell, W. W.

1858 Novum Organon Renovatum. *The Philosophy of the Inductive Sciences, 3rd Edition, Part II*. John W. Parker Publisher, London.

White, J. P., and D. H. Thomas

1972 What Mean These Stones? Ethno-Taxonomic Models and Archaeological Interpretation in the New Guinea Highlands. In *Models in Archaeology*, edited by D. Clarke, pp. 275-308. Methuen, London.

White, J. P., and H. L. Dibble

1986 Stone Tools: Small-Scale Variability. In *Stone Age Prehistory: Studies in Memory of Charles McBurney*, edited by G. N. Bailey and P. Callow, pp. 47-53. Cambridge University Press, Cambridge.

White, Leslie A.

1949 *The Science of Culture: A Study of Man and Civilization*. Farrar and Straus, New York.

1959 *The Evolution of Culture*. McGraw-Hill, New York.

White, Randall

1982 Rethinking the Middle/Upper Paleolithic Transition. *Current Anthropology* 23(2):169-192.

1983 Changing Land-Use Patterns Across the Middle/Upper Paleolithic Transition: The Complex Cas of the Perigord. In *The Mousterian Legacy: Human Biocultural Change in the Upper Pleistocene*, edited by E. Trinkaus, pp. 113-121. British Archaeological Reports International Series 164, Oxford.

1989 Production Complexity and Standardization in Early Aurignacian Bead and Pendant Manufacture: Evolutionary Implications. *The Human Revolution: Behavioural and Biological Perspectives in the Origins of Modern Humans*, edited by P. Mellars and C. Stringer, pp. 366-390. Edinburgh University Press, Edinburgh.

1993 Introduction: Gesture and Speech. In *Gesture and Speech*, edited by A. Leroi-Gourhan, pp. xiii-xxii. MIT Press, Cambridge.

2007 Systems of Personal Ornamentation in the Early Upper Paleolithic: Methodological Challenges and New Observations. In *Rethinking the Human Revolution: New Behavioural and Biological Perspectives on the Origins and Dispersal of Modern Humans*, edited by P. Mellars, K. Boyle, O. Bar-Yosef, and C. Stringer, pp. 287-302. MacDonald Institute Research Monograph, Cambridge.

White, Richard

1991 *The Middle Ground: Indians, Empires, and Republics in the Great Lakes Region, 1650-1815*. Cambridge Studies in North American Indian History. Cambridge University Press, Cambridge.

Whittaker, J. C.

1994 *Flintknapping: Making and Understanding Stone Tools*. University of Texas Press, Austin.

2004 *American Flintknappers : Stone Age Art in the Age of Computers*. University of Texas Press, Austin.

Whybrow, P. J.

1992 Land Movements and Species Dispersal. In *The Cambridge Encyclopedia of Human Evolution*, edited by S. Jones, R. Martin, and D. Pilbeam, pp. 169-173. Cambridge University Press, Cambridge.

Wiant, M., and H. Hassan

1985 The Role of Lithic Resource Accessibility in the Organization of Lithic Technology. In *Lithic Resource Procurement: Proceedings from the Second Conference on Prehistoric Chert Exploitation*, edited by S. Vehik, pp. 101-114. Centre for Archaeological Investigations, Occasional Papers 4. Southern Illinois University Press, Carbondale.

Wiessner, P.

1982 Risk, Reciprocity, and Social Influences on !Kung San Economics. In *Politics and History in Band Societies*, edited by E. Leacock and R. B. Lee, pp. 61-84. Cambridge University Press, Cambridge.

1983 Style and Social Information in Kalahari San Projectile Points. *American Antiquity* 48:253-276.

1984 Reconsidering the Behavioral Basis for Style: A Case Study among the Kalahari San. *Journal of Anthropological Archaeology* 3:190-234.

1985 Style or Isochrestic Variation? A Reply to Sackett. *American Antiquity* 50(1):160-166.

1990 Is There a Unity to Style? In *The Uses of Style in Archaeology*, edited by M. Conkey and C. Hastorf, pp. 105-112. Cambridge University Press, Cambridge.

2002 Hunting, Healing, and Hxaro Exchange: A Long-Term Perspective on !Kung (Ju/'hoansi) Large-Game Hunting. *Evolution and Human Behavior* 23:407-436.

Wilkie, L.

1996 Glassknapping at a Louisiana Plantation: Afro-American Tools. *Historical Archaeology* 30(4):37-49.

Willey, G. R., C. DiPeso, W. A. Ritchie, I. Rouse, J. Rowe, and D. Lathrap (alternately Lathrap et al. 1956)

1956 An Archaeological Classification of Culture Contact Situations. In *Seminars in Archaeology, 1955*, edited by R. Wauchope, *American Antiquity* 22(2ii):3-30.

Willey, G. R., and J.A. Sabloff

1993 *A History of American Archaeology, Third Edition*. W. H. Freeman and Co., New York.

Wilmsen, E. N.

1973 Interaction, Spacing Behavior and the Organization of Hunting Bands. *Journal of Anthropological Research* 29:1-31.

Wilson, M. L., W. R. Wallauer, and A. E. Pusey

2004 New Cases of Intergroup Violence among Chimpanzees in Gombe National Park, Tanzania. *International Journal of Primatology* 25(3):523-549.

Wilson, M. L., and R. W. Wrangham

2003 Intergroup Relations in Chimpanzees. *Annual Review of Anthropology* 32:363-392.

Wintle, A. G.

1990 A Review of Current Research on TL Dating of Loess. *Quaternary Science Reviews* 9:385-397.

Winterhalder, B.

1986 Diet Choice, Risk, and Food Sharing in a Stochastic Environment. *Journal of Anthropological Archaeology* 5:369-392.

1997 Gifts Given, Gifts Taken: The Behavioral Ecology of Nonmarket, Intragroup Exchange. *Journal of Archaeological Research* 5(2):121-168.

Winterhalder, B., and E. Smith

1981 *Hunter-Gatherer Foraging Strategies.* University of Chicago Press, Chicago.

Wobst, H. M.

1974 Boundary Conditions for Paleolithic Social Systems: A Simulation Approach. *American Antiquity* 39(2):147-178.

1977 Stylistic Behavior and Information Exchange. In *For the Director: Essays in honor of James B. Griffin*, edited by C. Cleland, pp. 317-342. Anthropological Papers of the University of Michigan 61. Regents of the University of Michigan, Ann Arbor.

1999 Style in Archaeology or Archaeologists in Style. In *Material Meanings: Critical Approaches to the Interpretation of Material Culture*, edited by E. Chilton, pp. 118-132. University of Utah Press, Salt Lake City.

Wolcott, Harry F.

1991 Propriospect and the Acquisition of Culture. *Anthropology & Education Quarterly* 22:251-273.

Wrangham, R., F. B. M. de Waal, and W. C. McGrew

1994 The Challenge of Behavioral Diversity. In *Chimpanzee Cultures*, edited by R. Wrangham, W. C. McGrew, F. B. M. de Waal, and P. G. Heltne, pp. 1-18. Harvard University Press, Cambridge.

Wurz, S., N. J. le Roux, S. Gardner, and H. J. Deacon

2003 Discriminating between the End Products of the Earlier Middle Stone Age Sub-Stages at Klasies River using Biplot Methodology. *Journal of Archaeological Science* 30:1107-1126.

Wurz, S., P. Van Peer, N. J. le Roux, S. Gardner, and H. J. Deacon

2005 Continental Patterns in Stone Tools: A Technological and Biplot-Based Comparison of Early Late Pleistocene Assemblages from Northern and Southern Africa. *African Archaeological Review* 22:1-24.

Wynne, T., and F. L. Coolidge

2004 The Expert Neandertal mind. *Journal of Human Evolution* 46:467-487.

2010 Beyond Symbolism and Language: An Introduction to Supplement 1, Working Memory. *Current Anthropology* 51(Supplement 1):S5-S16.

Young, D. E., and R. Bonnichsen

1985 Cognition, Behavior, and Material Culture. In *Stone Tool Analysis: Essays in Honor of Don E. Crabtree*, edited by M. G. Plew, J. C. Woods, and M. G. Pavesic, pp. 91-129. University of New Mexico Press, Albuquerque.

Zedeño, M. N.

1995 The Role of Population Movement and Technology Transfer in the Manufacture of Prehistoric Southwestern Ceramics. In *Ceramic Production in the American Southwest*, edited by B. J. Mills, and P. L Crown, pp. 115-141. University of Arizona Press, Tucson.

Zięba, A., V. Sitlivy, K. Sobczyk, and A. V. Kolesnik

2008 Raw Material Exploitation and Intra-site Spatial Distribution at Two Late Middle and Early Upper Paleolithic sites in the Krakow Region: Piekary IIa and Księcia Józefa. *Archaeology Ethnology and Anthropology of Eurasia* 33(1):46-57.

Zilhão, J.

2006 Neandertals and Moderns Mixed, and it Matters. *Evolutionary Anthropology* 15:183-195.

Zilhão, J., and F. d'Errico

1999 The Chronology and Taphonomy of the Earliest Aurignacian and Its Implications for the Understanding of Neandertal Extinction. *Journal of World Prehistory* 13(1):1-68.

2003 The Chronology of the Aurignacian and Transitional Technocomplexes. Where do we Stand? In *The Chronology of the Aurignacian and of the Transitional Technocomplexes: Dating, Stratigraphies, Cultural Implications*, edited by J. Zilhão and F. d'Errico, pp. 313-349. Trabalhos de Arquelogia 33. Instituto Português de Arqueologia, Lisboa.

Zilhão, J., E. Trinkaus, S. Constantin, Ş. Milota, M. Cherase, L. Sarcina, A. Danciu, H. Rougier, J. Quilès, and R. Rodrigo

2007 The Peştera cu Oase People, Europe's Earliest Modern Humans. In *Rethinking the Human Revolution: New Behavioural and Biological Perspectives on the Origins and Dispersal of Modern Humans*, edited by P. Mellars, K. Boyle, O. Bar-Yosef, and C. Stringer, pp. 249-262. MacDonald Institute Research Monograph, Cambridge.

Zilhão, J., F. d'Errico, J.-G. Bordes, R. Lenoble, J.-P. Texier, and J.-P. Rigaud

2008a Grotte des Fées (Châtelperron): History of Research, Stratigraphy, Dating, and Archaeology of the Châtelperronian Type-Site. *PaleoAnthropology* 2008:1-42.

2008b Like Hobbes' Chimney Birds. *PaleoAnthropology* 2008:65-67.

Zhou, W.-X., D. Sornette, R. A. Hill, and R. I. M. Dunbar

2005 Discrete Hierarchical Organization of Social Group Sizes. *Proceedings of the Royal Society B Series* 272:439-44.

Index